Unit Conversion Factors

TIME

1 min = 60 s
1 hr = 60 min = 3600 s
1 day = 24 hr = 86,400 s

LENGTH

1 m = 3.281 ft = 39.37 in.
1 km = 0.6214 mi
1 in. = 0.08333 ft = 0.02540 m
1 ft = 12 in. = 0.3048 m
1 mi = 5280 ft = 1.609 km
1 nautical mile = 1852 m = 6080 ft

ANGLE

1 rad = $180/\pi$ deg = 57.30 deg
1 deg = $\pi/180$ rad = 0.01745 rad
1 revolution = 2π rad = 360 deg
1 rev/min (rpm) = 0.1047 rad/s

AREA

1 mm^2 = 1.550 $\times$ 10^{-3} in^2 = 1.076 $\times$ 10^{-5} ft^2
1 m^2 = 10.76 ft^2
1 in^2 = 645.2 mm^2
1 ft^2 = 144 in^2 = 0.0929 m^2

VOLUME

1 mm^3 = 6.102 $\times$ 10^{-5} in^3 = 3.531 $\times$ 10^{-8} ft^3
1 m^3 = 6.102 $\times$ 10^4 in^3 = 35.31 ft^3
1 in^3 = 1.639 $\times$ 10^4 mm^3 = 1.639 $\times$ 10^{-5} m^3
1 ft^3 = 0.02832 m^3

VELOCITY

1 m/s = 3.281 ft/s
1 km/hr = 0.2778 m/s = 0.6214 mi/hr = 0.9113 ft/s
1 mi/hr = (88/60) ft/s = 1.609 km/hr = 0.4470 m/s
1 knot = 1 nautical mile/hr = 0.5144 m/s = 1.689 ft/s

ACCELERATION

1 m/s^2 = 3.281 ft/s^2 = 39.37 in/s^2
1 in/s^2 = 0.08333 ft/s^2 = 0.02540 m/s^2
1 ft/s^2 = 0.3048 m/s^2
1 g = 9.81 m/s^2 = 32.2 ft/s^2

MASS

1 kg = 0.0685 slug
1 slug = 14.59 kg
1 t (metric tonne) = 10^3 kg = 68.5 slug

FORCE

1 N = 0.2248 lb
1 lb = 4.448 N
1 kip = 1000 lb = 4448 N
1 ton = 2000 lb = 8896 N

WORK AND ENERGY

1 J = 1 N-m = 0.7376 ft-lb
1 ft-lb = 1.356 J

POWER

1 W = 1 N-m/s = 0.7376 ft-lb/s = 1.340 $\times$ 10^{-3} hp
1 ft-lb/s = 1.356 W
1 hp = 550 ft-lb/s = 746 W

PRESSURE

1 Pa = 1 N/m^2 = 0.0209 lb/ft^2 = 1.451 $\times$ 10^{-4} lb/in^2
1 bar = 10^5 Pa
1 lb/in^2 (psi) = 144 lb/ft^2 = 6891 Pa
1 lb/ft^2 = 6.944 $\times$ 10^{-3} lb/in^2 = 47.85 Pa

ENGINEERING MECHANICS

STATICS & DYNAMICS PRINCIPLES

Anthony Bedford • Wallace Fowler

University of Texas at Austin

Prentice Hall, Upper Saddle River, New Jersey 07458

Library of Congress Cataloging-in-Publication Data on file

Vice President and Editorial Director, ECS: *Marcia J. Horton*
Executive Editor: *Eric Svendsen*
Associate Editor: *Dee Bernhard*
Vice President and Director of Production and Manufacturing, ESM: *David W. Riccardi*
Executive Managing Editor: *Vince O'Brien*
Managing Editor: *David A. George*
Production Editor: *Sarah Parker*
Director of Creative Services: *Paul Belfanti*
Creative Director: *Carole Anson*
Art Director: *Jonathan Boylan*
Cover Designer: *John Christiana*
Interior Designer: *Circa 86*
Managing Editor, A/V Management and Production: *Patricia Burns*
Audio/Visual Editor: *Xiaohong Zhu*
Manufacturing Manager: *Trudy Pisciotti*
Manufacturing Buyer: *Lisa McDowell*
Marketing Manager: *Holly Stark*

Based on *Engineering Mechanics—Statics & Dynamics* by Anthony Bedford and Wallace Fowler, © 2002, 1999, and 1995 by Prentice Hall.

The author and publisher of this book have used their best efforts in preparing this book. These efforts include the development, research, and testing of the theories and programs to determine their effectiveness. The author and publisher make no warranty of any kind, expressed or implied, with regard to these programs or the documentation contained in this book. The author and publisher shall not be liable in any event for incidental or consequential damages in connection with, or arising out of, the furnishing, performance, or use of these programs.

MATLAB is a registered trademark of The MathWorks, Inc., 3 Apple Hill Drive, Natick, MA 01760-2098.

Mathcad is a registered trademark of MathSoft Engineering and Education, 101 Main St., Cambridge, MA 02142-1521.

Printed in the United States of America

10   9   8   7   6   5   4   3   2   1

ISBN 0-13-008209-0

Pearson Education Ltd., *London*
Pearson Education Australia Pty. Ltd., *Sydney*
Pearson Education *Singapore*, Pte. Ltd.
Pearson Education North Asia Ltd., *Hong Kong*
Pearson Education Canada, Inc., *Toronto*
Pearson Education de Mexico, S.A. de C.V.
Pearson Education—Japan, *Tokyo*
Pearson Education Malaysia, Pte. Ltd.
Pearson Education, Inc., Upper Saddle River, *New Jersey*

CONTENTS

STATICS

APPENDICES

DYNAMICS

About this Package

More than just a book, this text is part of a system to teach engineering mechanics, a system comprised of three components: (1) this core principles book, (2) algorithmic problem material available on-line, and (3) a course management system to track and monitor student progress. By using this system, we hope instructors and their students will benefit from increased flexibility in the ability to assign and grade problems, and the ability to make sure each student works a "unique" version of a problem, all coming at a lower price and in a smaller package.

The Components

Core Principles Text: This paperback text contains the complete set of descriptions, explanations and examples that appear in the hardback *Engineering Mechanics—Statics & Dynamics*, *Third Edition* text by Bedford and Fowler. No core theoretical or example material has been deleted; students and instructors will find a complete discussion of the subject. However, most all problem material has been deleted from the printed text, only a section of end-of-chapter review problems remain.

On-Line, Algorithmic, Assignments: Instructors assign homework problems on-line using PHGradeAssist—Prentice Hall's on-line, algorithmic homework generator. Using this system, professors can create assignments from over 1500 *Statics & Dynamics* problems keyed to this book by chapter and section. When students access these problems on-line, PHGradeAssist automatically generates a unique version of the problem for each student. Students may print out their problem sets and work them off line. Answers are then submitted and graded through the site and results are tracked in the instructor's on-line gradebook.

Course Management: Instructors have complete control over their course. They assign the homework they want, in the format they choose, and make it available for as long as they'd like. Instructors can change problems if they choose, as well as control the value of each problem. Complete help is available both in the system and to download at **http://www.prenhall.com/bedford**.

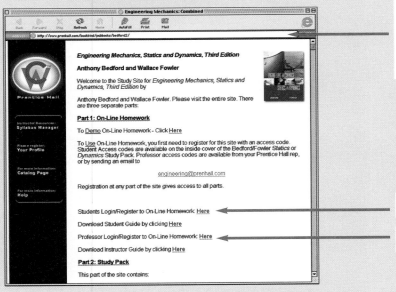

Home page for PH Grade Assist/On-Line Homework. Go to www.prenhall.com/bedford and choose Engineering Mechanics Statics/Dynamics.

Students register and login here.

Professors register and login here.

How to Get Started

For Instructors: Obtain a PHGradeAssist Access code. This will have either come bundled with the text, or is available through either your local Prentice Hall sales rep, or by emailing **engineering@prenhall.com**.

Go to **http://www.prenhall.com/bedford**, and choose the *Engineering Mechanics* Series. In part one of the next screen, you will see the **instructor registration/login** option. Choose this option and follow instructions to register for the site, create your own unique username and password, and your own unique course identifier.

Follow instructions to create assignments for your course. You can create them one at a time, or even all at once. You have complete access to the site any time you choose to create or change your assignments and view your students' results. Complete help is available both in the system and to download at **http://www.prenhall.com/bedford**. You may also contact the Prentice Hall help line at **1-800-677-6337**.

Pass out your *Course ID* to your students when you give them assignments. Students use this identifier to find your assignments at the site. These results flow directly into your gradebook letting you manage your homework on-line.

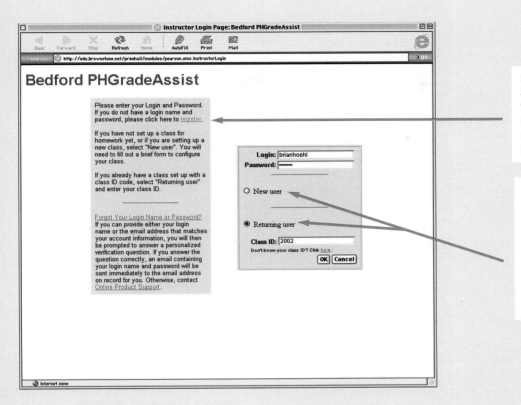

Initial Registration/Login Screen. Click register the first time you visit to redeem you access code and establish a username and a password.

If you are setting up a class for the first time, select new user after you have registered first and enter in your username and password. Otherwise select returning user and provide your class ID.

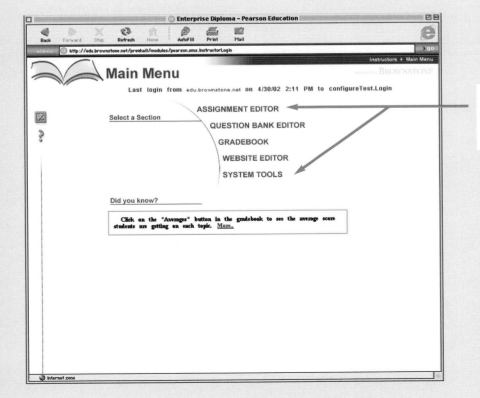

Create Assignments and manage your course on-line. Full help available at the site.

For Students: Keep the Access card that comes with new copies of the text—you need it to register at **http://www.prenhall.com/bedford** to do your homework. If you did not buy a new copy, you can purchase cards separately through your bookstore using ISBN **0-13-046043-5**.

Go to **http://www.prenhall.com**. Choose the *Engineering Mechanics* Series.

Choose **Student Login/Register** in Part One of the site. (*Note:* A student help file is located here to download.) Follow the instructions to process your access code and establish a username and password. Support is available by calling the Prentice Hall helpline at 1-800-677-6337.

Your instructor will have created assignments for you to work and submit answers to on-line. To access your instructor's assignments, you need to obtain your instructor's *Course ID* in class, or use the search function.

Once you have located your course, simply work the assignments as prescribed by your instructor. When you select an individual assignment, the PHGradeAssist system generates your own unique set of homework questions. If you choose, you can print these out, work your homework off line, and then come back on-line to answer these questions. The system will remember the assignment it created for you.

Submit your answers on-line for grading. Be careful to use correct notation and units. It is a good idea to consult the student help the first time you use the system. Press the grade button for results. Your instructor may give you the option to work incorrect questions again so follow the instructions they provided in class.

Go to www.prenhall.com/bedford Choose Engineering Mechanics Statics/Dynamics. Choose Student Login/Register. If this is your first time at the site, click register to redeem your access code and establish your username and password.

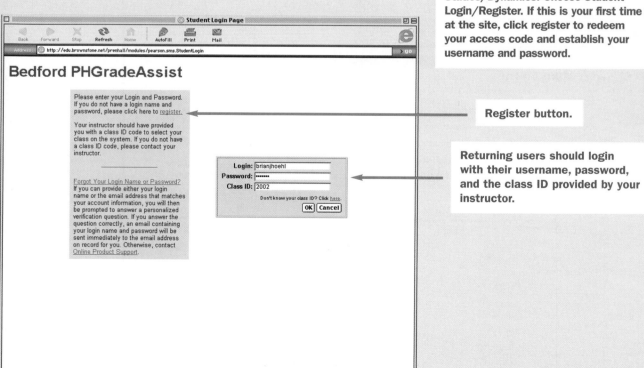

Register button.

Returning users should login with their username, password, and the class ID provided by your instructor.

Preface

Our original objective in writing this book was to present the foundations and applications of statics as we do in the classroom. We used many sequences of figures, emulating the gradual development of a figure by a teacher explaining a concept. We stressed the importance of visual analysis in gaining understanding, especially through the use of free-body diagrams. Because inspiration is so conducive to learning, we based many of our examples and problems on a variety of modern engineering applications. With encouragement and help from many students and fellow teachers who have used the book, we continue and expand upon these themes in this edition.

Examples that Teach

The Strategy/Solution/Discussion framework employed by most of our examples is designed to emphasize the critical importance of good problem-solving skills. Our objective is to teach students how to approach problems and critically judge the results.

"Strategy" sections show the preliminary planning needed to begin a solution. What principles and equations apply? What must be determined, and in what order?

The solution is then described in detail, using sequences of figures when needed to clarify the steps.

"Discussion" sections point out properties of the solution, or comment on alternative solution methods, or suggest out ways to check answers.

Example 9.3

Analyzing a Friction Brake

The motion of the disk in Fig. 9.11 is controlled by the friction force exerted at C by the brake ABC. The hydraulic actuator BE exerts a horizontal force of magnitude F on the brake at B. The coefficients of friction between the disk and the brake are μ_s and μ_k. What couple M is necessary to rotate the disk at a constant rate in the counterclockwise direction?

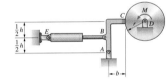

Figure 9.11

Strategy

We can use the free-body diagram of the disk to obtain a relation between M and the reaction exerted on the disk by the brake, then use the free-body diagram of the brake to determine the reaction in terms of F.

Solution

We draw the free-body diagram of the disk in Fig. a, representing the force exerted by the brake by a single force R. The force R opposes the counter-clockwise rotation of the disk, and the friction angle is the angle of kinetic friction $\theta_k = \arctan \mu_k$. Summing moments about D, we obtain

$$\Sigma M_{(\text{point } D)} = M - (R\sin\theta_k)r = 0.$$

Then, from the free-body diagram of the brake (Fig. b), we obtain

$$\Sigma M_{(\text{point } A)} = -F\left(\frac{1}{2}h\right) + (R\cos\theta_k)h - (R\sin\theta_k)b = 0.$$

We can solve these two equations for M and R. The solution for the couple M is

$$M = \frac{(1/2)hr\,F\sin\theta_k}{h\cos\theta_k - b\sin\theta_k} = \frac{(1/2)hr\,F\mu_k}{h - b\mu_k}.$$

(a) The free-body diagram of the disk.

(b) The free-body diagram of the brake.

Discussion

If μ_k is sufficiently small, then the denominator of the solution for the couple, $(h\cos\theta_k - b\sin\theta_k)$, is positive. As μ_k becomes larger, the denominator becomes smaller, because $\cos\theta_k$ decreases and $\sin\theta_k$ increases. As the denominator approaches zero, the couple required to rotate the disk approaches infinity. To understand this result, notice that the denominator equals zero when $\tan\theta_k = h/b$, which means that the line of action of R passes through point A (Fig. c). As μ_k becomes larger and the line of action of R approaches point A, the magnitude of R necessary to balance the moment of F about A approaches infinity and, as a result, M approaches infinity.

(c) The line of action of R passing through point A.

Engineering Design

We include simple design considerations in many examples and problems without compromising emphasis on fundamental mechanics. Design problems are marked with a $\mathscr{D}$ Icon. Optional examples titled "Application to Engineering" provide more detailed discussions of the uses of statics in engineering design:

Example 4.9

Application to Engineering:

Rotating Machines

The crewman in Fig. 4.25 exerts the forces shown on the handles of the coffee grinder winch, where $\mathbf{F} = 4\mathbf{j} + 32\mathbf{k}$ N. Determine the total moment he exerts (a) about point O, (b) about the axis of the winch, which coincides with the x axis.

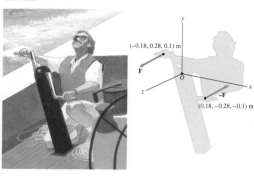

$(-0.18, 0.28, 0.1)$ m

$(0.18, -0.28, -0.1)$ m

Figure 4.25

> A specific engineering application is first described and analyzed.

Strategy

(a) To obtain the total moment about point O, we must sum the moments of the two forces about O. Let the sum be denoted by $\Sigma\mathbf{M}_O$. (b) Because point O is on the x axis, the total moment about the x axis is the component of $\Sigma\mathbf{M}_O$ parallel to the x axis, which is the x component of $\Sigma\mathbf{M}_O$.

$\mathscr{D}$esign Issues

The winch in this example is a simple representative of a class of rotating machines that includes hydrodynamic and aerodynamic power turbines, propellers, jet engines, and electric motors and generators. The ancestors of hydrodynamic and aerodynamic power turbines—water wheels and windmills—were among the earliest machines. These devices illustrate the importance of the concept of the moment of a force about a line. Their common feature is a part designed to rotate and perform some function when it is subjected to a moment about its axis of rotation. In the case of the winch, the forces exerted on the handles by the crewman exert a moment about the axis of rotation, causing the winch to rotate and wind a rope onto a drum, trimming the boat's sails. A hydrodynamic power turbine (Fig. 4.26) has turbine blades that are subjected to forces by flowing water, exerting a moment about the axis of rotation. This moment rotates the shaft to which the blades are attached, turning an electric generator that is connected to the same shaft.

> A "Design Issues" section then discusses design implications of the application and places it in a broader engineering context.

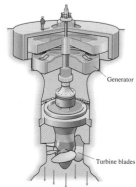

Generator

Turbine blades

Figure 4.26
A hydroelectric turbine. Water flowing through the turbine blades exerts a moment about the axis of the shaft, turning the generator.

Computational Mechanics

Some instructors prefer to teach statics without requiring the use of a computer. Others use statics as an opportunity to introduce students to the use of computers in engineering, having them either write their own programs in a lower level language or use higher level problem-solving software. Our book is suitable for each of these approaches. We provide optional, self-contained "Computational Mechanics" sections with examples and problems designed for solution by a programmable calculator or computer. In addition, tutorials on using Mathcad® and MATLAB® in engineering mechanics are available from our texts website. See supplements (p. xxi) for a further description.

Computational Example 9.11

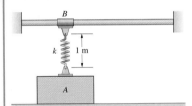

Figure 9.33

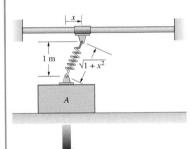

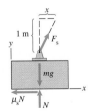

(a) Moving the slider to the right a distance x.

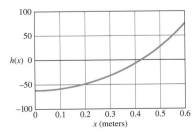

(b) Free-body diagram of the block when slip is impending.

The mass of the block A in Fig. 9.33 is 20 kg, and the coefficient of static friction between the block and the floor is $\mu_s = 0.3$. The spring constant $k = 1$ kN/m, and the spring is unstretched. How far can the slider B be moved to the right without causing the block to slip?

Solution

Suppose that moving the slider B a distance x to the right causes impending slip of the block (Fig. a). The resulting stretch of the spring is $\sqrt{1 + x^2} - 1$ m, so the magnitude of the force exerted on the block by the spring is

$$F_s = k(\sqrt{1 + x^2} - 1). \qquad (9.23)$$

From the free-body diagram of the block (Fig. b), we obtain the equilibrium equations

$$\Sigma F_x = \left(\frac{x}{\sqrt{1 + x^2}}\right)F_s - \mu_s N = 0,$$

$$\Sigma F_y = \left(\frac{1}{\sqrt{1 + x^2}}\right)F_s + N - mg = 0.$$

Substituting Eq. (9.23) into these two equations and then eliminating N, we can write the resulting equation in the form

$$h(x) = k(x + \mu_s)(\sqrt{1 + x^2} - 1) - \mu_s mg\sqrt{1 + x^2} = 0.$$

We must obtain the root of this function to determine the value of x corresponding to impending slip of the block. From the graph of $h(x)$ in Fig. 9.34, we estimate that $h(x) = 0$ at $x = 0.43$ m. By examining computed results near this value of x, we see that $h(x) = 0$, and slip is impending, when x is approximately 0.4284 m.

x (m)	$h(x)$
0.4281	−0.1128
0.4282	−0.0777
0.4283	−0.0425
0.4284	−0.0074
0.4285	0.0278
0.4286	0.0629
0.4287	0.0981

Figure 9.34
Graph of the function $h(x)$.

Consistent Use of Color

To help students recognize and interpret elements of figures, we use consistent indentifying colors:

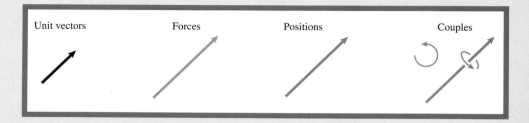

New to the Third Edition

Positive responses from users and reviewers have led us to retain the basic organization, and features of the first edition. During our preparation of this edition, we examined how we presented each concept, example, figure, summary statement, and problem. Where necessary, we made changes, additions, or deletions to simplify and clarify the presentation. In response to requests, we made the following notable changes:

- We have added new examples where users indicated more were needed. Many of the new examples continue our emphasis on realistic and motivational applications and engineering design.
- New sets of **Study Questions** appear after most sections to help students check their retention of key concepts.
- Each example is clearly labeled for its teaching purpose.
- We have redesigned the text and also added photographs throughout to help students connect the text to real world applications and situations.
- An extensive new supplement program includes web-based assessment software, visualization software, and much more. See the Supplements description for complete information.

Commitment to Students and Instructors

In revising the textbook and solutions manual, we have taken precautions to ensure accuracy to the best of our ability. We have each solved the new problems in an effort to be sure that their answers are correct and that they are of an appropriate level of difficulty. Karim Nohra of the University of South Florida also checked the text, examples, problems and solutions manual. Any errors that remain are the responsibility of the authors. We welcome communication from students and instructors concerning errors or areas for improvement. Our mailing address is Department of Aerospace Engineering and Engineering Mechanics, University of Texas at Austin, Austin, Texas 78712. Our electronic mail address is *abedford@mail.utexas.edu*.

Supplements

Student Supplements

PHGradeAssist lets students solve problems from the text with randomized variables so each student solves a slightly different problem. After students have submitted their answers, they receive the actual answers and can keep trying similar problems until they are successful. By integrating with an optional course management system, professors can have student results recorded electronically. Consult the About this Package section of this book, visit www.prenhall.com/bedford, or contact your PH sales representative for more information. Student Access Code Cards come bundled with each new copy of this text, and are also available by ordering with ISBN 0-13-0460435. Professor access codes are available through your PH sales rep or by emailing engineering@prenhall.com.

Statics and Dynamics Study Pack this optional supplement is designed to give students the tools to improve their study skills. It consists of three study components—a free body-diagram workbook, a Visualization CD based on Working Model Software, and an access code to a website with 500 sample Statics and Dynamics problems and solutions.

- **Free-Body Diagram Workbook** prepared by Peter Schiavone of the University of Alberta. This workbook begins with a tutorial on free body diagrams and then includes 50 practice problems of progressing difficulty with complete solutions. Further "strategies and tips" help students understand how to use the diagrams in solving the accompanying problems.

- **Working Model CD** contains 25 pre-set simulations of Statics examples in the text that include questions for further exploration. Simulations are powered by the Working Model Engine and were created with actual artwork from the text to enhance their correlation with the text.

- **Password-Protected Website** contains 500 sample Statics and Dynamics problems for students to study. Problems are keyed to each chapter of the text and contain complete solutions. All problems are supplemental and do not appear in the Third Edition. Student access codes are printed on the inside cover of the Free-Body Diagram Workbook. To access this site, students should go to http://www.prenhall.com/bedford and follow the on-line directions to register. Access Codes found in optional Statics Study Packs also provide access to PHGradeAssist.

Order stand-alone Study Packs with the ISBN 0-13-032473-6.

MATLAB® /Mathcad® Tutorials Twenty tutorials showing how to use computational software in engineering mechanics. Each tutorial discusses a basic mechanics concept, and then shows how to solve a specific problem related to this concept using MATLAB/Mathcad. There are twenty tutorials each for MATLAB and Mathcad, and are available in PDF format from the access codes protected area of the Bedford website. Worksheets were developed by Ronald Larsen and Stephen Hunt of Montana State University—Bozeman.

Website—http://www.prenhall.com/bedford contains multiple-choice and True/False quizzes keyed to each chapter in the book developed by Karim Nohra of the University of South Florida. PHGradeAssist, MATLAB/Mathcad tutorials, and supplemental questions and solutions are all available at the access code protected part of this website. Student access codes for protected portions come packaged for free with new books, or are available by purchasing the Statics Study Pack or the stand-alone student access card ISBN 0-13-046043-5. Professors access codes are available through your sales rep or by emailing engineering@prenhall.com.

ESource ACCESS Students may obtain a password to access to Prentice Hall's ESource, a more than 5000 page on-line database of Introductory Engineering titles. Topics in the database include mathematics review, MATLAB, Mathcad, Excel, programming languages, engineering design, and many more. This database is fully searchable and available 24 hours a day from the web. To learn more, visit *http://www.prenhall.com/esource*. Contact either your sales rep or *engineering@prenhall.com* for pricing and bundling options.

ADAMS Simulations for Statics and Dynamics—Mechanical Statics and Dynamics, Inc. has created over 100 simulations of problems from *Statics and Dynamics* using their ADAMS simulation/prototyping software. Professors and students can simulate and observe the effects of changing parameters in systems and gain deeper insight into their behavior. Simulations also come with an accompanying avi "movie" file. Files are located at the access code protected part of the website. Qualified adopters may also be able to obtain free site licenses. Contact *university@adams.com* for more information.

Instructor Supplements

Instructor's Solutions Manual and Presentation CD This supplement available to instructors contains problem statements with completely worked out solutions. The problem statements correspond by chapter and number to those at the PHGradeAssist site. Professors can thus see one sample solution to each PHGA problem. The separate CD contains PowerPoint slides of art from examples and text passages, as well as pdf files of art from the book.

PHGradeAssist lets students solve problems from the text with randomized variables so each student solves a slightly different problem. After students have submitted their answers, they receive the actual answers and can keep trying similar problems until they are successful. By integrating with an optional course management system, professors can have student results recorded electronically. Consult the About this Package section of this book, visit www.prenhall.com/bedford, or contact your PH sales representative for more information. Student Access Code Cards come bundled with each new copy of this text, and are also available by ordering with isbn#. Professor access codes are available through your PH sales rep or by emailing engineering@prenhall.com.

Acknowledgments

Many students and teachers have given us insightful comments on the earlier editions. The following academic colleagues critically reviewed the book and made valuable suggestions:

Edward E. Adams
Michigan Technological University

Raid S. Al-Akkad
University of Dayton

Jerry L. Anderson
Memphis State University

James G. Andrews
University of Iowa

Robert J. Asaro
University of California, San Diego

Leonard B. Baldwin
University of Wyoming

Gautam Batra
University of Nebraska

Mary Bergs
Marquette University

Spencer Brinkerhoff
Northern Arizona University

L.M. Brock
University of Kentucky

William (Randy) Burkett
Texas Tech University

Donald Carlson
University of Illinois

Major Robert M. Carpenter
U.S. Military Academy

Douglas Carroll
University of Missouri, Rolla

Paul C. Chan
New Jersey Institute of Technology

Namas Chandra
Florida State University

James Cheney
University of California, Davis

Ravinder Chona
Texas A & M University

Anthony DeLuzio
Merrimack College

Mitsunori Denda
Rutgers University

James F. Devine
University of South Florida

Craig Douglas
University of Massachusetts, Lowell

Marijan Dravinski
University of Southern California

S. Olani Durrant
Brigham Young University

Estelle Eke
California State University, Sacramento

William Ferrante
University of Rhode Island

Robert W. Fitzgerald
Worcester Polytechnic Institute

George T. Flowers
Auburn University

Mark Frisina
Wentworth Institute

Robert W. Fuessle
Bradley University

William Gurley
University of Tennessee, Chattanooga

John Hansberry
University of Massachusetts, Dartmouth

W. C. Hauser
California Polytechnic University Pomona

Linda Hayes
University of Texas - Austin

R. Craig Henderson
Tennessee Technological University

James Hill
University of Alabama

Allen Hoffman
Worcester Polytechnic Institute

Edward E. Hornsey
University of Missouri, Rolla

Robert A. Howland
University of Notre Dame

Joe Ianelli
University of Tennessee, Knoxville

Ali Iranmanesh
Gadsden State Community College

David B. Johnson
Southern Methodist University

E. O. Jones, Jr.
Auburn University

Serope Kalpakjian
Illinois Institute of Technology

Kathleen A. Keil
California Polytechnic University San Luis Obispo

Yohannes Ketema
University of Minnesota

Seyyed M. H. Khandani
Diablo Valley College

Charles M. Krousgrill
Purdue University

B. Kent Lall
Portland State University

Kenneth W. Lau
University of Massachusetts, Lowell

Norman Laws
University of Pittsburgh

William M. Lee
U.S. Naval Academy

Donald G. Lemke
University of Illinois, Chicago

Richard J. Leuba
North Carolina State University

Richard Lewis
Louisiana Technological University

Bertram Long
Northeastern University

V. J. Lopardo
U.S. Naval Academy

Frank K. Lu
University of Texas, Arlington

K. Madhaven
Christian Brothers College

Gary H. McDonald
University of Tennessee

James McDonald
Texas Technical University

Jim Meagher
California Polytechnic State University, San Luis Obispo

Lee Minardi
Tufts University

Norman Munroe
Florida International University

Shanti Nair
University of Massachusetts, Amherst

Saeed Niku
California Polytechnic State University, San Luis Obispo

Harinder Singh Oberoi
Western Washington University

James O'Connor
University of Texas, Austin

Samuel P. Owusu-Ofori
North Carolina A& T State University

Venkata Panchakarla
Florida State University

Assimina A Pelegri
Rutgers University

Noel C. Perkins
University of Michigan

David J. Purdy
Rose-Hulman Institute of Technology

Colin E Ratcliffe
U.S. Naval Academy

Daniel Riahi
University of Illinois

Charles Ritz
California Polytechnic State University Pomona

George Rosborough
University of Colorado, Boulder

Robert Schmidt
University of Detroit

Robert J. Schultz
Oregon State University

Patricia M. Shamamy
Lawrence Technological University

Sorin Siegler
Drexel University

L. N. Tao
Illinois Institute of Technology

Craig Thompson
Western Wyoming Community College

John Tomko
Cleveland State University

Kevin Z. Truman
Washington University

John Valasek
Texas A &M University

Dennis VandenBrink
Western Michigan University

Thomas J. Vasko
University of Hartford

Mark R. Virkler
University of Missouri, Columbia

William H. Walston, Jr.
University of Maryland

Reynolds Watkins
Utah State University

Charles White
Northeastern University

Norman Wittels
Worcester Polytechnic Institute

Julius P. Wong
University of Louisville

T. W. Wu
University of Kentucky

Constance Ziemian
Bucknell University

Books of this kind represent a collaborative effort by many individuals. It has been our good fortune to work with extremely talented and agreeable people at Prentice Hall. In large measure we owe the improvements in this edition to our editor and mentor Eric Svendsen. We have not met a more creative, hard-working and conscientious person in publishing. Eric's efforts and ours were overseen, inspired, and supported by Marcia Horton. Irwin Zucker and Vince O'Brien helped us through the day-to-day crises of the production process. David George was our Managing Editor. Jon Boylan and John Christiana were responsible for the book's design, and Xiaohong Zhu coordinated the art program. Karim Nohra of the University of South Florida carefully examined the text, examples and problems for accuracy. Among the many other people on whom we relied, we mention particularly Dee Bernhard, Kristen Blanco, Lisa McDowell, Trudy Pisciotti, Joe Russo, and Holly Stark.

Increasingly, a textbook is only part of an integrated set of pedagogical tools. Our website was created by Ryan Greene and Edward Cadillo. Peter Schiavone has written a free-body diagram workbook that supplements and expands upon the book's treatment. Ronald Larsen and Steven Hunt have developed MATLAB and Mathcad worksheets based on problems and examples in the book for optional use by instructors and students.

And we thank our wives, Nancy and Marsha, for their continued support, patience, and acceptance of years of lost weekends.

Anthony Bedford and Wallace Fowler
Austin, Texas

About the Authors

Anthony Bedford is Professor of Aerospace Engineering and Engineering Mechanics at the University of Texas at Austin. He received his B.S. degree at the University of Texas at Austin, his M.S. degree at the California Institute of Technology, and his Ph.D. degree at Rice University in 1967. He has industrial experience at Douglas Aircraft Company and at TRW, where he did structural dynamics and trajectory analyses for the Apollo program. He has been on the faculty of the University of Texas at Austin since 1968.

Dr. Bedford's main professional activity has been education and research in engineering mechanics. He has been principal investigator on grants from the National Science Foundation and the Office of Naval Research, and from 1973 until 1983 was a consultant to Sandia National Laboratories, Albuquerque, New Mexico. His other books include *Hamilton's Principle in Continuum Mechanics, Introduction to Elastic Wave Propagation* (with D.S. Drumheller), and *Mechanics of Materials* (with K.M. Liechti).

Wallace T. Fowler holds the Paul D. and Betty Robertson Meek Professorship in Engineering in the Department of Aerospace Engineering and Engineering Mechanics at the University of Texas at Austin. Dr. Fowler received his B.A., M.S., and Ph.D. degrees at the University of Texas at Austin, and has been on the faculty there since 1965. During Fall 1976, he was on the staff of the United States Air Force Test Pilot School, Edwards Air Force Base, California, and in 1981–1982 he was a visiting professor at the United States Air Force Academy. Since 1991 he has been Associate Director of the Texas Space Grant Consortium.

Dr. Fowler's areas of teaching and research are dynamics, orbital mechanics, and spacecraft mission design. He is author or coauthor of technical papers on trajectory optimization, attitude dynamics, and space mission planning and has also published papers on the theory and practice of engineering teaching. He has received numerous teaching awards including the Chancellor's Council Outstanding Teaching Award, the General Dynamics Teaching Excellence Award, the Halliburton Education Foundation Award of Excellence, the ASEE Fred Merryfield Design Award, and the AIAA-ASEE Distinguished Aerospace Educator Award. He is a member of the Academy of Distinguished Teachers at the University of Texas at Austin. He is a licensed professional engineer, a member of several technical societies, and a Fellow of both the American Institute of Aeronautics and Astronautics and the American Society for Engineering Education. In 2000–2001, he served as president of the American Society for Engineering Education.

Photo Credits

ENGINEERING MECHANICS

STATICS
PRINCIPLES

The architects and engineers are guided by the principles of statics during each step of the design and construction of a building. Statics is one of the sciences underlying the art of structural design.

Introduction

Engineers are responsible for the design, construction, and testing of the devices we use, from simple things such as chairs and pencil sharpeners to complicated ones such as dams, cars, airplanes, and spacecraft. They must have a deep understanding of the physics underlying these devices and must be familiar with the use of mathematical models to predict system behavior. Students of engineering begin to learn how to analyze and predict the behavior of physical systems by studying mechanics.

1.1 Engineering and Mechanics

How do engineers design complex systems and predict their characteristics before they are constructed? Engineers have always relied on their knowledge of previous designs, experiments, ingenuity, and creativity to develop new designs. Modern engineers add a powerful technique: They develop mathematical equations based on the physical characteristics of the devices they design. With these mathematical models, engineers predict the behavior of their designs, modify them, and test them prior to their actual construction. Aerospace engineers use mathematical models to predict the paths the space shuttle will follow in flight. Civil engineers use mathematical models to analyze the effects of loads on buildings and foundations.

At its most basic level, mechanics is the study of forces and their effects. Elementary mechanics is divided into *statics*, the study of objects in equilibrium, and *dynamics*, the study of objects in motion. The results obtained in elementary mechanics apply directly to many fields of engineering. Mechanical and civil engineers who design structures use the equilibrium equations derived in statics. Civil engineers who analyze the responses of buildings to earthquakes and aerospace engineers who determine the trajectories of satellites use the equations of motion derived in dynamics.

Mechanics was the first analytical science; consequently fundamental concepts, analytical methods, and analogies from mechanics are found in virtually every field of engineering. Students of chemical and electrical engineering gain a deeper appreciation for basic concepts in their fields such as equilibrium, energy, and stability by learning them in their original mechanical contexts. By studying mechanics, they retrace the historical development of these ideas.

1.2 Learning Mechanics

Mechanics consists of broad principles that govern the behavior of objects. In this book we describe these principles and provide examples that demonstrate some of their applications. Although it is essential that you practice working problems similar to these examples, and we include many problems of this kind, our objective is to help you understand the principles well enough to apply them to situations that are new to you. Each generation of engineers confronts new problems.

Problem Solving

In the study of mechanics you learn problem-solving procedures you will use in succeeding courses and throughout your career. Although different types of problems require different approaches, the following steps apply to many of them:

- Identify the information that is given and the information, or answer, you must determine. It's often helpful to restate the problem in your own words. When appropriate, make sure you understand the physical system or model involved.
- Develop a *strategy* for the problem. This means identifying the principles and equations that apply and deciding how you will use them to solve the

problem. Whenever possible, draw diagrams to help visualize and solve the problem.

- Whenever you can, try to predict the answer. This will develop your intuition and will often help you recognize an incorrect answer.

- Solve the equations and, whenever possible, interpret your results and compare them with your prediction. This last step is a *reality check*. Is your answer reasonable?

Calculators and Computers

Most of the problems in this book are designed to lead to an algebraic expression with which to calculate the answer in terms of given quantities. A calculator with trigonometric and logarithmic functions is sufficient to determine the numerical value of such answers. The use of a programmable calculator or a computer with problem-solving software such as *Mathcad* or MATLAB is convenient, but be careful not to become too reliant on tools you will not have during tests.

Sections headed "Computational Mechanics" contain examples and problems that are suitable for solution with a programmable calculator or a computer.

Engineering Applications

Although the problems are designed primarily to help you learn mechanics, many of them illustrate uses of mechanics in engineering. Sections headed "Application to Engineering" describe how mechanics is applied in various fields of engineering.

We also include problems that emphasize two essential aspects of engineering:

- *Design.* Some problems ask you to choose values of parameters to satisfy stated design criteria.

- *Safety.* Some problems ask you to evaluate the safety of devices and choose values of parameters to satisfy stated safety requirements.

Subsequent Use of This Text

This book contains tables and information you will find useful in subsequent engineering courses and throughout your engineering career. In addition, you will often want to review fundamental engineering subjects, both during the remainder of your formal education and when you are a practicing engineer. The most efficient way to do so is by using the textbooks with which you are familiar. Your engineering textbooks will form the core of your professional library.

1.3 Fundamental Concepts

Some topics in mechanics will be familiar to you from everyday experience or from previous exposure to them in mathematics and physics courses. In this section we briefly review the foundations of elementary mechanics.

Numbers

Engineering measurements, calculations, and results are expressed in numbers. You need to know how we express numbers in the examples and problems and how to express the results of your own calculations.

Significant Digits   This term refers to the number of meaningful (that is, accurate) digits in a number, counting to the right starting with the first nonzero digit. The two numbers 7.630 and 0.007630 are each stated to four significant digits. If only the first four digits in the number 7,630,000 are known to be accurate, this can be indicated by writing the number in scientific notation as 7.630×10^6.

If a number is the result of a measurement, the significant digits it contains are limited by the accuracy of the measurement. If the result of a measurement is stated to be 2.43, this means that the actual value is believed to be closer to 2.43 than to 2.42 or 2.44.

Numbers may be rounded off to a certain number of significant digits. For example, we can express the value of π to three significant digits, 3.14, or we can express it to six significant digits, 3.14159. When you use a calculator or computer, the number of significant digits is limited by the number of digits the machine is designed to carry.

Use of Numbers in This Book   You should treat numbers given in problems as exact values and not be concerned about how many significant digits they contain. If a problem states that a quantity equals 32.2, you can assume its value is 32.200. ... We express intermediate results and answers in the examples and the answers to the problems to at least three significant digits. If you use a calculator, your results should be that accurate. Be sure to avoid round-off errors that occur if you round off intermediate results when making a series of calculations. Instead, carry through your calculations with as much accuracy as you can by retaining values in your calculator.

Space and Time

Space simply refers to the three-dimensional universe in which we live. Our daily experiences give us an intuitive notion of space and the locations, or positions, of points in space. The distance between two points in space is the length of the straight line joining them.

Measuring the distance between points in space requires a unit of length. We use both the International System of units, or SI units, and U.S. Customary units. In SI units, the unit of length is the meter (m). In U.S. Customary units, the unit of length is the foot (ft).

Time is, of course, familiar—our lives are measured by it. The daily cycles of light and darkness and the hours, minutes, and seconds measured by our clocks and watches give us an intuitive notion of time. Time is measured by the intervals between repeatable events, such as the swings of a clock pendulum or the vibrations of a quartz crystal in a watch. In both SI units and U.S. Customary units, the unit of time is the second (s). The minute (min), hour (hr), and day are also frequently used.

If the position of a point in space relative to some reference point changes with time, the rate of change of its position is called its *velocity*, and the rate of change of its velocity is called its *acceleration*. In SI units, the velocity is expressed in meters per second (m/s) and the acceleration is expressed in meters per second per second, or meters per second squared (m/s^2). In U.S. Customary units, the velocity is expressed in feet per second (ft/s) and the acceleration is expressed in feet per second squared (ft/s^2).

Newton's Laws

Elementary mechanics was established on a firm basis with the publication in 1687 of *Philosophiae naturalis principia mathematica*, by Isaac Newton. Although highly original, it built on fundamental concepts developed by many others during a long and difficult struggle toward understanding (Fig. 1.1).

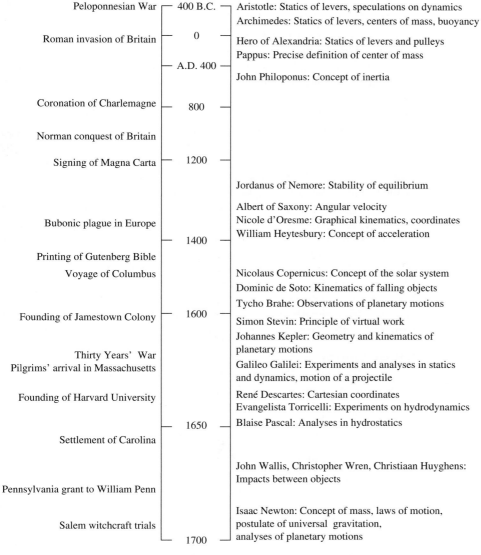

Figure 1.1
Chronology of developments in mechanics up to the publication of Newton's *Principia* in relation to other events in history.

Newton stated three "laws" of motion, which we express in modern terms:

1. *When the sum of the forces acting on a particle is zero, its velocity is constant. In particular, if the particle is initially stationary, it will remain stationary.*

2. *When the sum of the forces acting on a particle is not zero, the sum of the forces is equal to the rate of change of the linear momentum of the particle. If the mass is constant, the sum of the forces is equal to the product of the mass of the particle and its acceleration.*

3. *The forces exerted by two particles on each other are equal in magnitude and opposite in direction.*

Notice that we did not define force and mass before stating Newton's laws. The modern view is that these terms are defined by the second law. To demonstrate, suppose that we choose an arbitrary object and define it to have unit mass. Then we define a unit of force to be the force that gives our unit mass an acceleration of unit magnitude. In principle, we can then determine the mass of any object: We apply a unit force to it, measure the resulting acceleration, and use the second law to determine the mass. We can also determine the magnitude of any force: We apply it to our unit mass, measure the resulting acceleration, and use the second law to determine the force.

Thus Newton's second law gives precise meanings to the terms *mass* and *force*. In SI units, the unit of mass is the kilogram (kg). The unit of force is the newton (N), which is the force required to give a mass of one kilogram an acceleration of one meter per second squared. In U.S. Customary units, the unit of force is the pound (lb). The unit of mass is the slug, which is the amount of mass accelerated at one foot per second squared by a force of one pound.

Although the results we discuss in this book are applicable to many of the problems met in engineering practice, there are limits to the validity of Newton's laws. For example, they don't give accurate results if a problem involves velocities that are not small compared to the velocity of light $(3 \times 10^8$ m/s). Einstein's special theory of relativity applies to such problems. Elementary mechanics also fails in problems involving dimensions that are not large compared to atomic dimensions. Quantum mechanics must be used to describe phenomena on the atomic scale.

Study Questions

1. What is the definition of the significant digits of a number?
2. What are the units of length, mass, and force in the SI system?

1.4 Units

The SI system of units has become nearly standard throughout the world. In the United States, U.S. Customary units are also used. In this section we summarize these two systems of units and explain how to convert units from one system to another.

International System of Units

In SI units, length is measured in meters (m) and mass in kilograms (kg). Time is measured in seconds (s), although other familiar measures such as minutes (min), hours (hr), and days are also used when convenient. Meters, kilograms, and seconds are called the *base units* of the SI system. Force is measured in newtons (N). Recall that these units are related by Newton's second law: One newton is the force required to give an object of one kilogram mass an acceleration of one meter per second squared:

$$1 \text{ N} = (1 \text{ kg})(1 \text{ m/s}^2) = 1 \text{ kg-m/s}^2.$$

Because the newton can be expressed in terms of the base units, it is called a *derived unit.*

To express quantities by numbers of convenient size, multiples of units are indicated by prefixes. The most common prefixes, their abbreviations, and the multiples they represent are shown in Table 1.1. For example, 1 km is 1 kilometer, which is 1000 m, and 1 Mg is 1 megagram, which is 10^6 g, or 1000 kg. We frequently use kilonewtons (kN).

Table 1.1 The common prefixes used in SI units and the multiples they represent.

Prefix	Abbreviation	Multiple
nano-	n	10^{-9}
micro-	μ	10^{-6}
milli-	m	10^{-3}
kilo-	k	10^{3}
mega-	M	10^{6}
giga-	G	10^{9}

U.S. Customary Units

In U.S. Customary units, length is measured in feet (ft) and force is measured in pounds (lb). Time is measured in seconds (s). These are the base units of the U.S. Customary system. In this system of units, mass is a derived unit. The unit of mass is the slug, which is the mass of material accelerated at one foot per second squared by a force of one pound. Newton's second law states that

$$1 \text{ lb} = (1 \text{ slug})(1 \text{ ft/s}^2).$$

From this expression we obtain

$$1 \text{ slug} = 1 \text{ lb-s}^2/\text{ft}.$$

We use other U.S. Customary units such as the mile (1 mi = 5280 ft) and the inch (1 ft = 12 in.). We also use the kilopound (kip), which is 1000 lb.

Angular Units

In both SI and U.S. Customary units, angles are normally expressed in radians (rad). We show the value of an angle θ in radians in Fig. 1.2. It is defined to be the ratio of the part of the circumference subtended by θ to the radius of the circle. Angles are also expressed in degrees. Since there are 360 degrees

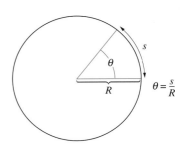

Figure 1.2
Definition of an angle in radians.

(360°) in a complete circle, and the complete circumference of the circle is $2\pi R$, 360° equals 2π rad.

Equations containing angles are nearly always derived under the assumption that angles are expressed in radians. Therefore when you want to substitute the value of an angle expressed in degrees into an equation, you should first convert it into radians. A notable exception to this rule is that many calculators are designed to accept angles expressed in either degrees or radians when you use them to evaluate functions such as $\sin\theta$.

Conversion of Units

Many situations arise in engineering practice that require you to convert values expressed in units of one kind into values in other units. If some data in a problem are given in terms of SI units and some are given in terms of U.S. Customary units, you must express all of the data in terms of one system of units. In problems expressed in terms of SI units, you will occasionally be given data in terms of units other than the base units of seconds, meters, kilograms, and newtons. You should convert these data into the base units before working the problem. Similarly, in problems involving U.S. Customary units, you should convert terms into the base units of seconds, feet, slugs, and pounds. After you gain some experience, you will recognize situations in which these rules can be relaxed, but for now the procedure we propose is the safest.

Converting units is straightforward, although you must do it with care. Suppose that we want to express 1 mi/hr in terms of ft/s. Since one mile equals 5280 ft and one hour equals 3600 seconds, we can treat the expressions

$$\left(\frac{5280\ \text{ft}}{1\ \text{mi}}\right) \quad \text{and} \quad \left(\frac{1\ \text{hr}}{3600\ \text{s}}\right)$$

as ratios whose values are 1. In this way we obtain

$$1\ \text{mi/hr} = 1\ \text{mi/hr} \times \left(\frac{5280\ \text{ft}}{1\ \text{mi}}\right) \times \left(\frac{1\ \text{hr}}{3600\ \text{s}}\right) = 1.47\ \text{ft/s}.$$

We give some useful unit conversions in Table 1.2.

Table 1.2 Unit conversions.

Time	1 minute	=	60 seconds
	1 hour	=	60 minutes
	1 day	=	24 hours
Length	1 foot	=	12 inches
	1 mile	=	5280 feet
	1 inch	=	25.4 millimeters
	1 foot	=	0.3048 meters
Angle	2π radians	=	360 degrees
Mass	1 slug	=	14.59 kilograms
Force	1 pound	=	4.448 newtons

Study Questions

1. What are the base units of the SI and U.S. Customary systems?
2. What is the definition of an angle in radians?

Example 1.1

Converting Units of Pressure

The pressure exerted at a point of the hull of the deep submersible in Fig. 1.3 is 3.00×10^6 Pa (pascals). A pascal is 1 newton per square meter. Determine the pressure in pounds per square foot.

Figure 1.3
Deep Submersible Vehicle.

Strategy

From Table 1.2, 1 pound = 4.448 newtons and 1 foot = 0.3048 meters. With these unit conversions we can calculate the pressure in pounds per square foot.

Solution

The pressure (to three significant digits) is

$$3.00 \times 10^6 \ \text{N/m}^2 = 3.00 \times 10^6 \ \text{N/m}^2 \times \left(\frac{1 \ \text{lb}}{4.448 \ \text{N}} \right) \times \left(\frac{0.3048 \ \text{m}}{1 \ \text{ft}} \right)^2$$

$$= 62{,}700 \ \text{lb/ft}^2.$$

Discussion

From the table of unit conversions in the inside front cover, 1 Pa = 0.0209 lb/ft^2. Therefore an alternative solution is

$$3.00 \times 10^6 \text{ N/m}^2 = 3.00 \times 10^6 \text{ N/m}^2 \times \left(\frac{0.0209 \text{ lb/ft}^2}{1 \text{ N/m}^2} \right)$$

$$= 62,700 \text{ lb/ft}^2.$$

Example 1.2

Determining Units from an Equation

Suppose that in Einstein's equation

$$E = mc^2,$$

the mass m is in kilograms and the velocity of light c is in meters per second.
(a) What are the SI units of E?
(b) If the value of E in SI units is 20, what is its value in U.S. Customary base units?

Strategy

(a) Since we know the units of the terms m and c, we can deduce the units of E from the given equation.
(b) We can use the unit conversions for mass and length from Table 1.2 to convert E from SI units to U.S. Customary units.

Solution

(a) From the equation for E,

$$E = (m \text{ kg})(c \text{ m/s})^2.$$

the SI units of E are kg-m^2/s^2.
(b) From Table 1.2, 1 slug = 14.59 kg and 1 ft = 0.3048 m. Therefore

$$1 \text{ kg-m}^2/\text{s}^2 = 1 \text{ kg-m}^2/\text{s}^2 \times \left(\frac{1 \text{ slug}}{14.59 \text{ kg}} \right) \times \left(\frac{1 \text{ ft}}{0.3048 \text{ m}} \right)^2$$

$$= 0.738 \text{ slug-ft}^2/\text{s}^2.$$

The value of E in U.S. Customary units is

$$E = (20)(0.738) = 14.8 \text{ slug-ft}^2/\text{s}^2.$$

Discussion

In part (a) we determined the units of E by using the fact that an equation must be dimensionally consistent. That is, the dimensions, or units, of each term must be the same.

1.5 Newtonian Gravitation

Newton postulated that the gravitational force between two particles of mass m_1 and m_2 that are separated by a distance r (Fig. 1.4) is

$$F = \frac{Gm_1m_2}{r^2},\qquad(1.1)$$

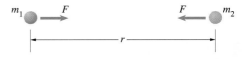

Figure 1.4
The gravitational forces between two particles are equal in magnitude and directed along the line between them.

where G is called the universal gravitational constant. Based on this postulate, he calculated the gravitational force between a particle of mass m_1 and a homogeneous sphere of mass m_2 and found that it is also given by Eq. (1.1), with r denoting the distance from the particle to the center of the sphere. Although the earth is not a homogeneous sphere, we can use this result to approximate the weight of an object of mass m due to the gravitational attraction of the earth.

$$W = \frac{Gmm_E}{r^2},\qquad(1.2)$$

where m_E is the mass of the earth and r is the distance from the center of the earth to the object. Notice that the weight of an object depends on its location relative to the center of the earth, whereas the mass of the object is a measure of the amount of matter it contains and doesn't depend on its position.

When an object's weight is the only force acting on it, the resulting acceleration is called the acceleration due to gravity. In this case, Newton's second law states that $W = ma$, and from Eq. (1.2) we see that the acceleration due to gravity is

$$a = \frac{Gm_E}{r^2}.\qquad(1.3)$$

The *acceleration due to gravity at sea level* is denoted by g. Denoting the radius of the earth by R_E, we see from Eq. (1.3) that $Gm_E = gR_E^2$. Substituting this result into Eq. (1.3), we obtain an expression for the acceleration due to gravity at a distance r from the center of the earth in terms of the acceleration due to gravity at sea level:

$$a = g\frac{R_E^2}{r^2}.\qquad(1.4)$$

Since the weight of the object $W = ma$, the weight of an object at a distance r from the center of the earth is

$$W = mg\frac{R_E^2}{r^2}.\qquad(1.5)$$

At sea level ($r = R_E$), the weight of an object is given in terms of its mass by the simple relation

$$W = mg.\qquad(1.6)$$

The value of g varies from location to location on the surface of the earth. The values we use in examples and problems are $g = 9.81$ m/s^2 in SI units and $g = 32.2$ ft/s^2 in U.S. Customary units.

Study Questions

1. Does the weight of an object depend on its location?
2. If you know an object's mass, how do you determine its weight at sea level?

Example 1.3

Determining an Object's Weight

In its final configuration, the International Space Station (Fig. 1.5) will have a mass of approximately 450,000 kg.

(a) What would be the weight of the ISS if it were at sea level?

(b) The orbit of the ISS is 354 km above the surface of the earth. The earth's radius is 6370 km. What is the weight of the ISS (the force exerted on it by gravity) when it is in orbit?

Figure 1.5
International Space Station.

Strategy

(a) The weight of an object at sea level is given by Eq. (1.6). Because the mass is given in kilograms, we will express g in SI units: $g = 9.81$ m/s^2.
(b) The weight of an object at a distance r from the center of the earth is given by Eq. (1.5).

Solution

(a) The weight at sea level is

$$W = mg$$
$$= (450,000)(9.81)$$
$$= 4.41 \times 10^6 \text{ N.}$$

(b) The weight in orbit is

$$W = mg\frac{R_\text{E}^2}{r^2}$$
$$= (450,000)(9.81)\frac{(6,370,000)^2}{(6,370,000 + 354,000)^2}$$
$$= 3.96 \times 10^6 \text{ N.}$$

Discussion

Notice that the force exerted on the ISS by gravity when it is in orbit is approximately 90% of its weight at sea level.

Vectors can specify the positions of points of a structure. Vectors are used to describe and analyze quantities that have magnitude and direction, including positions, forces, moments, velocities, and accelerations.

Vectors

To describe a force acting on a structural member, both the magnitude of the force and its direction must be specified. To describe the position of an airplane relative to an airport, both the distance and direction from the airport to the airplane must be specified. In engineering we deal with many quantities that have both magnitude and direction and can be expressed as vectors. In this chapter we review vector operations, resolve vectors into components, and give examples of engineering applications of vectors.

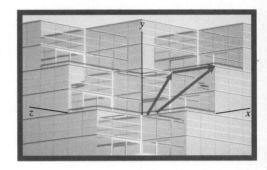

Vector Operations and Definitions

Engineers designing a structure must analyze the positions of its members and the forces acting on them. When designing a machine, they must analyze the velocities and accelerations of its moving parts. These and many other physical quantities important in engineering, can be represented by vectors and analyzed by vector operations. Here we review fundamental vector operations and definitions.

2.1 Scalars and Vectors

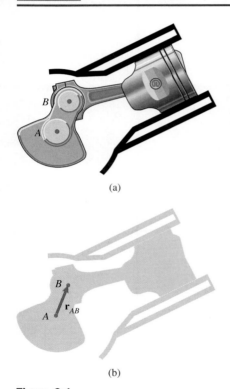

(a)

(b)

Figure 2.1
(a) Two points A and B of a mechanism.
(b) The vector $\mathbf{r}_{AB}$ from A to B.

A physical quantity that is completely described by a real number is called a *scalar*. Time is a scalar quantity. Mass is also a scalar quantity. For example, you completely describe the mass of a car by saying that its value is 1200 kg.

In contrast, you have to specify both a nonnegative real number, or *magnitude*, and a direction to describe a vector quantity. Two vector quantities are equal only if both their magnitudes and their directions are equal.

The position of a point in space relative to another point is a vector quantity. To describe the location of a city relative to your home, it is not enough to say that it is 100 miles away. You must say that it is 100 miles west of your home. Force is also a vector quantity. When you push a piece of furniture across the floor, you apply a force of magnitude sufficient to move the furniture and you apply it in the direction you want the furniture to move.

We will represent vectors by boldfaced letters, $\mathbf{U}$, $\mathbf{V}$, $\mathbf{W}$, ..., and will denote the magnitude of a vector $\mathbf{U}$ by $|\mathbf{U}|$. A vector is represented graphically by an arrow. The direction of the arrow indicates the direction of the vector, and the length of the arrow is defined to be proportional to the magnitude. For example, consider the points A and B of the mechanism in Fig. 2.1a. We can specify the position of point B relative to point A by the vector $\mathbf{r}_{AB}$ in Fig. 2.1b. The direction of $\mathbf{r}_{AB}$ indicates the direction from point A to point B. If the distance between the two points is 200 mm, the magnitude $|\mathbf{r}_{AB}|$ = 200 mm.

The cable AB in Fig. 2.2 helps support the television transmission tower. We can represent the force the cable exerts on the tower by a vector $\mathbf{F}$ as shown. If the cable exerts an 800-N force on the tower, $|\mathbf{F}|$ = 800 N.

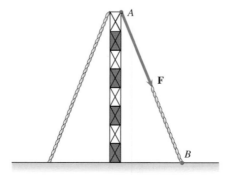

Figure 2.2
Representing the force cable AB exerts on the tower by a vector $\mathbf{F}$.

2.2 Rules for Manipulating Vectors

Vectors are a convenient means for representing physical quantities that have magnitude and direction, but that is only the beginning of their usefulness. Just as you manipulate real numbers with the familiar rules for addition, subtraction, multiplication, and so forth, there are rules for manipulating vectors. These rules provide you with powerful tools for engineering analysis.

Vector Addition

When an object moves from one location in space to another, we say it undergoes a *displacement*. If we move a book (or, speaking more precisely, some point of a book) from one location on a table to another, as shown in Fig. 2.3a, we can represent the displacement by the vector **U**. The direction of **U** indicates the direction of the displacement, and |**U**| is the distance the book moves.

Suppose that we give the book a second displacement **V**, as shown in Fig. 2.3b. The two displacements **U** and **V** are equivalent to a single displacement of the book from its initial position to its final position, which we represent by the vector **W** in Fig. 2.3c. Notice that the final position of the book is the same whether we first give it the displacement **U** and then the displacement **V** or we first give it the displacement **V** and then the displacement **U** (Fig. 2.3d). The displacement **W** is defined to be the sum of the displacements **U** and **V**:

$$\mathbf{U} + \mathbf{V} = \mathbf{W}.$$

The definition of vector addition is motivated by the addition of displacements. Consider the two vectors **U** and **V** shown in Fig. 2.4a. If we place them head to tail (Fig. 2.4b), their sum is defined to be the vector from the tail of **U** to the head of **V** (Fig. 2.4c). This is called the *triangle rule* for vector addition. Figure 2.4(d) demonstrates that the sum is independent of the order

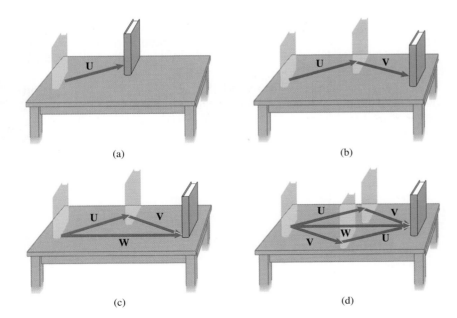

(a)

(b)

(c)

(d)

Figure 2.3
(a) A displacement represented by the vector **U**.
(b) The displacement **U** followed by the displacement **V**.
(c) The displacements **U** and **V** are equivalent to the displacement **W**.
(d) The final position of the book doesn't depend on the order of the displacements.

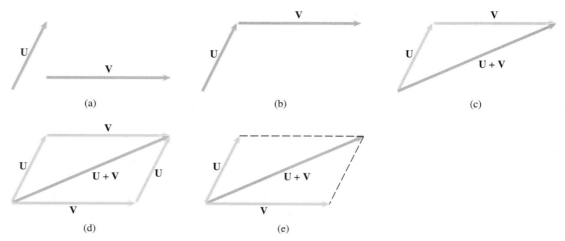

Figure 2.4
(a) Two vectors **U** and **V**.
(b) The head of **U** placed at the tail of **V**.
(c) The triangle rule for obtaining the sum of **U** and **V**.
(d) The sum is independent of the order in which the vectors are added.
(e) The parallelogram rule for obtaining the sum of **U** and **V**.

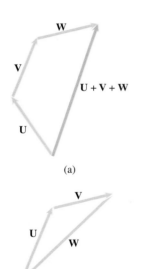

Figure 2.5
(a) The sum of three vectors.
(b) Three vectors whose sum is zero.

Figure 2.6
Arrows denoting the relative positions of
points are vectors.

in which the vectors are placed head to tail. From this figure we obtain the *parallelogram rule* for vector addition (Fig. 2.4e).

The definition of vector addition implies that

$$\mathbf{U} + \mathbf{V} = \mathbf{V} + \mathbf{U} \quad \text{Vector addition is commutative.} \tag{2.1}$$

and

$$(\mathbf{U} + \mathbf{V}) + \mathbf{W} = \mathbf{U} + (\mathbf{V} + \mathbf{W}) \quad \substack{\text{Vector addition} \\ \text{is associative.}} \tag{2.2}$$

for any vectors **U**, **V**, and **W**. These results mean that when you add two or more vectors, you don't need to worry about the order in which you add them. The sum is obtained by placing the vectors head to tail in any order. The vector from the tail of the first vector to the head of the last one is the sum (Fig. 2.5a). If the sum is zero, the vectors form a closed polygon when they are placed head to tail (Fig. 2.5b).

A physical quantity is called a vector if it has magnitude and direction and obeys the definition of vector addition. We have seen that a displacement is a vector. The position of a point in space relative to another point is also a vector quantity. In Fig. 2.6, the vector $\mathbf{r}_{AC}$ from A to C is the sum of $\mathbf{r}_{AB}$ and $\mathbf{r}_{BC}$.

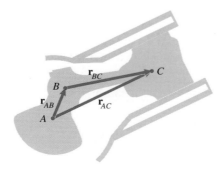

A force has direction and magnitude, but do forces obey the definition of vector addition? For now we will assume that they do. When we discuss dynamics we will show that Newton's second law implies that force is a vector.

Product of a Scalar and a Vector

The product of a scalar (real number) a and a vector $\mathbf{U}$ is a vector written as $a\mathbf{U}$. Its magnitude is $|a||\mathbf{U}|$, where $|a|$ is the absolute value of the scalar a. The direction of $a\mathbf{U}$ is the same as the direction of $\mathbf{U}$ when a is positive and is opposite to the direction of $\mathbf{U}$ when a is negative.

The product $(-1)\mathbf{U}$ is written as $-\mathbf{U}$ and is called "the negative of the vector $\mathbf{U}$." It has the same magnitude as $\mathbf{U}$ but the opposite direction. The division of a vector $\mathbf{U}$ by a scalar a is defined to be the product

$$\frac{\mathbf{U}}{a} = \left(\frac{1}{a}\right)\mathbf{U}.$$

Figure 2.7 shows a vector $\mathbf{U}$ and the products of $\mathbf{U}$ with the scalars 2, -1, and $1/2$.

The definitions of vector addition and the product of a scalar and a vector imply that

$$a(b\mathbf{U}) = (ab)\mathbf{U}, \quad \text{(2.3)}$$
The product is associative with respect to scalar multiplication.

$$(a + b)\mathbf{U} = a\mathbf{U} + b\mathbf{U} \quad \text{(2.4)}$$
The product is distributive with respect to scalar addition.

and

$$a(\mathbf{U} + \mathbf{V}) = a\mathbf{U} + a\mathbf{V} \quad \text{(2.5)}$$
The product is distributive with respect to vector addition.

for any scalars a and b and vectors $\mathbf{U}$ and $\mathbf{V}$. We will need these results when we discuss components of vectors.

Vector Subtraction

The difference of two vectors $\mathbf{U}$ and $\mathbf{V}$ is obtained by adding $\mathbf{U}$ to the vector $(-1)\mathbf{V}$:

$$\mathbf{U} - \mathbf{V} = \mathbf{U} + (-1)\mathbf{V}. \quad \text{(2.6)}$$

Consider the two vectors $\mathbf{U}$ and $\mathbf{V}$ shown in Fig. 2.8a. The vector $(-1)\mathbf{V}$ has the same magnitude as the vector $\mathbf{V}$ but is in the opposite direction (Fig. 2.8b). In Fig. 2.8c, we add the vector $\mathbf{U}$ to the vector $(-1)\mathbf{V}$ to obtain $\mathbf{U} - \mathbf{V}$.

Unit Vectors

A *unit vector* is simply a vector whose magnitude is 1. A unit vector specifies a direction and also provides a convenient way to express a vector that has a particular direction. If a unit vector $\mathbf{e}$ and a vector $\mathbf{U}$ have the same direction, we can write $\mathbf{U}$ as the product of its magnitude $|\mathbf{U}|$ and the unit vector $\mathbf{e}$ (Fig. 2.9),

$$\mathbf{U} = |\mathbf{U}|\mathbf{e}.$$

Figure 2.7
(a) A vector $\mathbf{U}$ and some of its scalar multiples.

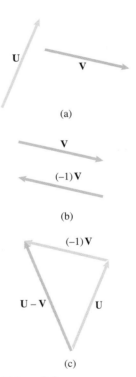

Figure 2.8
(a) Two vectors $\mathbf{U}$ and $\mathbf{V}$.
(b) The vectors $\mathbf{V}$ and $(-1)\,\mathbf{V}$.
(c) The sum of $\mathbf{U}$ and $(-1)\,\mathbf{V}$ is the vector difference $\mathbf{U} - \mathbf{V}$.

Figure 2.9
Since $\mathbf{U}$ and $\mathbf{e}$ have the same direction, the vector $\mathbf{U}$ equals the product of its magnitude with $\mathbf{e}$.

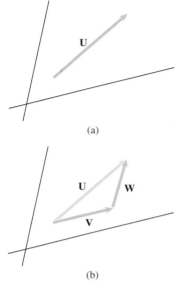

Figure 2.10
(a) A vector **U** and two intersecting lines.
(b) The vectors **V** and **W** are vector components of **U**.

Any vector **U** can be regarded as the product of its magnitude and a unit vector that has the same direction as **U**. Dividing both sides of this equation by |**U**|:

$$\frac{\mathbf{U}}{|\mathbf{U}|} = \mathbf{e},$$

we see that dividing any vector by its magnitude yields a unit vector that has the same direction.

Vector Components

When a vector **U** is expressed as the sum of a set of vectors, each vector of the set is called a *vector component* of **U**. Suppose that the vector **U** shown in Fig. 2.10a is parallel to the plane defined by the two intersecting lines. We can express **U** as the sum of vector components **V** and **W** that are parallel to the two lines, as shown in Fig. 2.10b. We say that **U** is *resolved* into the vector components **V** and **W**.

Study Questions

1. What is the triangle rule for vector addition?
2. Vector addition is commutative. What does that mean?
3. If you multiply a vector **U** by a number a, what do you know about the resulting vector $a\mathbf{U}$?
4. What is a unit vector?

Example 2.1

Adding Vectors

Figure 2.11 is an initial design sketch of part of the roof of a sports stadium that is to be supported by the cables *AB* and *AC*. The forces the cables exert on the pylon to which they are attached are represented by the vectors $\mathbf{F}_{AB}$ and $\mathbf{F}_{AC}$. The magnitudes of the forces are $|\mathbf{F}_{AB}| = 100$ kN and $|\mathbf{F}_{AC}| = 60$ kN. Determine the magnitude and direction of the sum of the forces exerted on the pylon by the cables (a) graphically and (b) by using trigonometry.

Strategy

(a) By drawing the parallelogram rule for adding the two forces *with the vectors drawn to scale*, we can measure the magnitude and direction of their sum. (b) We will calculate the magnitude and direction of the sum of the forces by applying the laws of sines and cosines (Appendix A, Section A.2) to the triangles formed by the parallelogram rule.

Figure 2.11

Solution

(a) We graphically construct the parallelogram rule for obtaining the sum of the two forces with the lengths of $\mathbf{F}_{AB}$ and $\mathbf{F}_{AC}$ proportional to their magnitudes (Fig. a). By measuring the figure, we estimate the magnitude of the vector $\mathbf{F}_{AB} + \mathbf{F}_{AC}$ to be 155 kN and its direction to be 19° above the horizontal.
(b) Consider the parallelogram rule for obtaining the sum of the two forces (Fig. b). Since $\alpha + 30° = 180°$, the angle $\alpha = 150°$. By applying the law of cosines to the shaded triangle,

$$\left|\mathbf{F}_{AB} + \mathbf{F}_{AC}\right|^2 = \left|\mathbf{F}_{AB}\right|^2 + \left|\mathbf{F}_{AC}\right|^2 - 2\left|\mathbf{F}_{AB}\right|\left|\mathbf{F}_{AC}\right|\cos\alpha$$

$$= (100)^2 + (60)^2 - 2(100)(60)\cos 150°,$$

we determine that the magnitude $\left|\mathbf{F}_{AB} + \mathbf{F}_{AC}\right| = 155$ kN.
    To determine the angle β between the vector $\mathbf{F}_{AB} + \mathbf{F}_{AC}$ and the horizontal, we apply the law of sines to the shaded triangle:

$$\frac{\sin\beta}{\left|\mathbf{F}_{AB}\right|} = \frac{\sin\alpha}{\left|\mathbf{F}_{AB} + \mathbf{F}_{AC}\right|}.$$

The solution is

$$\beta = \arcsin\left(\frac{\left|\mathbf{F}_{AB}\right|\sin\alpha}{\left|\mathbf{F}_{AB} + \mathbf{F}_{AC}\right|}\right) = \arcsin\left(\frac{100\sin 150°}{155}\right) = 18.8°$$

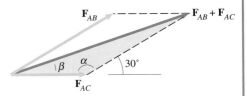

(a) Graphical solution.

(b) Trigonometric solution.

Discussion

Engineering applications of vectors usually require the precision of analytical solutions, but experience with graphical solutions can help you understand vector operations. Carrying out a graphical solution can also help you formulate an analytical solution.

Example 2.2

Resolving a Vector into Components

The force $\mathbf{F}$ in Fig. 2.12 lies in the plane defined by the intersecting lines L_A and L_B. Its magnitude is 400 lb. Suppose that you want to resolve $\mathbf{F}$ into vector components parallel to L_A and L_B. Determine the magnitudes of the vector components (a) graphically and (b) by using trigonometry.

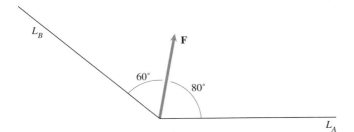

Figure 2.12

Strategy

The parallelogram rule (Fig. 2.4e) clearly indicates how we can resolve **F** into components parallel to L_A and L_B.

Solution

(a) We draw dashed lines from the head of **F** parallel to L_A and L_B to construct the vector components, which we denote $\mathbf{F}_A$ and $\mathbf{F}_B$ (Fig. a). By measuring the figure, we estimate their magnitudes to be $\left|\mathbf{F}_A\right| = 540$ lb and $\left|\mathbf{F}_B\right| = 610$ lb.

(b) Consider the force **F** and the vector components $\mathbf{F}_A$ and $\mathbf{F}_B$ (Fig. b). Since $\alpha + 80° + 60° = 180°$, the angle $\alpha = 40°$. By applying the law of sines to triangle 1,

$$\frac{\sin 60°}{\left|\mathbf{F}_A\right|} = \frac{\sin \alpha}{\left|\mathbf{F}\right|},$$

we obtain the magnitude of $\mathbf{F}_A$:

$$\left|\mathbf{F}_A\right| = \frac{\left|\mathbf{F}\right| \sin 60°}{\sin \alpha} = \frac{400 \sin 60°}{\sin 40°} = 539 \text{ lb}.$$

Then by applying the law of sines to triangle 2,

$$\frac{\sin 80°}{\left|\mathbf{F}_B\right|} = \frac{\sin \alpha}{\left|\mathbf{F}\right|},$$

we obtain the magnitude of $\mathbf{F}_B$:

$$\left|\mathbf{F}_B\right| = \frac{\left|\mathbf{F}\right| \sin 80°}{\sin \alpha} = \frac{400 \sin 80°}{\sin 40°} = 613 \text{ lb}.$$

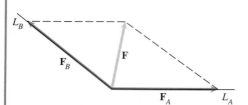

(a) Graphical solution.

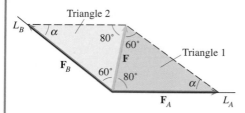

(b) Trigonometric solution.

Cartesian Components

Vectors are much easier to work with when they are expressed in terms of mutually perpendicular vector components. Here we explain how to resolve vectors into cartesian components in two and three dimensions and give examples of vector manipulations using components.

2.3 Components in Two Dimensions

Consider the vector **U** in Fig. 2.13a. By placing a cartesian coordinate system so that **U** is parallel to the x-y plane, we can resolve it into vector components $\mathbf{U}_x$ and $\mathbf{U}_y$ parallel to the x and y axes (Fig. 2.13b),

$$\mathbf{U} = \mathbf{U}_x + \mathbf{U}_y.$$

Then by introducing a unit vector **i** defined to point in the direction of the positive x axis and a unit vector **j** defined to point in the direction of the positive y axis (Fig. 2.13c), we can express the vector **U** in the form

$$\mathbf{U} = U_x\mathbf{i} + U_y\mathbf{j}. \tag{2.7}$$

The scalars U_x and U_y are called *scalar components of* **U**. *When we refer simply to the components of a vector, we will mean its scalar components.* We will call U_x and U_y the x and y components of **U**.

The components of a vector specify both its direction relative to the cartesian coordinate system and its magnitude. From the right triangle formed by the vector **U** and its vector components (Fig. 2.13c), we see that the magnitude of **U** is given in terms of its components by the Pythagorean theorem,

$$|\mathbf{U}| = \sqrt{U_x^2 + U_y^2}. \tag{2.8}$$

With this equation you can determine the magnitude of a vector when you know its components.

Manipulating Vectors in Terms of Components

The sum of two vectors **U** and **V** in terms of their components is

$$\mathbf{U} + \mathbf{V} = (U_x\mathbf{i} + U_y\mathbf{j}) + (V_x\mathbf{i} + V_y\mathbf{j})$$

$$= (U_x + V_x)\mathbf{i} + (U_y + V_y)\mathbf{j}. \tag{2.9}$$

The components of **U** + **V** are the sums of the components of the vectors **U** and **V**. Notice that in obtaining this result we used Eqs. (2.2), (2.4), and (2.5).

It is instructive to derive Eq. (2.9) graphically. The summation of **U** and **V** is shown in Fig. 2.14a. In Fig. 2.14b we introduce a coordinate system and resolve **U** and **V** into their components. In Fig. 2.14c we add the x and y components, obtaining Eq. (2.9).

The product of a number a and a vector **U** in terms of the components of **U** is

$$a\mathbf{U} = a(U_x\mathbf{i} + U_y\mathbf{j}) = aU_x\mathbf{i} + aU_y\mathbf{j}.$$

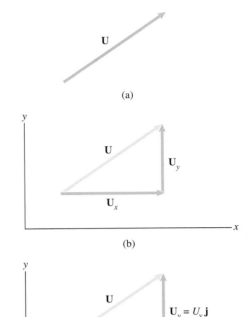

Figure 2.13
(a) A vector **U**.
(b) The vector components U_x and U_y.
(c) The vector components can be expressed in terms of **i** and **j**.

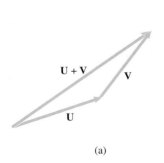

(a)

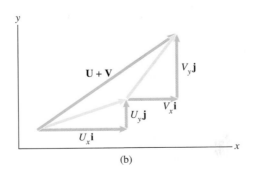

(b)

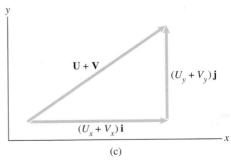

(c)

Figure 2.14
(a) The sum of **U** and **V**.
(b) The vector components of **U** and **V**.
(c) The sum of the components in each coordinate direction equals the component of **U** + **V** in that direction.

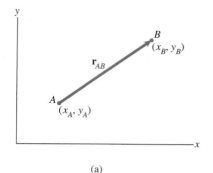

(a)

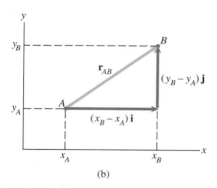

(b)

Figure 2.15
(a) Two points A and B and the position vector $\mathbf{r}_{AB}$ from A to B.
(b) The components of $\mathbf{r}_{AB}$ can be determined from the coordinates of points A and B.

The component of $a\mathbf{U}$ in each coordinate direction equals the product of a and the component of $\mathbf{U}$ in that direction. We used Eqs. (2.3) and (2.5) to obtain this result.

Position Vectors in Terms of Components

We can express the position vector of a point relative to another point in terms of the cartesian coordinates of the points. Consider point A with coordinates (x_A, y_A) and point B with coordinates (x_B, y_B). Let $\mathbf{r}_{AB}$ be the vector that specifies the position of B relative to A (Fig. 2.15a). That is, we denote the vector *from* a point A *to* a point B by $\mathbf{r}_{AB}$. We see from Fig. 2.15b that $\mathbf{r}_{AB}$ is given in terms of the coordinates of points A and B by

$$\mathbf{r}_{AB} = (x_B - x_A)\mathbf{i} + (y_B - y_A)\mathbf{j}. \tag{2.10}$$

Notice that the x component of the position vector from a point A to a point B is obtained by subtracting the x coordinate of A from the x coordinate of B, and the y component is obtained by subtracting the y coordinate of A from the y coordinate of B.

Study Questions

1. How are the scalar components of a vector defined in terms of a cartesian coordinate system?
2. If you know the scalar components of a vector, how can you determine its magnitude?
3. Suppose that you know the coordinates of two points A and B. How do you determine the scalar components of the position vector of point B relative to point A?

Example 2.3

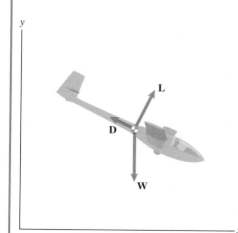

Figure 2.16

Adding Vectors in Terms of Components

The forces acting on the sailplane in Fig. 2.16 are its weight $\mathbf{W} = -600\mathbf{j}$ (lb), the drag $\mathbf{D} = -200\mathbf{i} + 100\mathbf{j}$ (lb), and the lift $\mathbf{L}$.
(a) If the sum of the forces on the sailplane is zero, what are the components of $\mathbf{L}$?
(b) If the lift $\mathbf{L}$ has the components determined in (a) and the drag $\mathbf{D}$ increases by a factor of 2, what is the magnitude of the sum of the forces on the sailplane?

Strategy

(a) By setting the sum of the forces equal to zero, we can determine the components of $\mathbf{L}$. (b) Using the value of $\mathbf{L}$ from (a), we can determine the components of the sum of the forces and use Eq. (2.8) to determine its magnitude.

Solution

(a) We set the sum of the forces equal to zero:

$$\mathbf{W} + \mathbf{D} + \mathbf{L} = \mathbf{0}.$$

$$(-600\mathbf{j}) + (-200\mathbf{i} + 100\mathbf{j}) + \mathbf{L} = \mathbf{0}.$$

Solving for the lift, we obtain

$$\mathbf{L} = 200\mathbf{i} + 500\mathbf{j} \text{ (lb).}$$

(b) If the drag increases by a factor of 2, the sum of the forces on the sailplane is

$$\mathbf{W} + 2\mathbf{D} + \mathbf{L} = (-600\mathbf{j}) + 2(-200\mathbf{i} + 100\mathbf{j}) + (200\mathbf{i} + 500\mathbf{j})$$
$$= -200\mathbf{i} + 100\mathbf{j} \text{ (lb).}$$

From Eq. (2.8), the magnitude of the sum is

$$|\mathbf{W} + 2\mathbf{D} + \mathbf{L}| = \sqrt{(-200)^2 + (100)^2} = 224 \text{ lb.}$$

Example 2.4

Determining Components in Terms of an Angle

Hydraulic cylinders are used to exert forces in many mechanical devices. The force is exerted by pressurized liquid (hydraulic fluid) pushing against a piston within the cylinder. The hydraulic cylinder AB in Fig. 2.17 exerts a 4000-lb force $\mathbf{F}$ on the bed of the dump truck at B. Express $\mathbf{F}$ in terms of components using the coordinate system shown.

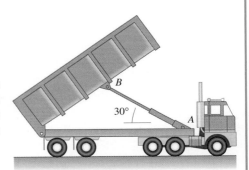

Strategy

When the direction of a vector is specified by an angle, as in this example, we can determine the values of the components from the right triangle formed by the vector and its components.

Solution

We draw the vector $\mathbf{F}$ and its vector components in Fig. a. From the resulting right triangle, we see that the magnitude of $\mathbf{F}_x$ is

$$|\mathbf{F}_x| = |\mathbf{F}| \cos 30° = (4000) \cos 30° = 3460 \text{ lb.}$$

$\mathbf{F}_x$ points in the negative x direction, so

$$\mathbf{F}_x = -3460\mathbf{i} \text{ (lb).}$$

The magnitude of $\mathbf{F}_y$ is

$$|\mathbf{F}_y| = |\mathbf{F}| \sin 30° = (4000) \sin 30° = 2000 \text{ lb.}$$

The vector component $\mathbf{F}_y$ points in the positive y direction, so

$$\mathbf{F}_y = 2000\mathbf{j} \text{ (lb).}$$

The vector $\mathbf{F}$ in terms of its components is

$$\mathbf{F} = \mathbf{F}_x + \mathbf{F}_y = -3460\mathbf{i} + 2000\mathbf{j} \text{ (lb).}$$

The x component of $\mathbf{F}$ is -3460 lb, and the y component is 2000 lb.

Discussion

When you determine the components of a vector, you should check to make sure they give you the correct magnitude. In this example,

$$|\mathbf{F}| = \sqrt{(-3460)^2 + (2000)^2} = 4000 \text{ lb.}$$

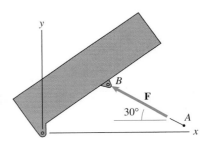

Figure 2.17

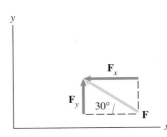

(a) The force $\mathbf{F}$ and its components form a right triangle.

Example 2.5

Determining Vector Components

The cable from point A to point B exerts an 800-N force $\mathbf{F}$ on the top of the television transmission tower in Fig. 2.18. Resolve $\mathbf{F}$ into components using the coordinate system shown.

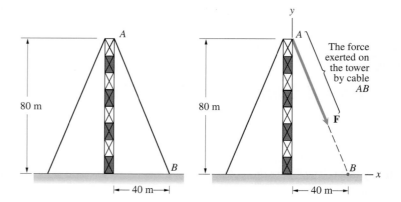

Figure 2.18

Strategy

We determine the components of $\mathbf{F}$ in three ways.

First Method From the given dimensions we can determine the angle α between $\mathbf{F}$ and the y axis (Fig. a), then determine the components from the right triangles formed by the vector $\mathbf{F}$ and its components.

Second Method The right triangles formed by $\mathbf{F}$ and its components are similar to the triangle OAB in Fig. a. We can determine the components of $\mathbf{F}$ by using the ratios of the sides of these similar triangles.

Third Method From the given dimensions we can determine the components of the position vector $\mathbf{r}_{AB}$ from point A to point B [Fig. b]. By dividing this vector by its magnitude, we will obtain a unit vector $\mathbf{e}_{AB}$ with the same direction as $\mathbf{F}$ (Fig. c), then obtain $\mathbf{F}$ in terms of its components by expressing it as the product of its magnitude and $\mathbf{e}_{AB}$.

Solution

First Method Consider the force $\mathbf{F}$ and its vector components (Fig. a). The tangent of the angle α between $\mathbf{F}$ and the y axis is $\tan\alpha = 40/80 = 0.5$, so $\alpha = \arctan(0.5) = 26.6°$. From the right triangles formed by $\mathbf{F}$ and its vector components, the magnitude of $\mathbf{F}_x$ is

$$|\mathbf{F}_x| = |\mathbf{F}|\sin 26.6° = (800)\sin 26.6° = 358\text{ N}$$

and the magnitude of $\mathbf{F}_y$ is

$$|\mathbf{F}_y| = |\mathbf{F}| \cos 26.6° = (800) \cos 26.6° = 716 \text{ N}.$$

Since $\mathbf{F}_x$ points in the positive x direction and $\mathbf{F}_y$ points in the negative y direction, the force $\mathbf{F}$ is

$$\mathbf{F} = 358\mathbf{i} - 716\mathbf{j} \text{ (N)}$$

Second Method The length of the cable AB is $\sqrt{(80)^2 + (40)^2} = 89.4$ m. Since the triangle OAB in Fig. a is similar to the triangle formed by $\mathbf{F}$ and its vector components,

$$\frac{|\mathbf{F}_x|}{|\mathbf{F}|} = \frac{OB}{AB} = \frac{40}{89.4}.$$

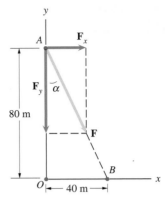

(a) Vector components of $\mathbf{F}$.

Thus the magnitude of $\mathbf{F}_x$ is

$$|\mathbf{F}_x| = \left(\frac{40}{89.4}\right)|\mathbf{F}| = \left(\frac{40}{89.4}\right)(800) = 358 \text{ N}.$$

We can also see from the similar triangles that

$$\frac{|\mathbf{F}_y|}{|\mathbf{F}|} = \frac{OA}{AB} = \frac{80}{89.4},$$

so the magnitude of $\mathbf{F}_y$ is

$$|\mathbf{F}_y| = \left(\frac{80}{89.4}\right)|\mathbf{F}| = \left(\frac{80}{89.4}\right)(800) = 716 \text{ N}.$$

Thus we again obtain the result

$$\mathbf{F} = 358\mathbf{i} - 716\mathbf{j} \text{ (N)}.$$

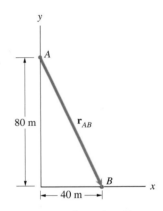

(b) The vector $\mathbf{r}_{AB}$ form A to B.

Third Method The vector $\mathbf{r}_{AB}$ in Fig. b is

$$\mathbf{r}_{AB} = (x_B - x_A)\mathbf{i} + (y_B - y_A)\mathbf{j} = (40 - 0)\mathbf{i} + (0 - 80)\mathbf{j}$$
$$= 40\mathbf{i} - 80\mathbf{j} \text{ (m)}.$$

We divide this vector by its magnitude to obtain a unit vector $\mathbf{e}_{AB}$ that has the same direction as the force $\mathbf{F}$ (Fig. c):

$$\mathbf{e}_{AB} = \frac{\mathbf{r}_{AB}}{|\mathbf{r}_{AB}|} = \frac{40\mathbf{i} - 80\mathbf{j}}{\sqrt{(40)^2 + (-80)^2}} = 0.447\mathbf{i} - 0.894\mathbf{j}.$$

The force $\mathbf{F}$ is equal to the product of its magnitude $|\mathbf{F}|$ and $\mathbf{e}_{AB}$:

$$\mathbf{F} = |\mathbf{F}|\mathbf{e}_{AB} = (800)(0.447\mathbf{i} - 0.894\mathbf{j}) = 358\mathbf{i} - 716\mathbf{j} \text{ (N)}.$$

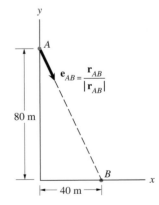

(c) The unit vector $\mathbf{e}_{AB}$ pointing from A toward B.

Example 2.6

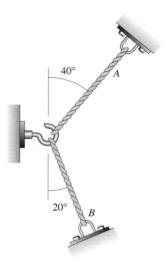

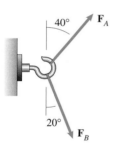

Determining an Unknown Vector Magnitude

The cables A and B in Fig. 2.19 exert forces $\mathbf{F}_A$ and $\mathbf{F}_B$ on the hook. The magnitude of $\mathbf{F}_A$ is 100 lb. The tension in cable B has been adjusted so that the total force $\mathbf{F}_A + \mathbf{F}_B$ is perpendicular to the wall to which the hook is attached.

(a) What is the magnitude of $\mathbf{F}_B$?

(b) What is the magnitude of the total force exerted on the hook by the two cables?

Strategy

The vector sum of the two forces is perpendicular to the wall, so the sum of the components parallel to the wall equals zero. From this condition we can obtain an equation for the magnitude of $\mathbf{F}_B$.

Solution

(a) In terms of the coordinate system shown in Fig. a, the components of $\mathbf{F}_A$ and $\mathbf{F}_B$ are

$$\mathbf{F}_A = |\mathbf{F}_A| \sin 40° \mathbf{i} + |\mathbf{F}_A| \cos 40° \mathbf{j},$$

$$\mathbf{F}_B = |\mathbf{F}_B| \sin 20° \mathbf{i} - |\mathbf{F}_B| \cos 20° \mathbf{j}.$$

The total force is

$$\mathbf{F}_A + \mathbf{F}_B = (|\mathbf{F}_A| \sin 40° + |\mathbf{F}_B| \sin 20°)\mathbf{i}$$
$$+ (|\mathbf{F}_A| \cos 40° - |\mathbf{F}_B| \cos 20°)\mathbf{j}.$$

By setting the component of the total force parallel to the wall (the y component) equal to zero,

$$|\mathbf{F}_A| \cos 40° - |\mathbf{F}_B| \cos 20° = 0,$$

we obtain an equation for the magnitude of $\mathbf{F}_B$:

$$|\mathbf{F}_B| = \frac{|\mathbf{F}_A| \cos 40°}{\cos 20°} = \frac{(100) \cos 40°}{\cos 20°} = 81.5 \text{ lb}.$$

(b) Since we now know the magnitude of $\mathbf{F}_B$, we can determine the total force acting on the hook:

$$\mathbf{F}_A + \mathbf{F}_B = (|\mathbf{F}_A| \sin 40° + |\mathbf{F}_B| \sin 20°)\mathbf{i}$$
$$= [(100) \sin 40° + (81.5) \sin 20°]\mathbf{i} = 92.2\mathbf{i} \text{ (lb)}.$$

The magnitude of the total force is 92.2 lb.

Figure 2.19

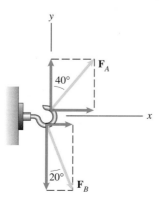

(a) Resolving $\mathbf{F}_A$ and $\mathbf{F}_B$ into components parallel and perpendicular to the wall.

Discussion

We can obtain the solution to (a) in a less formal way. If the component of the total force parallel to the wall is zero, we see in Fig. (a) that the magnitude of the vertical component of $\mathbf{F}_A$ must equal the magnitude of the vertical component of $\mathbf{F}_B$:

$$|\mathbf{F}_A| \cos 40° = |\mathbf{F}_B| \cos 20°.$$

Therefore the magnitude of $\mathbf{F}_B$ is

$$|\mathbf{F}_B| = \frac{|\mathbf{F}_A| \cos 40°}{\cos 20°} = \frac{(100) \cos 40°}{\cos 20°} = 81.5 \text{ lb}.$$

2.4 Components in Three Dimensions

Many engineering applications require you to resolve vectors into components in a three-dimensional coordinate system. In this section we explain this technique and demonstrate vector operations in three dimensions.

Let's first review how to draw objects in three dimensions. Consider a three-dimensional object such as a cube. If we draw the cube as it appears when your line of sight is perpendicular to one of its faces, we obtain the diagram shown in Fig. 2.20a. In this view the cube appears two-dimensional; you can't see the dimension perpendicular to the page. To remedy this, we can draw the cube as it appears if you move upward and to the right (Fig. 2.20b). In this oblique view you can see the third dimension. The hidden edges of the cube are shown as dashed lines.

We can use this method to draw three-dimensional coordinate systems. In Fig. 2.20c we align the x, y, and z axes of a three-dimensional cartesian coordinate system with the edges of the cube. The three-dimensional representation of the coordinate system is shown in Fig. 2.20d.

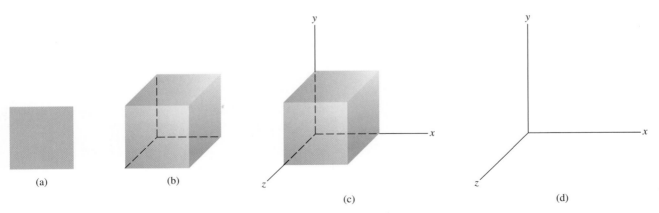

(a)   (b)   (c)   (d)

Figure 2.20
(a) A cube viewed with the line of sight perpendicular to a face.
(b) An oblique view of the cube.
(c) A cartesian coordinate system aligned with the edges of the cube.
(d) Three-dimensional representation of the coordinate system.

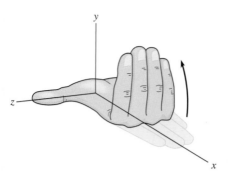

Figure 2.21
Recognizing a right-handed coordinate system.

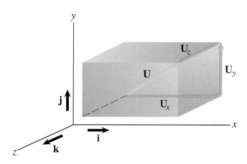

Figure 2.22
A vector **U** and its vector components.

The coordinate system in Fig. 2.20d is *right-handed*. If you point the fingers of your right hand in the direction of the positive x axis and bend them (as in preparing to make a fist) toward the positive y axis, your thumb will point in the direction of the positive z axis (Fig. 2.21). When the positive z axis points in the opposite direction, the coordinate system is left-handed. For some purposes, it doesn't matter which coordinate system you use. However, some equations we will derive do not give correct results with a left-handed coordinate system. For this reason we will use only right-handed coordinate systems.

We can resolve a vector **U** into vector components $\mathbf{U}_x$, $\mathbf{U}_y$, and $\mathbf{U}_z$ parallel to the x, y, and z axes (Fig. 2.22):

$$\mathbf{U} = \mathbf{U}_x + \mathbf{U}_y + \mathbf{U}_z. \tag{2.11}$$

(We have drawn a box around the vector to help you visualize the directions of the vector components.) By introducing unit vectors **i**, **j**, and **k** that point in the positive x, y, and z directions, we can express **U** in terms of scalar components as

$$\mathbf{U} = U_x\mathbf{i} + U_y\mathbf{j} + U_z\mathbf{k}. \tag{2.12}$$

We will refer to the scalars U_x, U_y, and U_z as the x, y, and z components of **U**.

Magnitude of a Vector in Terms of Components

Consider a vector **U** and its vector components (Fig. 2.23a). From the right triangle formed by the vectors $\mathbf{U}_y$, $\mathbf{U}_z$, and their sum $\mathbf{U}_y + \mathbf{U}_z$ (Fig. 2.23b), we can see that

$$\left|\mathbf{U}_y + \mathbf{U}_z\right|^2 = \left|\mathbf{U}_y\right|^2 + \left|\mathbf{U}_z\right|^2. \tag{2.13}$$

The vector **U** is the sum of the vectors $\mathbf{U}_x$ and $\mathbf{U}_y + \mathbf{U}_z$. These three vectors form a right triangle (Fig. 2.23c), from which we obtain

$$|\mathbf{U}|^2 = \left|\mathbf{U}_x\right|^2 + \left|\mathbf{U}_y + \mathbf{U}_z\right|^2.$$

Substituting Eq. (2.13) into this result yields the equation

$$|\mathbf{U}|^2 = \left|\mathbf{U}_x\right|^2 + \left|\mathbf{U}_y\right|^2 + \left|\mathbf{U}_z\right|^2 = U_x^2 + U_y^2 + U_z^2.$$

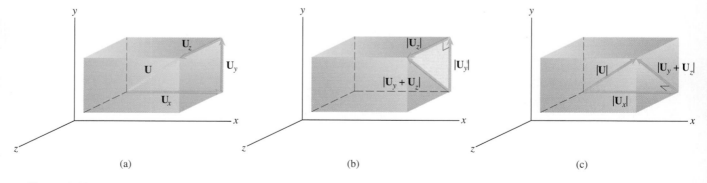

Figure 2.23
(a) A vector **U** and its vector components.
(b) The right triangle formed by the vectors $\mathbf{U}_y$, $\mathbf{U}_z$, and $\mathbf{U}_y + \mathbf{U}_z$.
(c) The right triangle formed by the vectors **U**, $\mathbf{U}_x$, and $\mathbf{U}_y + \mathbf{U}_z$.

Thus the magnitude of a vector $\mathbf{U}$ is given in terms of its components in three dimensions by

$$|\mathbf{U}| = \sqrt{U_x^2 + U_y^2 + U_z^2}. \tag{2.14}$$

Direction Cosines

We described the direction of a vector relative to a two-dimensional cartesian coordinate system by specifying the angle between the vector and one of the coordinate axes. One of the ways we can describe the direction of a vector in three dimensions is by specifying the angles θ_x, θ_y, and θ_z between the vector and the positive coordinate axes (Fig. 2.24a).

In Figs. 2.24b–d, we demonstrate that the components of the vector $\mathbf{U}$ are given in terms of the angles θ_x, θ_y, and θ_z. by

$$U_x = |\mathbf{U}| \cos\theta_x, \quad U_y = |\mathbf{U}| \cos\theta_y, \quad U_z = |\mathbf{U}| \cos\theta_z. \tag{2.15}$$

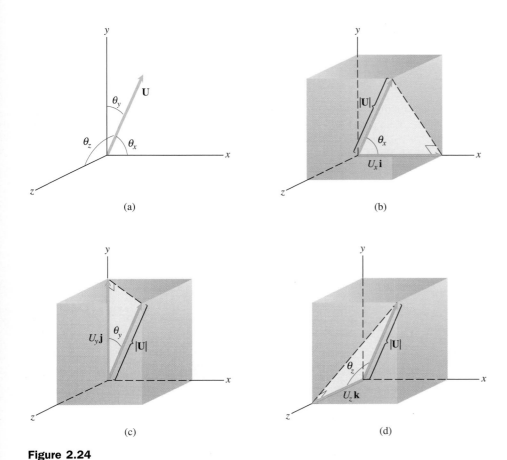

Figure 2.24
(a) A vector $\mathbf{U}$ and the angles θ_x, θ_y, and θ_z.
(b)–(d) The angles θ_x, θ_y, and θ_z and the vector components of $\mathbf{U}$.

The quantities $\cos\theta_x$, $\cos\theta_y$, and $\cos\theta_z$ are called the *direction cosines* of $\mathbf{U}$. The direction cosines of a vector are not independent. If we substitute Eqs. (2.15) into Eq. (2.14), we find that the direction cosines satisfy the relation

$$\cos^2\theta_x + \cos^2\theta_y + \cos^2\theta_z = 1. \tag{2.16}$$

Suppose that $\mathbf{e}$ is a unit vector with the same direction as $\mathbf{U}$, so that

$$\mathbf{U} = |\mathbf{U}|\mathbf{e}.$$

In terms of components, this equation is

$$U_x\mathbf{i} + U_y\mathbf{j} + U_z\mathbf{k} = |\mathbf{U}|(e_x\mathbf{i} + e_y\mathbf{j} + e_z\mathbf{k}).$$

Thus the relations between the components of $\mathbf{U}$ and $\mathbf{e}$ are

$$U_x = |\mathbf{U}|e_x, \quad U_y = |\mathbf{U}|e_y, \quad U_z = |\mathbf{U}|e_z.$$

By comparing these equations to Eqs. (2.15), we see that

$$\cos\theta_x = e_x, \quad \cos\theta_y = e_y, \quad \cos\theta_z = e_z.$$

The direction cosines of a vector $\mathbf{U}$ are the components of a unit vector with the same direction as $\mathbf{U}$.

Position Vectors in Terms of Components

Generalizing the two-dimensional case, let's consider a point A with coordinates (x_A, y_A, z_A) and a point B with coordinates (x_B, y_B, z_B). The position vector $\mathbf{r}_{AB}$ from A to B, shown in Fig. 2.25a, is given in terms of the coordinates of A and B by

$$\mathbf{r}_{AB} = (x_B - x_A)\mathbf{i} + (y_B - y_A)\mathbf{j} + (z_B - z_A)\mathbf{k}. \tag{2.17}$$

The components are obtained by subtracting the coordinates of point A from the coordinates of point B (Fig. 2.25b).

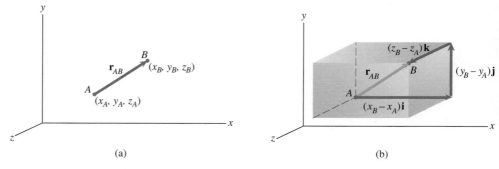

(a)   (b)

Figure 2.25
(a) The position vector from point A to point B.
(b) The components of $\mathbf{r}_{AB}$ can be determined from the coordinates of points A and B.

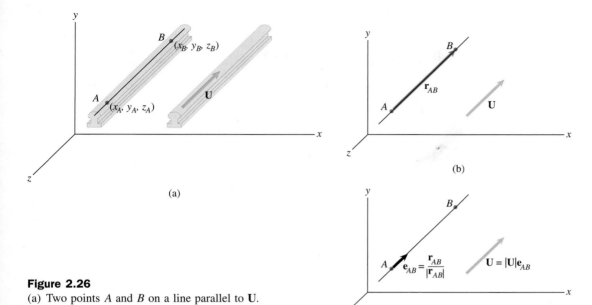

Figure 2.26
(a) Two points A and B on a line parallel to $\mathbf{U}$.
(b) The position vector from A to B.
(c) The unit vector $\mathbf{e}_{AB}$ that points from A toward B.

Components of a Vector Parallel to a Given Line

In three-dimensional applications, the direction of a vector is often defined
by specifying the coordinates of two points on a line that is parallel to the
vector. You can use this information to determine the components of the
vector.

Suppose that we know the coordinates of two points A and B on a line
parallel to a vector $\mathbf{U}$ (Fig. 2.26a). We can use Eq. (2.17) to determine the
position vector $\mathbf{r}_{AB}$ from A to B (Fig. 2.26b). We can divide $\mathbf{r}_{AB}$ by its magnitude to obtain a unit vector $\mathbf{e}_{AB}$ that points from A toward B (Fig. 2.26c).
Since $\mathbf{e}_{AB}$ has the same direction as $\mathbf{U}$, we can determine $\mathbf{U}$ in terms of its
scalar components by expressing it as the product of its magnitude and
$\mathbf{e}_{AB}$.

More generally, suppose that we know the magnitude of a vector $\mathbf{U}$ and
the components of any vector $\mathbf{V}$ that has the same direction as $\mathbf{U}$. Then $\mathbf{V}/|\mathbf{V}|$
is a unit vector with the same direction as $\mathbf{U}$, and we can determine the com-
ponents of $\mathbf{U}$ by expressing it as $\mathbf{U} = |\mathbf{U}|(\mathbf{V}/|\mathbf{V}|)$.

Study Questions

1. How do you identify a right-handed coordinate system?
2. If you know the scalar components of a vector in three dimensions, how can
   you determine its magnitude?
3. What are the direction cosines of a vector? If you know them, how do you
   determine the components of the vector?
4. Suppose that you know the coordinates of two points A and B in three
   dimensions. How do you determine the scalar components of the position vector
   of point B relative to point A?

Example 2.7

Magnitude and Direction Cosines of a Vector

An engineer designing a threshing machine determines that at a particular time the position vectors of the ends A and B of a shaft are $\mathbf{r}_A = 3\mathbf{i} - 4\mathbf{j} - 12\mathbf{k}$ (ft) and $\mathbf{r}_B = -\mathbf{i} + 7\mathbf{j} + 6\mathbf{k}$ (ft).
(a) What is the magnitude of $\mathbf{r}_A$?
(b) Determine the angles θ_x, θ_y, and θ_z between $\mathbf{r}_A$ and the positive coordinate axes.
(c) Determine the scalar components of the position vector of end B of the shaft relative to end A.

Strategy

(a) Since we know the components of $\mathbf{r}_A$, we can use Eq. (2.14) to determine its magnitude.
(b) We can obtain the angles θ_x, θ_y, and θ_z from Eqs. (2.15).
(c) The position vector of end B of the shaft relative to end A is $\mathbf{r}_B - \mathbf{r}_A$.

Solution

(a) The magnitude of $\mathbf{r}_A$ is

$$\left|\mathbf{r}_A\right| = \sqrt{r_{Ax}^2 + r_{Ay}^2 + r_{Az}^2} = \sqrt{(3)^2 + (-4)^2 + (-12)^2} = 13 \text{ ft.}$$

(b) The direction cosines of $\mathbf{r}_A$ are

$$\cos\theta_x = \frac{r_{Ax}}{\left|\mathbf{r}_A\right|} = \frac{3}{13},$$

$$\cos\theta_y = \frac{r_{Ay}}{\left|\mathbf{r}_A\right|} = \frac{-4}{13},$$

$$\cos\theta_z = \frac{r_{Az}}{\left|\mathbf{r}_A\right|} = \frac{-12}{13}.$$

From these equations we find that the angles between $\mathbf{r}_A$ and the positive coordinate axes are $\theta_x = 76.7°$, $\theta_y = 107.9°$, and $\theta_z = 157.4°$.
(c) The position vector of end B of the shaft relative to end A is

$$\mathbf{r}_B - \mathbf{r}_A = (-\mathbf{i} + 7\mathbf{j} + 6\mathbf{k}) - (3\mathbf{i} - 4\mathbf{j} - 12\mathbf{k})$$

$$= -4\mathbf{i} + 11\mathbf{j} + 18\mathbf{k} \text{ (ft).}$$

Example 2.8

Determining Scalar Components

The crane in Fig. 2.27 exerts a 600-lb force **F** on the caisson. The angle between **F** and the x axis is 54°, and the angle between **F** and the y axis is 40°. The z component of **F** is positive. Express **F** in terms of components.

Figure 2.27

Strategy

Only two of the angles between the vector and the positive coordinate axes are given, but we can use Eq. (2.16) to determine the third angle. Then we can determine the components of **F** by using Eqs. (2.15).

Solution

The angles between **F** and the positive coordinate axes are related by

$$\cos^2\theta_x + \cos^2\theta_y + \cos^2\theta_z = (\cos 54°)^2 + (\cos 40°)^2 + \cos^2\theta_z = 1.$$

Solving this equation for $\cos\theta_z$, we obtain the two solutions $\cos\theta_z = 0.260$ and $\cos\theta_z = -0.260$, which tells us that $\theta_z = 74.9°$ or $\theta_z = 105.1°$. The z component of the vector **F** is positive, so the angle between **F** and the positive z axis is less than 90°. Therefore $\theta_z = 74.9°$.
   The components of **F** are

$$F_x = |\mathbf{F}| \cos\theta_x = 600 \cos 54° \quad = 353 \text{ lb,}$$

$$F_y = |\mathbf{F}| \cos\theta_y = 600 \cos 40° \quad = 460 \text{ lb,}$$

$$F_z = |\mathbf{F}| \cos\theta_z = 600 \cos 74.9° = 156 \text{ lb.}$$

Example 2.9

Determining Scalar Components

The tether of the balloon in Fig. 2.28 exerts an 800-N force **F** on the hook at *O*. The vertical line *AB* intersects the *x-z* plane at point *A*. The angle between the *z* axis and the line *OA* is 60°, and the angle between the line *OA* and **F** is 45°. Express **F** in terms of components.

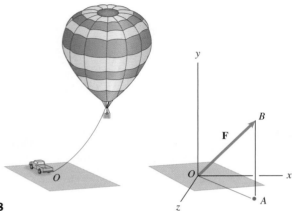

Figure 2.28

Strategy

We can determine the components of **F** from the given geometric information in two steps. First, we resolve **F** into two vector components parallel to the lines *OA* and *AB*. The component parallel to *AB* is the vector component **F**$_y$. Then we can resolve the component parallel to *OA* to determine the vector components **F**$_x$ and **F**$_z$.

Solution

In Fig. a, we resolve **F** into its *y* component **F**$_y$ and the component **F**$_h$ parallel to *OA*. The magnitude of **F**$_y$ is

$$|\mathbf{F}_y| = |\mathbf{F}| \sin 45° = 800 \sin 45° = 566 \text{ N},$$

and the magnitude of **F**$_h$ is

$$|\mathbf{F}_h| = |\mathbf{F}| \cos 45° = 800 \cos 45° = 566 \text{ N},$$

In Fig. b, we resolve **F**$_h$ into the vector components **F**$_x$ and **F**$_z$. The magnitude of **F**$_x$ is

$$|\mathbf{F}_x| = |\mathbf{F}_h| \sin 60° = 566 \sin 60° = 490 \text{ N},$$

and the magnitude of **F**$_z$ is

$$|\mathbf{F}_z| = |\mathbf{F}_h| \cos 60° = 566 \cos 60° = 283 \text{ N}.$$

The vector components **F**$_x$, **F**$_y$, and **F**$_z$ all point in the positive axis directions, so the scalar components of **F** are positive:

$$\mathbf{F} = 490\mathbf{i} + 566\mathbf{j} + 283\mathbf{k} \text{ (N)}.$$

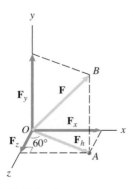

(a) Resolving **F** into vector components parallel to *OA* and *OB*.

(b) Resolving **F**$_h$ into vector components parallel to the *x* and *z* axes.

Example 2.10

Vector Whose Direction is Specified by Two Points

The bar AB in Fig. 2.29 exerts a 140-N force $\mathbf{F}$ on its support at A. The force is parallel to the bar and points toward B. Express $\mathbf{F}$ in terms of components.

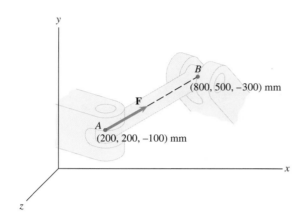

Figure 2.29

Strategy

Since we are given the coordinates of points A and B, we can determine the components of the position vector from A to B. By dividing the position vector by its magnitude, we can obtain a unit vector with the same direction as $\mathbf{F}$. Then by multiplying the unit vector by the magnitude of $\mathbf{F}$, we obtain $\mathbf{F}$ in terms of its components.

Solution

The position vector from A to B is (Fig. a)

$$\mathbf{r}_{AB} = (x_B - x_A)\mathbf{i} + (y_B - y_A)\mathbf{j} + (z_B - z_A)\mathbf{k}$$
$$= \left[(800) - (200)\right]\mathbf{i} + \left[(500) - (200)\right]\mathbf{j} + \left[(-300) - (-100)\right]\mathbf{k}$$
$$= 600\mathbf{i} + 300\mathbf{j} - 200\mathbf{k} \text{ mm,}$$

and its magnitude is

$$|\mathbf{r}_{AB}| = \sqrt{(600)^2 + (300)^2 + (-200)^2} = 700 \text{ mm.}$$

By dividing $\mathbf{r}_{AB}$ by its magnitude, we obtain a unit vector with the same direction as $\mathbf{F}$ (Fig. b),

$$\mathbf{e}_{AB} = \frac{\mathbf{r}_{AB}}{|\mathbf{r}_{AB}|} = \frac{6}{7}\mathbf{i} + \frac{3}{7}\mathbf{j} - \frac{2}{7}\mathbf{k}.$$

Then, in terms of its scalar components, $\mathbf{F}$ is

$$\mathbf{F} = |\mathbf{F}|\mathbf{e}_{AB} = (140)\left(\frac{6}{7}\mathbf{i} + \frac{3}{7}\mathbf{j} - \frac{2}{7}\mathbf{k}\right) = 120\mathbf{i} + 60\mathbf{j} - 40\mathbf{k} \text{ (N).}$$

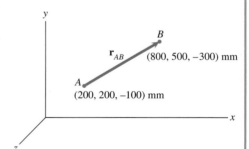

(a) The position vector $\mathbf{r}_{AB}$.

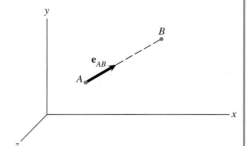

(b) The unit vector $\mathbf{e}_{AB}$ pointing from A toward B.

Example 2.11

Determining Components in Three Dimensions

The rope in Fig. 2.30 extends from point B through a metal loop attached to the wall at A to point C. The rope exerts forces $\mathbf{F}_{AB}$ and $\mathbf{F}_{AC}$ on the loop at A with magnitudes $|\mathbf{F}_{AB}| = |\mathbf{F}_{AC}| = 200$ lb. What is the magnitude of the total force $\mathbf{F} = \mathbf{F}_{AB} + \mathbf{F}_{AC}$ exerted on the loop by the rope?

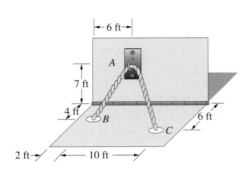

Figure 2.30

Strategy

The force $\mathbf{F}_{AB}$ is parallel to the line from A to B, and the force $\mathbf{F}_{AC}$ is parallel to the line from A to C. Since we can determine the coordinates of points A, B, and C from the given dimensions, we can determine the components of unit vectors that have the same directions as the two forces and use them to express the forces in terms of scalar components.

Solution

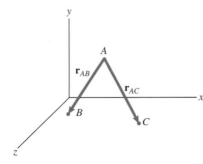

(a) The position vectors $\mathbf{r}_{AB}$ and $\mathbf{r}_{AC}$.

Let $\mathbf{r}_{AB}$ be the position vector from point A to point B and let $\mathbf{r}_{AC}$ be the position vector from point A to point C (Fig. a). From the given dimensions, the coordinates of points A, B, and C are

$$A: (6, 7, 0) \text{ ft}, \qquad B: (2, 0, 4) \text{ ft}, \qquad C: (12, 0, 6) \text{ ft}.$$

Therefore the components of $\mathbf{r}_{AB}$ and $\mathbf{r}_{AC}$ are

$$\mathbf{r}_{AB} = (x_B - x_A)\mathbf{i} + (y_B - y_A)\mathbf{j} + (z_B - z_A)\mathbf{k}$$
$$= (2 - 6)\mathbf{i} + (0 - 7)\mathbf{j} + (4 - 0)\mathbf{k}$$
$$= -4\mathbf{i} - 7\mathbf{j} + 4\mathbf{k} \text{ (ft)}$$

and

$$\mathbf{r}_{AC} = (x_C - x_A)\mathbf{i} + (y_C - y_A)\mathbf{j} + (z_C - z_A)\mathbf{k}$$
$$= (12 - 6)\mathbf{i} + (0 - 7)\mathbf{j} + (6 - 0)\mathbf{k}$$
$$= 6\mathbf{i} - 7\mathbf{j} + 6\mathbf{k} \text{ (ft)}.$$

Their magnitudes are $|\mathbf{r}_{AB}| = 9$ ft and $|\mathbf{r}_{AC}| = 11$ ft. By dividing $\mathbf{r}_{AB}$ and $\mathbf{r}_{AC}$ by their magnitudes, we obtain unit vectors $\mathbf{e}_{AB}$ and $\mathbf{e}_{AC}$ that point in the directions of $\mathbf{F}_{AB}$ and $\mathbf{F}_{AC}$ (Fig. b):

$$\mathbf{e}_{AB} = \frac{\mathbf{r}_{AB}}{|\mathbf{r}_{AB}|} = -0.444\mathbf{i} - 0.778\mathbf{j} + 0.444\mathbf{k},$$

$$\mathbf{e}_{AC} = \frac{\mathbf{r}_{AC}}{|\mathbf{r}_{AC}|} = 0.545\mathbf{i} - 0.636\mathbf{j} + 0.545\mathbf{k}.$$

The forces $\mathbf{F}_{AB}$ and $\mathbf{F}_{AC}$ are

$$\mathbf{F}_{AB} = 200\mathbf{e}_{AB} = -88.9\mathbf{i} - 155.6\mathbf{j} + 88.9\mathbf{k} \text{ (lb)},$$

$$\mathbf{F}_{AC} = 200\mathbf{e}_{AC} = 109.1\mathbf{i} - 127.3\mathbf{j} + 109.1\mathbf{k} \text{ (lb)}.$$

The total force exerted on the loop by the rope is

$$\mathbf{F} = \mathbf{F}_{AB} + \mathbf{F}_{AC} = 20.2\mathbf{i} - 282.8\mathbf{j} + 198.0\mathbf{k} \text{ (lb)},$$

and its magnitude is

$$|\mathbf{F}| = \sqrt{(20.2)^2 + (-282.8)^2 + (198.0)^2} = 346 \text{ lb}.$$

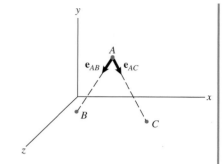

(b) The unit vector $\mathbf{e}_{AB}$ pointing and $\mathbf{e}_{AC}$.

Example 2.12

Determining Components of a Force

The cable AB in Fig. 2.31 exerts a 50-N force $\mathbf{T}$ on the collar at A. Express $\mathbf{T}$ in terms of components.

Strategy

Let $\mathbf{r}_{AB}$ be the position vector from A to B. We will divide $\mathbf{r}_{AB}$ by its magnitude to obtain a unit vector $\mathbf{e}_{AB}$ having the same direction as the force $\mathbf{T}$. Then we can obtain $\mathbf{T}$ in terms of scalar components by expressing it as the product of its magnitude and $\mathbf{e}_{AB}$. To begin this procedure, we must first determine the coordinates of the collar A. We will do so by obtaining a unit vector $\mathbf{e}_{CD}$ pointing from C toward D and multiplying it by 0.2 m to determine the position of the collar A relative to C.

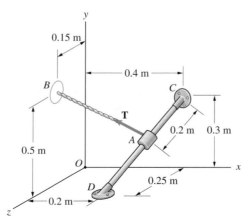

Figure 2.31

Solution

Determining the Coordinates of Point A   The position vector from C to D is

$$\mathbf{r}_{CD} = (0.2 - 0.4)\mathbf{i} + (0 - 0.3)\mathbf{j} + (0.25 - 0)\mathbf{k}$$
$$= -0.2\mathbf{i} - 0.3\mathbf{j} + 0.25\mathbf{k} \text{ (m)}.$$

Dividing this vector by its magnitude, we obtain the unit vector $\mathbf{e}_{CD}$ (Fig. a):

$$\mathbf{e}_{CD} = \frac{\mathbf{r}_{CD}}{|\mathbf{r}_{CD}|} = \frac{-0.2\mathbf{i} - 0.3\mathbf{j} + 0.25\mathbf{k}}{\sqrt{(-0.2)^2 + (-0.3)^2 + (0.25)^2}}$$
$$= -0.456\mathbf{i} - 0.684\mathbf{j} + 0.570\mathbf{k}.$$

Using this vector, we obtain the position vector from C to A:

$$\mathbf{r}_{CA} = (0.2 \text{ m})\mathbf{e}_{CD} = -0.091\mathbf{i} - 0.137\mathbf{j} + 0.114\mathbf{k} \text{ (m)}.$$

The position vector from the origin of the coordinate system to C is $\mathbf{r}_{OC} = 0.4\mathbf{i} + 0.3\mathbf{j}\text{(m)}$, so the position vector from the origin to A is

$$\mathbf{r}_{OA} = \mathbf{r}_{OC} + \mathbf{r}_{CA} = (0.4\mathbf{i} + 0.3\mathbf{j}) + (-0.091\mathbf{i} - 0.137\mathbf{j} + 0.114\mathbf{k})$$
$$= 0.309\mathbf{i} + 0.163\mathbf{j} + 0.114\mathbf{k} \text{ (m)}.$$

The coordinates of A are $(0.309, 0.163, 0.114)$ m.

Determining the Components of T   Using the coordinates of point A, the position vector from A to B is

$$\mathbf{r}_{AB} = (0 - 0.309)\mathbf{i} + (0.5 - 0.163)\mathbf{j} + (0.15 - 0.114)\mathbf{k}$$
$$= -0.309\mathbf{i} + 0.337\mathbf{j} + 0.036\mathbf{k} \text{ (m)}.$$

Dividing this vector by its magnitude, we obtain the unit vector $\mathbf{e}_{AB}$ [Fig. (a)]:

$$\mathbf{e}_{AB} = \frac{\mathbf{r}_{AB}}{|\mathbf{r}_{AB}|} = \frac{-0.309\mathbf{i} + 0.337\mathbf{j} + 0.036\mathbf{k}}{\sqrt{(-0.309)^2 + (0.337)^2 + (0.036)^2}}$$
$$= -0.674\mathbf{i} + 0.735\mathbf{j} + 0.079\mathbf{k}.$$

The force $\mathbf{T}$ is

$$\mathbf{T} = |\mathbf{T}|\mathbf{e}_{AB} = (50 \text{ N})(-0.674\mathbf{i} + 0.735\mathbf{j} + 0.079\mathbf{k})$$
$$= -33.7\mathbf{i} + 36.7\mathbf{j} + 3.9\mathbf{k} \text{ (N)}.$$

(a) The unit vectors $\mathbf{e}_{AB}$ and $\mathbf{e}_{CD}$.

Products of Vectors

Two kinds of products of vectors, the dot and cross products, have been found to have applications in science and engineering, especially in mechanics and electromagnetic field theory. We use both of these products in Chapter 4 to evaluate moments of forces about points and lines. We discuss them here so that you can concentrate on mechanics when we introduce moments and not be distracted by the details of vector operations.

2.5 Dot Products

The dot product of two vectors has many uses, including resolving a vector into components parallel and perpendicular to a given line and determining the angle between two lines in space.

Definition

Consider two vectors **U** and **V** (Fig. 2.32a). The *dot product* of **U** and **V**, denoted by **U · V** (hence the name "dot product"), is defined to be the product of the magnitude of **U**, the magnitude of **V**, and the cosine of the angle θ between **U** and **V** when they are placed tail to tail (Fig. 2.32b):

$$\mathbf{U \cdot V} = |\mathbf{U}||\mathbf{V}| \cos\theta. \tag{2.18}$$

Because the result of the dot product is a scalar, the dot product is sometimes called the scalar product. The units of the dot product are the product of the units of the two vectors. *Notice that the dot product of two nonzero vectors is equal to zero if and only if the vectors are perpendicular.*

The dot product has the properties

$$\mathbf{U \cdot V} = \mathbf{V \cdot U}, \quad \text{The dot product is commutative.} \tag{2.19}$$

$$a(\mathbf{U \cdot V}) = (a\mathbf{U}) \cdot \mathbf{V} = \mathbf{U} \cdot (a\mathbf{V}), \quad \text{The dot product is associative with respect to scalar multiplication.} \tag{2.20}$$

and

$$\mathbf{U \cdot (V + W)} = \mathbf{U \cdot V} + \mathbf{U \cdot W} \quad \text{The dot product is distributive with respect to vector addition.} \tag{2.21}$$

for any scalar a and vectors **U**, **V**, and **W**.

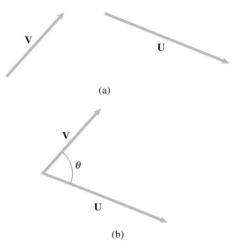

Figure 2.32
(a) The vectors **U** and **V**.
(b) The angle θ between **U** and **V** when the two vectors are placed tail to tail.

Dot Products in Terms of Components

In this section we derive an equation that allows you to determine the dot product of two vectors if you know their scalar components. The derivation also results in an equation for the angle between the vectors. The first step is to determine the dot products formed from the unit vectors **i**, **j**, and **k**. Let's evaluate the dot product **i · i**. The magnitude $|\mathbf{i}| = 1$, and the angle between two identical vectors placed tail to tail is zero, so we obtain

$$\mathbf{i \cdot i} = |\mathbf{i}||\mathbf{i}| \cos(0) = (1)(1)(1) = 1.$$

The dot product of **i** and **j** is

$$\mathbf{i \cdot j} = |\mathbf{i}||\mathbf{j}| \cos(90°) = (1)(1)(0) = 0.$$

Continuing in this way, we obtain

$$\begin{aligned}
\mathbf{i \cdot i} &= 1, & \mathbf{i \cdot j} &= 0, & \mathbf{i \cdot k} &= 0, \\
\mathbf{j \cdot i} &= 0, & \mathbf{j \cdot j} &= 1, & \mathbf{j \cdot k} &= 0, \\
\mathbf{k \cdot i} &= 0, & \mathbf{k \cdot j} &= 0, & \mathbf{k \cdot k} &= 1.
\end{aligned} \tag{2.22}$$

The dot product of two vectors **U** and **V** expressed in terms of their components is

$$\mathbf{U} \cdot \mathbf{V} = (U_x\mathbf{i} + U_y\mathbf{j} + U_z\mathbf{k}) \cdot (V_x\mathbf{i} + V_y\mathbf{j} + V_z\mathbf{k})$$

$$= U_xV_x(\mathbf{i} \cdot \mathbf{i}) + U_xV_y(\mathbf{i} \cdot \mathbf{j}) + U_xV_z(\mathbf{i} \cdot \mathbf{k})$$

$$+ U_yV_x(\mathbf{j} \cdot \mathbf{i}) + U_yV_y(\mathbf{j} \cdot \mathbf{j}) + U_yV_z(\mathbf{j} \cdot \mathbf{k})$$

$$+ U_zV_x(\mathbf{k} \cdot \mathbf{i}) + U_zV_y(\mathbf{k} \cdot \mathbf{j}) + U_zV_z(\mathbf{k} \cdot \mathbf{k}).$$

In obtaining this result, we used Eqs. (2.20) and (2.21). Substituting Eqs. (2.22) into this expression, we obtain an equation for the dot product in terms of the scalar components of the two vectors:

$$\mathbf{U} \cdot \mathbf{V} = U_xV_x + U_yV_y + U_zV_z. \tag{2.23}$$

To obtain an equation for the angle θ in terms of the components of the vectors, we equate the expression for the dot product given by Eq. (2.23) to the definition of the dot product, Eq. (2.18), and solve for $\cos\theta$:

$$\cos\theta = \frac{\mathbf{U} \cdot \mathbf{V}}{|\mathbf{U}||\mathbf{V}|} = \frac{U_xV_x + U_yV_y + U_zV_z}{|\mathbf{U}||\mathbf{V}|}. \tag{2.24}$$

Vector Components Parallel and Normal to a Line

In some engineering applications you must resolve a vector into components that are parallel and normal (perpendicular) to a given line. The component of a vector parallel to a line is called the *projection* of the vector onto the line. For example, when the vector represents a force, the projection of the force onto a line is the component of the force in the direction of the line.

We can determine the components of a vector parallel and normal to a line by using the dot product. Consider a vector **U** and a straight line L (Fig. 2.33a). We can resolve **U** into components $\mathbf{U}_p$ and $\mathbf{U}_n$ that are parallel and normal to L (Fig. 2.33b).

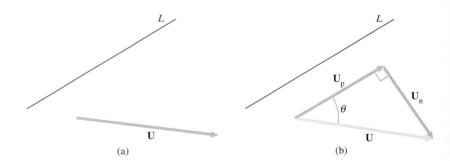

Figure 2.33
(a) A vector **U** and line L.
(b) Resolving **U** into components parallel and normal to L.

The Parallel Component   In terms of the angle θ between **U** and the component $\mathbf{U}_p$, the magnitude of $\mathbf{U}_p$ is

$$|\mathbf{U}_p| = |\mathbf{U}|\cos\theta. \tag{2.25}$$

Let **e** be a unit vector parallel to L (Fig. 2.34). The dot product of **e** and **U** is

$$\mathbf{e} \cdot \mathbf{U} = |\mathbf{e}||\mathbf{U}|\cos\theta = |\mathbf{U}|\cos\theta.$$

Comparing this result with Eq. (2.25), we see that the magnitude of $\mathbf{U_p}$ is

$$|\mathbf{U_p}| = \mathbf{e} \cdot \mathbf{U}.$$

Therefore the parallel component, or projection of $\mathbf{U}$ onto L, is

$$\mathbf{U_p} = (\mathbf{e} \cdot \mathbf{U})\mathbf{e}. \tag{2.26}$$

(This equation holds even if $\mathbf{e}$ doesn't point in the direction of $\mathbf{U_p}$. In that case, the angle $\theta > 90°$ and $\mathbf{e} \cdot \mathbf{U}$ is negative.) When the components of a vector and the components of a unit vector $\mathbf{e}$ parallel to a line L are known, we can use Eq. (2.26) to determine the component of the vector parallel to L.

The Normal Component Once the parallel component, has been determined, we can obtain the normal component from the relation $\mathbf{U} = \mathbf{U_p} + \mathbf{U_n}$:

$$\mathbf{U_n} = \mathbf{U} - \mathbf{U_p}. \tag{2.27}$$

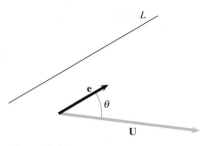

Figure 2.34
The unit vector $\mathbf{e}$ is parallel to L.

Study Questions

1. What is the definition of the dot product?
2. The dot product is commutative. What does that mean?
3. If you know the components of two vectors $\mathbf{U}$ and $\mathbf{V}$, how can you determine their dot product?
4. How can you use the dot product to determine the components of a vector parallel and normal to a line?

Example 2.13

Calculating a Dot Product

The magnitude of the force $\mathbf{F}$ in Fig. 2.35 is 100 lb. The magnitude of the vector $\mathbf{r}$ from point O to point A is 8 ft.
(a) Use the definition of the dot product to determine $\mathbf{r} \cdot \mathbf{F}$.
(b) Use Eq. (2.23) to determine $\mathbf{r} \cdot \mathbf{F}$.

Strategy

(a) Since we know the magnitudes of $\mathbf{r}$ and $\mathbf{F}$ and the angle between them when they are placed tail to tail, we can determine $\mathbf{r} \cdot \mathbf{F}$ directly from the definition.
(b) We can determine the components of $\mathbf{r}$ and $\mathbf{F}$ and use Eq. (2.23) to determine their dot product.

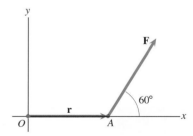

Figure 2.35

Solution

(a) Using the definition of the dot product,

$$\mathbf{r} \cdot \mathbf{F} = |\mathbf{r}||\mathbf{F}| \cos \theta = (8)(100) \cos 60° = 400 \text{ ft-lb}.$$

(b) The vector $\mathbf{r} = 8\mathbf{i}$ (ft). The vector $\mathbf{F}$ in terms of scalar components is

$$\mathbf{F} = 100 \cos 60° \, \mathbf{i} + 100 \sin 60° \, \mathbf{j} \text{ (lb)}.$$

Therefore the dot product of $\mathbf{r}$ and $\mathbf{F}$ is

$$\mathbf{r} \cdot \mathbf{F} = r_x F_x + r_y F_y + r_z F_z$$
$$= (8)(100 \cos 60°) + (0)(100 \sin 60°) + (0)(0) = 400 \text{ ft-lb}.$$

Example 2.14

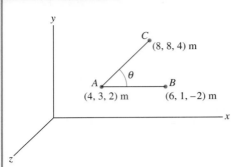

Figure 2.36

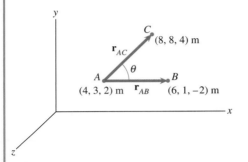

(a) The position vectors $\mathbf{r}_{AB}$ and $\mathbf{r}_{AC}$.

Using the Dot Product to Determine an Angle

What is the angle θ between the lines AB and AC in Fig. 2.36?

Strategy

We know the coordinates of the points A, B, and C, so we can determine the components of the vector $\mathbf{r}_{AB}$ from A to B and the vector $\mathbf{r}_{AC}$ from A to C (Fig. a). Then we can use Eq. (2.24) to determine θ.

Solution

The vectors $\mathbf{r}_{AB}$ and $\mathbf{r}_{AC}$ are

$$\mathbf{r}_{AB} = (6-4)\mathbf{i} + (1-3)\mathbf{j} + (-2-2)\mathbf{k} = 2\mathbf{i} - 2\mathbf{j} - 4\mathbf{k} \text{ (m)},$$

$$\mathbf{r}_{AC} = (8-4)\mathbf{i} + (8-3)\mathbf{j} + (4-2)\mathbf{k} = 4\mathbf{i} + 5\mathbf{j} + 2\mathbf{k} \text{ (m)}.$$

Their magnitudes are

$$|\mathbf{r}_{AB}| = \sqrt{(2)^2 + (-2)^2 + (-4)^2} = 4.90 \text{ m},$$

$$|\mathbf{r}_{AC}| = \sqrt{(4)^2 + (5)^2 + (2)^2} = 6.71 \text{ m}.$$

The dot product of $\mathbf{r}_{AB}$ and $\mathbf{r}_{AC}$ is

$$\mathbf{r}_{AB} \cdot \mathbf{r}_{AC} = (2)(4) + (-2)(5) + (-4)(2) = -10 \text{ m}^2.$$

Therefore

$$\cos\theta = \frac{\mathbf{r}_{AB} \cdot \mathbf{r}_{AC}}{|\mathbf{r}_{AB}||\mathbf{r}_{AC}|} = \frac{-10}{(4.90)(6.71)} = -0.304.$$

The angle $\theta = \arccos(-0.304) = 107.7°$.

Example 2.15

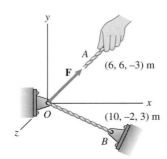

Figure 2.37

Components Parallel and Normal to a Line

Suppose that you pull on the cable OA in Fig. 2.37, exerting a 50-N force $\mathbf{F}$ at O. What are the components of $\mathbf{F}$ parallel and normal to the cable OB?

Strategy

Resolving $\mathbf{F}$ into components parallel and normal to OB (Fig. a), we can determine the components by using Eqs. (2.26) and (2.27). But to apply them, we must first express $\mathbf{F}$ in terms of scalar components and determine the components of a unit vector parallel to OB. We can obtain the components of

F by determining the components of the unit vector pointing from O toward A and multiplying them by $|\mathbf{F}|$.

Solution

The position vectors from O to A and from O to B are (Fig. b)

$$\mathbf{r}_{OA} = 6\mathbf{i} + 6\mathbf{j} - 3\mathbf{k} \text{ (m)},$$

$$\mathbf{r}_{OB} = 10\mathbf{i} - 2\mathbf{j} + 3\mathbf{k} \text{ (m)}.$$

Their magnitudes are $|\mathbf{r}_{OA}| = 9$ m and $|\mathbf{r}_{OB}| = 10.6$ m. Dividing these vectors by their magnitudes, we obtain unit vectors that point from the origin toward A and B (Fig. c):

$$\mathbf{e}_{OA} = \frac{\mathbf{r}_{OA}}{|\mathbf{r}_{OA}|} = \frac{6\mathbf{i} + 6\mathbf{j} - 3\mathbf{k}}{9} = 0.667\mathbf{i} + 0.667\mathbf{j} - 0.333\mathbf{k},$$

$$\mathbf{e}_{OB} = \frac{\mathbf{r}_{OB}}{|\mathbf{r}_{OB}|} = \frac{10\mathbf{i} - 2\mathbf{j} + 3\mathbf{k}}{10.6} = 0.941\mathbf{i} + 0.188\mathbf{j} - 0.282\mathbf{k}.$$

The force **F** in terms of scalar components is

$$\mathbf{F} = |\mathbf{F}|\mathbf{e}_{OA} = (50)(0.667\mathbf{i} + 0.667\mathbf{j} - 0.333\mathbf{k})$$

$$= 33.3\mathbf{i} + 33.3\mathbf{j} - 16.7\mathbf{k} \text{ (N)}.$$

Taking the dot product of $\mathbf{e}_{OB}$ and **F**, we obtain

$$\mathbf{e}_{OB} \cdot \mathbf{F} = (0.941)(33.3) + (-0.188)(33.3) + (0.282)(-16.7)$$

$$= 20.4 \text{ N}.$$

The parallel component of **F** is

$$\mathbf{F}_p = (\mathbf{e}_{OB} \cdot \mathbf{F})\mathbf{e}_{OB} = (20.4)(0.941\mathbf{i} - 0.188\mathbf{j} + 0.282\mathbf{k})$$

$$= 19.2\mathbf{i} - 3.8\mathbf{j} + 5.8\mathbf{k} \text{ (N)}.$$

and the normal component is

$$\mathbf{F}_n = \mathbf{F} - \mathbf{F}_p = 14.2\mathbf{i} + 37.2\mathbf{j} - 22.4\mathbf{k} \text{ (N)}.$$

Discussion

You can confirm that two vectors are perpendicular by making sure their dot product is zero. In this example,

$$\mathbf{F}_p \cdot \mathbf{F}_n = (19.2)(14.2) + (-3.8)(37.2) + (5.8)(-22.4) = 0.$$

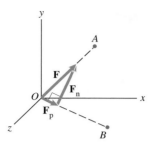

(a) The components of **F** parallel and normal to OB.

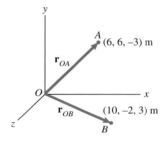

(b) The position vectors $\mathbf{r}_{OA}$ and $\mathbf{r}_{OB}$.

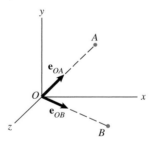

(c) The unit vectors $\mathbf{e}_{OA}$ and $\mathbf{e}_{OB}$.

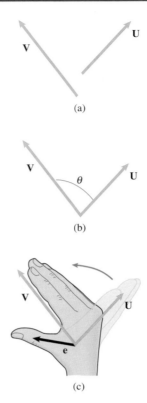

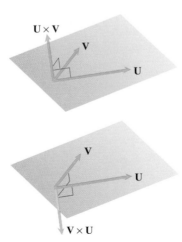

2.6 Cross Products

Figure 2.38
(a) The vectors **U** and **V**.
(b) The angle θ between the vectors when they are placed tail to tail.
(c) Determining the direction of **e** by the right-hand rule.

Like the dot product, the cross product of two vectors has many applications, including determining the rate of rotation of a fluid particle and calculating the force exerted on a charged particle by a magnetic field. Because of its usefulness for determining moments of forces, the cross product is an indispensable tool in mechanics. In this section we show you how to evaluate cross products and give examples of simple applications.

Definition

Consider two vectors **U** and **V** (Fig. 2.38a). The *cross product* of **U** and **V**, denoted **U** × **V**, is defined by

$$\mathbf{U} \times \mathbf{V} = |\mathbf{U}||\mathbf{V}| \sin\theta \, \mathbf{e}. \tag{2.28}$$

The angle θ is the angle between **U** and **V** when they are placed tail to tail (Fig. 2.38b). The vector **e** is a unit vector defined to be perpendicular to both **U** and **V**. Since this leaves two possibilities for the direction of **e**, the vectors **U**, **V**, and **e** are defined to be a right-handed system. The *right-hand rule* for determining the direction of **e** is shown in Fig. 2.38c. When you point the four fingers of your right hand in the direction of the vector **U** (the first vector in the cross product) and close your fingers toward the vector **V** (the second vector in the cross product), your thumb points in the direction of **e**.

Because the result of the cross product is a vector, it is sometimes called the vector product. The units of the cross product are the product of the units of the two vectors. Notice that the cross product of two nonzero vectors is equal to zero if and only if the two vectors are parallel.

An interesting property of the cross product is that it is *not* commutative. Eq. (2.28) implies that the magnitude of the vector **U** × **V** is equal to the magnitude of the vector **V** × **U**, but the right-hand rule indicates that they are opposite in direction (Fig. 2.39). That is,

$$\mathbf{U} \times \mathbf{V} = -\mathbf{V} \times \mathbf{U}. \quad \text{The cross product is } \textit{not} \text{ commutative.} \tag{2.29}$$

The cross product also satisfies the relations

$$a(\mathbf{U} \times \mathbf{V}) = (a\mathbf{U}) \times \mathbf{V} = \mathbf{U} \times (a\mathbf{V}) \quad \begin{array}{l}\text{The cross product is}\\\text{associative with}\\\text{respect to scalar}\\\text{multiplication.}\end{array} \tag{2.30}$$

and

$$\mathbf{U} \times (\mathbf{V} + \mathbf{W}) = (\mathbf{U} \times \mathbf{V}) + (\mathbf{U} \times \mathbf{W}) \quad \begin{array}{l}\text{The cross product}\\\text{is distributive with}\\\text{respect to vector}\\\text{addition.}\end{array} \tag{2.31}$$

for any scalar a and vectors **U**, **V**, and **W**.

Cross Products in Terms of Components

To obtain an equation for the cross product of two vectors in terms of their components, we must determine the cross products formed from the unit vectors **i**, **j**, and **k**. Since the angle between two identical vectors placed tail to tail is zero,

Figure 2.39
Directions of **U** × **V** and **V** × **U**.

$$\mathbf{i} \times \mathbf{i} = |\mathbf{i}||\mathbf{i}| \sin(0)\mathbf{e} = \mathbf{0}.$$

The cross product $\mathbf{i} \times \mathbf{j}$ is

$$\mathbf{i} \times \mathbf{j} = |\mathbf{i}||\mathbf{j}| \sin(90)°\mathbf{e} = \mathbf{e},$$

where $\mathbf{e}$ is a unit vector perpendicular to $\mathbf{i}$ and $\mathbf{j}$. Either $\mathbf{e} = \mathbf{k}$ or $\mathbf{e} = -\mathbf{k}$. Applying the right-hand rule, we find that $\mathbf{e} = \mathbf{k}$ (Fig. 2.40). Therefore

$$\mathbf{i} \times \mathbf{j} = \mathbf{k}.$$

Continuing in this way, we obtain

$$
\begin{array}{lll}
\mathbf{i} \times \mathbf{i} = \mathbf{0}, & \mathbf{i} \times \mathbf{j} = \mathbf{k}. & \mathbf{i} \times \mathbf{k} = -\mathbf{j}, \\
\mathbf{j} \times \mathbf{i} = -\mathbf{k}, & \mathbf{j} \times \mathbf{j} = \mathbf{0}, & \mathbf{j} \times \mathbf{k} = \mathbf{i}, \\
\mathbf{k} \times \mathbf{i} = \mathbf{j}, & \mathbf{k} \times \mathbf{j} = -\mathbf{i}, & \mathbf{k} \times \mathbf{k} = \mathbf{0}.
\end{array}
\tag{2.32}
$$

These results can be remembered easily by arranging the unit vectors in a circle, as shown in Fig. 2.41a. The cross product of adjacent vectors is equal to the third vector with a positive sign if the order of the vectors in the cross product is the order indicated by the arrows and a negative sign otherwise. For example, in Fig. 2.41b we see that $\mathbf{i} \times \mathbf{j} = \mathbf{k}$, but $\mathbf{i} \times \mathbf{k} = -\mathbf{j}$.

The cross product of two vectors $\mathbf{U}$ and $\mathbf{V}$ expressed in terms of their components is

$$
\begin{aligned}
\mathbf{U} \times \mathbf{V} &= (U_x\mathbf{i} + U_y\mathbf{j} + U_z\mathbf{k}) \times (V_x\mathbf{i} + V_y\mathbf{j} + V_z\mathbf{k}) \\
&= U_xV_x(\mathbf{i} \times \mathbf{i}) + U_xV_y(\mathbf{i} \times \mathbf{j}) + U_xV_z(\mathbf{i} \times \mathbf{k}) \\
&\quad + U_yV_x(\mathbf{j} \times \mathbf{i}) + U_yV_y(\mathbf{j} \times \mathbf{j}) + U_yV_z(\mathbf{j} \times \mathbf{k}) \\
&\quad + U_zV_x(\mathbf{k} \times \mathbf{i}) + U_zV_y(\mathbf{k} \times \mathbf{j}) + U_zV_z(\mathbf{k} \times \mathbf{k}).
\end{aligned}
$$

By substituting Eqs. (2.32) into this expression, we obtain the equation

$$
\begin{aligned}
\mathbf{U} \times \mathbf{V} &= (U_yV_z - U_zV_y)\mathbf{i} - (U_xV_z - U_zV_x)\mathbf{j} \\
&\quad + (U_xV_y - U_yV_x)\mathbf{k}.
\end{aligned}
\tag{2.33}
$$

This result can be compactly written as the determinant

$$
\mathbf{U} \times \mathbf{V} = \begin{vmatrix} \mathbf{i} & \mathbf{j} & \mathbf{k} \\ U_x & U_y & U_z \\ V_x & V_y & V_z \end{vmatrix}
\tag{2.34}
$$

This equation is based on Eqs. (2.32), which we obtained using a right-handed coordinate system. It gives the correct result for the cross product only if a right-handed coordinate system is used to determine the components of $\mathbf{U}$ and $\mathbf{V}$.

Evaluating a 3 × 3 Determinant

A 3 × 3 determinant can be evaluated by repeating its first two columns as shown and evaluating the products of the terms along the six diagonal lines.

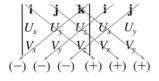

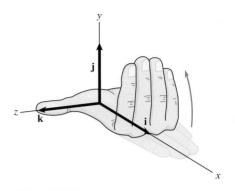

Figure 2.40
The right-hand rule indicates that $\mathbf{i} \times \mathbf{j} = \mathbf{k}$.

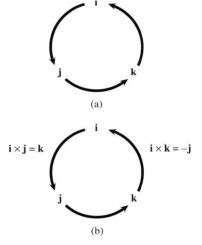

Figure 2.41
(a) Arrange the unit vectors in a circle with arrows to indicate their order.
(b) You can use the circle to determine their cross products.

Adding the terms obtained from the diagonals that run downward to the right (blue arrows) and subtracting the terms obtained from the diagonals that run downward to the left (red arrows) gives the value of the determinant:

$$\begin{vmatrix} \mathbf{i} & \mathbf{j} & \mathbf{k} \\ U_x & U_y & U_z \\ V_x & V_y & V_z \end{vmatrix} = \begin{array}{l} U_y V_z \mathbf{i} + U_z V_x \mathbf{j} + U_x V_y \mathbf{k} \\ - U_y V_x \mathbf{k} - U_z V_y \mathbf{i} - U_x V_z \mathbf{j}. \end{array}$$

A 3×3 determinant can also be evaluated by expressing it as

$$\begin{vmatrix} \mathbf{i} & \mathbf{j} & \mathbf{k} \\ U_x & U_y & U_z \\ V_x & V_y & V_z \end{vmatrix} = \mathbf{i} \begin{vmatrix} U_y & U_z \\ V_y & V_z \end{vmatrix} - \mathbf{j} \begin{vmatrix} U_x & U_z \\ V_x & V_z \end{vmatrix} + \mathbf{k} \begin{vmatrix} U_x & U_y \\ V_x & V_y \end{vmatrix}.$$

The terms on the right are obtained by multiplying each element of the first row of the 3×3 determinant by the 2×2 determinant obtained by crossing out that element's row and column. For example, the first element of the first row, $\mathbf{i}$, is multiplied by the 2×2 determinant

$$\begin{vmatrix} \cancel{\mathbf{i}} & \cancel{\mathbf{j}} & \cancel{\mathbf{k}} \\ \cancel{U}_x & U_y & U_z \\ \cancel{V}_x & V_y & V_z \end{vmatrix}$$

Be sure to remember that the second term is subtracted. Expanding the 2×2 determinants, we obtain the value of the determinant:

$$\begin{vmatrix} \mathbf{i} & \mathbf{j} & \mathbf{k} \\ U_x & U_y & U_z \\ V_x & V_y & V_z \end{vmatrix} = \begin{array}{l} (U_y V_z - U_z V_y)\mathbf{i} - (U_x V_z - U_z V_x)\mathbf{j} \\ + (U_x V_y - U_y V_x)\mathbf{k}. \end{array}$$

2.7 Mixed Triple Products

In Chapter 4, when we discuss the moment of a force about a line, we will use an operation called the *mixed triple product*, defined by

$$\mathbf{U} \cdot (\mathbf{V} \times \mathbf{W}). \tag{2.35}$$

In terms of the scalar components of the vectors,

$$\mathbf{U} \cdot (\mathbf{V} \times \mathbf{W}) = (U_x \mathbf{i} + U_y \mathbf{j} + U_z \mathbf{k}) \cdot \begin{vmatrix} \mathbf{i} & \mathbf{j} & \mathbf{k} \\ V_x & V_y & V_z \\ W_x & W_y & W_z \end{vmatrix}$$

$$= (U_x \mathbf{i} + U_y \mathbf{j} + U_z \mathbf{k}) \cdot [(V_y W_z - V_z W_y)\mathbf{i}$$

$$- (V_x W_z - V_z W_x)\mathbf{j} + (V_x W_y - V_y W_x)\mathbf{k}]$$

$$= U_x(V_y W_z - V_z W_y) - U_y(V_x W_z - V_z W_x)$$

$$+ U_z(V_x W_y - V_y W_x).$$

This result can be expressed as the determinant

$$\mathbf{U} \cdot (\mathbf{V} \times \mathbf{W}) = \begin{vmatrix} U_x & U_y & U_z \\ V_x & V_y & V_z \\ W_x & W_y & W_z \end{vmatrix}. \tag{2.36}$$

Interchanging any two of the vectors in the mixed triple product changes the sign but not the absolute value of the result. For example,

$$\mathbf{U} \cdot (\mathbf{V} \times \mathbf{W}) = -\mathbf{W} \cdot (\mathbf{V} \times \mathbf{U}).$$

If the vectors **U**, **V**, and **W** in Fig. 2.42 form a right-handed system, it can be shown that the volume of the parallelepiped equals $\mathbf{U} \cdot (\mathbf{V} \times \mathbf{W})$.

Study Questions

1. What is the definition of the cross product?
2. If you know the components of two vectors **U** and **V**, how can you determine their cross product?
3. If the cross product of two vectors is zero, what does that mean?

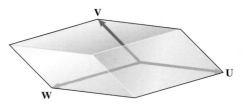

Figure 2.42
Parallelepiped defined by the vectors **U**, **V**, and **W**.

Example 2.16

Cross Product in Terms of Components

Determine the cross product $\mathbf{U} \times \mathbf{V}$ of the vectors $\mathbf{U} = -2\mathbf{i} + \mathbf{j}$ and $\mathbf{V} = 3\mathbf{i} - 4\mathbf{k}$.

Strategy

We can evaluate the cross product of the vectors in two ways: by evaluating the cross products of their components term by term and by using Eq. (2.34).

Solution

$$
\begin{aligned}
\mathbf{U} \times \mathbf{V} &= (-2\mathbf{i} + \mathbf{j}) \times (3\mathbf{i} - 4\mathbf{k}) \\
&= (-2)(3)(\mathbf{i} \times \mathbf{i}) + (-2)(-4)(\mathbf{i} \times \mathbf{k}) + (1)(3)(\mathbf{j} \times \mathbf{i}) \\
&\quad + (1)(-4)(\mathbf{j} \times \mathbf{k}) \\
&= (-6)(0) + (8)(-\mathbf{j}) + (3)(-\mathbf{k}) + (-4)(\mathbf{i}) \\
&= -4\mathbf{i} - 8\mathbf{j} - 3\mathbf{k}.
\end{aligned}
$$

Using Eq. (2.34), we obtain

$$\mathbf{U} \times \mathbf{V} = \begin{vmatrix} \mathbf{i} & \mathbf{j} & \mathbf{k} \\ U_x & U_y & U_z \\ V_x & V_y & V_z \end{vmatrix} = \begin{vmatrix} \mathbf{i} & \mathbf{j} & \mathbf{k} \\ -2 & 1 & 0 \\ 3 & 0 & -4 \end{vmatrix} = -4\mathbf{i} - 8\mathbf{j} - 3\mathbf{k}.$$

Example 2.17

Calculating the Cross Product

The magnitude of the force **F** in Fig. 2.43 is 100 lb. The magnitude of the vector **r** from point O to point A is 8 ft.
(a) Use the definition of the cross product to determine **r** × **F**.
(b) Use Eq. (2.34) to determine **r** × **F**.

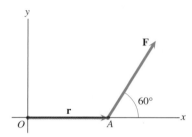

Figure 2.43

Strategy

(a) We know the magnitudes of **r** and **F** and the angle between them when they are placed tail to tail. Since both vectors lie in the *x*-*y* plane, the unit vector **k** is perpendicular to both **r** and **F**. We therefore have all the information we need to determine **r** × **F** directly from the definition.
(b) We can determine the components of **r** and **F** and use Eq. (2.34) to determine **r** × **F**.

Solution

(a) Using the definition of the cross product,

$$\mathbf{r} \times \mathbf{F} = |\mathbf{r}||\mathbf{F}| \sin\theta\,\mathbf{e} = (8)(100) \sin 60°\,\mathbf{e} = 693\,\mathbf{e}\ (\text{ft-lb}).$$

Since **e** is defined to be perpendicular to **r** and **F**, either **e** = **k** or **e** = −**k**. Pointing the fingers of the right hand in the direction of **r** and closing them toward **F**, the right-hand rule indicates that **e** = **k**. Therefore

$$\mathbf{r} \times \mathbf{F} = 693\mathbf{k}\ (\text{ft-lb}).$$

(b) The vector **r** = 8**i** (ft). The vector **F** in terms of scalar components is

$$\mathbf{F} = 100 \cos 60°\,\mathbf{i} + 100 \sin 60°\,\mathbf{j}\ (\text{lb}).$$

From Eq. (2.34),

$$\mathbf{r} \times \mathbf{F} = \begin{vmatrix} \mathbf{i} & \mathbf{j} & \mathbf{k} \\ r_x & r_y & r_z \\ F_x & F_y & F_z \end{vmatrix} = \begin{vmatrix} \mathbf{i} & \mathbf{j} & \mathbf{k} \\ 8 & 0 & 0 \\ 100 \cos 60° & 100 \sin 60° & 0 \end{vmatrix}$$

$$= (8)(100 \cos 60°)\mathbf{k} = 693\mathbf{k}\ (\text{ft-lb}).$$

Example 2.18

Minimum Distance from a Point to a Line

Consider the straight lines OA and OB in Fig. 2.44.
(a) Determine the components of a unit vector that is perpendicular to both OA and OB.
(b) What is the minimum distance from point A to the line OB?

Strategy

(a) Let $\mathbf{r}_{OA}$ and $\mathbf{r}_{OB}$ be the position vectors from O to A and from O to B (Fig. a). Since the cross product $\mathbf{r}_{OA} \times \mathbf{r}_{OB}$ is perpendicular to $\mathbf{r}_{OA}$ and $\mathbf{r}_{OB}$ we will determine it and divide it by its magnitude to obtain a unit vector perpendicular to the lines OA and OB.
(b) The minimum distance from A to the line OB is the length d of the straight line from A to OB that is perpendicular to OB (Fig. b). We can see that $d = |\mathbf{r}_{OA}| \sin\theta$, where θ is the angle between $\mathbf{r}_{OA}$ and $\mathbf{r}_{OB}$. From the definition of the cross product, the magnitude of $\mathbf{r}_{OA} \times \mathbf{r}_{OB}$ is $|\mathbf{r}_{OA}||\mathbf{r}_{OB}| \sin\theta$, so we can determine d by dividing the magnitude of $\mathbf{r}_{OA} \times \mathbf{r}_{OB}$ by the magnitude of $\mathbf{r}_{OB}$.

Solution

(a) The components of $\mathbf{r}_{OA}$ and $\mathbf{r}_{OB}$ are

$$\mathbf{r}_{OA} = 10\mathbf{i} - 2\mathbf{j} + 3\mathbf{k} \text{ (m)},$$
$$\mathbf{r}_{OB} = 6\mathbf{i} + 6\mathbf{j} - 3\mathbf{k} \text{ (m)}.$$

By using Eq. (2.34), we obtain $\mathbf{r}_{OA} \times \mathbf{r}_{OB}$:

$$\mathbf{r}_{OA} \times \mathbf{r}_{OB} = \begin{vmatrix} \mathbf{i} & \mathbf{j} & \mathbf{k} \\ 10 & -2 & 3 \\ 6 & 6 & -3 \end{vmatrix} = -12\mathbf{i} + 48\mathbf{j} + 72\mathbf{k} \ (\text{m}^2).$$

This vector is perpendicular to $\mathbf{r}_{OA}$ and $\mathbf{r}_{OB}$. Dividing it by its magnitude, we obtain a unit vector $\mathbf{e}$ that is perpendicular to the lines OA and OB:

$$\mathbf{e} = \frac{\mathbf{r}_{OA} \times \mathbf{r}_{OB}}{|\mathbf{r}_{OA} \times \mathbf{r}_{OB}|} = \frac{-12\mathbf{i} + 48\mathbf{j} + 72\mathbf{k}}{\sqrt{(-12)^2 + (48)^2 + (72)^2}}$$
$$= -0.137\mathbf{i} + 0.549\mathbf{j} + 0.824\mathbf{k}.$$

(b) From Fig. b, the minimum distance d is

$$d = |\mathbf{r}_{OA}| \sin\theta.$$

The magnitude of $\mathbf{r}_{OA} \times \mathbf{r}_{OB}$ is

$$|\mathbf{r}_{OA} \times \mathbf{r}_{OB}| = |\mathbf{r}_{OA}||\mathbf{r}_{OB}| \sin\theta.$$

Solving this equation for $\sin\theta$, the distance d is

$$d = |\mathbf{r}_{OA}|\left(\frac{|\mathbf{r}_{OA} \times \mathbf{r}_{OB}|}{|\mathbf{r}_{OA}||\mathbf{r}_{OB}|}\right) = \frac{|\mathbf{r}_{OA} \times \mathbf{r}_{OB}|}{|\mathbf{r}_{OB}|}$$
$$= \frac{\sqrt{(-12)^2 + (48)^2 + (72)^2}}{\sqrt{(6)^2 + (6)^2 + (-3)^2}} = 9.71 \text{ m}.$$

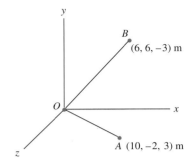

Figure 2.44

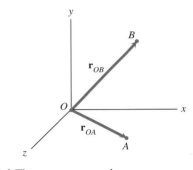

(a) The vectors $\mathbf{r}_{OA}$ and $\mathbf{r}_{OB}$.

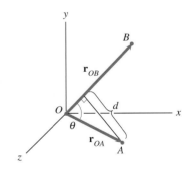

(b) The minimum distance d from A to the line OB.

Example 2.19

Component of a Vector Perpendicular to a Plane

The rope CE in Fig. 2.45 exerts a 500-N force $\mathbf{T}$ on the door $ABCD$. What is the magnitude of the component of $\mathbf{T}$ perpendicular to the door?

Strategy

We are given the coordinates of the corners A, B, and C of the door. By taking the cross product of the position vector $\mathbf{r}_{CB}$ from C to B and the position vector $\mathbf{r}_{CA}$ from C to A, we will obtain a vector that is perpendicular to the door. We can divide the resulting vector by its magnitude to obtain a unit vector perpendicular to the door and then apply Eq. (2.26) to determine the component of $\mathbf{T}$ perpendicular to the door.

Solution

The components of $\mathbf{r}_{CB}$ and $\mathbf{r}_{CA}$ are

$$\mathbf{r}_{CB} = 0.35\mathbf{i} - 0.2\mathbf{j} + 0.2\mathbf{k} \text{ (m)},$$
$$\mathbf{r}_{CA} = 0.5\mathbf{i} - 0.2\mathbf{j} \text{ (m)}.$$

Their cross product is

$$\mathbf{r}_{CB} \times \mathbf{r}_{CA} = \begin{vmatrix} \mathbf{i} & \mathbf{j} & \mathbf{k} \\ 0.35 & -0.2 & 0.2 \\ 0.5 & -0.2 & 0 \end{vmatrix} = 0.04\mathbf{i} + 0.1\mathbf{j} + 0.03\mathbf{k} \ (\text{m}^2).$$

Dividing this vector by its magnitude, we obtain a unit vector $\mathbf{e}$ that is perpendicular to the door (Fig. a):

$$\mathbf{e} = \frac{\mathbf{r}_{CB} \times \mathbf{r}_{CA}}{|\mathbf{r}_{CB} \times \mathbf{r}_{CA}|} = \frac{0.04\mathbf{i} + 0.1\mathbf{j} + 0.03\mathbf{k}}{\sqrt{(0.04)^2 + (0.1)^2 + (0.03)^2}}$$
$$= 0.358\mathbf{i} + 0.894\mathbf{j} + 0.268\mathbf{k}.$$

To use Eq. (2.26), we must express $\mathbf{T}$ in terms of its scalar components. The position vector from C to E is

$$\mathbf{r}_{CE} = 0.4\mathbf{i} + 0.05\mathbf{j} - 0.1\mathbf{k} \text{ (m)},$$

so we can express the force $\mathbf{T}$ as

$$\mathbf{T} = |\mathbf{T}| \frac{\mathbf{r}_{CE}}{|\mathbf{r}_{CE}|} = (500) \frac{0.4\mathbf{i} + 0.05\mathbf{j} - 0.1\mathbf{k}}{\sqrt{(0.4)^2 + (0.05)^2 + (-0.1)^2}}$$
$$= 481.5\mathbf{i} + 60.2\mathbf{j} - 120.4\mathbf{k} \text{ (N)}.$$

The component of $\mathbf{T}$ parallel to the unit vector $\mathbf{e}$, which is the component perpendicular to the door, is

$$\mathbf{T}_p = (\mathbf{e} \cdot \mathbf{T})\mathbf{e} = [(0.358)(481.5) + (0.894)(60.2) + (0.268)(-120.4)]\mathbf{e}$$
$$= 194\mathbf{e} \text{ (N)}.$$

The magnitude of $\mathbf{T}_p$ is 194 N.

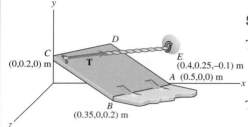

C (0,0.2,0) m, $\mathbf{T}$, D, E (0.4,0.25,-0.1) m, A (0.5,0,0) m, B (0.35,0,0.2) m

Figure 2.45

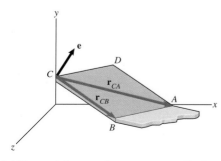

(a) Determining a unit vector perpendicular to the door.

Chapter Summary

In this chapter we have defined scalars, vectors, and vector operations. We showed how to express vectors in terms of cartesian components and carry out vector operations in terms of components. We introduced the definitions of the dot and cross products and the mixed triple product and demonstrated some applications of these operations, particularly the use of the dot product to resolve a vector into components parallel and perpendicular to a given direction. In Chapter 3 we will use vector operations to analyze forces acting on objects in equilibrium.

A physical quantity completely described by a real number is a *scalar*. A *vector* has both *magnitude* and *direction*. A vector is represented graphically by an arrow whose length is defined to be proportional to its magnitude.

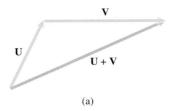

(a)

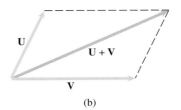

(b)

Rules for Manipulating Vectors

The sum of two vectors is defined by the *triangle rule* (Fig. a) or the equivalent *parallelogram rule* (Fig. b).

The product of a scalar a and a vector $\mathbf{U}$ is a vector $a\mathbf{U}$ with magnitude $|a||\mathbf{U}|$. Its direction is the same as $\mathbf{U}$ when a is positive and opposite to $\mathbf{U}$ when a is negative. The product $(-1)\mathbf{U}$ is written $-\mathbf{U}$ and is called the negative of $\mathbf{U}$. The division of $\mathbf{U}$ by a is the product $(1/a)\mathbf{U}$.

A *unit vector* is a vector whose magnitude is 1. A unit vector specifies a direction. Any vector $\mathbf{U}$ can be expressed as $|\mathbf{U}|\mathbf{e}$, where $\mathbf{e}$ is a unit vector with the same direction as $\mathbf{U}$. Dividing any vector by its magnitude yields a unit vector with the same direction as the vector.

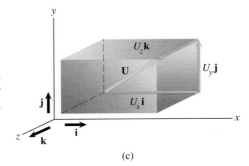

(c)

Cartesian Components

A vector $\mathbf{U}$ is expressed in terms of *scalar components* as

$$\mathbf{U} = U_x\mathbf{i} + U_y\mathbf{j} + U_z\mathbf{k} \quad \textbf{Eq. (2.12)}$$

(Fig. c). The coordinate system is *right-handed* (Fig. d): If the fingers of the right hand are pointed in the positive x direction and then closed toward the positive y direction, the thumb points in the z direction. The magnitude of $\mathbf{U}$ is

$$|\mathbf{U}| = \sqrt{U_x^2 + U_y^2 + U_z^2}. \quad \textbf{Eq. (2.14)}$$

Let θ_x, θ_y, and θ_z be the angles between $\mathbf{U}$ and the positive coordinate axes (Fig. e). Then the scalar components of $\mathbf{U}$ are

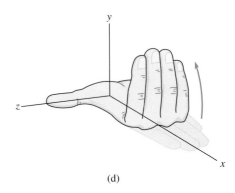

(d)

$$U_x = |\mathbf{U}|\cos\theta_x, \quad U_y = |\mathbf{U}|\cos\theta_y, \quad U_z = |\mathbf{U}|\cos\theta_z, \quad \textbf{Eq. (2.15)}$$

The quantities $\cos\theta_x$, $\cos\theta_y$, and $\cos\theta_z$ are the *direction cosines* of $\mathbf{U}$. They satisfy the relation

$$\cos^2\theta_x + \cos^2\theta_y + \cos^2\theta_z = 1. \quad \textbf{Eq. (2.16)}$$

The *position vector* $\mathbf{r}_{AB}$ from a point A with coordinates (x_A, y_A, z_A) to a point B with coordinates (x_B, y_B, z_B) is given by

$$\mathbf{r}_{AB} = (x_B - x_A)\mathbf{i} + (y_B - y_A)\mathbf{j} + (z_B - z_A)\mathbf{k}. \quad \textbf{Eq. (2.17)}$$

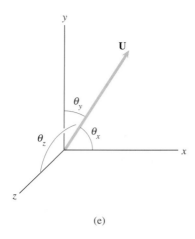

(e)

Dot Products

The dot product of two vectors **U** and **V** is

$$\mathbf{U} \cdot \mathbf{V} = |\mathbf{U}||\mathbf{V}| \cos \theta, \quad \textbf{Eq. (2.18)}$$

where θ is the angle between the vectors when they are placed tail to tail. The dot product of two nonzero vectors is equal to zero if and only if the two vectors are perpendicular.

In terms of scalar components,

$$\mathbf{U} \cdot \mathbf{V} = U_x V_x + U_y V_y + U_z V_z. \quad \textbf{Eq. (2.23)}$$

A vector **U** can be resolved into vector components $\mathbf{U}_p$ and $\mathbf{U}_n$ parallel and normal to a straight line L. In terms of a unit vector **e** that is parallel to L,

$$\mathbf{U}_p = (\mathbf{e} \cdot \mathbf{U})\mathbf{e}. \quad \textbf{Eq. (2.26)}$$

and

$$\mathbf{U}_n = \mathbf{U} - \mathbf{U}_p. \quad \textbf{Eq. (2.27)}$$

Cross Products

The cross product of two vectors **U** and **V** is

$$\mathbf{U} \times \mathbf{V} = |\mathbf{U}||\mathbf{V}| \sin \theta \, \mathbf{e}, \quad \textbf{Eq. (2.28)}$$

where θ is the angle between the vectors **U** and **V** when they are placed tail to tail and **e** is a unit vector perpendicular to **U** and **V**. The direction of **e** is specified by the *right-hand rule*: When the fingers of the right hand are pointed in the direction of **U** (the first vector in the cross product) and closed toward **V** (the second vector in the cross product), the thumb points in the direction of **e**. The cross product of two nonzero vectors is equal to zero if and only if the two vectors are parallel.

In terms of scalar components,

$$\mathbf{U} \times \mathbf{V} = \begin{vmatrix} \mathbf{i} & \mathbf{j} & \mathbf{k} \\ U_x & U_y & U_z \\ V_x & V_y & V_z \end{vmatrix} \quad \textbf{Eq. (2.34)}$$

Mixed Triple Products

The *mixed triple product* is the operation

$$\mathbf{U} \cdot (\mathbf{V} \times \mathbf{W}). \quad \textbf{Eq. (2.35)}$$

In terms of scalar components,

$$\mathbf{U} \cdot (\mathbf{V} \times \mathbf{W}) = \begin{vmatrix} U_x & U_y & U_z \\ V_x & V_y & V_z \\ W_x & W_y & W_z \end{vmatrix} \quad \textbf{Eq. (2.36)}$$

Review Problems

2.1 The magnitude of **F** is 8 kN. Express **F** in terms of scalar components.

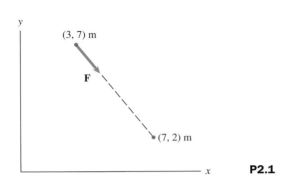

P2.1

2.2 The magnitude of the vertical force **W** is 600 lb, and the magnitude of the force **B** is 1500 lb. Given that **A** + **B** + **W** = **0**, determine the magnitude of the force **A** and the angle α.

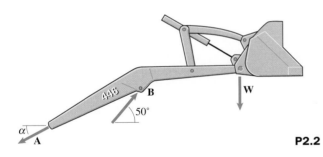

P2.2

2.3 The magnitude of the vertical force vector **A** is 200 lb. If **A** + **B** + **C** = **0**, what are the magnitudes of the force vectors **B** and **C**?

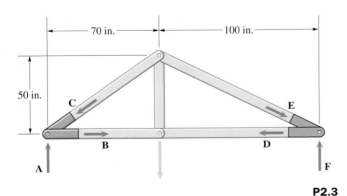

P2.3

2.4 The magnitude of the horizontal force vector **D** in Problem 2.3 is 280 lb. If **D** + **E** + **F** = **0**, what are the magnitudes of the force vectors **E** and **F**?

Refer to the following diagram when solving Problems 2.5 through 2.10.

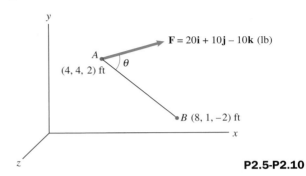

P2.5-P2.10

2.5 What are the direction cosines of **F**?

2.6 Determine the scalar components of a unit vector parallel to line AB that points from A toward B.

2.7 What is the angle θ between the line AB and the force **F**?

2.8 Determine the vector component of **F** that is parallel to the line AB.

2.9 Determine the vector component of **F** that is normal to the line AB.

2.10 Determine the vector $\mathbf{r}_{BA} \times \mathbf{F}$, where $\mathbf{r}_{BA}$ is the position vector from B to A.

2.11 (a) Write the position vector $\mathbf{r}_{AB}$ from point A to point B in terms of scalar components.
(b) The vector **F** has magnitude $|\mathbf{F}| = 200$ N and is parallel to the line from A to B. Write **F** in terms of scalar components.

2.12 The rope exerts a force of magnitude $|\mathbf{F}| = 200$ lb on the top of the pole at B.
(a) Determine the vector $\mathbf{r}_{AB} \times \mathbf{F}$, where $\mathbf{r}_{AB}$ is the position vector from A to B.
(b) Determine the vector $\mathbf{r}_{AC} \times \mathbf{F}$, where $\mathbf{r}_{AC}$ is the position vector from A to C.

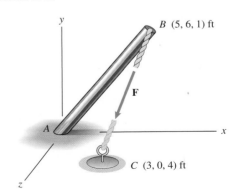

P2.12

2.13 The magnitude of $\mathbf{F}_B$ is 400 N and $|\mathbf{F}_A + \mathbf{F}_B| = 900$ N. Determine the components of $\mathbf{F}_A$.

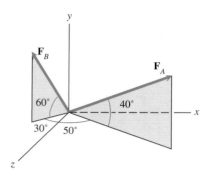

P2.13

2.14 Suppose that the forces $\mathbf{F}_A$ and $\mathbf{F}_B$ shown in Problem 2.13 have the same magnitude and $\mathbf{F}_A \cdot \mathbf{F}_B = 600$ N^2. What are $\mathbf{F}_A$ and $\mathbf{F}_B$?

2.15 The magnitude of the force vector $\mathbf{F}_B$ is 2 kN. Express it in terms of scalar components.

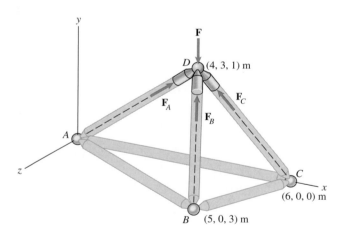

P2.15

2.16 The magnitude of the vertical force vector $\mathbf{F}$ in Problem 2.15 is 6 kN. Determine the vector components of $\mathbf{F}$ parallel and normal to the line from B to D.

2.17 The magnitude of the vertical force vector $\mathbf{F}$ in Problem 2.15 is 6 kN. Given that $\mathbf{F} + \mathbf{F}_A + \mathbf{F}_B + \mathbf{F}_C = \mathbf{0}$, what are the magnitudes of $\mathbf{F}_A$, $\mathbf{F}_B$, and $\mathbf{F}_C$?

2.18 The magnitude of the vertical force $\mathbf{W}$ is 160 N. The direction cosines of the position vector from A to B are $\cos\theta_x = 0.500$, $\cos\theta_y = 0.866$, and $\cos\theta_z = 0$, and the direction cosines of the position vector from B to C are $\cos\theta_x = 0.707$, $\cos\theta_y = 0.619$, and $\cos\theta_z = -0.342$. Point G is the midpoint of the line from B to C. Determine the vector $\mathbf{r}_{AG} \times \mathbf{W}$, where $\mathbf{r}_{AG}$ is the position vector from A to G.

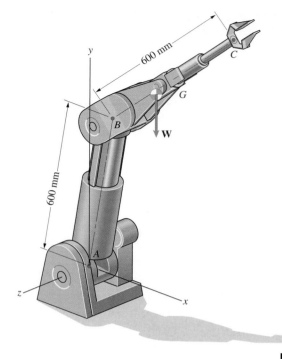

P2.18

2.19 The rope CE exerts a 500-N force $\mathbf{T}$ on the door $ABCD$. Determine the vector component of $\mathbf{T}$ in the direction parallel to the line from point A to point B.

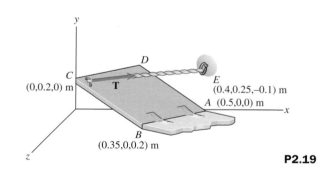

P2.19

2.20 In Problem 2.19, let $\mathbf{r}_{BC}$ be the position vector from point B to point C. Determine the cross product $\mathbf{r}_{BC} \times \mathbf{T}$.

2.21 In Problem 2.19, let $\mathbf{r}_{BC}$ be the position vector from point B to point C, and let $\mathbf{e}_{AB}$ be a unit vector that points from point A toward point B. Evaluate the mixed triple product $\mathbf{e}_{AB} \cdot (\mathbf{r}_{BC} \times \mathbf{T})$.

2.22 A structural engineer determines that the truss in Problem 2.10 will safely support the force $\mathbf{F}$ if the magnitudes of the vector components of $\mathbf{F}$ parallel to the bars do not exceed 20 kN. Based on this criterion, what is the largest safe magnitude of $\mathbf{F}$?

2.23 Consider the sling supporting the storage tank. The tension in the supporting cable is $|\mathbf{F}_A| = |\mathbf{F}_B|$. Suppose that you want a factor of safety of 1.5, which means the cable can support 1.5 times the tension to which it is expected to be subjected.

(a) What minimum tension must the cable used be able to support?

(b) Suppose that design constraints require you to increase the 40° angle. If the cable used will support a tension of 800 lb, what is the maximum acceptable value of the angle?

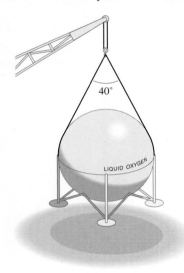

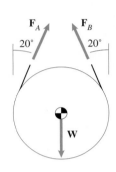

P2.23

2.24 By moving the block at B, the designer of the system supporting the lifeboat can increase the 20° angle between the vector $\mathbf{F}_{BC}$ and the horizontal, thereby decreasing the total force $|\mathbf{F}_{BA} + \mathbf{F}_{BC}|$ exerted on the block. (Assume that the support at A is also moved so that the vector $\mathbf{F}_{BA}$ remains vertical.) If the designer does not want the block to be subjected to a force greater than 740 N, what is the minimum acceptable value of the angle?

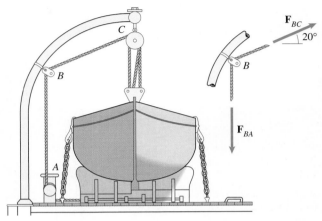

P2.24

2.25 Suppose that the bracket is to be subjected to forces $|\mathbf{F}_1| = |\mathbf{F}_2| = 3$ kN, and it will safely support a total force of 4-kN magnitude in any direction. What is the acceptable range of the angle α?

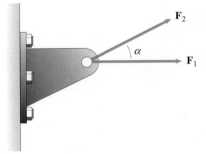

P2.25

The gravitational force on the climber is balanced
by the forces exerted by the rope suspending him.
In this chapter we use free-body diagrams to analyze
forces on objects in equilibrium.

Forces

I n Chapter 2 we represented forces by vectors and used vector addition to sum forces. In this chapter we discuss forces in more detail and introduce two of the most important concepts in mechanics, equilibrium and the free-body diagram. We will use free-body diagrams to identify the forces on objects and use equilibrium to determine unknown forces.

3.1 Types of Forces

Force is a familiar concept, as is evident from the words push, pull, and lift used in everyday conversation. In engineering we deal with different types of forces having a large range of magnitudes. In this section we introduce some terms used to describe forces and discuss particular forces that occur frequently in engineering applications.

Terminology

Line of Action When a force is represented by a vector, the straight line collinear with the vector is called the *line of action* of the force (Fig. 3.1).

Systems of Forces A *system of forces* is simply a particular set of forces. A system of forces is *coplanar*, or *two-dimensional,* if the lines of action of the forces lie in a plane. Otherwise it is *three-dimensional*. A system of forces is *concurrent* if the lines of action of the forces intersect at a point (Fig. 3.2a) and *parallel* if the lines of action are parallel (Fig. 3.2b).

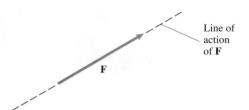

Figure 3.1
A force **F** and its line of action.

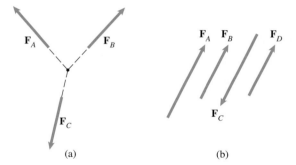

Figure 3.2
(a) Concurrent forces.
(b) Parallel forces.

External and Internal Forces We say that a given object is subjected to an *external force* if the force is exerted by a different object. When one part of a given object is subjected to a force by another part of the same object, we say it is subjected to an *internal force*. These definitions require that you clearly define the object you are considering. For example, suppose that you are the object. When you are standing, the floor—a different object——exerts an external force on your feet. If you press your hands together, your left hand exerts an internal force on your right hand. However, if your right hand is the object you are considering, the force exerted by your left hand is an external force.

Body and Surface Forces A force acting on an object is called a *body force* if it acts on the volume of the object and a *surface force* if it acts on its surface. The gravitational force on an object is a body force. A surface force can be exerted on an object by contact with another object. Both body and surface forces can result from electromagnetic effects.

Gravitational Forces

You are aware of the force exerted on an object by the earth's gravity whenever you pick up something heavy. We can represent the gravitational force, or weight, of an object by a vector (Fig. 3.3).

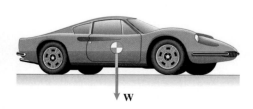

Figure 3.3
Representing an object's weight by a vector.

The magnitude of an object's weight is related to its mass m by

$$|\mathbf{W}| = mg,$$

where g is the acceleration due to gravity at sea level. We will use the values $g = 9.81 \text{ m/s}^2$ in SI units and $g = 32.2 \text{ ft/s}^2$ in U.S. Customary units.

Gravitational forces, and also electromagnetic forces, act at a distance. The objects they act on are not necessarily in contact with the objects exerting the forces. In the next section we discuss forces resulting from contacts between objects.

Contact Forces

Contact forces are the forces that result from contacts between objects. For example, you exert a contact force when you push on a wall (Fig. 3.4a). The surface of your hand exerts a force on the surface of the wall that can be represented by a vector $\mathbf{F}$ (Fig. 3.4b). The wall exerts an equal and opposite force $-\mathbf{F}$ on your hand (Fig. 3.4c). (Recall Newton's third law: The forces exerted on each other by any two particles are equal in magnitude and opposite in direction. If you have any doubt that the wall exerts a force on your hand, try pushing on the wall while standing on roller skates.)

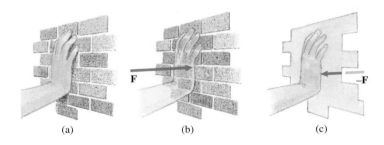

(a) (b) (c)

Figure 3.4
(a) Exerting a contact force on a wall by pushing on it.
(b) The vector $\mathbf{F}$ represents the force you exert on the wall.
(c) The wall exerts a force $-\mathbf{F}$ on your hand.

We will be concerned with contact forces exerted on objects by contact with the surfaces of other objects and by ropes, cables, and springs.

Surfaces   Consider two plane surfaces in contact (Fig. 3.5a). We represent the force exerted on the right surface by the left surface by the vector $\mathbf{F}$ in Fig. 3.5(b). We can resolve $\mathbf{F}$ into a component $\mathbf{N}$ that is normal to the surface and a component $\mathbf{f}$ that is parallel to the surface (Fig. 3.5c). The component $\mathbf{N}$ is called the *normal force*, and the component $\mathbf{f}$ is called the *friction force*. We sometimes assume that the friction force between two surfaces is negligible in comparison to the normal force, a condition we describe by saying that the surfaces are *smooth*. In this case we show only the normal force (Fig. 3.5d). When the friction force cannot be neglected, we say the surfaces are *rough*.

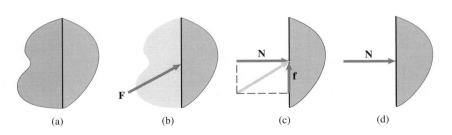

(a) (b) (c) (d)

Figure 3.5
(a) Two plane surfaces in contact.
(b) The force $\mathbf{F}$ exerted on the right surface.
(c) The force $\mathbf{F}$ resolved into components normal and parallel to the surface.
(d) Only the normal force is shown when friction is neglected.

If the contacting surfaces are curved (Fig. 3.6a), the normal force and the friction force are perpendicular and parallel to the plane tangent to the surfaces at their point of contact (Fig. 3.6b).

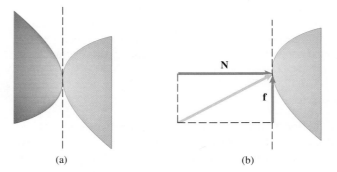

(a)   (b)

Figure 3.6
(a) Curved contacting surfaces. The dashed line indicates the plane tangent to the surfaces at their point of contact.
(b) The normal force and friction force on the right surface.

Ropes and Cables   You can exert a contact force on an object by attaching a rope or cable to the object and pulling on it. In Fig. 3.7a, the crane's cable is attached to a container of building materials. We can represent the force the cable exerts on the container by a vector **T** (Fig. 3.7b). The magnitude of **T** is called the *tension* in the cable, and the line of action of **T** is collinear with the cable. The cable exerts an equal and opposite force −**T** on the crane (Fig. 3.7c).

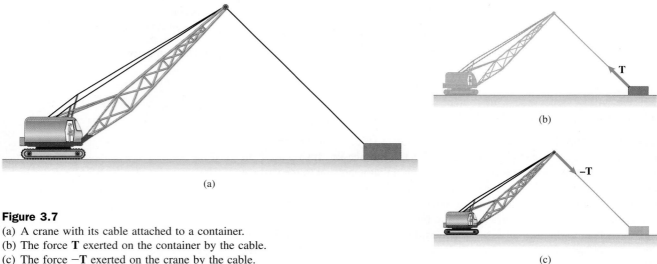

(a)

(b)

(c)

Figure 3.7
(a) A crane with its cable attached to a container.
(b) The force **T** exerted on the container by the cable.
(c) The force −**T** exerted on the crane by the cable.

Notice that we have assumed that the cable is straight and that the tension where the cable is connected to the container equals the tension near the crane. This is approximately true if the weight of the cable is small compared to the tension. Otherwise, the cable will sag significantly and the tension will vary along its length. In Chapter 9 we will discuss ropes and cables whose weights are not small in comparison to their tensions. For now, you should assume that ropes and cables are straight and that their tensions are constant along their lengths.

A *pulley* is a wheel with a grooved rim that can be used to change the direction of a rope or cable (Fig. 3.8a). For now, we assume that the tension is the same on both sides of a pulley (Fig. 3.8b). This is true, or at least approximately true, when the pulley can turn freely and the rope or cable either is stationary or turns the pulley at a constant rate.

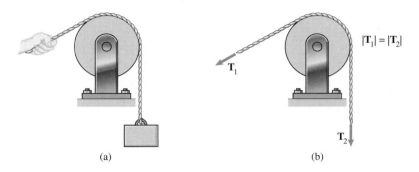

(a)   (b)

Figure 3.8
(a) A pulley changes the direction of a rope or cable.
(b) For now, you should assume that the tensions on each side of the pulley are equal.

Springs   Springs are used to exert contact forces in mechanical devices, for example, in the suspensions of cars (Fig. 3.9). Let's consider a coil spring whose unstretched length, the length of the spring when its ends are free, is L_0 (Fig. 3.10a). When the spring is stretched to a length L greater than L_0 (Fig. 3.10b), it pulls on the object to which it is attached with a force $\mathbf{F}$ (Fig. 3.10c). The object exerts an equal and opposite force $-\mathbf{F}$ on the spring (Fig. 3.10d).

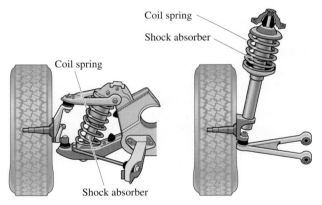

Figure 3.9
Coil springs in car suspensions. The arrangement on the right is called a MacPherson strut.

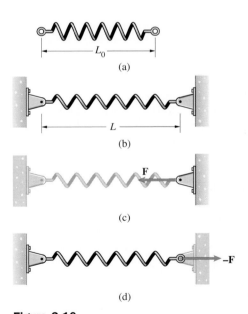

Figure 3.10
(a) A spring of unstretched length L_0.
(b) The spring stretched to a length $L > L_0$.
(c, d) The force $\mathbf{F}$ exerted by the spring and the force $-\mathbf{F}$ on the spring.

When the spring is compressed to a length L less than L_0 (Figs. 3.11a, b), the spring pushes on the object with a force $\mathbf{F}$ and the object exerts an equal and opposite force $-\mathbf{F}$ on the spring (Figs. 3.11c, d). If a spring is compressed too much, it may buckle (Fig. 3.11e). A spring designed to exert a force by being compressed is often provided with lateral support to prevent buckling, for example, by enclosing it in a cylindrical sleeve. In the car suspensions shown in Fig. 3.9, the shock absorbers within the coils prevent the springs from buckling.

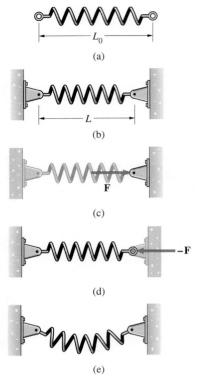

Figure 3.11
(a) A spring of length L_0.
(b) The spring compressed to a length $L < L_0$.
(c, d) The spring pushes on an object with a force **F**, and the object exerts a force $-$**F** on the spring.
(e) A coil spring will buckle if it is compressed too much.

The magnitude of the force exerted by a spring depends on the material it is made of, its design, and how much it is stretched or compressed relative to its unstretched length. When the change in length is not too large compared to the unstretched length, the coil springs commonly used in mechanical devices exert a force approximately proportional to the change in length:

$$|\mathbf{F}| = k|L - L_0|. \tag{3.1}$$

Because the force is a linear function of the change in length (Fig. 3.12), a spring that satisfies this relation is called a *linear spring*. The value of the *spring constant k* depends on the material and design of the spring. Its dimensions are (force)/(length). Notice from Eq. (3.1) that k equals the magnitude of the force required to stretch or compress the spring a unit of length.

Suppose that the unstretched length of a spring is $L_0 = 1$ m and $k = 3000$ N/m. If the spring is stretched to a length $L = 1.2$ m, the magnitude of the pull it exerts is

$$k|L - L_0| = 3000(1.2 - 1) = 600 \text{ N}.$$

Although coil springs are commonly used in mechanical devices, we are also interested in them for a different reason. Springs can be used to *model* situations in which forces depend on displacements. For example, the force necessary to bend the steel beam in Fig. 3.13a is a linear function of the displacement δ,

$$|\mathbf{F}| = k\delta,$$

if δ is not too large. Therefore we can model the force-deflection behavior of the beam with a linear spring (Fig. 3.13b).

Study Questions

1. What is a two-dimensional system of forces?
2. What are internal and external forces?
3. If a surface is said to be smooth, what does that mean?
4. What is the relation between the magnitude of the force exerted by a linear spring and the change in its length?

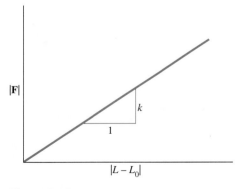

Figure 3.12
The graph of the force exerted by a linear spring as a function of its stretch or compression is a straight line with slope k.

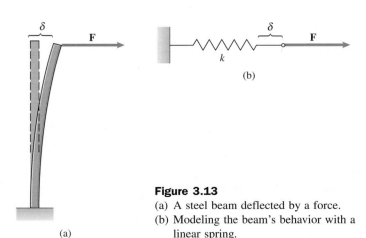

Figure 3.13
(a) A steel beam deflected by a force.
(b) Modeling the beam's behavior with a linear spring.

3.2 Equilibrium and Free-Body Diagrams

Statics is the study of objects in equilibrium. In everyday conversation, *equilibrium* means an unchanging state—a state of balance. Before we explain precisely what this term means in mechanics, let's consider some examples. Pieces of furniture sitting at rest in a room and a person standing stationary in the room are in equilibrium. If a train travels at constant speed on a straight track, objects that are at rest relative to the train, such as a person standing in the aisle, are in equilibrium (Fig. 3.14a). The person standing in the room and the person standing in the aisle of the train are not accelerating. If the train should start to increase or decrease its speed, however, the person standing in the aisle would no longer be in equilibrium and might lose his balance (Fig. 3.14b).

We say that an object is in *equilibrium* only if each point of the object has the same constant velocity, which is referred to as *steady translation*. The velocity must be measured relative to a frame of reference in which Newton's laws are valid, which is called an *inertial reference frame*. In most engineering applications, the velocity can be measured relative to the earth.

The vector sum of the external forces acting on an object in equilibrium is zero. We will use the symbol $\Sigma \mathbf{F}$ to denote the sum of the external forces. Thus when an object is in equilibrium,

$$\Sigma \mathbf{F} = \mathbf{0}. \tag{3.2}$$

In some situations we can use this *equilibrium equation* to determine unknown forces acting on an object in equilibrium. The first step will be to draw a *free-body diagram* of the object to identify the external forces acting on it. The free-body diagram is an essential tool in mechanics. It focuses attention on the object of interest and helps identify the external forces acting on it. Although in statics we will be concerned only with objects in equilibrium, free-body diagrams are also used in dynamics to analyze the motions of objects.

The free-body diagram is a simple concept. It is a drawing of an object and the external forces acting on it. Otherwise, nothing other than the object of interest is included. The drawing shows the object *isolated*, or *freed*, from its surroundings. Drawing a free-body diagram involves three steps:

1. *Identify the object you want to isolate.* As the following examples show, your choice is often dictated by particular forces you want to determine.

2. *Draw a sketch of the object isolated from its surroundings, and show relevant dimensions and angles.* Your drawing should be reasonably accurate, but it can omit irrelevant details.

3. *Draw vectors representing all of the external forces acting on the isolated object, and label them.* Don't forget to include the gravitational force if you are not intentionally neglecting it.

You will also need to choose a coordinate system so that you can express the forces on the isolated object in terms of components. Often you will find it convenient to choose the coordinate system before drawing the free-body diagram, but in some situations the best choice of coordinate system will not be apparent until after you have drawn it.

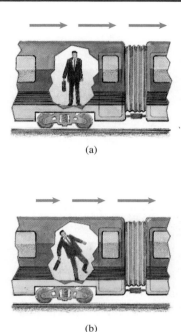

(a)

(b)

Figure 3.14
(a) While the train moves at a constant speed, a person standing in the aisle is in equilibrium.
(b) If the train starts to speed up, the person is no longer in equilibrium.

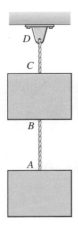

Figure 3.15

Stationary blocks suspended by cables.

A simple example demonstrates how you can choose free-body diagrams to determine particular forces and also that you must distinguish carefully between external and internal forces. Two stationary blocks of equal weight W are suspended by cables in Fig. 3.15. The system is in equilibrium. Suppose that we want to determine the tensions in the two cables.

To determine the tension in cable AB, we first isolate an "object" consisting of the lower block and part of cable AB (Fig. 3.16a). We then ask ourselves what forces can be exerted on our isolated object by objects not included in the diagram. The earth exerts a gravitational force of magnitude W on the block. Also, where we "cut" cable AB, the cable is subjected to a contact force equal to the tension in the cable (Fig. 3.16b). The arrows in this figure indicate the directions of the forces. The scalar W is the weight of the block and T_{AB} is the tension in cable AB. We assume that the weight of the part of cable AB included in the free-body diagram can be neglected in comparison to the weight of the block.

Since the free-body diagram is in equilibrium, the sum of the external forces equals zero. In terms of a coordinate system with the y axis upward (Fig. 3.16c), we obtain the equilibrium equation

$$\Sigma \mathbf{F} = T_{AB}\mathbf{j} - W\mathbf{j} = (T_{AB} - W)\mathbf{j} = \mathbf{0}.$$

Thus the tension in cable AB is $T_{AB} = W$.

Figure 3.16

(a) Isolating the lower block and part of cable AB.

(b) Indicating the external forces completes the free-body diagram.

(c) Introducing a coordinate system.

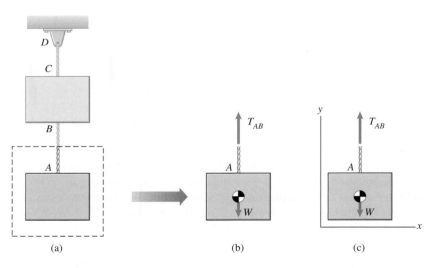

(a) (b) (c)

We can determine the tension in cable CD by isolating the upper block (Fig. 3.17a). The external forces are the weight of the upper block and the tensions in the two cables (Fig. 3.17b). In this case we obtain the equilibrium equation

$$\Sigma \mathbf{F} = T_{CD}\mathbf{j} - T_{AB}\mathbf{j} - W\mathbf{j} = (T_{CD} - T_{AB} - W)\mathbf{j} = \mathbf{0}.$$

Since $T_{AB} = W$, we find that $T_{CD} = 2W$.

We could also have determined the tension in cable CD by treating the two blocks and the cable AB as a single object (Figs. 3.18a, b). The equilibrium equation is

$$\Sigma \mathbf{F} = T_{CD}\mathbf{j} - W\mathbf{j} - W\mathbf{j} = (T_{CD} - 2W)\mathbf{j} = \mathbf{0},$$

and we again obtain $T_{CD} = 2W$.

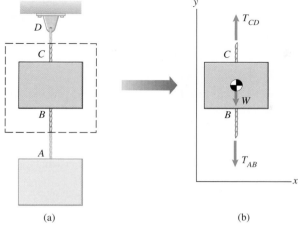

Figure 3.17
(a) Isolating the upper block to determine the tension in cable CD.
(b) Free-body diagram of the upper block.

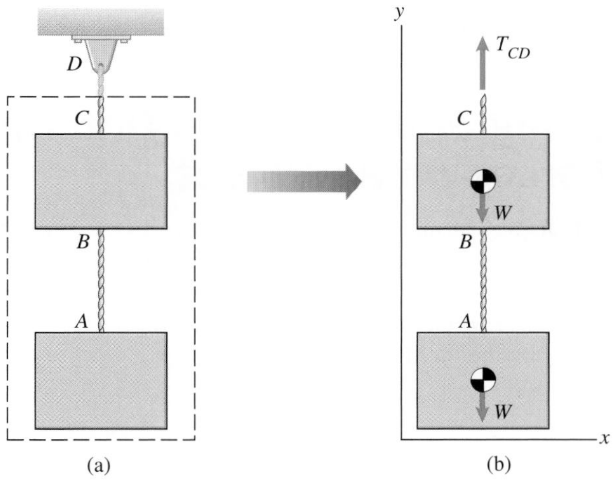

Figure 3.18
(a) An alternative choice for determining the tension in cable CD.
(b) Free-body diagram including both blocks and cable AB.

Why doesn't the tension in cable AB appear on the free-body diagram in Fig. 3.18b? Remember that only external forces are shown on free-body diagrams. Since cable AB is part of the free-body diagram in this case, the forces it exerts on the upper and lower blocks are internal forces.

We have described the procedure for drawing free-body diagrams. In the next section we will draw free-body diagrams of objects subjected to two-dimensional systems of forces and use them to determine unknown forces acting on objects in equilibrium.

3.3 Two-Dimensional Force Systems

Suppose that the system of external forces acting on an object in equilibrium is two-dimensional (coplanar). By orienting a coordinate system so that the forces lie in the x-y plane, we can express the sum of the external forces as

$$\Sigma \mathbf{F} = \left(\Sigma F_x\right)\mathbf{i} + \left(\Sigma F_y\right)\mathbf{j} = \mathbf{0},$$

where ΣF_x and ΣF_y are the sums of the x and y components of the forces. Since a vector is zero only if each of its components is zero, we obtain two scalar equilibrium equations:

$$\Sigma F_x = 0, \qquad \Sigma F_y = 0. \tag{3.3}$$

The sums of the x and y components of the external forces acting on an object in equilibrium must each equal zero.

Study Questions

1. What do you know about the sum of the external forces acting on an object in equilibrium?
2. Is a free-body diagram only useful when an object is in equilibrium?
3. What are the steps in drawing a free-body diagram?

Example 3.1

Using Equilibrium to Determine Forces on an Object

For display at an automobile show, the 1440-kg car in Fig. 3.19 is held in place on the inclined surface by the horizontal cable from A to B. Determine the tension that the cable (and the fixture to which it is connected at B) must support. The car's brakes are not engaged, so the tires exert only normal forces on the inclined surface.

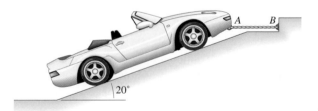

Figure 3.19

Strategy

Since the car is in equilibrium, we can draw its free-body diagram and use Eqs. (3.3) to determine the forces exerted on the car by the cable and the inclined surface.

Solution

Draw the Free-Body Diagram We first draw a diagram of the car isolated from its surrounding (Fig. a) and then complete the free-body diagram by showing the force exerted by the car's weight, the force T exerted by the cable, and the normal force N exerted by the inclined surface (Fig. b).

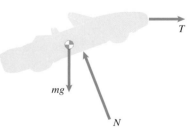

(a) Isolating the car.

(b) The completed free-body diagram shows the known and unknown external forces.

Apply the Equilibrium Equations In Fig. c, we introduce a coordinate system and resolve the normal force into x and y components. The equilibrium equations are

$$\Sigma F_x = T - N \sin 20° = 0,$$

$$\Sigma F_y = N \cos 20° - mg = 0.$$

We can solve the second equilibrium equation for N,

$$N = \frac{mg}{\cos 20°} = \frac{(1440)(9.81)}{\cos 20°} = 15.0 \text{ kN},$$

and then solve the first equilibrium equation for the tension T:

$$T = N \sin 20° = 5.14 \text{ kN}.$$

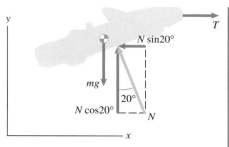

(c) Introducing a coordinate system and resolving N into its components.

Example 3.2

Choosing a Free-Body Diagram

The automobile engine block in Fig. 3.20 is suspended by a system of cables. The mass of the block is 200 kg. What are the tensions in cables AB and AC?

Strategy

We need a free-body diagram that is subjected to the forces we want to determine. By isolating part of the cable system near point A where the cables are joined, we can obtain a free-body diagram that is subjected to the weight of the block and the unknown tensions in cables AB and AC.

Solution

Draw the Free-Body Diagram Isolating part of the cable system near point A (Fig. a), we obtain a free-body diagram subjected to the weight of the block $W = mg = (200 \text{ kg})(9.81 \text{ m/s}^2) = 1962 \text{ N}$ and the tensions in cables AB and AC (Fig. b).

Apply the Equilibrium Equations We select the coordinate system shown in Fig. c and resolve the cable tensions into x and y components. The resulting equilibrium equations are

$$\Sigma F_x = T_{AC} \cos 45° - T_{AB} \cos 60° = 0,$$

$$\Sigma F_y = T_{AC} \sin 45° + T_{AB} \sin 60° - 1962 = 0.$$

Solving these equations, we find that the tensions in the cables are $T_{AB} = 1436 \text{ N}$ and $T_{AC} = 1016 \text{ N}$.

Alternative Solution: We can determine the tensions in the cables in another way that will also help you visualize the conditions for equilibrium. Since the sum of the three forces acting on our free-body diagram is zero, the vectors form a closed polygon when placed head to tail (Fig. d). You can see that the

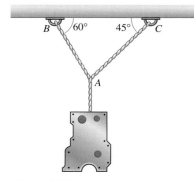

Figure 3.20

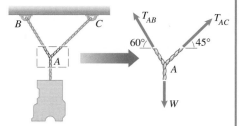

(a) Isolating part of the cable system.
(b) The completed free-body diagram.

(c) Selecting a coordinate system and resolving the forces into components.

(d) The triangle formed by the sum of the three forces.

sum of the vertical components of the tensions supports the weight and that the horizontal components of the tensions must balance each other. The angle of the triangle opposite the weight W is $180° - 30° - 45° = 105°$. By applying the law of sines,

$$\frac{\sin 45°}{T_{AB}} = \frac{\sin 30°}{T_{AC}} = \frac{\sin 105°}{1962},$$

we obtain $T_{AB} = 1436$ N and $T_{AC} = 1016$ N.

Discussion

How were we able to choose a free-body diagram that permitted us to determine the unknown tensions in the cables? There are no definite rules for choosing free-body diagrams. You will learn what to do in many cases from the examples we present, but you will also encounter new situations. It may be necessary to try several free-body diagrams before finding one that provides the information you need. Remember that forces you want to determine should appear as external forces on your free-body diagram, and your objective is to obtain a number of equilibrium equations equal to the number of unknown forces.

Example 3.3

Applying Equilibrium to a System of Pulleys

The mass of each pulley of the system in Fig. 3.21 is m, and the mass of the suspended object A is m_A. Determine the force T necessary for the system to be in equilibrium.

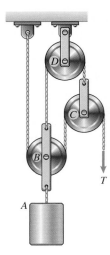

Figure 3.21

Strategy

By drawing free-body diagrams of the individual pulleys and applying equilibrium, we can relate the force T to the weights of the pulleys and the object A.

Solution

We first draw a free-body diagram of the pulley C to which the force T is applied (Fig. a). Notice that we assume the tension in the cable supported by the pulley to equal T on both sides (see Fig. 3.8). From the equilibrium equation

$$T_D - T - T - mg = 0,$$

we determine that the tension in the cable supported by pulley D is

$$T_D = 2T + mg.$$

We now know the tensions in the cables extending from pulleys C and D to pulley B in terms of T. Drawing the free-body diagram of pulley B (Fig. b), we obtain the equilibrium equation

$$T + T + 2T + mg - mg - m_A g = 0.$$

Solving, we obtain $T = m_A g/4$.

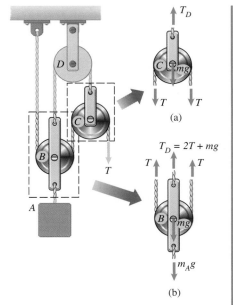

(a)

(a) Free-body diagram of pulley C.
(b) Free-body diagram of pulley B.

Example 3.4

Application to Engineering:

Steady Flight

Figure 3.22 shows an airplane flying in the vertical plane and its free-body diagram. The forces acting on the airplane are its weight W, the thrust T exerted by its engines, and aerodynamic forces. The dashed line indicates the path along which the airplane is moving. The aerodynamic forces are resolved into a component perpendicular to the path, the lift L, and a component parallel to

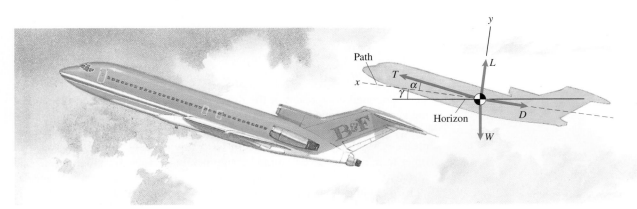

Figure 3.22
External forces on an airplane in flight.

the path, the drag D. The angle γ between the horizontal and the path is called the flight path angle, and α is the angle of attack. If the airplane remains in equilibrium for an interval of time, it is said to be in steady flight. If $\gamma = 6°$, $D = 125$ kN, $L = 680$ kN, and the mass of the airplane is 72 Mg (megagrams), what values of T and α are necessary to maintain steady flight?

Solution

In terms of the coordinate system in Fig. 3.22, the equilibrium equations are

$$\Sigma F_x = T \cos \alpha - D - W \sin \gamma = 0, \tag{3.4}$$

$$\Sigma F_y = T \sin \alpha + L - W \cos \gamma = 0. \tag{3.5}$$

We solve Eq. (3.5) for $\sin \alpha$, solve Eq. (3.4) for $\cos \alpha$, and divide to obtain an equation for $\tan \alpha$:

$$\tan a = \frac{\sin \alpha}{\cos \alpha} = \frac{W \cos \gamma - L}{W \sin \gamma + D}$$

$$= \frac{(72,000)(9.81) \cos 6° - 680,000}{(72,000)(9.81) \sin 6° + 125,000} = 0.113.$$

The angle of attack $\alpha = \arctan(0.113) = 6.44°$. Now we use Eq. (3.4) to determine the thrust:

$$T = \frac{W \sin \gamma + D}{\cos \alpha} = \frac{(72,000)(9.81) \sin 6° + 125,000}{\cos 6.44°} = 200,000 \text{ N}.$$

Notice that the thrust necessary for steady flight is 28% of the airplane's weight.

𝒟esign Issues

In the examples we have considered so far, the values of certain forces acting on an object in equilibrium were given, and our goal was simply to determine the unknown forces by setting the sum of the forces equal to zero. In many situations in engineering, an object in equilibrium is subjected to forces that have different values under different conditions, and this has a profound effect on its design.

When an airplane cruises at constant altitude ($\gamma = 0$), Eqs. (3.4) and (3.5) reduce to

$$T \cos \alpha = D,$$

$$T \sin \alpha + L = W.$$

The horizontal component of the thrust must equal the drag, and the sum of the vertical component of the thrust and the lift must equal the weight. For a fixed value of α, the lift and drag increase as the speed of the airplane increases. A principal design concern is to minimize D at cruising speed in order to minimize the thrust (and consequently the fuel consumption) needed to satisfy the first equilibrium equation. Much of the research on airplane design, including both theoretical analyses and model tests in wind tunnels (Fig. 3.23), is devoted to developing airplane shapes that minimize drag.

When an airplane cruises at low speed, satisfying the second equilibrium equation has the most serious implications for design. The airplane's wings

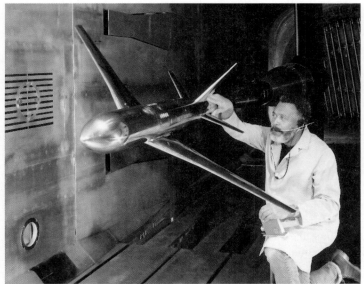

Figure 3.24
An F-15 being refueled by a KC-135 refueling plane.

Figure 3.23
Wind tunnels are used to measure the aerodynamic forces on airplane models.

must generate sufficient lift to balance its weight. This requirement is especially difficult to achieve in fast airplanes, because wings designed for low drag at high velocities do not generate as much lift at low speeds as wings that are designed for flight at lower velocities. For example, the F-15 in Fig. 3.24 must fly with a relatively large angle of attack (which increases both the lift and the vertical component of the thrust) in comparison to the refueling plane. In the case of the F-14 (Fig. 3.25), the engineers obtained both low drag at high velocities and good lift characteristics at low velocities by using variable sweep wings.

Figure 3.25
An F-14 with its wings in the takeoff and landing configuration and in the high-speed configuration.

3.4 Three-Dimensional Force Systems

The equilibrium situations we have considered so far have involved only coplanar forces. When the system of external forces acting on an object in equilibrium is three-dimensional, we can express the sum of the external forces as

$$\Sigma \mathbf{F} = \left(\Sigma F_x \right)\mathbf{i} + \left(\Sigma F_y \right)\mathbf{j} + \left(\Sigma F_z \right)\mathbf{k} = \mathbf{0}.$$

Each component of this equation must equal zero, resulting in three scalar equilibrium equations:

$$\Sigma F_x = 0, \qquad \Sigma F_y = 0, \qquad \Sigma F_z = 0. \tag{3.6}$$

The sums of the x, y, and z components of the external forces acting on an object in equilibrium must each equal zero.

Example 3.5

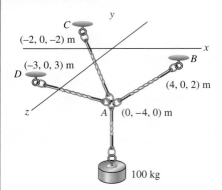

Figure 3.26

Applying Equilibrium in Three Dimensions

The 100-kg cylinder in Fig. 3.26 is suspended from the ceiling by cables attached at points B, C, and D. What are the tensions in cables AB, AC, and AD?

Strategy

We can determine the tensions by the same approach we used for similar two-dimensional problems. By isolating part of the cable system near point A, we can obtain a free-body diagram subjected to forces due to the tensions in the cables. Since the sums of the x, y, and z components of the external forces must each equal zero, we obtain three equations for the three unknown tensions.

Solution

Draw the Free-Body Diagram We isolate part of the cable system near point A (Fig. a) and complete the free-body diagram by showing the forces exerted by the tensions in the cables (Fig. b). The magnitudes of the vectors $\mathbf{T}_{AB}$, $\mathbf{T}_{AC}$, and $\mathbf{T}_{AD}$ are the tensions in cables AB, AC, and AD, respectively.

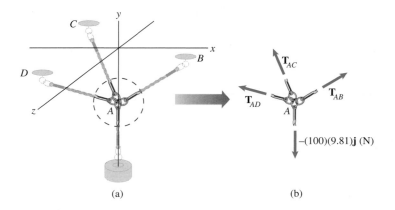

(a) Isolating part of the cable system.
(b) The completed free-body diagram showing the forces exerted by the tensions in the cables.

(a) (b)

Apply the Equilibrium Equations The sum of the external forces acting on the free-body diagram is

$$\Sigma \mathbf{F} = \mathbf{T}_{AB} + \mathbf{T}_{AC} + \mathbf{T}_{AD} - 981\mathbf{j} = \mathbf{0}.$$

To solve this equation for the tensions in the cables, we need to express the vectors $\mathbf{T}_{AB}$, $\mathbf{T}_{AC}$, and $\mathbf{T}_{AD}$ in terms of their components.

We first determine the components of a unit vector that points in the direction of the vector $\mathbf{T}_{AB}$. Let $\mathbf{r}_{AB}$ be the position vector from point A to point B (Fig. c):

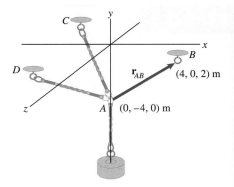

$$\mathbf{r}_{AB} = (x_B - x_A)\mathbf{i} + (y_B - y_A)\mathbf{j} + (z_B - z_A)\mathbf{k} = 4\mathbf{i} + 4\mathbf{j} + 2\mathbf{k} \text{ (m)}.$$

Dividing $\mathbf{r}_{AB}$ by its magnitude, we obtain a unit vector that has the same direction as $\mathbf{T}_{AB}$:

$$\mathbf{e}_{AB} = \frac{\mathbf{r}_{AB}}{|\mathbf{r}_{AB}|} = 0.667\mathbf{i} + 0.667\mathbf{j} + 0.333\mathbf{k}.$$

Now we can write the vector $\mathbf{T}_{AB}$ as the product of the tension T_{AB} in cable AB and $\mathbf{e}_{AB}$:

$$\mathbf{T}_{AB} = T_{AB}\mathbf{e}_{AB} = T_{AB}(0.667\mathbf{i} + 0.667\mathbf{j} + 0.333\mathbf{k}).$$

(c) The position vector $\mathbf{r}_{AB}$.

We now express the force vectors $\mathbf{T}_{AC}$ and $\mathbf{T}_{AD}$ in terms of the tensions T_{AC} and T_{AD} in cables AC and AD in the same way. The results are

$$\mathbf{T}_{AC} = T_{AC}(-0.408\mathbf{i} + 0.816\mathbf{j} - 0.408\mathbf{k}),$$

$$\mathbf{T}_{AD} = T_{AD}(-0.514\mathbf{i} + 0.686\mathbf{j} - 0.514\mathbf{k}).$$

We use these expressions to write the sum of the external forces in terms of the tensions T_{AB}, T_{AC}, and T_{AD}:

$$\begin{aligned}
\Sigma \mathbf{F} &= \mathbf{T}_{AB} + \mathbf{T}_{AC} + \mathbf{T}_{AD} - 981\mathbf{j} \\
&= (0.667T_{AB} - 0.408T_{AC} - 0.514T_{AD})\mathbf{i} \\
&\quad + (0.667T_{AB} + 0.816T_{AC} + 0.686T_{AD} - 981)\mathbf{j} \\
&\quad + (0.333T_{AB} - 0.408T_{AC} + 0.514T_{AD})\mathbf{k} \\
&= \mathbf{0}.
\end{aligned}$$

The sums of the forces in the x, y, and z directions must each equal zero:

$$\Sigma F_x = 0.667T_{AB} - 0.408T_{AC} - 0.514T_{AD} = 0,$$

$$\Sigma F_y = 0.667T_{AB} + 0.816T_{AC} + 0.686T_{AD} - 981 = 0,$$

$$\Sigma F_z = 0.333T_{AB} - 0.408T_{AC} + 0.514T_{AD} = 0.$$

Solving these equations, we find that the tensions are $T_{AB} = 519$ N, $T_{AC} = 636$ N, and $T_{AD} = 168$ N.

Discussion

Notice that this example required several of the techniques we covered in Chapter 2. In particular, we had to determine the components of a position vector, divide the position vector by its magnitude to obtain a unit vector with the same direction as a particular force, and express the force in terms of its components by writing it as the product of the unit vector and the magnitude of the force.

Example 3.6

Application of the Dot Product

The 100-lb "slider" C in Fig. 3.27 is held in place on the smooth bar by the cable AC. Determine the tension in the cable and the force exerted on the slider by the bar.

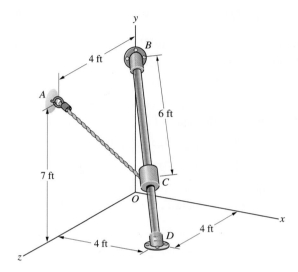

Figure 3.27

Strategy

Since we want to determine forces that act on the slider, we need to draw its free-body diagram. The external forces acting on the slider are its weight and the forces exerted on it by the cable and the bar. If we approached this example as we did the previous one, our next step would be to express the forces in terms of their components. However, we don't know the direction of the force exerted on the slider by the bar. Since the smooth bar exerts negligible friction force, we do know that the force exerted by the bar is normal to its axis. Therefore we can eliminate this force from the equation $\Sigma \mathbf{F} = \mathbf{0}$ by taking the dot product of the equation with a unit vector that is parallel to the bar.

Solution

Draw the Free-Body Diagram We isolate the slider (Fig. a) and complete the free-body diagram by showing the weight of the slider, the force $\mathbf{T}$ exerted by the tension in the cable, and the normal force $\mathbf{N}$ exerted by the bar (Fig. b).

Apply the Equilibrium Equations The sum of the external forces acting on the free-body diagram is

$$\Sigma \mathbf{F} = \mathbf{T} + \mathbf{N} - 100\mathbf{j} = \mathbf{0}. \tag{3.7}$$

Let $\mathbf{e}_{BD}$ be the unit vector pointing from point B toward point D. Since $\mathbf{N}$ is perpendicular to the bar, $\mathbf{e}_{BD} \cdot \mathbf{N} = 0$. Therefore

$$\mathbf{e}_{BD} \cdot (\Sigma \mathbf{F}) = \mathbf{e}_{BD} \cdot (\mathbf{T} - 100\mathbf{j}) = 0. \tag{3.8}$$

(a)

$\mathbf{T}$

$\mathbf{N}$

$-100\,\mathbf{j}$ (lb)

(b)

(a) Isolating the slider.
(b) Free-body diagram of the slider showing the forces exerted by its weight, the cable, and the bar.

This equation has a simple interpretation: The component of the slider's weight parallel to the bar is balanced by the component of **T** parallel to the bar. *Determining* $\mathbf{e}_{BD}$: We determine the vector from point B to point D,

$$\mathbf{r}_{BD} = (4 - 0)\mathbf{i} + (0 - 7)\mathbf{j} + (4 - 0)\mathbf{k} = 4\mathbf{i} - 7\mathbf{j} + 4\mathbf{k} \text{ (ft)},$$

and divide it by its magnitude to obtain the unit vector $\mathbf{e}_{BD}$:

$$\mathbf{e}_{BD} = \frac{\mathbf{r}_{BD}}{|\mathbf{r}_{BD}|} = \frac{4}{9}\mathbf{i} - \frac{7}{9}\mathbf{j} + \frac{4}{9}\mathbf{k}.$$

Resolving **T** *into components*: To express **T** in terms of its components, we need to determine the coordinates of the slider C. We can write the vector from B to C in terms of the unit vector $\mathbf{e}_{BD}$,

$$\mathbf{r}_{BC} = 6\mathbf{e}_{BD} = 2.67\mathbf{i} - 4.67\mathbf{j} + 2.67\mathbf{k} \text{ (ft)},$$

and then add it to the vector from the origin O to B to obtain the vector from O to C:

$$\mathbf{r}_{OC} = \mathbf{r}_{OB} + \mathbf{r}_{BC} = 7\mathbf{j} + (2.67\mathbf{i} - 4.67\mathbf{j} + 2.67\mathbf{k})$$
$$= 2.67\mathbf{i} + 2.33\mathbf{j} + 2.67\mathbf{k} \text{ (ft)}.$$

The components of this vector are the coordinates of point C.

Now we can determine a unit vector with the same direction as **T**. The vector from C to A is

$$\mathbf{r}_{CA} = (0 - 2.67)\mathbf{i} + (7 - 2.33)\mathbf{j} + (4 - 2.67)\mathbf{k}$$
$$= -2.67\mathbf{i} + 4.67\mathbf{j} + 1.33\mathbf{k} \text{ (ft)},$$

and the unit vector that points from point C toward point A is

$$\mathbf{e}_{CA} = \frac{\mathbf{r}_{CA}}{|\mathbf{r}_{CA}|} = -0.482\mathbf{i} + 0.843\mathbf{j} + 0.241\mathbf{k}.$$

Let T be the tension in the cable AC. Then we can write the vector **T** as

$$\mathbf{T} = T\mathbf{e}_{CA} = T(-0.482\mathbf{i} + 0.843\mathbf{j} + 0.241\mathbf{k}).$$

Determining **T** *and* **N**: Substituting our expressions for $\mathbf{e}_{BD}$ and **T** in terms of their components into Eq. (3.8),

$$\mathbf{e}_{BD} \cdot (\mathbf{T} - 100\mathbf{j})$$
$$= \left[\frac{4}{9}\mathbf{i} - \frac{7}{9}\mathbf{j} + \frac{4}{9}\mathbf{k}\right] \cdot \left[-0.482T\mathbf{i} + (0.843T - 100)\mathbf{j} + 0.241T\mathbf{k}\right]$$
$$= -0.762T + 77.8 = 0,$$

we obtain the tension $T = 102$ lb.

Now we can determine the force exerted on the slider by the bar by using Eq. (3.7):

$$\mathbf{N} = -\mathbf{T} + 100\mathbf{j} = -102(-0.482\mathbf{i} + 0.843\mathbf{j} + 0.241\mathbf{k}) + 100\mathbf{j}$$
$$= 49.1\mathbf{i} + 14.0\mathbf{j} - 24.6\mathbf{k} \text{ (lb)}.$$

⠿ *Computational Mechanics*

The following examples and problems are designed for the use of a programmable calculator or computer. Example 3.7 is similar to previous examples and problems except that the solution must be calculated for a range of input quantities. Example 3.8 leads to an algebraic equation that must be solved numerically.

Computational Example 3.7 ⠿

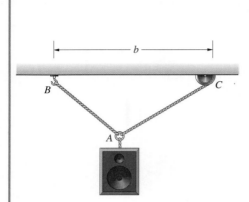

Figure 3.28

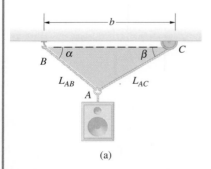

(a)

(a) Determining the angles α and β.

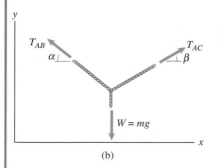

(b)

(b) Free-body diagram of part of the cable system.

Determining Tensions for a Range of Dimensions

The system of cables in Fig. 3.28 is designed to suspend a load with a mass of 1 Mg (megagram). The dimension $b = 2$ m, and the length of cable AB is 1 m. The height of the load can be adjusted by changing the length of cable AC.

(a) Plot the tensions in cables AB and AC for values of the length of cable AC from 1.2 m to 2.2 m.

(b) Cables AB and AC can each safely support a tension equal to the weight of the load. Use the results of (a) to estimate the allowable range of the length of cable AC.

Strategy

By drawing the free-body diagram of the part of the cable system where the cables join, we can determine the tensions in the cables in terms of the length of cable AC.

Solution

(a) Let the lengths of the cables be $L_{AB} = 1$ m and L_{AC}. We can apply the law of cosines to the triangle in Fig. a to determine α in terms of L_{AC}:

$$\alpha = \arccos\left(\frac{b^2 + L_{AB}^2 - L_{AC}^2}{2bL_{AB}}\right).$$

Then we can use the law of sines to determine β:

$$\beta = \arcsin\left(\frac{L_{AB}\sin\alpha}{L_{AC}}\right).$$

Draw the Free-Body Diagram   We draw the free-body diagram of the part of the cable system where the cables join in Fig. b, where T_{AB} and T_{AC} are the tensions in the cables.

Apply the Equilibrium Equations   Selecting the coordinate system shown in Fig. b, the equilibrium equations are

$$\Sigma F_x = -T_{AB}\cos\alpha + T_{AC}\cos\beta = 0,$$

$$\Sigma F_y = -T_{AB}\sin\alpha + T_{AC}\sin\beta - W = 0.$$

Solving these equations for the cable tensions, we obtain

$$T_{AB} = \frac{W \cos \beta}{\sin \alpha \cos \beta + \cos \alpha \sin \beta},$$

$$T_{AC} = \frac{W \cos \alpha}{\sin \alpha \cos \beta + \cos \alpha \sin \beta}.$$

To compute the results, we input a value of the length L_{AC} and calculate the angle α, then the angle β, and then the tensions T_{AB} and T_{AC}. The resulting values of T_{AB}/W and T_{AC}/W are plotted as functions of L_{AC} in Fig. 3.29.
(b) The allowable range of the length of cable AC is the range over which the tensions in both cables are less than W. From Fig. 3.29 we can see that the tension T_{AB} exceeds W for values of L_{AC} less than about 1.35 m, so the safe range is $L_{AC} > 1.35$ m.

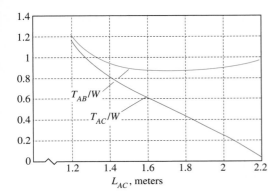

Figure 3.29
Ratios of the cable tensions to the suspended weight as functions of L_{AC}.

Computational Example 3.8

Equilibrium Position of an Object Supported by a Spring

The 12-lb collar A in Fig. 3.30 is held in equilibrium on the smooth vertical bar by the spring. The spring constant $k = 300$ lb/ft, the unstretched length of the spring is $L_0 = b$, and the distance $b = 1$ ft. What is the distance h?

Strategy

Both the direction and the magnitude of the force exerted on the collar by the spring depend on h. By drawing the free-body diagram of the collar and applying the equilibrium equations, we can obtain an equation for h.

Solution

Draw the Free-Body Diagram We isolate the collar (Fig. a) and complete the free-body diagram by showing its weight $W = 12$ lb, the force F exerted by the spring, and the normal force N exerted by the bar (Fig. b).

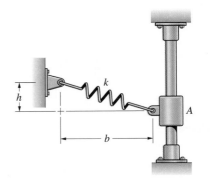

Figure 3.30

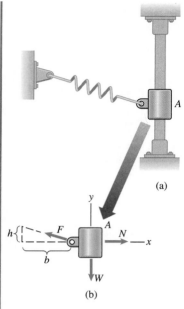

(a)

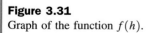

(b)

(a) Isolating the collar.
(b) The free-body diagram.

Apply the Equilibrium Equations   Selecting the coordinate system shown in Fig. b, we obtain the equilibrium equations

$$\Sigma F_x = N - \left(\frac{b}{\sqrt{h^2 + b^2}}\right)F = 0,$$

$$\Sigma F_y = \left(\frac{h}{\sqrt{h^2 + b^2}}\right)F - W = 0.$$

In terms of the length of the spring $L = \sqrt{h^2 + b^2}$, the force exerted by the spring is

$$F = k(L - L_0) = k(\sqrt{h^2 + b^2} - b).$$

Substituting this expression into the second equilibrium equation, we obtain the equation

$$\left(\frac{h}{\sqrt{h^2 + b^2}}\right)k(\sqrt{h^2 + b^2} - b) - W = 0.$$

Inserting the values of k, b, and W, we find that the distance h is a root of the equation

$$f(h) = \left(\frac{300h}{\sqrt{h^2 + 1}}\right)(\sqrt{h^2 + 1} - 1) - 12 = 0. \tag{3.9}$$

How can we solve this nonlinear algebraic equation for h? Some calculators and software are designed to obtain roots of such equations. Another approach is to calculate the value of $f(h)$ for a range of values of h and plot the results, as we have done in Fig. 3.31. From the graph we see that the solution is approximately $h = 0.45$ ft. By examining the computed results near $h = 0.45$ ft,

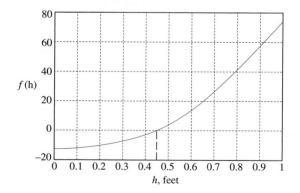

$h(ft)$	$f(h)$
0.449	−0.1818
0.450	−0.1094
0.451	−0.0368
0.452	0.0361
0.453	0.1092
0.454	0.1826

Figure 3.31
Graph of the function $f(h)$.

we see that the solution (to three significant digits) is $h = 0.452$ ft.

Chapter Summary

In this chapter we discussed the forces that occur frequently in engineering applications and introduced two of the most important concepts in mechanics: the free-body diagram and equilibrium. By drawing free-body diagrams and applying the vector techniques developed in Chapter 2, we showed how unknown forces acting on objects in equilibrium can be determined from the condition that the sum of the external forces must equal zero. The sum of the moments of the external forces on an object in equilibrium must also equal zero, and this condition can be used to obtain additional information about unknown forces on objects. We will discuss moments of forces in Chapter 4. We will then apply equilibrium to individual objects in Chapter 5 and to structures in Chapter 6.

The straight line coincident with a force vector is called the *line of action* of the force. A system of forces is *coplanar*, or *two-dimensional*, if the lines of action of the forces lie in a plane. Otherwise, it is *three-dimensional*. A system of forces is *concurrent* if the lines of action of the forces intersect at a point and *parallel* if the lines of action are parallel.

An object is subjected to an *external force* if the force is exerted by a different object. When one part of an object is subjected to a force by another part of the same object, the force is *internal*.

A *body force* acts on the volume of an object, and a *surface* or *contact force* acts on its surface.

Gravitational Forces

The weight of an object is related to its mass by $W = mg$, where $g = 9.81 \text{ m/s}^2$ in SI units and $g = 32.2 \text{ ft/s}^2$ in U.S. Customary units.

Surfaces

Two surfaces in contact exert forces on each other that are equal in magnitude and opposite in direction. Each force can be resolved into the *normal force* and the *friction force*. If the friction force is negligible in comparison to the normal force, the surfaces are said to be *smooth*. Otherwise, they are *rough*.

Ropes and Cables

A rope or cable attached to an object exerts a force on the object whose magnitude is equal to the tension and whose line of action is parallel to the rope or cable at the point of attachment.

A *pulley* is a wheel with a grooved rim that can be used to change the direction of a rope or cable. When a pulley can turn freely and the rope or cable either is stationary or turns the pulley at a constant rate, the tension is approximately the same on both sides of the pulley.

Springs

The force exerted by a *linear spring* is

$$|\mathbf{F}| = k|L - L_0|, \qquad \text{Eq. (3.1)}$$

where k is the *spring constant*, L is the length of the spring, and L_0 is its unstretched length.

1. Choose an object to isolate.

2. Draw the isolated object.

3. Show the external forces.

Free-Body Diagrams

A free-body diagram is a drawing of an object in which the object is isolated from its surroundings and the external forces acting on the object are shown. Drawing a free-body diagram requires the steps shown in Figs. 1–3. A coordinate system must be chosen to express the forces on the isolated object in terms of components.

Equilibrium

If an object is in equilibrium, the sum of the external forces acting on it is zero:

$$\Sigma \mathbf{F} = \mathbf{0}. \quad \text{Eq. (3.2)}$$

This implies that the sums of the external forces in the x, y, and z directions each equal zero:

$$\Sigma F_x = 0, \qquad \Sigma F_y = 0, \qquad \Sigma F_z = 0, \qquad\qquad \text{Eqs. (3.6)}$$

Review Problems

3.1 The 100-lb crate is held in place on the smooth surface by the rope AB. Determine the tension in the rope and the magnitude of the normal force exerted on the crate by the surface.

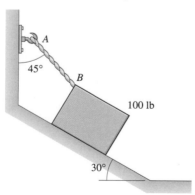

P3.1

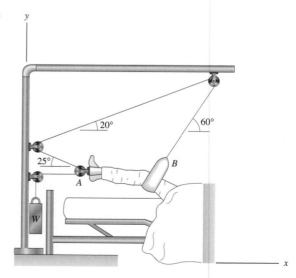

P3.2

3.2 The system shown is called Russell's traction. If the sum of the downward forces exerted at A and B by the patient's leg is 32.2 lb, what is the weight W?

3.3 A heavy rope used as a hawser for a cruise ship sags as shown. If it weighs 200 lb, what are the tensions in the rope at A and B?

P3.3

3.4 The cable AB is horizontal, and the box on the right weighs 100 lb. The surfaces are smooth.

(a) What is the tension in the cable?
(b) What is the weight of the box on the left?

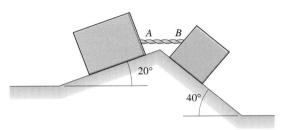

P3.4

3.5 A concrete bucket used at a construction site is supported by two cranes. The 100-kg bucket contains 500 kg of concrete. Determine the tensions in the cables AB and AC.

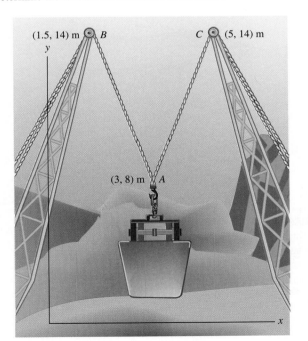

P3.5

3.6 The mass of the suspended object A is m_A and the masses of the pulleys are negligible. Determine the force T necessary for the system to be in equilibrium.

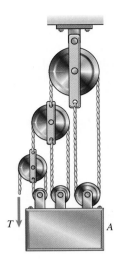

P3.6

3.7 The assembly A, including the pulley, weighs 60 lb. What force F is necessary for the system to be in equilibrium?

P3.7

3.8 The mass of block A is 42 kg, and the mass of block B is 50 kg. The surfaces are smooth. If the blocks are in equilibrium, what is the force F?

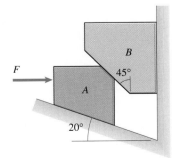

P3.8

3.9 The climber A is being helped up an icy slope by two friends. His mass is 80 kg, and the direction cosines of the force exerted on him by the slope are $\cos\theta_x = -0.286$, $\cos\theta_y = 0.429$, and $\cos\theta_z = 0.857$. The y axis is vertical. If the climber is in equilibrium in the position shown, what are the tensions in the ropes AB and AC and the magnitude of the force exerted on him by the slope?

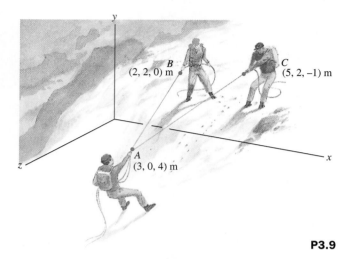

P3.9

3.10 Consider the climber A being helped by his friends in Problem 3.9 To try to make the tensions in the ropes more equal, the friend at B moves to the position (4, 2, 0) m. What are the new tensions in the ropes AB and AC and the magnitude of the force exerted on the climber by the slope?

3.11 A climber helps his friend up an icy slope. His friend is hauling a box of supplies. If the mass of the friend is 90 kg and the mass of the supplies is 22 kg, what are the tensions in the ropes AB and CD? Assume that the slope is smooth.

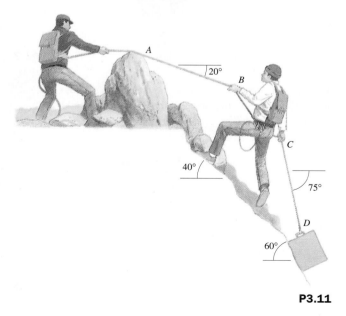

P3.11

3.12 The small sphere of mass m is attached to a string of length L and rests on the smooth surface of a sphere of radius R. Determine the tension in the string in terms of m, L, h, and R.

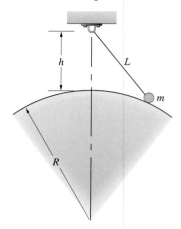

P3.12

3.13 An engineer doing preliminary design studies for a new radio telescope envisions a triangular receiving platform sus-pended by cables from three equally spaced 40-m towers. The receiving platform has a mass of 20 Mg (megagrams) and is 10 m below the tops of the towers. What tension would the cables be subjected to?

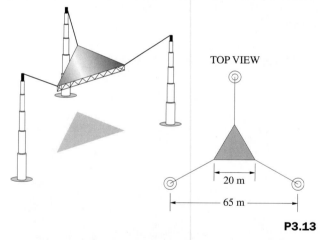

TOP VIEW

P3.13

3.14 The metal disk A weighs 10 lb. It is held in place at the center of the smooth inclined surface by the strings AB and AC. What are the tensions in the strings?

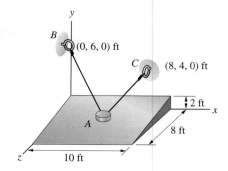

P3.14

3.15 Cable *AB* is attached to the top of the vertical 3-m post, and its tension is 50 kN. What are the tensions in cables *AO*, *AC*, and *AD*?

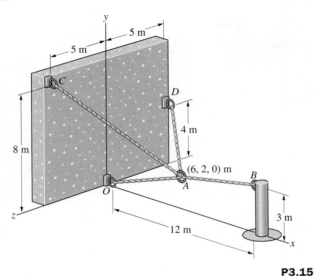

P3.15

3.16 The 1350-kg car is at rest on a plane surface with its brakes locked. The unit vector $\mathbf{e}_n = 0.231\mathbf{i} + 0.923\mathbf{j} + 0.308\mathbf{k}$ is perpendicular to the surface. The *y* axis points upward. The direction cosines of the cable from *A* to *B* are $\cos\theta_x = -0.816$, $\cos\theta_y = 0.408$, $\cos\theta_z = -0.408$, and the tension in the cable is 1.2 kN. Determine the magnitudes of the normal and friction forces the car's wheels exert on the surface.

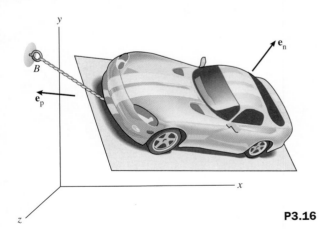

P3.16

3.17 The brakes of the car in Problem 3.16 are released, and the car is held in place on the plane surface by the cable *AB*. The car's front wheels are aligned so that the tires exert no friction forces parallel to the car's longitudinal axis. The unit vector $\mathbf{e}_p = -0.941\mathbf{i} + 0.131\mathbf{j} + 0.314\mathbf{k}$ is parallel to the plane surface and aligned with the car's longitudinal axis. What is the tension in the cable?

𝒟**esign Experience** A possible design for a simple scale to weigh objects is shown. The length of the string *AB* is 0.5 m. When an object is placed in the pan, the spring stretches and the string *AB* rotates. The object's weight can be determined by observing the change in the angle α.

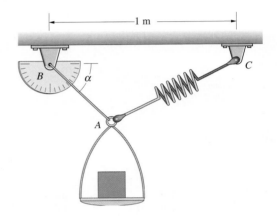

(a) Assume that objects with masses in the range 0.2–2 kg are to be weighed. Choose the unstretched length and spring constant of the spring in order to obtain accurate readings for weights in the desired range. (Neglect the weights of the pan and spring. Notice that a significant change in the angle α is needed to determine the weight accurately.)

(b) Suppose that you can use the same components—the pan, protractor, a spring, string—and also one or more pulleys. Suggest another possible configuration for the scale. Use statics to analyze your proposed configuration and compare its accuracy with that of the configuration shown for objects with masses in the range 0.2–2 kg.

Loads lifted by a building crane can exert large moments that the crane's structure must support. In this chapter we calculate moments of forces and analyze systems of forces and moments.

Systems of Forces and Moments

The effects of forces can depend not only on their magnitudes and directions but also on the moments, or torques, they exert. The rotations of objects such as the wheels of a vehicle, the crankshaft of an engine, and the rotor of an electric generator result from the moments of the forces exerted on them. If an object is in equilibrium, the moment about any point due to the forces acting on the object is zero. Before continuing our discussion of free-body diagrams and equilibrium, we must explain how to calculate moments and introduce the concept of equivalent systems of forces and moments.

4.1 Two-Dimensional Description of the Moment

Consider a force of magnitude F and a point P, and let's view them in the direction perpendicular to the plane containing the force vector and the point (Fig. 4.1a). The *magnitude of the moment* of the force about P is DF, where D is the perpendicular distance from P to the line of action of the force (Fig. 4.1b). In this example, the force would tend to cause counterclockwise rotation about point P. That is, if we imagine the force acts on an object that can rotate about point P, the force would tend to cause counterclockwise rotation (Fig. 4.1c). We say that the *sense of the moment* is counterclockwise. *We define counterclockwise moments to be positive and clockwise moments to be negative.* (This is the usual convention, although we occasionally encounter situations in which it is more convenient to define clockwise moments to be positive.) Thus the moment of the force about P is

$$M_P = DF. \tag{4.1}$$

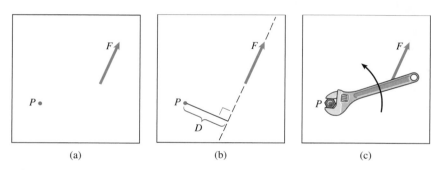

(a) (b) (c)

Figure 4.1
(a) The force and point P.
(b) The perpendicular distance D from point P to the line of action of F.
(c) The sense of the moment is counterclockwise.

Notice that if the line of action of F passes through P, the perpendicular distance $D = 0$ and the moment of F about P is zero.

The dimensions of the moment are (distance) $\times$ (force). For example, moments can be expressed in newton-meters in SI units and in foot-pounds in U.S. Customary units.

Suppose that you want to place a television set on a shelf, and you aren't certain the attachment of the shelf to the wall is strong enough to support it. Instinctively, you place it near the wall (Fig. 4.2a), knowing that the attachment is more likely to fail if you place it away from the wall (Fig. 4.2b). What is the difference in the two cases? The magnitude and direction of the force exerted on the shelf by the weight of the television are the same in each case, but the moments exerted on the attachment are different. The moment exerted about P by its weight when it is near the wall, $M_P = -D_1 W$, is smaller in magnitude than the moment about P when it is placed away from the wall, $M_P = -D_2 W$.

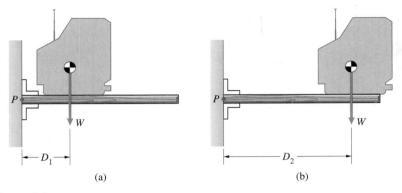

Figure 4.2
It is better to place the television near the wall (a) instead of away from it
(b) because the moment exerted on the support at P is smaller.

The method we describe in this section can be used to determine the sum
of the moments of a system of forces about a point if the forces are two-dimensional (coplanar) and the point lies in the same plane. For example, consider the construction crane shown in Fig. 4.3. The sum of the moments
exerted about point P by the load W_1 and the counterweight W_2 is

$$\Sigma M_P = D_1 W_1 - D_2 W_2.$$

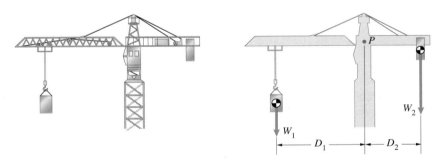

Figure 4.3
A tower crane used in the construction of high-rise buildings.

This moment tends to cause the top of the vertical tower to rotate and could
cause it to collapse. If the distance D_2 is adjusted so that $D_1 W_1 = D_2 W_2$, the
moment about point P due to the load and the counterweight is zero.

If a force is resolved into components, the moment of the force about a
point P is equal to the sum of the moments of its components about P. We
prove this very useful result in the next section.

Study Questions

1. How do you determine the magnitude of the moment of a force about a point?
2. The moment of a force about a point is defined to be positive if its sense is
   counterclockwise. What does that mean?
3. If the line of action of a force passes through a point P, what do you know
   about the moment of the force about P?

Example 4.1

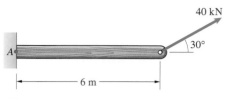

Determining the Moment of a Force

What is the moment of the 40-kN force in Fig. 4.4 about point A?

Figure 4.4

Strategy

We can calculate the moment in two ways: by determining the perpendicular distance from point A to the line of action of the force or by resolving the force into components and determining the sum of the moments of the components about A.

Solution

First Method From Fig. a, the perpendicular distance from A to the line of action of the force is

$$D = 6\sin 30° = 3 \text{ m}.$$

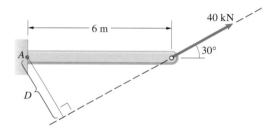

(a) Determining the perpendicular distance D.

The magnitude of the moment of the force about A is $(3 \text{ m})(40 \text{ kN}) = 120$ kN-m, and the sense of the moment about A is counterclockwise. Therefore the moment is

$$M_A = 120 \text{ kN-m.}$$

Second Method In Fig. b, we resolve the force into horizontal and vertical components. The perpendicular distance from A to the line of action of the horizontal component is zero, so the horizontal component exerts no moment about A. The magnitude of the moment of the vertical component about A is $(6 \text{ m})(40\sin 30° \text{ kN}) = 120$ kN-m, and the sense of its moment about A is counterclockwise. The moment is

$$M_A = 120 \text{ kN-m.}$$

(b) Resolving the force into components.

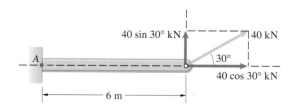

Example 4.2

Moment of a System of Forces

Four forces act on the machine part in Fig. 4.5. What is the sum of the moments of the forces about the origin O?

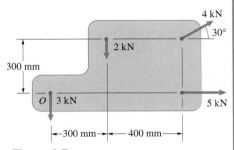

Figure 4.5

Strategy

We can determine the moments of the forces about point O directly from the given information except for the 4-kN force. We will determine its moment by resolving it into components and summing the moments of the components.

Solution

Moment of the 3-kN Force The line of action of the 3-kN force passes through O. It exerts no moment about O.

Moment of the 5-kN Force The line of action of the 5-kN force also passes through O. It too exerts no moment about O.

Moment of the 2-kN Force The perpendicular distance from O to the line of action of the 2-kN force is 0.3 m, and the sense of the moment about O is clockwise. The moment of the 2-kN force about O is

$$-(0.3 \text{ m})(2 \text{ kN}) = -0.600 \text{ kN-m}.$$

(Notice that we converted the perpendicular distance from millimeters into meters, obtaining the result in terms of kilonewton-meters.)

Moment of the 4-kN Force In Fig. a, we introduce a coordinate system and resolve the 4-kN force into x and y components. The perpendicular distance from O to the line of action of the x component is 0.3 m, and the sense of the moment about O is clockwise. The moment of the x component about O is

$$-(0.3 \text{ m})(4 \cos 30° \text{ kN}) = -1.039 \text{ kN-m}.$$

The perpendicular distance from point O to the line of action of the y component is 0.7 m, and the sense of the moment about O is counterclockwise. The moment of the y component about O is

$$(0.7 \text{ m})(4 \sin 30° \text{ kN}) = 1.400 \text{ kN-m}.$$

The sum of the moments of the four forces about point O is

$$\Sigma M_0 = -0.600 - 1.039 + 1.400 = -0.239 \text{ kN-m}.$$

The four forces exert a 0.239 kN-m clockwise moment about point O.

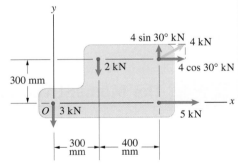

(a) Resolving the 4-kN force into components.

Example 4.3

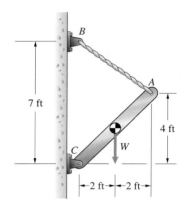

Figure 4.6

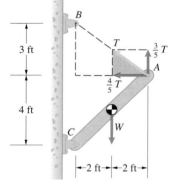

(a) Resolving the force exerted by the cable into horizontal and vertical components.

Summing Moments to Determine an Unknown Force

The weight $W = 300$ lb (Fig. 4.6). The sum of the moments about C due to the weight W and the force exerted on the bar CA by the cable AB is zero. What is the tension in the cable?

Strategy

Let T be the tension in cable AB. Using the given dimensions, we can express the horizontal and vertical components of the force exerted on the bar by the cable in terms of T. Then by setting the sum of the moments about C due to the weight of the bar and the force exerted by the cable equal to zero, we can obtain an equation for T.

Solution

Using similar triangles, we resolve the force exerted on the bar by the cable into horizontal and vertical components (Fig. a). The sum of the moments about C due to the weight of the bar and the force exerted by the cable AB is

$$\Sigma M_C = 4\left(\frac{4}{5}T\right) + 4\left(\frac{3}{5}T\right) - 2W = 0.$$

Solving for T, we obtain

$$T = 0.357W = 107.1 \text{ lb.}$$

4.2 The Moment Vector

The moment of a force about a point is a vector. In this section we define this vector and explain how it is evaluated. We then show that when we use the two-dimensional description of the moment described in Section 4.1, we are specifying the magnitude and direction of the moment vector.

Consider a force vector $\mathbf{F}$ and point P (Fig. 4.7a). The *moment* of $\mathbf{F}$ about P is the vector

$$\mathbf{M}_P = \mathbf{r} \times \mathbf{F}. \tag{4.2}$$

where **r** is a position vector from P to *any* point on the line of action of **F** (Fig. 4.7b).

Magnitude of the Moment

From the definition of the cross product, the magnitude of $\mathbf{M}_P$ is

$$|\mathbf{M}_P| = |\mathbf{r}||\mathbf{F}| \sin\theta,$$

where θ is the angle between the vectors **r** and **F** when they are placed tail to tail. The perpendicular distance from P to the line of action of **F** is $D = |\mathbf{r}| \sin\theta$ (Fig. 4.7c). Therefore the magnitude of the moment $\mathbf{M}_P$ equals the product of the perpendicular distance from P to the line of action of **F** and the magnitude of **F**:

$$|\mathbf{M}_P| = D|\mathbf{F}|. \tag{4.3}$$

Notice that if you know the vectors $\mathbf{M}_P$ and **F**, you can solve this equation for the perpendicular distance D.

Sense of the Moment

We know from the definition of the cross product that $\mathbf{M}_P$ is perpendicular to both **r** and **F**. That means that $\mathbf{M}_P$ is perpendicular to the plane containing P and **F** (Fig. 4.8a). Notice in this figure that we denote a moment by a circular arrow around the vector.

The direction of $\mathbf{M}_P$ also indicates the sense of the moment: If you point the thumb of your right hand in the direction of $\mathbf{M}_P$, the "arc" of your fingers indicates the sense of the rotation that **F** tends to cause about P (Fig. 4.8b).

The result obtained from Eq. (4.2) doesn't depend on where the vector **r** intersects the line of action of **F**. Instead of using the vector **r** in

(a)

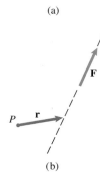

(b)

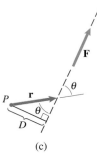

(c)

Figure 4.7
(a) The force **F** and point P.
(b) A vector **r** from P to a point on the line of action of **F**.
(c) The angle θ and the perpendicular distance D.

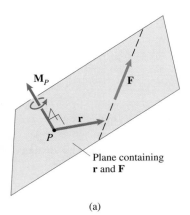

(a)

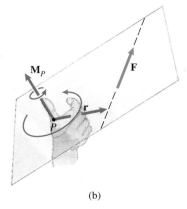

(b)

Figure 4.8
(a) $\mathbf{M}_P$ is perpendicular to the plane containing P and **F**.
(b) The direction of $\mathbf{M}_P$ indicates the sense of the moment.

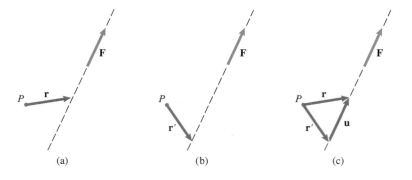

Figure 4.9
(a) A vector **r** from P to the line of action of **F**.
(b) A different vector **r'**.
(c) **r** = **r'** + **u**.

Fig. 4.9a, we could use the vector **r'** in Fig. 4.9b. The vector **r** = **r'** + **u**, where **u** is parallel to **F** (Fig. 4.9c). Therefore

$$\mathbf{r} \times \mathbf{F} = (\mathbf{r'} + \mathbf{u}) \times \mathbf{F} = \mathbf{r'} \times \mathbf{F}$$

because the cross product of the parallel vectors **u** and **F** is zero.

In summary, the moment of a force **F** about a point P has three properties:

1. The magnitude of $\mathbf{M}_P$ is equal to the product of the magnitude of **F** and the perpendicular distance from P to the line of action of **F**. If the line of action of **F** passes through P, $\mathbf{M}_P = \mathbf{0}$.

2. $\mathbf{M}_P$ is perpendicular to the plane containing P and **F**.

3. The direction of $\mathbf{M}_P$ indicates the sense of the moment through a right-hand rule (Fig. 4.8b). Since the cross product is not commutative, you must be careful to maintain the correct sequence of the vectors in the equation $\mathbf{M}_P = \mathbf{r} \times \mathbf{F}$.

Let us determine the moment of the force **F** in Fig. 4.10a about the point P. Since the vector **r** in Eq. (4.2) can be a position vector to any point on the line of action of **F**, we can use the vector from P to the point of application of **F** (Fig. 4.10b):

$$\mathbf{r} = (12 - 3)\mathbf{i} + (6 - 4)\mathbf{j} + (-5 - 1)\mathbf{k} = 9\mathbf{i} + 2\mathbf{j} - 6\mathbf{k} \text{ (ft)}.$$

The moment is

$$\mathbf{M}_P = \mathbf{r} \times \mathbf{F} = \begin{vmatrix} \mathbf{i} & \mathbf{j} & \mathbf{k} \\ 9 & 2 & -6 \\ 4 & 4 & 7 \end{vmatrix} = 38\mathbf{i} - 87\mathbf{j} + 28\mathbf{k} \text{ (ft-lb)}.$$

The magnitude of $\mathbf{M}_P$,

$$|\mathbf{M}_P| = \sqrt{(38)^2 + (-87)^2 + (28)^2} = 99.0 \text{ ft-lb},$$

equals the product of the magnitude of **F** and the perpendicular distance D from point P to the line of action of **F**. Therefore

$$D = \frac{|\mathbf{M}_P|}{|\mathbf{F}|} = \frac{99.0 \text{ ft-lb}}{9 \text{ lb}} = 11.0 \text{ ft}.$$

The direction of $\mathbf{M}_P$ tells us both the orientation of the plane containing P and **F** and the sense of the moment (Fig. 4.10c).

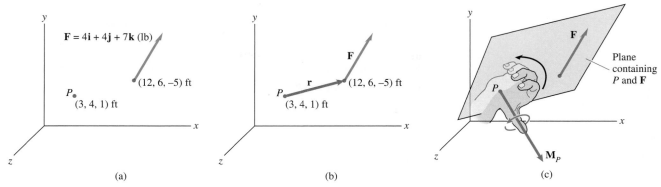

Figure 4.10
(a) A force **F** and point P.
(b) The vector **r** from P to the point of application of **F**.
(c) $\mathbf{M}_P$ is perpendicular to the plane containing P and **F**. The right-hand rule
indicates the sense of the moment.

Relation to the Two-Dimensional Description

If our view is perpendicular to the plane containing the point P and the force
F, the two-dimensional description of the moment we used in Section 4.1
specifies both the magnitude and direction of $\mathbf{M}_P$. In this situation, $\mathbf{M}_P$ is perpendicular to the page, and the right-hand rule indicates whether it points out
of or into the page.

For example, in Fig. 4.11a, the view is perpendicular to the x–y plane
and the 10-N force is contained in the x–y plane. Suppose that we want to determine the moment of the force about the origin O. The perpendicular distance from O to the line of action of the force is 4 m. The two-dimensional
description of the moment of the force about O is that its magnitude is
(4 m)(10 N) = 40 N-m and its sense is counterclockwise, or

$$M_O = 40 \text{ N-m.}$$

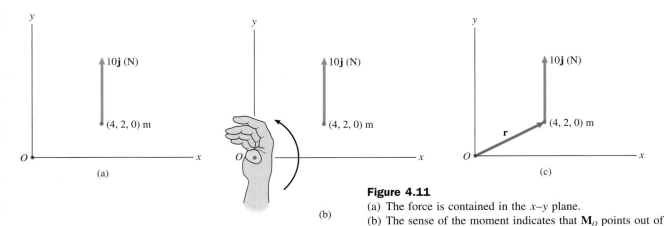

Figure 4.11
(a) The force is contained in the x–y plane.
(b) The sense of the moment indicates that $\mathbf{M}_O$ points out of
the page.
(c) The vector **r** from O to the point of application of **F**.

That tells us that the magnitude of the vector $\mathbf{M}_O$ is 40 N-m, and the right-hand rule (Fig. 4.11b) indicates that it points out of the page. Therefore

$$\mathbf{M}_O = 40\mathbf{k} \ (\text{N-m}).$$

We can confirm this result by using Eq. (4.2). If we let $\mathbf{r}$ be the vector from O to the point of application of the force (Fig. 4.11c),

$$\mathbf{M}_O = \mathbf{r} \times \mathbf{F} = (4\mathbf{i} + 2\mathbf{j}) \times 10\mathbf{j} = 40\mathbf{k} \ (\text{N-m}).$$

As this example illustrates, the two-dimensional description of the moment determines the moment vector. The converse is also true. The magnitude of $\mathbf{M}_O$ equals the product of the magnitude of the force and the perpendicular distance from O to the line of action of the force, 40 N-m, and the direction of $\mathbf{M}_O$ indicates that the sense of the moment is counterclockwise (Fig. 4.11b).

Varignon's Theorem

Let $\mathbf{F}_1, \mathbf{F}_2, \ldots, \mathbf{F}_N$ be a concurrent system of forces whose lines of action intersect at a point Q. The moment of the system about a point P is

$$\left(\mathbf{r}_{PQ} \times \mathbf{F}_1\right) + \left(\mathbf{r}_{PQ} \times \mathbf{F}_2\right) + \cdots + \left(\mathbf{r}_{PQ} \times \mathbf{F}_N\right)$$

$$= \mathbf{r}_{PQ} \times \left(\mathbf{F}_1 + \mathbf{F}_2 + \cdots + \mathbf{F}_N\right),$$

where $\mathbf{r}_{PQ}$ is the vector from P to Q (Fig. 4.12). This result, known as *Varignon's theorem*, follows from the distributive property of the cross product, Eq. (2.31). It confirms that the moment of a force about a point P is equal to the sum of the moments of its components about P.

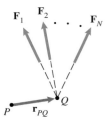

Figure 4.12
A system of concurrent forces and a point P.

Study Questions

1. When you use the equation $\mathbf{M}_P = \mathbf{r} \times \mathbf{F}$ to determine the moment of a force $\mathbf{F}$ about a point P, how do you choose the vector $\mathbf{r}$?
2. If you know the components of the vector $\mathbf{M}_P = \mathbf{r} \times \mathbf{F}$, how can you determine the product of the magnitude of $\mathbf{F}$ and the perpendicular distance from P to the line of action of $\mathbf{F}$?
3. How does the direction of the vector $\mathbf{M}_P = \mathbf{r} \times \mathbf{F}$ indicate the sense of the moment of $\mathbf{F}$ about P?

Example 4.4

Two-Dimensional Description and the Moment Vector

Determine the moment of the 400-N force in Fig. 4.13 about O.
(a) What is the two-dimensional description of the moment?
(b) Express the moment as a vector without using Eq. (4.2).
(c) Use Eq. (4.2) to determine the moment.

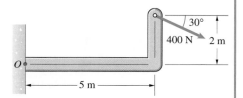

Figure 4.13

Solution

(a) Resolving the force into horizontal and vertical components (Fig. a), the two-dimensional description of the moment is

$$M_O = -(2 \text{ m})(400 \cos 30° \text{ N}) - (5 \text{ m})(400 \sin 30° \text{ N})$$

$$= -1.69 \text{ kN-m.}$$

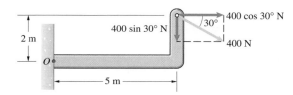

(a) Resolving the force into components.

(b) To express the moment as a vector, we introduce the coordinate system shown in Fig. b. The magnitude of the moment is 1.69 kN-m, and its sense is clockwise. Pointing the arc of the fingers of the right- hand clockwise, the thumb points into the page. Therefore

$$\mathbf{M}_O = -1.69\mathbf{k} \text{ (kN-m).}$$

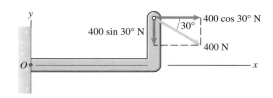

(b) Introducing a coordinate system.

(c) We apply Eq. (4.2):

Choose the Vector r   We can let $\mathbf{r}$ be the vector from O to the point of application of the force (Fig. c):

$$\mathbf{r} = 5\mathbf{i} + 2\mathbf{j} \text{ (m).}$$

Evaluate r $\times$ F   The moment is

$$\mathbf{M}_O = \mathbf{r} \times \mathbf{F} = (5\mathbf{i} + 2\mathbf{j}) \times (400 \cos 30°\mathbf{i} - 400 \sin 30°\mathbf{j})$$

$$= -1.69\mathbf{k} \text{ (kN-m).}$$

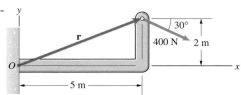

(c) The vector $\mathbf{r}$ from 0 to the point of application of the force.

Example 4.5

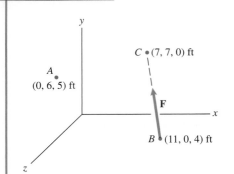

Figure 4.14

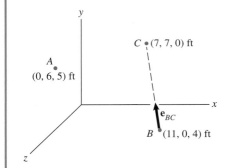

(a) The unit vector $\mathbf{e}_{BC}$.

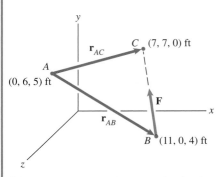

(b) The moment can be determined using either $\mathbf{r}_{AB}$ or $\mathbf{r}_{AC}$.

Determining the Moment and the Perpendicular Distance to the Line of Action

The line of action of the 90-lb force $\mathbf{F}$ in Fig. 4.14 passes through points B and C.
(a) What is the moment of $\mathbf{F}$ about point A?
(b) What is the perpendicular distance from point A to the line of action of $\mathbf{F}$?

Strategy

(a) We must use Eq. (4.2) to determine the moment. Since $\mathbf{r}$ is a vector from A to any point on the line of action of $\mathbf{F}$, we can use either the vector from A to B or the vector from A to C. To demonstrate that we obtain the same result, we will determine the moment using both.
(b) Since the magnitude of the moment is equal to the product of the magnitude of $\mathbf{F}$ and the perpendicular distance from A to the line of action of $\mathbf{F}$, we can use the result of (a) to determine the perpendicular distance.

Solution

(a) To evaluate the cross product in Eq. (4.2), we need the components of $\mathbf{F}$. The vector from B to C is

$$(7 - 11)\mathbf{i} + (7 - 0)\mathbf{j} + (0 - 4)\mathbf{k} = -4\mathbf{i} + 7\mathbf{j} - 4\mathbf{k} \text{ (ft)}.$$

Dividing this vector by its magnitude, we obtain a unit vector $\mathbf{e}_{BC}$ that has the same direction as $\mathbf{F}$ (Fig. a):

$$\mathbf{e}_{BC} = -\frac{4}{9}\mathbf{i} + \frac{7}{9}\mathbf{j} - \frac{4}{9}\mathbf{k}.$$

Now we express $\mathbf{F}$ as the product of its magnitude and $\mathbf{e}_{BC}$:

$$\mathbf{F} = 90\mathbf{e}_{BC} = -40\mathbf{i} + 70\mathbf{j} - 40\mathbf{k} \text{ (lb)}.$$

Choose the Vector r The position vector from A to B (Fig. b) is

$$\mathbf{r}_{AB} = (11 - 0)\mathbf{i} + (0 - 6)\mathbf{j} + (4 - 5)\mathbf{k} = 11\mathbf{i} - 6\mathbf{j} - \mathbf{k} \text{ (ft)}.$$

Evaluate r × F The moment of $\mathbf{F}$ about A is

$$\mathbf{M}_A = \mathbf{r}_{AB} \times \mathbf{F} = \begin{vmatrix} \mathbf{i} & \mathbf{j} & \mathbf{k} \\ 11 & -6 & -1 \\ -40 & 70 & -40 \end{vmatrix}$$

$$= 310\mathbf{i} + 480\mathbf{j} + 530\mathbf{k} \text{ (ft-lb)}.$$

Alternative Choice of Position Vector If we use the vector from A to C instead,

$$\mathbf{r}_{AC} = (7 - 0)\mathbf{i} + (7 - 6)\mathbf{j} + (0 - 5)\mathbf{k} = 7\mathbf{i} + \mathbf{j} - 5\mathbf{k} \text{ (ft)},$$

we obtain the same result:

$$\mathbf{M}_A = \mathbf{r}_{AC} \times \mathbf{F} = \begin{vmatrix} \mathbf{i} & \mathbf{j} & \mathbf{k} \\ 7 & 1 & -5 \\ -40 & 70 & -40 \end{vmatrix}$$

$$= 310\mathbf{i} + 480\mathbf{j} + 530\mathbf{k} \text{ (ft-lb)}.$$

(b) The perpendicular distance is

$$\frac{|\mathbf{M}_A|}{|\mathbf{F}|} = \frac{\sqrt{(310)^2 + (480)^2 + (530)^2}}{\sqrt{(-40)^2 + (70)^2 + (-40)^2}} = 8.66 \text{ ft}.$$

Example 4.6

Applying the Moment Vector

The cables *AB* and *AC* in Fig. 4.15 extend from an attachment point *A* on the floor to attachment points *B* and *C* in the walls. The tension in cable *AB* is 10 kN, and the tension in cable *AC* is 20 kN. What is the sum of the moments about *O* due to the forces exerted on *A* by the two cables?

Solution

Let $\mathbf{F}_{AB}$ and $\mathbf{F}_{AC}$ be the forces exerted on the attachment point *A* by the two cables (Fig. a). To express $\mathbf{F}_{AB}$ in terms of its components, we determine the position vector from *A* to *B*,

$$(0 - 4)\mathbf{i} + (4 - 0)\mathbf{j} + (8 - 6)\mathbf{k} = -4\mathbf{i} + 4\mathbf{j} + 2\mathbf{k} \text{ (m)},$$

and divide it by its magnitude to obtain a unit vector $\mathbf{e}_{AB}$ with the same direction as $\mathbf{F}_{AB}$ (Fig. b):

$$\mathbf{e}_{AB} = \frac{-4\mathbf{i} + 4\mathbf{j} + 2\mathbf{k}}{6} = -\frac{2}{3}\mathbf{i} + \frac{2}{3}\mathbf{j} + \frac{1}{3}\mathbf{k}.$$

Now we write $\mathbf{F}_{AB}$ as

$$\mathbf{F}_{AB} = 10\mathbf{e}_{AB} = -6.67\mathbf{i} + 6.67\mathbf{j} + 3.33\mathbf{k} \text{ (kN)}.$$

We express the force $\mathbf{F}_{AC}$ in terms of its components in the same way:

$$\mathbf{F}_{AC} = 5.71\mathbf{i} + 8.57\mathbf{j} - 17.14\mathbf{k} \text{ (kN)}.$$

Choose the Vector r Since the lines of action of both forces pass through point *A*, we can use the vector from *O* to *A* to determine the moments of both forces about point *O* (Fig. a):

$$\mathbf{r} = 4\mathbf{i} + 6\mathbf{k} \text{ (m)}.$$

Evaluate r × F The sum of the moments is

$$\Sigma \mathbf{M}_O = (\mathbf{r} \times \mathbf{F}_{AB}) + (\mathbf{r} \times \mathbf{F}_{AC})$$

$$= \begin{vmatrix} \mathbf{i} & \mathbf{j} & \mathbf{k} \\ 4 & 0 & 6 \\ -6.67 & 6.67 & 3.33 \end{vmatrix} + \begin{vmatrix} \mathbf{i} & \mathbf{j} & \mathbf{k} \\ 4 & 0 & 6 \\ 5.71 & 8.57 & -17.14 \end{vmatrix}$$

$$= -91.4\mathbf{i} + 49.5\mathbf{j} + 61.0\mathbf{k} \text{ (kN-m)}.$$

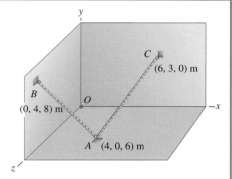

Figure 4.15

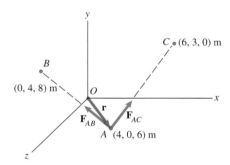

(a) The forces $\mathbf{F}_{AB}$ and $\mathbf{F}_{AC}$ exerted at *A* by the cables.

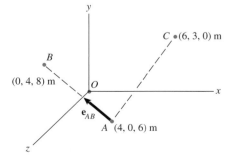

(b) The unit vector $\mathbf{e}_{AB}$ has the same direction as $\mathbf{F}_{AB}$.

4.3 Moment of a Force About a Line

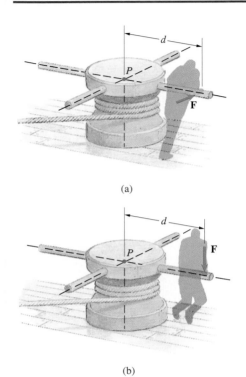

(a)

(b)

Figure 4.16
(a) Turning a capstan.
(b) A vertical force does not turn the capstan.

The device in Fig. 4.16, called a *capstan*, was used in the days of square-rigged sailing ships. Crewmen turned it by pushing on the handles as shown in Fig. 4.16a, providing power for such tasks as raising anchors and hoisting yards. A vertical force **F** applied to one of the handles as shown in Fig. 4.16b does not cause the capstan to turn, even though the magnitude of the moment about point P is $d|\mathbf{F}|$ in both cases.

The measure of the tendency of a force to cause rotation about a line, or axis, is called the moment of the force about the line. Suppose that a force **F** acts on an object such as a turbine that rotates about an axis L, and we resolve **F** into components in terms of the coordinate system shown in Fig. 4.17. The components F_x and F_z do not tend to rotate the turbine, just as the force parallel to the axis of the capstan did not cause it to turn. It is the component F_y that tends to cause rotation, by exerting a moment of magnitude aF_y about the turbine's axis. In this example we can determine the moment of **F** about L easily because the coordinate system is conveniently placed. We now introduce an expression that determines the moment of a force about any line.

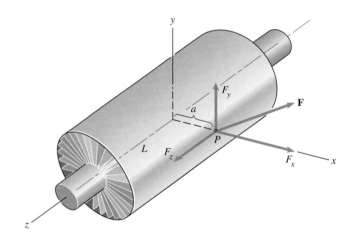

Figure 4.17
Applying a force to a turbine with axis of rotation L.

Definition

Consider a line L and force **F** (Fig. 4.18a). Let $\mathbf{M}_P$ be the moment of **F** about an arbitrary point P on L (Fig. 4.18b). The moment of **F** about L is the component of $\mathbf{M}_P$ parallel to L, which we denote by $\mathbf{M}_L$ (Fig. 4.18c). The magnitude of the moment of **F** about L is $|\mathbf{M}_L|$, and when the thumb of the right hand is pointed in the direction of $\mathbf{M}_L$, the arc of the fingers indicates the sense of the moment about L.

In terms of a unit vector **e** along L (Fig. 4.18d), $\mathbf{M}_L$ is given by

$$\mathbf{M}_L = (\mathbf{e} \cdot \mathbf{M}_P)\mathbf{e}. \tag{4.4}$$

(The unit vector **e** can point in either direction. See our discussion of vector components parallel and normal to a line in Section 2.5.) The moment $\mathbf{M}_P = \mathbf{r} \times \mathbf{F}$, so we can also express $\mathbf{M}_L$ as

$$\mathbf{M}_L = \big[\mathbf{e} \cdot (\mathbf{r} \times \mathbf{F})\big]\mathbf{e}. \tag{4.5}$$

The mixed triple product in this expression is given in terms of the components of the three vectors by

$$\mathbf{e} \cdot (\mathbf{r} \times \mathbf{F}) = \begin{vmatrix} e_x & e_y & e_z \\ r_x & r_y & r_z \\ F_x & F_y & F_z \end{vmatrix}. \tag{4.6}$$

Notice that the value of the scalar $\mathbf{e} \cdot \mathbf{M}_P = \mathbf{e} \cdot (\mathbf{r} \times \mathbf{F})$ determines both the magnitude and direction of $\mathbf{M}_L$. The absolute value of $\mathbf{e} \cdot \mathbf{M}_P$ is the magnitude of $\mathbf{M}_L$. If $\mathbf{e} \cdot \mathbf{M}_P$ is positive, $\mathbf{M}_L$ points in the direction of **e**, and if $\mathbf{e} \cdot \mathbf{M}_P$ is negative, $\mathbf{M}_L$ points in the direction opposite to **e**.

The result obtained with Eq. (4.4) or (4.5) doesn't depend on which point on L is chosen to determine $\mathbf{M}_P = \mathbf{r} \times \mathbf{F}$. If we use point P in Fig. 4.19 to determine the moment of **F** about L, we get the result given by Eq. (4.5). If we use P' instead, we obtain the same result,

$$\big[\mathbf{e} \cdot (\mathbf{r}' \times \mathbf{F})\big]\mathbf{e} = \big\{\mathbf{e} \cdot \big[(\mathbf{r} + \mathbf{u}) \times \mathbf{F}\big]\big\}\mathbf{e}$$

$$= \big[\mathbf{e} \cdot (\mathbf{r} \times \mathbf{F}) + \mathbf{e} \cdot (\mathbf{u} \times \mathbf{F})\big]\mathbf{e}$$

$$= \big[\mathbf{e} \cdot (\mathbf{r} \times \mathbf{F})\big]\mathbf{e},$$

because $\mathbf{u} \times \mathbf{F}$ is perpendicular to **e**.

Applying the Definition

To demonstrate that $\mathbf{M}_L$ is the measure of the tendency of **F** to cause rotation about L, we return to the turbine in Fig. 4.17. Let Q be a point on L at an arbitrary distance b from the origin (Fig. 4.20a). The vector **r** from Q to P is $\mathbf{r} = a\mathbf{i} - b\mathbf{k}$, so the moment of **F** about Q is

$$\mathbf{M}_Q = \mathbf{r} \times \mathbf{F} = \begin{vmatrix} \mathbf{i} & \mathbf{j} & \mathbf{k} \\ a & 0 & -b \\ F_x & F_y & F_z \end{vmatrix} = bF_y\mathbf{i} - (aF_z + bF_x)\mathbf{j} + aF_y\mathbf{k}.$$

Since the z axis is coincident with L, the unit vector **k** is along L. Therefore the moment of **F** about L is

$$\mathbf{M}_L = (\mathbf{k} \cdot \mathbf{M}_Q)\mathbf{k} = aF_y\mathbf{k}.$$

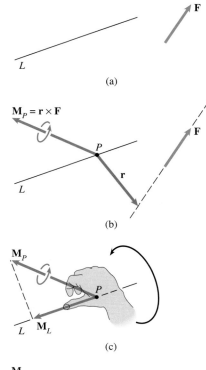

Figure 4.18
(a) The line L and force **F**.
(b) $\mathbf{M}_P$ is the moment of **F** about any point P on L.
(c) The component $\mathbf{M}_L$ is the moment of **F** about L.
(d) A unit vector **e** along L.

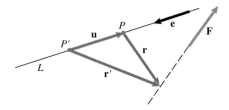

Figure 4.19
Using different points P and P' to determine the moment of **F** about L.

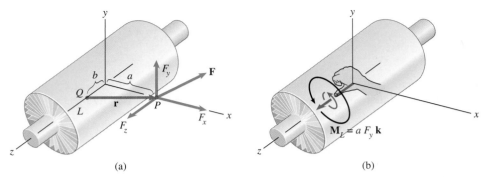

Figure 4.20
(a) An arbitrary point Q on L and the vector $\mathbf{r}$ from Q to P.
(b) $\mathbf{M}_L$ and the sense of the moment about L.

The components F_x and F_z exert no moment about L. If we assume that F_y is positive, it exerts a moment of magnitude aF_y about the turbine's axis in the direction shown in Fig. 4.20b.

Now let's determine the moment of a force about an arbitrary line L (Fig. 4.21a). The first step is to choose a point on the line. If we choose point A (Fig. 4.21b), the vector $\mathbf{r}$ from A to the point of application of $\mathbf{F}$ is

$$\mathbf{r} = (8 - 2)\mathbf{i} + (6 - 0)\mathbf{j} + (4 - 4)\mathbf{k} = 6\mathbf{i} + 6\mathbf{j} \text{ (m).}$$

The moment of $\mathbf{F}$ about A is

$$\mathbf{M}_A = \mathbf{r} \times \mathbf{F} = \begin{vmatrix} \mathbf{i} & \mathbf{j} & \mathbf{k} \\ 6 & 6 & 0 \\ 10 & 60 & -20 \end{vmatrix}$$

$$= -120\mathbf{i} + 120\mathbf{j} + 300\mathbf{k} \text{ (N-m).}$$

The next step is to determine a unit vector along L. The vector from A to B is

$$(-7 - 2)\mathbf{i} + (6 - 0)\mathbf{j} + (2 - 4)\mathbf{k} = -9\mathbf{i} + 6\mathbf{j} - 2\mathbf{k} \text{ (m).}$$

Dividing this vector by its magnitude, we obtain a unit vector $\mathbf{e}_{AB}$ that points from A toward B (Fig. 4.21c):

$$\mathbf{e}_{AB} = -\frac{9}{11}\mathbf{i} + \frac{6}{11}\mathbf{j} - \frac{2}{11}\mathbf{k}.$$

The moment of $\mathbf{F}$ about L is

$$\mathbf{M}_L = (\mathbf{e}_{AB} \cdot \mathbf{M}_A)\mathbf{e}_{AB}$$

$$= \left[\left(-\frac{9}{11}\right)(-120) + \left(\frac{6}{11}\right)(120) + \left(-\frac{2}{11}\right)(300)\right]\mathbf{e}_{AB}$$

$$= 109\mathbf{e}_{AB} \text{ (N-m).}$$

The magnitude of $\mathbf{M}_L$ is 109 N-m; pointing the thumb of the right hand in the direction of $\mathbf{e}_{AB}$ indicates the direction.

If we calculate $\mathbf{M}_L$ using the unit vector $\mathbf{e}_{BA}$ that points from B toward A instead, we obtain

$$\mathbf{M}_L = -109\mathbf{e}_{BA} \text{ (N-m).}$$

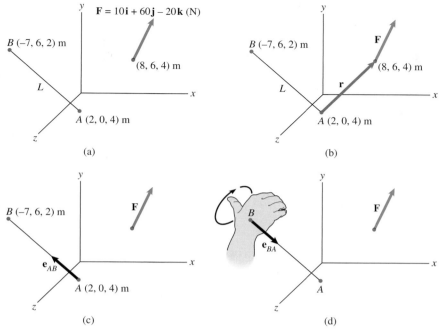

Figure 4.21
(a) A force **F** and line L.
(b) The vector **r** from A to the point of application of **F**.
(c) $\mathbf{e}_{AB}$ points from A toward B.
(d) The right-hand rule indicates the sense of the moment.

We obtain the same magnitude, and the minus sign indicates that $\mathbf{M}_L$ points in the direction opposite to $\mathbf{e}_{BA}$, so the direction of $\mathbf{M}_L$ is the same. Therefore the right-hand rule indicates the same sense (Fig. 4.21d).

The preceding examples demonstrate three useful results that we can state in more general terms:

- When the line of action of **F** is perpendicular to a plane containing L (Fig. 4.22a), the magnitude of the moment of **F** about L is equal to the product of the magnitude of **F** and the perpendicular distance D from L to the point where the line of action intersects the plane: $|\mathbf{M}_L| = |\mathbf{F}|D$.

- When the line of action of **F** is parallel to L (Fig. 4.22b), the moment of **F** about L is zero: $\mathbf{M}_L = 0$. Since $\mathbf{M}_P = \mathbf{r} \times \mathbf{F}$ is perpendicular to **F**, $\mathbf{M}_P$ is perpendicular to L and the vector component of $\mathbf{M}_P$ parallel to L is zero.

- When the line of action of **F** intersects L (Fig. 4.22c), the moment of **F** about L is zero. Since we can choose any point on L to evaluate $\mathbf{M}_P$, we can use the point where the line of action of **F** intersects L. The moment $\mathbf{M}_P$ about that point is zero, so its vector component parallel to L is zero.

In summary, determining the moment of a force **F** about a point P using Eqs. (4.4)–(4.6) requires three steps:

1. **Determine a vector r**—Choose any point P on L, and determine the components of a vector **r** from P to any point on the line of action of **F**.

2. **Determine a vector e**—Determine the components of a unit vector along L. It doesn't matter in which direction along L it points.

3. **Evaluate $\mathbf{M}_L$**—You can calculate $\mathbf{M}_P = \mathbf{r} \times \mathbf{F}$ and determine $\mathbf{M}_L$ by using Eq. (4.4), or you can use Eq. (4.6) to evaluate the mixed triple product and substitute the result into Eq. (4.5).

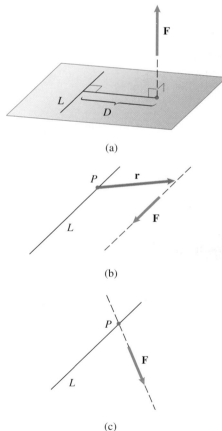

Figure 4.22
(a) **F** is perpendicular to a plane containing L.
(b) **F** is parallel to L.
(c) The line of action of **F** intersects L at P.

Study Questions

1. When you use Eq. (4.5) to determine the moment of a force **F** about a line L, how do you choose the vector **r**? What is the definition of the vector **e**?
2. Explain how the direction of the vector $\mathbf{M}_L$ in Eq. (4.5) indicates the sense of the moment of **F** about L.
3. What is the moment of a force **F** about a line L if the line of action of **F** passes through L? What is the moment if the line of action of **F** is parallel to L?

Example 4.7

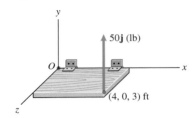

Figure 4.23

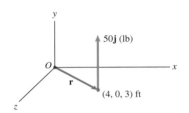

(a) The vector **r** from O to the point of application of the force.

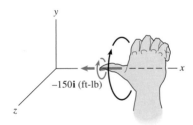

(b) The sense of the moment.

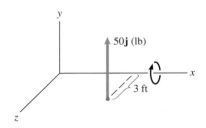

(c) The distance from the x axis to the point where the line of action of the force intersects the x–z plane is 3 ft. The arrow indicates the sense of the moment about the x axis.

Moment of a Force About the x Axis

What is the moment of the 50-lb force in Fig. 4.23 about the x axis?

Strategy

We can determine the moment in two ways.

First Method We can use Eqs. (4.5) and (4.6). Since **r** can extend from any point on the x axis to the line of action of the force, we can use the vector from O to the point of application of the force. The vector **e** must be a unit vector along the x axis, so we can use either **i** or $-\mathbf{i}$.

Second Method This example is the first of the special cases we just discussed, because the 50-lb force is perpendicular to the x–z plane. We can determine the magnitude and direction of the moment directly from the given information.

Solution

First Method *Determine a vector* **r**. The vector from O to the point of application of the force is (Fig. a)

$$\mathbf{r} = 4\mathbf{i} + 3\mathbf{k} \text{ (ft)}.$$

Determine a vector **e**. We can use the unit vector **i**.
Evaluate $\mathbf{M}_L$. Using Eq. (4.6), the mixed triple product is

$$\mathbf{i} \cdot (\mathbf{r} \times \mathbf{F}) = \begin{vmatrix} 1 & 0 & 0 \\ 4 & 0 & 3 \\ 0 & 50 & 0 \end{vmatrix} = -150 \text{ ft-lb}.$$

Then from Eq. (4.5), the moment of the force about the x axis is

$$\mathbf{M}_{(x \text{ axis})} = \left[\mathbf{i} \cdot (\mathbf{r} \times \mathbf{F}) \right]\mathbf{i} = -150\mathbf{i} \text{ (ft-lb)}.$$

The magnitude of the moment is 150 ft-lb, and its sense is as shown in Fig. b.

Second Method Since the 50-lb force is perpendicular to a plane (the x–z plane) containing the x axis, the magnitude of the moment about the x axis is equal to the perpendicular distance from the x axis to the point where the line of action of the force intersects the x–z plane (Fig. c):

$$\left| \mathbf{M}_{(x \text{ axis})} \right| = (3 \text{ ft})(50 \text{ lb}) = 150 \text{ ft-lb}.$$

Pointing the arc of the fingers in the direction of the sense of the moment about the x axis (Fig. c), we find that the right-hand rule indicates that $\mathbf{M}_{(x\,\text{axis})}$ points in the negative x-axis direction. Therefore

$$\mathbf{M}_{(x\,\text{axis})} = -150\mathbf{i} \text{ (ft-lb)}.$$

Example 4.8

Moment of a Force About a Line

What is the moment of the force $\mathbf{F}$ in Fig. 4.24 about the bar BC?

Strategy

We can use Eqs. (4.5) and (4.6) to determine the moment. Since we know the coordinates of points B and C, we can determine the components of a vector $\mathbf{r}$ that extends either from B to the point of application of the force or from C to the point of application. We can also use the coordinates of points B and C to determine a unit vector along the line BC.

Solution

Determine a Vector r   We need a vector from any point on the line BC to any point on the line of action of the force. We can let $\mathbf{r}$ be the vector from B to the point of application of $\mathbf{F}$ (Fig. a):

$$\mathbf{r} = (4 - 0)\mathbf{i} + (2 - 0)\mathbf{j} + (2 - 3)\mathbf{k} = 4\mathbf{i} + 2\mathbf{j} - \mathbf{k} \text{ (m)}.$$

Determine a Vector e   To obtain a unit vector along the bar BC, we determine the vector from B to C,

$$(0 - 0)\mathbf{i} + (4 - 0)\mathbf{j} + (0 - 3)\mathbf{k} = 4\mathbf{j} - 3\mathbf{k} \text{ (m)},$$

and divide it by its magnitude (Fig. a):

$$\mathbf{e}_{BC} = \frac{4\mathbf{j} - 3\mathbf{k}}{5} = 0.8\mathbf{j} - 0.6\mathbf{k}.$$

Evaluate $\mathbf{M_L}$ Using Eq. (4.6), the mixed triple product is

$$\mathbf{e}_{BC} \cdot (\mathbf{r} \times \mathbf{F}) = \begin{vmatrix} 0 & 0.8 & -0.6 \\ 4 & 2 & -1 \\ -2 & 6 & 3 \end{vmatrix} = -24.8 \text{ kN-m}.$$

Substituting this result into Eq. (4.5), the moment of $\mathbf{F}$ about the bar BC is

$$\mathbf{M}_{BC} = \left[[\mathbf{e}_{BC} \cdot (\mathbf{r} \times \mathbf{F})]\mathbf{e}_{BC} = -24.8\mathbf{e}_{BC} \text{ (kN-m)}.$$

The magnitude of $\mathbf{M}_{BC}$ is 24.8 kN-m, and its direction is opposite to that of $\mathbf{e}_{BC}$. The sense of the moment is shown in Fig. b.

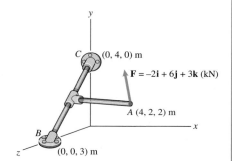

Figure 4.24

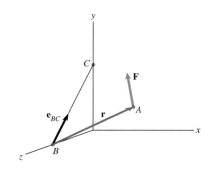

(a) The vectors $\mathbf{r}$ and $\mathbf{e}_{BC}$.

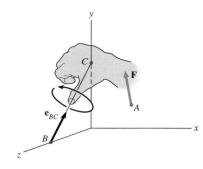

(b) The right-hand rule indicates the sense of the moment about BC.

Example 4.9

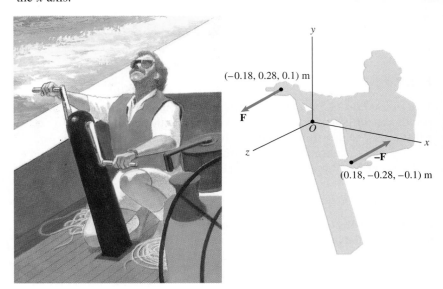

Application to Engineering:

Rotating Machines

The crewman in Fig. 4.25 exerts the forces shown on the handles of the coffee grinder winch, where $\mathbf{F} = 4\mathbf{j} + 32\mathbf{k}$ N. Determine the total moment he exerts (a) about point O, (b) about the axis of the winch, which coincides with the x axis.

Figure 4.25

Strategy

(a) To obtain the total moment about point O, we must sum the moments of the two forces about O. Let the sum be denoted by $\Sigma\mathbf{M}_O$. (b) Because point O is on the x axis, the total moment about the x axis is the component of $\Sigma\mathbf{M}_O$ parallel to the x axis, which is the x component of $\Sigma\mathbf{M}_O$.

Solution

(a) The total moment about point O is

$$\Sigma\mathbf{M}_O = \begin{vmatrix} \mathbf{i} & \mathbf{j} & \mathbf{k} \\ -0.18 & 0.28 & 0.1 \\ 0 & 4 & 32 \end{vmatrix} + \begin{vmatrix} \mathbf{i} & \mathbf{j} & \mathbf{k} \\ 0.18 & -0.28 & -0.1 \\ 0 & -4 & -32 \end{vmatrix}$$

$$= 17.1\mathbf{i} + 11.5\mathbf{j} - 1.4\mathbf{k} \text{ (N-m)}.$$

(b) The total moment about the x axis is the x component of $\Sigma\mathbf{M}_O$ (Fig. a):

$$\Sigma\mathbf{M}_{(x\text{ axis})} = 17.1 \text{ (N-m)}.$$

Notice that this is the result given by Eq. (4.4): Since $\mathbf{i}$ is a unit vector parallel to the x axis,

$$\Sigma\mathbf{M}_{(x\text{ axis})} = (\mathbf{i} \cdot \Sigma\mathbf{M}_O)\mathbf{i} = 17.1 \text{ (N-m)}.$$

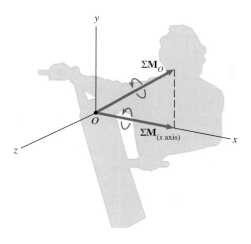

(a) The total moment about the x axis.

$\mathscr{D}$esign Issues

The winch in this example is a simple representative of a class of rotating machines that includes hydrodynamic and aerodynamic power turbines, propellers, jet engines, and electric motors and generators. The ancestors of hydrodynamic and aerodynamic power turbines—water wheels and windmills—were among the earliest machines. These devices illustrate the importance of the concept of the moment of a force about a line. Their common feature is a part designed to rotate and perform some function when it is subjected to a moment about its axis of rotation. In the case of the winch, the forces exerted on the handles by the crewman exert a moment about the axis of rotation, causing the winch to rotate and wind a rope onto a drum, trimming the boat's sails. A hydrodynamic power turbine (Fig. 4.26) has turbine blades that are subjected to forces by flowing water, exerting a moment about the axis of rotation. This moment rotates the shaft to which the blades are attached, turning an electric generator that is connected to the same shaft.

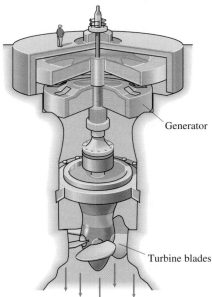

Generator

Turbine blades

Figure 4.26
A hydroelectric turbine. Water flowing through the turbine blades exerts a moment about the axis of the shaft, turning the generator.

4.4 Couples

Now that we have described how to calculate the moment due to a force, consider this question: Is it possible to exert a moment on an object without subjecting it to a net force? The answer is yes, and it occurs when a compact disk begins rotating or a screw is turned by a screwdriver. Forces are exerted on these objects, but in such a way that the net force is zero while the net moment is not zero.

Two forces that have equal magnitudes, opposite directions, and different lines of action are called a *couple* (Fig. 4.27a). A couple tends to cause rotation of an object even though the vector sum of the forces is zero, and it has the remarkable property that *the moment it exerts is the same about any point.*

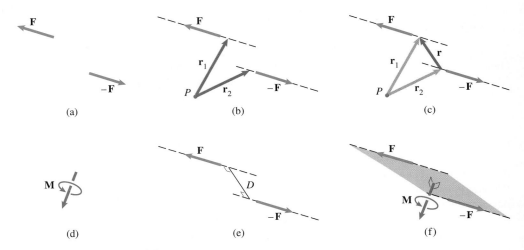

(a)   (b)   (c)

(d)   (e)   (f)

Figure 4.27
(a) A couple.
(b) Determining the moment about *P*.
(c) The vector $\mathbf{r} = \mathbf{r}_1 - \mathbf{r}_2$.
(d) Representing the moment of the couple.
(e) The distance *D* between the lines of action.
(f) $\mathbf{M}$ is perpendicular to the plane containing $\mathbf{F}$ and $-\mathbf{F}$.

The moment of a couple is simply the sum of the moments of the forces about a point *P* (Fig. 4.27b):

$$\mathbf{M} = \left[\mathbf{r}_1 \times \mathbf{F}\right] + \left[\mathbf{r}_2 \times (-\mathbf{F})\right] = (\mathbf{r}_1 - \mathbf{r}_2) \times \mathbf{F}.$$

The vector $\mathbf{r}_1 - \mathbf{r}_2$ is equal to the vector $\mathbf{r}$ shown in Fig. 4.27c, so we can express the moment as

$$\mathbf{M} = \mathbf{r} \times \mathbf{F}.$$

Since $\mathbf{r}$ doesn't depend on the position of *P*, the moment $\mathbf{M}$ is the same for *any* point *P*.

Because a couple exerts a moment but the sum of the forces is zero, it is often represented in diagrams simply by showing the moment (Fig. 4.27d). Like the Cheshire cat in *Alice's Adventures in Wonderland*, which vanished except for its grin, the forces don't appear; you see only the moment they exert. But we recognize the origin of the moment by referring to it as a *moment of a couple*, or simply a *couple*.

Notice in Fig. 4.27c that $\mathbf{M} = \mathbf{r} \times \mathbf{F}$ is the moment of $\mathbf{F}$ about a point on the line of action of the force $-\mathbf{F}$. The magnitude of the moment of a force about a point equals the product of the magnitude of the force and the perpendicular distance from the point to the line of action of the force, so $|\mathbf{M}| = D|\mathbf{F}|$, where D is the perpendicular distance between the lines of action of the two forces (Fig. 4.27e). The cross product $\mathbf{r} \times \mathbf{F}$ is perpendicular to $\mathbf{r}$ and $\mathbf{F}$, which means that $\mathbf{M}$ is perpendicular to the plane containing $\mathbf{F}$ and $-\mathbf{F}$ (Fig. 4.27f). Pointing the thumb of the right hand in the direction of $\mathbf{M}$, the arc of the fingers indicates the sense of the moment.

In Fig. 4.28a, our view is perpendicular to the plane containing the two forces. The distance between the lines of action of the forces is 4 m, so the magnitude of the moment of the couple is $|\mathbf{M}| = (4 \text{ m})(2 \text{ kN}) = 8$ kN-m. The moment $\mathbf{M}$ is perpendicular to the plane containing the two forces. Pointing the arc of the fingers of the right hand counterclockwise, we find that the right-hand rule indicates that $\mathbf{M}$ points out of the page. Therefore the moment of the couple is

$$\mathbf{M} = 8\mathbf{k} \text{ (kN-m)}.$$

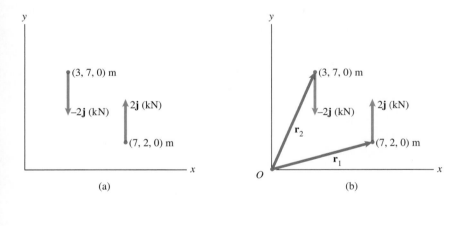

(a)

(b)

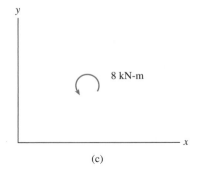

(c)

Figure 4.28
(a) A couple consisting of 2-kN forces.
(b) Determining the sum of the moments of the forces about O.
(c) Representing a couple in two dimensions.

We can also determine the moment of the couple by calculating the sum of the moments of the two forces about *any* point. The sum of the moments of the forces about the origin O is (Fig. 4.28b)

$$\mathbf{M} = \left[\mathbf{r}_1 \times (2\mathbf{j})\right] + \left[\mathbf{r}_2 \times (-2\mathbf{j})\right]$$

$$= \left[(7\mathbf{i} + 2\mathbf{j}) \times (2\mathbf{j})\right] + \left[(3\mathbf{i} + 7\mathbf{j}) \times (-2\mathbf{j})\right]$$

$$= 8\mathbf{k} \ (\text{kN-m}).$$

In a two-dimensional situation like this example, it isn't convenient to represent a couple by showing the moment vector, because the vector is perpendicular to the page. Instead, we represent the couple by showing its magnitude and a circular arrow that indicates its sense (Fig. 4.28c).

By grasping a bar and twisting it (Fig. 4.29a), a moment can be exerted about its axis (Fig. 4.29b). Although the system of forces exerted is distributed over the surface of the bar in a complicated way, the effect is the same as if two equal and opposite forces are exerted (Fig. 4.29c). When we represent a couple as in Fig. 4.29b, or by showing the moment vector $\mathbf{M}$, we imply that some system of forces exerts that moment. The system of forces (such as the forces exerted in twisting the bar, or the forces on the crankshaft that exert a moment on the drive shaft of a car) is nearly always more complicated than two equal and opposite forces, but the effect is the same. For this reason, we can *model* the actual system as a simple system of two forces.

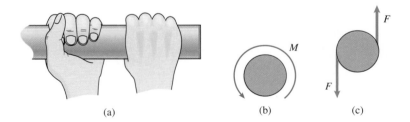

(a)   (b)   (c)

Figure 4.29
(a) Twisting a bar.
(b) The moment about the axis of the bar.
(c) The same effect is obtained by applying two equal and opposite forces.

Study Questions

1. How do you determine the moment exerted about a point P by a couple consisting of forces $\mathbf{F}$ and $-\mathbf{F}$?
2. If you know the moment of a couple about a point P, what do you know about the moment of the couple about a different point P'?
3. A couple consists of forces $\mathbf{F}$ and $-\mathbf{F}$. The perpendicular distance between the lines of action of the forces is D. What is the magnitude of the moment of the couple?

Example 4.10

Determining the Moment of a Couple

The force **F** in Fig. 4.30 is $10\mathbf{i} - 4\mathbf{j}$ (N). Determine the moment of the couple and represent it as shown in Fig. 4.29b.

Strategy

We can determine the moment in two ways: We can calculate the sum of the moments of the forces about a point, or we can sum the moments of the two couples formed by the x and y components of the forces.

Solution

First Method   If we calculate the sum of the moments of the forces about a point on the line of action of one of the forces, the moment of that force is zero and we only need to calculate the moment of the other force. Choosing the point of application of **F** (Fig. a), we calculate the moment as

$$\mathbf{M} = \mathbf{r} \times (-\mathbf{F}) = (-2\mathbf{i} + 3\mathbf{j}) \times (-10\mathbf{i} + 4\mathbf{j}) = 22\mathbf{k} \text{ (N-m)}.$$

We would obtain the same result by calculating the sum of the moments about any point. For example, the sum of the moments about the point P in Fig. b is

$$\mathbf{M} = \left[\mathbf{r}_1 \times \mathbf{F}\right] + \left[\mathbf{r}_2 \times (-\mathbf{F})\right]$$

$$= \begin{vmatrix} \mathbf{i} & \mathbf{j} & \mathbf{k} \\ -2 & -4 & -3 \\ 10 & -4 & 0 \end{vmatrix} + \begin{vmatrix} \mathbf{i} & \mathbf{j} & \mathbf{k} \\ -4 & -1 & -3 \\ -10 & 4 & 0 \end{vmatrix}$$

$$= 22\mathbf{k} \text{ (N-m)}.$$

Second Method   The x and y components of the forces form two couples (Fig. c). We determine the moment of the original couple by summing the moments of the couples formed by the components.

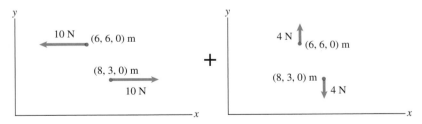

Consider the 10-N couple. The magnitude of its moment is (3 m)(10 N) = 30 N-m, and its sense is counterclockwise, indicating that the moment vector points out of the page. Therefore the moment is $30\mathbf{k}$ N-m.

The 4-N couple causes a moment of magnitude (2 m)(4 N) = 8 N-m and its sense is clockwise, so the moment is $-8\mathbf{k}$ N-m. The moment of the original couple is

$$\mathbf{M} = 30\mathbf{k} - 8\mathbf{k} = 22\mathbf{k} \text{ (N-m)}.$$

Its magnitude is 22 N-m, and its sense is counterclockwise (Fig. d).

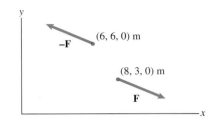

Figure 4.30

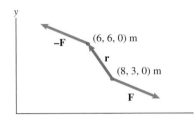

(a) Determining the moment about the point of application of **F**.

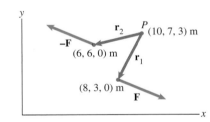

(b) Determining the moment about P.

(c) The x and y components form two couples.

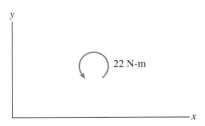

(d) Representing the moment.

Example 4.11

Determining Unknown Forces

Two forces A and B and a 200 ft-lb couple act on the beam in Fig. 4.31. The sum of the forces is zero, and the sum of the moments about the left end of the beam is zero. What are the forces A and B?

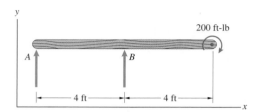

Figure 4.31

Solution

The sum of the forces is

$$\Sigma F_y = A + B = 0.$$

The moment of the couple (200 ft-lb clockwise) is the same about any point, so the sum of the moments about the left end of the beam is

$$\Sigma M_{(\text{left end})} = 4B - 200 = 0.$$

The forces are $B = 50$ lb and $A = -50$ lb.

Discussion

Notice that A and B form a couple (Fig. a). It causes a moment of magnitude (4 ft)(50 lb) = 200 ft-lb, and its sense is counterclockwise, so the sum of the moments of the couple formed by A and B and the 200 ft-lb clockwise couple is zero.

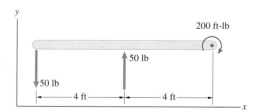

(a) The forces on the beam form a couple.

Example 4.12

Sum of the Moments Due to Two Couples

Determine the sum of the moments exerted on the pipe in Fig. 4.32 by the two couples.

Solution

Consider the 20-N couple. The magnitude of the moment of the couple is (2 m)(20 N) = 40 N-m. The direction of the moment vector is perpendicular to the y–z plane, and the right-hand rule indicates that it points in the positive x-axis direction. The moment of the 20-N couple is 40$\mathbf{i}$ (N-m).

By resolving the 30-N forces into y and z components, we obtain the two couples in Fig. a. The moment of the couple formed by the y components is $-(30 \sin 60°)(4)\mathbf{k}$ (N-m), and the moment of the couple formed by the z components is $(30 \cos 60°)(4)\mathbf{j}$ (N-m).

The sum of the moments is

$$\Sigma\,\mathbf{M} = 40\mathbf{i} + (30 \cos 60°)(4)\mathbf{j} - (30 \sin 60°)(4)\mathbf{k}$$

$$= 40\mathbf{i} + 60\mathbf{j} - 103.9\mathbf{k} \ (\text{N-m}).$$

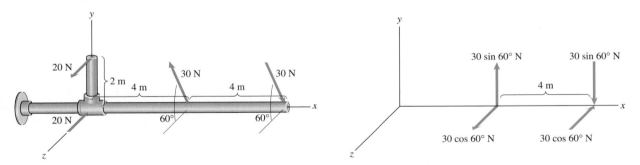

Figure 4.32

(a) Resolving the 30-N forces into y and z components.

Discussion

Although the method we used in this example helps you recognize the contributions of the individual couples to the sum of the moments, it is convenient only when the orientations of the forces and their points of application relative to the coordinate system are fairly simple. When that is not the case, you can determine the sum of the moments by choosing any point and calculating the sum of the moments of the forces about that point.

Example 4.13

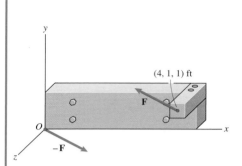

Figure 4.33

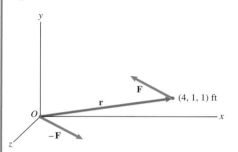

(a) Determining the sum of the moments about O.

Distance Between the Lines of Action

The force **F** in Fig. 4.33 is $-20\mathbf{i} + 20\mathbf{j} + 10\mathbf{k}$ (lb).
(a) What moment does the couple exert on the bracket?
(b) What is the perpendicular distance D between the lines of action of the two forces?

Strategy

(a) We can choose a point and determine the sum of the moments of the forces about that point.
(b) The magnitude of the moment of the couple equals $D\,|\mathbf{F}|$, so we can use the result of (a) to determine D.

Solution

(a) If we determine the sum of the moments of the forces about the origin O, the moment of the force $-\mathbf{F}$ is zero. The moment of the couple is (Fig. a)

$$\mathbf{M} = \mathbf{r} \times \mathbf{F} = \begin{vmatrix} \mathbf{i} & \mathbf{j} & \mathbf{k} \\ 4 & 1 & 1 \\ -20 & 20 & 10 \end{vmatrix} = -10\mathbf{i} - 60\mathbf{j} + 100\mathbf{k} \text{ (ft-lb)}.$$

(b) The perpendicular distance is

$$D = \frac{|\mathbf{M}|}{|\mathbf{F}|} = \frac{\sqrt{(-10)^2 + (-60)^2 + (100)^2}}{\sqrt{(-20)^2 + (20)^2 + (10)^2}} = 3.90 \text{ ft}.$$

4.5 Equivalent Systems

A *system of forces and moments* is simply a particular set of forces and moments of couples. The systems of forces and moments dealt with in engineering can be complicated. This is especially true in the case of distributed forces, such as the pressure forces exerted by water on a dam. Fortunately, if we are concerned only with the total force and moment exerted, we can represent complicated systems of forces and moments by much simpler systems.

Conditions for Equivalence

We define two systems of forces and moments, designated as system 1 and system 2, to be *equivalent* if the sums of the forces are equal,

$$(\Sigma\mathbf{F})_1 = (\Sigma\mathbf{F})_2, \tag{4.7}$$

and the sums of the moments about a point P are equal,

$$(\Sigma\mathbf{M}_P)_1 = (\Sigma\mathbf{M}_P)_2. \tag{4.8}$$

Demonstration of Equivalence

To see what the conditions for equivalence mean, consider the systems of forces and moments in Fig. 4.34a. In system 1, an object is subjected to two forces $\mathbf{F}_A$ and $\mathbf{F}_B$ and a couple $\mathbf{M}_C$. In system 2, the object is subjected to a force $\mathbf{F}_D$ and two couples $\mathbf{M}_E$ and $\mathbf{M}_F$. The first condition for equivalence is

System 1

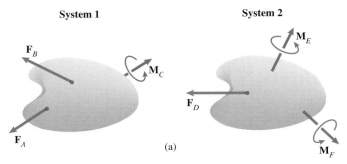

System 2

(a)

Figure 4.34
(a) Different systems of forces and moments applied to an object.
(b) Determining the sum of the moments about a point P for each system.

$$(\Sigma \mathbf{F})_1 = (\Sigma \mathbf{F})_2:$$

$$\mathbf{F}_A + \mathbf{F}_B = \mathbf{F}_D. \tag{4.9}$$

If we determine the sums of the moments about the point P in Fig. 4.34b, the second condition for equivalence is

$$(\Sigma \mathbf{M}_P)_1 = (\Sigma \mathbf{M}_P)_2:$$

$$(\mathbf{r}_A \times \mathbf{F}_A) + (\mathbf{r}_B \times \mathbf{F}_B) + \mathbf{M}_C = (\mathbf{r}_D \times \mathbf{F}_D) + \mathbf{M}_E + \mathbf{M}_F. \tag{4.10}$$

If these conditions are satisfied, systems 1 and 2 are equivalent.

We will use this example to demonstrate that *if the sums of the forces are equal for two systems of forces and moments and the sums of the moments about one point P are equal, then the sums of the moments about any point are equal.* Suppose that Eq. (4.9) is satisfied, and Eq. (4.10) is satisfied for the point P in Fig. 4.34b. For a different point P' (Fig. 4.35), we will show that

$$(\Sigma \mathbf{M}_{P'})_1 = (\Sigma \mathbf{M}_{P'})_2:$$

$$(\mathbf{r}'_A \times \mathbf{F}_A) + (\mathbf{r}'_B \times \mathbf{F}_B) + \mathbf{M}_C = (\mathbf{r}'_D \times \mathbf{F}_D) + \mathbf{M}_E + \mathbf{M}_F. \tag{4.11}$$

In terms of the vector $\mathbf{r}$ from P' to P, the relations between the vectors $\mathbf{r}'_A$, $\mathbf{r}'_B$, and $\mathbf{r}'_D$ in Fig. 4.35 and the vectors $\mathbf{r}_A$, $\mathbf{r}_B$, and $\mathbf{r}_D$ in Fig. 4.34b are

$$\mathbf{r}'_A = \mathbf{r} + \mathbf{r}_A, \qquad \mathbf{r}'_B = \mathbf{r} + \mathbf{r}_B, \qquad \mathbf{r}'_D = \mathbf{r} + \mathbf{r}_D.$$

Substituting these expressions into Eq. (4.11), we obtain

$$[(\mathbf{r} + \mathbf{r}_A) \times \mathbf{F}_A] + [(\mathbf{r} + \mathbf{r}_B) \times \mathbf{F}_B] + \mathbf{M}_C$$
$$= [(\mathbf{r} + \mathbf{r}_D) \times \mathbf{F}_D] + \mathbf{M}_E + \mathbf{M}_F.$$

Rearranging terms, we can write this equation as

$$[\mathbf{r} \times (\Sigma \mathbf{F})_1] + (\Sigma \mathbf{M}_P)_1 = [\mathbf{r} \times (\Sigma \mathbf{F})_2] + (\Sigma \mathbf{M}_P)_2,$$

which holds in view of Eqs. (4.9) and (4.10). The sums of the moments of the two systems about any point are equal.

Study Questions

1. What conditions must be satisfied for two systems of forces and moments to be equivalent?
2. If the sums of the forces in two systems of forces and moments are the same, and the sums of the moments about a point P are the same, what do you know about the sums of the moments about a different point P'?

System 1

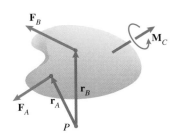

System 2

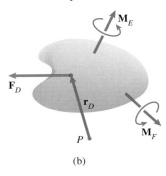

(b)

System 1

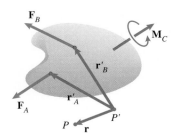

System 2

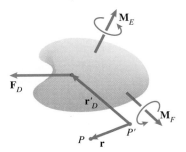

Figure 4.35
Determining the sum of the moments about a different point P' for each system.

Example 4.14

Determining Whether Systems are Equivalent

Three systems of forces and moments act on the beam in Fig. 4.36. Are they equivalent?

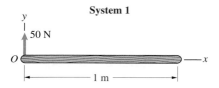

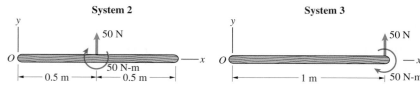

Figure 4.36

Solution

Are the Sums of the Forces Equal?   The sums of the forces are

$$(\Sigma \mathbf{F})_1 = 50\mathbf{j} \ (\text{N}),$$

$$(\Sigma \mathbf{F})_2 = 50\mathbf{j} \ (\text{N}),$$

$$(\Sigma \mathbf{F})_3 = 50\mathbf{j} \ (\text{N}).$$

Are the Sums of the Moments About an Arbitrary Point Equal?   The sums of the moments about the origin O are

$$(\Sigma M_O)_1 = 0,$$

$$(\Sigma M_O)_2 = (50 \text{ N})(0.5 \text{ m}) - (50 \text{ N-m}) = -25 \text{ N-m},$$

$$(\Sigma M_O)_3 = (50 \text{ N})(1 \text{ m}) - (50 \text{ N-m}) = 0.$$

Systems 1 and 3 are equivalent.

Discussion

Remember that you can choose any convenient point to determine whether the sums of the moments are equal. For example, the sums of the moments about the right end of the beam are

$$(\Sigma M_{\text{right end}})_1 = -(50 \text{ N})(1 \text{ m}) = 50 \text{ N-m},$$

$$(\Sigma M_{\text{right end}})_2 = -(50 \text{ N})(0.5 \text{ m}) - (50 \text{ N-m}) = -75 \text{ N-m},$$

$$(\Sigma M_{\text{right end}})_3 = -50 \text{ N-m}.$$

Example 4.15

Determining Whether Systems are Equivalent

Two systems of forces and moments act on the rectangular plate in Fig. 4.37. Are they equivalent?

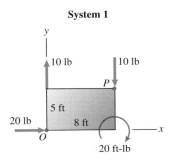

System 1

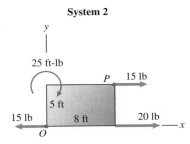

System 2

Figure 4.37

Solution

Are the Sums of the Forces Equal? The sums of the forces are

$$(\Sigma \mathbf{F})_1 = 20\mathbf{i} + 10\mathbf{j} - 10\mathbf{j} = 20\mathbf{i} \text{ (lb)},$$

$$(\Sigma \mathbf{F})_2 = 20\mathbf{i} + 15\mathbf{i} - 15\mathbf{i} = 20\mathbf{i} \text{ (lb)}.$$

Are the Sums of the Moments About an Arbitrary Point Equal? The sums of the moments about the origin O are

$$(\Sigma M_O)_1 = -(8 \text{ ft})(10 \text{ lb}) - (20 \text{ ft-lb}) = -100 \text{ ft-lb},$$

$$(\Sigma M_O)_2 = -(5 \text{ ft})(15 \text{ lb}) - (25 \text{ ft-lb}) = -100 \text{ ft-lb}.$$

The systems are equivalent.

Discussion

Let's confirm that the sums of the moments of the two systems about a different point are equal. The sums of the moments about P are

$$(\Sigma M_P)_1 = -(8 \text{ ft})(10 \text{ lb}) + (5 \text{ ft})(20 \text{ lb}) - (20 \text{ ft-lb}) = 0,$$

$$(\Sigma M_P)_2 = -(5 \text{ ft})(15 \text{ lb}) + (5 \text{ ft})(20 \text{ lb}) - (25 \text{ ft-lb}) = 0.$$

Example 4.16

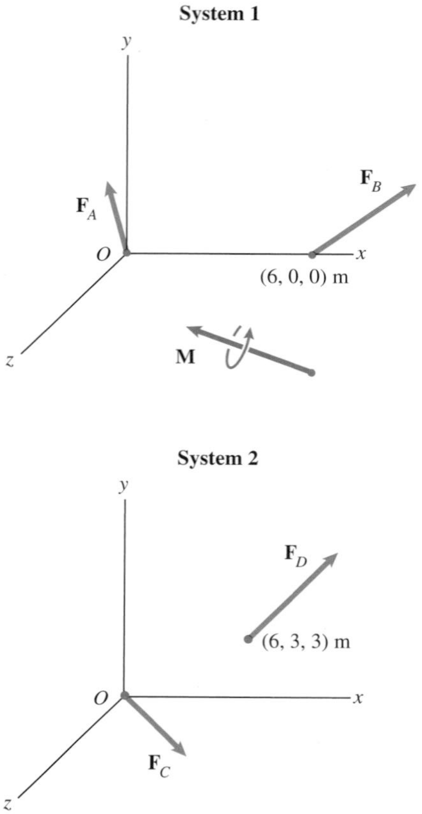

Determining Whether Systems are Equivalent

Two systems of forces and moments are shown in Fig. 4.38, where

$$\mathbf{F}_A = -10\mathbf{i} + 10\mathbf{j} - 15\mathbf{k} \ (\text{kN}),$$
$$\mathbf{F}_B = 30\mathbf{i} + 5\mathbf{j} + 10\mathbf{k} \ (\text{kN}),$$
$$\mathbf{M} = -90\mathbf{i} + 150\mathbf{j} + 60\mathbf{k} \ (\text{kN-m}),$$
$$\mathbf{F}_C = 10\mathbf{i} - 5\mathbf{j} + 5\mathbf{k} \ (\text{kN}),$$
$$\mathbf{F}_D = 10\mathbf{i} + 20\mathbf{j} - 10\mathbf{k} \ (\text{kN}).$$

Are they equivalent?

Solution

Are the Sums of the Forces Equal? The sums of the forces are

$$(\Sigma \mathbf{F})_1 = \mathbf{F}_A + \mathbf{F}_B = 20\mathbf{i} + 15\mathbf{j} - 5\mathbf{k} \ (\text{kN}).$$
$$(\Sigma \mathbf{F})_2 = \mathbf{F}_C + \mathbf{F}_D = 20\mathbf{i} + 15\mathbf{j} - 5\mathbf{k} \ (\text{kN}).$$

Are the Sums of the Moments About an Arbitrary Point Equal? The sum of the moments about the origin O in system 1 is

$$(\Sigma \mathbf{M}_O)_1 = (6\mathbf{i} \times \mathbf{F}_B) + \mathbf{M}$$

$$= \begin{vmatrix} \mathbf{i} & \mathbf{j} & \mathbf{k} \\ 6 & 0 & 0 \\ 30 & 5 & 10 \end{vmatrix} + (-90\mathbf{i} + 150\mathbf{j} + 60\mathbf{k})$$

$$= -90\mathbf{i} + 90\mathbf{j} + 90\mathbf{k} \ (\text{kN-m}).$$

The sum of the moments about O in system 2 is

$$(\Sigma \mathbf{M}_O)_2 = (6\mathbf{i} + 3\mathbf{j} + 3\mathbf{k}) \times \mathbf{F}_D = \begin{vmatrix} \mathbf{i} & \mathbf{j} & \mathbf{k} \\ 6 & 3 & 3 \\ 10 & 20 & -10 \end{vmatrix}$$

$$= -90\mathbf{i} + 90\mathbf{j} + 90\mathbf{k} \ (\text{kN-m}).$$

The systems are equivalent.

Figure 4.38

4.6 Representing Systems by Equivalent Systems

If we are concerned only with the total force and total moment exerted on an object by a given system of forces and moments, we can *represent* the system by an equivalent one. By this we mean that instead of showing the actual forces and couples acting on an object, we would show a different system that exerts the same total force and moment. In this way, we can replace a given system by a less complicated one to simplify the analysis of the forces and moments acting on an object and to gain a better intuitive understanding of their effects on the object.

Representing a System by a Force and a Couple

Let's consider an arbitrary system of forces and moments and a point P (system 1 in Fig. 4.39). We can represent this system by one consisting of a single force acting at P and a single couple (system 2). The conditions for equivalence are

$$(\Sigma \mathbf{F})_2 = (\Sigma \mathbf{F})_1:$$
$$\mathbf{F} = (\Sigma \mathbf{F})_1$$

and

$$(\Sigma \mathbf{M}_P)_2 = (\Sigma \mathbf{M}_P)_1:$$
$$\mathbf{M} = (\Sigma \mathbf{M}_P)_1.$$

These conditions are satisfied if $\mathbf{F}$ equals the sum of the forces in system 1 and $\mathbf{M}$ equals the sum of the moments about P in system 1.

Thus *no matter how complicated a system of forces and moments may be, we can represent it by a single force acting at a given point and a single couple*. Three particular cases occur frequently in practice:

Representing a Force by a Force and a Couple We can represent a force $\mathbf{F}_P$ acting at a point P (system 1 in Fig. 4.40a) by a force $\mathbf{F}$ acting at a different point Q and a couple $\mathbf{M}$ (system 2). The moment of system 1 about point Q is $\mathbf{r} \times \mathbf{F}_P$, where $\mathbf{r}$ is the vector from Q to P (Fig. 4.40b). The conditions for equivalence are

$$(\Sigma \mathbf{F})_2 = (\Sigma \mathbf{F})_1:$$
$$\mathbf{F} = \mathbf{F}_P$$

and

$$(\Sigma \mathbf{M}_Q)_2 = (\Sigma \mathbf{M}_Q)_1:$$
$$\mathbf{M} = \mathbf{r} \times \mathbf{F}_P.$$

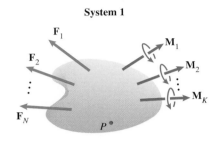

System 1

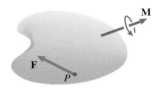

System 2

Figure 4.39
(a) An arbitrary system of forces and moments.
(b) A force acting at P and a couple.

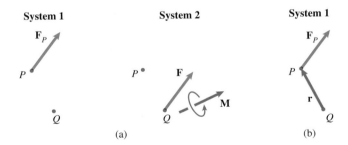

(a)   (b)

Figure 4.40
(a) System 1 is a force $\mathbf{F}_P$ acting at point P. System 2 consists of a force $\mathbf{F}$ acting at point Q and a couple $\mathbf{M}$.
(b) Determining the moment of system 1 about point Q.

The systems are equivalent if the force $\mathbf{F}$ equals the force $\mathbf{F}_P$ and the couple $\mathbf{M}$ equals the moment of $\mathbf{F}_P$ about Q.

Concurrent Forces Represented by a Force We can represent a system of concurrent forces whose lines of action intersect at a point P (system 1 in Fig. 4.41) by a single force whose line of action intersects P (system 2). The sums of the forces in the two systems are equal if

$$\mathbf{F} = \mathbf{F}_1 + \mathbf{F}_2 + \cdots + \mathbf{F}_N.$$

The sum of the moments about P equals zero for each system, so the systems are equivalent if the force $\mathbf{F}$ equals the sum of the forces in system 1.

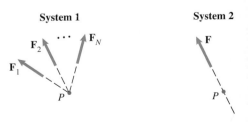

Figure 4.41
A system of concurrent forces and a system consisting of a single force $\mathbf{F}$.

Parallel Forces Represented by a Force We can represent a system of parallel forces whose sum is not zero by a single force **F** (Fig. 4.42). We demonstrate this result in Example 4.20.

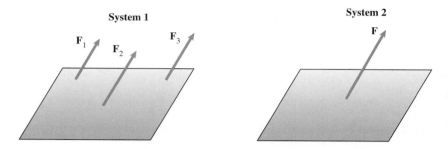

Figure 4.42
A system of parallel forces and a system consisting of a single force **F**.

Study Questions

1. If you represent a system of forces and moments by a force **F** acting at a point *P* and a couple **M**, how do you determine **F** and **M**?
2. If you represent a system of concurrent forces by a single force **F**, what condition must be satisfied by the line of action of **F**?

Example 4.17

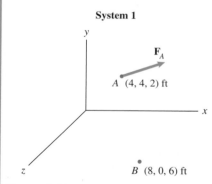

Figure 4.43

(a) A force acting at *B* and a couple.

Representing a Force by a Force and Couple

System 1 in Fig. 4.43 consists of a force $\mathbf{F}_A = 10\mathbf{i} + 4\mathbf{j} - 3\mathbf{k}$ (lb) acting at *A*. Represent it by a force acting at *B* and a couple.

Strategy

We want to represent the force $\mathbf{F}_A$ by a force **F** acting at *B* and a couple **M** (system 2 in Fig. a). We can determine **F** and **M** by using the two conditions for equivalence.

Solution

The sums of the forces must be equal:

$$(\Sigma \mathbf{F})_2 = (\Sigma \mathbf{F})_1:$$

$$\mathbf{F} = \mathbf{F}_A = 10\mathbf{i} + 4\mathbf{j} - 3\mathbf{k} \text{ (lb)}.$$

The sums of the moments about an arbitrary point must be equal: The vector from *B* to *A* is

$$\mathbf{r}_{BA} = (4 - 8)\mathbf{i} + (4 - 0)\mathbf{j} + (2 - 6)\mathbf{k} = -4\mathbf{i} + 4\mathbf{j} - 4\mathbf{k} \text{ (ft)},$$

so the moment about *B* in system 1 is

$$\mathbf{r}_{BA} \times \mathbf{F}_A = \begin{vmatrix} \mathbf{i} & \mathbf{j} & \mathbf{k} \\ -4 & 4 & -4 \\ 10 & 4 & -3 \end{vmatrix} = 4\mathbf{i} - 52\mathbf{j} - 56\mathbf{k} \text{ (ft-lb)}.$$

The sums of the moments about B must be equal:

$$(\mathbf{M}_B)_2 = (\mathbf{M}_B)_1:$$

$$\mathbf{M} = 4\mathbf{i} - 52\mathbf{j} - 56\mathbf{k} \text{ (ft-lb)}.$$

Example 4.18

Representing a System by a Simpler Equivalent System

System 1 in Fig. 4.44 consists of two forces and a couple acting on a pipe. Represent system 1 by (a) a single force acting at the origin O of the coordinate system and a single couple and (b) a single force.

Strategy

(a) We can represent system 1 by a force $\mathbf{F}$ acting at the origin and a couple M (system 2 in Fig. a) and use the conditions for equivalence to determine $\mathbf{F}$ and $\mathbf{M}$.

(b) Suppose that we place the force $\mathbf{F}$ with its point of application a distance D along the x axis (system 3 in Fig. b). The sums of the forces in systems 2 and 3 are equal. If we can choose the distance D so that the moment about O in system 3 equals $\mathbf{M}$, system 3 will be equivalent to system 2 and therefore equivalent to system 1.

Solution

(a) The conditions for equivalence are

$$(\Sigma \mathbf{F})_2 = (\Sigma \mathbf{F})_1:$$

$$\mathbf{F} = 30\mathbf{j} + (20\mathbf{i} + 20\mathbf{j}) = 20\mathbf{i} + 50\mathbf{j} \text{ (kN)},$$

and

$$(\Sigma M_O)_2 = (\Sigma M_O)_1:$$

$$M = (30 \text{ kN})(3 \text{ m}) + (20 \text{ kN})(5 \text{ m}) + 210 \text{ kN-m}$$

$$= 400 \text{ kN-m}.$$

(b) The sums of the forces in systems 2 and 3 are equal. Equating the sums of the moments about O,

$$(\Sigma M_O)_3 = (\Sigma M_O)_2:$$

$$(50 \text{ kN})D = 400 \text{ kN-m}.$$

we find that system 3 is equivalent to system 2 if $D = 8$ m.

Discussion

To represent the system by a single force in (b), we needed to place the line of action of the force so that the force exerted a 400 kN-m counterclockwise moment about O. Placing the point of application of the force a distance D along the x axis was simply a convenient way to accomplish that.

System 1

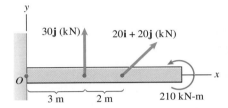

Figure 4.44

System 2

(a) A force $\mathbf{F}$ acting at O and a couple M.

System 3

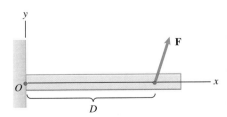

(b) A system consisting of the force $\mathbf{F}$ acting at a point on the x axis.

Example 4.19

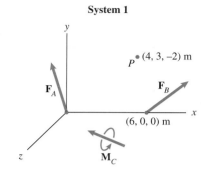

System 1

System 2

Figure 4.45

Representing a System by a Force and Couple

System 1 in Fig. 4.45 consists of the following forces and couple:

$$\mathbf{F}_A = -10\mathbf{i} + 10\mathbf{j} - 15\mathbf{k} \text{ (kN)},$$

$$\mathbf{F}_B = 30\mathbf{i} + 5\mathbf{j} + 10\mathbf{k} \text{ (kN)},$$

$$\mathbf{M}_C = -90\mathbf{i} + 150\mathbf{j} + 60\mathbf{k} \text{ (kN-m)}.$$

Suppose you want to represent it by a force $\mathbf{F}$ acting at P and a couple $\mathbf{M}$ (system 2). Determine $\mathbf{F}$ and $\mathbf{M}$.

Solution

The sums of the forces must be equal:

$$(\Sigma \mathbf{F})_2 = (\Sigma \mathbf{F})_1:$$

$$\mathbf{F} = \mathbf{F}_A + \mathbf{F}_B = 20\mathbf{i} + 15\mathbf{j} - 5\mathbf{k} \text{ (kN)}.$$

The sums of the moments about an arbitrary point must be equal: The sums of the moments about point P must be equal:

$$(\Sigma \mathbf{M}_P)_2 = (\Sigma \mathbf{M}_P)_1:$$

$$\mathbf{M} = \begin{vmatrix} \mathbf{i} & \mathbf{j} & \mathbf{k} \\ -4 & -3 & 2 \\ -10 & 10 & -15 \end{vmatrix} + \begin{vmatrix} \mathbf{i} & \mathbf{j} & \mathbf{k} \\ 2 & -3 & 2 \\ 30 & 5 & 10 \end{vmatrix}$$

$$+ (-90\mathbf{i} + 150\mathbf{j} + 60\mathbf{k})$$

$$= -105\mathbf{i} + 110\mathbf{j} + 90\mathbf{k} \text{ (kN-m)}.$$

Example 4.20

Representing Parallel Forces by a Single Force

System 1 in Fig. 4.46 consists of parallel forces. Suppose you want to represent it by a force $\mathbf{F}$ (system 2). What is $\mathbf{F}$, and where does its line of action intersect the x–z plane?

Strategy

We can determine $\mathbf{F}$ from the condition that the sums of the forces in the two systems must be equal. For the two systems to be equivalent, we must choose the point of application P so that the sums of the moments about a point are equal. This condition will tell us where the line of action intersects the x–z plane.

Solution

The sums of the forces must be equal:

$$(\Sigma \mathbf{F})_2 = (\Sigma \mathbf{F})_1:$$

$$\mathbf{F} = 30\mathbf{j} + 20\mathbf{j} - 10\mathbf{j} = 40\mathbf{j} \text{ (lb)}.$$

The sums of the moments about an arbitrary point must be equal: Let the co-ordinates of point P be (x, y, z). The sums of the moments about the origin O must be equal.

$$(\Sigma M_O)_2 = (\Sigma M_O)_1:$$

$$\begin{vmatrix} \mathbf{i} & \mathbf{j} & \mathbf{k} \\ x & y & z \\ 0 & 40 & 0 \end{vmatrix} = \begin{vmatrix} \mathbf{i} & \mathbf{j} & \mathbf{k} \\ 6 & 0 & 2 \\ 0 & 30 & 0 \end{vmatrix} + \begin{vmatrix} \mathbf{i} & \mathbf{j} & \mathbf{k} \\ 2 & 0 & 4 \\ 0 & -10 & 0 \end{vmatrix}$$

$$+ \begin{vmatrix} \mathbf{i} & \mathbf{j} & \mathbf{k} \\ -3 & 0 & -2 \\ 0 & 20 & 0 \end{vmatrix}.$$

Expanding the determinants, we obtain

$$(20 + 40z)\mathbf{i} + (100 - 40x)\mathbf{k} = \mathbf{0}.$$

The sums of the moments about the origin are equal if

$$x = 2.5 \text{ ft,}$$

$$z = -0.5 \text{ ft.}$$

The systems are equivalent if $\mathbf{F} = 40\mathbf{j}$ (lb) and its line of action inter-sects the x–z plane at $x = 2.5$ ft and $z = -0.5$ ft. Notice that we did not ob-tain an equation for the y coordinate of P. The systems are equivalent if $\mathbf{F}$ is applied at any point along the line of action.

Discussion

We could have determined the x and z coordinates of point P in a simpler way. Since the sums of the moments about any point must be equal for the systems to be equivalent, the sums of the moments about any *line* must also be equal. Equating the sums of the moments about the x axis,

$$\left(\Sigma M_{x \text{ axis}}\right)_2 = \left(\Sigma M_{x \text{ axis}}\right)_1:$$

$$-40z = -(30)(2) + (10)(4) + (20)(2),$$

we obtain $z = -0.5$ ft, and equating the sums of the moments about the z axis,

$$\left(\Sigma M_{z \text{ axis}}\right)_2 = \left(\Sigma M_{z \text{ axis}}\right)_1:$$

$$40x = (30)(6) - (10)(2) - (20)(3),$$

we obtain $x = 2.5$ ft.

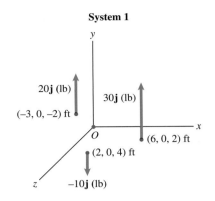

System 1

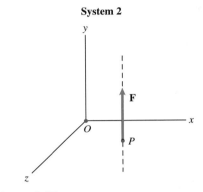

System 2

Figure 4.46

Representing a System by a Wrench

We have shown that any system of forces and moments can be represented by a single force acting at a given point and a single couple. This raises an interesting question: What is the simplest system that can be equivalent to any system of forces and moments?

To consider this question, let's begin with an arbitrary force **F** acting at a point P and an arbitrary couple **M** (system 1 in Fig. 4.47a) and see whether we can represent this system by a simpler one. For example, can we represent it by the force **F** acting at a different point Q and no couple (Fig 4.47b)? The sum of the forces is the same as in system 1. If we can choose the point Q so that $\mathbf{r} \times \mathbf{F} = \mathbf{M}$, where **r** is the vector from P to Q (Fig. 4.47c), the sum of the moments about P is the same as in system 1 and the systems are equivalent. But the vector $\mathbf{r} \times \mathbf{F}$ is perpendicular to **F**, so it can equal **M** only if **M** is perpendicular to **F**. That means that, in general, we can't represent system 1 by the force **F** alone.

However, we can represent system 1 by the force **F** acting at a point Q and the component of **M** that is parallel to **F**. Figure 4.47d shows system 1 with a coordinate system placed so that **F** is along the y axis and **M** is contained in the x–y plane. In terms of this coordinate system, we can express

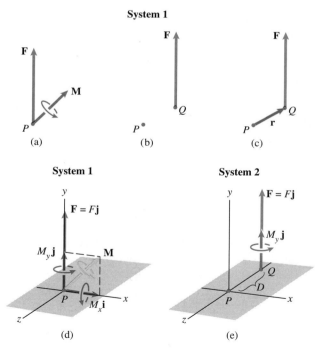

Figure 4.47
(a) System 1 is a single force and a single couple.
(b) Can system 1 be represented by a single force and no couple?
(c) The moment of **F** about P is $\mathbf{r} \times \mathbf{F}$.
(d) **F** is along the y axis, and **M** is contained in the x–y plane.
(e) System 2 is the force **F** and the component of **M** parallel to **F**.

the force and couple as $\mathbf{F} = F\mathbf{j}$ and $\mathbf{M} = M_x\mathbf{i} + M_y\mathbf{j}$. System 2 in Fig. 4.47e consists of the force $\mathbf{F}$ acting at a point on the z axis and the component of $\mathbf{M}$ parallel to $\mathbf{F}$. If we choose the distance D so that $D = M_x/F$, system 2 is equivalent to system 1. The sum of the forces in each system is $\mathbf{F}$. The sum of the moments about P in system 1 is $\mathbf{M}$, and the sum of the moments about P in system 2 is

$$(\Sigma\mathbf{M}_P)_2 = [(-D\mathbf{k}) \times (F\mathbf{j})] + M_y\mathbf{j} = M_x\mathbf{i} + M_y\mathbf{j} = \mathbf{M}.$$

A force $\mathbf{F}$ and a couple $\mathbf{M}_p$ that is parallel to $\mathbf{F}$ is called a *wrench*; it is the simplest system that can be equivalent to an arbitrary system of forces and moments.

How can you represent a given system of forces and moments by a wrench? If the system is a single force or a single couple or if it consists of a force $\mathbf{F}$ and a couple that is parallel to $\mathbf{F}$, it is a wrench, and you can't simplify it further. If the system is more complicated than a single force and a single couple, begin by choosing a convenient point P and representing the system by a force $\mathbf{F}$ acting at P and a couple $\mathbf{M}$ (Fig. 4.48a). Then representing this system by a wrench requires two steps:

1. Determine the components of $\mathbf{M}$ parallel and normal to $\mathbf{F}$ (Fig. 4.48b).
2. The wrench consists of the force $\mathbf{F}$ acting at a point Q and the parallel component $\mathbf{M}_P$ (Fig. 4.48c). To achieve equivalence, you must choose the point Q so that the moment of $\mathbf{F}$ about P equals the normal component $\mathbf{M}_n$ (Fig. 4.48d)—that is, so that $\mathbf{r}_{PQ} \times \mathbf{F} = \mathbf{M}_n$.

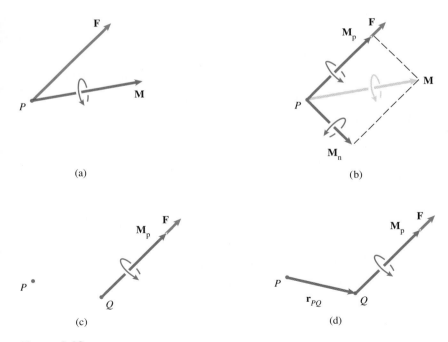

(a)

(b)

(c)

(d)

Figure 4.48
(a) If necessary, first represent the system by a single force and a single couple.
(b) The components of $\mathbf{M}$ parallel and normal to $\mathbf{F}$.
(c) The wrench.
(d) Choose Q so that the moment of $\mathbf{F}$ about P equals the normal component of $\mathbf{M}$.

Example 4.21

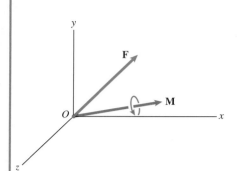

Figure 4.49

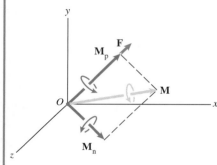

(a) Resolving **M** into components parallel and normal to **F**.

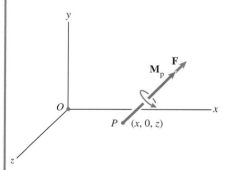

(b) The wrench acting at a point in the x–z plane.

Representing a Force and Couple by a Wrench

The system in Fig. 4.49 consists of the force and couple

$$\mathbf{F} = 3\mathbf{i} + 6\mathbf{j} + 2\mathbf{k} \ (\text{N}),$$

$$\mathbf{M} = 12\mathbf{i} + 4\mathbf{j} + 6\mathbf{k} \ (\text{N-m}).$$

Represent it by a wrench, and determine where the line of action of the wrench's force intersects the x–z plane.

Strategy

The wrench is the force **F** and the component of **M** parallel to **F** (Figs. a, b). We must choose the point of application P so that the moment of **F** about O equals the normal component $\mathbf{M}_n$. By letting P be an arbitrary point of the x–z plane, we can determine where the line of action of **F** intersects that plane.

Solution

Dividing **F** by its magnitude, we obtain a unit vector **e** with the same direction as **F**:

$$\mathbf{e} = \frac{\mathbf{F}}{|\mathbf{F}|} = \frac{3\mathbf{i} + 6\mathbf{j} + 2\mathbf{k}}{\sqrt{(3)^2 + (6)^2 + (2)^2}} = 0.429\mathbf{i} + 0.857\mathbf{j} + 0.286\mathbf{k}.$$

We can use **e** to calculate the component of **M** parallel to **F**:

$$\mathbf{M}_p = (\mathbf{e} \cdot \mathbf{M})\mathbf{e} = \big[(0.429)(12) + (0.857)(4) + (0.286)(6)\big]\mathbf{e}$$

$$= 4.408\mathbf{i} + 8.816\mathbf{j} + 2.939\mathbf{k} \ (\text{N-m}).$$

The component of **M** normal to **F** is

$$\mathbf{M}_n = \mathbf{M} - \mathbf{M}_p = 7.592\mathbf{i} - 4.816\mathbf{j} + 3.061\mathbf{k} \ (\text{N-m}).$$

The wrench is shown in Fig. b. Let the coordinates of P be $(x, 0, z)$. The moment of **F** about O is

$$\mathbf{r}_{OP} \times \mathbf{F} = \begin{vmatrix} \mathbf{i} & \mathbf{j} & \mathbf{k} \\ x & 0 & z \\ 3 & 6 & 2 \end{vmatrix} = -6z\mathbf{i} - (2x - 3z)\mathbf{j} + 6x\mathbf{k}.$$

By equating this moment to $\mathbf{M}_n$,

$$-6z\mathbf{i} - (2x - 3z)\mathbf{j} + 6x\mathbf{k} = 7.592\mathbf{i} - 4.816\mathbf{j} + 3.061\mathbf{k},$$

we obtain the equations

$$-6z = 7.592,$$

$$-2x + 3z = -4.816,$$

$$6x = 3.061.$$

Solving these equations, we find the coordinates of point P are $x = 0.510$ m, $z = -1.265$ m.

Computational Mechanics

The following example and problems are designed for the use of a programmable calculator or computer.

Computational Example 4.22

The radius R of the steering wheel in Fig. 4.50 is 200 mm. The distance from O to C is 1 m. The center C of the steering wheel lies in the x–y plane. The force $\mathbf{F} = \sin\alpha(10\mathbf{i} + 10\mathbf{j} - 5\mathbf{k})$ N. Determine the value of α at which the magnitude of the moment of $\mathbf{F}$ about the shaft OC of the steering wheel is a maximum. What is the maximum magnitude?

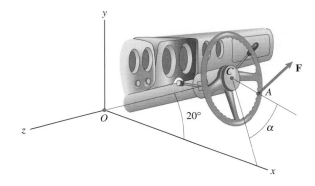

Figure 4.50

Strategy

We will determine the moment of $\mathbf{F}$ about OC in terms of the angle α and obtain a graph of the moment as a function of α.

Solution

In terms of the vector $\mathbf{r}_{CA}$ from point C on the shaft to the point of application of the force, and the unit vector $\mathbf{e}_{OC}$ that points along the shaft from point O toward point C (Fig. a), the moment of $\mathbf{F}$ about the shaft is

$$\mathbf{M}_{OC} = \left[\mathbf{e}_{OC} \cdot (\mathbf{r}_{CA} \times \mathbf{F})\right]\mathbf{e}_{OC}.$$

From Fig. a, the unit vector $\mathbf{e}_{OC}$ is

$$\mathbf{e}_{OC} = \cos 20°\,\mathbf{i} + \sin 20°\,\mathbf{j},$$

and the z component of $\mathbf{r}_{CA}$ is $-R\sin\alpha$. By viewing the steering wheel with the z axis perpendicular to the page (Fig. b), we can see that the x component of $\mathbf{r}_{CA}$ is $R\cos\alpha\sin 20°$ and the y component is $-R\cos\alpha\cos 20°$, so

$$\mathbf{r}_{CA} = R(\cos\alpha\sin 20°\,\mathbf{i} - \cos\alpha\cos 20°\,\mathbf{j} - \sin\alpha\,\mathbf{k}).$$

The magnitude of $\mathbf{M}_{OC}$ is the absolute value of the scalar

$$\mathbf{e}_{OC} \cdot (\mathbf{r}_{CA} \times \mathbf{F}) = \begin{vmatrix} \cos 20° & \sin 20° & 0 \\ R\cos\alpha\sin 20° & -R\cos\alpha\cos 20° & -R\sin\alpha \\ 10\sin\alpha & 10\sin\alpha & -5\sin\alpha \end{vmatrix}$$

$$= R[5\sin\alpha\cos\alpha + 10(\cos 20° - \sin 20°)\sin^2\alpha].$$

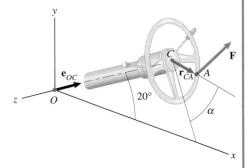

(a) The position vector $\mathbf{r}_{CA}$ and the unit vector $\mathbf{e}_{OC}$.

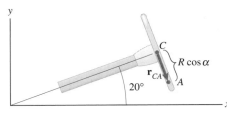

(b) Determining the x and y components of $\mathbf{r}_{CA}$.

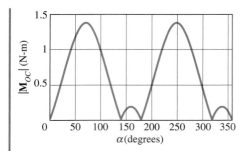

Figure 4.51
Magnitude of the moment as a function of α.

Computing the absolute value of this expression as a function of α, we obtain the graph shown in Fig. 4.51. The magnitude of the moment is an extremum at values of α of approximately 70° and 250°. By examining the computed results near 70°,

α	$|\mathbf{M}_{OC}|$ (N-m)
67°	1.3725
68°	1.3749
69°	1.3764
70°	1.3769
71°	1.3765
72°	1.3751
73°	1.3728

we can see that the maximum value is approximately 1.38 N-m. The value of the moment at $\alpha = 250°$ is also 1.38 N-m.

Chapter Summary

In this chapter we have defined the moment of a force about a point and about a line and explained how to evaluate them. We introduced the concept of a couple and defined equivalent systems of forces and moments. We can now apply two consequences of equilibrium: The sum of the forces equals zero, and the sum of the moments about any point equals zero. We will consider individual objects in Chapter 5 and structures in Chapter 6.

Moment of a Force About a Point

The moment of a force about a point is the measure of the tendency of the force to cause rotation about the point. The *moment* of a force $\mathbf{F}$ about a point P is the vector

$$\mathbf{M}_P = \mathbf{r} \times \mathbf{F}, \qquad \text{Eq. (4.2)}$$

where $\mathbf{r}$ is a position vector from P to *any* point on the line of action of $\mathbf{F}$. The magnitude of $\mathbf{M}_P$ is equal to the product of the perpendicular distance D from P to the line of action of $\mathbf{F}$ and the magnitude of $\mathbf{F}$:

$$|\mathbf{M}_P| = D|\mathbf{F}|. \qquad \text{Eq. (4.3)}$$

The vector $\mathbf{M}_P$ is perpendicular to the plane containing P and $\mathbf{F}$. When the thumb of the right hand points in the direction of $\mathbf{M}_P$, the arc of the fingers indicates the sense of the rotation that $\mathbf{F}$ tends to cause about P. The dimensions of the moment are (distance) × (force).

If a force is resolved into components, the moment of the force about a point P is equal to the sum of the moments of its components about P. If the line of action of a force passes through a point P, the moment of the force about P is zero.

When the view is perpendicular to the plane containing the force and the point (Fig. a), the two-dimensional description of the moment is

$$M_p = DF. \qquad \text{Eq. (4.1)}$$

Moment of a Force About a Line

The moment of a force about a line is the measure of the tendency of the force to cause rotation about the line. Let P be any point on a line L and let $\mathbf{M}_P$ be the moment about P of a force $\mathbf{F}$ (Fig. b). The moment $\mathbf{M}_L$ of $\mathbf{F}$ about L is the vector component of $\mathbf{M}_P$ parallel to L. If $\mathbf{e}$ is a unit vector along L,

$$\mathbf{M}_L = (\mathbf{e} \cdot \mathbf{M}_P)\mathbf{e} = \big[\mathbf{e} \cdot (\mathbf{r} \times \mathbf{F})\big]\mathbf{e}. \qquad \text{Eq. (4.4), (4.5)}$$

When the line of action of $\mathbf{F}$ is perpendicular to a plane containing L, $|\mathbf{M}_L|$ is equal to the product of the magnitude of $\mathbf{F}$ and the perpendicular distance D from L to the point where the line of action intersects the plane. When the line of action of $\mathbf{F}$ is parallel to L or intersects L, $\mathbf{M}_L = 0$.

Couples

Two forces that have equal magnitudes, opposite directions, and do not have the same line of action are called a *couple*. The moment $\mathbf{M}$ of a couple is the same about any point. The magnitude of $\mathbf{M}$ is equal to the product of the magnitude of one of the forces and the perpendicular distance between the lines of action, and its direction is perpendicular to the plane containing the lines of action.

Because a couple exerts a moment but no net force, it can be represented by showing the moment vector (Fig. c), or it can be represented in two dimensions by showing the magnitude of the moment and a circular arrow to indicate the sense (Fig. d). The moment represented in this way is called the *moment of a couple*, or simply a *couple*.

Equivalent Systems

Two systems of forces and moments are defined to be *equivalent* if the sums of the forces are equal,

$$(\Sigma \mathbf{F})_1 = (\Sigma \mathbf{F})_2, \qquad \text{Eq. (4.7)}$$

and the sums of the moments about a point P are equal,

$$(\Sigma \mathbf{M}_P)_1 = (\Sigma \mathbf{M}_P)_2. \qquad \text{Eq. (4.8)}$$

If the sums of the forces are equal and the sums of the moments about one point are equal, the sums of the moments about any point are equal.

Representing Systems by Equivalent Systems

If the system of forces and moments acting on an object is represented by an equivalent system, the equivalent system exerts the same total force and total moment on the object.

Any system can be represented by an equivalent system consisting of a force $\mathbf{F}$ acting at a given point P and a couple $\mathbf{M}$ (Fig. e). The simplest system that can be equivalent to any system of forces and moments is the *wrench*, which is a force $\mathbf{F}$ and a couple $\mathbf{M}_p$ that is parallel to $\mathbf{F}$ (Fig. f).

A system of concurrent forces can be represented by a single force. A system of parallel forces whose sum is not zero can be represented by a single force.

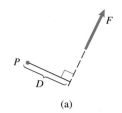

(a)

Figure (a)

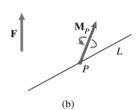

(b)

Figure (b)

(c)

Figure (c)

(d)

Figure (d)

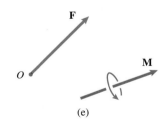

(e)

Figure (e)

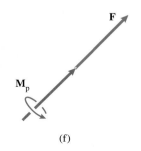

(f)

Figure (f)

Review Problems

4.1 Determine the moment of the 50-N force about (a) point A, (b) point B.

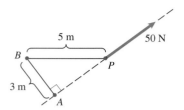

P4.1

4.2 Determine the moment of the 200-N force about A.

(a) What is the two-dimensional description of the moment?
(b) Express the moment as a vector.

P4.2

4.3 The Leaning Tower of Pisa is approximately 55 m tall and 7 m in diameter. The horizontal displacement of the top of the tower from the vertical is approximately 5 m. Its mass is approximately 3.2×10^6 kg. If you model the tower as a cylinder and assume that its weight acts at the center, what is the magnitude of the moment exerted by the weight about the point at the center of the tower's base?

P4.3

4.4 The device shown has been suggested as a design for a perpetual motion machine. Determine the moment about the axis of rotation due to the four masses as a function of the angle as the device rotates 90° clockwise from the position shown, and indicate whether gravity could cause rotation in that direction.

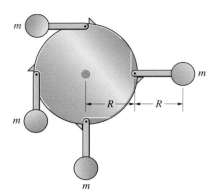

P4.4

4.5 In Problem 4.4, determine whether gravity could cause rotation in the counterclockwise direction.

4.6 Determine the moment of the 400-N force (a) about A, (b) about B.

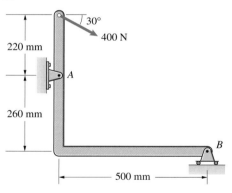

P4.6

4.7 Determine the sum of the moments exerted about A by the three forces and the couple.

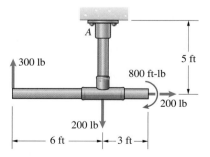

P4.7

4.8 In Problem 4.7, if you represent the three forces and the couple by an equivalent system consisting of a force **F** acting at A and a couple **M**, what are the magnitudes of **F** and **M**?

4.9 The vector sum of the forces acting on the beam is zero, and the sum of the moments about A is zero.

(a) What are the forces A_x, A_y, and B?
(b) What is the sum of the moments about B?

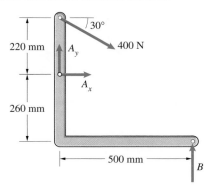

P4.9

4.10 To support the ladder, the force exerted at B by the hydraulic piston AB must exert a moment about C equal in magnitude to the moment about C due to the ladder's 450-lb weight. What is the magnitude of the force exerted at B?

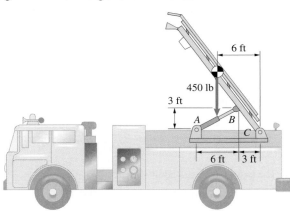

P4.10

4.11 The force $\mathbf{F} = -60\mathbf{i} + 60\mathbf{j}$ (lb).

(a) Determine the moment of $\mathbf{F}$ about point A.
(b) What is the perpendicular distance from point A to the line of action of $\mathbf{F}$?

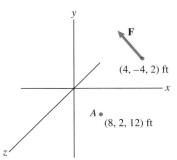

P4.11

4.12 The 20-kg mass is suspended by cables attached to three vertical 2-m posts. Point A is at (0, 1.2, 0) m. Determine the moment about the base E due to the force exerted on the post BE by the cable AB.

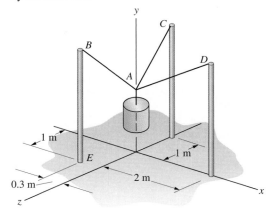

P4.12

4.13 Three forces of equal magnitude are applied parallel to the sides of an equilateral triangle.

(a) Show that the sum of the moments of the forces is the same about any point.
(b) Determine the magnitude of the moment.

    Strategy: To do (a), resolve one of the forces into vector components parallel to the other two forces.

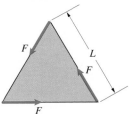

P4.13

4.14 The bar AB supporting the lid of the grand piano exerts a force $\mathbf{F} = -6\mathbf{i} + 35\mathbf{j} - 12\mathbf{k}$ (lb) at B. The coordinates of B are (3, 4, 3) ft. What is the moment of the force about the hinge line of the lid (the x axis)?

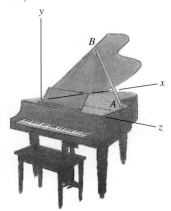

P4.14

4.15 Determine the moment of the vertical 800-lb force about point C.

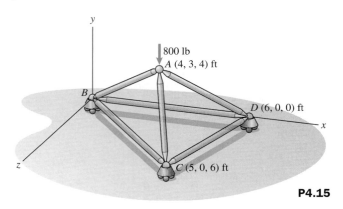

P4.15

4.16 In Problem 4.15, determine the moment of the vertical 800-lb force about the straight line through points C and D.

4.17 The system of cables and pulleys supports the 300-lb weight of the work platform. If you represent the upward force exerted at E by cable EF and the upward force exerted at G by cable GH by a single equivalent force $\mathbf{F}$, what is $\mathbf{F}$, and where does its line of action intersect the x axis?

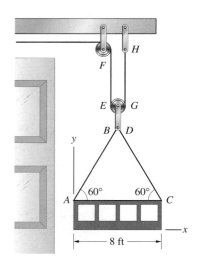

P4.17

4.18 Consider the system in Problem 4.17.
(a) What are the tensions in cables AB and CD?
(b) If you represent the forces exerted by the cables at A and C by a single equivalent force $\mathbf{F}$, what is $\mathbf{F}$, and where does its line of action intersect the x axis?

4.19 The two systems are equivalent. Determine the forces A_x and A_y, and the couple M_A.

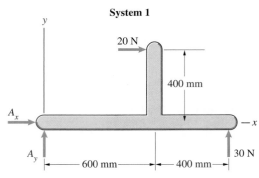

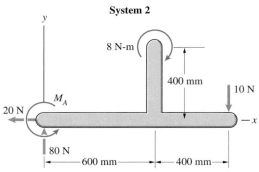

P4.19

4.20 If you represent the equivalent systems in Problem 4.19 by a force $\mathbf{F}$ acting at the origin and a couple M, what are $\mathbf{F}$ and M?

4.21 If you represent the equivalent systems in Problem 4.19 by a force $\mathbf{F}$, what is $\mathbf{F}$, and where does its line of action intersect the x axis?

4.22 The two systems are equivalent. If

$$\mathbf{F} = -100\mathbf{i} + 40\mathbf{j} + 30\mathbf{k} \text{ (lb)},$$

$$\mathbf{M}' = -80\mathbf{i} + 120\mathbf{j} + 40\mathbf{k} \text{ (in-lb)},$$

determine $\mathbf{F}'$ and $\mathbf{M}$.

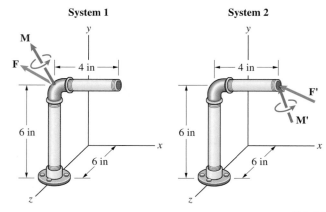

P4.22

4.23 The tugboats A and B exert forces $F_A = 1$ kN and $F_B = 1.2$ kN on the ship. The angle $\theta = 30°$. If you represent the two forces by a force $\mathbf{F}$ acting at the origin O and a couple M, what are $\mathbf{F}$ and M?

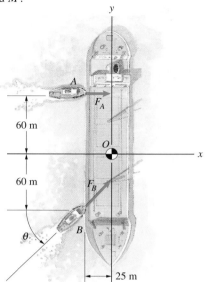

P4.23

4.24 The tugboats A and B in Problem 4.23 exert forces $F_A = 600$ N and $F_B = 800$ N on the ship. The angle $\theta = 45°$. If you represent the two forces by a force $\mathbf{F}$, what is $\mathbf{F}$, and where does its line of action intersect the y axis?

4.25 The tugboats A and B in Problem 4.23 want to exert two forces on the ship that are equivalent to a force $\mathbf{F}$ acting at the origin O of 2-kN magnitude. If $F_A = 800$ N, determine the necessary values of F_B and θ.

4.26 If you represent the forces exerted by the floor on the table legs by a force $\mathbf{F}$ acting at the origin O and a couple $\mathbf{M}$, what are $\mathbf{F}$ and $\mathbf{M}$?

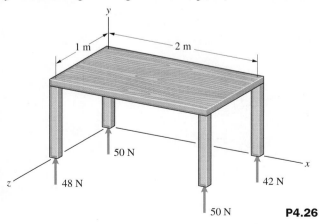

P4.26

4.27 If you represent the forces exerted by the floor on the table legs in Problem 4.26 by a force $\mathbf{F}$, what is $\mathbf{F}$, and where does its line of action intersect the x–z plane?

4.28 Two forces are exerted on the crankshaft by the connecting rods. The direction cosines of $\mathbf{F}_A$ are $\cos\theta_x = -0.182$, $\cos\theta_y = 0.818$, and $\cos\theta_z = 0.545$, and its magnitude is 4 kN. The direction cosines of $\mathbf{F}_B$ are $\cos\theta_x = 0.182$, $\cos\theta_y = 0.818$, and $\cos\theta_z = -0.545$, and its magnitude is 2 kN. If you represent the two forces by a force $\mathbf{F}$ acting at the origin O and a couple $\mathbf{M}$, what are $\mathbf{F}$ and $\mathbf{M}$?

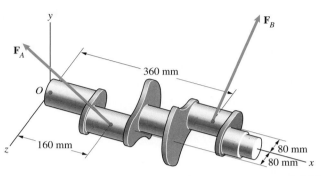

P4.28

4.29 If you represent the two forces exerted on the crankshaft in Problem 4.28 by a wrench consisting of a force $\mathbf{F}$ and a parallel couple $\mathbf{M}_p$, what are $\mathbf{F}$ and $\mathbf{M}_p$, and where does the line of action of $\mathbf{F}$ intersect the x–z plane?

𝒟esign Experience A relatively primitive device for exercising the biceps muscle is shown. Suggest an improved configuration for the device. You can use elastic cords (which behave like linear springs), weights, and pulleys. Seek a design such that the variation of the moment about the elbow joint as the device is used is small in comparison to the design shown. Give consideration to the safety of your device, its reliability, and the requirement to accommodate users having a range of dimensions and strengths. Choosing specific dimensions, determine the range of the magnitude of the moment exerted about the elbow joint as your device is used.

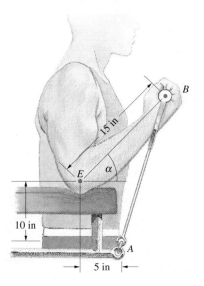

The Space Shuttle main engine being held in equilibrium by a support. In this chapter we use the equilibrium equations to determine forces and couples exerted on objects by their supports.

Objects in Equilibrium

By applying the techniques developed in Chapters 3 and 4, we can now analyze many of the equilibrium problems that arise in engineering applications. After stating the equilibrium equations, we describe the various types of supports that are used. We then show how free-body diagrams and equilibrium are used to determine unknown forces and couples acting on objects.

5.1   The Equilibrium Equations

In Chapter 3 we defined an object to be in equilibrium when it is stationary or in steady translation relative to an inertial reference frame. When an object acted upon by a system of forces and moments is in equilibrium, the following conditions are satisfied.

1. The sum of the forces is zero:

$$\Sigma \mathbf{F} = \mathbf{0}. \tag{5.1}$$

2. The sum of the moments about any point is zero:

$$\Sigma \mathbf{M}_{(\text{any point})} = \mathbf{0}. \tag{5.2}$$

Before we consider specific applications, some general observations about these equations are in order.

From our discussion of equivalent systems of forces and moments in Chapter 4, Eqs. (5.1) and (5.2) imply that the system of forces and moments acting on an object in equilibrium is equivalent to a system consisting of no forces and no couples. This provides insight into the nature of equilibrium. From the standpoint of the total force and total moment exerted on an object in equilibrium, the effects are the same as if no forces or couples acted on the object. This observation also makes it clear that if the sum of the forces on an object is zero and the sum of the moments about one point is zero, then the sum of the moments about every point is zero.

Figure 5.1 shows an object subjected to concurrent forces $\mathbf{F}_1, \mathbf{F}_2, \ldots, \mathbf{F}_N$ and no couples. If the sum of these forces is zero,

$$\mathbf{F}_1 + \mathbf{F}_2 + \cdots + \mathbf{F}_N = \mathbf{0}, \tag{5.3}$$

the conditions for equilibrium are satisfied, because the moment about point P is zero. The only condition imposed by equilibrium on a set of concurrent forces is that their sum is zero.

To determine the sum of the moments about a line L due to a system of forces and moments acting on an object, we choose any point P on the line and determine the sum of the moments $\Sigma \mathbf{M}_P$ about P (Fig. 5.2). Then the sum of the moments about the line is the component of $\Sigma \mathbf{M}_P$ parallel to the line. If the object is in equilibrium, $\Sigma \mathbf{M}_P = \mathbf{0}$. We see that the sum of the moments about any line due to the forces and couples acting on an object in equilibrium is zero. This result is useful in certain types of problems.

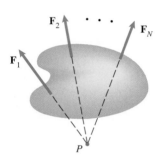

Figure 5.1
An object subjected to concurrent forces.

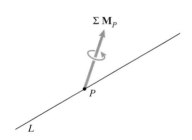

Figure 5.2
The sum of the moments $\Sigma \mathbf{M}_P$ about a point P on the line L.

5.2   Two-Dimensional Applications

Many engineering applications involve two-dimensional systems of forces and moments. These include the forces and moments exerted on many beams and planar structures, pliers, some cranes and other machines, and some types of bridges and dams. In this section we discuss supports, free-body diagrams, and the equilibrium equations for two-dimensional applications.

Supports

When you are standing, the floor supports you. When you sit in a chair, the chair supports you. In this section we are concerned with the ways objects are held in place or are attached to other objects. Forces and couples exerted on an object by its supports are called *reactions*, expressing the fact that the supports "react" to the other forces and couples, or *loads*, acting on the object. For example, a bridge is held up by the reactions exerted by its supports, and the loads are the forces exerted by the weight of the bridge itself, the traffic crossing it, and the wind.

Some very common kinds of supports are represented by stylized models called support conventions. Actual supports often closely resemble the support conventions, but even when they don't, we represent them by these conventions if the actual supports exert the same (or approximately the same) reactions as the models.

The Pin Support   Figure 5.3a shows a *pin support*. The diagram represents a bracket to which an object (such as a beam) is attached by a smooth pin that passes through the bracket and the object. The side view is shown in Fig. 5.3b.

To understand the reactions that a pin support can exert, it's helpful to imagine holding a bar attached to a pin support (Fig. 5.3c). If you try to move the bar without rotating it (that is, translate the bar), the support exerts a reactive force that prevents this movement. However, you can rotate the bar about the axis of the pin. The support cannot exert a couple about the pin axis to prevent rotation. Thus a pin support can't exert a couple about the pin axis, but it can exert a force on an object in any direction, which is usually expressed by representing the force in terms of components (Fig. 5.3d). The arrows indicate the directions of the reactions if A_x and A_y are positive. If you determine A_x or A_y to be negative, the reaction is in the direction opposite to that of the arrow.

The pin support is used to represent any real support capable of exerting a force in any direction but not exerting a couple. Pin supports are used in many common devices, particularly those designed to allow connected parts to rotate relative to each other (Fig. 5.4).

The Roller Support   The convention called a *roller support* (Fig. 5.5a) represents a pin support mounted on wheels. Like the pin support, it cannot exert

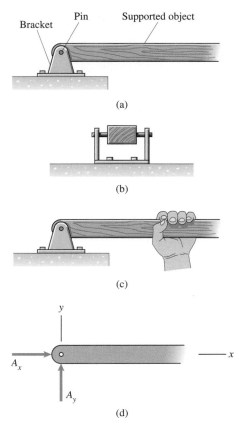

Figure 5.3
(a) A pin support.
(b) Side view showing the pin passing through the beam.
(c) Holding a supported bar.
(d) The pin support is capable of exerting two components of force.

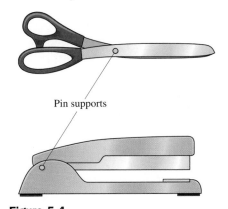

Figure 5.4
Pin supports in a pair of scissors and a stapler.

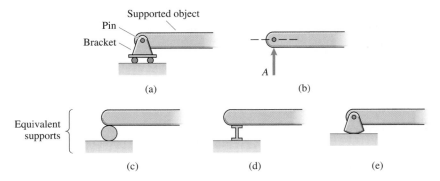

Figure 5.5
(a) A roller support.
(b) The reaction consists of a force normal to the surface.
(c)–(e) Supports equivalent to the roller support.

Figure 5.6
Supporting an object with a plane smooth surface.

a couple about the axis of the pin. Since it can move freely in the direction parallel to the surface on which it rolls, it can't exert a force parallel to the surface but can only exert a force normal (perpendicular) to this surface (Fig. 5.5b). Figures 5.5c–e are other commonly used conventions equivalent to the roller support. The wheels of vehicles and wheels supporting parts of machines are roller supports if the friction forces exerted on them are negligible in comparison to the normal forces. A plane smooth surface can also be modeled by a roller support (Fig. 5.6). Beams and bridges are sometimes supported in this way so that they will be free to undergo thermal expansion and contraction.

The supports shown in Fig. 5.7 are similar to the roller support in that they cannot exert a couple and can only exert a force normal to a particular direction. (Friction is neglected.) In these supports, the supported object is attached to a pin or slider that can move freely in one direction but is constrained in the perpendicular direction. Unlike the roller support, these supports can exert a normal force in either direction.

Figure 5.7
Supports similar to the roller support except that the normal force can be exerted in either direction.
(a) Pin in a slot.
(b) Slider in a slot.
(c) Slider on a shaft.

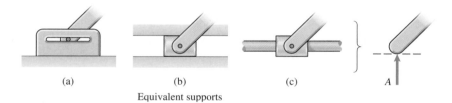

(a) (b) (c) A

Equivalent supports

The Built-In Support The *built-in support* shows the supported object literally built into a wall (Fig. 5.8a). This convention is also called a *fixed support*. To understand the reactions, imagine holding a bar attached to a built-in support (Fig. 5.8b). If you try to translate the bar, the support exerts a reactive force that prevents translation, and if you try to rotate the bar, the support exerts a reactive couple that prevents rotation. A built-in support can exert two components of force and a couple (Fig. 5.8c). The term M_A is the couple exerted by the support, and the curved arrow indicates its direction. Fence posts and lampposts have built-in supports. The attachments of parts connected so that they cannot move or rotate relative to each other, such as the head of a hammer and its handle, can be modeled as built-in supports.

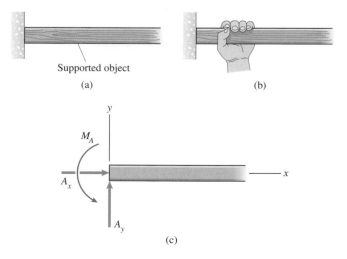

Supported object

(a) (b)

y

M_A

A_x x

A_y

(c)

Figure 5.8
(a) Built-in support.
(b) Holding a supported bar.
(c) The reactions a built-in support is capable of exerting.

Table 5.1 summarizes the support conventions commonly used in two-dimensional applications, including those we discussed in Chapter 3. Although the number of conventions may appear daunting, the examples and problems

Table 5.1 Supports used in two-dimensional applications.

Supports	Reactions
Rope or Cable Spring	One Collinear Force
Contact with a Smooth Surface	One Force Normal to the Supporting Surface
Contact with a Rough Surface	Two Force Components
Pin Support	Two Force Components
Roller Support Equivalents	One Force Normal to the Supporting Surface
Constrained Pin or Slider	One Normal Force
Built-in (Fixed) Support	Two Force Components and One Couple

will help you become familiar with them. You should also observe how various objects you see in your everyday experience are supported and think about whether each support could be represented by one of the conventions.

Free-Body Diagrams

We introduced free-body diagrams in Chapter 3 and used them to determine forces acting on simple objects in equilibrium. By using the support conventions, we can model more elaborate objects and construct their free-body diagrams in a systematic way.

For example, the beam in Fig. 5.9a has a pin support at the left end and a roller support at the right end and is loaded by a force F. The roller support rests on a surface inclined at 30° to the horizontal. To obtain the free-body diagram of the beam, we first isolate it from its supports (Fig. 5.9b), since the free-body diagram must contain no object other than the beam. We complete the free-body diagram by showing the reactions that may be exerted on the beam by the supports (Fig. 5.9c). Notice that the reaction B exerted by the roller support is normal to the surface on which the support rests.

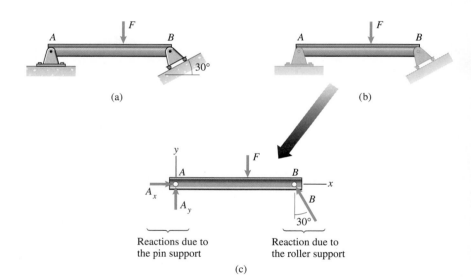

Figure 5.9
(a) A beam with pin and roller supports.
(b) Isolating the beam from its supports.
(c) The completed free-body diagram.

The object in Fig. 5.10a has a fixed support at the left end. A cable passing over a pulley is attached to the object at two points. We isolate it from its supports (Fig. 5.10b) and complete the free-body diagram by showing the reactions at the built-in support and the forces exerted by the cable (Fig. 5.10c). *Don't forget the couple at a built-in support.* Since we assume the tension in the cable is the same on both sides of the pulley, the two forces exerted by the cable have the same magnitude T.

Once you have obtained the free-body diagram of an object in equilibrium to identify the loads and reactions acting on it, you can apply the equilibrium equations.

(a)

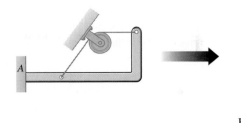

(b)

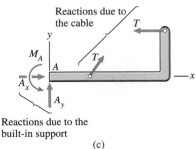

(c)

Figure 5.10
(a) An object with a built-in support.
(b) Isolating the object.
(c) The completed free-body diagram.

The Scalar Equilibrium Equations

When the loads and reactions on an object in equilibrium form a two-dimensional system of forces and moments, they are related by three scalar equilibrium equations:

$$\Sigma F_x = 0, \tag{5.4}$$
$$\Sigma F_y = 0, \tag{5.5}$$
$$\Sigma M_{(\text{any point})} = 0. \tag{5.6}$$

A natural question is whether more than one equation can be obtained from Eq. (5.6) by evaluating the sum of the moments about more than one point. The answer is yes, and in some cases it is convenient to do so. But there is a catch—the additional equations will not be independent of Eqs. (5.4)–(5.6). In other words, *more than three independent equilibrium equations cannot be obtained from a two-dimensional free-body diagram, which means we can solve for at most three unknown forces or couples.* We discuss this point further in Section 5.3.

The seesaw found on playgrounds, consisting of a board with a pin support at the center that allows it to rotate, is a simple and familiar example that illustrates the role of Eq. (5.6). If two people of unequal weight sit at the seesaw's ends, the heavier person sinks to the ground (Fig. 5.11a). To obtain equilibrium, that person must move closer to the center (Fig. 5.11b).

We draw the free-body diagram of the seesaw in Fig. 5.11c, showing the weights of the people W_1 and W_2 and the reactions at the pin support. Evaluating the sum of the moments about A, the equilibrium equations are

$$\Sigma F_x = A_x = 0, \tag{5.7}$$
$$\Sigma F_y = A_y - W_1 - W_2 = 0, \tag{5.8}$$
$$\Sigma M_{(\text{point } A)} = D_1 W_1 - D_2 W_2 = 0. \tag{5.9}$$

Thus $A_x = 0$, $A_y = W_1 + W_2$, and $D_1 W_1 = D_2 W_2$. The last condition indicates the relation between the positions of the two persons necessary for equilibrium.

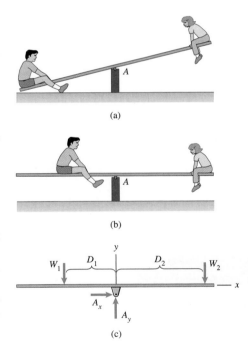

(a)

(b)

(c)

Figure 5.11
(a) If both people sit at the ends of the seesaw, the heavier one sinks.
(b) The seesaw and people in equilibrium.
(c) The free-body diagram of the seesaw, showing the weights of the people and the reactions at the pin support.

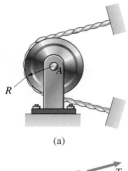

(a)

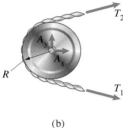

(b)

Figure 5.12
(a) A pulley of radius R.
(b) Free-body diagram of the pulley and
   part of the cable.

To demonstrate that an additional independent equation is not obtained by evaluating the sum of the moments about a different point, we can sum the moments about the right end of the seesaw:

$$\Sigma M_{(\text{right end})} = (D_1 + D_2)W_1 - D_2 A_y = 0.$$

This equation is a linear combination of Eqs. (5.8) and (5.9):

$$(D_1 + D_2)W_1 - D_2 A_y = \underbrace{-D_2\left(A_y - W_1 - W_2\right)}_{\textbf{Eq. (5.8)}}$$

$$+ \underbrace{\left(D_1 W_1 - D_2 W_2\right)}_{\textbf{Eq. (5.9)}} = 0.$$

Until now we have assumed in examples and problems that the tension in a rope or cable is the same on both sides of a pulley. Consider the pulley in Fig. 5.12a. In its free-body diagram in Fig. 5.12b, we do not assume that the tensions are equal. Summing the moments about the center of the pulley, we obtain the equilibrium equation

$$\Sigma M_{(\text{point } A)} = RT_1 - RT_2 = 0.$$

The tensions must be equal if the pulley is in equilibrium. However, notice that we have assumed that the pulley's support behaves like a pin support and cannot exert a couple on the pulley. When that is not true—for example, due to friction between the pulley and the support—the tensions are not necessarily equal.

Study Questions

1. What is a pin support? What reactions can it exert on an object subjected to a two-dimensional system of forces and moments?
2. What is a roller support? What reactions can it exert on an object subjected to a two-dimensional system of forces and moments?
3. How many independent equilibrium equations can you obtain from a two-dimensional free-body diagram?

Example 5.1

Reactions at Pin and Roller Supports

The beam in Fig. 5.13 has pin and roller supports and is subjected to a 2-kN force. What are the reactions at the supports?

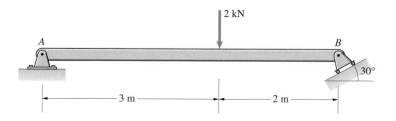

Figure 5.13

Solution

Draw the Free-Body Diagram We isolate the beam from its supports and show the loads and the reactions that may be exerted by the pin and roller supports (Fig. a). There are three unknown reactions: two components of force A_x and A_y at the pin support and a force B at the roller support.

Apply the Equilibrium Equations Summing the moments about point A, the equilibrium equations are

$$\Sigma F_x = A_x - B\sin 30° = 0,$$

$$\Sigma F_y = A_y + B\cos 30° - 2 = 0,$$

$$\Sigma M_{(\text{point } A)} = (5)(B\cos 30°) - (3)(2) = 0.$$

Solving these equations, the reactions are $A_x = 0.69$ kN, $A_y = 0.80$ kN, and $B = 1.39$ kN. The load and reactions are shown in Fig. b. It is good practice to show your answers in this way and confirm that the equilibrium equations are satisfied:

$$\Sigma F_x = 0.69 - 1.39\sin 30° = 0,$$

$$\Sigma F_y = 0.80 + 1.39\cos 30° - 2 = 0,$$

$$\Sigma M_{(\text{point } A)} = (5)(1.39\cos 30°) - (3)(2) = 0.$$

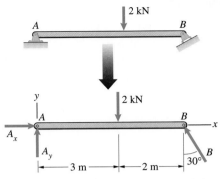

(a) Drawing the free-body diagram of the beam.

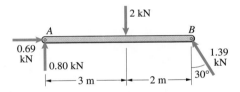

(b) The load and reaction.

Discussion

We drew the arrows indicating the directions of the reactions A_x and A_y in the positive x and y axis directions, but we could have drawn them in either direction. In Fig. c we draw the free-body diagram of the beam with the component A_y pointed downward. From this free-body diagram we obtain the equilibrium equations

$$\Sigma F_x = A_x - B\sin 30° = 0,$$

$$\Sigma F_y = -A_y + B\cos 30° - 2 = 0,$$

$$\Sigma M_{(\text{point } A)} = (5)(B\cos 30°) - (2)(3) = 0.$$

The solutions are $A_x = 0.69$ kN, $A_y = -0.80$ kN, and $B = 1.39$ kN. The negative value of A_y indicates that the vertical force exerted on the beam by the pin support is in the direction opposite to that of the arrow in Fig. c; that is, the force is 0.80 kN upward. Thus we again obtain the reactions shown in Fig. b.

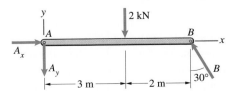

(c) An alternative free-body diagram.

Example 5.2

Reactions at a Built-In Support

The object in Fig. 5.14 has a built-in support and is subjected to two forces and a couple. What are the reactions at the support?

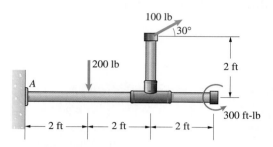

Figure 5.14

Solution

Draw the Free-Body Diagram We isolate the object from its support and show the reactions at the built-in support (Fig. a). There are three unknown reactions: two force components A_x and A_y and a couple M_A. (Remember that we can choose the directions of these arrows arbitrarily.) We also resolve the 100-lb force into its components.

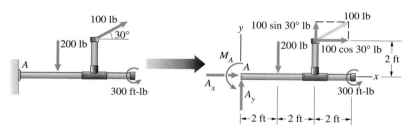

(a) Drawing the free-body diagram.

Apply the Equilibrium Equations Summing the moments about point A, the equilibrium equations are

$$\Sigma F_x = A_x + 100\cos 30° = 0,$$

$$\Sigma F_y = A_y - 200 + 100\sin 30° = 0,$$

$$\Sigma M_{(\text{point } A)} = M_A + 300 - (200)(2) - (100\cos 30°)(2)$$
$$+ (100\sin 30°)(4) = 0.$$

Solving these equations, we obtain the reactions $A_x = -86.6$ lb, $A_y = 150.0$ lb, and $M_A = 73.2$ ft-lb.

Discussion

Notice that the 300-ft-lb couple and the couple M_A exerted by the built-in support don't appear in the first two equilibrium equations because a couple exerts no net force. Also, since the moment due to a couple is the same about any point, the moment about point A due to the 300-ft-lb counterclockwise couple is 300 ft-lb counterclockwise.

Example 5.3

Reactions on a Car's Tires

The 2800-lb car in Fig. 5.15 is stationary. Determine the normal forces exerted on the front and rear tires by the road.

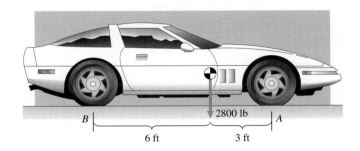

B 6 ft   2800 lb   3 ft A

Figure 5.15

Solution

Draw the Free-Body Diagram   In Fig. a we isolate the car and show its weight and the reactions exerted by the road. There are two unknown reactions: the forces A and B exerted on the front and rear tires.

Apply the Equilibrium Equations   The forces have no x components. Summing the moments about point B, the equilibrium equations are

$$\Sigma F_y = A + B - 2800 = 0,$$

$$\Sigma M_{(\text{point } B)} = (6)(2800) - 9A = 0.$$

Solving these equations, the reactions are $A = 1867$ lb and $B = 933$ lb.

Discussion

This example doesn't fall within our definition of a two-dimensional system of forces and moments because the forces acting on the car are not coplanar. Let's examine why you can analyze problems of this kind as if they were two-dimensional.

In Fig. b we show an oblique view of the free-body diagram of the car. In this view you can see the forces acting on the individual tires. The total normal force on the front tires is $A_L + A_R = A$, and the total normal force on the rear tires is $B_L + B_R = B$. The sum of the forces in the y direction is

$$\Sigma F_y = A_L + A_R + B_L + B_R - 2800 = A + B - 2800 = 0.$$

Since the sum of the moments about any line due to the forces and couples acting on an object in equilibrium is zero, the sum of the moments about the z axis due to the forces acting on the car is zero:

$$\Sigma M_{(z \text{ axis})} = (9)(A_L + A_R) - (6)(2800) = 9A - (6)(2800) = 0.$$

Thus we obtain the same equilibrium equations we did when we solved the problem using a two-dimensional analysis.

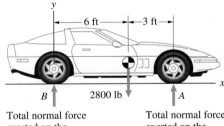

y 6 ft   3 ft x

B 2800 lb A

Total normal force exerted on the two rear tires

Total normal force exerted on the two front tires

(a) The free-body diagram.

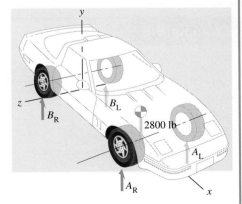

y

z B_R B_L

2800 lb

A_L

A_R x

(b) An oblique view showing the forces on the individual tires.

Example 5.4

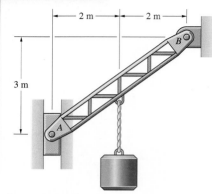

Figure 5.16

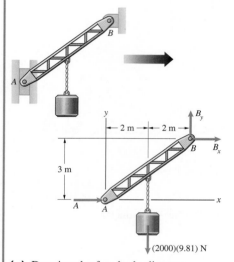

(a) Drawing the free-body diagram.

Choosing the Point About Which to Evaluate Moments

The structure AB in Fig. 5.16 supports a suspended 2-Mg (megagram) mass. The structure is attached to a slider in a vertical slot at A and has a pin support at B. What are the reactions at A and B?

Solution

Draw the Free-Body Diagram We isolate the structure and mass from the supports and show the reactions at the supports and the force exerted by the weight of the 2000-kg mass (Fig. a). The slot at A can exert only a horizontal force on the slider.

Apply the Equilibrium Equations Notice that if we sum the moments about point B, we obtain an equation containing only one unknown reaction, the force A. The equilibrium equations are

$$\Sigma F_x = A + B_x = 0,$$

$$\Sigma F_y = B_y - (2000)(9.81) = 0,$$

$$\Sigma M_{(\text{point } B)} = A(3) + (2000)(9.81)(2) = 0.$$

The reactions are $A = -13.1$ kN, $B_x = 13.1$ kN, and $B_y = 19.6$ kN.

Discussion

You can often simplify equilibrium equations by a careful choice of the point about which you sum moments. For example, when you can choose a point where the lines of action of unknown forces intersect, those forces will not appear in your moment equation.

Example 5.5

 # Application to Engineering:

Design for Human Factors

Figure 5.17 shows an airport luggage carrier and its free-body diagram when it is held in equilibrium in the tilted position. If the luggage carrier supports a weight $W = 50$ lb, the angle $\alpha = 30°$, $a = 8$ in., $b = 16$ in., and $d = 48$ in., what force F must the user exert?

Strategy

The unknown reactions on the free-body diagram are the force F and the normal force N exerted by the floor. If we sum moments about the center of the wheel C, we obtain an equation in which F is the only unknown reaction.

Solution

Summing moments about C,

$$\Sigma M_{(point\ C)} = d(F \cos \alpha) + a(W \sin \alpha) - b(W \cos \alpha) = 0,$$

and solving for F, we obtain

$$F = \frac{(b - a \tan \alpha)W}{d}. \tag{5.10}$$

Substituting the values of W, α, a, b, and d, the solution is $F = 11.9$ lb.

𝒟esign Issues

Design that accounts for human physical dimensions, capabilities, and characteristics is a special challenge. This art is called design for human factors. Here we consider a simple device, the airport luggage carrier in Fig. 5.17, and show how consideration of its potential users *and the constraints imposed by the equilibrium equations* affect its design.

The user moves the carrier by grasping the bar at the top, tilting it, and walking while pulling the carrier. The height of the handle (the dimension h) needs to be comfortable. Since $h = R + d \sin \alpha$, if we choose values of h and the wheel radius R, we obtain a relation between the length of the carrier's handle d and the tilt angle α:

$$d = \frac{h - R}{\sin \alpha}. \tag{5.11}$$

Substituting this expression for d into Eq. (5.10), we obtain

$$F = \frac{\sin \alpha (b - a \tan \alpha)W}{h - R}. \tag{5.12}$$

Suppose that based on statistical data on human dimensions, we decide to design the carrier for convenient use by persons up to 6 ft 2 in. tall, which corresponds to a dimension h of approximately 36 in. Let $R = 3$ in., $a = 6$ in., and $b = 12$ in. The resulting value of F/W as a function of α is shown in Fig. 5.18. At $\alpha = 63°$, the force the user must exert is zero, which means the weight of the luggage acts at a point directly above the wheels. This would be the optimum solution if the user could maintain exactly that value of α. However, α inevitably varies, and the resulting changes in F make it difficult to control the carrier. In addition, the relatively steep angle would make the carrier awkward to pull. From this point of view, it is desirable to choose a design within the range of values of α in which F varies slowly, say, $30° \leq \alpha \leq 45°$. (Even though the force the user must exert is large in this range of α in comparison with larger values of α, it is only about 13% of the weight.) Over this range of α, the dimension d varies from 5.5 ft to 3.9 ft. A smaller carrier is desirable for lightness and ease of storage, so we choose $d = 4$ ft for our preliminary design.

We have chosen the dimension d based on particular values of the dimensions R, a, and b. In an actual design study, we would carry out the analysis for expected ranges of values of these parameters. Our final design would also reflect decisions based on safety (for example, there must be adequate means to secure the luggage and no sharp projections), reliability (the frame must be sufficiently strong and the wheels must have adequate and reliable bearings), and the cost of manufacture.

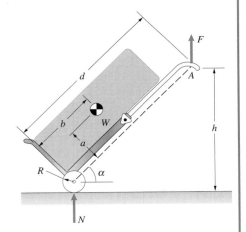

Figure 5.17

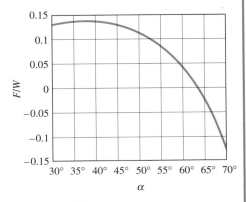

Figure 5.18
Graph of the ratio F/W as a function of α.

In Section 5.2 we discussed examples in which we were able to use the equilibrium equations to determine unknown forces and couples acting on objects in equilibrium. You need to be aware of two common situations in which this procedure doesn't lead to a solution.

First, the free-body diagram of an object can have more unknown forces or couples than the number of independent equilibrium equations you can obtain. Since you can write no more than three such equations for a given free-body diagram in a two-dimensional problem, when there are more than three unknowns you can't determine them from the equilibrium equations alone. This occurs, for example, when an object has more supports than the minimum number necessary to maintain it in equilibrium. Such an object is said to have *redundant supports*. The second situation is when the supports of an object are improperly designed such that they cannot maintain equilibrium under the loads acting on it. The object is said to have *improper supports*. In either situation, the object is said to be *statically indeterminate*.

Engineers use redundant supports whenever possible for strength and safety. Some designs, however, require that the object be incompletely supported so that it is free to undergo certain motions. These two situations—more supports than necessary for equilibrium or not enough—are so common that we consider them in detail.

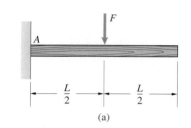

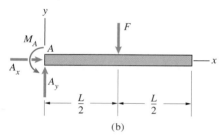

Figure 5.19
(a) A beam with a built-in support.
(b) The free-body diagram has three unknown reactions.

Redundant Supports

Let's consider a beam with a built-in support (Fig. 5.19a). From its free-body diagram (Fig. 5.19b), we obtain the equilibrium equations

$$\Sigma F_x = A_x = 0,$$

$$\Sigma F_y = A_y - F = 0,$$

$$\Sigma M_{(\text{point } A)} = M_A - \left(\frac{L}{2}\right)F = 0.$$

Assuming we know the load F, we have three equations and three unknown reactions, for which we obtain the solutions $A_x = 0$, $A_y = F$, and $M_A = FL/2$.

Now suppose we add a roller support at the right end of the beam (Fig. 5.20a). From the new free-body diagram (Fig. 5.20b), we obtain the equilibrium equations

$$\Sigma F_x = A_x = 0, \tag{5.13}$$

$$\Sigma F_y = A_y - F + B = 0, \tag{5.14}$$

$$\Sigma M_{(\text{point } A)} = M_A - \left(\frac{L}{2}\right)F + LB = 0. \tag{5.15}$$

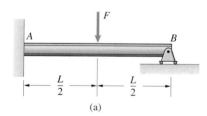

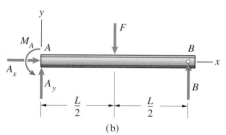

Figure 5.20
(a) A beam with built-in and roller supports.
(b) The free-body diagram has four unknown reactions.

Now we have three equations and four unknown reactions. Although the first equation tells us that $A_x = 0$, we can't solve the two equations (5.14) and (5.15) for the three reactions A_y, B, and M_A.

When faced with this situation, students often attempt to sum the moments about another point, such as point B, to obtain an additional equation:

$$\Sigma M_{(\text{point } B)} = M_A + \left(\frac{L}{2}\right)F - LA_y = 0.$$

Unfortunately, this doesn't help. This is not an independent equation but is a linear combination of Eqs. (5.14) and (5.15):

$$\Sigma M_{(\text{point } B)} = M_A + \left(\frac{L}{2}\right)F - LA_y$$

$$= \underbrace{M_A - \left(\frac{L}{2}\right)F + LB}_{\text{Eq. (5.15)}} - \underbrace{L\left(A_y - F + B\right)}_{\text{Eq. (5.14)}}.$$

As this example demonstrates, each support added to an object results in additional reactions. The difference between the number of reactions and the number of independent equilibrium equations is called the *degree of redundancy*.

Even if an object is statically indeterminate due to redundant supports, it may be possible to determine some of the reactions from the equilibrium equations. Notice that in our previous example we were able to determine the reaction A_x even though we could not determine the other reactions.

Since redundant supports are so ubiquitous, you may wonder why we devote so much effort to teaching you how to analyze objects whose reactions can be determined with the equilibrium equations. We want to develop your understanding of equilibrium and give you practice writing equilibrium equations. The reactions on an object with redundant supports *can* be determined by supplementing the equilibrium equations with additional equations that relate the forces and couples acting on the object to its deformation, or change in shape. Thus obtaining the equilibrium equations is the first step of the solution.

Example 5.6

Recognizing a Statically Indeterminate Object

The beam in Fig. 5.21 has two pin supports and is loaded by a 2-kN force.
(a) Show that the beam is statically indeterminate.
(b) Determine as many reactions as possible.

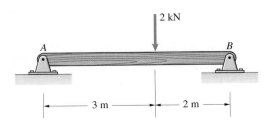

Figure 5.21

Strategy

The beam is statically indeterminate if its free-body diagram has more unknown reactions than the number of independent equilibrium equations we can obtain. But even if this is the case, we may be able to solve the equilibrium equations for some of the reactions.

Solution

Draw the Free-Body Diagram We draw the free-body diagram of the beam in Fig. a. There are four unknown reactions—A_x, A_y, B_x, and B_y—and we can write only three independent equilibrium equations. Therefore the beam is statically indeterminate.

Apply the Equilibrium Equations Summing the moments about point A, the equilibrium equations are

$$\Sigma F_x = A_x + B_x = 0,$$

$$\Sigma F_y = A_y + B_y - 2 = 0,$$

$$\Sigma M_{(\text{point } A)} = 5B_y - (2)(3) = 0.$$

We can solve the third equation for B_y and then solve the second equation for A_y. The results are $A_y = 0.8$ kN and $B_y = 1.2$ kN. The first equation tells us that $B_x = -A_x$, but we can't solve for their values.

Discussion

This example can give you insight into why the reactions on objects with redundant constraints can't be determined from the equilibrium equations alone. The two pin supports can exert horizontal reactions on the beam even in the absence of loads (Fig. b), and these reactions satisfy the equilibrium equations for any value of T ($\Sigma F_x = -T + T = 0$).

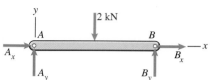

(a) The free-body diagram of the beam.

(b) The supports can exert reactions on the beam.

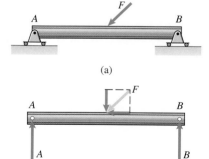

Figure 5.22

(a) A beam with two roller supports is not in equilibrium when subjected to the load shown.

(b) The sum of the forces in the horizontal direction is not zero.

Improper Supports

We say that an object has improper supports if it will not remain in equilibrium under the action of the loads exerted on it. Thus an object with improper supports will move when the loads are applied. In two-dimensional problems, this can occur in two ways:

1. *The supports can exert only parallel forces.* This leaves the object free to move in the direction perpendicular to the support forces. If the loads exert a component of force in that direction, the object is not in equilibrium. Figure 5.22a shows an example of this situation. The two roller supports can exert only vertical forces, while the force F has a horizontal component. The beam will move horizontally when F is applied. This is particularly apparent from the free-body diagram (Fig. 5.22b). The sum of the forces in the horizontal direction cannot be zero because the roller supports can exert only vertical reactions.

2. *The supports can exert only concurrent forces*. If the loads exert a moment about the point where the lines of action of the support forces intersect, the object is not in equilibrium. For example, consider the beam in Fig. 5.23a. From its free-body diagram (Fig. 5.23b) we see that the reactions A and B exert no moment about the point P, where their lines of action intersect, but the load F does. The sum of the moments about point P is not zero, and the beam will rotate when the load is applied.

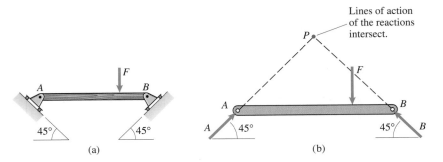

(a)

(b)

Figure 5.23
(a) A beam with roller supports on sloped surfaces.
(b) The sum of the moments about point P is not zero.

Except for problems that deal explicitly with improper supports, objects in our examples and problems have proper supports. You should develop the habit of examining objects in equilibrium and thinking about why they are properly supported for the loads acting on them.

Study Questions

1. What does it mean when an object is said to have redundant supports?
2. How can you recognize if an object is statically indeterminate due to redundant supports?
3. What is the "degree of redundancy" of an object?

Example 5.7

Proper and Improper Supports

State whether each L-shaped bar in Fig. 5.24 is properly or improperly supported. If a bar is properly supported, determine the reactions at its supports.

Solution

We draw the free-body diagrams of the bars in Fig. 5.25.

Bar (a) The lines of action of the reactions due to the two roller supports intersect at P, and the load F exerts a moment about P. This bar is improperly supported.

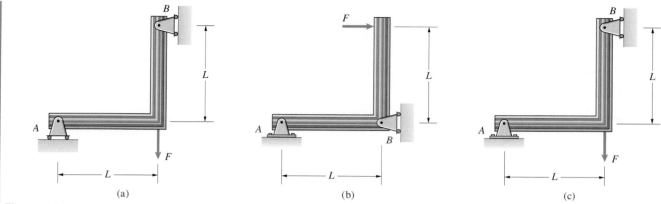

Figure 5.24

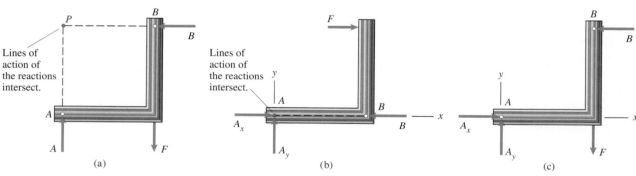

Figure 5.25
Free-body diagrams of the three bars.

Bar (b) The lines of action of the reactions intersect at A, and the load F exerts a moment about A. This bar is also improperly supported.

Bar (c) The three support forces are neither parallel nor concurrent. This bar is properly supported. The equilibrium equations are

$$\Sigma F_x = A_x - B = 0,$$

$$\Sigma F_y = A_y - F = 0,$$

$$\Sigma M_{(\text{point } A)} = BL - FL = 0.$$

Solving these equations, the reactions are $A_x = F$, $A_y = F$, and $B = F$.

5.4 Three-Dimensional Applications

You have seen that when an object in equilibrium is subjected to a two-dimensional system of forces and moments, you can obtain no more than three independent equilibrium equations. In the case of a three-dimensional system of forces and moments, you can obtain up to six independent equilibrium equations: The three components of the sum of the forces must equal zero, and the three components of the sum of the moments about any point must equal zero. Your procedure for determining the reactions on objects subjected to three-dimensional systems of forces and moments—drawing the free-body diagram and applying the equilibrium equations—is the same as in two-dimensions. You just need to become familiar with the support conventions used in three-dimensional applications.

Supports

We present five conventions frequently used in three-dimensional problems. Again, even when actual supports do not physically resemble these models, we represent them by the models if they exert the same (or approximately the same) reactions.

The Ball and Socket Support In the *ball and socket support*, the supported object is attached to a ball enclosed within a spherical socket (Fig. 5.26a). The socket permits the ball to rotate freely (friction is neglected) but prevents it from translating in any direction.

Imagine holding a bar attached to a ball and socket support (Fig. 5.26b). If you try to translate the bar (move it without rotating it) in any direction, the support exerts a reactive force to prevent the motion. However, you can rotate the bar about the support. The support cannot exert a couple to prevent rotation. Thus a ball and socket support can't exert a couple but can exert three components of force (Fig. 5.26c). It is the three-dimensional analog of the two-dimensional pin support.

The human hip joint is an example of a ball and socket support (Fig. 5.27). The support of the gear shift lever of a car can be modeled as a ball and socket support within the lever's range of motion.

The Roller Support The *roller support* (Fig. 5.28a) is a ball and socket support that can roll freely on a supporting surface. A roller support can exert only a force normal to the supporting surface (Fig. 5.28b). The rolling "casters" sometimes used to support furniture legs are supports of this type.

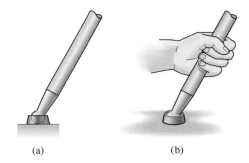

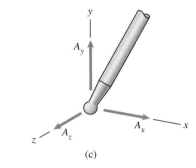

Figure 5.26
(a) A ball and socket support.
(b) Holding a supported bar.
(c) The ball and socket support can exert three components of force.

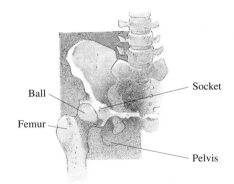

Figure 5.27
The human femur is attached to the pelvis by a ball and socket support.

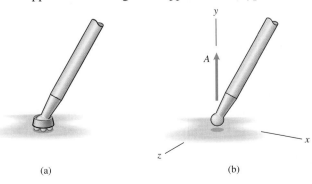

Figure 5.28
(a) A roller support.
(b) The reaction is normal to the supporting surface.

The Hinge   The hinge support is the familiar device used to support doors. It permits the supported object to rotate freely about a line, the *hinge axis*. An object is attached to a hinge in Fig. 5.29a. The z axis of the coordinate system is aligned with the hinge axis.

If you imagine holding a bar attached to a hinge (Fig. 5.29b), notice that you can rotate the bar about the hinge axis. The hinge cannot exert a couple about the hinge axis (the z axis) to prevent rotation. However, you can't rotate the bar about the x or y axis because the hinge can exert couples about those axes to resist the motion. In addition, you can't translate the bar in any direction. The reactions a hinge can exert on an object are shown in Fig. 5.29c. There are three components of force, A_x, A_y, and A_z, and couples about the x and y axes, M_{Ax} and M_{Ay}.

In some situations, either a hinge exerts no couples on the object it supports, or they are sufficiently small to neglect. An example of the latter case is when the axes of the hinges supporting a door are properly aligned (the axes of the individual hinges coincide). In these situations the hinge exerts only forces on an object (Fig. 5.29d). Situations also arise in which a hinge exerts no couples on an object and exerts no force in the direction of the hinge axis. (The hinge may actually be designed so that it cannot support a force parallel to the hinge axis.) Then the hinge exerts forces only in the directions perpendicular to the hinge axis (Fig. 5.29e). In examples and problems, we indicate when a hinge does not exert all five of the reactions in Fig. 5.29c.

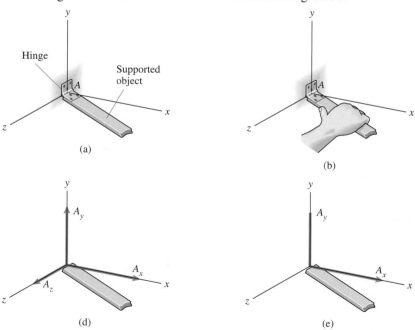

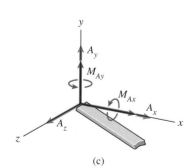

Figure 5.29
(a) A hinge. The z axis is aligned with the hinge axis.
(b) Holding a supported bar.
(c) In general, a hinge can exert five reactions: three force components and two couple components.
(d) The reactions when the hinge exerts no couples.
(e) The reactions when the hinge exerts neither couples nor a force parallel to the hinge axis.

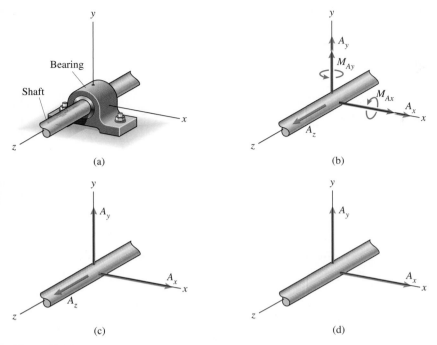

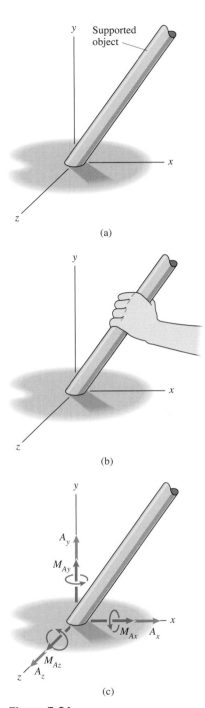

Figure 5.30
(a) A bearing. The z axis is aligned with the axis of the shaft.
(b) In general, a bearing can exert five reactions: three force components and two couple components.
(c) The reactions when the bearing exerts no couples.
(d) The reactions when the bearing exerts neither couples nor a force parallel to the axis of the shaft.

The Bearing   The type of bearing shown in Fig. 5.30a supports a circular shaft while permitting it to rotate about its axis. The reactions are identical to those exerted by a hinge. In the most general case (Fig. 5.30b), the bearing can exert a force on the supported shaft in each coordinate direction and can exert couples about axes perpendicular to the shaft but cannot exert a couple about the axis of the shaft.

As in the case of the hinge, situations can occur in which the bearing exerts no couples (Fig. 5.30c) or exerts no couples and no force parallel to the shaft axis (Fig. 5.30d). Some bearings are designed in this way for specific applications. In examples and problems, we indicate when a bearing does not exert all of the reactions in Fig. 5.30b.

The Built-In Support   You are already familiar with the built-in, or fixed, support (Fig. 5.31a). Imagine holding a bar with a built-in support (Fig. 5.31b). You cannot translate it in any direction, and you cannot rotate it about any axis. The support is capable of exerting forces A_x, A_y, and A_z in each coordinate direction and couples M_{Ax}, M_{Ay}, and M_{Az} about each coordinate axis (Fig. 5.31c).

Table 5.2 summarizes the support conventions commonly used in three-dimensional applications.

Figure 5.31
(a) A built-in support.
(b) Holding a supported bar.
(c) A built-in support can exert six reactions: three force components and three couple components.

Table 5.2 Supports used in three-dimensional applications.

Supports	Reactions
Rope or Cable	One Collinear Force
Contact with a Smooth Surface	One Normal Force
Contact with a Rough Surface	Three Force Components
Ball and Socket Support	Three Force Components
Roller Support	One Normal Force

Table 5.2 *(continued)*

Supports	Reactions
  Hinge (The *z* axis is parallel to the hinge axis.)	 Three Force Components, Two Couple Components
  Bearing (The *z* axis is parallel to the axis of the supported shaft.)	 (When no couples are exerted) (When no couples and no axial force are exerted)
  Built-in (Fixed) Support	 Three Force Components, Three Couple Components

The Scalar Equilibrium Equations

The loads and reactions on an object in equilibrium satisfy the six scalar equilibrium equations

$$\Sigma F_x = 0, \tag{5.16}$$
$$\Sigma F_y = 0, \tag{5.17}$$
$$\Sigma F_z = 0, \tag{5.18}$$
$$\Sigma M_x = 0, \tag{5.19}$$
$$\Sigma M_y = 0, \tag{5.20}$$
$$\Sigma M_z = 0. \tag{5.21}$$

You can evaluate the sums of the moments about any point. Although you can obtain other equations by summing the moments about additional points, they will not be independent of these equations. *More than six independent equilibrium equations cannot be obtained from a given free-body diagram, so we can solve for at most six unknown forces or couples.*

The steps required to determine reactions in three dimensions are familiar from your experience with two-dimensional applications. You must first obtain a free-body diagram by isolating an object and showing the loads and reactions acting on it, then use Eqs. (5.16)–(5.21) to determine the reactions.

Study Questions

1. What is a ball and socket support? What reactions can it exert on an object?
2. In general, a hinge support can exert five reactions on an object. What are they?
3. If an object has a built-in support and any additional supports, it is statically indeterminate. Why is this true?

Example 5.8

Determining Reactions in Three Dimensions

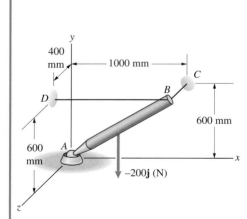

Figure 5.32

The bar AB in Fig. 5.32 is supported by the cables BC and BD and a ball and socket support at A. Cable BC is parallel to the z axis, and cable BD is parallel to the x axis. The 200-N weight of the bar acts at its midpoint. What are the tensions in the cables and the reactions at A?

Strategy

We must obtain the free-body diagram of the bar AB by isolating it from the support at A and the two cables. Then we can use the equilibrium equations to determine the reactions at A and the tensions in the cables.

Solution

Draw the Free-Body Diagram In Fig. a we isolate the bar and show the reactions that may be exerted on it. The ball and socket support can exert three components of force, A_x, A_y, and A_z. The terms T_{BC} and T_{BD} represent the tensions in the cables.

(a) Obtaining the free-body diagram of the bar.

Apply the Equilibrium Equations   The sums of the forces in each coordinate direction equal zero:

$$\Sigma F_x = A_x - T_{BD} = 0,$$

$$\Sigma F_y = A_y - 200 = 0,$$

$$\Sigma F_z = A_z - T_{BC} = 0. \tag{5.22}$$

Let $\mathbf{r}_{AB}$ be the position vector from A to B. The sum of the moments about A is

$$\Sigma \mathbf{M}_{(\text{point } A)} = \left[\mathbf{r}_{AB} \times (-T_{BC}\mathbf{k})\right] + \left[\mathbf{r}_{AB} \times (-T_{BD}\mathbf{i})\right]$$

$$+ \left[\tfrac{1}{2}\mathbf{r}_{AB} \times (-200\mathbf{j})\right]$$

$$= \begin{vmatrix} \mathbf{i} & \mathbf{j} & \mathbf{k} \\ 1 & 0.6 & 0.4 \\ 0 & 0 & -T_{BC} \end{vmatrix} + \begin{vmatrix} \mathbf{i} & \mathbf{j} & \mathbf{k} \\ 1 & 0.6 & 0.4 \\ -T_{BD} & 0 & 0 \end{vmatrix}$$

$$+ \begin{vmatrix} \mathbf{i} & \mathbf{j} & \mathbf{k} \\ 0.5 & 0.3 & 0.2 \\ 0 & -200 & 0 \end{vmatrix}$$

$$= \left(-0.6T_{BC} + 40\right)\mathbf{i} + \left(T_{BC} - 0.4T_{BD}\right)\mathbf{j} + \left(0.6T_{BD} - 100\right)\mathbf{k}.$$

The components of this vector (the sums of the moments about the three coordinate axes) each equal zero:

$$\Sigma M_x = -0.6T_{BC} + 40 = 0,$$

$$\Sigma M_y = T_{BC} - 0.4T_{BD} = 0,$$

$$\Sigma M_z = 0.6T_{BD} - 100 = 0.$$

Solving these equations, we obtain the tensions in the cables:

$$T_{BC} = 66.7 \text{ N}, \qquad T_{BD} = 166.7 \text{ N}.$$

(Notice that we needed only two of the three equations to obtain the two tensions. The third equation is redundant.)

Then from Eqs. (5.22) we obtain the reactions at the ball and socket support:

$$A_x = 166.7 \text{ N}, \qquad A_y = 200 \text{ N}, \qquad A_z = 66.7 \text{ N}.$$

Discussion

Notice that by summing moments about A we obtained equations in which the unknown reactions A_x, A_y, and A_z did not appear. You can often simplify your solutions in this way.

Example 5.9

Reactions at a Hinge Support

The bar AC in Fig. 5.33 is 4 ft long and is supported by a hinge at A and the cable BD. The hinge axis is along the z axis. The centerline of the bar lies in the x–y plane, and the cable attachment point B is the midpoint of the bar. Determine the tension in the cable and the reactions exerted on the bar by the hinge.

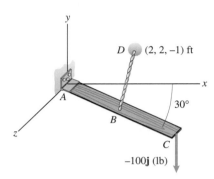

Figure 5.33

Solution

Draw the Free-Body Diagram   We isolate the bar from the hinge support and the cable and show the reactions they exert (Fig. a). The terms A_x, A_y, and A_z are the components of force exerted by the hinge, and the terms M_{Ax} and M_{Ay} are the couples exerted by the hinge about the x and y axes. (Remember that the hinge cannot exert a couple on the bar about the hinge axis.) The term T is the tension in the cable.

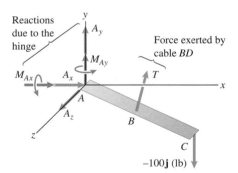

(a) The free-body diagram of the bar.

Apply the Equilibrium Equations   To write the equilibrium equations, we must first express the cable force in terms of its components. The coordinates of point B are $(2 \cos 30°, -2 \sin 30°, 0)$ ft, so the position vector from B to D is

$$\mathbf{r}_{BD} = (2 - 2 \cos 30°)\mathbf{i} + \left[2 - (-2 \sin 30°)\right]\mathbf{j} + (-1 - 0)\mathbf{k}$$

$$= 0.268\mathbf{i} + 3\mathbf{j} - \mathbf{k}.$$

We divide this vector by its magnitude to obtain a unit vector $\mathbf{e}_{BD}$ that points from point B toward point D:

$$\mathbf{e}_{BD} = \frac{\mathbf{r}_{BD}}{|\mathbf{r}_{BD}|} = 0.084\mathbf{i} + 0.945\mathbf{j} - 0.315\mathbf{k}.$$

Now we can write the cable force as the product of its magnitude and $\mathbf{e}_{BD}$:

$$T\mathbf{e}_{BD} = T(0.084\mathbf{i} + 0.945\mathbf{j} - 0.315\mathbf{k}).$$

The sums of the forces in each coordinate direction must equal zero:

$$\Sigma F_x = A_x + 0.084T = 0,$$

$$\Sigma F_y = A_y + 0.945T - 100 = 0,$$

$$\Sigma F_z = A_z - 0.315T = 0. \tag{5.23}$$

If we sum moments about A, the resulting equations do not contain the unknown reactions A_x, A_y, and A_z. The position vectors from A to B and from A to C are

$$\mathbf{r}_{AB} = 2\cos 30°\mathbf{i} - 2\sin 30°\mathbf{j}.$$

$$\mathbf{r}_{AC} = 4\cos 30°\mathbf{i} - 4\sin 30°\mathbf{j}.$$

The sum of the moments about A is

$$\Sigma \mathbf{M}_{(\text{point } A)} = M_{Ax}\mathbf{i} + M_{Ay}\mathbf{j} + \left[\mathbf{r}_{AB} \times (T\mathbf{e}_{BD})\right] + \left[\mathbf{r}_{AC} \times (-100\mathbf{j})\right]$$

$$= M_{Ax}\mathbf{i} + M_{Ay}\mathbf{j} + \begin{vmatrix} \mathbf{i} & \mathbf{j} & \mathbf{k} \\ 1.732 & -1 & 0 \\ 0.084T & 0.945T & -0.315T \end{vmatrix}$$

$$+ \begin{vmatrix} \mathbf{i} & \mathbf{j} & \mathbf{k} \\ 3.464 & -2 & 0 \\ 0 & -100 & 0 \end{vmatrix}$$

$$= (M_{Ax} + 0.315T)\mathbf{i} + (M_{Ay} + 0.546T)\mathbf{j}$$

$$+ (1.72T - 346)\mathbf{k} = 0.$$

From this vector equation we obtain the scalar equations

$$\Sigma M_x = M_{Ax} + 0.315T = 0,$$

$$\Sigma M_y = M_{Ay} + 0.546T = 0,$$

$$\Sigma M_z = 1.72T - 346 = 0.$$

Solving these equations, we obtain the reactions

$$T = 201 \text{ lb}, \qquad M_{Ax} = -63.4 \text{ ft-lb}, \qquad M_{Ay} = -109.8 \text{ ft-lb}.$$

Then from Eqs. (5.23) we obtain the forces exerted on the bar by the hinge:

$$A_x = -17.0 \text{ lb}, \qquad A_y = -90.2 \text{ lb}, \qquad A_z = 63.4 \text{ lb}.$$

Example 5.10

Reactions at Properly Aligned Hinges

The plate in Fig. 5.34 is supported by hinges at A and B and the cable CE. The properly aligned hinges do not exert couples on the plate, and the hinge at A does not exert a force on the plate in the direction of the hinge axis. Determine the reactions at the hinges and the tension in the cable.

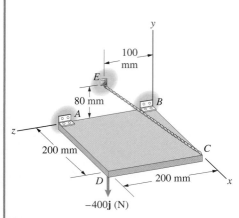

Figure 5.34

Solution

Draw the Free-Body Diagram We isolate the plate and show the reactions at the hinges and the force exerted by the cable (Fig. a). The term T is the force exerted on the plate by cable CE.

Apply the Equilibrium Equations Since we know the coordinates of points C and E, we can express the cable force as the product of its magnitude T and a unit vector directed from C toward E. The result is

$$T(-0.842\mathbf{i} + 0.337\mathbf{j} + 0.421\mathbf{k}).$$

The sums of the forces in each coordinate direction equal zero:

$$\Sigma F_x = A_x + B_x - 0.842T = 0,$$
$$\Sigma F_y = A_y + B_y + 0.337T - 400 = 0,$$
$$\Sigma F_z = B_z + 0.421T = 0.$$

If we sum the moments about B, the resulting equations will not contain the three unknown reactions at B. The sum of the moments about B is

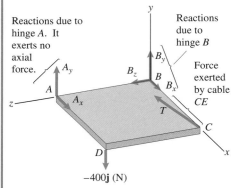

(a) The free-body diagram of the plate.

$$\Sigma\mathbf{M}_{(\text{point } B)} = \begin{vmatrix} \mathbf{i} & \mathbf{j} & \mathbf{k} \\ 0.2 & 0 & 0 \\ -0.842T & 0.337T & 0.421T \end{vmatrix} + \begin{vmatrix} \mathbf{i} & \mathbf{j} & \mathbf{k} \\ 0 & 0 & 0.2 \\ A_x & A_y & 0 \end{vmatrix}$$

$$+ \begin{vmatrix} \mathbf{i} & \mathbf{j} & \mathbf{k} \\ 0.2 & 0 & 0.2 \\ 0 & -400 & 0 \end{vmatrix}$$

$$= (-0.2A_y + 80)\mathbf{i} + (-0.0842T + 0.2A_x)\mathbf{j}$$
$$+ (0.0674T - 80)\mathbf{k} = 0.$$

The scalar equations are

$$\Sigma M_x = -0.2A_y + 80 = 0,$$
$$\Sigma M_y = -0.0842T + 0.2A_x = 0,$$
$$\Sigma M_z = 0.0674T - 80 = 0.$$

Solving these equations, we obtain the reactions

$$T = 1187 \text{ N}, \qquad A_x = 500 \text{ N}, \qquad A_y = 400 \text{ N}.$$

Then from Eqs. (5.24), the reactions at B are

$$B_x = 500 \text{ N}, \qquad B_y = -400 \text{ N}, \qquad B_z = -500 \text{ N}.$$

Discussion

If our only objective had been to determine the tension T, we could have done so easily by setting the sum of the moments about the line AB (the z axis) equal to zero. Since the reactions at the hinges exert no moment about the z axis, we obtain the equation

$$(0.2)(0.337T) - (0.2)(400) = 0,$$

which yields the result $T = 1187$ N.

5.5 Two-Force and Three-Force Members

You have seen how the equilibrium equations are used to analyze objects supported and loaded in different ways. Here we discuss two particular cases that occur so frequently you need to be familiar with them. The first one is especially important and plays a central role in our analysis of structures in the next chapter.

Two-Force Members

If the system of forces and moments acting on an object is equivalent to two forces acting at different points, we refer to the object as a *two-force member*. For example, the object in Fig. 5.35a is subjected to two sets of concurrent forces whose lines of action intersect at A and B. Since we can represent them

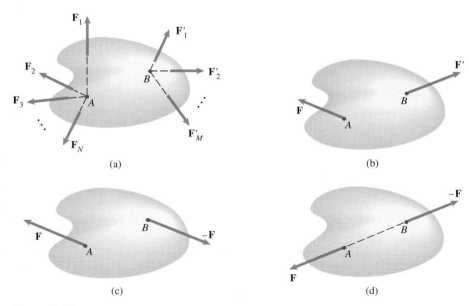

Figure 5.35
(a) An object subjected to two sets of concurrent forces.
(b) Representing the concurrent forces by two forces **F** and **F′**.
(c) If the object is in equilibrium, the forces must be equal and opposite.
(d) The forces form a couple unless they have the same line of action.

by single forces acting at A and B (Fig. 5.35b), where $\mathbf{F} = \mathbf{F}_1 + \mathbf{F}_2 + \cdots + \mathbf{F}_N$ and $\mathbf{F}' = \mathbf{F}'_1 + \mathbf{F}'_2 + \cdots + \mathbf{F}'_M$, this object is a two-force member.

If the object is in equilibrium, what can we infer about the forces $\mathbf{F}$ and $\mathbf{F}'$? The sum of the forces equals zero only if $\mathbf{F}' = -\mathbf{F}$ (Fig. 5.35c). Furthermore, the forces $\mathbf{F}$ and $-\mathbf{F}$ form a couple, so the sum of the moments is not zero unless the lines of action of the forces lie along the line through the points A and B (Fig. 5.35d). Thus equilibrium tells us that *the two forces are equal in magnitude, are opposite in direction, and have the same line of action.* However, without additional information, we cannot determine their magnitude.

A cable attached at two points (Fig. 5.36a) is a familiar example of a two-force member (Fig. 5.36b). The cable exerts forces on the attachment points that are directed along the line between them (Fig. 5.36c).

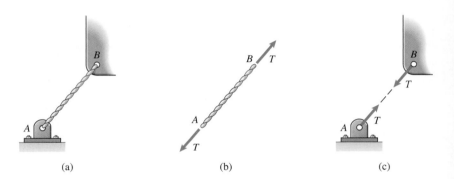

Figure 5.36
(a) A cable attached at A and B.
(b) The cable is a two-force member.
(c) The forces exerted by the cable.

(a)   (b)   (c)

A bar that has two supports that exert only forces on it (no couples) and is not subjected to any loads is a two-force member (Fig. 5.37a). Such bars are often used as supports for other objects. Because the bar is a two-force member, the lines of action of the forces exerted on the bar must lie along the line between the supports (Fig. 5.37b). Notice that, unlike the cable, the bar can exert forces at A and B either in the directions shown in Fig. 5.37c or in the opposite directions. (In other words, the cable can only pull on its supports, while the bar can either pull or push.)

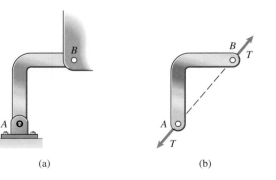

Figure 5.37
(a) The bar AB attaches the object to the pin support.
(b) The bar AB is a two-force member.
(c) The force exerted on the supported object by the bar AB.

(a)   (b)   (c)

In these examples we assumed that the weights of the cable and the bar could be neglected in comparison with the forces exerted on them by their supports. When that is not the case, they are clearly not two-force members.

Three-Force Members

If the system of forces and moments acting on an object is equivalent to three forces acting at different points, we call it a *three-force member*. We can show that if a three-force member is in equilibrium, the three forces are coplanar and are either parallel or concurrent.

We first prove that the forces are coplanar. Let them be called F_1, F_2, and F_3, and let P be the plane containing the three points of application (Fig. 5.38a). Let L be the line through the points of application of F_1 and F_2. Since the moments due to F_1 and F_2 about L are zero, the moment due to F_3 about L must equal zero (Fig. 5.38b):

$$[e \cdot (r \times F_3)]e = [F_3 \cdot (e \times r)]e = 0.$$

This equation requires that F_3 be perpendicular to $e \times r$, which means that F_3 is contained in P. The same procedure can be used to show that F_1 and F_2 are contained in P, so the forces are coplanar. (A different proof is required if the points of application lie on a straight line, but the result is the same.)

If the three coplanar forces are not parallel, there will be points where their lines of action intersect. Suppose that the lines of action of two of the forces intersect at a point Q. Then the moments of those two forces about Q are zero, and the sum of the moments about Q is zero only if the line of action of the third force also passes through Q. Therefore either the forces are parallel or they are concurrent (Fig. 5.38c).

You can often simplify the analysis of an object in equilibrium by recognizing that it is a two-force or three-force member. However, you are not getting something for nothing. Once you have drawn the free-body diagram of a two-force member as shown in Figs. 5.36b and 5.37b, you cannot obtain any further information about the forces from the equilibrium equations. When you require that the lines of action of nonparallel forces acting on a three-force member be coincident, you have used the fact that the sum of the moments about a point must be zero and cannot obtain any further information from that condition.

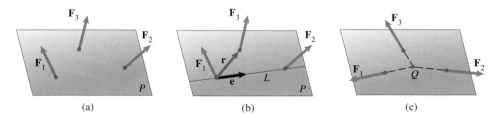

(a)　　　　　　　　(b)　　　　　　　　(c)

Figure 5.38
(a) The three forces and the plane P.
(b) Determining the moment due to force F_3 about L.
(c) If the forces are not parallel, they must be concurrent.

Example 5.11

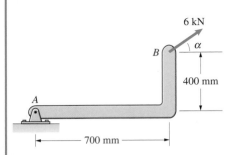

Figure 5.39

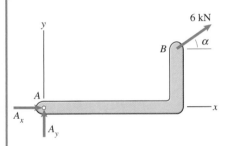

(a) The free-body diagram of the bar.

A Two-Force Member

The L-shaped bar in Fig. 5.39 has a pin support at A and is loaded by a 6-kN force at B. Neglect the weight of the bar. Determine the angle α and the reactions at A.

Strategy

The bar is a two-force member because it is subjected only to the 6-kN force at B and the force exerted by the pin support. (If we could not neglect the weight of the bar, it would not be a two-force member.) We will determine the angle α and the reactions at A in two ways, first by applying the equilibrium equations in the usual way and then by using the fact that the bar is a two-force member.

Solution

Applying the Equilibrium Equations We draw the free-body diagram of the bar in Fig. a, showing the reactions at the pin support. Summing moments about point A, the equilibrium equations are

$$\Sigma F_x = A_x + 6\cos\alpha = 0,$$

$$\Sigma F_y = A_y + 6\sin\alpha = 0,$$

$$\Sigma M_{(\text{point } A)} = (6\sin\alpha)(0.7) - (6\cos\alpha)(0.4) = 0.$$

From the third equation we see that $\alpha = \arctan(0.4/0.7)$. In the range $0 \le \alpha \le 360°$, this equation has the two solutions $\alpha = 29.7°$ and $\alpha = 209.7°$. Knowing α, we can determine A_x and A_y from the first two equilibrium equations. The solutions for the two values of α are

$$\alpha = 29.7°, \qquad A_x = -5.21 \text{ kN}, \qquad A_y = -2.98 \text{ kN},$$

and

$$\alpha = 209.7°, \qquad A_x = 5.21 \text{ kN}, \qquad A_y = 2.98 \text{ kN}.$$

Treating the Bar as a Two-Force Member We know that the 6-kN force at B and the force exerted by the pin support must be equal in magnitude, opposite in direction, and directed along the line between points A and B. The two possibilities are shown in Figs. b and c. Thus by recognizing that the bar is a two-force member, we immediately know the possible directions of the forces and the magnitude of the reaction at A.

In Fig. b we can see that $\tan\alpha = 0.4/0.7$, so $\alpha = 29.7°$ and the components of the reaction at A are

$$A_x = -6\cos 29.7° = -5.21 \text{ kN}$$

$$A_y = -6\sin 29.7° = -2.98 \text{ kN}$$

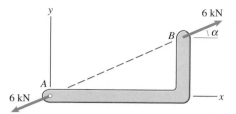
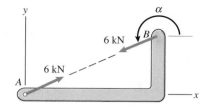

(b), (c) The possible directions of the forces.

In Fig. c, $\alpha = 180° + 29.7° = 209.7°$, and the components of the reaction at A are

$$A_x = 6 \cos 29.7° = 5.21 \text{ kN}$$

$$A_y = 6 \sin 29.7° = 2.98 \text{ kN}$$

Example 5.12

Two- and Three-Force Members

The 100-lb weight of the rectangular plate in Fig. 5.40 acts at its midpoint. Determine the reactions exerted on the plate at B and C.

Strategy

The plate is subjected to its weight and the reactions exerted by the pin supports at B and C, so it is a three-force member. Furthermore, the bar AB is a two-force member, so we know that the line of action of the reaction it exerts on the plate at B is directed along the line between A and B. We can use this information to simplify the free-body diagram of the plate.

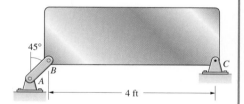

Figure 5.40

Solution

The reaction exerted on the plate by the two-force member AB must be directed along the line between A and B, and the line of action of the weight is vertical. Since the three forces on the plate must be either parallel or concurrent, their lines of action must intersect at the point P shown in Fig. a. From the equilibrium equations

$$\Sigma F_x = B \sin 45° - C \sin 45° = 0,$$

$$\Sigma F_y = B \cos 45° + C \cos 45° - 100 = 0,$$

we obtain the reactions $B = C = 70.7$ lb.

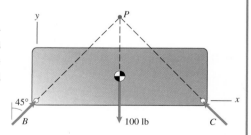

(a) The free-body diagram of the plate. The three forces must be concurrent.

Computational Mechanics

The following example and problems are designed for the use of a programmable calculator or computer.

Computational Example 5.13

The beam in Fig. 5.41 weighs 200 lb and is supported by a pin support at A and the wire BC. The wire behaves like a linear spring with spring constant $k = 60$ lb/ft and is unstretched when the beam is in the position shown. Determine the reactions at A and the tension in the wire when the beam is in equilibrium.

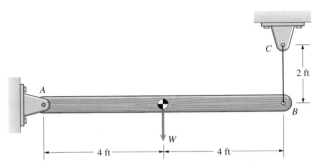

Figure 5.41

Strategy

When the beam is in equilibrium, the sum of the moments about A due to the beam's weight and the force exerted by the wire equals zero. We will obtain a graph of the sum of the moments as a function of the angle of rotation of the beam relative to the horizontal to determine the position of the beam when it is in equilibrium. Once we know the position, we can determine the tension in the wire and the reactions at A.

Solution

Let α be the angle from the horizontal to the centerline of the beam (Fig. a). The distances b and h are

$$b = 8(1 - \cos\alpha),$$
$$h = 2 + 8\sin\alpha,$$

and the length of the stretched wire is

$$L = \sqrt{b^2 + h^2}.$$

The tension in the wire is

$$T = k(L - 2).$$

We draw the free-body diagram of the beam in Fig. b. In terms of the components of the force exerted by the wire,

$$T_x = \frac{b}{L}T, \qquad T_y = \frac{h}{L}T,$$

the sum of the moments about A is

$$\Sigma M_A = (8\sin\alpha)T_x + (8\cos\alpha)T_y - (4\cos\alpha)W.$$

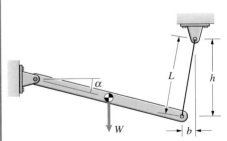

(a) Rotating the beam through an angle α.

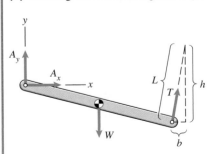

(b) The free-body diagram of the beam.

If we choose a value of α, we can sequentially evaluate these quantities. Computing ΣM_A as a function of α, we obtain the graph shown in Fig. 5.42. From the graph we estimate that $\Sigma M_A = 0$ when $\alpha = 12°$. By examining computed results near 12°, we estimate that the beam is in equilibrium when $\alpha = 11.89°$. The corresponding value of the tension in the wire is $T = 99.1$ lb.

α	ΣM_A (ft-lb)
11.87°	−1.2600
11.88°	−0.5925
11.89°	0.0750
11.90°	0.7424
11.91°	1.4099

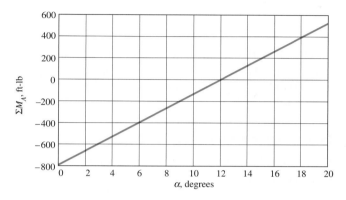

Figure 5.42
The sum of the moments as a function of α.

To determine the reactions at A, we use the equilibrium equations

$$\Sigma F_x = A_x + T_x = 0,$$
$$\Sigma F_y = A_y + T_y - W = 0,$$

obtaining $A_x = -4.7$ lb and $A_y = 101.0$ lb.

Chapter Summary

Building on our discussions of forces in Chapter 3 and moments in Chapter 4, in this chapter we have used the equilibrium equations to analyze the forces and couples acting on many types of objects. We defined the support conventions commonly used in engineering and presented examples of their use. We discussed situations that can result in an object's being statically indeterminate. Finally, we defined two-force and three-force members. In Chapter 6 we will use the concepts and methods developed in this chapter to analyze the individual members of structures, beginning with structures consisting entirely of two-force members.

When an object is in equilibrium, the following conditions are satisfied:

1. The sum of the forces is zero,

$$\Sigma \mathbf{F} = 0. \qquad \text{Eq. (5.1)}$$

2. The sum of the moments about any point is zero,

$$\Sigma \mathbf{M}_{(\text{any point})} = 0. \qquad \text{Eq. (5.2)}$$

Forces and couples exerted on an object by its supports are called *reactions*. The other forces and couples on the object are the *loads*. Common supports are represented by models called *support conventions*.

Two-Dimensional Applications

When the loads and reactions on an object in equilibrium form a two-dimensional system of forces and moments, they are related by three scalar equilibrium equations:

$$\Sigma F_x = 0,$$
$$\Sigma F_y = 0,$$
$$\Sigma M_{(\text{any point})} = 0.$$

Eqs. (5.4)–(5.6)

No more than three independent equilibrium equations can be obtained from a given two-dimensional free-body diagram.

Support conventions commonly used in two-dimensional applications are summarized in Table 5.1.

Three-Dimensional Applications

The loads and reactions on an object in equilibrium satisfy the six scalar equilibrium equations

$$\Sigma F_x = 0, \qquad \Sigma F_y = 0, \qquad \Sigma F_z = 0,$$
$$\Sigma M_x = 0, \qquad \Sigma M_y = 0, \qquad \Sigma M_z = 0.$$

Eqs. (5.16)–(5.21)

No more than six independent equilibrium equations can be obtained from a given free-body diagram.

Support conventions commonly used in three-dimensional applications are summarized in Table 5.2.

Statically Indeterminate Objects

An object has *redundant supports* when it has more supports than the minimum number necessary to maintain it in equilibrium and *improper supports* when its supports are improperly designed to maintain equilibrium under the applied loads. In either situation, the object is *statically indeterminate*. The difference between the number of reactions and the number of independent equilibrium equations is called the *degree of redundancy*. Even if an object is statically indeterminate due to redundant supports, it may be possible to determine some of the reactions from the equilibrium equations.

Two-Force and Three-Force Members

If the system of forces and moments acting on an object is equivalent to two forces acting at different points, the object is a *two-force member*. If the object is in equilibrium, the two forces are equal in magnitude, opposite in direction, and directed along the line through their points of application. If the system of forces and moments acting on an object is equivalent to three forces acting at different points, it is a *three-force member*. If the object is in equilibrium, the three forces are coplanar and either parallel or concurrent.

Review Problems

5.1 The beam has a built-in support and is loaded by a 2-kN force and a 6 kN-m couple.
(a) Draw the free-body diagram of the beam.
(b) Determine the reactions at the supports.

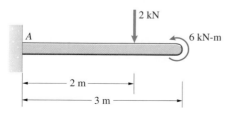

P5.1

5.2 Determine the reactions at A and B.

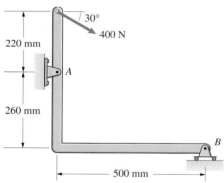

P5.2

5.3 Paleontologists speculate that the stegosaur could stand on its hind limbs for short periods to feed. Based on the free-body diagram shown and assuming that $m = 2000$ kg, determine the magnitudes of the forces B and C exerted by the ligament–muscle brace and vertebral column, and determine the angle α.

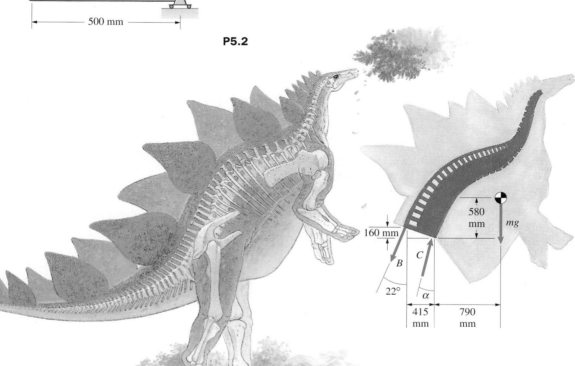

P5.3

5.4 (a) Draw the free-body diagram of the 50-lb plate, and explain why it is statically indeterminate.
(b) Determine as many of the reactions at A and B as possible.

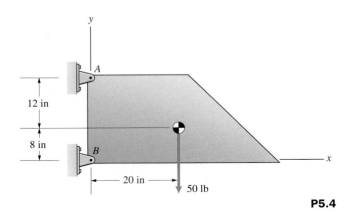

P5.4

5.5 The mass of the truck is 4 Mg. Its wheels are locked, and the tension in its cable is $T = 10$ kN.

(a) Draw the free-body diagram of the truck.
(b) Determine the normal forces exerted on the truck's wheels by the road.

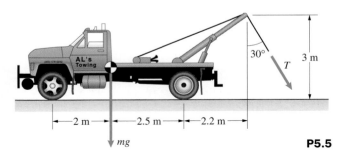

P5.5

5.6 Assume that the force exerted on the head of the nail by the hammer is vertical, and neglect the hammer's weight.

(a) Draw the free-body diagram of the hammer.
(b) If $F = 10$ lb, what are the magnitudes of the force exerted on the nail by the hammer and the normal and friction forces exerted on the floor by the hammer?

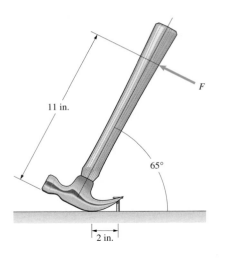

P5.6

5.7 (a) Draw the free-body diagram of the beam.
(b) Determine the reactions at the supports.

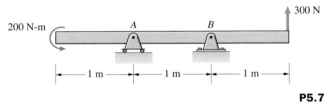

P5.7

5.8 Consider the beam shown in Problem 5.7. First represent the loads (the 300-N force and the 200-N-m couple) by a single equivalent force; then determine the reactions at the supports.

5.9 The truss supports a 90-kg suspended object. What are the reactions at the supports A and B?

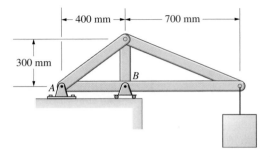

P5.9

5.10 The trailer is parked on a 15° slope. Its wheels are free to turn. The hitch H behaves like a pin support. Determine the reactions at A and H.

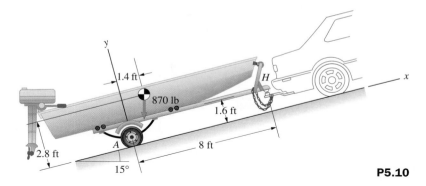

P5.10

5.11 To determine the location of the point where the weight of a car acts (the *center of mass*), an engineer places the car on scales and measures the normal reactions at the wheels for two values of α, obtaining the following results.

α	A_y (kN)	B (kN)
10°	10.134	4.357
20°	10.150	3.677

What are the distances b and h?

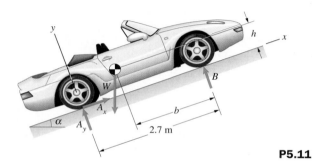

P5.11

5.12 The bar is attached by pin supports to collars that slide on the two fixed bars. Its mass is 10 kg, it is 1 m in length, and its weight acts at its midpoint. Neglect friction and the masses of the collars. The spring is unstretched when the bar is vertical ($\alpha = 0$), and the spring constant is $k = 100$ N/m. Determine the values of α in the range $0 \le \alpha \le 60°$ at which the bar is in equilibrium.

5.13 The 450-lb ladder is supported by the hydraulic cylinder AB and the pin support at C. The reaction at B is parallel to the hydraulic cylinder. Determine the reactions on the ladder.

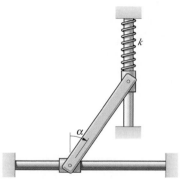

P5.12

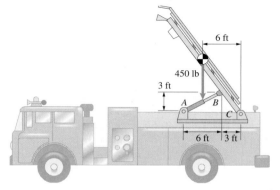

P5.13

5.14 Consider the crane. The hydraulic actuator *BC* exerts a force at *C* that points along the line from *B* to *C*. Treat *A* as a pin support. The mass of the suspended load is 6000 kg. If the angle $\alpha = 35°$, what are the reactions at *A*?

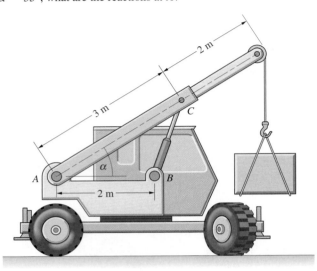

P5.14

5.15 The horizontal rectangular plate weighs 800 N and is suspended by three vertical cables. The weight of the plate acts at its midpoint. What are the tensions in the cables?

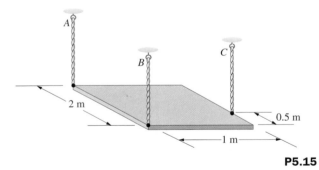

P5.15

5.16 Consider the suspended 800-N plate in Problem 5.15. The weight of the plate acts at its midpoint. If you represent the reactions exerted on the plate by the three cables by a single equivalent force, what is the force, and where does its line of action intersect the plate?

5.17 The 20-kg mass is suspended by cables attached to three vertical 2-m posts. Point *A* is at $(0, 1.2, 0)$ m. Determine the reactions at the built-in support at *E*.

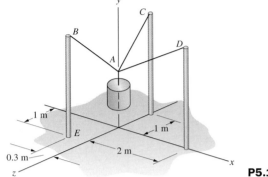

P5.17

5.18 In Problem 5.17, the built-in support of each vertical post will safely support a couple of 800 N-m magnitude. Based on this criterion, what is the maximum safe value of the suspended mass?

5.19 The 80-lb bar is supported by a ball and socket support at *A*, the smooth wall it leans against, and the cable *BC*. The weight of the bar acts at its midpoint.

(a) Draw the free-body diagram of the bar.
(b) Determine the tension in cable *BC* and the reactions at *A*.

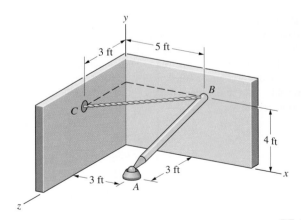

P5.19

5.20 The horizontal bar of weight W is supported by a roller support at A and the cable BC. Use the fact that the bar is a three-force member to determine the angle α, the tension in the cable, and the magnitude of the reaction at A.

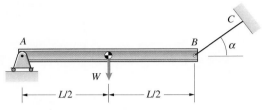

P5.20

5.21 The bicycle brake on the right is pinned to the bicycle's frame at A. Determine the force exerted by the brake pad on the wheel rim at B in terms of the cable tension T.

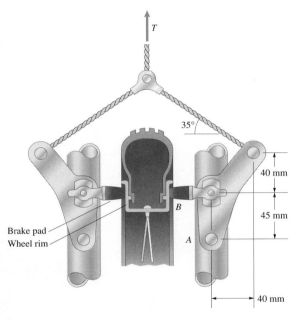

P5.21

$\mathscr{D}$**esign Experience**   The traditional wheelbarrow shown is designed to transport a load W while being supported by an upward force F applied to the handles by the user. (a) Use statics to analyze the effects of a range of choices of the dimensions a and b on the size of load that could be carried. Also consider the implications of these dimensions on the wheelbarrow's ease and practicality of use. (b) Suggest a different design for this classic device that achieves the same function. Use statics to compare your design to the wheelbarrow with respect to load-carrying ability and ease of use.

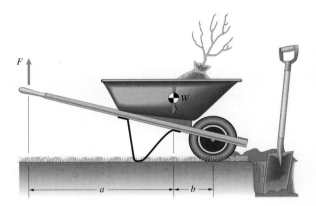

The highway bridge is supported by a truss structure. In this chapter we describe techniques for determining the forces and couples acting on the individual members of structures.

CHAPTER 6

Structures in Equilibrium

I n engineering, the term *structure* can refer to any object that has the capacity to support and exert loads. In this chapter we consider structures composed of interconnected parts, or *members*. To design such a structure, or to determine whether an existing one is adequate, you must determine the forces and couples acting on the structure as a whole as well as on its individual members. We first demonstrate how this is done for the structures called trusses, which are composed entirely of two-force members. The familiar frameworks of steel members that support some highway bridges are trusses. We then consider other structures, called *frames* if they are designed to remain stationary and support loads and *machines* if they are designed to move and exert loads.

6.1 Trusses

Figure 6.1
A typical house is supported by trusses made of wood beams.

We can explain the nature of truss structures such as the beams supporting a house (Fig. 6.1) by starting with very simple examples. Suppose we pin three bars together at their ends to form a triangle. If we add supports as shown in Fig. 6.2a, we obtain a structure that will support a load F. We can construct more elaborate structures by adding more triangles (Figs. 6.2b and c). The bars are the members of these structures, and the places where the bars are pinned together are called the *joints*. Even though these examples are quite simple, you can see that Fig. 6.2c, which is called a Warren truss, begins to resemble the structures used to support bridges and the roofs of houses (Fig. 6.3). If these structures are supported and loaded at their joints and we neglect the weights of the bars, each bar is a two-force member. We call such a structure a *truss*.

We draw the free-body diagram of a member of a truss in Fig. 6.4a. Because it is a two-force member, the forces at the ends, which are the sums of the forces exerted on the member at its joints, must be equal in magnitude, opposite in direction, and directed along the line between the joints. We call the force T the *axial force* in the member. When T is positive in the direction shown (that is, when the forces are directed away from each other), the member is in *tension*. When the forces are directed toward each other, the member is in *compression*.

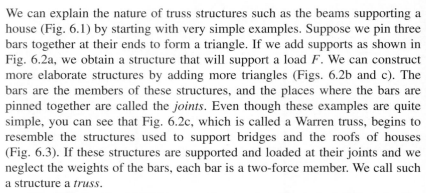

(a)   (b)   (c)

Figure 6.2
Making structures by pinning bars together to form triangles.

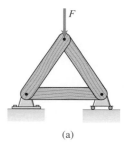

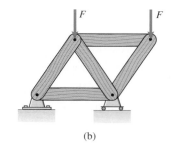

Howe Bridge Truss

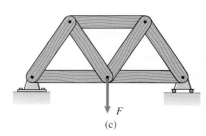

Pratt Bridge Truss

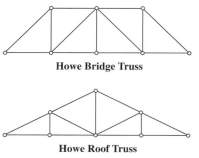

Howe Roof Truss

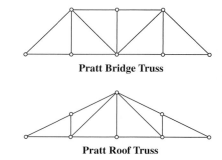

Pratt Roof Truss

Figure 6.3
Simple examples of bridge and roof structures. (The lines represent members, and the circles represent joints.)

In Fig. 6.4b, we "cut" the member by a plane and draw the free-body diagram of the part of the member on one side of the plane. We represent the system of internal forces and moments exerted by the part not included in the free-body diagram by a force **F** acting at the point P where the plane intersects the axis of the member and a couple **M**. The sum of the moments about P must equal zero, so $\mathbf{M} = \mathbf{0}$. Therefore we have a two-force member, which means that **F** must be equal in magnitude and opposite in direction to the force T acting at the joint (Fig. 6.4c). The internal force is a tension or compression equal to the tension or compression exerted at the joint. Notice the similarity to a rope or cable, in which the internal force is a tension equal to the tension applied at the ends.

Although many actual structures, including "roof trusses" and "bridge trusses," consist of bars connected at the ends, very few have pinned joints. For example, if you examine a joint of a bridge truss, you will see that the members are bolted or riveted together so that they are not free to rotate at the joint (Fig. 6.5). It is obvious that such a joint can exert couples on the members. Why are these structures called trusses?

The reason is that they are designed to function as trusses, meaning that they support loads primarily by subjecting their members to axial forces. They can usually be *modeled* as trusses, treating the joints as pinned connections under the assumption that couples they exert on the members are small in comparison to axial forces. When we refer to structures with riveted joints as trusses in problems, we mean that you can model them as trusses.

In the following sections we describe two methods for determining the axial forces in the members of trusses. The method of joints is usually the preferred approach when you need to determine the axial forces in all members of a truss. When you only need to determine the axial forces in a few members, the method of sections often results in a faster solution than the method of joints.

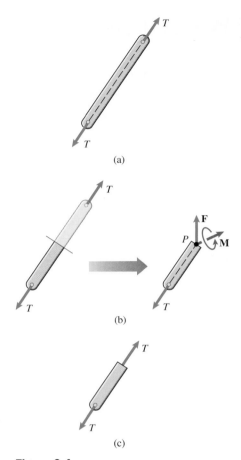

(a)

(b)

(c)

Figure 6.4
(a) Each member of a truss is a two-force member.
(b) Obtaining the free-body diagram of part of the member.
(c) The internal force is equal and opposite to the force acting at the joint, and the internal couple is zero.

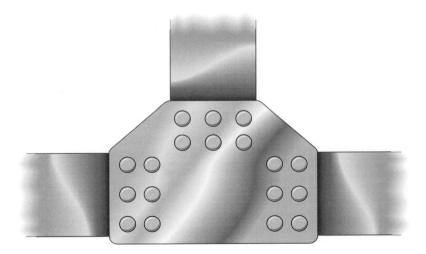

Figure 6.5
A joint of a bridge truss.

6.2 The Method of Joints

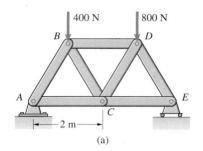

(a)

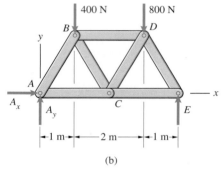

(b)

Figure 6.6
(a) A Warren truss supporting two loads.
(b) Free-body diagram of the truss.

The method of joints involves drawing free-body diagrams of the joints of a truss one by one and using the equilibrium equations to determine the axial forces in the members. Before beginning, it is usually necessary to draw a free-body diagram of the entire truss (that is, treat the truss as a single object) and determine the reactions at its supports. For example, let's consider the Warren truss in Fig. 6.6a, which has members 2 m in length and supports loads at B and D. We draw its free-body diagram in Fig. 6.6b. From the equilibrium equations,

$$\Sigma F_x = A_x = 0,$$

$$\Sigma F_y = A_y + E - 400 - 800 = 0,$$

$$\Sigma M_{(\text{point } A)} = -(1)(400) - (3)(800) + 4E = 0,$$

we obtain the reactions $A_x = 0$, $A_y = 500$ N, and $E = 700$ N.

Our next step is to choose a joint and draw its free-body diagram. In Fig. 6.7a, we isolate joint A by cutting members AB and AC. The terms T_{AB} and T_{AC} are the axial forces in members AB and AC, respectively. Although the directions of the arrows representing the unknown axial forces can be chosen arbitrarily, notice that we have chosen them so that a member is in tension if we obtain a positive value for the axial force. We feel that consistently choosing the directions in this way helps avoid errors.

Figure 6.7
(a) Obtaining the free-body diagram of joint A.
(b) The axial forces on members AB and AC.
(c) Realistic and simple free-body diagrams of joint A.

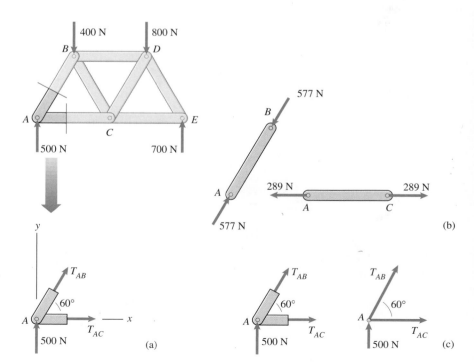

The equilibrium equations for joint A are

$$\Sigma F_x = T_{AC} + T_{AB}\cos 60° = 0,$$

$$\Sigma F_y = T_{AB}\sin 60° + 500 = 0.$$

Solving these equations, we obtain the axial forces $T_{AB} = -577\,\text{N}$ and $T_{AC} = 289\,\text{N}$. Member AB is in compression, and member AC is in tension (Fig. 6.7b).

Although we use a realistic figure for the joint in Fig. 6.7a to help you understand the free-body diagram, in your own work you can use a simple figure showing only the forces acting on the joint (Fig. 6.7c).

We next obtain a free-body diagram of joint B by cutting members AB, BC, and BD (Fig. 6.8a). From the equilibrium equations for joint B,

$$\Sigma F_x = T_{BD} + T_{BC}\cos 60° + 577\cos 60° = 0,$$

$$\Sigma F_y = -400 + 577\sin 60° - T_{BC}\sin 60° = 0,$$

we obtain $T_{BC} = 115\,\text{N}$ and $T_{BD} = -346\,\text{N}$. Member BC is in tension, and member BD is in compression (Fig. 6.8b). By continuing to draw free-body diagrams of the joints, we can determine the axial forces in all of the members.

In two dimensions, you can obtain only two independent equilibrium equations from the free-body diagram of a joint. Summing the moments about a point does not result in an additional independent equation because the forces are concurrent. Therefore when applying the method of joints, you should choose joints to analyze that are subjected to no more than two unknown forces. In our example, we analyzed joint A first because it was subjected to the known reaction exerted by the pin support and two unknown forces, the axial forces T_{AB} and T_{AC} (Fig. 6.7a). We could then analyze joint B because it was subjected to two known forces and two unknown forces, T_{BC} and T_{BD} (Fig. 6.8a). If we had attempted to analyze joint B first, there would have been three unknown forces.

When you determine the axial forces in the members of a truss, your task will often be simpler if you are familiar with three particular types of joints.

- **Truss joints with two collinear members and no load** (Fig. 6.9). The sum of the forces must equal zero, $T_1 = T_2$. The axial forces are equal.

- **Truss joints with two noncollinear members and no load** (Fig. 6.10). Because the sum of the forces in the x direction must equal zero, $T_2 = 0$. Therefore T_1 must also equal zero. The axial forces are zero.

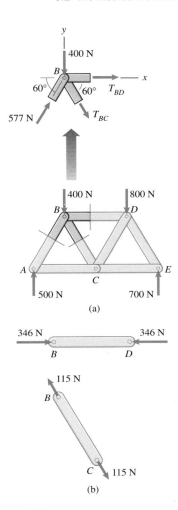

(a)

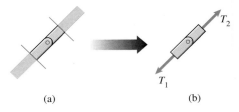

(b)

Figure 6.8
(a) Obtaining the free-body diagram of joint B.
(b) Axial forces in members BD and BC.

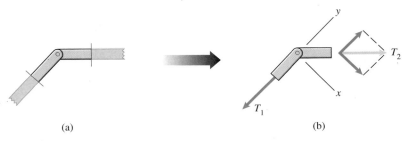

(a)   (b)

Figure 6.9
(a) A joint with two collinear members and no load.
(b) Free-body diagram of the joint.

Figure 6.10
(a) A joint with two noncollinear members and no load.
(b) Free-body diagram of the joint.

• **Truss joints with three members, two of which are collinear, and no
load** (Fig. 6.11). Because the sum of the forces in the x direction must
equal zero, $T_3 = 0$. The sum of the forces in the y direction must equal
zero, so $T_1 = T_2$. The axial forces in the collinear members are equal, and
the axial force in the third member is zero.

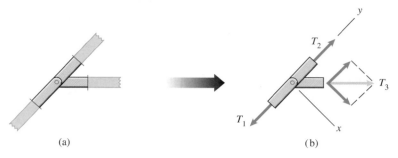

Figure 6.11
(a) A joint with three members, two of
   which are collinear, and no load.
(b) Free-body diagram of the joint.

(a)

(b)

Study Questions

1. What is a truss?
2. What is the method of joints?
3. How many independent equilibrium equations can you obtain from the free-
    body diagram of a joint?

Example 6.1

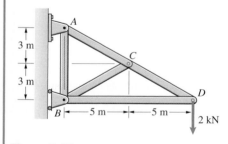

Figure 6.12

Applying the Method of Joints

Determine the axial forces in the members of the truss in Fig. 6.12.

Solution

Determine the Reactions at the Supports   We first draw the free-body
diagram of the entire truss (Fig. a). From the equilibrium equations,

$$\Sigma F_x = A_x + B = 0,$$

$$\Sigma F_y = A_y - 2 = 0,$$

$$\Sigma M_{(\text{point } B)} = -6A_x - (10)(2) = 0,$$

we obtain the reactions $A_x = -3.33$ kN, $A_y = 2$ kN, and $B = 3.33$ kN.

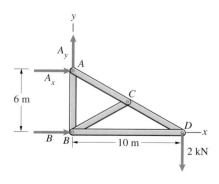

(a) Free-body
diagram of the entire
truss.

Identify Special Joints   Because joint C has three members, two of which are collinear, and no load, the axial force in member BC is zero, $T_{BC} = 0$, and the axial forces in the collinear members AC and CD are equal, $T_{AC} = T_{CD}$.

Draw Free-Body Diagrams of the Joints   We know the reaction exerted on joint A by the support, and joint A is subjected to only two unknown forces, the axial forces in members AB and AC. We draw its free-body diagram in Fig. b. The angle $\alpha = \arctan(5/3) = 59.0°$. The equilibrium equations for joint A are

$$\Sigma F_x = T_{AC} \sin \alpha - 3.33 = 0,$$

$$\Sigma F_y = 2 - T_{AB} - T_{AC} \cos \alpha = 0.$$

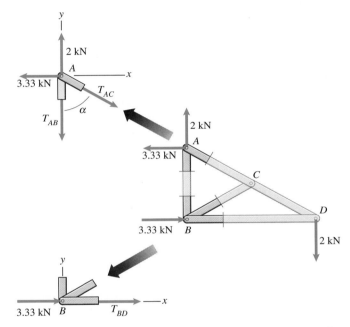

(b) Free-body diagram of joint A.

(c) Free-body diagram of joint B.

Solving these equations, we obtain $T_{AB} = 0$ and $T_{AC} = 3.89$ kN. Because the axial forces in members AC and CD are equal, $T_{CD} = 3.89$ kN.

Now we draw the free-body diagram of joint B in Fig. c. (We already know that the axial forces in members AB and BC are zero.) From the equilibrium equation

$$\Sigma F_x = T_{BD} + 3.33 = 0,$$

we obtain $T_{BD} = -3.33$ kN. The negative sign indicates that member BD is in compression.

The axial forces in the members are

AB: 0,

AC: 3.89 kN in tension (T),

BC: 0,

BD: 3.33 kN in compression (C),

CD: 3.89 kN in tension (T).

Example 6.2

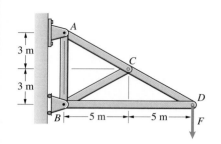

Figure 6.13

Determining the Largest Force a Truss Will Support

Each member of the truss in Fig. 6.13 will safely support a tensile force of 10 kN and a compressive force of 2 kN. What is the largest downward load F that the truss will safely support?

Strategy

This truss is identical to the one we analyzed in Example 6.1. By applying the method of joints in the same way, the axial forces in the members can be determined in terms of the load F. The smallest value of F that will cause a tensile force of 10 kN or a compressive force of 2 kN in any of the members is the largest value of F that the truss will support.

Solution

By using the method of joints in the same way as in Example 6.1, we obtain the axial forces

$$AB: 0,$$
$$AC: 1.94F \text{ (T)},$$
$$BC: 0,$$
$$BD: 1.67F \text{ (C)},$$
$$CD: 1.94F \text{ (T)}.$$

For a given load F, the largest tensile force is $1.94F$ (in members AC and CD) and the largest compressive force is $1.67F$ (in member BD). The largest safe tensile force would occur when $1.94F = 10$ kN or when $F = 5.14$ kN. The largest safe compressive force would occur when $1.67F = 2$ kN or when $F = 1.20$ kN. Therefore the largest load F that the truss will safely support is 1.20 kN.

Example 6.3

Application to Engineering:

Bridge Design

The loads a bridge structure must support and pin supports where the structure is to be attached are shown in Fig. 6.14(1). Assigned to design the structure, a civil engineering student proposes the structure shown in Fig. 6.14(2). What are the axial forces in the members?

Solution

The vertical members AG, BH, CI, DJ, and EK are subjected to compressive forces of magnitude F. From the free-body diagram of joint C, we obtain $T_{BC} = T_{CD} = -1.93F$. We draw the free-body diagram of joint B in Fig. a.

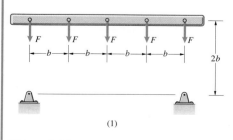

Figure 6.14

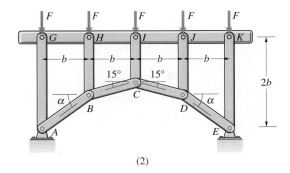

(2)

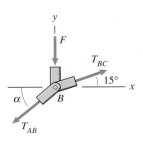

(a) Free-body diagram of joint B.

From the equilibrium equations

$$\Sigma F_x = -T_{AB} \cos\alpha + T_{BC} \cos 15° = 0,$$

$$\Sigma F_y = -T_{AB} \sin\alpha + T_{BC} \sin 15° - F = 0,$$

we obtain $T_{AB} = -2.39F$ and $\alpha = 38.8°$. By symmetry, $T_{DE} = T_{AB}$. The axial forces in the members are shown in Table 6.1.

Table 6.1 Axial forces in the members of the bridge structure.

Members	Axial Force	
AG, BH, CI, DJ, EK	F	(C)
AB, DE	$2.39F$	(C)
BC, CD	$1.93F$	(C)

𝒟esign Issues

The bridge was an early application of engineering. Although initially the solution was as primitive as laying a log between the banks, engineers constructed surprisingly elaborate bridges in the remote past. For example, archaeologists have identified foundations of the seven piers of a 120-m (400-ft) highway bridge over the Euphrates that existed in Babylon at the time of Nebuchadnezzar II (reigned 605–562 B.C.).

The basic difficulty in bridge design is that a single beam extended between the banks will fail if the distance between banks, or *span*, is too large. To meet the need for bridges of increasing strength and span, civil engineers created ingenious and aesthetic designs in antiquity and continue to do so today.

The bridge structure proposed by the student in Example 6.3, called an *arch*, is an ancient design. Notice in Table 6.1 that all the members of the structure are in compression. Because masonry (stone, brick, or concrete) is weak in tension but very strong in compression, many bridges made of these materials were designed with arched spans in the past. For the same reason, modern concrete bridges are often built with arched spans (Fig. 6.15).

Figure 6.15
A bridge along Highway 1 in California that is supported by concrete arches anchored in rock.

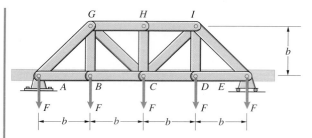

Figure 6.16
A Pratt truss supporting a bridge.

Figure 6.17
The Forth Bridge (Scotland, 1890) is an example of a large truss bridge. Each main span is 520 m long.

Table 6.2 Axial forces in the members of the Pratt truss.

Members	Axial Force	
AB, BC, CD, DE	1.5F	(T)
AG, EI	2.12F	(C)
CG, CI	0.71F	(T)
GH, HI	2F	(C)
BG, DI	F	(T)
CH	0	

Table 6.3 Axial forces in the members of the suspension structure.

Members	Axial Force	
BH, CI, DJ	F	(T)
AB, DE	2.39F	(T)
BC, CD	1.93F	(T)

Unlike masonry, wood and steel can support substantial forces in both compression and tension. Beginning with the wooden truss bridges designed by the architect Andrea Palladio (1518–1580), both of these materials have been used to construct a large variety of trusses to support bridges. For example, the forces in Fig. 6.14(1) can be supported by the Pratt truss shown in Fig. 6.16. Its members are subjected to both tension and compression (Table 6.2). The Forth Bridge (Fig. 6.17) has a truss structure.

Truss structures are too heavy for the largest bridges. (The Forth Bridge contains 58,000 tons of steel.) By taking advantage of the ability of relatively light cables to support large tensile forces, civil engineers use suspension structures to bridge very large spans. The system of five forces we are using as an example can be supported by the simple suspension structure in Fig. 6.18. In effect, the compression arch used since antiquity is inverted. (Compare Figs. 6.14(2) and 6.18.) The loads in Fig. 6.18 are "suspended" from members AB, BC, CD, and DE. Every member of this structure except the towers AG and EK is in tension (Table 6.3). The largest existing bridges, such as the Golden Gate Bridge (Fig. 6.19), consist of cable-suspended spans supported by towers.

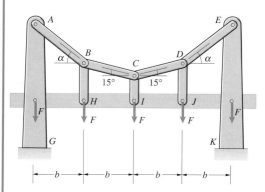

Figure 6.18
A suspension structure supporting a bridge.

Figure 6.19
The Golden Gate Bridge (California) has a central suspended span 1280 m (4200 ft) in length.

6.3 The Method of Sections

When we need to know the axial forces only in certain members of a truss, we often can determine them more quickly using the method of sections than using the method of joints. For example, let's reconsider the Warren truss we used to introduce the method of joints (Fig. 6.20a). It supports loads at B and D, and each member is 2 m in length. Suppose that we need to determine only the axial force in member BC.

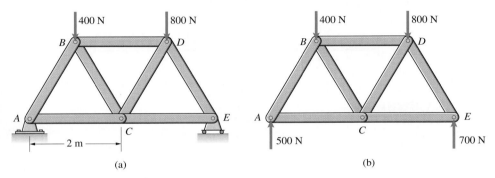

Figure 6.20
(a) A Warren truss supporting two loads.
(b) Free-body diagram of the truss, showing the reactions at the supports.

Just as in the method of joints, we begin by drawing a free-body diagram of the entire truss and determining the reactions at the supports. The results of this step are shown in Fig. 6.20b. Our next step is to cut the members AC, BC, and BD to obtain a free-body diagram of a part, or **section**, of the truss (Fig. 6.21). Summing moments about point B, the equilibrium equations for the section are

$$\Sigma F_x = T_{AC} + T_{BD} + T_{BC} \cos 60° = 0,$$

$$\Sigma F_y = 500 - 400 - T_{BC} \sin 60° = 0,$$

$$\Sigma M_{(\text{point } B)} = T_{AC}(2 \sin 60°) - (500)(2 \cos 60°) = 0.$$

Solving them, we obtain $T_{AC} = 289$ N, $T_{BC} = 115$ N, and $T_{BD} = -346$ N.

Notice how similar this method is to the method of joints. Both methods involve cutting members to obtain free-body diagrams of parts of a truss. In the method of joints, we move from joint to joint, drawing free-body diagrams of the joints and determining the axial forces in the members as we go. In the method of sections, we try to obtain a single free-body diagram that allows us to determine the axial forces in specific members. In our example, we obtained a free-body diagram by cutting three members, including the one (member BC) whose axial force we wanted to determine.

In contrast to the free-body diagrams of joints, the forces on the free-body diagrams used in the method of sections are not usually concurrent, and as in our example, we can obtain three independent equilibrium equations. Although there are exceptions, it is usually necessary to choose a section that requires cutting no more than three members, or there will be more unknown axial forces than equilibrium equations.

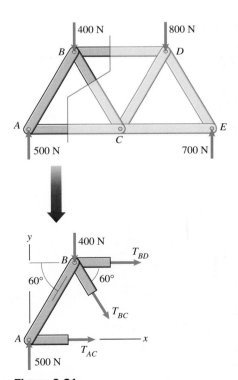

Figure 6.21
Obtaining a free-body diagram of a section of the truss.

Example 6.4

Applying the Method of Sections

The truss in Fig. 6.22 supports a 100-kN load. The horizontal members are each 1 m in length. Determine the axial force in member CJ, and state whether it is in tension or compression.

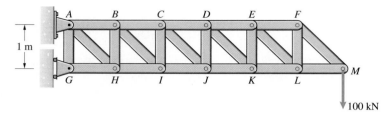

Figure 6.22

100 kN

Strategy

We need to obtain a section by cutting members that include member CJ. By cutting members CD, CJ, and IJ, we will obtain a free-body diagram with three unknown axial forces.

Solution

To obtain a section (Fig. a), we cut members CD, CJ, and IJ and draw the free-body diagram of the part of the truss on the right side of the cuts. From the equilibrium equation

$$\Sigma F_y = T_{CJ} \sin 45° - 100 = 0,$$

we obtain $T_{CJ} = 141.4$ kN. The axial force in member CJ is 141.4 kN (T).

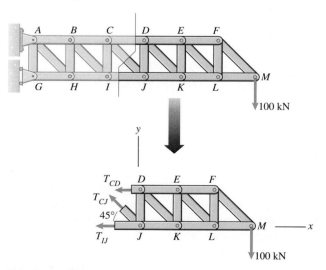

(a) Obtaining the section.

Discussion

Notice that by using the section on the right side of the cuts, we did not need to determine the reactions at the supports A and G.

Example 6.5

Choosing an Appropriate Section

Determine the axial forces in members DG and BE of the truss in Fig. 6.23.

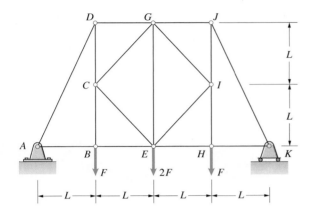

Figure 6.23

Strategy

An appropriate choice of section is not obvious, and it isn't clear beforehand that we can determine the requested information by the method of sections. We can't obtain a section that involves cutting members DG and BE without cutting more than three members. However, cutting members DG, BE, CD, and BC results in a section with which we can determine the axial forces in members DG and BE even though the resulting free-body diagram is statically indeterminate.

Solution

Determine the Reactions at the Supports   We draw the free-body diagram of the entire truss in Fig. a. From the equilibrium equations,

$$\Sigma F_x = A_x = 0,$$
$$\Sigma F_y = A_y + K - F - 2F - F = 0,$$
$$\Sigma M_{(\text{point } A)} = -FL - 2F(2L) - F(3L) + K(4L) = 0,$$

we obtain the reactions $A_x = 0, A_y = 2F$, and $K = 2F$.

Choose a Section   In Fig. b, we obtain a section by cutting members DG, CD, BC, and BE. Because the lines of action of T_{BE}, T_{BC}, and T_{CD} pass through point B, we can determine T_{DG} by summing moments about B:

$$\Sigma M_{(\text{point } B)} = -2FL - T_{DG}(2L) = 0.$$

The axial force $T_{DG} = -F$. Then from the equilibrium equation

$$\Sigma F_x = T_{DG} + T_{BE} = 0,$$

we see that $T_{BE} = -T_{DG} = F$. Member DG is in compression, and member BE is in tension.

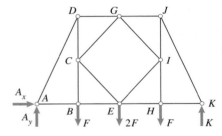

(a) Free-body diagram of the entire truss.

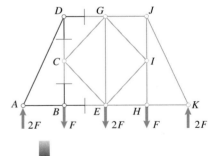

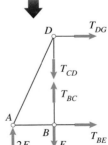

(b) A section of the truss obtained by passing planes through members DG, CD, BC, and BE.

6.4 Space Trusses

We can form a simple three-dimensional structure by connecting six bars at their ends to obtain a tetrahedron, as shown in Fig. 6.24a. By adding members, we can obtain more elaborate structures (Figs. 6.24b and c). Three-dimensional structures such as these are called *space trusses* if they have joints that do not exert couples on the members (that is, the joints behave like ball and socket supports) and they are loaded and supported at their joints. Space trusses are analyzed by the same methods we described for two-dimensional trusses. The only difference is the need to cope with the more complicated geometry.

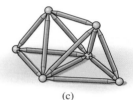

(a) (b) (c)

Figure 6.24
Space trusses with 6, 9, and 12 members.

Consider the space truss in Fig. 6.25a. Suppose that the load $\mathbf{F} = -2\mathbf{i} - 6\mathbf{j} - \mathbf{k}$ (kN). The joints A, B, and C rest on the smooth floor. Joint A is supported by the corner where the smooth walls meet, and joint C rests against the back wall. We can apply the method of joints to this truss.

First we must determine the reactions exerted by the supports (the floor and walls). We draw the free-body diagram of the entire truss in Fig. 6.25b. The corner can exert three components of force at A, the floor and wall can exert two components of force at C, and the floor can exert a normal force at B. Summing moments about A, the equilibrium equations are

$$\Sigma F_x = A_x - 2 = 0,$$

$$\Sigma F_y = A_y + B_y + C_y - 6 = 0,$$

$$\Sigma F_z = A_z + C_z - 1 = 0,$$

$$\Sigma M_{(\text{point } A)} = (\mathbf{r}_{AB} \times B_y\mathbf{j}) + \left[\mathbf{r}_{AC} \times (C_y\mathbf{j} + C_z\mathbf{k})\right] + (\mathbf{r}_{AD} \times \mathbf{F})$$

$$= \begin{vmatrix} \mathbf{i} & \mathbf{j} & \mathbf{k} \\ 2 & 0 & 3 \\ 0 & B_y & 0 \end{vmatrix} + \begin{vmatrix} \mathbf{i} & \mathbf{j} & \mathbf{k} \\ 4 & 0 & 0 \\ 0 & C_y & C_z \end{vmatrix}$$

$$+ \begin{vmatrix} \mathbf{i} & \mathbf{j} & \mathbf{k} \\ 2 & 3 & 1 \\ -2 & -6 & -1 \end{vmatrix}$$

$$= (-3B_y + 3)\mathbf{i} + (-4C_z)\mathbf{j}$$

$$+ (2B_y + 4C_y - 6)\mathbf{k} = 0.$$

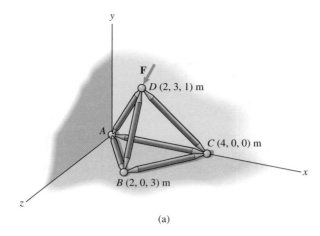

(a)

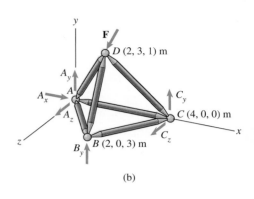

(b)

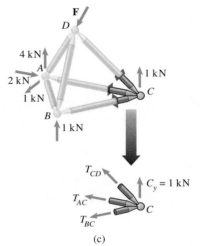

(c)

Figure 6.25
(a) A space truss supporting a load **F**.
(b) Free-body diagram of the entire truss.
(c) Obtaining the free-body diagram of joint C.

Solving these equations, we obtain the reactions $A_x = 2$ kN, $A_y = 4$ kN, $A_z = 1$ kN, $B_y = 1$ kN, $C_y = 1$ kN, and $C_z = 0$.

In this example, we can determine the axial forces in members AC, BC, and CD from the free-body diagram of joint C (Fig. 6.25c). To write the equilibrium equations for the joint, we must express the three axial forces in terms of their components. Because member AC lies along the x axis, we express the force exerted on joint C by the axial force T_{AC} as the vector $-T_{AC}\mathbf{i}$. Let $\mathbf{r}_{CB}$ be the position vector from C to B:

$$\mathbf{r}_{CB} = (2 - 4)\mathbf{i} + (0 - 0)\mathbf{j} + (3 - 0)\mathbf{k} = -2\mathbf{i} + 3\mathbf{k} \text{ (m)}.$$

Dividing this vector by its magnitude to obtain a unit vector that points from C toward B,

$$\mathbf{e}_{CB} = \frac{\mathbf{r}_{CB}}{|\mathbf{r}_{CB}|} = -0.555\mathbf{i} + 0.832\mathbf{k},$$

we express the force exerted on joint C by the axial force T_{BC} as the vector

$$T_{BC}\mathbf{e}_{CB} = T_{BC}(-0.555\mathbf{i} + 0.832\mathbf{k}).$$

In the same way, we express the force exerted on joint C by the axial force T_{CD} as the vector

$$T_{CD}(-0.535\mathbf{i} + 0.802\mathbf{j} + 0.267\mathbf{k}).$$

Setting the sum of the forces on the joint equal to zero,

$$-T_{AC}\mathbf{i} + T_{BC}(-0.555\mathbf{i} + 0.832\mathbf{k})$$

$$+ T_{CD}(-0.535\mathbf{i} + 0.802\mathbf{j} + 0.267\mathbf{k}) + (1)\mathbf{j} = 0,$$

we obtain the three equilibrium equations

$$\Sigma F_x = -T_{AC} - 0.555T_{BC} - 0.535T_{CD} = 0,$$

$$\Sigma F_y = 0.802T_{CD} + 1 = 0,$$

$$\Sigma F_z = 0.832T_{BC} + 0.267T_{CD} = 0.$$

Solving these equations, the axial forces are $T_{AC} = 0.444$ kN, $T_{BC} = 0.401$ kN, and $T_{CD} = -1.247$ kN. Members AC and BC are in tension, and member CD is in compression. By continuing to draw free-body diagrams of the joints, we can determine the axial forces in all the members.

As our example demonstrates, three equilibrium equations can be obtained from the free-body diagram of a joint in three dimensions, so it is usually necessary to choose joints to analyze that are subjected to known forces and no more than three unknown forces.

6.5 Frames and Machines

Many structures, such as the frame of a car and the human structure of bones, tendons, and muscles (Fig. 6.26), are not composed entirely of two-force members and thus cannot be modeled as trusses. In this section we consider structures of interconnected members that do not satisfy the definition of a truss. Such structures are called *frames* if they are designed to remain stationary and support loads and *machines* if they are designed to move and apply loads.

When trusses are analyzed by cutting members to obtain free-body diagrams of joints or sections, the internal forces acting at the "cuts" are simple axial forces (see Fig. 6.4). This is not generally true for frames or machines, and a different method of analysis is necessary. Instead of cutting members, you isolate entire members, or in some cases combinations of members, from the structure.

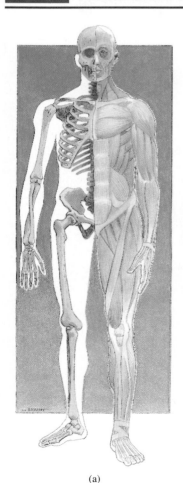

(a)

(b)

Figure 6.26
The internal structure of a person (a) and a car's frame (b) are not trusses.

To begin analyzing a frame or machine, we draw a free-body diagram of the entire structure (that is, treat the structure as a single object) and determine the reactions at its supports. In some cases the entire structure will be statically indeterminate, but it is helpful to determine as many of the reactions as possible. We then draw free-body diagrams of individual members, or selected combinations of members, and apply the equilibrium equations to determine the forces and couples acting on them. For example, let's consider the stationary structure in Fig. 6.27. Member BE is a two-force member, but the other three members—ABC, CD, and DEG—are not. This structure is a frame. Our objective is to determine the forces on its members.

Analyzing the Entire Structure

We draw the free-body diagram of the entire frame in Fig. 6.28. It is statically indeterminate: There are four unknown reactions, A_x, A_y, G_x, and G_y, whereas we can write only three independent equilibrium equations. However, notice that the lines of action of three of the unknown reactions intersect at A. By summing moments about A,

$$\Sigma M_{(\text{point } A)} = 2G_x + (1)(8) - (3)(6) = 0,$$

we obtain the reaction $G_x = 5$ kN. Then from the equilibrium equation

$$\Sigma F_x = A_x + G_x + 8 = 0,$$

we obtain the reaction $A_x = -13$ kN. Although we cannot determine A_y or G_y from the free-body diagram of the entire structure, we can do so by analyzing the individual members.

Analyzing the Members

Our next step is to draw free-body diagrams of the members. To do so, we treat the attachment of a member to another member just as if it were a support. Looked at in this way, we can think of each member as a supported object of the kind analyzed in Chapter 5. Furthermore, the forces and couples the members exert on one another are *equal in magnitude and opposite in direction*. A simple demonstration is instructive. If you clasp your hands as shown in Fig. 6.29a and exert a force on your left hand with your right hand, your left hand exerts an equal and opposite force on your right hand (Fig. 6.29b). Similarly, if you exert a couple on your left hand, your left hand exerts an equal and opposite couple on your right hand.

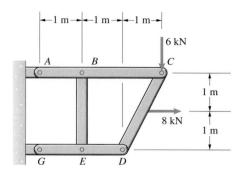

Figure 6.27
A frame supporting two loads.

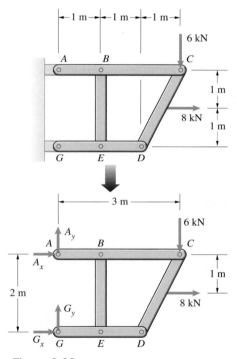

Figure 6.28
Obtaining the free-body diagram of the entire frame.

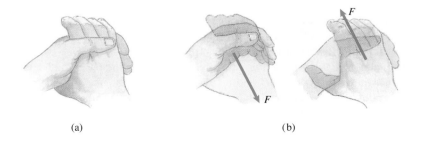

Figure 6.29
Demonstrating Newton's third law:
(a) Clasp your hands and pull on your left hand.
(b) Your hands exert equal and opposite forces.

In Fig. 6.30 we "disassemble" the frame and draw free-body diagrams of its members. Observe that the forces exerted on one another by the members are equal and opposite. For example, at point C on the free-body diagram of member ABC, the force exerted by member CD is denoted by the components C_x, and C_y. We can choose the directions of these unknown forces arbitrarily, but once we have done so, the forces exerted by member ABC on member CD at point C must be equal and opposite, as shown.

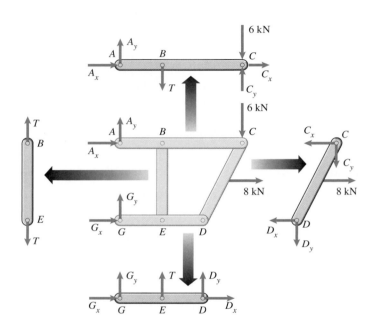

Figure 6.30
Obtaining the free-body diagrams of the members.

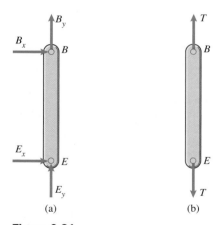

Figure 6.31
Free-body diagram of member BE:
(a) Not treating it as a two-force member.
(b) Treating it as a two-force member.

We need to discuss two important aspects of these free-body diagrams before completing the analysis.

Two-Force Members Member BE is a two-force member, and we have taken this into account in drawing its free-body diagram in Fig. 6.30. The force T is the axial force in member BE, and an equal and opposite force is subjected on member ABC at B and on member GED at E.

Recognizing two-force members in frames and machines and drawing their free-body diagrams as we have done will reduce the number of un-knowns and will greatly simplify the analysis. In our example, if we did not treat member BE as a two-force member, its free-body diagram would have four unknown forces (Fig. 6.31a). By treating it as a two-force member (Fig. 6.31b), we reduce the number of unknown forces by three.

Loads Applied at Joints A question arises when a load is applied at a joint: Where does the load appear on the free-body diagrams of the individual members? The answer is that you can place the load on *any one* of the members attached at the joint. For example, in Fig. 6.27, the 6-kN load acts at the joint where members ABC and CD are connected. In drawing the free-body diagrams of the individual members (Fig. 6.30), we assumed that the 6-kN load acted on member ABC. The force components C_x and C_y on the free-body diagram of member ABC are the forces exerted by the member CD.

To explain why we can draw the free-body diagrams in this way, let us assume that the 6-kN force acts on the pin connecting members ABC and CD, and draw separate free-body diagrams of the pin and the two members (Fig. 6.32a). The force components C'_x and C'_y are the forces exerted by the pin on member ABC, and C_x and C_y are the forces exerted by the pin on member CD. If we superimpose the free-body diagrams of the pin and member ABC, we obtain the two free-body diagrams in Fig. 6.32b, which is the way we drew them in Fig. 6.30. Alternatively, by superimposing the free-body diagrams of the pin and member CD, we obtain the two free-body diagrams in Fig. 6.32c.

Thus if a load acts at a joint, it can be placed on any one of the members attached at the joint when drawing the free-body diagrams of the individual members. Just make sure not to place it on more than one member.

To detect errors in the free-body diagrams of the members, it is helpful to "reassemble" them (Fig. 6.33a). The forces at the connections between the members cancel (they are internal forces once the members are reassembled), and the free-body diagram of the entire structure is recovered (Fig. 6.33b).

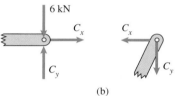

(a)

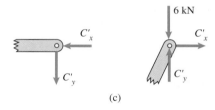

(b)

(c)

Figure 6.32

(a) Drawing free-body diagrams of the pin and the two members.

(b) Superimposing the pin on member ABC.

(c) Superimposing the pin on member CD.

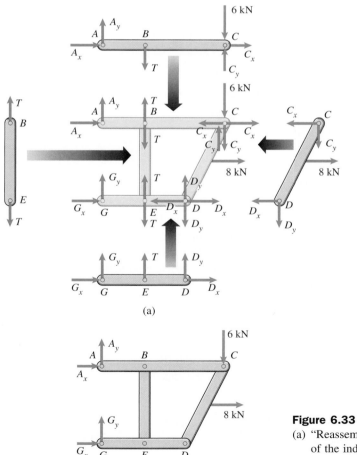

(a)

(b)

Figure 6.33

(a) "Reassembling" the free-body diagrams of the individual members.

(b) The free-body diagram of the entire frame is recovered.

Our final step is to apply the equilibrium equations to the free-body diagrams of the members (Fig. 6.34). In two dimensions, we can obtain three independent equilibrium equations from the free-body diagram of each member of a structure that we do not treat as a two-force member. (By assuming that the forces on a two-force member are equal and opposite axial forces, we have already used the three equilibrium equations for that member.) In this example, there are three members in addition to the two-force member, so we can write $(3)(3) = 9$ independent equilibrium equations, and there are 9 unknown forces: $A_x, A_y, C_x, C_y, D_x, D_y, G_x, G_y,$ and T.

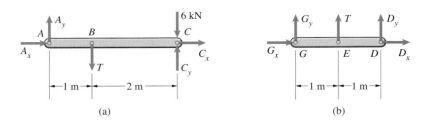

(a)   (b)

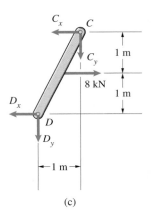

(c)

Figure 6.34
Free-body diagrams of the members.

Recall that we determined that $A_x = -13$ kN and $G_x = 5$ kN from our analysis of the entire structure. The equilibrium equations we obtained from the free-body diagram of the entire structure are not independent of the equilibrium equations obtained from the free-body diagrams of the members, but by using them to determine A_x and G_x, we get a head start on solving the equations for the members. Consider the free-body diagram of member ABC (Fig. 6.34a). Because we know A_x, we can determine C_x from the equation

$$\Sigma F_x = A_x + C_x = 0,$$

obtaining $C_x = -A_x = 13$ kN. Now consider the free-body diagram of GED (Fig. 6.34b). We can determine D_x from the equation

$$\Sigma F_x = G_x + D_x = 0,$$

obtaining $D_x = -G_x = -5$ kN. Now consider the free-body diagram of member CD (Fig. 6.34c). Because we know C_x, we can determine C_y by summing moments about D:

$$\Sigma M_{(\text{point } D)} = (2)C_x - (1)C_y - (1)(8) = 0.$$

We obtain $C_y = 18$ kN. Then from the equation

$$\Sigma F_y = -C_y - D_y = 0,$$

we find that $D_y = -C_y = -18$ kN. Now we can return to the free-body diagrams of members ABC and GED to determine A_y and G_y. Summing moments about point B of member ABC,

$$\Sigma M_{(\text{point } B)} = -(1)A_y + (2)C_y - (2)(6) = 0,$$

we obtain $A_y = 2C_y - 12 = 24$ kN. Then by summing moments about point E of member GED,

$$\Sigma M_{(\text{point } E)} = (1)D_y - (1)G_y = 0,$$

we obtain $G_y = D_y = -18$ kN. Finally, from the free-body diagram of member GED, we use the equilibrium equation

$$\Sigma F_y = D_y + G_y + T = 0,$$

which gives us the result $T = -D_y - G_y = 36$ kN. The forces on the members are shown in Fig. 6.35. As this example demonstrates, determination of the forces on the members can often be simplified by carefully choosing the order in which the equations are solved.

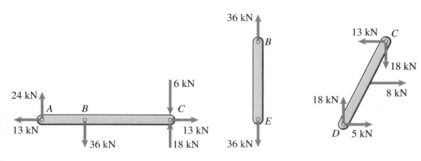

Figure 6.35
Forces on the members of the frame.

We see that determining the forces and couples on the members of frames and machines requires two steps:

1. Determine the reactions at the supports—Draw the free-body diagram of the entire structure, and determine the reactions at its supports. This step can greatly simplify your analysis of the members. If the free-body diagram is statically indeterminant, determine as many of the reactions as possible.

2. Analyze the members—Draw free-body diagrams of the members, and apply the equilibrium equations to determine the forces acting on them. You can simplify this step by identifying two-force members. If a load acts at a joint of the structure, you can place the load on the free-body diagram of any one of the members attached at that joint.

Example 6.6

Analyzing a Frame

The frame in Fig. 6.36 is subjected to a 200-N-m couple. Determine the forces and couples on its members.

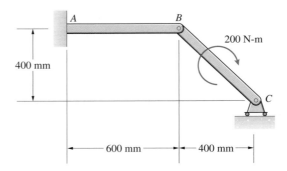

Figure 6.36

Solution

Determine the Reactions at the Supports   We draw the free-body diagram of the entire frame in Fig. a. The term M_A is the couple exerted by the built-in support. From the equilibrium equations

$$\Sigma F_x = A_x = 0,$$

$$\Sigma F_y = A_y + C = 0,$$

$$\Sigma M_{(\text{point } A)} = M_A - 200 + (1)C = 0,$$

we obtain the reaction $A_x = 0$. We can't determine A_y, M_A, or C from this free-body diagram.

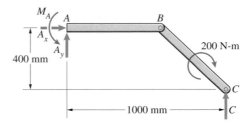

(a) Free-body diagram of the entire frame.

Analyze the Members   We "disassemble" the frame to obtain the free-body diagrams of the members in Fig. b. The equilibrium equations for member BC are

$$\Sigma F_x = -B_x = 0,$$

$$\Sigma F_y = -B_y + C = 0,$$

$$\Sigma M_{(\text{point } B)} = -200 + (0.4)C = 0.$$

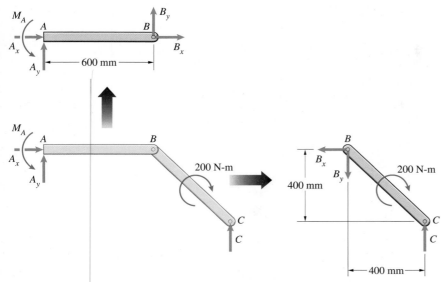

(b) Obtaining the free-body diagrams of the members.

Solving these equations, we obtain $B_x = 0$, $B_y = 500$ N, and $C = 500$ N. The equilibrium equations for member AB are

$$\Sigma F_x = A_x + B_x = 0,$$
$$\Sigma F_y = A_y + B_y = 0,$$
$$\Sigma M_{(\text{point } A)} = M_A + (0.6)B_y = 0.$$

Because we already know A_x, B_x, and B_y, we can solve these equations for A_y and M_A. The results are $A_y = -500$ N and $M_A = -300$ N-m. This completes the solution (Fig. c).

Discussion

We were able to solve the equilibrium equations for member BC without having to consider the free-body diagram of member AB. We were then able to solve the equilibrium equations for member AB. By choosing the members with the fewest unknowns to analyze first, you will often be able to solve them sequentially, but in some cases you will have to solve the equilibrium equations for the members simultaneously.

Even though we were unable to determine the four reactions A_x, A_y, M_A, and C with the three equilibrium equations obtained from the free-body diagram of the entire frame, we were able to determine them from the free-body diagrams of the individual members. By drawing free-body diagrams of the members, we gained three equations because we obtained three equilibrium equations from each member but only two new unknowns, B_x and B_y.

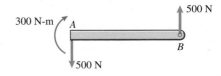

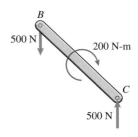

(c) Forces and couples on the members.

Example 6.7

Determining Forces on Members of a Frame

The frame in Fig. 6.37 supports a suspended weight $W = 40$ lb. Determine the forces on members $ABCD$ and CEG.

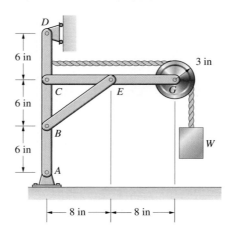

Figure 6.37

Solution

Determine the Reactions at the Supports We draw the free-body diagram of the entire frame in Fig. a. From the equilibrium equations

$$\Sigma F_x = A_x - D = 0,$$

$$\Sigma F_y = A_y - 40 = 0,$$

$$\Sigma M_{(\text{point } A)} = (18)D - (19)(40) = 0,$$

we obtain the reactions $A_x = 42.2$ lb, $A_y = 40$ lb, and $D = 42.2$ lb.

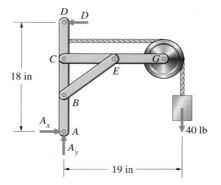

(a) Free-body diagram of the entire frame.

Analyze the Members We obtain the free-body diagrams of the members in Fig. b. Notice that BE is a two-force member. The angle $\alpha = \arctan(6/8) = 36.9°$.

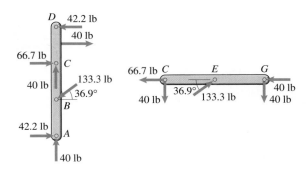

(b) Obtaining the free-body diagrams of the members.

The free-body diagram of the pulley has only two unknown forces. From the equilibrium equations

$$\Sigma F_x = G_x - 40 = 0,$$

$$\Sigma F_y = G_y - 40 = 0,$$

we obtain $G_x = 40$ lb and $G_y = 40$ lb. There are now only three unknown forces on the free-body diagram of member CEG. From the equilibrium equations

$$\Sigma F_x = -C_x - R\cos\alpha - 40 = 0,$$

$$\Sigma F_y = -C_y - R\sin\alpha - 40 = 0,$$

$$\Sigma M_{(\text{point } C)} = -(8)R\sin\alpha - (16)(40) = 0,$$

we obtain $C_x = 66.7$ lb, $C_y = 40$ lb, and $R = -133.3$ lb, completing the solution (Fig. c).

(c) Forces on members $ABCD$ and CEG.

Example 6.8

Free-Body Diagrams for Three Joined Members

Determine the forces on the members of the frame in Fig. 6.38.

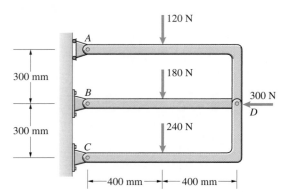

Figure 6.38

Strategy

You can confirm that no information can be obtained from the free-body diagram of the entire frame. To analyze the members, we must deal with an interesting challenge at joint D, where a load acts and three members are connected. We will obtain the free-body diagrams of the members by first isolating member AD, then separating members BD and CD.

Solution

We first isolate member AD from the rest of the structure, introducing the reactions D_x and D_y (Fig. a). We then separate members BD and CD, introducing equal and opposite forces E_x and E_y (Fig. b). In this step we could have placed the 300-N load and the forces D_x and D_y on either free-body diagram.

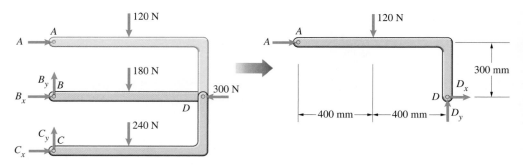

(a) Isolating member AD.

Only three unknown forces act on member AD. From the equilibrium equations

$$\Sigma F_x = A + D_x = 0,$$

$$\Sigma F_y = D_y - 120 = 0,$$

$$\Sigma M_{(\text{point } D)} = -(0.3)A + (0.4)(120) = 0,$$

we obtain $A = 160$ N, $D_x = -160$ N, and $D_y = 120$ N. Now we consider the free-body diagram of member BD. From the equation

$$\Sigma M_{(\text{point } D)} = -(0.8)B_y + (0.4)(180) = 0,$$

we obtain $B_y = 90$ N. Now we use the equation

$$\Sigma F_y = B_y - D_y + E_y - 180 = 90 - 120 + E_y - 180 = 0,$$

obtaining $E_y = 210$ N. Now that we know E_y, there are only three unknown forces on the free-body diagram of member CD. From the equilibrium equations

$$\Sigma F_x = C_x - E_x = 0,$$

$$\Sigma F_y = C_y - E_y - 240 = C_y - 210 - 240 = 0,$$

$$\Sigma M_{(\text{point } C)} = (0.3)E_x - (0.8)E_y - (0.4)(240)$$

$$= (0.3)E_x - (0.8)(210) - (0.4)(240) = 0,$$

we obtain $C_x = 880$ N, $C_y = 450$ N, and $E_x = 880$ N. Finally, we return to the free-body diagram of member BD and use the equation

$$\Sigma F_x = B_x + E_x - D_x - 300 = B_x + 880 + 160 - 300 = 0$$

to obtain $B_x = -740$ N, completing the solution (Fig. c).

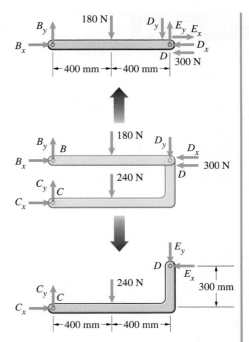

(b) Separating members BD and CD.

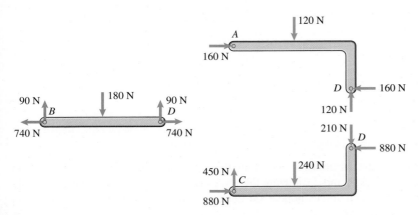

(c) Solutions for the forces on the members.

Example 6.9

Analyzing a Truck and Trailer as a Frame

The truck in Fig. 6.39 is parked on a 10° slope. Its brakes prevent the wheels at B from turning, but the wheels at C and the wheels of the trailer at A can turn freely. The trailer hitch at D behaves like a pin support. Determine the forces exerted on the truck at B, C, and D.

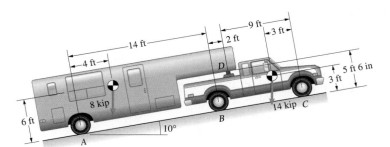

Figure 6.39

Strategy

We can treat this example as a structure whose "members" are the truck and trailer. We must isolate the truck and trailer and draw their individual free-body diagrams to determine the forces acting on the truck.

Solution

Determine the Reactions at the Supports   The reactions in this example are the forces exerted on the truck and trailer by the road. We draw the free-body diagram of the connected truck and trailer in Fig. a. Because the tires at B are locked, the road can exert both a normal force and a friction force, but only normal forces are exerted at A and C. The equilibrium equations are

$$\Sigma F_x = B_x - 8 \sin 10° - 14 \sin 10° = 0,$$

$$\Sigma F_y = A + B_y + C - 8 \cos 10° - 14 \cos 10° = 0,$$

$$\Sigma M_{(\text{point } A)} = 14B_y + 25C + (6)(8 \sin 10°)$$
$$- (4)(8 \cos 10°) + (3)(14 \sin 10°)$$
$$- (22)(14 \cos 10°) = 0.$$

From the first equation we obtain the reaction $B_x = 3.82$ kip, but we can't solve the other two equations for the three reactions A, B_y, and C.

Analyze the Members   We draw the free-body diagrams of the trailer and truck in Figs. b and c, showing the forces D_x and D_y exerted at the hitch. Only three unknown forces appear on the free-body diagram of the trailer. From the equilibrium equations for the trailer,

$$\Sigma F_x = D_x - 8 \sin 10° = 0,$$

$$\Sigma F_y = A + D_y - 8 \cos 10° = 0,$$

$$\Sigma M_{(\text{point } D)} = (0.5)(8 \sin 10°) + (12)(8 \cos 10°) - 16A = 0.$$

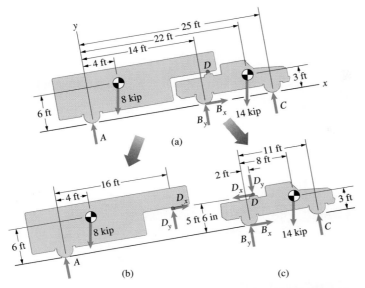

(a) Free-body diagram of the combined truck and trailer.

(b), (c) The individual free-body diagrams.

we obtain $A = 5.95$ kip, $D_x = 1.39$ kip, and $D_y = 1.93$ kip. (Notice that by summing moments about D, we obtained an equation containing only one unknown force.)

The equilibrium equations for the truck are

$$\Sigma F_x = B_x - D_x - 14 \sin 10° = 0,$$

$$\Sigma F_y = B_y + C - D_y - 14 \cos 10° = 0,$$

$$\Sigma M_{(\text{point } B)} = 11C + 5.5D_x - 2D_y + (3)(14 \sin 10°)$$
$$- (8)(14 \cos 10°) = 0.$$

Using the known values of D_x and D_y, we can solve these equations, obtaining $B_x = 3.82$ kip, $B_y = 6.69$ kip, and $C = 9.02$ kip.

Discussion

We were unable to solve two of the equilibrium equations for the connected truck and trailer. When that happens, you can use the equilibrium equations for the entire structure to check your results:

$$\Sigma F_x = B_x - 8 \sin 10° - 14 \sin 10°$$
$$= 3.82 - 8 \sin 10° - 14 \sin 10° = 0,$$

$$\Sigma F_y = A + B_y + C - 8 \cos 10° - 14 \cos 10°$$
$$= 5.95 + 6.69 + 9.02 - 8 \cos 10° - 14 \cos 10° = 0,$$

$$\Sigma M_{(\text{point } A)} = 14B_y + 25C + (6)(8 \sin 10°)$$
$$- (4)(8 \cos 10°) + (3)(14 \sin 10°) - (22)(14 \cos 10°)$$
$$= (14)(6.69) + (25)(9.02) + (6)(8 \sin 10°)$$
$$- (4)(8 \cos 10°) + (3)(14 \sin 10°)$$
$$- (22)(14 \cos 10°) = 0.$$

Example 6.10

Analyzing a Machine

What forces are exerted on the bolt at E in Fig. 6.40 as a result of the 150-N forces on the pliers?

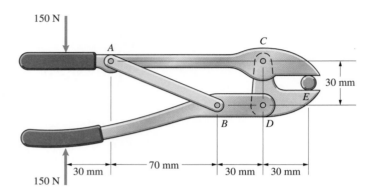

Figure 6.40

Strategy

A pair of pliers is a simple example of a machine, a structure designed to move and exert forces. The interconnections of the members are designed to create a mechanical advantage, subjecting an object to forces greater than the forces exerted by the user.

In this case there is no information to be gained from the free-body diagram of the entire structure. We must determine the forces exerted on the bolt by drawing free-body diagrams of the members.

Solution

We "disassemble" the pliers in Fig. a to obtain the free-body diagrams of the members, labeled (1), (2), and (3). The force R on free-body diagrams (1) and (3) is exerted by the two-force member AB. The angle $\alpha = $ arctan $(30/70) = 23.2°$. Our objective is to determine the force E exerted by the bolt.

The free-body diagram of member (3) has only three unknown forces and the 150-N load, so we can determine R, D_x, and D_y from this free-body diagram alone. The equilibrium equations are

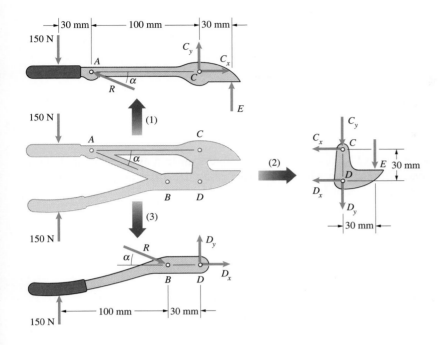

(a) Obtaining the free-body diagrams of the members.

$$\Sigma F_x = D_x + R\cos\alpha = 0,$$

$$\Sigma F_y = D_y - R\sin\alpha + 150 = 0,$$

$$\Sigma M_{(\text{point } B)} = 30D_y - (100)(150) = 0.$$

Solving these equations, we obtain $D_x = -1517$ N, $D_y = 500$ N, and $R = 1650$ N. Knowing D_x, we can determine E from the free-body diagram of member (2) by summing moments about C,

$$\Sigma M_{(\text{point } C)} = -30E - 30D_x = 0.$$

The force exerted on the bolt by the pliers is $E = -D_x = 1517$ N. The mechanical advantage of the pliers is $(1517 \text{ N})/(150 \text{ N}) = 10.1$.

Discussion

Notice that we did not need to use the free-body diagram of member (1) to determine E. When this happens, you can use the "leftover" free-body diagram to check your work. Using our results for R and E, we can confirm that the sum of the moments about point C of member (1) is zero:

$$\Sigma M_{(\text{point } C)} = (130)(150) - 100R\sin\alpha + 30E$$

$$= (130)(150) - (100)(1650)\sin 23.2° + (30)(1517) = 0.$$

Computational Mechanics

The following example and problems are designed for the use of a programmable calculator or computer.

Computational Example 6.11

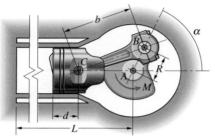

Figure 6.41

The device in Fig. 6.41 is used to compress air in a cylinder by applying a couple M to the arm AB. The pressure p in the cylinder and the net force F exerted on the piston by pressure are

$$p = p_{atm}\left(\frac{V_0}{V}\right),$$

$$F = Ap_{atm}\left(\frac{V_0}{V} - 1\right),$$

where $A = 0.02$ m^2 is the cross-sectional area of the piston, $p_{atm} = 10^5$ Pa (Pascals, or N/m^2) is atmospheric pressure, V is the volume of air in the cylinder, and V_0 is the value of V when $\alpha = 0$. The dimensions $R = 150$ mm, $b = 350$ mm, $d = 150$ mm, and $L = 1050$ mm. If M and α are initially zero and M is slowly increased until its value is 40 N-m, what are the resulting values of α and p?

Strategy

By expressing the volume of air in the cylinder in terms of α, we will determine the force exerted on the cylinder by pressure in terms of α. From the free-body diagram of the piston we will determine the axial force in the two-force member BC in terms of the pressure force on the cylinder. Then from the free-body diagram of the arm AB we will obtain a relation between M and α.

Solution

From the geometry of the arms AB and BC (Fig. a), the volume of air in the cylinder is

$$V = A\left(L - d - \sqrt{b^2 - R^2 \sin^2\alpha} + R\cos\alpha\right).$$

When $\alpha = 0$, the volume is

$$V_0 = A(L - d - b + R).$$

Therefore the force exerted on the piston by pressure is

$$F = Ap_{atm}\left(\frac{V_0}{V} - 1\right)$$

$$= Ap_{atm}\left(\frac{L - d - b + R}{L - d - \sqrt{b^2 - R^2 \sin^2\alpha} + R\cos\alpha} - 1\right).$$

We draw the free-body diagrams of the piston and the arm AB in Figs. b and c, where N is the force exerted on the piston by the cylinder (friction is

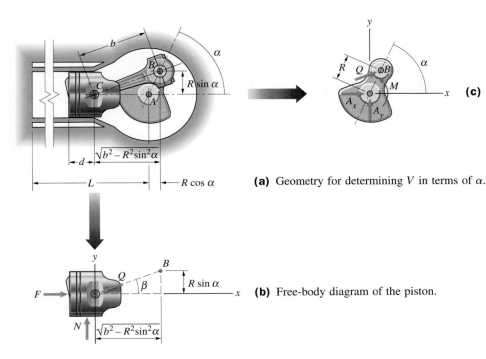

(c) Free-body diagram of the arm AB.

(a) Geometry for determining V in terms of α.

(b) Free-body diagram of the piston.

neglected), Q is the axial force in the two-force member BC, and A_x and A_y are the reactions due to the pin support A. From Fig. b, we obtain the equilibrium equation

$$\Sigma F_x = F - Q\cos\beta = 0,$$

where

$$\beta = \arctan\left(\frac{R\sin\alpha}{\sqrt{b^2 - R^2\sin^2\alpha}}\right).$$

The force exerted on the arm AB at B is

$$Q\cos\beta\,\mathbf{i} + Q\sin\beta\,\mathbf{j}.$$

The moment of this force about A is

$$\mathbf{r}_{AB} \times (Q\cos\beta\,\mathbf{i} + Q\sin\beta\,\mathbf{j}) = \begin{vmatrix} \mathbf{i} & \mathbf{j} & \mathbf{k} \\ R\cos\alpha & R\sin\alpha & 0 \\ Q\cos\beta & Q\sin\beta & 0 \end{vmatrix}$$

$$= QR(\cos\alpha\sin\beta - \sin\alpha\cos\beta)\,\mathbf{k}.$$

Using this result, the sum of the moments about A is

$$\Sigma M_{(\text{point } A)} = M + QR(\cos\alpha\sin\beta - \sin\alpha\cos\beta) = 0.$$

If we choose a value of α, we can sequentially calculate V, F, β, Q, and M. Computing M as a function of α, we obtain the graph shown in Fig. 6.42. The moment $M = 40$ N-m at approximately $\alpha = 80°$. By examining computed results near $80°$ (see table), we estimate that $\alpha = 79.61°$ when $M = 40$ N-m. Once we know α, we can calculate V and then p, obtaining

$$p = 1.148 p_{\text{atm}} = 1.148 \times 10^5 \text{ Pa}.$$

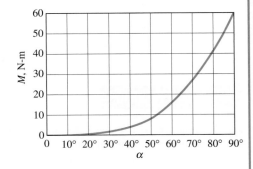

Figure 6.42
The moment M as a function of α.

α	M **(N-m)**
79.59°	39.9601
79.60°	39.9769
79.61°	39.9937
79.62°	40.0105
79.63°	40.0272

Chapter Summary

A structure of *members* interconnected at *joints* is a *truss* if it is composed entirely of two-force members. Otherwise, it is a *frame* if it is designed to remain stationary and support loads and a *machine* if it is designed to move and exert loads.

Trusses

A member of a truss is in *tension* if the *axial forces* at the ends are directed away from each other and is in *compression* if the axial forces are directed toward each other. Before beginning to determine the axial forces in the members of a truss, it is usually necessary to draw a free-body diagram of the entire truss and determine the reactions at its supports. The axial forces in the members can be determined by two methods. The *method of joints* involves drawing free-body diagrams of the joints of a truss one by one and using the equilibrium equations to determine the axial forces in the members. In two dimensions, choose joints to analyze that are subjected to known forces and no more than two unknown forces. The *method of sections* involves drawing free-body diagrams of parts, or *sections*, of a truss and using the equilibrium equations to determine the axial forces in selected members.

A *space truss* is a three-dimensional truss. Space trusses are analyzed by the same methods used for two-dimensional trusses. Choose joints to analyze that are subjected to known forces and no more than three unknown forces.

Frames and Machines

Begin analyzing a frame or machine by drawing a free-body diagram of the entire structure and determining the reactions at its supports. If the entire structure is statically indeterminate, determine as many reactions as possible. Then draw free-body diagrams of individual members, or selected combinations of members, and apply the equilibrium equations to determine the forces and couples acting on them. Recognizing two-force members will reduce the number of unknown forces that must be determined. If a load is applied at a joint, it can be placed on the free-body diagram of *any one* of the members attached at the joint.

Review Problems

6.1 Determine the axial forces in the members of the truss and indicate whether they are in tension (T) or compression (C).

   Strategy: Draw a free-body diagram of joint A. By writing the equilibrium equations for the joint, you can determine the axial forces in the two members.

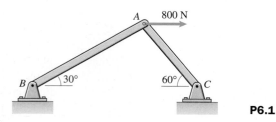

P6.1

6.2 The loads $F_1 = 60$ N and $F_2 = 40$ N.

(a) Draw the free-body diagram of the entire truss, and determine the reactions at its supports.

(b) Determine the axial forces in the members. Indicate whether they are in tension (T) or compression (C).

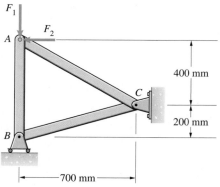

P6.2

6.3 Consider the truss in Problem 6.2. The loads $F_1 = 440$ N and $F_2 = 160$ N. Determine the axial forces in the members. Indicate whether they are in tension (T) or compression (C).

6.4 The truss supports a load $F = 10$ kN. Determine the axial forces in members AB, AC, and BC.

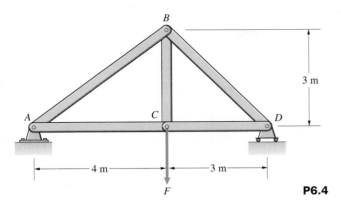

P6.4

✐ **6.5** Each member of the truss shown in Problem 6.4 will safely support a tensile force of 40 kN and a compressive force of 32 kN. Based on this criterion, what is the largest downward load F that can safely be applied at C?

6.6 The Pratt bridge truss supports loads at F, G, and H. Determine the axial forces in members BC, BG, and FG.

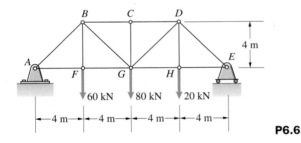

P6.6

6.7 Consider the truss in Problem 6.6. Determine the axial forces in members CD, GD, and GH.

6.8 The truss supports loads at F and H. Determine the axial forces in members AB, AC, BC, BD, CD, and CE.

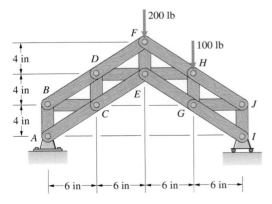

P6.8

6.9 Consider the truss in Problem 6.8. Determine the axial forces in members EH and FH.

6.10 Determine the axial forces in members BD, CD, and CE.

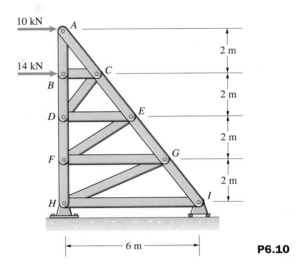

P6.10

6.11 For the truss in Problem 6.10, determine the axial forces in members DF, EF, and EG.

6.12 The truss supports a 400-N load at G. Determine the axial forces in members AC, CD, and CF.

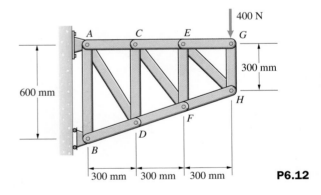

P6.12

6.13 Consider the truss in Problem 6.12. Determine the axial forces in members CE, EF, and EH.

6.14 Consider the truss in Problem 6.12. Which members have the largest tensile and compressive forces, and what are their values?

6.15 The Howe truss helps support a roof. Model the supports at *A* and *G* as roller supports. Use the method of joints to determine the axial forces in members *BC*, *CD*, *CI*, and *CJ*.

(b) When you draw the free-body diagrams of the individual members, place the 400-lb load on the free-body diagram of member *CD*.

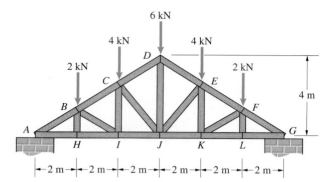

P6.15

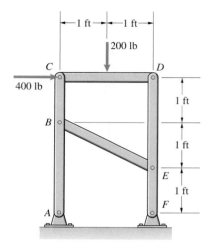

P6.19

6.16 For the roof truss in Problem 6.15, use the method of sections to determine the axial forces in members *CD*, *CJ*, and *IJ*.

6.17 A speaker system is suspended from the truss by cables attached at *D* and *E*. The mass of the speaker system is 130 kg, and its weight acts at *G*. Determine the axial forces in members *BC* and *CD*.

6.20 The mass *m* = 120 kg. Determine the forces on member *ABC*.

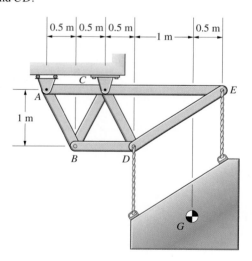

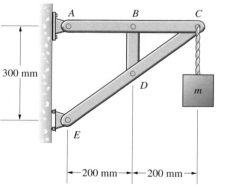

P6.20

6.21 Determine the forces on member *ABC*, presenting your answers as shown in Fig. 6.35.

P6.17

⑨ **6.18** Consider the system described in Problem 6.17. If each member of the truss will safely support a tensile force of 5 kN and a compressive force of 3 kN, what is the maximum safe value of the mass of the speaker system?

6.19 Determine the forces on member *ABC*, presenting your answers as shown in Fig. 6.35. Obtain the answers in two ways:

(a) When you draw the free-body diagrams of the individual members, place the 400-lb load on the free-body diagram of member *ABC*.

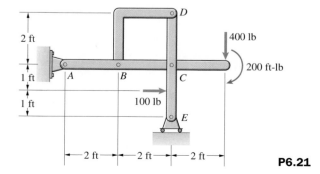

P6.21

6.22 Determine the force exerted on the bolt by the bolt cutters and the magnitude of the force the members exert on each other at the pin connection *A*.

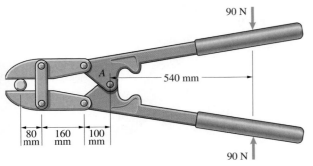

P6.22

6.23 The 600-lb weight of the scoop acts at a point 1 ft 6 in. to the right of the vertical line *CE*. The line *ADE* is horizontal. The hydraulic actuator *AB* can be treated as a two-force member. Determine the axial force in the hydraulic actuator *AB* and the forces exerted on the scoop at *C* and *E*.

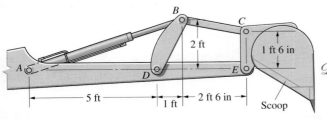

P6.23

6.24 This structure supports a conveyer belt used in a lignite mining operation. The cables connected to the belt exert the force *F* at *J*. As a result of the counterweight *W* = 8 kip, the reaction at *E* and the vertical reaction at *D* are equal. Determine *F* and the axial forces in members *BG* and *EF*.

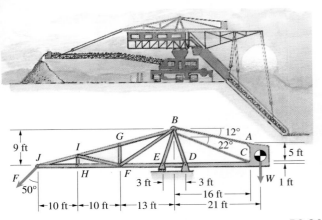

P6.24

6.25 Consider the structure described in Problem 6.24. The counterweight *W* = 8 kip is pinned at *D* and is supported by the cable *ABC*, which passes over a pulley at *A*. What is the tension in the cable, and what forces are exerted on the counterweight at *D*?

6.26 The weights W_1 = 4 kN and W_2 = 10 kN. Determine the forces on member *ACDE* at points *A* and *E*.

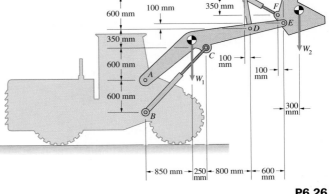

P6.26

D **Design Experience** Design a truss structure to support a foot bridge with an unsupported span (width) of 8 m. Make conservative estimates of the loads the structure will need to support if the pathway supported by the truss is made of wood. Consider two options: (1) Your client wants the bridge to be supported by a truss below the bridge so that the upper surface will be unencumbered by structure. (2) The client wants the truss to be above the bridge and designed so that it can serve as handrails. For each option, use statics to estimate the maximum axial forces to which the members of the structure will be subjected. Investigate alternative designs and compare the resulting axial loads.

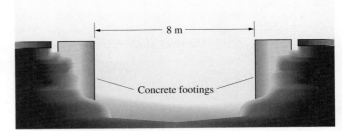

The loads that the legs of the piano must be designed to support depend not only on the piano's weight but also on the position of its center of mass—the point at which the weight effectively acts.

7

Centroids and Centers of Mass

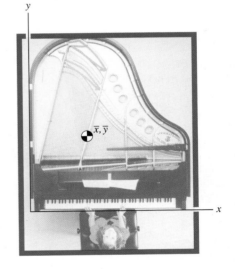

An object's weight does not act at a single point—it is distributed over the entire volume of the object. But we can represent the weight by a single equivalent force acting at a point called the center of mass. In this chapter we define the center of mass and show how it is determined for various kinds of objects. Along the way, we also introduce definitions that can be interpreted as the average positions of areas, volumes, and lines. These average positions are called centroids. Centroids coincide with the centers of mass of particular classes of objects, but they also arise in many other applications.

Centroids

Because centroids have such varied applications, we first define them using the general concept of a *weighted average*. Let's begin with the familiar idea of an average position. Suppose we want to determine the average position of a group of students sitting in a room. First, we introduce a coordinate system so that we can specify the position of each student. For example, we can align the axes with the walls of the room (Fig. 7.1 a). We number the students from 1 to N and denote the position of student 1 by x_1, y_1, the position of student 2 by x_2, y_2, and so on. The average x coordinate $\bar{x}$ is the sum of their x coordinates divided by N,

$$\bar{x} = \frac{x_1 + x_2 + \cdots + x_N}{N} = \frac{\sum_i x_i}{N}, \tag{7.1}$$

where the symbol $\sum_i$ stands for "sum over the range of i". The average y coordinate is

$$\bar{y} = \frac{\sum_i y_i}{N}. \tag{7.2}$$

We indicate the average position by the symbol shown in Fig. 7.1b.

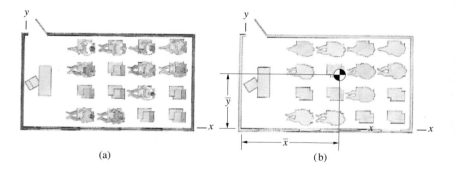

Figure 7.1
(a) A group of students in a classroom.
(b) Their average position.

(a)

(b)

Now suppose that we pass out some pennies to the students. Let the number of coins given to student 1 be c_1, the number given to student 2 be c_2, and so on. What is the average position of the coins in the room? Clearly, the average position of the coins may not be the same as the average position of the students. For example, if the students in the front of the room have more coins, the average position of the coins will be closer to the front of the room than the average position of the students.

To determine the x coordinate of the average position of the coins, we need to sum the x coordinates of the coins and divide by the number of coins. We can obtain the sum of the x coordinates of the coins by multiplying the number of coins each student has by his or her x coordinate and summing.

We can obtain the number of coins by summing the numbers $c_1, c_2, \ldots$. Thus the average x coordinate of the coins is

$$\bar{x} = \frac{\sum_i x_i c_i}{\sum_i c_i}.$$ (7.3)

We can determine the average y coordinate of the coins in the same way:

$$\bar{y} = \frac{\sum_i y_i c_i}{\sum_i c_i}.$$ (7.4)

By assigning other meanings to $c_1, c_2, \ldots$, we can determine the average positions of other measures associated with the students. For example, we could determine the average position of their age or the average position of their height.

More generally, we can use Eqs. (7.3) and (7.4) to determine the average position of any set of quantities with which we can associate positions. An average position obtained from these equations is called a *weighted average position*, or *centroid*. The "weight" associated with position x_1, y_1, is c_1, the weight associated with position x_2, y_2 is c_2, and so on. In Eqs. (7.1) and (7.2), the weight associated with the position of each student is 1. When the census is taken, the centroid of the population of the United States—the average position of the population—is determined in this way. In the next section we use Eqs. (7.3) and (7.4) to determine centroids of areas.

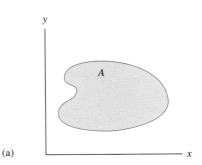

(a)

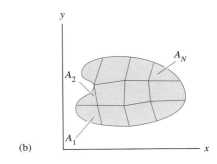

(b)

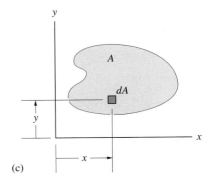

(c)

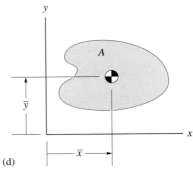

(d)

Figure 7.2
(a) The area A.
(b) Dividing A into N parts.
(c) A differential element of area dA with coordinates x, y.
(d) The centroid of the area.

7.1 Centroids of Areas

Consider an arbitrary area A in the x–y plane (Fig. 7.2a). Let us divide the area into parts $A_1, A_2, \ldots, A_N$ (Fig. 7.2b) and denote the positions of the parts by $(x_1, y_1), (x_2, y_2), \ldots, (x_N, y_N)$. We can obtain the centroid, or average position of the area, by using Eqs. (7.3) and (7.4) with the areas of the parts as the weights:

$$\bar{x} = \frac{\sum_i x_i A_i}{\sum_i A_i}, \qquad \bar{y} = \frac{\sum_i y_i A_i}{\sum_i A_i}.$$ (7.5)

A question arises if we try to carry out this procedure: What are the exact positions of the areas $A_1, A_2, \ldots, A_N$? We could reduce the uncertainty in their positions by dividing A into smaller parts, but we would still obtain only approximate values for $\bar{x}$ and $\bar{y}$. To determine the exact location of the centroid, we must take the limit as the sizes of the parts approach zero. We obtain this limit by replacing Eqs. (7.5) by the integrals

$$\bar{x} = \frac{\int_A x \, dA}{\int_A dA}, \tag{7.6}$$

$$\bar{y} = \frac{\int_A y \, dA}{\int_A dA}, \tag{7.7}$$

where x and y are the coordinates of the differential element of area dA (Fig. 7.2c). The subscript A on the integral signs means the integration is carried out over the entire area. The centroid of the area is shown in Fig. 7.2d.

Keeping in mind that the centroid of an area is its average position will often help you locate it. For example, the centroid of a circular area or a rectangular area obviously lies at the center of the area. If an area has "mirror image" symmetry about an axis, the centroid lies on the axis (Fig. 7.3a), and if an area is symmetric about two axes, the centroid lies at their intersection (Fig. 7.3b).

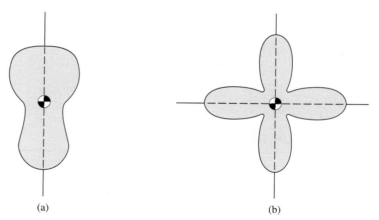

(a) (b)

Figure 7.3
(a) An area that is symmetric about an axis.
(b) An area with two axes of symmetry.

Study Questions

1. How is a weighted average position defined?
2. How is the concept of a weighted average used to define the centroid of a plane area?
3. Why is integration generally needed to determine the exact position of the centroid of an area?

Example 7.1

Centroid of an Area by Integration

Determine the centroid of the triangular area in Fig. 7.4.

Strategy

We will determine the coordinates of the centroid by using an element of area dA in the form of a "strip" of width dx.

Solution

Let dA be the vertical strip in Fig. a. The height of the strip is $(h/b)x$, so $dA = (h/b)x\,dx$. To integrate over the entire area, we must integrate with respect to x from $x = 0$ to $x = b$. The x coordinate of the centroid is

$$\bar{x} = \frac{\displaystyle\int_A x\,dA}{\displaystyle\int_A dA} = \frac{\displaystyle\int_0^b x\left(\frac{h}{b}x\,dx\right)}{\displaystyle\int_0^b \frac{h}{b}x\,dx} = \frac{\dfrac{h}{b}\left[\dfrac{x^3}{3}\right]_0^b}{\dfrac{h}{b}\left[\dfrac{x^2}{2}\right]_0^b} = \frac{2}{3}b.$$

To determine $\bar{y}$, we let y in Eq. (7.7) be the y coordinate of the midpoint of the strip (Fig. b):

$$\bar{y} = \frac{\displaystyle\int_A y\,dA}{\displaystyle\int_A dA} = \frac{\displaystyle\int_0^b \frac{1}{2}\left(\frac{h}{b}x\right)\left(\frac{h}{b}x\,dx\right)}{\displaystyle\int_0^b \frac{h}{b}x\,dx} = \frac{\dfrac{1}{2}\left(\dfrac{h}{b}\right)^2\left[\dfrac{x^3}{3}\right]_0^b}{\dfrac{h}{b}\left[\dfrac{x^2}{2}\right]_0^b} = \frac{1}{3}h.$$

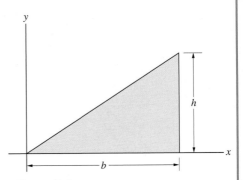

Figure 7.4

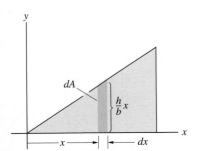

(a) An element dA in the form of a strip.

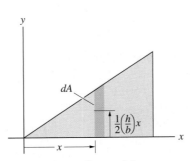

(b) The y coordinate of the midpoint of the strip is $\frac{1}{2}(h/b)x$.

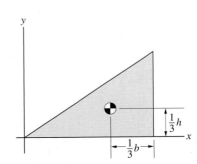

(c) Centroid of the area.

The centroid is shown in Fig. c.

Discussion

You should always be alert for opportunities to check your results. In this example we should make sure that our integration procedure gives the correct result for the area of the triangle:

$$\int_A dA = \int_0^b \frac{h}{b}x\,dx = \frac{h}{b}\left[\frac{x^2}{2}\right]_0^b = \frac{1}{2}bh.$$

Example 7.2

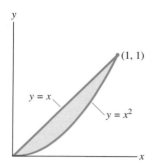

Figure 7.5

Area Defined by Two Equations

Determine the centroid of the area in Fig. 7.5.

Solution

Let dA be the vertical strip in Fig. a. The height of the strip is $x - x^2$, so $dA = (x - x^2)\,dx$. The x coordinate of the centroid is

$$\bar{x} = \frac{\displaystyle\int_A x\,dA}{\displaystyle\int_A dA} = \frac{\displaystyle\int_0^1 x(x - x^2)\,dx}{\displaystyle\int_0^1 (x - x^2)\,dx} = \frac{\left[\dfrac{x^3}{3} - \dfrac{x^4}{4}\right]_0^1}{\left[\dfrac{x^2}{2} - \dfrac{x^3}{3}\right]_0^1} = \frac{1}{2}.$$

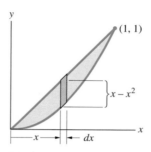

(a) A vertical strip of width dx. The height of the strip is equal to the difference in the two functions.

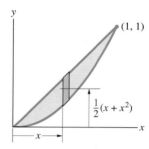

(b) The y coordinate of the midpoint of the strip.

The y coordinate of the midpoint of the strip is $x^2 + \frac{1}{2}(x - x^2) = \frac{1}{2}(x + x^2)$ (Fig. b). Substituting this expression for y in Eq. (7.7), we obtain the y coordinate of the centroid:

$$\bar{y} = \frac{\displaystyle\int_A y\,dA}{\displaystyle\int_A dA} = \frac{\displaystyle\int_0^1 \left[\frac{1}{2}(x + x^2)\right](x - x^2)\,dx}{\displaystyle\int_0^1 (x - x^2)\,dx} = \frac{\dfrac{1}{2}\left[\dfrac{x^3}{3} - \dfrac{x^5}{5}\right]_0^1}{\left[\dfrac{x^2}{2} - \dfrac{x^3}{3}\right]_0^1} = \frac{2}{5}.$$

7.2 Centroids of Composite Areas

Although centroids of areas can be determined by integration, the process becomes difficult and tedious for complicated areas. In this section we describe a much easier approach that can be used if an area consists of a combination of simple areas, which we call a *composite area*. We can determine the centroid of a composite area without integration if the centroids of its parts are known.

The area in Fig. 7.6a consists of a triangle, a rectangle, and a semicircle, which we call parts 1, 2, and 3. The x coordinate of the centroid of the composite area is

$$\bar{x} = \frac{\displaystyle\int_A x\, dA}{\displaystyle\int_A dA} = \frac{\displaystyle\int_{A_1} x\, dA + \int_{A_2} x\, dA + \int_{A_3} x\, dA}{\displaystyle\int_{A_1} dA + \int_{A_2} dA + \int_{A_3} dA}. \tag{7.8}$$

The x coordinates of the centroids of the parts are shown in Fig. 7.6b. From the equation for the x coordinate of the centroid of part 1,

$$\bar{x}_1 = \frac{\displaystyle\int_{A_1} x\, dA}{\displaystyle\int_{A_1} dA},$$

we obtain

$$\int_{A_1} x\, dA = \bar{x}_1 A_1.$$

Using this equation and equivalent equations for parts 2 and 3, we can write Eq. (7.8) as

$$\bar{x} = \frac{\bar{x}_1 A_1 + \bar{x}_2 A_2 + \bar{x}_3 A_3}{A_1 + A_2 + A_3}.$$

We have obtained an equation for the x coordinate of the composite area in terms of those of its parts. The coordinates of the centroid of a composite area with an arbitrary number of parts are

$$\bar{x} = \frac{\displaystyle\sum_i \bar{x}_i A_i}{\displaystyle\sum_i A_i}, \qquad \bar{y} = \frac{\displaystyle\sum_i \bar{y}_i A_i}{\displaystyle\sum_i A_i}. \tag{7.9}$$

When you can divide an area into parts whose centroids are known, you can use these expressions to determine its centroid. The centroids of some simple areas are tabulated in Appendix B.

We began our discussion of the centroid of an area by dividing an area into finite parts and writing equations for its weighted average position. The results, Eqs. (7.5), are approximate because of the uncertainty in the positions of the parts of the area. The exact Eqs. (7.9) are identical except that the positions of the parts are their centroids.

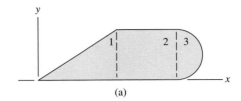

(a)

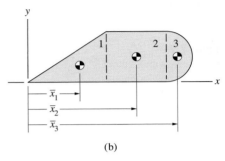

(b)

Figure 7.6
(a) A composite area composed of three simple areas.
(b) The centroids of the parts.

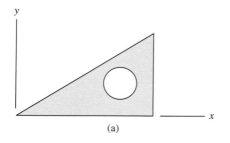

(a)

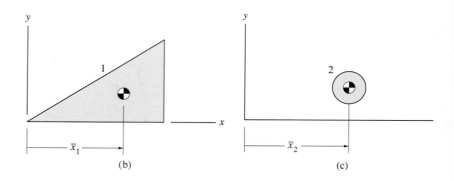

(b)   (c)

Figure 7.7
(a) An area with a cutout.
(b) The triangular area.
(c) The area of the cutout.

The area in Fig. 7.7a consists of a triangular area with a circular hole, or cutout. Designating the triangular area (without the cutout) as part 1 of the composite area (Fig. 7.7b) and the area of the cutout as part 2 (Fig. 7.7c), we obtain the x coordinate of the centroid of the composite area:

$$\bar{x} = \frac{\int_{A_1} x\, dA - \int_{A_2} x\, dA}{\int_{A_1} dA - \int_{A_2} dA} = \frac{\bar{x}_1 A_1 - \bar{x}_2 A_2}{A_1 - A_2}.$$

This equation is identical in form to the first of Eqs. (7.9) except that the terms corresponding to the cutout are negative. As this example demonstrates, you can use Eqs. (7.9) to determine the centroids of composite areas containing cutouts by treating the cutouts as negative areas.

We see that determining the centroid of a composite area requires three steps:

1. Choose the parts—Try to divide the composite area into parts whose centroids you know or can easily determine.

2. Determine the values for the parts—Determine the centroid and the area of each part. Watch for instances of symmetry that can simplify your task.

3. Calculate the centroid—Use Eqs. (7.9) to determine the centroid of the composite area.

Example 7.3

Centroid of a Composite Area

Determine the centroid of the area in Fig. 7.8.

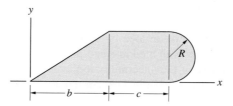

Figure 7.8

Solution

Choose the Parts   We can divide the area into a triangle, a rectangle, and a semicircle, which we call parts 1, 2, and 3, respectively.

Determine the Values for the Parts   The x coordinates of the centroids of the parts are shown in Fig. a. The x coordinates, the areas of the parts, and their products are summarized in Table 7.1.

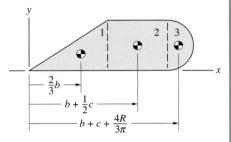

(a) The x coordinates of the centroids of the parts.

Table 7.1 Information for determining the x coordinate of the centroid

	$\bar{x}_i$	A_i	$\bar{x}_i A_i$
Part 1 (triangle)	$\frac{2}{3}b$	$\frac{1}{2}b(2R)$	$\left(\frac{2}{3}b\right)\left[\frac{1}{2}b(2R)\right]$
Part 2 (rectangle)	$b + \frac{1}{2}c$	$c(2R)$	$\left(b + \frac{1}{2}c\right)\left[c(2R)\right]$
Part 3 (semicircle)	$b + c + \dfrac{4R}{3\pi}$	$\frac{1}{2}\pi R^2$	$\left(b + c + \dfrac{4R}{3\pi}\right)\left(\frac{1}{2}\pi R^2\right)$

Calculate the Centroid   The x coordinate of the centroid of the composite area is

$$\bar{x} = \frac{\bar{x}_1 A_1 + \bar{x}_2 A_2 + \bar{x}_3 A_3}{A_1 + A_2 + A_3}$$

$$= \frac{\left(\frac{2}{3}b\right)\left[\frac{1}{2}b(2R)\right] + \left(b + \frac{1}{2}c\right)\left[c(2R)\right] + \left(b + c + \dfrac{4R}{3\pi}\right)\left(\frac{1}{2}\pi R^2\right)}{\frac{1}{2}b(2R) + c(2R) + \frac{1}{2}\pi R^2}.$$

We repeat the last two steps to determine the y coordinate of the centroid. The y coordinates of the centroids of the parts are shown in Fig. b. Using the information summarized in Table 7.2, we obtain

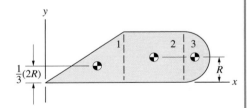

(b) The y coordinates of the centroids of the parts.

$$\bar{y} = \frac{\bar{y}_1 A_1 + \bar{y}_2 A_2 + \bar{y}_3 A_3}{A_1 + A_2 + A_3}$$

$$= \frac{\left[\frac{1}{3}(2R)\right]\left[\frac{1}{2}b(2R)\right] + R\left[c(2R)\right] + R\left(\frac{1}{2}\pi R^2\right)}{\frac{1}{2}b(2R) + c(2R) + \frac{1}{2}\pi R^2}.$$

Table 7.2 Information for determining the y coordinate of the centroid

	$\bar{y}_i$	A_i	$\bar{y}_i A_i$
Part 1 (triangle)	$\frac{1}{3}(2R)$	$\frac{1}{2}b(2R)$	$\left[\frac{1}{3}(2R)\right]\left[\frac{1}{2}b(2R)\right]$
Part 2 (rectangle)	R	$c(2R)$	$R\left[c(2R)\right]$
Part 3 (semicircle)	R	$\frac{1}{2}\pi R^2$	$R\left(\frac{1}{2}\pi R^2\right)$

Example 7.4

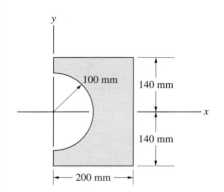

Figure 7.9

Centroid of an Area with a Cutout

Determine the centroid of the area in Fig. 7.9.

Solution

Choose the Parts We will treat the area as a composite area consisting of the rectangle without the semicircular cutout and the area of the cutout, which we call parts 1 and 2, respectively (Fig. a).

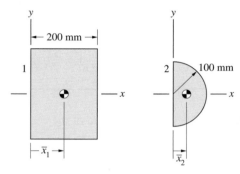

(a) The rectangle and the semicircular cutout.

Determine the Values for the Parts From Appendix B, the x coordinate of the centroid of the cutout is

$$\bar{x}_2 = \frac{4R}{3\pi} = \frac{4(100)}{3\pi} \text{ mm.}$$

The information for determining the x coordinate of the centroid is summarized in Table 7.3. Notice that we treat the cutout as a negative area.

Table 7.3 Information for determining $\bar{x}$

	$\bar{x}_i\text{(mm)}$	$A_i\text{(mm}^2)$	$\bar{x}_i A_i\text{(mm}^3)$
Part 1 (rectangle)	100	$(200)(280)$	$(100)\big[(200)(280)\big]$
Part 2 (cutout)	$\dfrac{4(100)}{3\pi}$	$-\tfrac{1}{2}\pi\,(100)^2$	$-\dfrac{4(100)}{3\pi}\big[\tfrac{1}{2}\pi\,(100)^2\big]$

Calculate the Centroid The x coordinate of the centroid is

$$\bar{x} = \frac{\bar{x}_1 A_1 + \bar{x}_2 A_2}{A_1 + A_2} = \frac{(100)\big[(200)(280)\big] - \dfrac{4(100)}{3\pi}\big[\tfrac{1}{2}\pi(100)^2\big]}{(200)(280) - \tfrac{1}{2}\pi(100)^2} = 122\text{mm}$$

Because of the symmetry of the area, $\bar{y} = 0$.

7.3 Distributed Loads

The load exerted on a beam (stringer) supporting a floor of a building is distributed over the beam's length (Fig. 7.10a). The load exerted by wind on a television transmission tower is distributed along the tower's height (Fig. 7.10b). In many engineering applications, loads are continuously distributed along lines. We will show that the concept of the centroid of an area can be useful in the analysis of objects subjected to such loads.

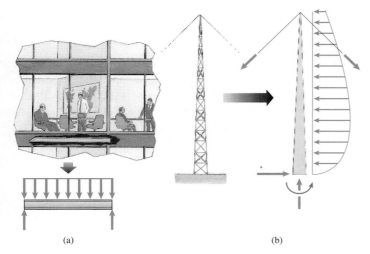

Figure 7.10
Examples of distributed forces: (a) Uniformly distributed load exerted on a beam of a building's frame by the floor. (b) Wind load distributed along the height of a tower.

Describing a Distributed Load

We can use a simple example to demonstrate how such loads are expressed analytically. Suppose that we pile bags of sand on a beam, as shown in Fig. 7.11a. You can see that the load exerted by the bags is distributed over the length of the beam and that its magnitude at a given position x depends on how high the bags are piled at that position. To describe the load, we define a function w such that the *downward* force exerted on an infinitesimal element dx of the beam is $w\,dx$. With this function we can model the varying magnitude of the load exerted by the sand bags (Fig. 7.11b). The arrows in the figure indicate that the load acts in the downward direction. Loads distributed along lines, from simple examples such as a beam's own weight to complicated ones such as the lift distributed along the length of an airplane's wing, are modeled by the function w. Since the product of w, and dx is a force, the dimensions of w are (force)/(length). For example, w, can be expressed in newtons per meter in SI units or in pounds per foot in U.S. Customary units.

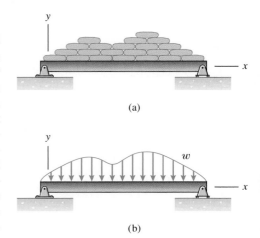

Figure 7.11
(a) Loading a beam with bags of sand.
(b) The distributed load w models the load exerted by the bags.

Determining Force and Moment

Let's assume that the function w describing a particular distributed load is known (Fig. 7.12a). The graph of w, is called the *loading curve*. Since the force acting on an element dx of the line is $w\,dx$, we can determine the total

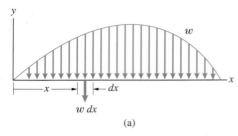

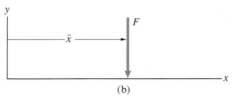

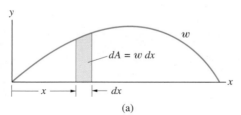

Figure 7.12
(a) A distributed load and the force exerted on a differential element dx.
(b) The equivalent force.

force F exerted by the distributed load by integrating the loading curve with respect to x:

$$F = \int_L w\, dx. \tag{7.10}$$

We can also integrate to determine the moment about a point exerted by the distributed load. For example, the moment about the origin due to the force exerted on the element dx is $xw\,dx$, so the total moment about the origin due to the distributed load is

$$M = \int_L xw\, dx. \tag{7.11}$$

When you are concerned only with the total force and moment exerted by a distributed load, you can represent it by a single equivalent force F (Fig. 7.12b). For equivalence, the force must act at a position $\bar{x}$ on the x axis such that the moment of F about the origin is equal to the moment of the distributed load about the origin:

$$\bar{x}F = \int_L xw\, dx.$$

Therefore the force F is equivalent to the distributed load if we place it at the position

$$\bar{x} = \frac{\displaystyle\int_L xw\, dx}{\displaystyle\int_L w\, dx}. \tag{7.12}$$

The Area Analogy

Notice that the term $w\,dx$ is equal to an element of "area" dA between the loading curve and the x axis (Fig. 7.13a). (We use quotation marks because $w\,dx$ is actually a force and not an area.) Interpreted in this way, Eq. (7.10) states that the total force exerted by the distributed load is equal to the "area" A between the loading curve and the x axis:

$$F = \int_L w\, dx = \int_A dA = A. \tag{7.13}$$

Substituting $w\,dx = dA$ into Eq. (7.12), we obtain

$$\bar{x} = \frac{\displaystyle\int_L xw\, dx}{\displaystyle\int_L w\, dx} = \frac{\displaystyle\int_A x\, dA}{\displaystyle\int_A dA}. \tag{7.14}$$

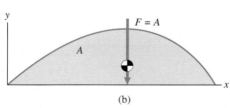

Figure 7.13
(a) Determining the "area" between the function w and the x axis.
(b) The equivalent force is equal to the "area," and the line of action passes through its centroid.

The force F is equivalent to the distributed load if it acts at the centroid of the "area" between the loading curve and the x axis (Fig. 7.13b). Using this analogy to represent a distributed load by an equivalent force can be very useful when the loading curve is relatively simple (see Example 7.5).

Study Questions

1. What is the definition of the function w?
2. How is the force exerted by a distributed load determined from the loading curve?
3. How is the moment exerted by a distributed load determined from the loading curve?

Example 7.5

Beam with a Triangular Distributed Load

The beam in Fig. 7.14 is subjected to a "triangular" distributed load whose value at B is 100 N/m.
(a) Represent the distributed load by a single equivalent force.
(b) Determine the reactions at A and B.

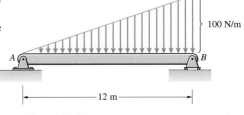

Figure 7.14

Strategy

(a) The magnitude of the force is equal to the "area" under the triangular loading curve, and the equivalent force acts at the centroid of the triangular "area."
(b) Once the distributed load is represented by a single equivalent force, we can apply the equilibrium equations to determine the reactions.

Solution

(a) The "area" of the triangular distributed load is one-half its base times its height, or $\frac{1}{2}(12 \text{ m}) \times (100 \text{ N/m}) = 600 \text{ N}$. The centroid of the triangular "area" is located at $\bar{x} = \frac{2}{3}(12 \text{ m}) = 8 \text{ m}$. We can therefore represent the distributed load by an equivalent downward force of 600-N magnitude acting at $x = 8 \text{ m}$ (Fig. a).

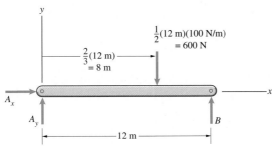

(a) Representing the distributed load by an equivalent force.

(b) From the equilibrium equations

$$\Sigma F_x = A_x = 0,$$

$$\Sigma F_y = A_y + B - 600 = 0,$$

$$\Sigma M_{(\text{point } A)} = 12B - (8)(600) = 0,$$

we obtain $A_x = 0, A_y = 200 \text{ N}$, and $B = 400 \text{ N}$.

Discussion

The loading curve in this example was sufficiently simple that we did not need to integrate to determine its area and centroid. In the following example we must integrate to determine the area and centroid.

Example 7.6

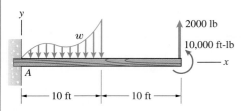

Figure 7.15

Beam with a Distributed Load

The beam in Fig. 7.15 is subjected to a distributed load, a force, and a couple. The distributed load is $w = 300x - 50x^2 + 0.3x^4$ lb/ft.
(a) Represent the distributed load by a single equivalent force.
(b) Determine the reactions at the built-in support A.

Strategy

(a) Since we know the function w, we can use Eq. (7.13) to determine the "area" under the loading curve, which is equal to the total force exerted by the distributed load. The x coordinate of the centroid is given by Eq. (7.14).
(b) Once the distributed load is represented by a single equivalent force, we can apply the equilibrium equations to determine the reactions at the built-in support.

Solution

(a) The downward force exerted by the distributed load is

$$F = \int_L w\,dx = \int_0^{10} \left(300x - 50x^2 + 0.3x^4\right)dx = 4330 \text{ lb.}$$

The x coordinate of the centroid of the distributed load is

$$\bar{x} = \frac{\displaystyle\int_L xw\,dx}{\displaystyle\int_L w\,dx} = \frac{\displaystyle\int_0^{10} x\left(300x - 50x^2 + 0.3x^4\right)dx}{\displaystyle\int_0^{10}\left(300x - 50x^2 + 0.3x^4\right)dx}$$

$$= \frac{25{,}000}{4330} = 5.77 \text{ ft.}$$

The distributed load is equivalent to a downward force of 4330-lb magnitude acting at $x = 5.77$ ft.
(b) In Fig. a, we draw the free-body diagram of the beam with the distributed force represented by the single equivalent force. From the equilibrium equations

$$\Sigma F_x = A_x = 0,$$

$$\Sigma F_y = A_y + 2000 - 4330 = 0,$$

$$\Sigma M_{\text{(point } A)} = (20)(2000) + 10{,}000 - (5.77)(4330) + M_A = 0,$$

we obtain $A_x = 0$, $A_y = 2330$ lb, and $M_A = -25{,}000$ ft-lb.

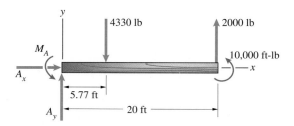

(a) Free-body
diagram of the beam.

Example 7.7

Beam Subjected to Distributed Loads

The beam in Fig. 7.16 is subjected to two distributed loads. Determine the reactions at A and B.

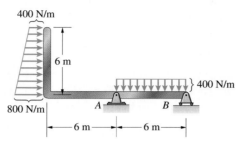

Figure 7.16

Strategy

We can easily represent the uniform distributed load on the right by an equivalent force. We can treat the distributed load on the left as the sum of uniform and triangular distributed loads and represent each load by an equivalent force.

Solution

We draw the free-body diagram of the beam in Fig. a, expressing the left distributed load as the sum of uniform and triangular loads. In Fig. b, we represent the three distributed loads by equivalent forces. The "area" of the uniform distributed load on the right is $(6 \text{ m}) \times (400 \text{ N/m}) = 2400 \text{ N}$, and its centroid is 3 m from B. The area of the uniform distributed load on the vertical part of the beam is $(6 \text{ m}) \times (400 \text{ N/m}) = 2400 \text{ N}$, and its centroid is located at $y = 3$ m. The area of the triangular distributed load is $\frac{1}{2}(6 \text{ m}) \times (400 \text{ N/m}) = 1200 \text{ N}$, and its centroid is located at $y = \frac{1}{3}(6 \text{ m}) = 2$ m.

From the equilibrium equations

$$\Sigma F_x = A_x + 1200 + 2400 = 0,$$

$$\Sigma F_y = A_y + B - 2400 = 0,$$

$$\Sigma M_{(\text{point } A)} = 6B - (3)(2400) - (2)(1200) - (3)(2400) = 0,$$

we obtain $A_x = -3600 \text{ N}$, $A_y = -400 \text{ N}$, and $B = 2800 \text{ N}$.

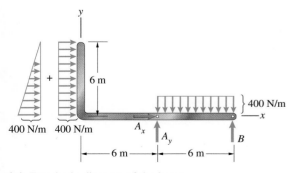

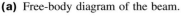

(a) Free-body diagram of the beam.

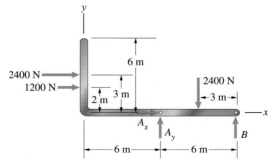

(b) Representing the distributed loads by equivalent forces.

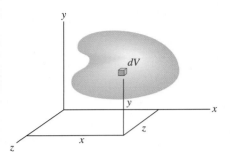

Figure 7.17
A volume V and differential element dV.

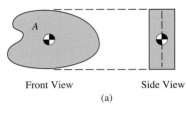

Front View   Side View

(a)

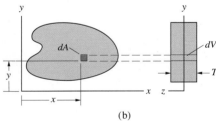

(b)

Figure 7.18
(a) A volume of uniform thickness.
(b) Obtaining dV by projecting dA through the volume.

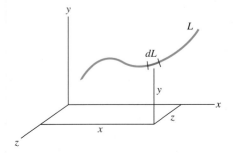

Figure 7.19
A line L and differential element dL.

7.4 Centroids of Volumes and Lines

Here we define the centroids, or average positions, of volumes and lines, and show how to determine the centroids of composite volumes and lines. We will show in Section 7.7 that knowing the centroids of volumes and lines allows you to determine the centers of mass of certain types of objects, which tells you where their weights effectively act.

Definitions

Volumes Consider a volume V, and let dV be a differential element of V with coordinates x, y, and z (Fig. 7.17). By analogy with Eqs. (7.6) and (7.7), the coordinates of the centroid of V are

$$\bar{x} = \frac{\int_V x \, dV}{\int_V dV}, \qquad \bar{y} = \frac{\int_V y \, dV}{\int_V dV}, \qquad \bar{z} = \frac{\int_V z \, dV}{\int_V dV}. \qquad (7.15)$$

The subscript V on the integral signs means that the integration is carried out over the entire volume.

If a volume has the form of a plate with uniform thickness and cross-sectional area A (Fig. 7.18a), its centroid coincides with the centroid of A and lies at the midpoint between the two faces. To show that this is true, we obtain a volume element dV by projecting an element dA of the cross-sectional area through the thickness T of the volume, so that $dV = T \, dA$ (Fig. 7.18b). Then the x and y coordinates of the centroid of the volume are

$$\bar{x} = \frac{\int_V x \, dV}{\int_V dV} = \frac{\int_A xT \, dA}{\int_A T \, dA} = \frac{\int_A x \, dA}{\int_A dA},$$

$$\bar{y} = \frac{\int_V y \, dV}{\int_V dV} = \frac{\int_A yT \, dA}{\int_A T \, dA} = \frac{\int_A y \, dA}{\int_A dA}.$$

The coordinate $\bar{z} = 0$ by symmetry. Thus you know the centroid of this type of volume if you know (or can determine) the centroid of its cross-sectional area.

Lines The coordinates of the centroid of a line L are

$$\bar{x} = \frac{\int_L x \, dL}{\int_L dL}, \qquad \bar{y} = \frac{\int_L y \, dL}{\int_L dL}, \qquad \bar{z} = \frac{\int_L z \, dL}{\int_L dL}, \qquad (7.16)$$

where dL is a differential length of the line with coordinates x, y, and z. (Fig. 7.19).

Example 7.8

Centroid of a Cone by Integration

Determine the centroid of the cone in Fig. 7.20.

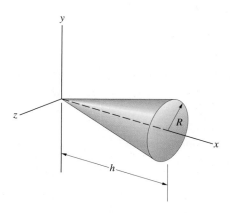

Figure 7.20

Strategy

The centroid must lie on the x axis because of symmetry. We will determine its x coordinate by using an element of volume dV in the form of a "disk" of width dx.

Solution

Let dV be the disk in Fig. a. The radius of the disk is $(R/h)x$ (Fig. b), and its volume equals the product of the area of the disk and its thickness, $dV = \pi\left[(R/h)x\right]^2 dx$. To integrate over the entire volume, we must integrate with respect to x from $x = 0$ to $x = h$. The x coordinate of the centroid is

$$\bar{x} = \frac{\displaystyle\int_V x\, dV}{\displaystyle\int_V dV} = \frac{\displaystyle\int_0^h x\pi\, \frac{R^2}{h^2}\, x^2\, dx}{\displaystyle\int_0^h \pi\, \frac{R^2}{h^2}\, x^2\, dx} = \frac{3}{4}\, h.$$

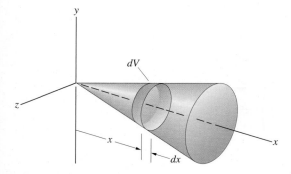

(a) An element dV in the form of a disk.

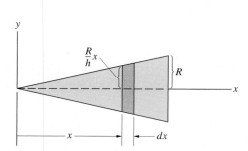

(b) The radius of the element is $(R/h)x$.

Example 7.9

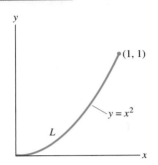

Figure 7.21

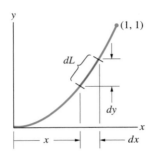

(a) A differential line element *dL*.

Centroid of a Line by Integration

The line *L* in Fig. 7.21 is defined by the function $y = x^2$. Determine the *x* coordinate of its centroid.

Solution

We can express a differential element *dL* of the line (Fig. a) in terms of *dx* and *dy*:

$$dL = \sqrt{dx^2 + dy^2} = \sqrt{1 + \left(\frac{dy}{dx}\right)^2}\, dx.$$

From the equation describing the line, the derivative $dy/dx = 2x$, so we obtain an expression for *dL* in terms of *x*:

$$dL = \sqrt{1 + 4x^2}\, dx.$$

To integrate over the entire line, we must integrate from $x = 0$ to $x = 1$. The *x* coordinate of the centroid is

$$\bar{x} = \frac{\int_L x\, dL}{\int_L dL} = \frac{\int_0^1 x\sqrt{1 + 4x^2}\, dx}{\int_0^1 \sqrt{1 + 4x^2}\, dx} = 0.574.$$

Example 7.10

Centroid of a Semicircular Line by Integration

Determine the centroid of the semicircular line in Fig. 7.22.

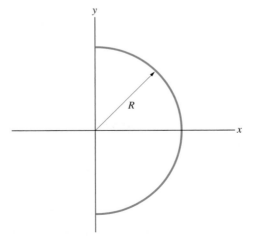

Figure 7.22

Strategy

Because of the symmetry of the line, the centroid lies on the x axis. To determine $\bar{x}$, we will integrate in terms of polar coordinates.

Solution

By letting θ change by an amount $d\theta$, we obtain a differential line element of length $dL = R\,d\theta$ (Fig. a). The x coordinate of dL is $x = R\cos\theta$. To integrate over the entire line, we must integrate with respect to θ from $\theta = -\pi/2$ to $\theta = +\pi/2$:

$$\bar{x} = \frac{\displaystyle\int_L x\,dL}{\displaystyle\int_L dL} = \frac{\displaystyle\int_{-\pi/2}^{\pi/2} (R\cos\theta)R\,d\theta}{\displaystyle\int_{-\pi/2}^{\pi/2} R\,d\theta} = \frac{R^2[\sin\theta]_{-\pi/2}^{\pi/2}}{R[\theta]_{-\pi/2}^{\pi/2}} = \frac{2R}{\pi}.$$

Discussion

Notice that our integration procedure gives the correct length of the line:

$$\int_L dL = \int_{-\pi/2}^{\pi/2} R\,d\theta = R[\theta]_{-\pi/2}^{\pi/2} = \pi R.$$

(a) A differential line element $dL = R\,d\theta$.

Centroids of Composite Volumes and Lines

The centroids of composite volumes and lines can be derived using the same approach we applied to areas. The coordinates of the centroid of a composite volume are

$$\bar{x} = \frac{\displaystyle\sum_i \bar{x}_i V_i}{\displaystyle\sum_i V_i}, \qquad \bar{y} = \frac{\displaystyle\sum_i \bar{y}_i V_i}{\displaystyle\sum_i V_i}, \qquad \bar{z} = \frac{\displaystyle\sum_i \bar{z}_i V_i}{\displaystyle\sum_i V_i}, \qquad (7.17)$$

and the coordinates of the centroid of a composite line are

$$\bar{x} = \frac{\displaystyle\sum_i \bar{x}_i L_i}{\displaystyle\sum_i L_i}, \qquad \bar{y} = \frac{\displaystyle\sum_i \bar{y}_i L_i}{\displaystyle\sum_i L_i}, \qquad \bar{z} = \frac{\displaystyle\sum_i \bar{z}_i L_i}{\displaystyle\sum_i L_i}. \qquad (7.18)$$

The centroids of some simple volumes and lines are tabulated in Appendixes B and C.

Determining the centroid of a composite volume or line requires three steps:

1. **Choose the parts**—Try to divide the composite into parts whose centroids you know or can easily determine.

2. **Determine the values for the parts**—Determine the centroid and the volume or length of each part. Watch for instances of symmetry that can simplify your task.

3. **Calculate the centroid**—Use Eqs. (7.17) or (7.18) to determine the centroid of the composite volume or line.

Example 7.11

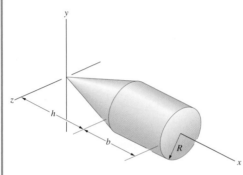

Figure 7.23

Centroid of a Composite Volume

Determine the centroid of the volume in Fig. 7.23.

Solution

Choose the Parts The volume consists of a cone and a cylinder, which we call parts 1 and 2, respectively.

Determine the Values for the Parts The centroid and volume of the cone are given in Appendix C. The x coordinates of the centroids of the parts are shown in Fig. a, and the information for determining the x coordinate of the centroid is summarized in Table 7.4.

Table 7.4 Information for determining $\bar{x}$

	$\bar{x}_i$	V_i	$\bar{x}_i V_i$
Part 1 (cone)	$\frac{3}{4}h$	$\frac{1}{3}\pi R^2 h$	$\left(\frac{4}{3}h\right)\left(\frac{1}{3}\pi R^2 h\right)$
Part 2 (cylinder)	$h + \frac{1}{2}b$	$\pi R^2 b$	$\left(h + \frac{1}{2}b\right)\left(\pi R^2 b\right)$

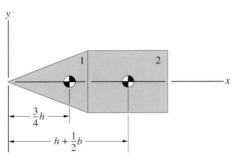

(a) The x coordinates of the centroids of the cone and cylinder.

Calculate the Centroid The x coordinate of the centroid of the composite volume is

$$\bar{x} = \frac{\bar{x}_1 V_1 + \bar{x}_2 V_2}{V_1 + V_2} = \frac{\left(\frac{3}{4}h\right)\left(\frac{1}{3}\pi R^2 h\right) + \left(h + \frac{1}{2}b\right)\left(\pi R^2 b\right)}{\frac{1}{3}\pi R^2 h + \pi R^2 b}.$$

Because of symmetry, $\bar{y} = 0$ and $\bar{z} = 0$.

Example 7.12

Centroid of a Volume Containing a Cutout

Determine the centroid of the volume in Fig. 7.24.

Solution

Choose the Parts We can divide the volume into the five simple parts shown in Fig. a. Part 5 is the volume of the 20-mm-diameter hole.

Determine the Values for the Parts The centroids of parts 1 and 3 are located at the centroids of their semicircular cross sections (Fig. b). The information for determining the x coordinate of the centroid is summarized in Table 7.5. Part 5 is a negative volume.

Table 7.5 Information for determining $\bar{x}$.

	$\bar{x}_i\,(\text{mm})$	$V_i\,(\text{mm}^3)$	$\bar{x}_i V_i\,(\text{mm}^4)$
Part 1	$-\dfrac{4(25)}{3\pi}$	$\dfrac{\pi(25)^2}{2}(20)$	$\left[-\dfrac{4(25)}{3\pi}\right]\left[\dfrac{\pi(25)^2}{2}(20)\right]$
Part 2	100	$(200)(50)(20)$	$(100)\left[(200)(50)(20)\right]$
Part 3	$200+\dfrac{4(25)}{3\pi}$	$\dfrac{\pi(25)^2}{2}(20)$	$\left[200+\dfrac{4(25)}{3\pi}\right]\left[\dfrac{\pi(25)^2}{2}(20)\right]$
Part 4	0	$\pi(25)^2(40)$	0
Part 5	200	$-\pi(10)^2(20)$	$-(200)\left[\pi(10)^2(20)\right]$

Calculate the Centroid The x coordinate of the centroid of the composite volume is

$$\bar{x} = \frac{\bar{x}_1 V_1 + \bar{x}_2 V_2 + \bar{x}_3 V_3 + \bar{x}_4 V_4 + \bar{x}_5 V_5}{V_1 + V_2 + V_3 + V_4 + V_5}$$

$$= \frac{\begin{array}{l}\left[-\dfrac{4(25)}{3\pi}\right]\left[\dfrac{\pi(25)^2}{2}(20)\right] + (100)\left[(200)(50)(20)\right] \\[2mm] \quad + \left[200+\dfrac{4(25)}{3\pi}\right]\left[\dfrac{\pi(25)^2}{2}(20)\right] + 0 - (200)\left[\pi(10)^2(20)\right]\end{array}}{\dfrac{\pi(25)^2}{2}(20) + (200)(50)(20) + \dfrac{\pi(25)^2}{2}(20) + \pi(25)^2(40) - \pi(10)^2(20)}$$

$$= 72.77 \text{ mm.}$$

The z coordinates of the centroids of the parts are zero except $\bar{z}_4 = 30$ mm. Therefore the z coordinate of the centroid of the composite volume is

$$\bar{z} = \frac{\bar{z}_4 V_4}{V_1 + V_2 + V_3 + V_4 + V_5}$$

$$= \frac{30\left[\pi(25)^2(40)\right]}{\dfrac{\pi(25)^2}{2}(20) + (200)(50)(20) + \dfrac{\pi(25)^2}{2}(20) + \pi(25)^2(40) - \pi(10)^2(20)}$$

$$= 7.56 \text{ mm.}$$

Because of symmetry, $\bar{y} = 0$.

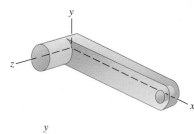

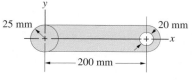

Side View

End View

Figure 7.24

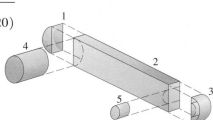

(a) Dividing the volume into five parts.

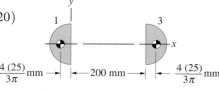

(b) Positions of the centroids of parts 1 and 3.

Example 7.13

Centroid of a Composite Line

Determine the centroid of the line in Fig. 7.25. The quarter-circular arc lies in the y–z plane.

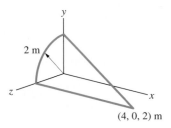

Figure 7.25

Solution

Choose the Parts The line consists of a quarter-circular arc and two straight segments, which we call parts 1, 2, and 3 (Fig. a).

Determine the Values for the Parts From Appendix B, the coordinates of the centroid of the quarter-circular arc are $\bar{x}_1 = 0$, $\bar{y}_1 = \bar{z}_1 = 2(2)/\pi$ m. The centroids of the straight segments lie at their midpoints. For segment 2, $\bar{x}_2 = 2$ m, $\bar{y}_2 = 0$, and $\bar{z}_2 = 2$ m, and for segment 3, $\bar{x}_3 = 2$ m, $\bar{y}_3 = 1$ m, and $\bar{z}_3 = 1$ m. The length of segment 3 is $L_3 = \sqrt{(4)^2 + (2)^2 + (2)^2} = 4.90$ m. This information is summarized in Table 7.6.

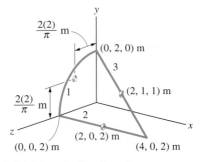

(a) Dividing the line into three parts.

Table 7.6 Information for determining the centroid.

	$\bar{x}_i$	$\bar{y}_i$	$\bar{z}_i$	L_i
Part 1	0	$2(2)/\pi$	$2(2)/\pi$	$\pi(2)/2$
Part 2	2	0	2	4
Part 3	2	1	1	4.90

Calculate the Centroid The coordinates of the centroid of the composite line are

$$\bar{x} = \frac{\bar{x}_1 L_1 + \bar{x}_2 L_2 + \bar{x}_3 L_3}{L_1 + L_2 + L_3} = \frac{0 + (2)(4) + (2)(4.90)}{\pi + 4 + 4.90} = 1.478 \text{ m},$$

$$\bar{y} = \frac{\bar{y}_1 L_1 + \bar{y}_2 L_2 + \bar{y}_3 L_3}{L_1 + L_2 + L_3} = \frac{[2(2)/\pi][\pi(2)/2] + 0 + (1)(4.90)}{\pi + 4 + 4.90} = 0.739 \text{ m},$$

$$\bar{z} = \frac{\bar{z}_1 L_1 + \bar{z}_2 L_2 + \bar{z}_3 L_3}{L_1 + L_2 + L_3} = \frac{[2(2)/\pi][\pi(2)/2] + (2)(4) + (1)(4.90)}{\pi + 4 + 4.90} = 1.404 \text{ m}.$$

The Pappus–Guldinus Theorems

In this section we discuss two simple and useful theorems relating surfaces and volumes of revolution to the centroids of the lines and areas that generate them.

First Theorem

Consider a line L in the x–y plane that does not intersect the x axis (Fig. 7.26a). Let the coordinates of the centroid of the line be $\bar{x}, \bar{y}$. We can generate a surface by revolving the line about the x axis (Fig. 7.26b). As the line revolves about the x axis, the centroid of the line moves in a circular path of radius $\bar{y}$.

The first Pappus-Guldinus theorem states that the area of the surface of revolution is equal to the product of the distance through which the centroid of the line moves and the length of the line:

$$A = 2\pi \bar{y} L. \tag{7.19}$$

To prove this result, we observe that as the line revolves about the x axis, the area dA generated by an element dL of the line is $dA = 2\pi y \, dL$, where y is the y coordinate of the element dL (Fig. 7.26c). Therefore the total area of the surface of revolution is

$$A = 2\pi \int_L y \, dL. \tag{7.20}$$

From the definition of the y coordinate of the centroid of the line,

$$\bar{y} = \frac{\displaystyle\int_L y \, dL}{\displaystyle\int_L dL},$$

we obtain

$$\int_L y \, dL = \bar{y} L.$$

Substituting this result into Eq. (7.20), we obtain Eq. (7.19).

Second Theorem

Consider an area A in the x–y plane that does not intersect the x axis (Fig. 7.27a). Let the coordinates of the centroid of the area be $\bar{x}, \bar{y}$. We can generate a volume by revolving the area about the x axis (Fig. 7.27b). As the area revolves about the x axis, the centroid of the area moves in a circular path of length $2\pi \bar{y}$.

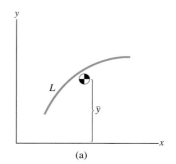

(a)

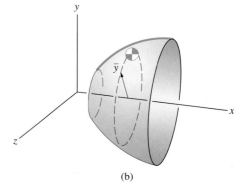

(b)

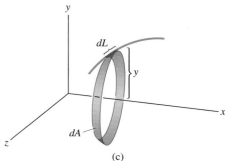

(c)

Figure 7.26
(a) A line L and the y coordinate of its centroid.
(b) The surface generated by revolving the line L about the x axis and the path followed by the centroid of the line.
(c) An element dL of the line and the element of area dA it generates.

The second Pappus-Guldinus theorem states that the volume V of the volume of revolution is equal to the product of the distance through which the centroid of the area moves and the area:

$$V = 2\pi \bar{y} A. \tag{7.21}$$

As the area revolves about the x axis, the volume dV generated by an element dA of the area is $dV = 2\pi \, y \, dA$, where y is the y coordinate of the element dA (Fig. 7.27c). Therefore the total volume is

$$V = 2\pi \int_A y \, dA. \tag{7.22}$$

From the definition of the y coordinate of the centroid of the area,

$$\bar{y} = \frac{\displaystyle\int_A y \, dA}{\displaystyle\int_A dA},$$

we obtain

$$\int_A y \, dA = \bar{y} A.$$

Substituting this result into Eq. (7.22), we obtain Eq. (7.21).

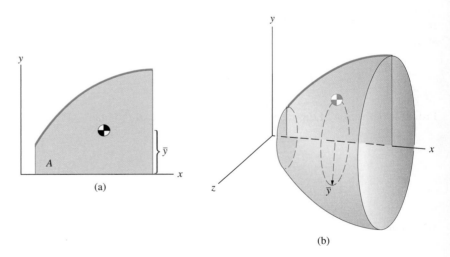

(a)

(b)

Figure 7.27
(a) An area A and the y coordinate of its centroid.
(b) The volume generated by revolving the area A about the x axis and the path followed by the centroid of the area.
(c) An element dA of the area and the element of volume dV it generates.

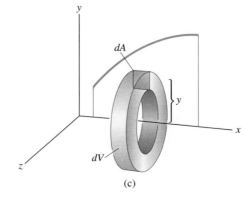

(c)

Example 7.14

Use the Pappus-Guldinus theorems to determine the surface area A and volume V of the cone in Fig. 7.28.

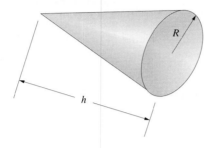

Figure 7.28

Strategy

We can generate the curved surface of the cone by revolving a straight line about an axis, and we can generate its volume by revolving a right triangular area about the axis. Since we know the centroids of the straight line and the triangular area, we can use the Pappus-Guldinus theorems to determine the area and volume of the cone.

Solution

Revolving the straight line in Fig. (a) about the x axis generates the curved surface of the cone. The y coordinate of the centroid of the line is $\bar{y}_L = \frac{1}{2}R$, and its length is $L = \sqrt{h^2 + R^2}$. The centroid of the line moves a distance $2\pi\bar{y}_L$ as the line revolves about the x axis, so the area of the curved surface is

$$(2\pi\bar{y})L = \pi R\sqrt{h^2 + R^2}.$$

We obtain the total surface area A of the cone by adding the area of the base,

$$A = \pi R\sqrt{h^2 + R^2} + \pi R^2.$$

Revolving the triangular area in Fig. (b) about the x axis generates the volume V. The y coordinate of its centroid is $\bar{y}_T = \frac{1}{3}R$, and its area is $A = \frac{1}{2}hR$, so the volume of the cone is

$$V = (2\pi\bar{y}_T)A = \frac{1}{3}\pi hR^2.$$

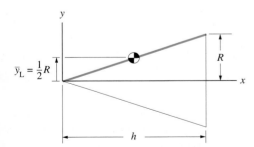

(a) The straight line that generates the curved surface of the cone.

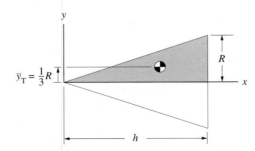

(b) The area that generates the volume of the cone.

Example 7.15

Determining Centroids with Pappus–Guldinus Theorems

The circumference of a sphere of radius R is $2\pi R$, its surface area is $4\pi R^2$, and its volume is $\frac{4}{3}\pi R^3$. Use this information to determine (a) the centroid of a semicircular line; (b) the centroid of a semicircular area.

Strategy

Revolving a semicircular line about an axis generates a spherical area, and revolving a semicircular area around an axis generates a spherical volume. Knowing the area and volume, we can use the Pappus-Guldinus theorems to determine the centroids of the generating line and area.

Solution

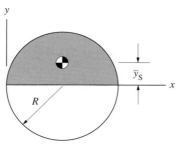

(a) Revolving a semicircular line about the x axis.

(a) Revolving the semicircular line in Fig. a about the x axis generates the surface area of a sphere. The length of the line is $L = \pi R$, and $\bar{y}_L$ is the y coordinate of its centroid. The centroid of the line moves a distance $2\pi\bar{y}_L$, so the surface area of the sphere is

$$(2\pi\bar{y}_L)L = 2\pi^2 R\bar{y}_L.$$

By equating this expression to the surface area $4\pi R^2$, we determine $\bar{y}_L$:

$$\bar{y}_L = \frac{2R}{\pi}.$$

(b) Revolving a semicircular area about the x axis.

(b) Revolving the semicircular area in Fig. b generates the sphere's volume. The area of the semicircle is $A = \frac{1}{2}\pi R^2$, and $\bar{y}_S$ is the y coordinate of its centroid. The centroid moves a distance $2\pi\bar{y}_S$, so the volume of the sphere is

$$(2\pi\bar{y}_S)A = \pi^2 R^2\bar{y}_S.$$

Equating this expression to the volume $\frac{4}{3}\pi R^3$, we obtain

$$\bar{y}_S = \frac{4R}{3\pi}.$$

Discussion

If you can obtain a result by using the Pappus-Guldinus theorems, you will often save time and effort in comparison with other approaches. Compare this example with Example 7.10, in which we use integration to determine the centroid of a semicircular line.

Centers of Mass

The *center of mass* of an object is the centroid, or average position, of its mass. In the following section we give the analytical definition of the center of mass and demonstrate one of its most important properties: An object's weight can be represented by a single equivalent force acting at its center of mass. We then discuss how to locate centers of mass and show that for particular classes of objects, the center of mass coincides with the centroid of a volume, area, or line. Finally, we show how to locate centers of mass of composite objects.

7.6 Definition of the Center of Mass

The center of mass of an object is defined by

$$\bar{x} = \frac{\displaystyle\int_m x \, dm}{\displaystyle\int_m dm}, \qquad \bar{y} = \frac{\displaystyle\int_m y \, dm}{\displaystyle\int_m dm}, \qquad \bar{z} = \frac{\displaystyle\int_m z \, dm}{\displaystyle\int_m dm}, \qquad (7.23)$$

where x, y, and z are the coordinates of the differential element of mass dm (Fig. 7.29). The subscripts m indicate that the integration must be carried out over the entire mass of the object.

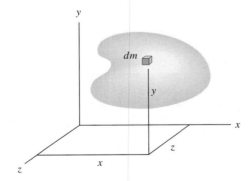

Figure 7.29
An object and differential element of mass dm.

Before considering how to determine the center of mass of an object, we will demonstrate that the weight of an object can be represented by a single equivalent force acting at its center of mass. Consider an element of mass dm of an object (Fig. 7.30a). If the y axis of the coordinate system points upward, the weight of dm is $-dm \, g\mathbf{j}$. Integrating this expression over the mass m, we obtain the total weight of the object,

$$\int_m -g\mathbf{j} \, dm = -mg\mathbf{j} = -W\mathbf{j}.$$

The moment of the weight of the element dm about the origin is

$$(x\mathbf{i} + y\mathbf{j} + z\mathbf{k}) \times (-dm \, g\mathbf{j}) = gz\mathbf{i} \, dm - gx\mathbf{k} \, dm.$$

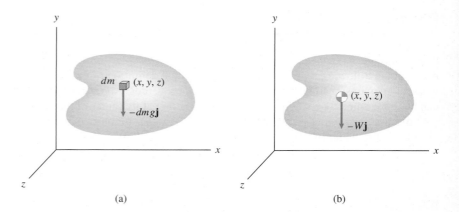

Figure 7.30
(a) Weight of the element dm.
(b) Representing the weight by a single force at the center of mass.

Integrating this expression over m, we obtain the total moment about the origin due to the weight of the object:

$$\int_m (gz\mathbf{i}\,dm - gx\mathbf{k}\,dm) = mg\bar{z}\mathbf{i} - mg\bar{x}\mathbf{k} = W\bar{z}\mathbf{i} - W\bar{x}\mathbf{k}.$$

If we represent the weight of the object by the force $-W\mathbf{j}$ acting at the center of mass (Fig. 7.30b), the moment of this force about the origin is equal to the total moment due to the weight:

$$(\bar{x}\mathbf{i} + \bar{y}\mathbf{j} + \bar{z}\mathbf{k}) \times (-W\mathbf{j}) = W\bar{z}\mathbf{i} - W\bar{x}\mathbf{k}.$$

This result shows that when you are concerned only with the total force and total moment exerted by the weight of an object, you can assume that its weight acts at the center of mass.

7.7 Centers of Mass of Objects

To apply Eqs. (7.23) to specific objects, we will change the variable of integration from mass to volume by introducing the mass density.

The *mass density* ρ of an object is defined such that the mass of a differential element of its volume is $dm = \rho\,dV$. The dimensions of ρ are therefore (mass)/(volume). For example, it can be expressed in kg/m^3 in SI units or in $slug/ft^3$ in U.S. Customary units. The total mass of an object is

$$m = \int_m dm = \int_V \rho\,dV. \tag{7.24}$$

An object whose mass density is uniform throughout its volume is said to be *homogeneous*. In this case, the total mass equals the product of the mass density and the volume:

$$m = \rho \int_V dV = \rho V. \qquad \textbf{Homogeneous object} \tag{7.25}$$

The *weight density* $\gamma = g\rho$. It can be expressed in N/m^3 in SI units or in lb/ft^3 in U.S. Customary units. The weight of an element of volume dV of an object is $dW = \gamma\,dV$, and the total weight of a homogeneous object equals γV.

By substituting $dm = \rho\,dV$ into Eqs. (7.23), we can express the coordinates of the center of mass in terms of volume integrals:

$$\bar{x} = \frac{\displaystyle\int_V \rho x\,dV}{\displaystyle\int_V \rho\,dV}, \qquad \bar{y} = \frac{\displaystyle\int_V \rho y\,dV}{\displaystyle\int_V \rho\,dV}, \qquad \bar{z} = \frac{\displaystyle\int_V \rho z\,dV}{\displaystyle\int_V \rho\,dV}. \qquad (7.26)$$

If ρ is known as a function of position in an object, these integrals determine its center of mass. Furthermore, we can use them to show that the centers of mass of particular classes of objects coincide with centroids of volumes, areas, and lines:

- **The center of mass of a homogeneous object coincides with the centroid of its volume.** If an object is homogeneous, ρ = constant and Eqs. (7.26) become the equations for the centroid of the volume,

$$\bar{x} = \frac{\displaystyle\int_V x\,dV}{\displaystyle\int_V dV}, \qquad \bar{y} = \frac{\displaystyle\int_V y\,dV}{\displaystyle\int_V dV}, \qquad \bar{z} = \frac{\displaystyle\int_V z\,dV}{\displaystyle\int_V dV}.$$

- **The center of mass of a homogeneous plate of uniform thickness coincides with the centroid of its cross-sectional area** (Fig. 7.31). The center of mass of the plate coincides with the centroid of its volume, and we showed in Section 7.4 that the centroid of the volume of a plate of uniform thickness coincides with the centroid of its cross-sectional area.

- **The center of mass of a homogeneous slender bar of uniform cross-sectional area coincides approximately with the centroid of the axis of the bar** (Fig. 7.32a). The axis of the bar is defined to be the line through the centroid of its cross section. Let $dm = \rho A\,dL$, where A is the cross-sectional area of the bar and dL is a differential element of length of its axis (Fig. 7.32b). If we substitute this expression into Eqs. (7.26), they become the equations for the centroid of the axis:

$$\bar{x} = \frac{\displaystyle\int_L x\,dL}{\displaystyle\int_L dL}, \qquad \bar{y} = \frac{\displaystyle\int_L y\,dL}{\displaystyle\int_L dL}, \qquad \bar{z} = \frac{\displaystyle\int_L z\,dL}{\displaystyle\int_L dL}.$$

This result is approximate because the center of mass of the element dm does not coincide with the centroid of the cross section in regions where the bar is curved.

Study Questions

1. If you want to represent the weight of an object as a single equivalent force, at what point must the force act?
2. How is the mass density of an object defined?
3. What is the relationship between the mass density ρ and the weight density γ?
4. If an object is homogeneous, what do you know about the position of its center of mass?

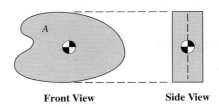

Front View   Side View

Figure 7.31
A plate of uniform thickness.

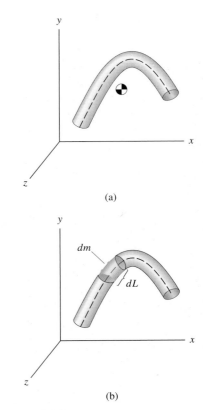

(a)

(b)

Figure 7.32
(a) A slender bar and the centroid of its axis.
(b) The element dm.

Example 7.16

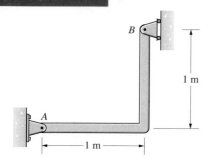

Figure 7.33

Representing the Weight of an L-Shaped Bar

The mass of the homogeneous slender bar in Fig. 7.33 is 80 kg. What are the reactions at A and B?

Strategy

We determine the reactions in two ways.

First Method We represent the weight of each straight segment of the bar by a force acting at the center of mass of the segment.

Second Method We determine the center of mass of the bar by determining the centroid of its axis and represent the weight of the bar by a single force acting at the center of mass.

Solution

First Method In the free-body diagram in Fig. a, we place half of the weight of the bar at the center of mass of each straight segment. From the equilibrium equations

$$\Sigma F_x = A_x - B = 0,$$

$$\Sigma F_y = A_y - (40)(9.81) - (40)(9.81) = 0,$$

$$\Sigma M_{(point\ A)} = (1)B - (1)(40)(9.81) - (0.5)(40)(9.81) = 0,$$

we obtain $A_x = 589$ N, $A_y = 785$ N, and $B = 589$ N.

Second Method We can treat the centerline of the bar as a composite line composed of two straight segments (Fig. b). The coordinates of the centroid of the composite line are

$$\bar{x} = \frac{\bar{x}_1 L_1 + \bar{x}_2 L_2}{L_1 + L_2} = \frac{(0.5)(1) + (1)(1)}{1 + 1} = 0.75\ \text{m},$$

$$\bar{y} = \frac{\bar{y}_1 L_1 + \bar{y}_2 L_2}{L_1 + L_2} = \frac{(0)(1) + (0.5)(1)}{1 + 1} = 0.25\ \text{m}.$$

In the free-body diagram in Fig. c, we place the weight of the bar at its center of mass. From the equilibrium equations

$$\Sigma F_x = A_x - B = 0,$$

$$\Sigma F_y = A_y - (80)(9.81) = 0$$

$$\Sigma M_{(point\ A)} = (1)B - (0.75)(80)(9.81) = 0,$$

we again obtain $A_x = 589$ N, $A_y = 785$ N, and $B = 589$ N.

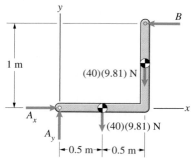

(a) Placing the weights of the straight segments at their centers of mass.

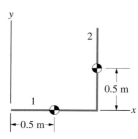

(b) Centroids of the straight segments of the axis.

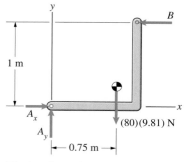

(c) Placing the weight of the bar at its center of mass.

Example 7.17

Cylinder with Nonuniform Density

Determine the mass of the cylinder in Fig. 7.34 and the position of its center of mass if (a) it is homogeneous with mass density ρ_0; (b) its density is given by the equation $\rho = \rho_0(1 + x/L)$.

Strategy

In (a), the mass of the cylinder is simply the product of its mass density and its volume and the center of mass is located at the centroid of its volume. In (b), the cylinder is nonhomogeneous and we must use Eqs. (7.24) and (7.26) to determine its mass and center of mass.

Solution

(a) The volume of the cylinder is LA, so its mass is $\rho_0 LA$. Since the center of mass is coincident with the centroid of the volume of the cylinder, the coordinates of the center of mass are $\bar{x} = \frac{1}{2}L$, $\bar{y} = 0$, $\bar{z} = 0$.

(b) We can determine the mass of the cylinder by using an element of volume dV in the form of a disk of thickness dx (Fig. a). The volume $dV = A\,dx$. The mass of the cylinder is

$$m = \int_V \rho\,dV = \int_0^L \rho_0\left(1 + \frac{x}{L}\right)A\,dx = \frac{3}{2}\rho_0 AL.$$

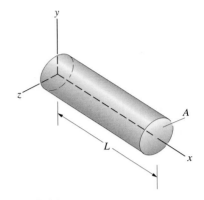

Figure 7.34

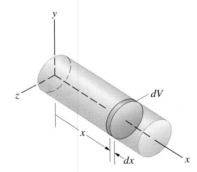

(a) An element of volume dV in the form of a disk.

The x coordinate of the center of mass is

$$\bar{x} = \frac{\displaystyle\int_V x\rho\,dV}{\displaystyle\int_V \rho\,dV} = \frac{\displaystyle\int_0^L \rho_0\left(x + \frac{x^2}{L}\right)A\,dx}{\dfrac{3}{2}\rho_0 AL} = \frac{5}{9}L.$$

Because the density does not depend on y or z, we know from symmetry that $\bar{y} = 0$ and $\bar{z} = 0$.

Discussion

Notice that the center of mass of the nonhomogeneous cylinder is *not* located at the centroid of its volume.

7.8 Centers of Mass of Composite Objects

You can easily determine the center of mass of an object consisting of a combination of parts if you know the centers of mass of its parts. The coordinates of the center of mass of a composite object composed of parts with masses $m_1, m_2, \ldots,$ are

$$\bar{x} = \frac{\sum_i \bar{x}_i m_i}{\sum_i m_i}, \qquad \bar{y} = \frac{\sum_i \bar{y}_i m_i}{\sum_i m_i}, \qquad \bar{z} = \frac{\sum_i \bar{z}_i m_i}{\sum_i m_i}, \qquad (7.27)$$

where $\bar{x}_i, \bar{y}_i, \bar{z}_i$ are the coordinates of the centers of mass of the parts. Because the weights of the parts are related to their masses by $W_i = gm_i$, Eqs. (7.27) can also be expressed as

$$\bar{x} = \frac{\sum_i \bar{x}_i W_i}{\sum_i W_i}, \qquad \bar{y} = \frac{\sum_i \bar{y}_i W_i}{\sum_i W_i}, \qquad \bar{z} = \frac{\sum_i \bar{z}_i W_i}{\sum_i W_i}. \qquad (7.28)$$

When you know the masses or weights and the centers of mass of the parts of a composite object, you can use these equations to determine its center of mass.

Determining the center of mass of a composite object requires three steps:

1. Choose the parts—Try to divide the object into parts whose centers of mass you know or can easily determine.

2. Determine the values for the parts—Determine the center of mass and the mass or weight of each part. Watch for instances of symmetry that can simplify your task.

3. Calculate the center of mass—Use Eqs. (7.27) or (7.28) to determine the center of mass of the composite object.

Example 7.18

Center of Mass of a Composite Object

The L-shaped machine part in Fig. 7.35 is composed of two homogeneous bars. Bar 1 is tungsten alloy with mass density 14,000 kg/m³, and bar 2 is steel with mass density 7800 kg/m³. Determine the center of mass of the machine part.

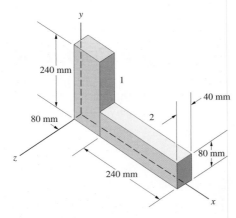

Figure 7.35

Solution

The volume of bar 1 is

$$(80)(240)(40) = 7.68 \times 10^5 \text{ mm}^3 = 7.68 \times 10^{-4} \text{ m}^3,$$

so its mass is $(7.68 \times 10^{-4})(1.4 \times 10^4) = 10.75 \ kg$. The center of mass of bar 1 coincides with the centroid of its volume: $\bar{x}_1 = 40$ mm, $\bar{y}_1 = 120$ mm, $\bar{z}_1 = 0$.

Bar 2 has the same volume as bar 1, so its mass is (7.68×10^{-4}) $(7.8 \times 10^3) = 5.99$ kg. The coordinates of its center of mass are $\bar{x}_2 = 200$ mm, $\bar{y}_2 = 40$ mm, $\bar{z}_2 = 0$. Using the information summarized in Table 7.7, we obtain the x coordinate of the center of mass,

$$\bar{x} = \frac{\bar{x}_1 m_1 + \bar{x}_2 m_2}{m_1 + m_2} = \frac{(40)(10.75) + (200)(5.99)}{10.75 + 5.99} = 97.2 \text{ mm},$$

and the y coordinate,

$$\bar{y} = \frac{\bar{y}_1 m_1 + \bar{y}_2 m_2}{m_1 + m_2} = \frac{(120)(10.75) + (40)(5.99)}{10.75 + 5.99} = 91.4 \text{ mm}.$$

Because of the symmetry of the object, $\bar{z} = 0$.

Table 7.7 Information for determining the center of mass

	m_1 (kg)	$\bar{x}_i$ (mm)	$\bar{x}_i m_i$ (mm-kg)	$\bar{y}_i$ (mm)	$\bar{y}_i m_i$ (mm-kg)
Bar 1	10.75	40	(40)(10.75)	120	(120)(10.75)
Bar 2	5.99	200	(200)(5.99)	40	(40)(5.99)

Example 7.19

Center of Mass of a Composite Object

The composite object in Fig. 7.36 consists of a bar welded to a cylinder. The homogeneous bar is aluminum (weight density 168 lb/ft^3), and the homogeneous cylinder is bronze (weight density 530 lb/ft^3). Determine the center of mass of the object.

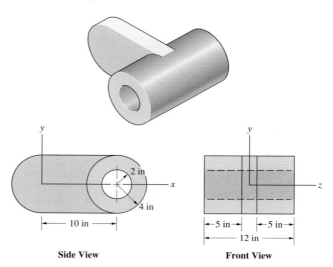

Figure 7.36   Side View   Front View

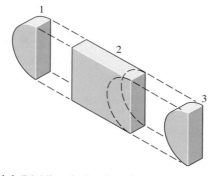

(a) Dividing the bar into three parts.

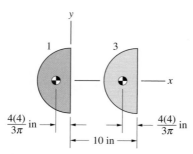

(b) The centroids of the two semicircular parts.

Strategy

We can determine the weight of each homogeneous part by multiplying its volume by its weight density. We also know that the center of mass of each part coincides with the centroid of its volume. The centroid of the cylinder is located at its center, but we must determine the location of the centroid of the bar by treating it as a composite volume.

Solution

The volume of the cylinder is $12[\pi(4)^2 - \pi(2)^2] = 452$ in.$^3 = 0.262$ ft^3, so its weight is

$$W_{(cylinder)} = (0.262)(530) = 138.8 \text{ lb.}$$

The x coordinate of its center of mass is $\bar{x}_{(cylinder)} = 10$ in.

The volume of the bar is $(10)(8)(2) + \frac{1}{2}\pi(4)^2(2) - \frac{1}{2}\pi(4)^2(2) = 160$ in.$^3 = 0.0926$ ft^3, and its weight is

$$W_{(bar)} = (0.0926)(168) = 15.6 \text{ lb.}$$

We can determine the centroid of the volume of the bar by treating it as a composite volume consisting of three parts (Fig. a). Part 3 is a semicircular "cutout." The centroids of part 1 and the semicircular cutout 3 are located at the centroids of their semicircular cross sections (Fig b). Using the information summarized in Table 7.8, we have

Table 7.8 Information for determining the x coordinate of the centroid of the bar

	$\bar{x}_i$ (in.)	V_i (in^3)	$\bar{x}_i V_i$ (in^4)
Part 1	$-\dfrac{4(4)}{3\pi}$	$\frac{1}{2}\pi(4)^2(2)$	$-\dfrac{4(4)}{3\pi}\left[\frac{1}{2}\pi(4)^2(2)\right]$
Part 2	5	$(10)(8)(2)$	$5\big[(10)(8)(2)\big]$
Part 3	$10 - \dfrac{4(4)}{3\pi}$	$-\frac{1}{2}\pi(4)^2(2)$	$-\left[10 - \dfrac{4(4)}{3\pi}\right]\left[\frac{1}{2}\pi(4)^2(2)\right]$

$$\bar{x}_{(bar)} = \frac{\bar{x}_1 V_1 + \bar{x}_2 V_2 + \bar{x}_3 V_3}{V_1 + V_2 + V_3}$$

$$= \frac{-\dfrac{4(4)}{3\pi}\left[\frac{1}{2}\pi(4)^2(2)\right] + 5\big[(10)(8)(2)\big] - \left[10 - \dfrac{4(4)}{3\pi}\right]\left[\frac{1}{2}\pi(4)^2(2)\right]}{\frac{1}{2}\pi(4)^2(2) + (10)(8)(2) - \frac{1}{2}\pi(4)^2(2)}$$

$$= 1.86 \text{ in.}$$

Therefore the x coordinate of the center of mass of the composite object is

$$\bar{x} = \frac{\bar{x}_{(bar)} W_{(bar)} + \bar{x}_{(cylinder)} W_{(cylinder)}}{W_{(bar)} + W_{(cylinder)}}$$

$$= \frac{(1.86)(15.6) + (10)(138.8)}{15.6 + 138.8} = 9.18 \text{ in.}$$

Because of the symmetry of the bar, the y and z coordinates of its center of mass are $\bar{y} = 0$ and $\bar{z} = 0$.

Example 7.20

Application to Engineering:

Centers of Mass of Vehicles

A car is placed on a platform that measures the normal force exerted by each tire independently (Fig. 7.37). Measurements made with the platform horizontal and with the platform tilted at $\alpha = 15°$ are shown in Table 7.9. Determine the position of the car's center of mass.

Table 7.9 Measurements of the normal forces exerted by the tires

Wheelbase = 2.82 m		
Track = 1.55 m	**Measured Loads (N)**	
	$\alpha = 0$	$\alpha = 15°$
Left front wheel, N_{LF}	5104	4463
Right front wheel, N_{RF}	5027	4396
Left rear wheel, N_{LR}	3613	3956
Right rear wheel, N_{RR}	3559	3898

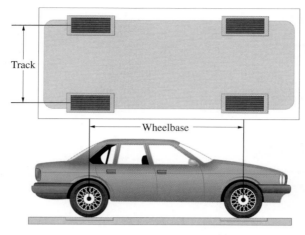

Figure 7.37

Solution

We draw the free-body diagram of the car when the platform is in the horizontal position in Figs. a and b. The car's weight is

$$W = N_{LF} + N_{RF} + N_{LR} + N_{RR}$$

$$= 5104 + 5027 + 3613 + 3559$$

$$= 17{,}303 \text{ N}$$

From Fig. a, we obtain the equilibrium equation

$$\Sigma M_{(z \text{ axis})} = (\text{wheelbase})(N_{LF} + N_{RF}) - \bar{x}W = 0,$$

which we can solve for $\bar{x}$:

$$\bar{x} = \frac{(\text{wheelbase})(N_{LF} + N_{RF})}{W}$$

$$= \frac{(2.82)(5104 + 5027)}{17{,}303}$$

$$= 1.651 \text{ m}.$$

From Fig. b,

$$\Sigma M_{(x \text{ axis})} = \bar{z}W - (\text{track})(N_{RF} + N_{RR}) = 0,$$

which we can solve for $\bar{z}$:

$$\bar{z} = \frac{(\text{track})(N_{RF} + N_{RR})}{W}$$

$$= \frac{(1.55)(5027 + 3559)}{17{,}303}$$

$$= 0.769 \text{ m}.$$

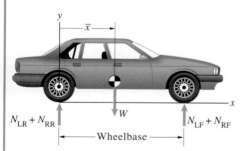

(a) Side view of the free-body diagram with the platform horizontal.

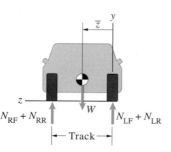

(b) Front view of the free-body diagram with the platform horizontal.

Now that we know $\bar{x}$, we can determine $\bar{y}$ from the free-body diagram of the car when the platform is in the tilted position (Fig. c). From the equilibrium equation

$$\Sigma M_{(z\,\text{axis})} = (\text{wheelbase})(N_{\text{LF}} + N_{\text{RF}}) + \bar{y}W\sin 15° - \bar{x}W\cos 15°$$
$$= 0,$$

we obtain

$$\bar{y} = \frac{\bar{x}W\cos 15° - (\text{wheelbase})(N_{\text{LF}} + N_{\text{RF}})}{W\sin 15°}$$

$$= \frac{(1.651)(17{,}303)\cos 15° - (2.82)(4463 + 4396)}{17{,}303\sin 15°}$$

$$= 0.584 \text{ m.}$$

Notice that we could not have determined $\bar{y}$ without the measurements made with the car in the tilted position.

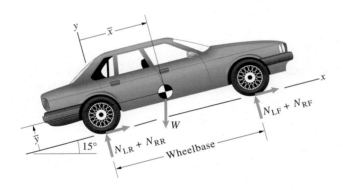

(c) Side view of the free-body diagram with the platform tilted.

$\mathcal{D}$esign Issues

The location of the center of mass of a vehicle affects its operation and performance. The forces exerted on the suspensions and wheels of cars and train coaches, the tractions their wheels create, and their dynamic behaviors are affected by the locations of their centers of mass. Not only are the performances of airplanes affected by the locations of their centers of mass, they cannot fly unless their centers of mass lie within prescribed bounds. For engineers who design vehicles, the position of the center of mass is one of the principal parameters governing decisions about the configuration of the vehicle and the layout of its contents. In testing new designs of both land vehicles and airplanes, the position of the center of mass is affected by the configuration of the particular vehicle and the weights and locations of stowage and passengers. It is often necessary to locate the center of mass experimentally by a technique such as the one we have described. Such experimental measurements are also used to confirm center of mass locations predicted by calculations made during design.

Chapter Summary

Centroids

A *centroid* is a weighted average position. The coordinates of the centroid of an area A in the x–y plane are

$$\bar{x} = \frac{\displaystyle\int_A x\,dA}{\displaystyle\int_A dA}, \qquad \bar{y} = \frac{\displaystyle\int_A y\,dA}{\displaystyle\int_A dA}. \qquad\qquad \text{Eqs. (7.6), (7.7)}$$

The coordinates of the centroid of a *composite area* composed of parts $A_1, A_2, \ldots$, are

$$\bar{x} = \frac{\displaystyle\sum_i \bar{x}_i A_i}{\displaystyle\sum_i A_i}, \qquad \bar{y} = \frac{\displaystyle\sum_i \bar{y}_i A_i}{\displaystyle\sum_i A_i}. \qquad\qquad \text{Eq. (7.9)}$$

Similar equations define the centroids of volumes [Eqs. (7.15) and (7.17)] and lines [(Eqs. (7.16) and (7.18)].

Distributed Forces

A force distributed along a line is described by a function w, defined such that the force on a differential element dx of the line is $w\,dx$. The force exerted by a distributed load is

$$F = \int_L w\,dx, \qquad\qquad \text{Eq. (7.10)}$$

and the moment about the origin is

$$M = \int_L xw\,dx. \qquad\qquad \text{Eq. (7.11)}$$

The force F is equal to the "area" between the function w, and the x axis and is equivalent to the distributed load if it is placed at the centroid of the "area."

The Pappus–Guldinus Theorems

First Theorem Consider a line of length L in the x–y plane with centroid $\bar{x}, \bar{y}$. The area A of the surface generated by revolving the line about the x axis is

$$A = 2\pi\bar{y}L. \qquad\qquad \text{Eq. (7.19)}$$

Second Theorem Let A be an area in the x–y plane with centroid $\bar{x}, \bar{y}$. The volume V generated by revolving A about the x axis is

$$V = 2\pi\bar{y}A. \qquad\qquad \text{Eq. (7.21)}$$

Centers of Mass

The *center of mass* of an object is the centroid of its mass. The weight of an object can be represented by a single equivalent force acting at its center of mass.

The *mass density* ρ is defined such that the mass of a differential element of volume is $dm = \rho \, dV$. An object whose mass density is uniform throughout its volume is said to be *homogeneous*. The *weight density* $\gamma = g\rho$.

The coordinates of the center of mass of an object are

$$\bar{x} = \frac{\int_V \rho x \, dV}{\int_V \rho \, dV}, \qquad \bar{y} = \frac{\int_V \rho y \, dV}{\int_V \rho \, dV}, \qquad \bar{z} = \frac{\int_V \rho z \, dV}{\int_V \rho \, dV}. \qquad \text{Eq. (7.26)}$$

The center of mass of a homogeneous object coincides with the centroid of its volume. The center of mass of a homogeneous plate of uniform thickness coincides with the centroid of its cross-sectional area. The center of mass of a homogeneous slender bar of uniform cross-sectional area coincides approximately with the centroid of the axis of the bar.

Review Problems

7.1 Determine the centroid of the area by letting dA be a vertical strip of width dx.

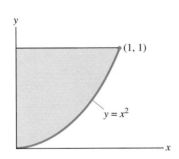

P7.1

7.2 Determine the centroid of the area in Problem 7.1 by letting dA be a horizontal strip of height dy.

7.3 Determine the centroid of the area.

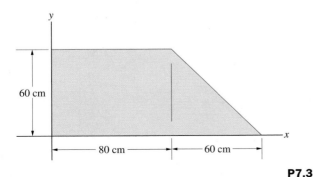

P7.3

7.4 Determine the centroid of the area.

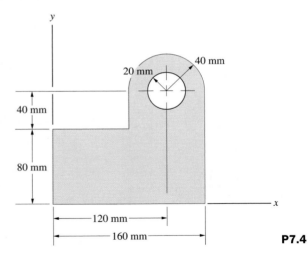

P7.4

7.5 The cantilever beam is subjected to a triangular distributed load. What are the reactions at A?

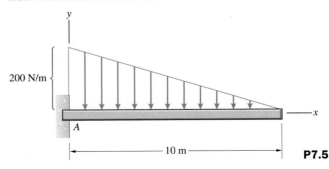

P7.5

7.6 What is the axial load in member *BD* of the frame?

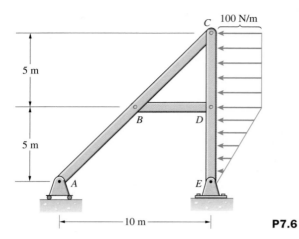

P7.6

7.7 An engineer estimates that the maximum wind load on the 40-m tower in Fig. a is described by the distributed load in Fig. b. The tower is supported by three cables, *A*, *B*, and *C*, from the top of the tower to equally spaced points 15 m from the bottom of the tower (Fig. c). If the wind blows from the west and cables *B* and *C* are slack, what is the tension in cable *A*? (Model the base of the tower as a ball and socket support.)

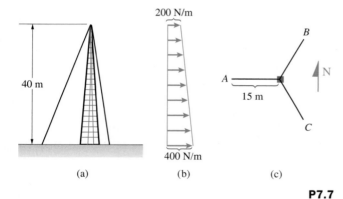

(a)   (b)   (c)

P7.7

7.8 If the wind in Problem 7.7 blows from the east and cable *A* is slack, what are the tensions in cables *B* and *C*?

7.9 Estimate the centroid of the volume of the *Apollo* lunar return configuration (not including its rocket nozzle) by treating it as a cone and a cylinder.

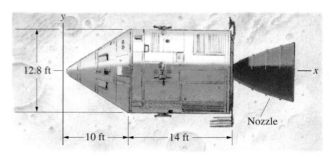

P7.9

7.10 The shape of the rocket nozzle of the *Apollo* lunar return configuration is approximated by revolving the curve shown around the *x* axis. In terms of the coordinate system shown, determine the centroid of the volume of the nozzle.

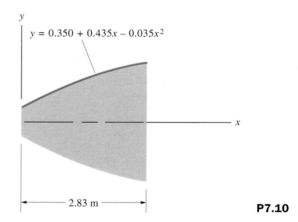

$y = 0.350 + 0.435x - 0.035x^2$

P7.10

7.11 Determine the volume of the volume of revolution.

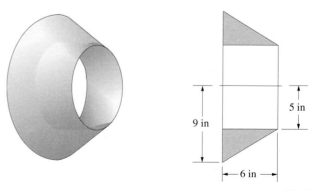

P7.11

7.12 Determine the surface area of the volume of revolution in Problem 7.11.

7.13 Determine the *y* coordinate of the center of mass of the homogeneous steel plate.

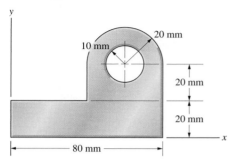

P7.13

7.14 Determine the *x* coordinate of the center of mass of the homogeneous steel plate.

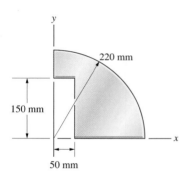

P7.14

7.15 The area of the homogeneous plate is 10 ft². The vertical reactions on the plate at *A* and *B* are 80 lb and 84 lb, respectively. Suppose that you want to equalize the reactions at *A* and *B* by drilling a 1-ft-diameter hole in the plate. What horizontal distance from *A* should the center of the hole be? What are the resulting reactions at *A* and *B*?

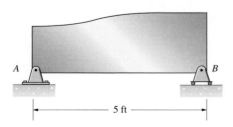

P7.15

7.16 The plate is of uniform thickness and is made of homogeneous material whose mass per unit area of the plate is 2 kg/m². The vertical reactions at *A* and *B* are 6 N and 10 N, respectively. What is the *x* coordinate of the centroid of the hole?

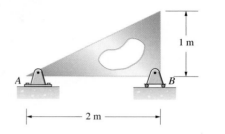

P7.16

7.17 Determine the center of mass of the homogeneous sheet of metal.

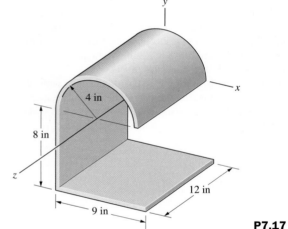

P7.17

7.18 Determine the center of mass of the homogeneous object.

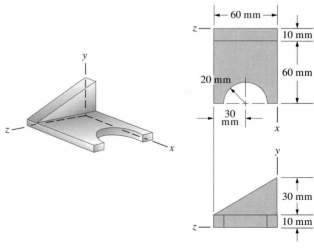

P7.18

7.19 Determine the center of mass of the homogeneous object.

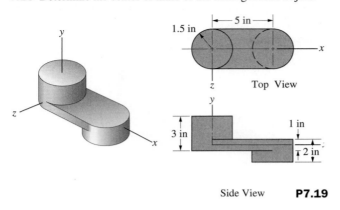

Top View

Side View   **P7.19**

7.20 The arrangement shown can be used to determine the location of the center of mass of a person. A horizontal board has a pin support at A and rests on a scale that measures weight at B. The distance from A to B is 2.3 m. When the person is not on the board, the scale at B measures 90 N.

(a) When a 63-kg person is in position (1), the scale at B measures 496 N. What is the x coordinate of the person's center of mass?

(b) When the same person is in position (2), the scale measures 523 N. What is the x coordinate of his center of mass?

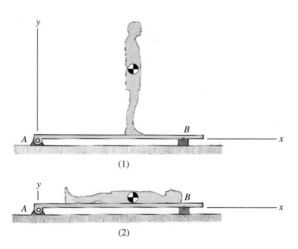

(1)

(2)   **P7.20**

7.21 If a string is tied to the slender bar at A and the bar is allowed to hang freely, what will be the angle between AB and the vertical?

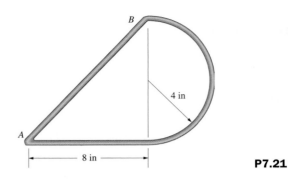

P7.21

7.22 The positions of the centers of three homogeneous spheres of equal radii are shown. The mass density of sphere 1 is ρ_0, the mass density of sphere 2 is $1.2\rho_0$, and the mass density of sphere 3 is $1.4\rho_0$.

(a) Determine the centroid of the volume of the three spheres.

(b) Determine the center of mass of the three spheres.

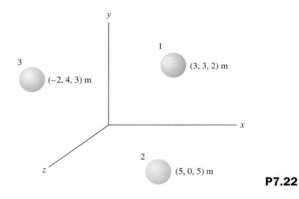

P7.22

7.23 The mass of the moon is 0.0123 times the mass of the earth. If the moon's center of mass is 383,000 km from the center of mass of the earth, what is the distance from the center of mass of the earth to the center of mass of the earth–moon system?

Project   Construct a homogeneous thin flat plate with the shape shown. (Use the cardboard back of a pad of paper to construct the plate. Choose your dimensions so that the plate is as large as possible.) Calculate the location of the center of mass of the plate. Measuring as carefully as possible, mark the center of mass clearly on both sides of the plate. Then carry out the following experiments.

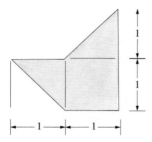

(a) Balance the plate on your finger (Fig. a) and observe that it balances at its center of mass. Explain the result of this experiment by drawing a free-body diagram of the plate.
(b) This experiment requires a needle or slender nail, a length of string, and a small weight. Tie the weight to one end of the string and make a small loop at the other end. Stick the needle through the plate at any point other than its center of mass. Hold the needle horizontal so that the plate hangs freely from it (Fig. b). Use the loop to hang the weight from the needle, and let the weight hang freely so that the string lies along the face of the plate. Observe that the string passes through the center of mass of the plate. Repeat this experiment several times, sticking the needle through various points on the plate. Explain the results of this experiment by drawing a free-body diagram of the plate.
(c) Hold the plate so that the plane of the plate is vertical, and throw the plate upward, spinning it like a Frisbee. Observe that the plate spins about its center of mass.

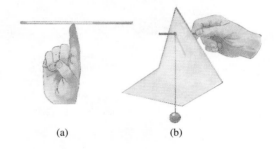

(a)   (b)

A beam's resistance to bending and ability to support loads depend on a property of its cross section called the moment of inertia. In this chapter we define and calculate moments of inertia of areas.

Moments of Inertia

The quantities called moments of inertia arise repeatedly in analyses of engineering problems. Moments of inertia of areas are used in the study of distributed forces and in calculating deflections of beams. The moment exerted by the pressure on a submerged flat plate can be expressed in terms of the moment of inertia of the plate's area. In dynamics, mass moments of inertia are used in calculating the rotational motions of objects. We show how to calculate the moments of inertia of simple areas and objects and then use results called parallel-axis theorems to calculate moments of inertia of more complex areas and objects.

Areas

8.1 Definitions

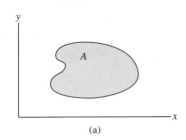

(a)

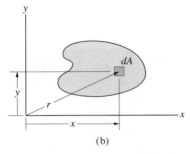

(b)

Figure 8.1
(a) An area A in the x–y plane.
(b) A differential element of A.

The moments of inertia of an area are integrals similar in form to those used to determine the centroid of an area. Consider an area A in the x–y plane (Fig. 8.1a). Four moments of inertia of A are defined:

1. **Moment of inertia about the x axis:**

$$I_x = \int_A y^2 \, dA, \tag{8.1}$$

where y is the y coordinate of the differential element of area dA (Fig. 8.1b). This moment of inertia is sometimes expressed in terms of the *radius of gyration* about the x axis, k_x, defined by

$$I_x = k_x^2 A. \tag{8.2}$$

2. **Moment of inertia about the y axis:**

$$I_y = \int_A x^2 \, dA, \tag{8.3}$$

where x is the x coordinate of the element dA (Fig. 8.1b). The radius of gyration about the y axis, k_y, is defined by

$$I_y = k_y^2 A. \tag{8.4}$$

3. **Product of inertia:**

$$I_{xy} = \int_A xy \, dA. \tag{8.5}$$

4. **Polar moment of inertia:**

$$J_O = \int_A r^2 \, dA, \tag{8.6}$$

where r is the radial distance from the origin of the coordinate system to dA (Fig. 8.1b). The radius of gyration about the origin, k_O, is defined by

$$J_O = k_O^2 A. \tag{8.7}$$

The polar moment of inertia is equal to the sum of the moments of inertia about the x and y axes:

$$J_O = \int_A r^2 \, dA = \int_A (y^2 + x^2) \, dA = I_x + I_y.$$

Substituting the expressions for the moments of inertia in terms of the radii of gyration into this equation, we obtain

$$k_O^2 = k_x^2 + k_y^2.$$

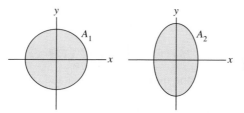

Figure 8.2
The areas $A_1 = A_2$. Based on their shapes, conclusions can be made about the relative sizes of their moments of inertia:
$$(I_x)_2 > (I_x)_1, \quad (I_y)_2 < (I_y)_1, \quad (J_O)_2 > (J_O)_1.$$

The dimensions of the moments of inertia of an area are $(\text{length})^4$, and the radii of gyration have dimensions of length. You can see that the definitions of the moments of inertia I_x, I_y, and J_O and the radii of gyration imply that they have positive values for any area. They cannot be negative or zero.

We can also make some qualitative deductions about the values of these quantities based on their definitions. In Fig. 8.2, $A_1 = A_2$. But because the contribution of a given element dA to the integral for I_x is proportional to the square of its perpendicular distance from the x axis, the value of I_x is larger for A_2 than for A_1: $(I_x)_2 > (I_x)_1$. For the same reason, $(I_y)_2 < (I_y)_1$ and $(J_O)_2 > (J_O)_1$. The circular areas in Fig. 8.3 are identical, but because of their positions relative to the coordinate system, $(I_x)_2 > (I_x)_1$, $(I_y)_2 > (I_y)_1$, and $(J_O)_2 > (J_O)_1$.

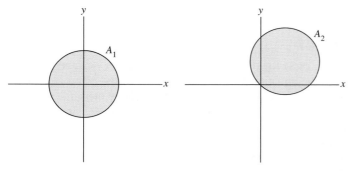

Figure 8.3
These identical areas have different moments of inertia depending on the position of the xy coordinate system:
$$(I_x)_2 > (I_x)_1, \quad (I_y)_2 > (I_y)_1, \quad (J_O)_2 > (J_O)_1.$$

The areas A_2 and A_3 in Fig. 8.4 are obtained by rotating A_1 about the y and x axes, respectively. We can see from the definitions that the moments of inertia I_x, I_y, and J_O of these areas are equal. The products of inertia $(I_{xy})_2 = -(I_{xy})_1$ and $(I_{xy})_3 = -(I_{xy})_1$: For each element dA of A_1 with coordinates (x, y), there is a corresponding element of A_2 with coordinates $(-x, y)$ and a corresponding element of A_3 with coordinates $(x, -y)$. These results also imply that if an area is symmetric about either the x axis or the y axis, its product of inertia is zero.

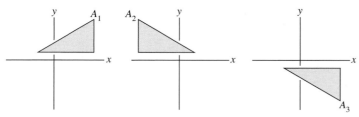

Figure 8.4
The areas A_2 and A_3 are obtained by rotating the area A_1 about the y and x axes, respectively. The moments of inertia I_x, I_y, and J_O of these areas are equal. The products of inertia are related by
$$(I_{xy})_2 = -(I_{xy})_1, \quad (I_{xy})_3 = -(I_{xy})_1.$$

Example 8.1

Moments of Inertia of a Triangular Area

Determine the moments of inertia and radii of gyration of the triangular area in Fig. 8.5.

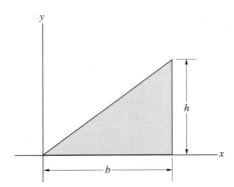

Figure 8.5

Strategy

Equation (8.3) for the moment of inertia about the y axis is very similar in form to the equation for the x coordinate of the centroid of an area, and we can evaluate it for this triangular area in exactly the same way: by using a differential element of area dA in the form of a vertical strip of width dx. We can then show that I_x and I_{xy} can be evaluated using the same element of area. The polar moment of inertia J_O is the sum of I_x and I_y.

Solution

Let dA be the vertical strip in Fig. a. The height of the strip is $(h/b)x$, so $dA = (h/b)x\,dx$. To integrate over the entire area, we must integrate with respect to x from $x = 0$ to $x = b$.

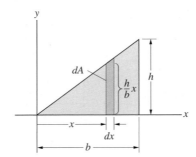

(a) An element dA in the form of a strip.

Moment of Inertia About the y Axis

$$I_y = \int_A x^2 \, dA = \int_0^b x^2 \left(\frac{h}{b} x\right) dx = \frac{h}{b}\left[\frac{x^4}{4}\right]_0^b = \frac{1}{4} hb^3.$$

The radius of gyration k_y is

$$k_y = \sqrt{\frac{I_y}{A}} = \sqrt{\frac{(1/4)hb^3}{(1/2)bh}} = \frac{1}{\sqrt{2}}\,b.$$

Moment of Inertia About the _x_ Axis   We will first determine the moment of inertia of the strip dA about the x axis while holding x and dx fixed. In terms of the element of area $dA_s = dx\,dy$ shown in Fig. b,

$$\left(I_x\right)_{\text{strip}} = \int_{\text{strip}} y^2\,dA_s = \int_0^{(h/b)x} \left(y^2\,dx\right)dy$$

$$= \left[\frac{y^3}{3}\right]_0^{(h/b)x} dx = \frac{h^3}{3b^3}\,x^3\,dx.$$

Integrating this expression with respect to x from $x = 0$ to $x = b$, we obtain the value of I_x for the entire area:

$$I_x = \int_0^b \frac{h^3}{3b^3}\,x^3\,dx = \frac{1}{12}\,bh^3.$$

The radius of gyration k_x is

$$k_x = \sqrt{\frac{I_x}{A}} = \sqrt{\frac{(1/12)bh^3}{(1/2)bh}} = \frac{1}{\sqrt{6}}\,h.$$

Product of Inertia   We can determine I_{xy} the same way we determined I_x. We first evaluate the product of inertia of the strip dA, holding x and dx fixed (Fig. b):

$$\left(I_{xy}\right)_{\text{strip}} = \int_{\text{strip}} xy\,dA_s = \int_0^{(h/b)x} \left(xy\,dx\right)dy$$

$$= \left[\frac{y^2}{2}\right]_0^{(h/b)x} x\,dx = \frac{h^2}{2b^2}\,x^3\,dx.$$

Integrating this expression with respect to x from $x = 0$ to $x = b$, we obtain the value of I_{xy} for the entire area:

$$I_{xy} = \int_0^b \frac{h^2}{2b^2}\,x^3\,dx = \frac{1}{8}\,b^2h^2.$$

Polar Moment of Inertia

$$J_O = I_x + I_y = \frac{1}{12}\,bh^3 + \frac{1}{4}\,hb^3.$$

The radius of gyration k_O is

$$k_O = \sqrt{k_x^2 + k_y^2} = \sqrt{\frac{1}{6}\,h^2 + \frac{1}{2}\,b^2}.$$

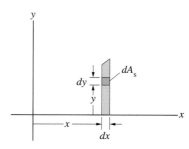

(b) An element of the strip element dA.

Discussion

As this example demonstrates, the integrals defining the moments and products of inertia are so similar in form to the integrals used to determine centroids of areas (Section 7.1) that you can use the same methods to evaluate them.

Example 8.2

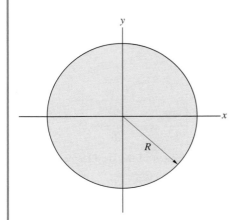

Figure 8.6

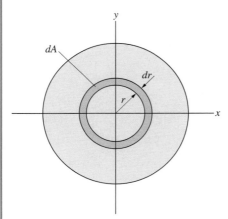

(a) An annular element dA.

Moments of Inertia of a Circular Area

Determine the moments of inertia and radii of gyration of the circular area in Fig. 8.6.

Strategy

We will first determine the polar moment of inertia J_O by integrating in terms of polar coordinates. We know from the symmetry of the area that $I_x = I_y$, and since $I_x + I_y = J_O$, the moments of inertia I_x and I_y are each equal to $\frac{1}{2} J_O$. We also know from the symmetry of the area that $I_{xy} = 0$.

Solution

By letting r change by an amount dr, we obtain an annular element of area $dA = 2\pi r \, dr$ (Fig. a). The polar moment of inertia is

$$J_O = \int_A r^2 \, dA = \int_0^R 2\pi r^3 \, dr = 2\pi \left[\frac{r^4}{4} \right]_0^R = \frac{1}{2} \pi R^4,$$

and the radius of gyration about O is

$$k_O = \sqrt{\frac{J_O}{A}} = \sqrt{\frac{(1/2)\pi R^4}{\pi R^2}} = \frac{1}{\sqrt{2}} R.$$

The moments of inertia about the x and y axes are

$$I_x = I_y = \frac{1}{2} J_O = \frac{1}{4} \pi R^4,$$

and the radii of gyration about the x and y axes are

$$k_x = k_y = \sqrt{\frac{I_x}{A}} = \sqrt{\frac{(1/4)\pi R^4}{\pi R^2}} = \frac{1}{2} R.$$

The product of inertia is zero:

$$I_{xy} = 0.$$

Discussion

The symmetry of this example saved us from having to integrate to determine I_x, I_y, and I_{xy}. Be alert for symmetry that can shorten your work. In particular, remember that $I_{xy} = 0$ if the area is symmetric about either the x or the y axis.

8.2 Parallel-Axis Theorems

In some situations the moments of inertia of an area are known in terms of a particular coordinate system but we need their values in terms of a different coordinate system. When the coordinate systems are parallel, the desired moments of inertia can be obtained by using the theorems we describe in this section. Furthermore, these theorems make it possible for us to determine the moments of inertia of a composite area when the moments of inertia of its parts are known.

Suppose that we know the moments of inertia of an area A in terms of a coordinate system $x'y'$ with its origin at the centroid of the area, and we wish to determine the moments of inertia in terms of a parallel coordinate system xy (Fig. 8.7a). We denote the coordinates of the centroid of A in the xy coordinate system by (d_x, d_y), and $d = \sqrt{d_x^2 + d_y^2}$ is the distance from the origin of the xy coordinate system to the centroid (Fig. 8.7b).

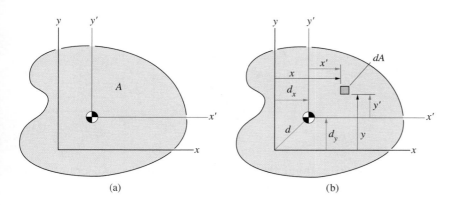

(a) (b)

Figure 8.7
(a) The area A and the coordinate systems $x'y'$ and xy.
(b) The differential element dA.

We need to obtain two preliminary results before deriving the parallel-axis theorems. In terms of the $x'y'$ coordinate system, the coordinates of the centroid of A are

$$\bar{x}' = \frac{\displaystyle\int_A x' \, dA}{\displaystyle\int_A dA}, \qquad \bar{y}' = \frac{\displaystyle\int_A y' \, dA}{\displaystyle\int_A dA}.$$

But the origin of the $x'y'$ coordinate system is located at the centroid of A, so $\bar{x}' = 0$ and $\bar{y}' = 0$. Therefore

$$\int_A x' \, dA = 0, \qquad \int_A y' \, dA = 0. \tag{8.8}$$

Moment of Inertia About the x Axis   In terms of the xy coordinate system, the moment of inertia of A about the x axis is

$$I_x = \int_A y^2 \, dA, \tag{8.9}$$

where y is the coordinate of the element of area dA relative to the xy coordinate system. From Fig. 8.7b, you can see that $y = y' + d_y$, where y' is the coordinate of dA relative to the $x'y'$ coordinate system. Substituting this expression into Eq. (8.9), we obtain

$$I_x = \int_A (y' + d_y)^2 \, dA = \int_A (y')^2 \, dA + 2d_y \int_A y' \, dA + d_y^2 \int_A dA.$$

The first integral on the right is the moment of inertia of A about the x' axis. From Eq. (8.8), the second integral on the right equals zero. Therefore we obtain

$$I_x = I_{x'} + d_y^2 A. \tag{8.10}$$

This is a *parallel-axis theorem*. It relates the moment of inertia of A about the x' axis through the centroid to the moment of inertia about the parallel axis x (Fig. 8.8).

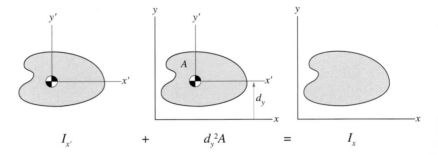

Figure 8.8
The parallel-axis theorem for the moment of inertia about the x axis.

Moment of Inertia About the y Axis   In terms of the xy coordinate system, the moment of inertia of A about the y axis is

$$I_y = \int_A x^2 \, dA = \int_A (x' + d_x)^2 \, dA$$

$$= \int_A (x')^2 \, dA + 2d_x \int_A x' \, dA + d_x^2 \int_A dA.$$

From Eq. (8.8), the second integral on the right equals zero. Therefore the parallel-axis theorem that relates the moment of inertia of A about the y' axis through the centroid to the moment of inertia about the parallel axis y is

$$I_y = I_{y'} + d_x^2 A. \tag{8.11}$$

Product of Inertia   The parallel-axis theorem for the product of inertia is

$$I_{xy} = I_{x'y'} + d_x d_y A. \tag{8.12}$$

Polar Moment of Inertia   The parallel-axis theorem for the polar moment of inertia is

$$J_O = J'_O + (d_x^2 + d_y^2)A = J'_O + d^2 A, \tag{8.13}$$

where d is the distance from the origin of the $x'y'$ coordinate system to the origin of the xy coordinate system.

How can the parallel-axis theorems be used to determine the moments of inertia of a composite area? Suppose that we want to determine the moment of inertia about the y axis of the area in Fig. 8.9a. We can divide it into a triangle, a semicircle, and a circular cutout, denoted parts 1, 2, and 3 (Fig. 8.9b). By using the parallel-axis theorem for I_y, can determine the moment of inertia of each part about the y axis. For example, the moment of inertia of part 2 (the semicircle) about the y axis is (Fig. 8.9c)

$$\left(I_y\right)_2 = \left(I_{y'}\right)_2 + \left(d_x\right)_2^2 A_2.$$

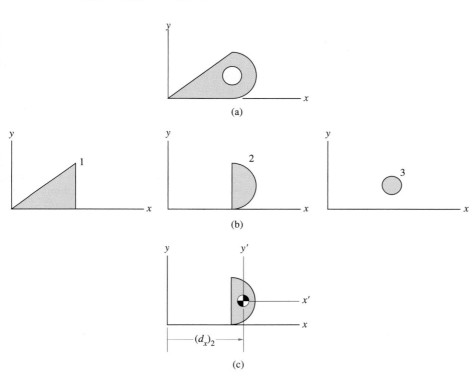

Figure 8.9
(a) A composite area.
(b) The three parts of the area.
(c) Determining $\left(I_y\right)_2$.

We must determine the values of $\left(I_{y'}\right)_2$ and $\left(d_x\right)_2$. Moments of inertia and centroid locations for some simple areas are tabulated in Appendix B. Once this procedure is carried out for each part, the moment of inertia of the composite area is

$$I_y = \left(I_y\right)_1 + \left(I_y\right)_2 - \left(I_y\right)_3.$$

Notice that the moment of inertia of the circular cutout is subtracted.

We see that determining a moment of inertia of a composite area in terms of a given coordinate system involves three steps:

1. Choose the parts—Try to divide the composite area into parts whose moments of inertia you know or can easily determine.
2. Determine the moments of inertia of the parts—Determine the moment of inertia of each part in terms of a parallel coordinate system with its

origin at the centroid of the part, and then use the parallel-axis theorem to determine the moment of inertia in terms of the given coordinate system.

3. Sum the results—Sum the moments of inertia of the parts (or subtract in the case of a cutout) to obtain the moment of inertia of the composite area.

Example 8.3

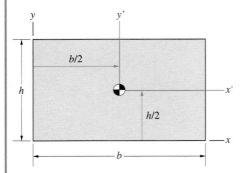

Figure 8.10

Demonstration of the Parallel Axis Theorems

The moments of inertia of the rectangular area in Fig. 8.10 in terms of the $x'y'$ coordinate system are $I_{x'} = \frac{1}{12}bh^3$, $I_{y'} = \frac{1}{12}hb^3$, $I_{x'y'} = 0$, and $J'_O = \frac{1}{12}(bh^3 + hb^3)$ (see Appendix B). Determine its moments of inertia in terms of the xy coordinate system.

Strategy

The $x'y'$ coordinate system has its origin at the centroid of the area and is parallel to the xy coordinate system. We can use the parallel-axis theorems to determine the moments of inertia of A in terms of the xy coordinate system.

Solution

The coordinates of the centroid in terms of the xy coordinate system are $d_x = b/2$, $d_y = h/2$. The moment of inertia about the x axis is

$$I_x = I_{x'} + d_y^2 A = \frac{1}{12}bh^3 + \left(\frac{1}{2}h\right)^2 bh = \frac{1}{3}bh^3.$$

The moment of inertia about the y axis is

$$I_y = I_{y'} + d_x^2 A = \frac{1}{12}hb^3 + \left(\frac{1}{2}b\right)^2 bh = \frac{1}{3}hb^3.$$

The product of inertia is

$$I_{xy} = I_{x'y'} + d_x d_y A = 0 + \left(\frac{1}{2}b\right)\left(\frac{1}{2}h\right)bh = \frac{1}{4}b^2h^2.$$

The polar moment of inertia is

$$J_O = J'_O + d^2 A = \frac{1}{12}(bh^3 + hb^3) + \left[\left(\frac{1}{2}b\right)^2 + \left(\frac{1}{2}h\right)^2\right]bh$$

$$= \frac{1}{3}(bh^3 + hb^3).$$

Discussion

Notice that we could also have determined J_O by using the relation

$$J_O = I_x + I_y = \frac{1}{3}bh^3 + \frac{1}{3}hb^3.$$

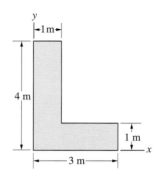

Figure 8.11

Example 8.4

Moments of Inertia of a Composite Area

Determine I_x, k_x, and I_{xy} for the composite area in Fig. 8.11.

Solution

Choose the Parts   We can determine the moments of inertia by dividing the area into the rectangular parts 1 and 2 shown in Fig. a.

Determine the Moments of Inertia of the Parts   For each part, we introduce a coordinate system $x'y'$ with its origin at the centroid of the part (Fig. b). The moments of inertia of the rectangular parts in terms of these coordinate systems are given in Appendix B. We then use the parallel-axis theorem to determine the moment of inertia of each part about the x axis (Table 8.1).

Table 8.1   Determining the moments of inertia of the parts about the x axis.

	d_y (m)	A (m²)	$I_{x'}$ (m⁴)	$I_x = I_{x'} + d_y^2$ (m⁴)
Part 1	2	(1)(4)	$\frac{1}{12}(1)(4)^3$	21.33
Part 2	0.5	(2)(1)	$\frac{1}{12}(2)(1)^3$	0.67

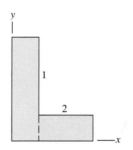

(a) Dividing the area into rectangles 1 and 2.

Sum the Results   The moment of inertia of the composite area about the x axis is

$$I_x = (I_x)_1 + (I_x)_2 = 21.33 + 0.67 = 22.00 \text{ m}^4.$$

The sum of the areas is $A = A_1 + A_2 = 6 \text{ m}^2$, so the radius of gyration about the x axis is

$$k_x = \sqrt{\frac{I_x}{A}} = \sqrt{\frac{22}{6}} = 1.91 \text{ m}.$$

Repeating this procedure, we determine I_{xy} for each part in Table 8.2. The product of inertia of the composite area is

$$I_{xy} = (I_{xy})_1 + (I_{xy})_2 = 4 + 2 = 6 \text{ m}^4.$$

Table 8.2   Determining the products of inertia of the parts in terms of the xy coordinate system

	d_x (m)	d_y (m)	A (m²)	$I_{x'y'}$	$I_{xy} = I_{x'y'} + d_x d_y A$ (m⁴)
Part 1	0.5	2	(1)(4)	0	4
Part 2	2	0.5	(2)(1)	0	2

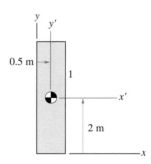

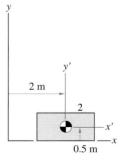

(b) Parallel coordinate systems $x'y'$ with origins at the centroids of the parts.

Discussion

The moments of inertia you obtain do not depend on how you divide a composite area into parts, and you will often have a choice of convenient ways to divide a given area. See Problem 8.28, in which we divide the composite area in this example in a different way.

Example 8.5

Moments of Inertia of a Composite Area

Determine I_y and k_y for the composite area in Fig. 8.12.

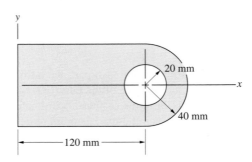

Figure 8.12

Solution

Choose the Parts We divide the area into a rectangle, a semicircle, and the circular cutout, calling them parts 1, 2, and 3, respectively (Fig. a).

Determine the Moments of Inertia of the Parts The moments of inertia of the parts in terms of the $x'y'$ coordinate systems and the location of the centroid of the semicircular part are given in Appendix B. In Table 8.3 we use the parallel-axis theorem to determine the moment of inertia of each part about the y axis.

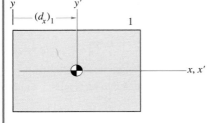

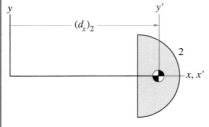

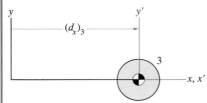

(a) Parts 1, 2, and 3.

Table 8.3 Determining the moments of inertia of the parts.

	d_x (mm)	A (mm^2)	$I_{y'}$ (mm^4)	$I_y = I_{y'} + d_x^2 A$ (mm^4)
Part 1	60	$(120)(80)$	$\frac{1}{12}(80)(120)^3$	4.608×10^7
Part 2	$120 + \dfrac{4(40)}{3\pi}$	$\frac{1}{2}\pi(40)^2$	$\left(\dfrac{\pi}{8} - \dfrac{8}{9\pi}\right)(40)^4$	4.744×10^7
Part 3	120	$\pi(20)^2$	$\frac{1}{4}\pi(20)^4$	1.822×10^7

Sum the Results The moment of inertia of the composite area about the y axis is

$$I_y = (I_y)_1 + (I_y)_2 - (I_y)_3 = (4.608 + 4.744 - 1.822) \times 10^7$$
$$= 7.530 \times 10^7 \text{ mm}^4.$$

The total area is

$$A = A_1 + A_2 - A_3 = (120)(80) + \frac{1}{2}\pi(40)^2 - \pi(20)^2$$
$$= 1.086 \times 10^4 \text{ mm}^2,$$

so the radius of gyration about the y axis is

$$k_y = \sqrt{\frac{I_y}{A}} = \sqrt{\frac{7.530 \times 10^7}{1.086 \times 10^4}} = 83.3 \text{ mm}.$$

Example 8.6

Application to Engineering:

Beam Design

The equal areas in Fig. 8.13 are candidates for the cross section of a beam. (A beam with the second cross section shown is called an I-beam.) Compare their moments of inertia about the x axis.

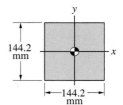

Solution

Square Cross Section From Appendix B, the moment of inertia of the square cross section about the x axis is

$$I_x = \frac{1}{12}(144.2)(144.2)^3 = 3.60 \times 10^7 \text{ mm}^4.$$

I-Beam Cross Section We can divide the area into the rectangular parts shown in Fig. a. Introducing coordinate systems $x'y'$ with their origins at the centroids of the parts (Fig. b), we use the parallel-axis theorem to determine the moments of inertia about the x axis (Table 8.4). Their sum is

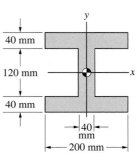

$$I_x = (I_x)_1 + (I_x)_2 + (I_x)_3 = (5.23 + 0.58 + 5.23) \times 10^7$$

$$= 11.03 \times 10^7 \text{ mm}^4.$$

The moment of inertia of the I-beam about the x axis is 3.06 times that of the square cross section of equal area.

Figure 8.13

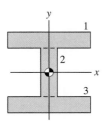

(a) Dividing the I-beam cross section into parts.

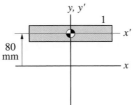

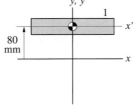

(b) Parallel coordinate systems $x'y'$ with origins at the centroids of the parts.

Table 8.4 Determining the moments of inertia of the parts about the x axis.

	d_y (mm)	A (mm^2)	$I_{x'}$ (mm^4)	$I_x = I_{x'} + d_y^2 A$ (mm^4)
Part 1	80	$(200)(40)$	$\frac{1}{12}(200)(40)^3$	5.23×10^7
Part 2	0	$(40)(120)$	$\frac{1}{12}(40)(120)^3$	0.58×10^7
Part 3	−80	$(200)(40)$	$\frac{1}{12}(200)(40)^3$	5.23×10^7

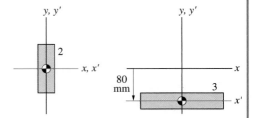

A simply supported beam.

Design Issues

A *beam* is a bar of material that supports lateral loads, meaning loads perpendicular to the axis of the bar. Two common types of beams are shown in Fig. 8.14 supporting a lateral load F. A beam with pinned ends is called a simply supported beam, and a beam with a single, built-in support is called a cantilever beam.

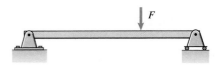

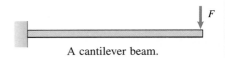

A cantilever beam.

Figure 8.14

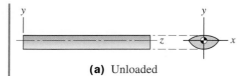

(a) Unloaded

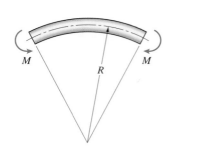

(b) Subjected to couples at the ends.

Figure 8.15
A beam with symmetrical cross section.

The lateral loads on a beam cause it to bend, and it must be stiff, or resistant to bending, to support them. A beam's resistance to bending depends directly on the moment of inertia of its cross-sectional area. Let's consider the beam in Fig. 8.15a. The cross section is symmetric about the y axis and the origin of the coordinate system is placed at its centroid. If the beam consists of a homogeneous structural material such as steel and it is subjected to couples at the ends, as shown in Fig. 8.15b, it bends into a circular arc of radius R. It can be shown that

$$R = \frac{EI_x}{M},$$

where I_x is the moment of inertia of the beam cross section about the x axis. The "modulus of elasticity" or "Young's modulus" E has different values for different materials. (This equation holds only if M is small enough so that the beam returns to its original shape when the couples are removed. The bending in Fig. 8.15b is exaggerated.) Thus the amount the beam bends for a given value of M depends on the material and the moment of inertia of its cross section. Increasing I_x increases the value of R, which means the resistance of the beam to bending is increased.

This explains in large part the cross sections of many of the beams you see in use—for example, in highway overpasses and in the frames of buildings. They are configured to increase their moments of inertia. The cross sections in Fig. 8.16 all have the same area. The numbers are the ratios of the moment of inertia I_x to the value of I_x for the solid square cross section.

(a) A box beam with thin walls.

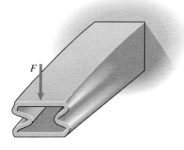

(b) Failure by buckling.

(c) Stabilizing the walls with a filler.

Figure 8.17

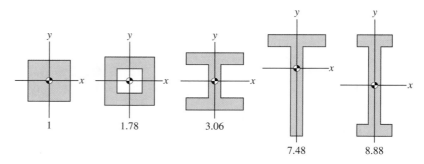

Figure 8.16
Typical beam cross sections and the ratio of I_x to the value for a solid square beam of equal cross-sectional area.

However, configuring the cross section of a beam to increase its moment of inertia can be carried too far. The "box" beam in Fig. 8.17a has a value of I_x that is four times as large as a solid square beam of the same cross-sectional area, but its walls are so thin they may "buckle," as shown in Fig. 8.17b. The stiffness implied by the beam's large moment of inertia is not realized because it becomes geometrically unstable. One solution used by engineers to achieve a large moment of inertia in a relatively light beam while avoiding failure due to buckling is to stabilize its walls by filling the beam with a light material such as honeycombed metal or foamed plastic (Fig. 8.17c).

8.3 Rotated and Principal Axes

Suppose that Fig. 8.18(a) is the cross section of a cantilever beam. If you apply a vertical force to the end of the beam, a larger vertical deflection results if the cross section is oriented as shown in Fig. 8.18(b) than if it is oriented as shown in Fig. 8.18(c). The minimum vertical deflection results when the beam's cross section is oriented so that the moment of inertia I_x is a maximum (Fig. 8.18d).

In many engineering applications you must determine moments of inertia of areas with various angular orientations relative to a coordinate system and also determine the orientation for which the value of a moment of inertia is a maximum or minimum. We discuss these procedures in this section.

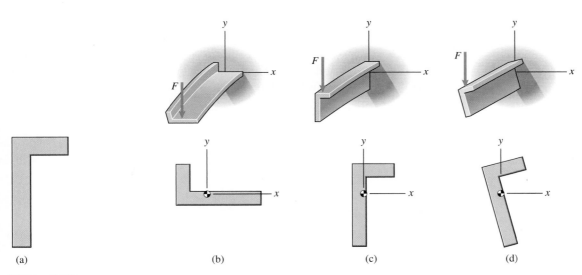

(a) (b) (c) (d)

Figure 8.18
(a) A beam cross section.
(b)–(d) Applying a lateral load with different orientations of the cross section.

Rotated Axes

Let's consider an area A, a coordinate system xy, and a second coordinate system $x'y'$ that is rotated through an angle θ relative to the xy coordinate system (Fig. 8.19a). Suppose that we know the moments of inertia of A in terms of the xy coordinate system. Our objective is to determine the moments of inertia in terms of the $x'y'$ coordinate system.

In terms of the radial distance r to a differential element of area dA and the angle α in Fig. 8.19b, the coordinates of dA in the xy coordinate system are

$$x = r \cos \alpha, \tag{8.14}$$

$$y = r \sin \alpha. \tag{8.15}$$

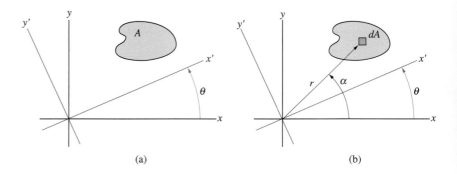

Figure 8.19
(a) The $x'y'$ coordinate system is rotated through an angle θ relative to the xy coordinate system.
(b) A differential element of area dA.

The coordinates of dA in the $x'y'$ coordinate system are

$$x' = r\cos(\alpha - \theta) = r(\cos\alpha\cos\theta + \sin\alpha\sin\theta), \quad (8.16)$$

$$y' = r\sin(\alpha - \theta) = r(\sin\alpha\cos\theta - \cos\alpha\sin\theta). \quad (8.17)$$

In Eqs. (8.16) and (8.17), we use identities for the cosine and sine of the difference of two angles (Appendix A). By substituting Eqs. (8.14) and (8.15) into Eqs. (8.16) and (8.17), we obtain equations relating the coordinates of dA in the two coordinate systems:

$$x' = x\cos\theta + y\sin\theta, \quad (8.18)$$

$$y' = -x\sin\theta + y\cos\theta. \quad (8.19)$$

We can use these expressions to derive relations between the moments of inertia of A in terms of the xy and $x'y'$ coordinate systems:

Moment of Inertia About the x′ Axis

$$I_{x'} = \int_A (y')^2\, dA = \int_A (-x\sin\theta + y\cos\theta)^2\, dA$$

$$= \cos^2\theta \int_A y^2\, dA - 2\sin\theta\cos\theta \int_A xy\, dA + \sin^2\theta \int_A x^2\, dA.$$

From this equation we obtain

$$I_{x'} = I_x\cos^2\theta - 2I_{xy}\sin\theta\cos\theta + I_y\sin^2\theta. \quad (8.20)$$

Moment of Inertia About the y′ Axis

$$I_{y'} = \int_A (x')^2\, dA = \int_A (x\cos\theta + y\sin\theta)^2\, dA$$

$$= \sin^2\theta \int_A y^2\, dA + 2\sin\theta\cos\theta \int_A xy\, dA + \cos^2\theta \int_A x^2\, dA.$$

This equation gives us the result

$$I_{y'} = I_x\sin^2\theta + 2I_{xy}\sin\theta\cos\theta + I_y\cos^2\theta. \quad (8.21)$$

Product of Inertia   In terms of the $x'y'$ coordinate system, the product of inertia of A is

$$I_{x'y'} = (I_x - I_y) \sin\theta \cos\theta + (\cos^2\theta - \sin^2\theta) I_{xy}. \qquad (8.22)$$

Polar Moment of Inertia   From Eqs. (8.20) and (8.21), the polar moment of inertia in terms of the $x'y'$ coordinate system is

$$J'_O = I_{x'} + I_{y'} = I_x + I_y = J_O.$$

Thus the value of the polar moment of inertia is unchanged by a rotation of the coordinate system.

Principal Axes

You have seen that the moments of inertia of A in terms of the $x'y'$ coordinate system depend on the angle θ in Fig. 8.19a. Let's consider the following question: For what values of θ is the moment of inertia $I_{x'}$ a maximum or minimum?

To consider this question, it is convenient to use the identities

$$\sin 2\theta = 2\sin\theta\cos\theta,$$

$$\cos 2\theta = \cos^2\theta - \sin^2\theta = 1 - 2\sin^2\theta = 2\cos^2\theta - 1.$$

With these expressions, we can write Eqs. (8.20)–(8.22) in the forms

$$I_{x'} = \frac{I_x + I_y}{2} + \frac{I_x - I_y}{2}\cos 2\theta - I_{xy}\sin 2\theta, \qquad (8.23)$$

$$I_{y'} = \frac{I_x + I_y}{2} - \frac{I_x - I_y}{2}\cos 2\theta + I_{xy}\sin 2\theta, \qquad (8.24)$$

$$I_{x'y'} = \frac{I_x - I_y}{2}\sin 2\theta + I_{xy}\cos 2\theta. \qquad (8.25)$$

We will denote a value of θ at which $I_{x'}$ is a maximum or minimum by θ_p. To determine θ_p, we evaluate the derivative of Eq. (8.23) with respect to 2θ and equate it to zero, obtaining

$$\tan 2\theta_p = \frac{2I_{xy}}{I_y - I_x}. \qquad (8.26)$$

If we set the derivative of Eq. (8.24) with respect to 2θ equal to zero to determine a value of θ for which $I_{y'}$ is a maximum or minimum, we again obtain Eq. (8.26). The second derivatives of $I_{x'}$ and $I_{y'}$ with respect to 2θ are opposite in sign,

$$\frac{d^2 I_{x'}}{d(2\theta)^2} = -\frac{d^2 I_{y'}}{d(2\theta)^2},$$

which means that at an angle θ_p for which $I_{x'}$ is a maximum, $I_{y'}$ is a minimum, and at an angle θ_p for which $I_{x'}$ is a minimum, $I_{y'}$ is a maximum.

A rotated coordinate system $x'y'$ that is oriented so that $I_{x'}$ and $I_{y'}$ have maximum or minimum values is called a set of *principal axes* of the area A.

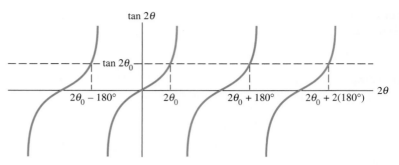

Figure 8.20
For a given value of $\tan 2\theta_0$, there are multiple roots $2\theta_0 + n(180°)$.

The corresponding moments of inertia $I_{x'}$ and $I_{y'}$ are called the *principal moments of inertia*. In the next section we can show that the product of inertia $I_{x'y'}$ corresponding to a set of principal axes equals zero.

Because the tangent is a periodic function, Eq. (8.26) does not yield a unique solution for the angle θ_p. We can show, however, that it does determine the orientation of the principal axes within an arbitrary multiple of 90°. Observe in Fig. 8.20 that if $2\theta_0$ is a solution of Eq. (8.26), then $2\theta_0 + n(180°)$ is also a solution for any integer n. The resulting orientations of the $x'y'$ coordinate system are shown in Fig. 8.21.

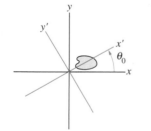 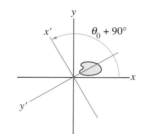 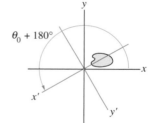 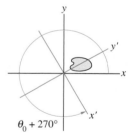

Figure 8.21
The orientation of the $x'y'$ coordinate system is determined within a multiple of 90°.

Determining principal axes and principal moments of inertia of an area involves three steps:

1. Determine I_x, I_y, and I_{xy}—You must determine the moments of inertia of the area in terms of the xy coordinate system.

2. Determine θ_p—Solve Eq. (8.26) to determine the orientation of the principal axes within an arbitrary multiple of 90°.

3. Calculate $I_{x'}$ and $I_{y'}$—Once you have chosen the orientation of the principal axes, you can use Eqs. (8.20) and (8.21) or Eqs. (8.23) and (8.24) to determine the principal moments of inertia.

Example 8.7

Determining Principal Axes and Moments of Inertia

Determine a set of principal axes and the corresponding principal moments of inertia for the triangular area in Fig. 8.22.

Strategy

We can obtain the moments of inertia of the triangular area from Appendix B. Then we can use Eq. (8.26) to determine the orientation of the principal axes and evaluate the principal moments of inertia with Eqs. (8.23) and (8.24).

Solution

Determine I_x, I_y, and I_{xy} The moments of inertia of the triangular area are

$$I_x = \frac{1}{12}(4)(3)^3 = 9 \text{ m}^4,$$

$$I_y = \frac{1}{4}(4)^3(3) = 48 \text{ m}^4,$$

$$I_{xy} = \frac{1}{8}(4)^2(3)^2 = 18 \text{ m}^4.$$

Determine θ_p From Eq. (8.26),

$$\tan 2\theta_p = \frac{2I_{xy}}{I_y - I_x} = \frac{2(18)}{48 - 9} = 0.923,$$

and we obtain $\theta_p = 21.4°$. The principal axes corresponding to this value of θ_p are shown in Fig. a.

Calculate $I_{x'}$, and $I_{y'}$ Substituting $\theta_p = 21.4°$ into Eqs. (8.23) and (8.24), we obtain

$$I_{x'} = \frac{I_x + I_y}{2} + \frac{I_x - I_y}{2}\cos 2\theta - I_{xy}\sin 2\theta$$

$$= \left(\frac{9 + 48}{2}\right) + \left(\frac{9 - 48}{2}\right)\cos\left[2(21.4°)\right] - (18)\sin\left[2(21.4°)\right] = 1.96 \text{ m}^4,$$

$$I_{y'} = \frac{I_x + I_y}{2} - \frac{I_x - I_y}{2}\cos 2\theta + I_{xy}\sin 2\theta$$

$$= \left(\frac{9 + 48}{2}\right) - \left(\frac{9 - 48}{2}\right)\cos\left[2(21.4°)\right] + (18)\sin\left[2(21.4°)\right] = 55.0 \text{ m}^4.$$

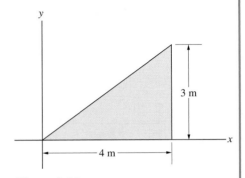

Figure 8.22

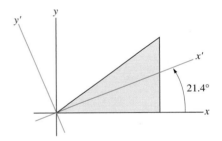

(a) The principal axes corresponding to $\theta = 21.4°$.

Discussion

The product of inertia corresponding to a set of principal axes is zero. In this example, substituting $\theta_p = 21.4°$ into Eq. (8.25) confirms that $I_{x'y'} = 0$.

Example 8.8

Rotated and Principal Axes

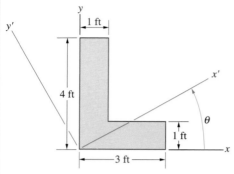

Figure 8.23

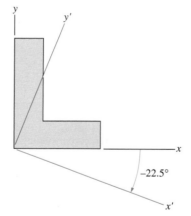

(a) The set of principal axes corresponding to $\theta_p = -22.5°$.

The moments of inertia of the area in Fig. 8.23 in terms of the xy coordinate system shown are $I_x = 22$ ft^4, $I_y = 10$ ft^4, and $I_{xy} = 6$ ft^4. (a) Determine $I_{x'}$, $I_{y'}$, and $I_{x'y'}$ for $\theta = 30°$. (b) Determine a set of principal axes and the corresponding principal moments of inertia.

Solution

(a) Determine $I_{x'}$, $I_{y'}$, and $I_{x'y'}$ By setting $\theta = 30°$ in Eqs. (8.23)–(8.25), we obtain the moments of inertia:

$$I_{x'} = \frac{I_x + I_y}{2} + \frac{I_x - I_y}{2} \cos 2\theta - I_{xy} \sin 2\theta$$

$$= \left(\frac{22 + 10}{2}\right) + \left(\frac{22 - 10}{2}\right) \cos\left[2(30°)\right] - (6) \sin\left[2(30°)\right] = 13.8 \text{ ft}^4,$$

$$I_{y'} = \frac{I_x + I_y}{2} - \frac{I_x - I_y}{2} \cos 2\theta + I_{xy} \sin 2\theta$$

$$= \left(\frac{22 + 10}{2}\right) - \left(\frac{22 - 10}{2}\right) \cos\left[2(30°)\right] + (6) \sin\left[2(30°)\right] = 18.2 \text{ ft}^4,$$

$$I_{x'y'} = \frac{I_x - I_y}{2} \sin 2\theta + I_{xy} \cos 2\theta$$

$$= \left(\frac{22 - 10}{2}\right) \sin\left[2(30°)\right] + (6) \cos\left[2(30°)\right] = 8.2 \text{ ft}^4.$$

(b) Determine θ_p Substituting the moments of inertia in terms of the xy coordinate system into Eq. (8.26),

$$\tan 2\theta_p = \frac{2I_{xy}}{I_y - I_x} = \frac{2(6)}{10 - 22} = -1,$$

we obtain $\theta_p = -22.5°$. The principal axes corresponding to this value of θ_p are shown in Fig. a.

Calculate $I_{x'}$ and $I_{y'}$ We substitute $\theta_p = -22.5°$ into Eqs. (8.23) and (8.24), obtaining the principal moments of inertia:

$$I_{x'} = 24.5 \text{ ft}^4, \qquad I_{y'} = 7.5 \text{ ft}^4.$$

Mohr's Circle

Given the moments of inertia of an area in terms of a particular coordinate system, we have presented equations with which you can determine the moments of inertia in terms of a rotated coordinate system, the orientation of the principal axes, and the principal moments of inertia. You can also obtain this information by using a graphical method called *Mohr's circle*, which is very useful for visualizing the solutions of Eqs. (8.23)–(8.25).

Determining $I_{x'}$, $I_{y'}$, and $I_{x'y'}$ We first describe how to construct Mohr's circle and then explain why it works. Suppose we know the moments of inertia I_x, I_y, and I_{xy} of an area in terms of a coordinate system xy and we want to determine the moments of inertia for a rotated coordinate system $x'y'$ (Fig. 8.24). Constructing Mohr's circle involves three steps:

1. Establish a set of horizontal and vertical axes and plot two points: point 1 with coordinates (I_x, I_{xy}) and point 2 with coordinates $(I_y, -I_{xy})$ as shown in Fig. 8.25a.
2. Draw a straight line connecting points 1 and 2. Using the intersection of the straight line with the horizontal axis as the center, draw a circle that passes through the two points (Fig. 8.25b).
3. Draw a straight line through the center of the circle at an angle 2θ measured counterclockwise from point 1. This line intersects the circle at point 1' with coordinates $(I_{x'}, I_{x'y'})$ and point 2' with coordinates $(I_{y'}, -I_{x'y'})$, as shown in Fig. 8.25c.

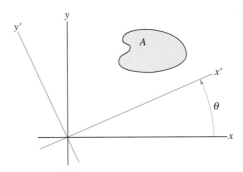

Figure 8.24
The xy coordinate system and the rotated $x'y'$ coordinate system.

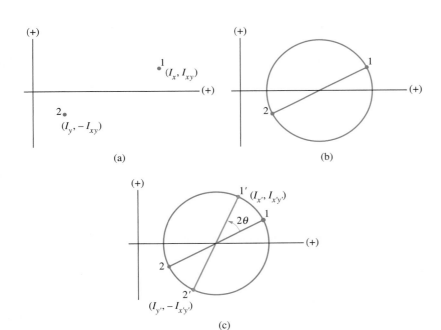

(a)

(b)

(c)

Figure 8.25
(a) Plotting the points 1 and 2.
(b) Drawing Mohr's circle. The center of the circle is the intersection of the line from 1 to 2 with the horizontal axis.
(c) Finding the points 1' and 2'.

Thus for a given angle θ, the coordinates of points 1' and 2' determine the moments of inertia in terms of the rotated coordinate system. Why does this graphical construction work? In Fig. 8.26a, we show the points 1 and 2 and Mohr's circle. Notice that the horizontal coordinate of the center of the circle is $(I_x + I_y)/2$. The sine and cosine of the angle β are

$$\sin\beta = \frac{I_{xy}}{R}, \qquad \cos\beta = \frac{I_x - I_y}{2R},$$

where R, the radius of the circle, is given by

$$R = \sqrt{\left(\frac{I_x - I_y}{2}\right)^2 + (I_{xy})^2}.$$

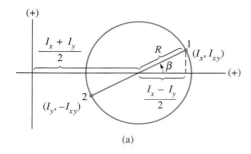

(a)

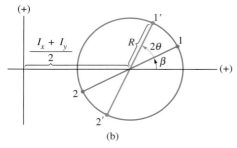

(b)

Figure 8.26
(a) The points 1 and 2 and Mohr's circle.
(b) The points 1' and 2'.

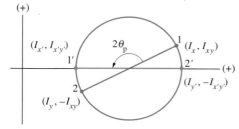

Figure 8.27
To determine the orientation of a set of principal axes, let points 1' and 2' be the points where the circle intersects the horizontal axis.

From Fig. 8.26b, the horizontal coordinate of point 1' is

$$\frac{I_x + I_y}{2} + R\cos(\beta + 2\theta)$$

$$= \frac{I_x + I_y}{2} + R(\cos\beta\cos 2\theta - \sin\beta\sin 2\theta)$$

$$= \frac{I_x + I_y}{2} + \frac{I_x - I_y}{2}\cos 2\theta - I_{xy}\sin 2\theta = I_{x'},$$

and the horizontal coordinate of point 2' is

$$\frac{I_x + I_y}{2} - R\cos(\beta + 2\theta)$$

$$= \frac{I_x + I_y}{2} - R(\cos\beta\cos 2\theta - \sin\beta\sin 2\theta)$$

$$= \frac{I_x + I_y}{2} - \frac{I_x - I_y}{2}\cos 2\theta + I_{xy}\sin 2\theta = I_{y'}.$$

The vertical coordinate of point 1' is

$$R\sin(\beta + 2\theta) = R(\sin\beta\cos 2\theta + \cos\beta\sin 2\theta)$$

$$= I_{xy}\cos 2\theta + \frac{I_x - I_y}{2}\sin 2\theta = I_{x'y'},$$

and the vertical coordinate of point 2' is

$$-R\sin(\beta + 2\theta) = -I_{x'y'}.$$

We have shown that the coordinates of point 1' are $\left(I_{x'}, I_{x'y'}\right)$ and the coordinates of point 2' are $\left(I_{y'}, -I_{x'y'}\right)$.

Determining Principal Axes and Principal Moments of Inertia Because the moments of inertia $I_{x'}$ and $I_{y'}$ are the horizontal coordinates of points 1' and 2' of Mohr's circle, their maximum and minimum values occur when points 1' and 2' coincide with the intersections of the circle with the horizontal axis (Fig. 8.27). (Which intersection you designate as 1' is arbitrary. In Fig. 8.27, we have designated the minimum moment of inertia as point 1'.) You can determine the orientation of the principal axes by measuring the angle $2\theta_p$ from point 1 to point 1', and the coordinates of points 1' and 2' are the principal moments of inertia.

Notice that Mohr's circle demonstrates that the product of inertia $I_{x'y'}$ corresponding to a set of principal axes (the vertical coordinate of point 1' in Fig. 8.27) is always zero. Furthermore, we can use Fig. 8.26a to obtain an analytical expression for the horizontal coordinates of the points where the circle intersects the horizontal axis, which are the principal moments of inertia:

$$\text{Principal moments of inertia} = \frac{I_x + I_y}{2} \pm R$$

$$= \frac{I_x + I_y}{2} \pm \sqrt{\left(\frac{I_x - I_y}{2}\right)^2 + \left(I_{xy}\right)^2}.$$

Example 8.9

Moments of Inertia by Mohr's Circle

The moments of inertia of the area in Fig. 8.28 in terms of the xy coordinate system are $I_x = 22 \text{ ft}^4$, $I_y = 10 \text{ ft}^4$, and $I_{xy} = 6 \text{ ft}^4$. Use Mohr's circle to determine (a) the moments of inertia $I_{x'}$, $I_{y'}$, and $I_{x'y'}$ for $\theta = 30°$; (b) a set of principal axes and the corresponding principal moments of inertia.

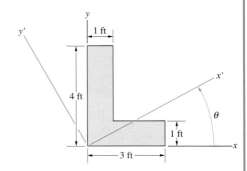

Figure 8.28

Solution

(a) First we plot point 1 with coordinates $(I_x, I_{xy}) = (22, 6) \text{ ft}^4$ and point 2 with coordinates $(I_y, -I_{xy}) = (10, -6) \text{ ft}^4$ (Fig. a). Then we draw a straight line between points 1 and 2 and, using the intersection of the line with the horizontal axis as the center, draw a circle that passes through the points (Fig. b).

To determine the moments of inertia for $\theta = 30°$, we measure an angle $2\theta = 60°$ counterclockwise from point 1 (Fig. c). From the coordinates of points 1′ and 2′, we obtain

$$I_{x'} = 14 \text{ ft}^4, \qquad I_{x'y'} = 8 \text{ ft}^4, \qquad I_{y'} = 18 \text{ ft}^4.$$

(b) To determine the principal axes, we let the points 1′ and 2′ be the points where the circle intersects the horizontal axis (Fig. d). Measuring the angle from point 1 to point 1′, we determine that $2\theta_p = 135°$. From the coordinates of points 1′ and 2′, we obtain the principal moments of inertia:

$$I_{x'} = 7.5 \text{ ft}^4, \qquad I_{y'} = 24.5 \text{ ft}^4.$$

The principal axes are shown in Fig. e.

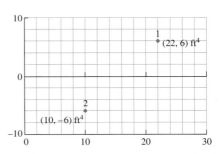

(a) Plot point 1 with coordinates (I_x, I_{xy}) and point 2 with coordinates $(I_y, -I_{xy})$.

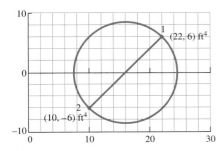

(b) Draw a line from point 1 to point 2 and construct the circle.

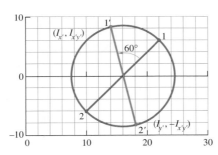

(c) Measure the angle $2\theta = 60°$ counterclockwise from point 1 to determine the points 1′ and 2′.

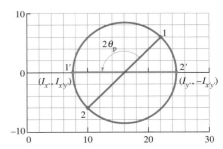

(d) Determine the principal axes by letting points 1′ and 2′ correspond to the points where the circle intersects the horizontal axis.

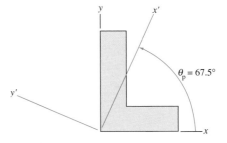

(e) The principal axes corresponding to $\theta_p = 67.5°$.

Discussion

In Example 8.8 we solved this problem by using Eqs. (8.23)–(8.26). For $\theta = 30°$, we obtained $I_{x'} = 13.8 \text{ ft}^4$, $I_{x'y'} = 8.2 \text{ ft}^4$, and $I_{y'} = 18.2 \text{ ft}^4$. The differences between these results and the ones we obtained using Mohr's circle are due to the errors inherent in measuring the answer graphically. By using Eq. (8.26) to determine the orientation of the principal axes, we obtained the principal axes shown in Fig. a of Example 8.8 and the principal moments of inertia $I_{x'} = 24.5 \text{ ft}^4$ and $I_{y'} = 7.5 \text{ ft}^4$. The difference between those results and the ones we obtained using Mohr's circle simply reflects the fact that the orientation of the principal axes can be determined only within a multiple of 90°.

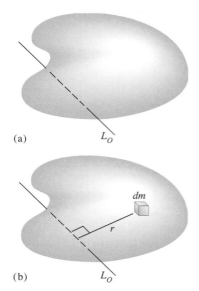

(a)

(b)

Figure 8.29
(a) An object and axis L_O.
(b) A differential element of mass dm.

Masses

The acceleration of an object that results from the forces acting on it depends on its mass. The angular acceleration, or rotational acceleration, that results from the forces and couples acting on an object depends on quantities called the mass moments of inertia of the object. In this section we discuss methods for determining mass moments of inertia of particular objects. We show that for special classes of objects, their mass moments of inertia can be expressed in terms of moments of inertia of areas, which explains how the names of those area integrals originated.

An object and a line or "axis" L_O are shown in Fig. 8.29a. The *mass moment of inertia* of the object about the axis L_O is defined by

$$I_O = \int_m r^2 \, dm, \tag{8.27}$$

where r is the perpendicular distance from the axis to the differential element of mass dm (Fig. 8.29b). Often L_O is an axis about which the object rotates, and the value of I_O is required to determine the angular acceleration, or the rate of change of the rate of rotation, caused by a given couple about L_O.

8.4 Simple Objects

The mass moments of inertia of complicated objects can be determined by summing the mass moments of inertia of their individual parts. We therefore begin by determining mass moments of inertia of some simple objects. Then in the next section we describe the parallel-axis theorem, which makes it possible to determine mass moments of inertia of objects composed of combinations of parts.

Slender Bars

Let us determine the mass moment of inertia of a straight, slender bar about a perpendicular axis L through the center of mass of the bar (Fig. 8.30a). "Slender" means that we assume that the bar's length is much greater than its width. Let the bar have length l, cross-sectional area A, and mass m. We assume that A is uniform along the length of the bar and that the material is homogeneous.

Consider a differential element of the bar of length dr at a distance r from the center of mass (Fig. 8.30b). The element's mass is equal to the product of its volume and the mass density: $dm = \rho A \, dr$. Substituting this expression into Eq. (8.27), we obtain the mass moment of inertia of the bar about a perpendicular axis through its center of mass:

$$I = \int_m r^2 \, dm = \int_{-l/2}^{l/2} \rho A r^2 \, dr = \frac{1}{12} \rho A l^3.$$

The mass of the bar equals the product of the mass density and the volume of the bar, $m = \rho A l$, so we can express the mass moment of inertia as

$$I = \frac{1}{12} m l^2. \tag{8.28}$$

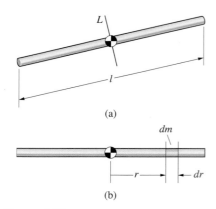

(a)

(b)

Figure 8.30
(a) A slender bar.
(b) A differential element of length dr.

We have neglected the lateral dimensions of the bar in obtaining this result. That is, we treated the differential element of mass dm as if it were concentrated on the axis of the bar. As a consequence, Eq. (8.28) is an approximation for the mass moment of inertia of a bar. Later in this section we determine the mass moments of inertia for a bar of finite lateral dimension and show that Eq. (8.28) is a good approximation when the width of the bar is small in comparison to its length.

Thin Plates

Consider a homogeneous flat plate that has mass m and uniform thickness T. We will leave the shape of the cross-sectional area of the plate unspecified. Let a cartesian coordinate system be oriented so that the plate lies in the x–y plane (Fig. 8.31a). Our objective is to determine the mass moments of inertia of the plate about the x, y, and z axes.

We can obtain a differential element of volume of the plate by projecting an element of area dA through the thickness T of the plate (Fig. 8.31b). The resulting volume is $T\,dA$. The mass of this element of volume is equal to the product of the mass density and the volume: $dm = \rho T\,dA$. Substituting this expression into Eq. (8.27), we obtain the mass moment of inertia of the plate about the z axis in the form

$$I_{(z\ \text{axis})} = \int_m r^2\,dm = \rho T \int_A r^2\,dA,$$

where r is the distance from the z axis to dA. Since the mass of the plate is $m = \rho T\,A$, where A is the cross-sectional area of the plate, $\rho T = m/A$. The integral on the right is the polar moment of inertia J_O of the cross-sectional area of the plate. We can therefore write the mass moment of inertia of the plate about the z axis as

$$I_{(z\ \text{axis})} = \frac{m}{A} J_O. \qquad (8.29)$$

From Fig 8.31b, we see that the perpendicular distance from the x axis to the element of area dA is the y coordinate of dA. Therefore the mass moment of inertia of the plate about the x axis is

$$I_{(x\ \text{axis})} = \int_m y^2\,dm = \rho T \int_A y^2\,dA = \frac{m}{A} I_x, \qquad (8.30)$$

where I_x is the moment of inertia of the cross-sectional area of the plate about the x axis. The mass moment of inertia of the plate about the y axis is

$$I_{(y\ \text{axis})} = \int_m x^2\,dm = \rho T \int_A x^2\,dA = \frac{m}{A} I_y, \qquad (8.31)$$

where I_y is the moment of inertia of the cross-sectional area of the plate about the y axis.

Thus we have expressed the mass moments of inertia of a thin homogeneous plate of uniform thickness in terms of the moments of inertia of the cross-sectional area of the plate. In fact, these results explain why the area integrals I_x, I_y, and J_O are called moments of inertia.

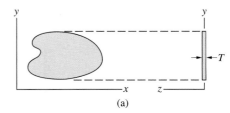

(a)

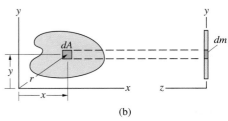

(b)

Figure 8.31
(a) A plate of arbitrary shape and uniform thickness T.
(b) An element of volume obtained by projecting an element of area dA through the plate.

Since the sum of the area moments of inertia I_x and I_y is equal to the polar moment of inertia J_O, the mass moment of inertia of the thin plate about the z axis is equal to the sum of its moments of inertia about the x and y axes:

$$I_{(z\text{ axis})} = I_{(x\text{ axis})} + I_{(y\text{ axis})}. \qquad \textbf{Thin plate} \qquad (8.32)$$

Example 8.10

Moments of Inertia of a Slender Bar

Two homogeneous slender bars, each of length l, mass m, and cross-sectional area A, are welded together to form the L-shaped object in Fig. 8.32. Determine the mass moment of inertia of the object about the axis L_O through point O. (The axis L_O is perpendicular to the two bars.)

Strategy

Using the same integration procedure we used for a single bar, we will determine the mass moment of inertia of each bar about L_O and sum the results.

Solution

Our first step is to introduce a coordinate system with the z axis along L_O and the x axis collinear with bar 1 (Fig. a). The mass of the differential element of bar 1 of length dx is $dm = \rho A\, dx$. The mass moment of inertia of bar 1 about L_O is

$$(I_O)_1 = \int_m r^2\, dm = \int_0^l \rho A x^2\, dx = \frac{1}{3}\rho A l^3.$$

In terms of the mass of the bar, $m = \rho A l$, we can write this result as

$$(I_O)_1 = \frac{1}{3}ml^2.$$

The mass of an element of bar 2 of length dy, shown in Fig. b, is $dm = \rho A\, dy$. From the figure we see that the perpendicular distance from L_O to the element is $r = \sqrt{l^2 + y^2}$. Therefore the mass moment of inertia of bar 2 about L_O is

$$(I_O)_2 = \int_m r^2\, dm = \int_0^l \rho A(l^2 + y^2)\, dy = \frac{4}{3}\rho A l^3.$$

In terms of the mass of the bar, we obtain

$$(I_O)_2 = \frac{4}{3}ml^2.$$

The mass moment of inertia of the L-shaped object about L_O is

$$I_O = (I_O)_1 + (I_O)_2 = \frac{1}{3}ml^2 + \frac{4}{3}ml^2 = \frac{5}{3}ml^2.$$

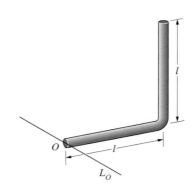

Figure 8.32

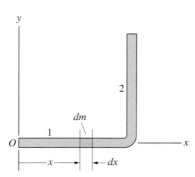

(a) Differential element of bar 1.

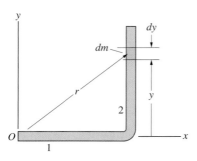

(b) Differential element of bar 2.

Example 8.11

Moments of Inertia of a Triangular Plate

The thin homogeneous plate in Fig. 8.33 is of uniform thickness and mass m.
Determine its mass moments of inertia about the x, y, and z axes.

Strategy

The mass moments of inertia about the x and y axes are given by Eqs. (8.30)
and (8.31) in terms of the moments of inertia of the cross-sectional area of
the plate. We can determine the mass moment of inertia of the plate about the
z axis from Eq. (8.32).

Solution

From Appendix B, the moments of inertia of the triangular area about the x
and y axes are $I_x = \frac{1}{12} bh^3$ and $I_y = \frac{1}{4} hb^3$. Therefore the mass moments of inertia of the plate about the x and y axes are

$$I_{(x \text{ axis})} = \frac{m}{A} I_x = \left(\frac{m}{\frac{1}{2} bh} \right)\left(\frac{1}{12} bh^3 \right) = \frac{1}{6} mh^2,$$

$$I_{(y \text{ axis})} = \frac{m}{A} I_y = \left(\frac{m}{\frac{1}{2} bh} \right)\left(\frac{1}{4} hb^3 \right) = \frac{1}{2} mb^2.$$

The mass moment of inertia about the z axis is

$$I_{(z \text{ axis})} = I_{(x \text{ axis})} + I_{(y \text{ axis})} = m\left(\frac{1}{6} h^2 + \frac{1}{2} b^2 \right).$$

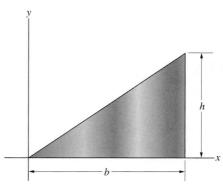

Figure 8.33

8.5 Parallel-Axis Theorem

The parallel-axis theorem allows us to determine the mass moment of inertia of an object about any axis when the mass moment of inertia about a
parallel axis through the center of mass is known. This theorem can be
used to calculate the mass moment of inertia of a composite object about
an axis given the mass moments of inertia of each of its parts about parallel axes.

Suppose that we know the mass moment of inertia I about an axis L
through the center of mass of an object, and we wish to determine its mass
moment of inertia I_O about a parallel axis L_O (Fig. 8.34a). To determine I_O,
we introduce parallel coordinate systems xyz and $x'y'z'$ with the z axis along
L_O and the z' axis along L, as shown in Fig. 8.34b. (In this figure the axes L_O
and L are perpendicular to the page.) The origin O of the xyz coordinate system is contained in the x'–y' plane. The terms d_x and d_y are the coordinates of
the center of mass relative to the xyz coordinate system.

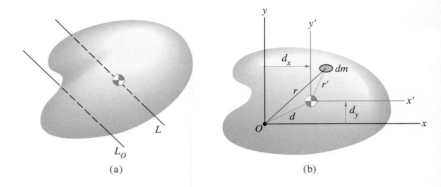

Figure 8.34
(a) An axis L through the center of mass of an object and a parallel axis L_O.
(b) The xyz and $x'y'z'$ coordinate systems.

The mass moment of inertia of the object about L_O is

$$I_O = \int_m r^2 \, dm = \int_m (x^2 + y^2) \, dm, \tag{8.33}$$

where r is the perpendicular distance from L_O to the differential element of mass dm, and x, y are the coordinates of dm in the x–y plane. The coordinates of dm in the two coordinate systems are related by

$$x = x' + d_x, \qquad y = y' + d_y.$$

By substituting these expressions into Eq. (8.33), we can write it as

$$I_O = \int_m \left[(x')^2 + (y')^2 \right] dm + 2d_x \int_m x' \, dm + 2d_y \int_m y' \, dm$$

$$+ \int_m (d_x^2 + d_y^2) \, dm. \tag{8.34}$$

Since $(x')^2 + (y')^2 = (r')^2$, where r' is the perpendicular distance from L to dm, the first integral on the right side of this equation is the mass moment of inertia I of the object about L. Recall that the x' and y' coordinates of the center of mass of the object relative to the $x'y'z'$ coordinate system are defined by

$$\bar{x}' = \frac{\int_m x' \, dm}{\int_m dm}, \qquad \bar{y}' = \frac{\int_m y' \, dm}{\int_m dm}.$$

Because the center of mass of the object is at the origin of the $x'y'z'$ system, $\bar{x}' = 0$ and $\bar{y}' = 0$. Therefore the integrals in the second and third terms on the right side of Eq. (8.34) are equal to zero. From Fig. 8.34b, we see that $d_x^2 + d_y^2 = d^2$, where d is the perpendicular distance between the axes L and L_O. Therefore we obtain

$$I_O = I + d^2 m, \tag{8.35}$$

where m is the mass of the object. This is the parallel-axis theorem. If the mass moment of inertia of an object is known about a given axis, we can use this theorem to determine its mass moment of inertia about any parallel axis.

Determining the mass moment of inertia about a given axis L_O typically requires three steps:

1. Choose the parts—Try to divide the object into parts whose mass moments of inertia you know or can easily determine.

2. Determine the mass moments of inertia of the parts—You must first determine the mass moment of inertia of each part about the axis through its center of mass parallel to L_O. Then you can use the parallel-axis theorem to determine its mass moment of inertia about L_O.

3. Sum the results—Sum the mass moments of inertia of the parts (or subtract in the case of a hole or cutout) to obtain the mass moment of inertia of the composite object.

Example 8.12

Moment of Inertia of a Composite Bar

Two homogeneous slender bars, each of length l and mass m, are welded together to form the L-shaped object in Fig. 8.35. Determine the mass moment of inertia of the object about the axis L_O through point O. (The axis L_O is perpendicular to the two bars.)

Solution

Choose the Parts The parts are the two bars, which we call bar 1 and bar 2 (Fig. a).

Determine the Mass Moments of Inertia of the Parts From Eq. (8.28), the mass moment of inertia of each bar about a perpendicular axis through its center of mass is $I = \frac{1}{12}ml^2$. The distance from L_O to the parallel axis through the center of mass of bar 1 is $\frac{1}{2}l$ (Fig. a). Therefore the mass moment of inertia of bar 1 about L_O is

$$(I_O)_1 = I + d^2m = \frac{1}{12}ml^2 + \left(\frac{1}{2}l\right)^2 m = \frac{1}{3}ml^2.$$

The distance from L_O to the parallel axis through the center of mass of bar 2 is $\left[l^2 + \left(\frac{1}{2}l\right)^2\right]^{1/2}$. The mass moment of inertia of bar 2 about L_O is

$$(I_O)_2 = I + d^2m = \frac{1}{12}ml^2 + \left[l^2 + \left(\frac{1}{2}l\right)^2\right]m = \frac{4}{3}ml^2.$$

Sum the Results The mass moment of inertia of the L-shaped object about L_O is

$$I_O = (I_O)_1 + (I_O)_2 = \frac{1}{3}ml^2 + \frac{4}{3}ml^2 = \frac{5}{3}ml^2.$$

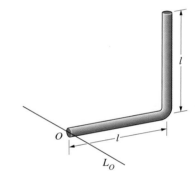

Figure 8.35

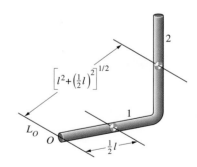

(a) The distances from L_O to parallel axes through the centers of mass of bars 1 and 2.

Discussion

Compare this solution to Example 8.10, in which we used integration to determine the mass moment of inertia of this object about L_O. We obtained the result much more easily with the parallel-axis theorem, but of course we needed to know the mass moments of inertia of the bars about the axes through their centers of mass.

Example 8.13

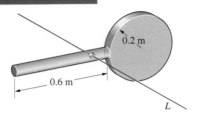

Figure 8.36

Moment of Inertia of a Composite Object

The object in Fig. 8.36 consists of a slender, 3-kg bar welded to a thin, circular 2-kg disk. Determine its mass moment of inertia about the axis L through its center of mass. (The axis L is perpendicular to the bar and disk.)

Strategy

We must first locate the center of mass of the composite object and then apply the parallel-axis theorem to the parts separately and sum the results.

Solution

Choose the Parts The parts are the bar and the disk. Introducing the coordinate system in Fig. a, the x coordinate of the center of mass of the composite object is

$$\bar{x} = \frac{\bar{x}_{(bar)} m_{(bar)} + \bar{x}_{(disk)} m_{(disk)}}{m_{(bar)} + m_{(disk)}} = \frac{(0.3)(3) + (0.6 + 0.2)(2)}{3 + 2} = 0.5 \text{ m}.$$

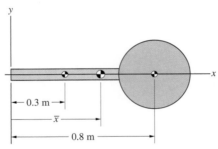

(a) The coordinate $\bar{x}$ of the center of mass of the object.

Determine the Mass Moments of Inertia of the Parts The distance from the center of mass of the bar to the center of mass of the composite object is 0.2 m (Fig. b). Therefore the mass moment of inertia of the bar about L is

$$I_{(bar)} = \frac{1}{12}(3)(0.6)^2 + (0.2)^2(3) = 0.210 \text{ kg-m}^2.$$

The distance from the center of mass of the disk to the center of mass of the composite object is 0.3 m (Fig. c). The mass moment of inertia of the disk about L is

$$I_{(disk)} = \frac{1}{2}(2)(0.2)^2 + (0.3)^2(2) = 0.220 \text{ kg-m}^2.$$

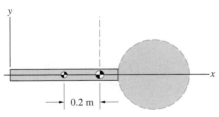

(b) Distance from L to the center of mass of the bar.

Sum the Results The mass moment of inertia of the composite object about L is

$$I = I_{(bar)} + I_{(disk)} = 0.430 \text{ kg-m}^2.$$

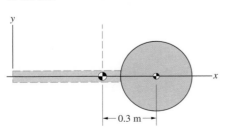

(c) Distance from L to the center of mass of the disk.

Example 8.14

Moments of Inertia of a Cylinder

The homogeneous cylinder in Fig. 8.37 has mass m, length l, and radius R. Determine its mass moments of inertia about the x, y, and z axes.

Strategy

We first determine the mass moments of inertia about the x, y, and z axes of an infinitesimal element of the cylinder consisting of a disk of thickness dz. We then integrate the results with respect to z to obtain the mass moments of inertia of the cylinder. We must apply the parallel-axis theorem to determine the mass moments of inertia of the disk about the x and y axes.

Solution

Consider an element of the cylinder of thickness dz at a distance z from the center of the cylinder (Fig. a). (You can imagine obtaining this element by "slicing" the cylinder perpendicular to its axis.) The mass of the element is equal to the product of the mass density and the volume of the element, $dm = \rho(\pi R^2 \, dz)$. We obtain the mass moments of inertia of the element by using the values for a thin circular plate given in Appendix C. The mass moment of inertia about the z axis is

$$dI_{(z\,\text{axis})} = \frac{1}{2} \, dm \, R^2 = \frac{1}{2}(\rho\pi R^2 \, dz)R^2.$$

By integrating this result with respect to z from $-l/2$ to $l/2$, we sum the mass moments of inertia of the infinitesimal disk elements that make up the cylinder. The result is the mass moment of inertia of the cylinder about the z axis:

$$I_{(z\,\text{axis})} = \int_{-l/2}^{l/2} \frac{1}{2} \rho\pi \, R^4 \, dz = \frac{1}{2} \rho\pi \, R^4 l.$$

We can write this result in terms of the mass of the cylinder, $m = \rho(\pi R^2 l)$, as

$$I_{(z\,\text{axis})} = \frac{1}{2} mR^2.$$

The mass moment of inertia of the disk element about the x' axis is

$$dI_{(x'\,\text{axis})} = \frac{1}{4} \, dm \, R^2 = \frac{1}{4}(\rho\pi R^2 \, dz)R^2.$$

We can use this result and the parallel-axis theorem to determine the mass moment of inertia of the element about the x axis:

$$dI_{(x\,\text{axis})} = dI_{(x'\,\text{axis})} + z^2 \, dm = \frac{1}{4}(\rho\pi R^2 \, dz)R^2 + z^2(\rho\pi R^2 \, dz).$$

Integrating this expression with respect to z from $-l/2$ to $l/2$, we obtain the mass moment of inertia of the cylinder about the x axis:

$$I_{(x\,\text{axis})} = \int_{-l/2}^{l/2} \left(\frac{1}{4} \rho\pi R^4 + \rho\pi R^2 z^2 \right) dz = \frac{1}{4} \rho\pi R^4 l + \frac{1}{12} \rho\pi \, R^2 l^3.$$

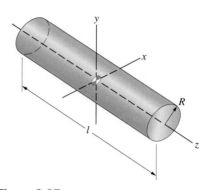

Figure 8.37

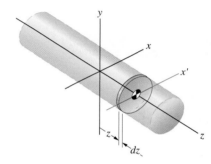

(a) A differential element of the cylinder in the form of a disk.

In terms of the mass of the cylinder,

$$I_{(x \text{ axis})} = \frac{1}{4} mR^2 + \frac{1}{12} ml^2.$$

Due to the symmetry of the cylinder,

$$I_{(y \text{ axis})} = I_{(x \text{ axis})}.$$

Discussion

When the cylinder is very long in comparison to its width, $l \gg R$, the first term in the equation for $I_{(x \text{ axis})}$ can be neglected, and we obtain the mass moment of inertia of a slender bar about a perpendicular axis, Eq. (8.28). Conversely, when the radius of the cylinder is much greater than its length, $R \gg l$, the second term in the equation for $I_{(x \text{ axis})}$ can be neglected, and we obtain the mass moment of inertia for a thin circular disk about an axis parallel to the disk. This indicates the sizes of the terms you neglect when you use the approximate expressions for the mass moments of inertia of a "slender" bar and a "thin" disk.

Chapter Summary

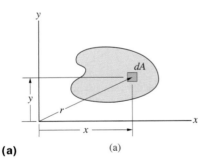

(a)

Areas

Four *area moments of inertia* are defined (Fig. a):

1. The moment of inertia about the x axis:

$$I_x = \int_A y^2 \, dA. \qquad \text{Eq. (8.1)}$$

2. The moment of inertia about the y axis:

$$I_y = \int_A x^2 \, dA. \qquad \text{Eq. (8.3)}$$

3. The product of inertia:

$$I_{xy} = \int_A xy \, dA. \qquad \text{Eq. (8.5)}$$

4. The polar moment of inertia:

$$J_O = \int_A r^2 \, dA. \qquad \text{Eq. (8.6)}$$

The *radii of gyration* about the x and y axes are defined by $k_x = \sqrt{I_x/A}$ and $k_y = \sqrt{I_y/A}$, respectively, and the radius of gyration about the origin O is defined by $k_O = \sqrt{J_O/A}$.

The polar moment of inertia is equal to the sum of the moments of inertia about the x and y axes: $J_O = I_x + I_y$. If an area is symmetric about either the x axis or the y axis, its product of inertia is zero.

Let $x'y'$ be a coordinate system with its origin at the centroid of an area A, and let xy be a parallel coordinate system. The moments of inertia of A in terms of the two systems are related by the *parallel-axis theorems* [Eqs. (8.10)–(8.13)]:

$$I_x = I_{x'} + d_y^2 A,$$

$$I_y = I_{y'} + d_x^2 A,$$

$$I_{xy} = I_{x'y'} + d_x d_y A,$$

$$J_O = J'_O + (d_x^2 + d_y^2)A = J'_O + d^2 A,$$

where d_x and d_y are the coordinates of the centroid of A in the xy coordinate system.

Masses

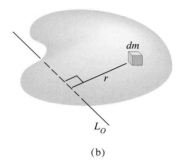

(b)

The mass moment of inertia of an object about an axis L_O is (Fig. b)

$$I_O = \int_m r^2 \, dm, \qquad \text{Eq. (8.27)}$$

where r is the perpendicular distance from L_O to the differential element of mass dm.

Let L be an axis through the center of mass of an object, and let L_O be a parallel axis (Fig. c). The moment of inertia I_O to about L_O is given in terms of the moment of inertia I about L by the parallel-axis theorem,

$$I_O = I + d^2 m, \qquad \text{Eq. (8.35)}$$

where m is the mass of the object and d is the perpendicular distance between L and L_O.

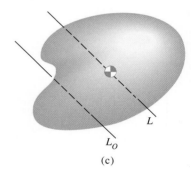

(c)

Review Problems

8.1 Determine I_y and k_y.

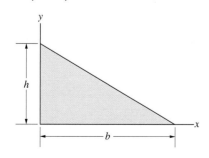

P8.1

Refer to P8.2 for Problems 8.2–8.5.

8.2 Determine I_y and k_y.

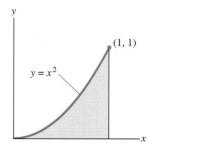

P8.2

8.3 Determine I_x and k_x.

8.4 Determine J_O and k_O.

8.5 Determine I_{xy}.

Refer to P8.6 for Problems 8.6–8.8.

8.6 Determine I_y and k_y.

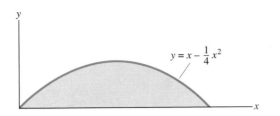

P8.6

8.7 Determine I_x and k_x.

8.8 Determine I_{xy}.

Refer to P8.9 for Problems 8.9–8.11. The origin of the $x'y'$ coordinate system is at the centroid of the area.

8.9 Determine $I_{y'}$ and $k_{y'}$.

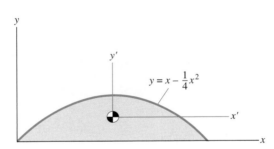

P8.9

8.10 Determine $I_{x'}$ and $k_{x'}$.

8.11 Determine $I_{x'y'}$.

8.12 Determine I_y and k_y.

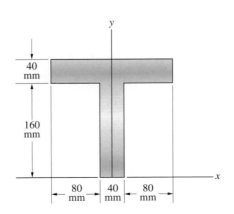

P8.12

8.13 Determine I_x and k_x for the area in Problem 8.12.

8.14 Determine I_x and k_x.

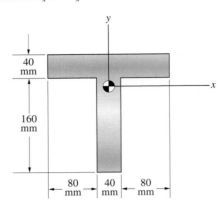

P8.14

8.15 Determine J_O and k_O for the area in Problem 8.14.

8.16 Determine I_y and k_y.

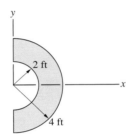

P8.16

8.17 Determine J_O and k_O for the area in Problem 8.16.

8.18 Determine I_x and k_x.

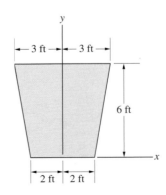

P8.18

8.19 Determine I_y and k_y for the area in Problem 8.18.

8.20 The moments of inertia of the area are $I_x = 36$ m^4, $I_y = 145$ m^4, and $I_{xy} = 44.25$ m^4. Determine a set of principal axes and the principal moments of inertia.

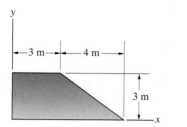

P8.20

8.21 The mass moment of inertia of the 31-oz bat about a perpendicular axis through point B is 0.093 slug-ft^2. What is the bat's mass moment of inertia about a perpendicular axis through point A? (Point A is the bat's "instantaneous center," or center of rotation, at the instant shown.)

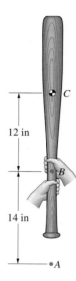

P8.21

8.22 The mass of the thin homogeneous plate is 4 kg. Determine its mass moment of inertia about the y axis.

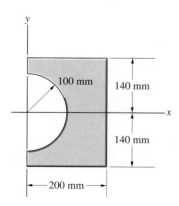

P8.22

8.23 Determine the mass moment of inertia of the plate in Problem 8.22 about the z axis.

8.24 The homogeneous pyramid is of mass m. Determine its mass moment of inertia about the z axis.

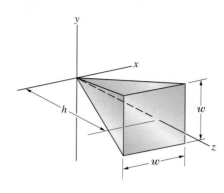

P8.24

8.25 Determine the mass moments of inertia of the homogeneous pyramid in Problem 8.24 about the x and y axes.

8.26 The homogeneous object weighs 400 lb. Determine its mass moment of inertia about the x axis.

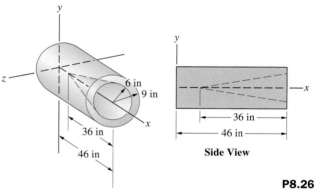

Side View

P8.26

8.27 Determine the mass moments of inertia of the object in Problem 8.26 about the y and z axes.

8.28 Determine the mass moment of inertia of the 14-kg flywheel about the axis L.

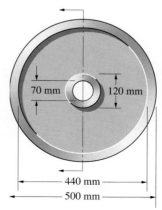

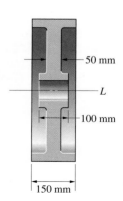

P8.28

Shoe soles are designed to support the friction forces necessary to prevent slipping. In this chapter we analyze friction forces between surfaces in contact.

Friction

Friction forces have many important effects, both desirable and undesirable, in engineering applications. The Coulomb theory of friction
allows us to estimate the maximum friction forces that can be exerted
by contacting surfaces and the friction forces exerted by sliding surfaces. This
opens the path to the analysis of important new classes of supports and machines, including wedges (shims), threaded connections, bearings, and belts.

9.1 Theory of Dry Friction

When you climb a ladder, it remains in place because of the friction force exerted on it by the floor (Fig. 9.1a). If you remain stationary on the ladder, the equilibrium equations determine the friction force. But an important question cannot be answered by the equilibrium equations alone: Will the ladder remain in place, or will it slip on the floor? If a truck is parked on an incline, the friction force exerted on it by the road prevents it from sliding down the incline (Fig. 9.1b). Here too there is another question: What is the steepest incline on which the truck can be parked?

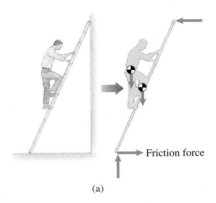

(a)

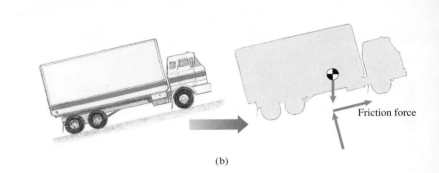

Friction force

(b)

Figure 9.1
Objects supported by friction forces.

To answer these questions, we must examine the nature of friction forces in more detail. Place a book on a table and push it with a small horizontal force, as shown in Fig. 9.2a. If the force you exert is sufficiently small, the book does not move. The free-body diagram of the book is shown in Fig. 9.2b. The force W is the book's weight, and N is the normal force exerted by the table. The force F is the horizontal force you apply, and f is the friction force exerted by the table. Because the book is in equilibrium, $f = F$.

(a)

(b)

Figure 9.2
(a) Exerting a horizontal force on a book.
(b) The free-body diagram of the book.

Now slowly increase the force you apply to the book. As long as the book remains in equilibrium, the friction force must increase correspondingly, since it equals the force you apply. When the force you apply becomes too large, the book moves. It slips on the table. After reaching some maximum value, the friction force can no longer maintain the book in equilibrium. Also, notice that the force you must apply to keep the book moving on the table is smaller than the force required to cause it to slip. (You are familiar with this phenomenon if you've ever pushed a piece of furniture across a floor.)

How does the table exert a friction force on the book? Why does the book slip? Why is less force required to slide the book across the table than is required to start it moving? If the surfaces of the table and the book are magnified sufficiently, they will appear rough (Fig. 9.3). Friction forces arise in part from the interactions of the roughnesses, or *asperities*, of the contacting surfaces. On a still smaller scale, contacting surfaces tend to form atomic bonds that "glue" them together (Fig. 9.4). The fact that more force is required to start an object sliding on a surface than to keep it sliding is explained in part by the necessity to break these bonds before sliding can begin.

In the following sections we present a theory that predicts the basic phenomena we have described and has been found useful for approximating friction forces between dry surfaces in engineering applications. (Friction between lubricated surfaces is a hydrodynamic phenomenon and must be analyzed in the context of fluid mechanics.)

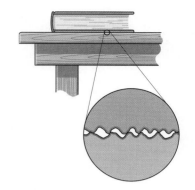

Figure 9.3
The roughnesses of the surfaces can be seen in a magnified view.

Coefficients of Friction

The theory of dry friction, or *Coulomb friction*, predicts the maximum friction forces that can be exerted by dry, contacting surfaces that are stationary relative to each other. It also predicts the friction forces exerted by the surfaces when they are in relative motion, or sliding. We first consider surfaces that are not in relative motion.

The Static Coefficient The magnitude of the *maximum* friction force that can be exerted between two plane dry surfaces in contact is

$$f = \mu_s N, \tag{9.1}$$

where N is the normal component of the contact force between the surfaces and μ_s is a constant called the *coefficient of static friction*.

The value of μ_s is assumed to depend only on the materials of the contacting surfaces and the conditions (smoothness and degree of contamination by other materials) of the surfaces. Typical values of μ_s for various materials are shown in Table 9.1. The relatively large range of values for each pair of materials reflects the sensitivity of μ_s to the conditions of the surfaces. In engineering applications it is usually necessary to measure the value of μ_s for the actual surfaces used.

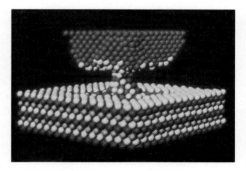

Figure 9.4
Computer simulation of a bond or "neck" of atoms formed between a nickel tip (red) and a gold surface.

Table 9.1 Typical values of the coefficient of static friction.

Materials	Coefficient of Static Friction μ_s
Metal on metal	0.15–0.20
Masonry on masonry	0.60–0.70
Wood on wood	0.25–0.50
Metal on masonry	0.30–0.70
Metal on wood	0.20–0.60
Rubber on concrete	0.50–0.90

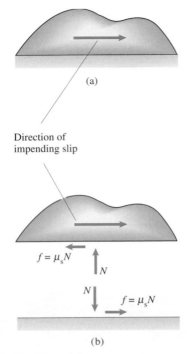

Figure 9.5
(a) The upper surface is on the verge of slipping to the right.
(b) Directions of the friction forces.

Let's return to the example of the book on the table (Fig. 9.2). If the force F exerted on the book is small enough that the book does not move, the condition for equilibrium requires that the friction force $f = F$. Why do we need the theory of dry friction? If we begin to increase F, the friction force f will increase until the book slips. Equation (9.1) gives the *maximum* friction force that the two surfaces can exert and thus tells us the largest force F that can be applied to the book without causing it to slip. Suppose that we know the coefficient of static friction μ_s between the book and the table and the weight W of the book. Since the normal force $N = W$, the largest value of F that can be applied to the book without causing it to slip is $F = f = \mu_s W$.

Equation (9.1) determines the magnitude of the maximum friction force but not its direction. The friction force is a maximum, and Eq. (9.1) is applicable, when two surfaces are on the verge of slipping relative to each other. We say that slip is *impending*, and the friction forces resist the impending motion. In Fig. 9.5a, suppose that the lower surface is fixed and slip of the upper surface toward the right is impending. The friction force on the upper surface resists its impending motion (Fig. 9.5b). The friction force on the lower surface is in the opposite direction.

The Kinetic Coefficient According to the theory of dry friction, the magnitude of the friction force between two plane dry contacting surfaces that are in motion (sliding) relative to each other is

$$f = \mu_k N, \tag{9.2}$$

where N is the normal force between the surfaces and μ_k is the *coefficient of kinetic friction*. The value of μ_k is assumed to depend only on the compositions of the surfaces and their conditions. For a given pair of surfaces, its value is generally smaller than that of μ_s.

Once you have caused the book in Fig. 9.2 to begin sliding on the table, the friction force $f = \mu_k N = \mu_k W$. Therefore the force you must exert to keep the book in uniform motion is $F = f = \mu_k W$.

When two surfaces are sliding relative to each other, the friction forces resist the relative motion. In Fig. 9.6a, suppose that the lower surface is fixed and the upper surface is moving to the right. The friction force on the upper surface acts in the direction opposite to its motion (Fig. 9.6b). The friction force on the lower surface is in the opposite direction.

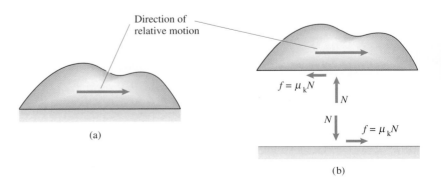

Figure 9.6
(a) The upper surface is moving to the right relative to the lower surface.
(b) Directions of the friction forces.

Angles of Friction

Instead of resolving the reaction exerted on a surface due to its contact with another surface into the normal force N and friction force f (Fig. 9.7a), we can express it in terms of its magnitude R and the *angle of friction* θ between the force and the normal to the surface (Fig. 9.7b). The normal force and friction force are related to R and θ by

$$f = R\sin\theta, \tag{9.3}$$

$$N = R\cos\theta. \tag{9.4}$$

The value of θ when slip is impending is called the *angle of static friction* θ_s, and its value when the surfaces are sliding relative to each other is called the *angle of kinetic friction* θ_k. By using Eqs. (9.1)–(9.4), we can express the angles of static and kinetic friction in terms of the coefficients of friction:

$$\tan\theta_s = \mu_s, \tag{9.5}$$

$$\tan\theta_k = \mu_k. \tag{9.6}$$

In summary, if slip is impending, the magnitude of the friction force is given by Eq. (9.1) and the angle of friction by Eq. (9.5). If surfaces are sliding relative to each other, the magnitude of the friction force is given by Eq. (9.2) and the angle of friction by Eq. (9.6). Otherwise, the friction force must be determined from the equilibrium equations. The sequence of decisions in evaluating the friction force and angle of friction is summarized in Fig. 9.8.

Study Questions

1. How is the coefficient of static friction defined?
2. How is the coefficient of kinetic friction defined?
3. If relative slip of two dry surfaces in contact is impending, what do you know about the friction forces they exert on each other?
4. If two dry surfaces in contact are sliding relative to each other, what do you know about the resulting friction forces?

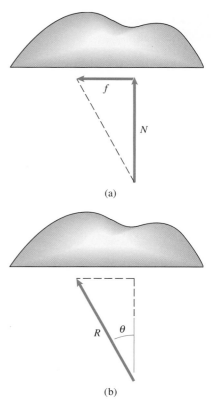

(a)

(b)

Figure 9.7
(a) The normal force N and the friction force f.
(b) The magnitude R and the angle of friction θ.

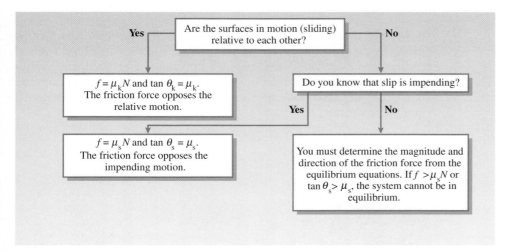

Figure 9.8
Evaluating the friction force.

Example 9.1

Determining the Friction Force

The arrangement in Fig. 9.9 exerts a horizontal force on the stationary 80-kg crate. The coefficient of static friction between the crate and the ramp is $\mu_s = 0.4$.

(a) If the rope exerts a 400-N force on the crate, what is the friction force exerted on the crate by the ramp?

(b) What is the largest force the rope can exert on the crate without causing it to slide up the ramp?

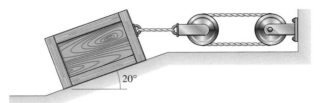

Figure 9.9

Strategy

(a) We can follow the logic in Fig. 9.8 to decide how to evaluate the friction force. The crate is not sliding on the ramp, and we don't know whether slip is impending, so we must determine the friction force by using the equilibrium equations.

(b) We want to determine the value of the force exerted by the rope that causes the crate to be on the verge of slipping up the ramp. When slip is impending, the magnitude of the friction force is $f = \mu_s N$ and the friction force opposes the impending slip. We can use the equilibrium equations to determine the force exerted by the rope.

Solution

(a) We draw the free-body diagram of the crate in Fig. a, showing the force T exerted by the rope, the weight mg of the crate, and the normal force N and friction force f exerted by the ramp. We can choose the direction of f arbitrarily, and our solution will indicate the actual direction of the friction force. By aligning the coordinate system with the ramp as shown, we obtain the equilibrium equation

$$\Sigma F_x = f + T\cos 20° - mg\sin 20° = 0.$$

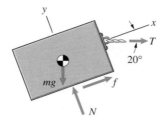

(a) Free-body
diagram of the crate.

Solving for the friction force, we obtain

$$f = -T \cos 20° + mg \sin 20° = -(400) \cos 20° + (80)(9.81) \sin 20°$$

$$= -107 \text{ N}.$$

The minus sign indicates that the direction of the friction force on the crate is down the ramp.

(b) The friction force is $f = \mu_s N$, and it opposes the impending slip. To simplify our solution for T, we align the coordinate system as shown in Fig. b, obtaining the equilibrium equations

$$\Sigma F_x = T - N \sin 20° - \mu_s N \cos 20° = 0,$$

$$\Sigma F_y = N \cos 20° - \mu_s N \sin 20° - mg = 0.$$

Solving the second equilibrium equation for N, we obtain

$$N = \frac{mg}{\cos 20° - \mu_s \sin 20°} = \frac{(80)(9.81)}{\cos 20° - (0.4) \sin 20°} = 977 \text{ N}.$$

Then from the first equilibrium equation, T is

$$T = N(\sin 20° + \mu_s \cos 20°) = (977)\left[\sin 20° + (0.4) \cos 20° \right]$$

$$= 702 \text{ N}.$$

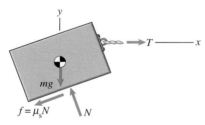

(b) The free-body diagram when slip up the ramp is impending.

Alternative Solution We can also determine T by representing the reaction exerted on the crate by the ramp as a single force (Fig. c). Because slip of the crate up the ramp is impending, R opposes the impending motion and the friction angle is $\theta_s = \arctan \mu_s = \arctan (0.4) = 21.8°$. From the triangle formed by the sum of the forces acting on the crate (Fig. d), we obtain

$$T = mg \tan(20° + \theta_s) = (80)(9.81) \tan(20° + 21.8°) = 702 \text{ N}.$$

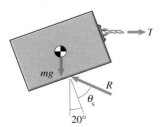

(c) Representing the reaction exerted by the ramp as a single force.

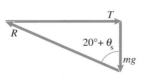

(d) The forces on the crate.

Example 9.2

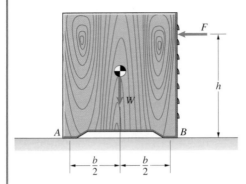

Figure 9.10

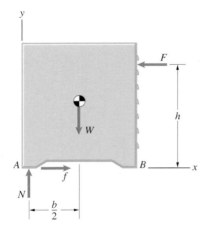

(a) The free-body diagram when the chest is on the verge of tipping over.

Determining Whether an Object Will Tip Over

Suppose that we want to push the tool chest in Fig. 9.10 across the floor by applying the horizontal force F. If we apply the force at too great a height h, the chest will tip over before it slips. If the coefficient of static friction between the floor and the chest is μ_s, what is the largest value of h for which the chest will slip before it tips over?

Strategy

When the chest is on the verge of tipping over, it is in equilibrium with no reaction at B. We can use this condition to determine F in terms of h. Then, by determining the value of F that will cause the chest to slip, we will obtain the value of h that causes the chest to be on the verge of tipping over *and* on the verge of slipping.

Solution

We draw the free-body diagram of the chest when it is on the verge of tipping over in Fig. a. Summing moments about A, we obtain

$$\Sigma M_{(\text{point } A)} = Fh - W\left(\frac{1}{2}b\right) = 0.$$

Equilibrium also requires that $f = F$ and $N = W$.

When the chest is on the verge of slipping,

$$f = \mu_s N,$$

so

$$F = f = \mu_s N = \mu_s W.$$

Substituting this expression into the moment equation, we obtain

$$\mu_s Wh - W\left(\frac{1}{2}b\right) = 0.$$

Solving this equation for h, we find that the chest is on the verge of tipping over *and* on the verge of slipping when

$$h = \frac{b}{2\mu_s}.$$

If h is smaller than this value, the chest will begin sliding before it tips over.

Discussion

Notice that the largest value of h for which the chest will slip before it tips over is independent of F. Whether the chest will tip over depends only on where the force is applied, not how large it is.

Example 9.3

Analyzing a Friction Brake

The motion of the disk in Fig. 9.11 is controlled by the friction force exerted at C by the brake ABC. The hydraulic actuator BE exerts a horizontal force of magnitude F on the brake at B. The coefficients of friction between the disk and the brake are μ_s and μ_k. What couple M is necessary to rotate the disk at a constant rate in the counterclockwise direction?

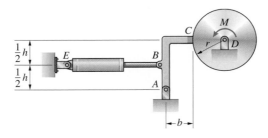

Figure 9.11

Strategy

We can use the free-body diagram of the disk to obtain a relation between M and the reaction exerted on the disk by the brake, then use the free-body diagram of the brake to determine the reaction in terms of F.

Solution

We draw the free-body diagram of the disk in Fig. a, representing the force exerted by the brake by a single force R. The force R opposes the counterclockwise rotation of the disk, and the friction angle is the angle of kinetic friction $\theta_k = \arctan \mu_k$. Summing moments about D, we obtain

$$\Sigma M_{(\text{point } D)} = M - (R \sin \theta_k) r = 0.$$

Then, from the free-body diagram of the brake (Fig. b), we obtain

$$\Sigma M_{(\text{point } A)} = -F\left(\frac{1}{2}h\right) + (R \cos \theta_k)h - (R \sin \theta_k)b = 0.$$

We can solve these two equations for M and R. The solution for the couple M is

$$M = \frac{(1/2)hr\, F \sin \theta_k}{h \cos \theta_k - b \sin \theta_k} = \frac{(1/2)hr\, F \mu_k}{h - b\mu_k}.$$

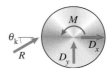

(a) The free-body diagram of the disk.

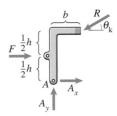

(b) The free-body diagram of the brake.

Discussion

If μ_k is sufficiently small, then the denominator of the solution for the couple, $(h \cos \theta_k - b \sin \theta_k)$, is positive. As μ_k becomes larger, the denominator becomes smaller, because $\cos \theta_k$ decreases and $\sin \theta_k$ increases. As the denominator approaches zero, the couple required to rotate the disk approaches infinity. To understand this result, notice that the denominator equals zero when $\tan \theta_k = h/b$, which means that the line of action of R passes through point A (Fig. c). As μ_k becomes larger and the line of action of R approaches point A, the magnitude of R necessary to balance the moment of F about A approaches infinity and, as a result, M approaches infinity.

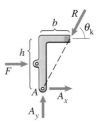

(c) The line of action of R passing through point A.

Example 9.4

A Friction Problem in Three Dimensions

The 80-kg climber at A in Fig. 9.12 is being helped up an icy slope by friends. The tensions in ropes AB and AC are 130 N and 220 N, respectively. The y axis is vertical, and the unit vector $\mathbf{e} = -0.182\mathbf{i} + 0.818\mathbf{j} + 0.545\mathbf{k}$ is perpendicular to the ground where the climber stands. What minimum coefficient of static friction between the climber's shoes and the ground is necessary to prevent him from slipping?

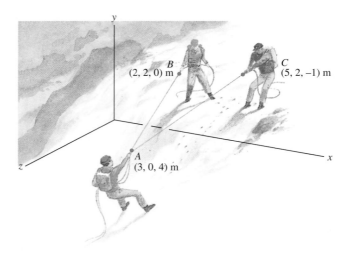

Figure 9.12

Strategy

We know the forces exerted on the climber by the two ropes and by his weight, so we can use equilibrium to determine the force $\mathbf{R}$ exerted on him by the ground. When slip is impending, the angle between $\mathbf{R}$ and the unit vector $\mathbf{e}$ is equal to the angle of static friction θ_s. We can use this condition to calculate the coefficient of static friction for impending slip.

Solution

We draw the free-body diagram of the climber in Fig. a, showing the forces $\mathbf{T}_{AB}$ and $\mathbf{T}_{AC}$ exerted by the ropes, the force $\mathbf{R}$ exerted by the ground, and his weight. The sum of the forces equals zero:

$$\mathbf{R} + \mathbf{T}_{AB} + \mathbf{T}_{AC} - mg\mathbf{j} = \mathbf{0}.$$

By expressing $\mathbf{T}_{AB}$ and $\mathbf{T}_{AC}$ in terms of their components, we can solve this equation for the components of $\mathbf{R}$. The force $\mathbf{T}_{AB}$ is

$$\mathbf{T}_{AB} = |\mathbf{T}_{AB}| \left[\frac{(2-3)\mathbf{i} + (2-0)\mathbf{j} + (0-4)\mathbf{k}}{\sqrt{(2-3)^2 + (2-0)^2 + (0-4)^2}} \right]$$

$$= (130)(-0.218\mathbf{i} + 0.436\mathbf{j} - 0.873\mathbf{k})$$

$$= -28.4\mathbf{i} + 56.7\mathbf{j} - 113.5\mathbf{k} \ (\text{N}),$$

and the force $\mathbf{T}_{AC}$ is

$$\mathbf{T}_{AC} = |\mathbf{T}_{AC}| \left[\frac{(5-3)\mathbf{i} + (2-0)\mathbf{j} + (-1-4)\mathbf{k}}{\sqrt{(5-3)^2 + (2-0)^2 + (-1-4)^2}} \right]$$

$$= (220)(0.348\mathbf{i} + 0.348\mathbf{j} - 0.870\mathbf{k})$$

$$= 76.6\mathbf{i} + 76.6\mathbf{j} - 191.5\mathbf{k} \ (\text{N}).$$

Substituting these expressions into the equilibrium equation and solving for $\mathbf{R}$, we obtain

$$\mathbf{R} = -48.2\mathbf{i} + 651.5\mathbf{j} + 305.0\mathbf{k} \ (\text{N}).$$

To determine the angle θ between $\mathbf{R}$ and the unit vector $\mathbf{e}$ that is normal to the surface on which the climber stands (Fig. b), we use the dot product. From the definition $\mathbf{R} \cdot \mathbf{e} = |\mathbf{R}| |\mathbf{e}| \cos\theta$, we obtain

$$\cos\theta = \frac{\mathbf{R} \cdot \mathbf{e}}{|\mathbf{R}| |\mathbf{e}|} = \frac{(-48.2)(-0.182) + (651.5)(0.818) + (305.0)(0.545)}{\sqrt{(-48.2)^2 + (651.5)^2 + (305.0)^2}}$$

$$= 0.982.$$

The angle $\theta = 10.9°$. Setting this angle equal to the angle of static friction, we obtain the coefficient of static friction for impending slip:

$$\mu_\text{s} = \tan\theta_\text{s} = \tan 10.9° = 0.193.$$

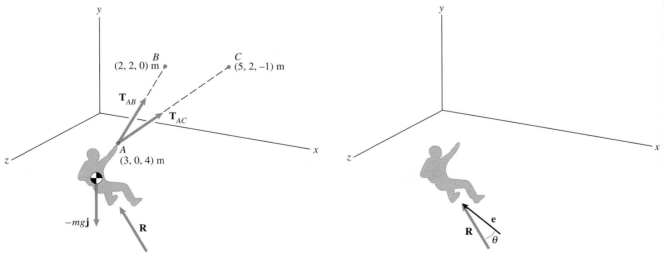

(a) Free-body diagram of the climber.

(b) The angle θ.

9.2   Applications

Effects of friction forces, such as wear, loss of energy, and generation of heat, are often undesirable. But many devices cannot function properly without friction forces and may actually be designed to create them. A car's brakes work by exerting friction forces on the rotating wheels, and its tires are designed to maximize the friction forces they exert on the road under various weather conditions. In this section we analyze several types of devices in which friction forces play important roles.

Wedges

A *wedge* is a bifacial tool with the faces set at a small acute angle (Figs. 9.13a and b). When a wedge is pushed forward, the faces exert large lateral forces as a result of the small angle between them (Fig. 9.13c). In various forms, wedges are used in many engineering applications.

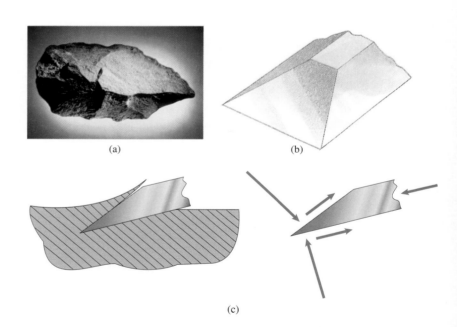

(a)

(b)

(c)

Figure 9.13
(a) An early wedge tool—a bifacial "hand axe" from Olduvai Gorge, Tanzania.
(b) A modern chisel blade.
(c) The faces of a wedge can exert large lateral forces.

The large lateral force generated by a wedge can be used to lift a load (Fig. 9.14a). Let W_L be the weight of the load and W_W the weight of the wedge. To determine the force F necessary to start raising the load, we assume that slip of the load and wedge are impending (Fig. 9.14b). From the free-body diagram of the load, we obtain the equilibrium equations

$$\Sigma F_x = Q - N \sin \alpha - \mu_s N \cos \alpha = 0,$$

$$\Sigma F_y = N \cos \alpha - \mu_s N \sin \alpha - \mu_s Q - W_L = 0.$$

From the free-body diagram of the wedge, we obtain the equations

$$\Sigma F_x = N \sin\alpha + \mu_s N \cos\alpha + \mu_s P - F = 0,$$

$$\Sigma F_y = P - N \cos\alpha + \mu_s N \sin\alpha - W_W = 0.$$

These four equations determine the three normal forces Q, N, and P and the force F. The solution for F is

$$F = \mu_s W_W + \left[\frac{(1 - \mu_s^2) \tan\alpha + 2\mu_s}{(1 - \mu_s^2) - 2\mu_s \tan\alpha} \right] W_L.$$

Suppose that $W_W = 0.2W_L$ and $\alpha = 10°$. If $\mu_s = 0$, the force necessary to lift the load is only $0.176W_L$. But if $\mu_s = 0.2$, the force becomes $0.680W_L$, and if $\mu_s = 0.4$, it becomes $1.44W_L$. From this standpoint, friction is undesirable. But if there were no friction, the wedge would not remain in place when the force F is removed.

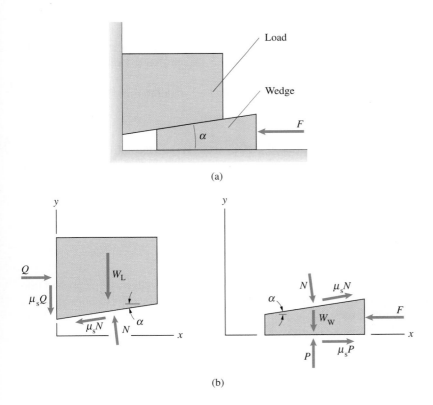

(a)

(b)

Figure 9.14
(a) Raising a load with a wedge.
(b) Free-body diagrams of the load and the wedge when slip is impending.

Example 9.5

Forces on a Wedge

Splitting a log must have been among the first applications of the wedge (Fig. 9.15). Although it is a dynamic process—the wedge is hammered into the wood—you can get an idea of the forces involved from a static analysis. Suppose that $\alpha = 10°$ and the coefficients of friction between the surfaces of the wedge and the log are $\mu_s = 0.22$ and $\mu_k = 0.20$. Neglect the weight of the wedge.

Figure 9.15

(a) If the wedge is driven into the log at a constant rate by a vertical force F, what are the magnitudes of the normal forces exerted on the log by the wedge?
(b) Will the wedge remain in place in the log when the force is removed?

Strategy

(a) The friction forces resist the motion of the wedge into the log and are equal to $\mu_k N$, where N is the normal force the log exerts on the faces. We can use equilibrium to determine N in terms of F.
(b) By assuming that the wedge is on the verge of slipping out of the log, we can determine the minimum value of μ_s necessary for the wedge to stay in place.

Solution

(a) In Fig. a we draw the free-body diagram of the wedge as it is pushed into the log by a force F. The faces of the wedge are subjected to normal forces and friction forces by the log. The friction forces resist the motion of the wedge. From the equilibrium equation

$$2N \sin\left(\frac{\alpha}{2}\right) + 2\mu_k N \cos\left(\frac{\alpha}{2}\right) - F = 0,$$

we obtain the normal force N:

$$N = \frac{F}{2\left[\sin(\alpha/2) + \mu_k \cos(\alpha/2)\right]} = \frac{F}{2\left[\sin(10°/2) + (0.20)\cos(10°/2)\right]}$$

$$= 1.75F.$$

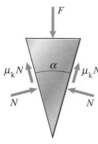

(a) Free-body diagram of the wedge with a vertical force F applied to it.

(b) In Fig. b we draw the free-body diagram when $F = 0$ and the wedge is on the verge of slipping out. From the equilibrium equation

$$2N \sin\left(\frac{\alpha}{2}\right) - 2\mu_s N \cos\left(\frac{\alpha}{2}\right) = 0,$$

we obtain the minimum coefficient of friction necessary for the wedge to remain in place:

$$\mu_s = \tan\left(\frac{\alpha}{2}\right) = \tan\left(\frac{10°}{2}\right) = 0.087.$$

We can also obtain this result by representing the reaction exerted on the wedge by the log as a single force (Fig. c). When the wedge is on the verge of slipping out, the friction angle is the angle of static friction θ_s. The sum of the forces in the vertical direction is zero only if

$$\theta_s = \arctan\left(\mu_s\right) = \frac{\alpha}{2} = 5°,$$

so $\mu_s = \tan 5° = 0.087$. Thus we conclude that the wedge will remain in place.

(b) Free-body diagram of the wedge when it is on the verge of slipping out.

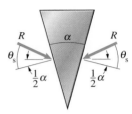

(c) Representing the reactions by a single force.

Threads

Threads are familiar from their use on wood screws, machine screws, and other machine elements. We show a shaft with square threads in Fig. 9.16a. The axial distance p from one thread to the next is called the *pitch* of the thread, and the angle α is its *slope*. We will consider only the case in which the shaft has a single continuous thread, so the relation between the pitch and slope is

$$\tan\alpha = \frac{p}{2\pi r}, \tag{9.7}$$

where r is the mean radius of the thread.

    Suppose that the threaded shaft is enclosed in a fixed sleeve with a mating groove and is subjected to an axial load F (Fig. 9.16b). Applying a couple M in the direction shown will tend to cause the shaft to start rotating and moving in the axial direction opposite to F. Our objective is to determine the couple M necessary to cause the shaft to start rotating.

    We draw the free-body diagram of a differential element of the thread of length dL in Fig. 9.16c, representing the reaction exerted by the mating groove by the force dR. If the shaft is on the verge of rotating, dR resists the impending motion and the friction angle is the angle of static friction θ_s. The vertical component of the reaction on the element is $dR\cos(\theta_s + \alpha)$. To determine the total vertical force on the thread, we must integrate this expression over the length L of the thread. For equilibrium, the result must equal the axial force F acting on the shaft:

$$\cos(\theta_s + \alpha) \int_L dR = F. \tag{9.8}$$

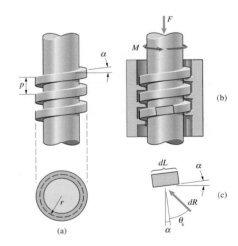

Figure 9.16
(a) A shaft with a square thread.
(b) The shaft within a sleeve with a mating groove and the direction of M that can cause the shaft to start moving in the axial direction opposite to F.
(c) A differential element of the thread when slip is impending.

The moment about the center of the shaft due to the reaction on the element is $r \, dR \sin(\theta_s + \alpha)$. The total moment must equal the couple M exerted on the shaft:

$$r \sin(\theta_s + \alpha) \int_L dR = M.$$

Dividing this equation by Eq. (9.8), we obtain the couple M necessary for the shaft to be on the verge of rotating and moving in the axial direction opposite to F:

$$M = rF \tan(\theta_s + \alpha). \qquad (9.9)$$

Replacing the angle of static friction θ_s in this expression with the angle of kinetic friction θ_k gives the couple required to cause the shaft to rotate at a constant rate.

If the couple M is applied to the shaft in the opposite direction (Fig. 9.17a), the shaft tends to start rotating and moving in the axial direction of the load F. Figure 9.17b shows the reaction on a differential element of the thread of length dL when slip is impending. The direction of the reaction opposes the rotation of the shaft. In this case, the vertical component of the reaction on the element is $dR \cos(\theta_s - \alpha)$. Equilibrium requires that

$$\cos(\theta_s - \alpha) \int_L dR = F. \qquad (9.10)$$

The moment about the center of the shaft due to the reaction is $r \, dR \sin(\theta_s - \alpha)$, so

$$r \sin(\theta_s - \alpha) \int_L dR = M.$$

Dividing this equation by Eq. (9.10), we obtain the couple M necessary for the shaft to be on the verge of rotating and moving in the direction of the force F:

$$M = rF \tan(\theta_s - \alpha). \qquad (9.11)$$

Replacing θ_s with θ_k in this expression gives the couple necessary to rotate the shaft at a constant rate.

Notice in Eq. (9.11) that the couple required for impending motion is zero when $\theta_s = \alpha$. When the angle of static friction is less than this value, the shaft will rotate and move in the direction of the force F with no couple applied.

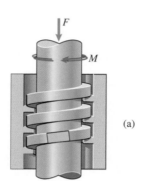

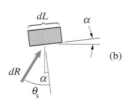

Figure 9.17
(a) The direction of M that can cause the shaft to move in the axial direction of F.
(b) A differential element of the thread when slip is impending.

Study Questions

1. How is the slope α of a thread defined?
2. If you know the pitch and mean radius of a thread, how do you determine its slope?
3. If a threaded shaft is subjected to an axial load, how do you determine the couple necessary to rotate the shaft at a constant rate and cause it to move in the direction opposite to the direction of the axial load?

Rotating a Threaded Collar

The right end of bar AB in Fig. 9.18 is pinned to an unthreaded collar B that rests on the threaded collar C. The mean radius of the thread is $r = 40$ mm and its pitch is $p = 5$ mm. The coefficients of static and kinetic friction between the threads of the collar C and those of the threaded shaft are $\mu_s = 0.25$ and $\mu_k = 0.22$. The 180-kg suspended object can be raised or lowered by turning the collar C.
(a) When the system is in the position shown, what couple must be applied to the collar C to rotate it at a constant rate and cause the suspended object to move upward?
(b) Will the system remain in equilibrium in the position shown if no couple is applied to the collar C?

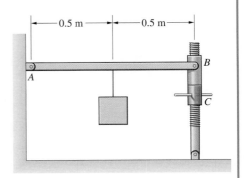

Figure 9.18

Strategy

(a) By drawing the free-body diagram of the bar and collar B, we can determine the axial force exerted on collar C. Then we can use Eq. (9.9), with θ_s replaced by θ_k, to determine the required couple.
(b) From Eq. (9.11), the collar C is on the verge of rotating and moving in the direction of the axial load when no couple is exerted on it if $\theta_s = \alpha$. If the angle of static friction θ_s is greater than or equal to the slope α, the system will remain in equilibrium with no couple applied.

Solution

(a) We draw the free-body diagram of the bar and collar B in Fig. a, where F is the force exerted on the collar B by the collar C. From the equilibrium equation

$$\Sigma M_{(\text{point } A)} = (1.0)F - (0.5)mg = 0,$$

we obtain $F = \frac{1}{2}mg = \frac{1}{2}(180)(9.81) = 883$ N. This is the axial force exerted on collar C (Fig. b). Replacing θ_s by θ_k in Eq. (9.9), the couple necessary to rotate the collar at a constant rate is

$$M = rF \tan(\theta_k + \alpha).$$

The slope α is related to the pitch and mean radius of the thread by Eq. (9.7):

$$\tan\alpha = \frac{p}{2\pi r} = \frac{0.005}{2\pi(0.04)} = 0.0199.$$

We obtain $\alpha = \arctan(0.0199) = 1.14°$. The angle of kinetic friction is

$$\theta_k = \arctan(\mu_k) = \arctan(0.22) = 12.41°.$$

Using these values, the required couple is

$$M = rF \tan(\theta_k + \alpha)$$
$$= (0.04)(883) \tan(12.41° + 1.14°)$$
$$= 8.51 \text{ N-m.}$$

(b) The angle of static friction is

$$\theta_s = \arctan(\mu_s) = \arctan(0.25) = 14.04°.$$

Therefore θ_s is greater than the slope α, and we conclude from Eq. (9.11) that the system will remain in equilibrium with no couple applied to collar C.

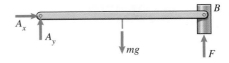

(a) Free-body diagram of bar AB and the collar B.

(b) The threaded shaft and the collar C.

Journal Bearings

A *bearing* is a support. This term usually refers to supports designed to allow the supported object to move. For example, in Fig. 9.19a, a horizontal shaft is supported by two *journal bearings*, which allow the shaft to rotate. The shaft can then be used to support a load perpendicular to its axis, such as that subjected by a pulley (Fig. 9.19b).

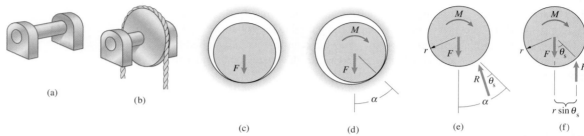

(a)

(b)

(c)

(d)

(e)

(f)

Figure 9.19
(a) A shaft supported by journal bearings.
(b) A pulley supported by the shaft.
(c) The shaft and bearing when no couple is applied to the shaft.
(d) A couple causes the shaft to roll within the bearing.
(e) Free-body diagram of the shaft.
(f) The two forces on the shaft must be equal and opposite.

(a)

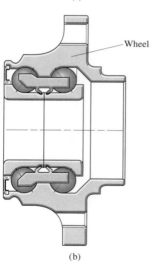

Wheel

(b)

Figure 9.20
(a) A journal bearing with one row of balls.
(b) Journal bearing assembly of the wheel of a car. There are two rows of balls between the rotating wheel and the fixed inner cylinder.

Here we analyze journal bearings consisting of brackets with holes through which the shaft passes. The radius of the shaft is slightly smaller than the radius of the holes in the bearings. Our objective is to determine the couple that must be applied to the shaft to cause it to rotate in the bearings. Let F be the total load supported by the shaft including the weight of the shaft itself. When no couple is exerted on the shaft, the force F presses it against the bearings as shown in Fig. 9.19c. When a couple M is exerted on the shaft, it rolls up the surfaces of the bearings (Fig. 9.19d). The term α is the angle from the original point of contact of the shaft to its point of contact when M is applied.

In Fig. 9.19e, we draw the free-body diagram of the shaft when M is sufficiently large that slip is impending. The force R is the total reaction exerted on the shaft by the two bearings. Since R and F are the only forces acting on the shaft, equilibrium requires that $\alpha = \theta_s$ and $R = F$ (Fig. 9.19f). The reaction exerted on the shaft by the bearings is displaced a distance $r \sin \theta_s$ from the vertical line through the center of the shaft. By summing moments about the center of the shaft, we obtain the couple M that causes the shaft to be on the verge of slipping:

$$M = rF \sin \theta_s. \tag{9.12}$$

This is the largest couple that can be exerted on the shaft without causing it to start rotating. Replacing θ_s in this expression by the angle of kinetic friction θ_k gives the couple necessary to rotate the shaft at a constant rate.

The simple type of journal bearing we have described is too primitive for most applications. The surfaces where the shaft and bearing are in contact would quickly become worn. Designers usually incorporate "ball" or "roller" bearings in journal bearings to minimize friction (Fig. 9.20).

Example 9.7

Pulley Supported by Journal Bearings

The mass of the suspended load in Fig. 9.21 is 450 kg. The pulley P has a 150-mm radius and is rigidly attached to a horizontal shaft supported by journal bearings. The radius of the horizontal shaft is 12 mm and the coefficient of kinetic friction between the shaft and the bearings is 0.2. The masses of the pulley and shaft are negligible. What tension must the winch A exert on the cable to raise the load at a constant rate?

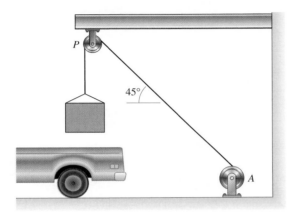

Figure 9.21

Strategy

Equation (9.12) with θ_s replaced by θ_k relates the couple M required to turn the pulley at a constant rate to the total force F on the shaft. By expressing M and F in terms of the load and the tension exerted by the winch, we can obtain an equation for the required tension.

Solution

Let T be the tension exerted by the winch (Fig. a). By calculating the magnitude of the sum of the forces exerted by the tension and the load (Fig. b), we obtain an expression for the total force F on the shaft supporting the pulley:

$$F = \sqrt{(mg + T \sin 45°)^2 + (T \cos 45°)^2}.$$

The (clockwise) couple exerted on the pulley by the tension and the load is

$$M = 0.15(T - mg).$$

The radius of the shaft is $r = 0.012$ m and the angle of kinetic friction is $\theta_k = \arctan(0.2) = 11.3°$. We substitute our expressions for F and M into Eq. (9.12).

$$M = rF \sin \theta_k:$$

$$0.15\left[T - (450)(9.81)\right] = 0.012\sqrt{\left[(450)(9.81) + T \sin 45°\right]^2 + (T \cos 45°)^2} \sin(11.3°).$$

Solving for the tension, we obtain $T = 4.54$ kN.

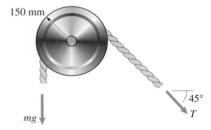

150 mm

mg

$45°$

T

(a) Free-body diagram of the pulley.

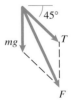

$45°$

mg

T

F

(b) The total force F on the shaft.

Thrust Bearings and Clutches

A *thrust bearing* supports a rotating shaft that is subjected to an axial load. In the type shown in Figs. 9.22a and 9.22b, the conical end of the shaft is pressed against the mating conical cavity by an axial load F. Let us determine the couple M necessary to rotate the shaft.

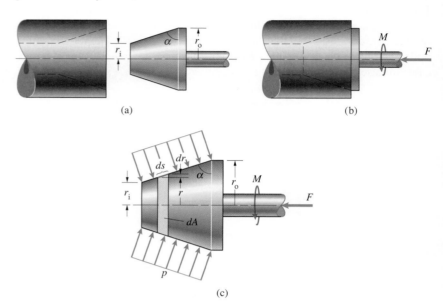

Figure 9.22

(a), (b) A thrust bearing supports a shaft subjected to an axial load.

(c) The differential element dA and the uniform pressure p exerted by the cavity.

The differential element of area dA in Fig. 9.22(c) is

$$dA = 2\pi r \, ds = 2\pi r \left(\frac{dr}{\cos \alpha} \right).$$

Integrating this expression from $r = r_i$ to $r = r_o$, we obtain the area of contact:

$$A = \frac{\pi \left(r_o^2 - r_i^2 \right)}{\cos \alpha}.$$

If we assume that the mating surface exerts a uniform pressure p, the axial component of the total force due to p must equal F: $pA \cos \alpha = F$. Therefore the pressure is

$$p = \frac{F}{A \cos \alpha} = \frac{F}{\pi \left(r_o^2 - r_i^2 \right)}.$$

As the shaft rotates about its axis, the moment about the axis due to the friction force on the element dA is $r \mu_k \, (p \, dA)$. The total moment equals M:

$$M = \int_A \mu_k r p \, dA = \int_{r_i}^{r_o} \mu_k r \left[\frac{F}{\pi \left(r_o^2 - r_i^2 \right)} \right] \left(\frac{2\pi r \, dr}{\cos \alpha} \right).$$

Integrating, we obtain the couple M necessary to rotate the shaft at a constant rate:

$$M = \frac{2\mu_k F}{3 \cos \alpha} \left(\frac{r_o^3 - r_i^3}{r_o^2 - r_i^2} \right). \tag{9.13}$$

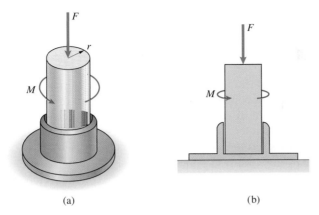

(a)   (b)

Figure 9.23
A thrust bearing that supports a flat-ended shaft.

A simpler thrust bearing is shown in Figs. 9.23a and 9.23b. The bracket supports the flat end of a shaft of radius r that is subjected to an axial load F. We can obtain the couple necessary to rotate the shaft at a constant rate from Eqs. (9.13) by setting $\alpha = 0$, $r_i = 0$, and $r_o = r$:

$$M = \frac{2}{3} \mu_k F r. \tag{9.14}$$

Although they are good examples of the analysis of friction forces, the thrust bearings we have described would become worn too quickly to be used in most applications. The designer of the thrust bearing in Fig. 9.24 minimizes friction by incorporating "roller" bearings.

A *clutch* is a device used to connect and disconnect two coaxial rotating shafts. The type shown in Figs. 9.25a and 9.25b consists of disks of radius r attached to the ends of the shafts. When the disks are separated (Fig. 9.25a), the clutch is *disengaged*, and the shafts can rotate freely relative to each other. When the clutch is engaged by pressing the disks together with axial forces F (Fig. 9.25b), the shafts can support a couple M due to the friction forces between the disks. If the couple M becomes too large. the clutch slips.

The friction forces exerted on one face of the clutch by the other face are identical to the friction forces exerted on the flat-ended shaft by the bracket in Fig. 9.23. We can therefore determine the largest couple the clutch can support without slipping by replacing μ_k by μ_s in Eqs. (9.14):

$$M = \frac{2}{3} \mu_s F r. \tag{9.15}$$

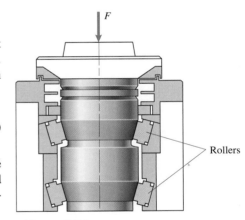

Figure 9.24
A thrust bearing with two rows of cylindrical rollers between the shaft and the fixed support.

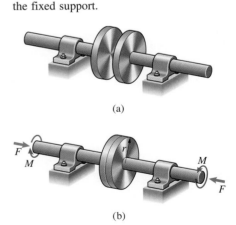

(a)

(b)

Figure 9.25
A clutch.
(a) Disengaged position.
(b) Engaged position.

Study Questions

1. What is a journal bearing?
2. If the shaft of a journal bearing is subjected to a lateral force F, how do you determine the couple M necessary to rotate the shaft at a constant rate?
3. When the axis of a clutch is subjected to an axial force F (Fig. 9.25b), how do you determine the largest couple M the clutch can support without slipping?

Example 9.8

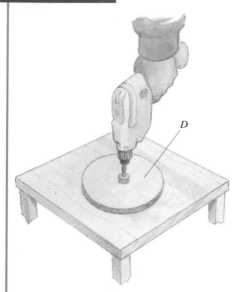

Figure 9.26

Friction on a Disk Sander

The handheld sander in Fig. 9.26 has a rotating disk D of 4-in. radius with sandpaper bonded to it. The total downward force exerted by the operator and the weight of the sander is 15 lb. The coefficient of kinetic friction between the sandpaper and the surface is $\mu_k = 0.6$. What couple (torque) M must the motor exert to turn the sander at a constant rate?

Strategy

As the disk D rotates, it is subjected to friction forces analogous to the friction forces exerted on the flat-ended shaft by the bracket in Fig. 9.23. We can determine the couple required to turn the disk D at a constant rate from Eq. (9.14).

Solution

The couple required to turn the disk at a constant rate is

$$M = \frac{2}{3} \mu_k rF = \frac{2}{3}(0.6)\left(\frac{4}{12}\right)(15) = 2 \text{ ft-lb.}$$

T_2 T_1

Figure 9.27
A rope wrapped around a post.

Belt Friction

If a rope is wrapped around a fixed post as shown in Fig. 9.27, a large force T_2 exerted on one end can be supported by a relatively small force T_1 applied to the other end. In this section we analyze this familiar phenomenon. It is referred to as *belt friction* because a similar approach can be used to analyze belts used in machines, such as the belts that drive alternators and other devices in a car.

Let's consider a rope wrapped through an angle β around a fixed cylinder (Fig. 9.28a). We will assume that the tension T_1 is known. Our objective is to determine the largest force T_2 that can be applied to the other end of the rope without causing the rope to slip.

We begin by drawing the free-body diagram of an element of the rope whose boundaries are at angles α and $\alpha + \Delta\alpha$ from the point where the rope comes into contact with the cylinder (Figs. 9.28b and 9.28c). The force T is the tension in the rope at the position defined by the angle α. We know that the tension in the rope varies with position, because it increases from T_1 at $\alpha = 0$ to T_2 at $\alpha = \beta$. We therefore write the tension in the rope at the position $\alpha + \Delta\alpha$ as $T + \Delta T$. The force ΔN is the normal force exerted on the element by the cylinder. Because we want to determine the largest value of T_2 that will not cause the rope to slip, we assume that the friction force is equal to its maximum possible value $\mu_s \Delta N$, where μ_s is the coefficient of static friction between the rope and the cylinder.

The equilibrium equations in the directions tangential to and normal to the centerline of the rope are

$$\Sigma F_{(\text{tangential})} = \mu_s \Delta N + T \cos\left(\frac{\Delta\alpha}{2}\right) - (T + \Delta T)\cos\left(\frac{\Delta\alpha}{2}\right) = 0,$$

$$\Sigma F_{(\text{normal})} = \Delta N - (T + \Delta T)\sin\left(\frac{\Delta\alpha}{2}\right) - T\sin\left(\frac{\Delta\alpha}{2}\right) = 0. \qquad (9.16)$$

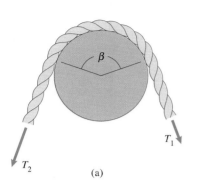
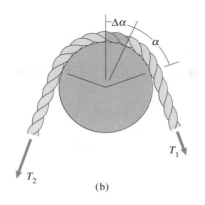
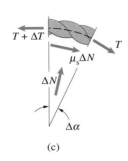

Figure 9.28
(a) A rope wrapped around a fixed cylinder.
(b) A differential element with boundaries at angles α and $\alpha + \Delta\alpha$.
(c) Free-body diagram of the element.

Eliminating ΔN, we can write the resulting equation as

$$\left[\cos\left(\frac{\Delta\alpha}{2}\right) - \mu_s \sin\left(\frac{\Delta\alpha}{2}\right)\right]\frac{\Delta T}{\Delta\alpha} - \mu_s T \frac{\sin(\Delta\alpha/2)}{(\Delta\alpha/2)} = 0.$$

Evaluating the limit of this equation as $\Delta\alpha \to 0$ and observing that

$$\frac{\sin(\Delta\alpha/2)}{(\Delta\alpha/2)} \to 1,$$

we obtain

$$\frac{dT}{d\alpha} - \mu_s T = 0.$$

This differential equation governs the variation of the tension in the rope. By separating variables,

$$\frac{dT}{T} = \mu_s\, d\alpha,$$

we can integrate to determine the tension T_2 in terms of the tension T_1 and the angle β:

$$\int_{T_1}^{T_2} \frac{dT}{T} = \int_0^\beta \mu_s\, d\alpha.$$

Thus we obtain the largest force T_2 that can be applied without causing the rope to slip when the force on the other end is T_1:

$$T_2 = T_1 e^{\mu_s\beta}. \tag{9.17}$$

The angle β in this equation must be expressed in radians. Replacing μ_s by the coefficient of kinetic friction μ_k gives the force T_2 required to cause the rope to slide at a constant rate.

Equation (9.17) explains why a large force can be supported by a relatively small force when a rope is wrapped around a fixed support. The force required to cause the rope to slip increases exponentially as a function of the angle through which the rope is wrapped. Suppose that $\mu_s = 0.3$. When the rope is wrapped one complete turn around the post ($\beta = 2\pi$), the ratio $T_2/T_1 = 6.59$. When the rope is wrapped four complete turns around the post ($\beta = 8\pi$), the ratio $T_2/T_1 = 1880$.

Study Questions

1. What is the definition of the term β in Eq. (9.17)?
2. If a rope is wrapped through a given angle around a fixed post and one end is subjected to a given tension T_1, how can you determine the tension T_2 necessary to cause the rope to be on the verge of slipping in the direction of T_2? How can you determine the smallest value of T_2 that will prevent the rope from slipping in the direction of T_1?

Example 9.9

Rope Wrapped Around Two Cylinders

The 50-kg crate in Fig. 9.29 is suspended from a rope that passes over two fixed cylinders. The coefficient of static friction is 0.2 between the rope and the left cylinder and 0.4 between the rope and the right cylinder. What is the smallest force the woman can exert and support the crate?

Strategy

She exerts the smallest possible force when slip of the rope is impending on both cylinders. Because we know the weight of the crate, we can use Eq. (9.17) to determine the tension in the rope between the two cylinders and then use Eq. (9.17) again to determine the force she exerts.

Solution

The weight of the crate is $W = (50)(9.81) = 491$ N. Let T be the tension in the rope between the two cylinders (Fig. a). The rope is wrapped around the left cylinder through an angle $\beta = \pi/2$ rad. The tension T necessary to prevent the rope from slipping on the left cylinder is related to W by

$$W = Te^{\mu_s\beta} = Te^{(0.2)(\pi/2)}.$$

Solving for T, we obtain

$$T = We^{-(0.2)(\pi/2)} = (491)e^{-(0.2)(\pi/2)} = 358 \text{ N}.$$

The rope is also wrapped around the right cylinder through an angle $\beta = \pi/2$ rad. The force F the woman must exert to prevent the rope from slipping on the right cylinder is related to T by

$$T = Fe^{\mu_s\beta} = Fe^{(0.4)(\pi/2)}.$$

The solution for F is

$$F = Te^{-(0.4)(\pi/2)} = (358)e^{-(0.4)(\pi/2)} = 191 \text{ N}.$$

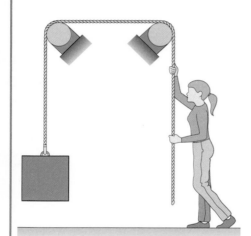

Figure 9.29

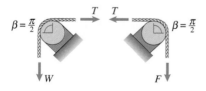

(a) The tensions in the rope.

Example 9.10

Application to Engineering:

Belts and Pulleys

The pulleys in Fig. 9.30 turn at a constant rate. The large pulley is attached to a fixed support. The small pulley is supported by a smooth horizontal slot and is pulled to the right by the force $F = 200$ N. The coefficient of static friction between the pulleys and the belt is $\mu_s = 0.8$, the dimension $b = 500$ mm, and the radii of the pulleys are $R_A = 200$ mm and $R_B = 100$ mm. What are the largest values of the couples M_A and M_B for which the belt will not slip?

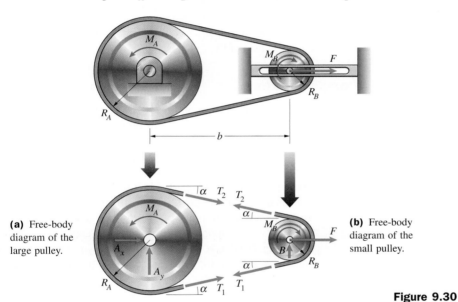

(a) Free-body diagram of the large pulley.

(b) Free-body diagram of the small pulley.

Figure 9.30

Strategy

By drawing free-body diagrams of the pulleys, we can use the equilibrium equations to relate the tensions in the belt to M_A and M_B and obtain a relation between the tensions in the belt and the force F. When slip is impending, the tensions are also related by Eq. (9.17). From these equations we can determine M_A and M_B.

Solution

From the free-body diagram of the large pulley (Fig. 9.30a). we obtain the equilibrium equation

$$M_A = R_A(T_2 - T_1), \tag{9.18}$$

and from the free-body diagram of the small pulley (Fig. 9.30b), we obtain

$$F = (T_1 + T_2)\cos\alpha, \tag{9.19}$$

$$M_B = R_B(T_2 - T_1). \tag{9.20}$$

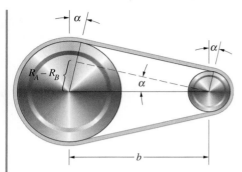

(c) Determining the angle α.

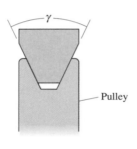

(a) Cross-sectional view of a V-belt and pulley.

(b) V-belt wrapped around a pulley.

The belt is in contact with the small pulley through the angle $\pi - 2\alpha$ (Fig. 9.30c). From the dashed line parallel to the belt, we see that the angle α satisfies the relation

$$\sin \alpha = \frac{R_A - R_B}{b} = \frac{200 - 100}{500} = 0.2.$$

Therefore $\alpha = 11.5° = 0.201$ rad. If we assume that slip impends between the small pulley and the belt, Eq. (9.17) states that

$$T_2 = T_1 e^{\mu_s \beta} = T_1 e^{0.8(\pi - 2\alpha)} = 8.95 T_1.$$

We solve this equation together with Eq. (9.19) for the two tensions, obtaining $T_1 = 20.5$ N and $T_2 = 183.6$ N. Then from Eqs. (9.18) and (9.20), the couples are $M_A = 32.6$ N-m and $M_B = 16.3$ N-m.

If we assume that slip impends between the large pulley and the belt, we obtain $M_A = 36.3$ N-m and $M_B = 18.1$ N-m, so the belt slips on the small pulley at smaller values of the couples.

𝒟esign Issues

Belts and pulleys are used to transfer power in cars and many other types of machines, including printing presses, farming equipment, and industrial robots. Because two pulleys of different diameters connected by a belt are subjected to different torques and have different rates of rotation, they can be used as a mechanical "transformer" to alter torque or rotation rate.

In this example we assumed that the belt was flat, but "V-belts" that fit into matching grooves in the pulleys are often used in applications (Fig. 9.31a). This configuration keeps the belt in place on the pulleys and also decreases the tendency of the belt to slip. Suppose that a V-belt is wrapped through an angle β around a pulley (Fig. 9.31b). If the tension T_1 is known, what is the largest tension T_2 that can be applied to the other end of the belt without causing it to slip relative to the pulley?

In Fig. 9.31c, we draw the free-body diagram of an element of the belt whose boundaries are at angles α and $\alpha + \Delta\alpha$ from the point where the belt comes into contact with the pulley. (Compare this figure with Fig. 9.28c.) The equilibrium equations in the directions tangential to and normal to the centerline of the belt are

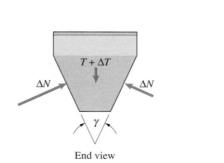

Figure 9.31

(c) Free-body diagram of an element of the belt.

$$\Sigma F_{(\text{tangential})} = 2\mu_s \Delta N + T\cos\left(\frac{\Delta\alpha}{2}\right) - (T + \Delta T)\cos\left(\frac{\Delta\alpha}{2}\right) = 0,$$

$$\Sigma F_{(\text{normal})} = 2\Delta N \sin\left(\frac{\gamma}{2}\right) - (T + \Delta T)\sin\left(\frac{\Delta\alpha}{2}\right) - T\sin\left(\frac{\Delta\alpha}{2}\right) = 0.$$

$$(9.21)$$

By the same steps leading from Eqs. (9.16) to Eq. (9.17), it can be shown that

$$T_2 = T_1 e^{\mu_s \beta / \sin(\gamma/2)}. \tag{9.22}$$

Thus using a V-belt effectively increases the coefficient of friction between the belt and pulley by the factor $1/\sin(\gamma/2)$.

When it is essential that the belt not slip relative to the pulley, a belt with cogs and a matching pulley (Fig. 9.32a) or a chain and sprocket wheel (Fig. 9.32b) can be used. The chains and sprocket wheels in bicycles and motorcycles are examples.

Figure 9.32
Designs that prevent slip of the belt relative to the pulley.

Computational Mechanics

The following example and problems are designed for the use of a programmable calculator or computer.

Computational Example 9.11

The mass of the block A in Fig. 9.33 is 20 kg, and the coefficient of static friction between the block and the floor is $\mu_s = 0.3$. The spring constant $k = 1$ kN/m, and the spring is unstretched. How far can the slider B be moved to the right without causing the block to slip?

Solution

Suppose that moving the slider B a distance x to the right causes impending slip of the block (Fig. a). The resulting stretch of the spring is $\sqrt{1 + x^2} - 1$ m, so the magnitude of the force exerted on the block by the spring is

$$F_s = k\left(\sqrt{1 + x^2} - 1\right). \tag{9.23}$$

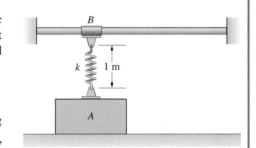

Figure 9.33

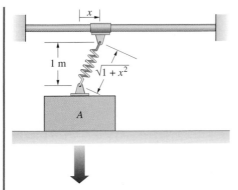

(a) Moving the slider to the right a distance x.

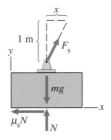

(b) Free-body diagram of the block when slip is impending.

From the free-body diagram of the block (Fig. b), we obtain the equilibrium equations

$$\Sigma F_x = \left(\frac{x}{\sqrt{1 + x^2}}\right) F_s - \mu_s N = 0,$$

$$\Sigma F_y = \left(\frac{1}{\sqrt{1 + x^2}}\right) F_s + N - mg = 0.$$

Substituting Eq. (9.23) into these two equations and then eliminating N, we can write the resulting equation in the form

$$h(x) = k(x + \mu_s)(\sqrt{1 + x^2} - 1) - \mu_s mg \sqrt{1 + x^2} = 0.$$

We must obtain the root of this function to determine the value of x corresponding to impending slip of the block. From the graph of $h(x)$ in Fig. 9.34, we estimate that $h(x) = 0$ at $x = 0{,}43$ m. By examining computed results near this value of x, we see that $h(x) = 0$, and slip is impending, when x is approximately 0.4284 m.

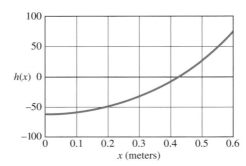

x (m)	$h(x)$
0.4281	−0.1128
0.4282	−0.0777
0.4283	−0.0425
0.4284	−0.0074
0.4285	0.0278
0.4286	0.0629
0.4287	0.0981

Figure 9.34
Graph of the function $h(x)$.

Chapter Summary

Dry Friction

The forces resulting from the contact of two plane surfaces can be expressed in terms of the normal force N and friction force f (Fig. a) or the magnitude R and angle of friction θ (Fig. b).

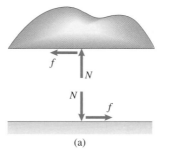

(a)

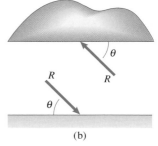

(b)

If slip is impending, the magnitude of the friction force is

$$f = \mu_s N,$$

Eq. (9.1)

and its direction opposes the impending slip. The angle of friction equals the angle of static friction $\theta_s = \arctan(\mu_s)$.

   If the surfaces are sliding, the magnitude of the friction force is

$$f = \mu_k N,$$

Eq. (9.2)

and its direction opposes the relative motion. The angle of friction equals the angle of kinetic friction $\theta_k = \arctan(\mu_k)$.

Threads

The slope α of the thread (Fig. c) is related to its pitch p by

$$\tan\alpha = \frac{p}{2\pi r}.$$

Eq. (9.7)

The couple required for impending rotation and axial motion opposite to the direction of F is

$$M = rF\tan(\theta_s + \alpha),$$

Eq. (9.9)

and the couple required for impending rotation and axial motion of the shaft in the direction of F is

$$M = rF\tan(\theta_s - \alpha).$$

Eq. (9.11)

When $\theta_s < \alpha$, the shaft will rotate and move in the direction of the force F with no couple applied.

Journal Bearings

The couple required for impending slip of the circular shaft (Fig. d) is

$$M = rF\sin\theta_s,$$

Eq. (9.12)

where F is the total load on the shaft.

Thrust Bearings and Clutches

The couple required to rotate the shaft at a constant rate (Fig. e) is

$$M = \frac{2\mu_k F}{3\cos\alpha}\left(\frac{r_o^3 - r_i^3}{r_o^2 - r_i^2}\right).$$

Eq. (9.13)

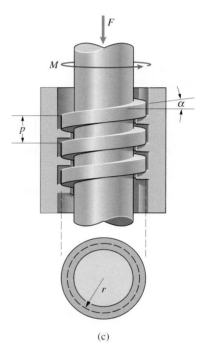

(c)

(d)

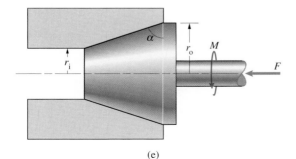

(e)

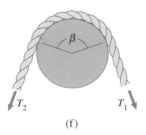

(f)

Belt Friction

The force T_2 required for impending slip in the direction of T_2 (Fig. f) is

$$T_2 = T_1 e^{\mu_s \beta},$$

Eq. (9.17)

where β is in radians.

Review Problems

9.1 The coefficient of static friction between the tires of the 8000-kg truck and the road is $\mu_s = 0.6$.
(a) If the truck is stationary on the incline and $\alpha = 15°$, what is the magnitude of the total friction force exerted on the tires by the road?
(b) What is the largest value of α for which the truck will not slip?

P9.1

9.2 The weight of the box is $W = 30$ lb, and the force F is perpendicular to the inclined surface. The coefficient of static friction between the box and the inclined surface is $\mu_s = 0.2$.

(a) If $F = 30$ lb, what is the magnitude of the friction force exerted on the stationary box?
(b) If $F = 10$ lb, show that the box cannot remain at rest on the inclined surface.

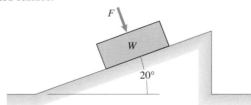

P9.2

9.3 In Problem 9.2, what is the smallest force F necessary to hold the box stationary on the inclined surface?

9.4 Blocks A and B are connected by a horizontal bar. The coefficient of static friction between the inclined surface and the 400-lb block A is 0.3. The coefficient of static friction between the surface and the 300-lb block B is 0.5. What is the smallest force F that will prevent the blocks from slipping down the surface?

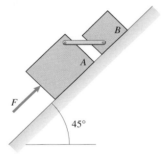

P9.4

9.5 What force F is necessary to cause the blocks in Problem 9.4 to start sliding up the plane?

9.6 The masses of crates A and B are 25 kg and 30 kg, respectively. The coefficient of static friction between the contacting surfaces is $\mu_s = 0.34$. What is the largest value of α for which the crates will remain in equilibrium?

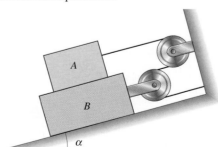

P9.6

9.7 The side of a soil embankment has a 45° slope (Fig. a). If the coefficient of static friction of soil on soil is $\mu_s = 0.6$, will the embankment be stable or will it collapse? If it will collapse, what is the smallest slope that can be stable?

Strategy: Draw a free-body diagram by isolating part of the embankment as shown in Fig. b.

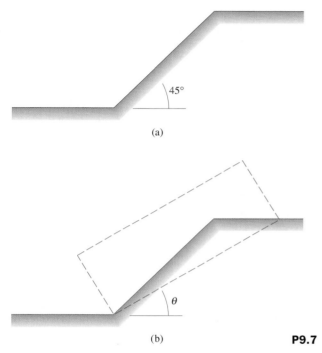

(a)

(b)　　**P9.7**

9.8 The mass of the van is 2250 kg, and the coefficient of static friction between its tires and the road is 0.6. If its front wheels are locked and its rear wheels can turn freely, what is the largest value of α for which it can remain in equilibrium?

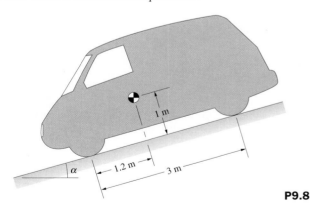

P9.8

9.9 In Problem 9.8, what is the largest value of α for which the van can remain in equilibrium if it points up the slope?

9.10 The shelf is designed so that it can be placed at any height on the vertical beam. The shelf is supported by friction between the two horizontal cylinders and the vertical beam. The combined weight of the shelf and camera is W. If the coefficient of static friction between the vertical beam and the horizontal cylinders is μ_s, what is the minimum distance b necessary for the shelf to stay in place?

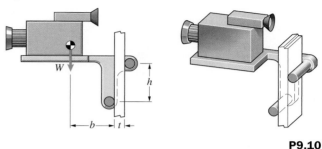

P9.10

9.11 The 20-lb homogeneous object is supported at A and B. The distance $h = 4$ in., friction can be neglected at B, and the coefficient of static friction at A is 0.4. Determine the largest force F that can be exerted without causing the object to slip.

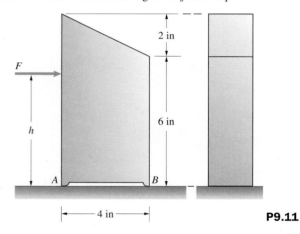

P9.11

9.12 In Problem 9.11, suppose that the coefficient of static friction at B is 0.36. What is the largest value of h for which the object will slip before it tips over?

9.13 The 180-lb climber is supported in the "chimney" by the normal and friction forces exerted on his shoes and back. The static coefficients of friction between his shoes and the wall and between his back and the wall are 0.8 and 0.6, respectively. What is the minimum normal force his shoes must exert?

P9.13

9.14 The sides of the 200-lb door fit loosely into grooves in the walls. Cables at A and B raise the door at a constant rate. The coefficient of kinetic friction between the door and the grooves is $\mu_k = 0.3$. What force must the cable at A exert to continue raising the door at a constant rate if the cable at B breaks?

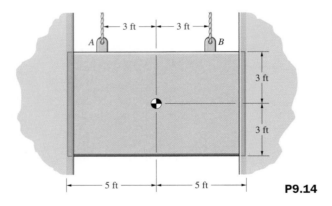

P9.14

9.15 The coefficients of static friction between the tires of the 1000-kg tractor and the ground and between the 450-kg crate and the ground are 0.8 and 0.3, respectively. Starting from rest, what torque must the tractor's engine exert on the rear wheels to cause the crate to move? (The front wheels can turn freely.)

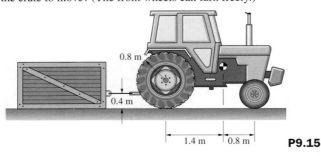

P9.15

9.16 In Problem 9.15, what is the most massive crate the tractor can cause to move from rest if its engine can exert sufficient torque? What torque is necessary?

9.17 The mass of the vehicle is 900 kg, it has rear-wheel drive, and the coefficient of static friction between its tires and the surface is 0.65. The coefficient of static friction between the crate and the surface is 0.4. If the vehicle attempts to pull the crate up the incline, what is the largest value of the mass of the crate for which it will slip up the incline before the vehicle's tires slip?

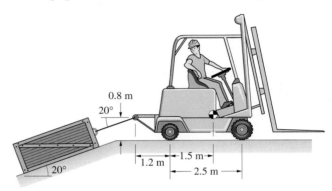

P9.17

9.18 Each of the uniform 1-m bars has a mass of 4 kg. The coefficient of static friction between the bar and the surface at B is 0.2. If the system is in equilibrium, what is the magnitude of the friction force exerted on the bar at B?

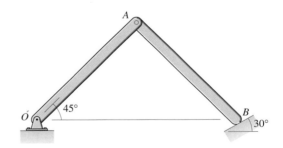

P9.18

9.19 In Problem 9.18, what is the minimum coefficient of static friction between the bar and the surface at B necessary for the system to be in equilibrium?

9.20 The collars A and B each have a mass of 2 kg. If friction between collar B and the bar can be neglected, what minimum coefficient of static friction between collar A and the bar is necessary for the collars to remain in equilibrium in the position shown?

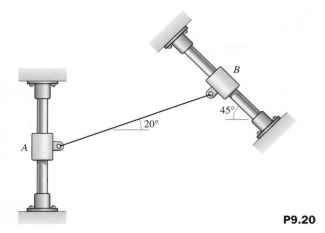

P9.20

and the coefficient of kinetic friction between the axles and the bearings is $\mu_k = 0.14$. The mass of the tram and its load is 160 kg. If the weight of the tram and its load is evenly divided between the axles, what force P is necessary to push the tram at a constant speed?

9.25 The two pulleys have a radius of 6 in. and are mounted on shafts of 1-in. radius supported by journal bearings. Neglect the weights of the pulleys and shafts. The coefficient of kinetic friction between the shafts and the bearings is $\mu_k = 0.2$. If a force $T = 200$ lb is required to raise the man at a constant rate, what is his weight?

9.21 In Problem 9.20, if the coefficient of static friction has the same value μ_s between collars A and B and the bars, what minimum value of μ_s, is necessary for the collars to remain in equilibrium in the position shown? (Assume that slip impends at A and B.)

9.22 The clamp presses two pieces of wood together. The pitch of the threads is $p = 2$ mm, the mean radius of the thread is $r = 8$ mm, and the coefficient of kinetic friction between the thread and the mating groove is 0.24. What couple must be exerted on the threaded shaft to press the pieces of wood together with a force of 200 N?

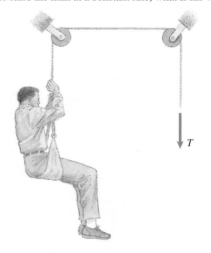

P9.25

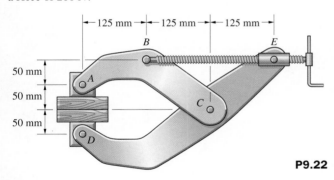

P9.22

9.26 If the man in Problem 9.25 weighs 160 lb, what force T is necessary to lower him at a constant rate?

9.27 If the two cylinders are held fixed, what is the range of W for which the two weights will remain stationary?

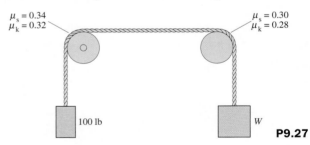

P9.27

9.23 In Problem 9.22, the coefficient of static friction between the thread and the mating groove is 0.28. After the threaded shaft is rotated sufficiently to press the pieces of wood together with a force of 200 N, what couple must be exerted on the shaft to loosen it?

9.24 The axles of the tram are supported by journal bearing. The radius of the wheels is 75 mm, the radius of the axles is 15 mm,

P9.24

9.28 In Problem 9.27, if the system is initially stationary and the left cylinder is slowly rotated, determine the largest weight W that can be (a) raised; (b) lowered.

*𝒟*esign Experience   Design and build a device to measure the coefficient of static friction μ_s between two materials. Use it to measure μ_s for several of the materials listed in Table 9.1 and compare your results with the values in the table. Discuss possible sources of error in your device and determine how closely your values agree when you perform repeated experiments with the same two materials.

The system of cables supporting the roadway subjects the bridge's arch to a distribution of internal forces and couples.

Internal Forces and Moments

W e began our study of equilibrium by drawing free-body diagrams of individual objects to determine unknown forces and
moments acting on them. In this chapter we carry this process
one step further and draw free-body diagrams of parts of individual objects to
determine internal forces and moments. In doing so, we arrive at the central
concern of the design engineer: It is the forces within an object that determine
whether it will support the external loads to which it is subjected.

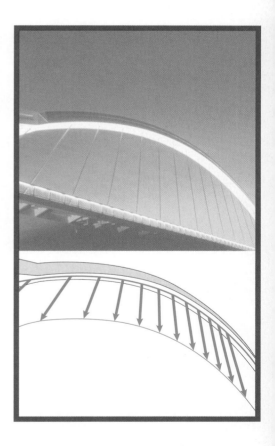

Beams

10.1 Axial Force, Shear Force, and Bending Moment

To ensure that a structural member will not fail (break or collapse) due to the forces and moments acting on it, the design engineer must know not only the external loads and reactions acting on it but also the forces and moments acting *within* the member.

Consider a beam subjected to an external load and reactions (Fig. 10.1a). How can we determine the forces and moments within the beam? In Fig. 10.1b, we "cut" the beam by a plane at an arbitrary cross section and isolate part of it. You can see that the isolated part cannot be in equilibrium unless it is subjected to some system of forces and moments at the plane where it joins the other part of the beam. These are the internal forces and moments we seek.

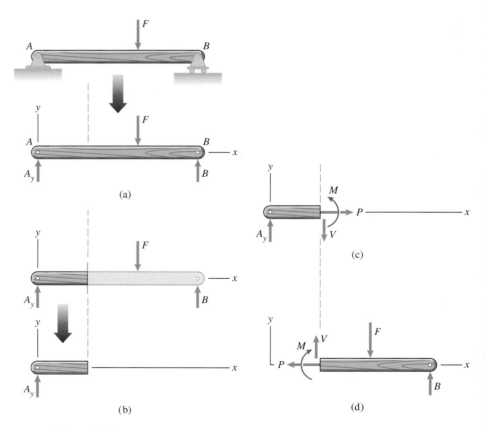

Figure 10.1
(a) A beam subjected to a load and reactions.
(b) Isolating a part of the beam.
(c), (d) The axial force, shear force, and bending moment.

In Chapter 4 we demonstrated that *any* system of forces and moments can be represented by an equivalent system consisting of a force and a couple. Since the system of external loads and reactions on the beam is two-dimensional, we can represent the internal forces and moments by an equivalent system consisting of two components of force and a couple (Fig. 10.1c). The component P parallel to the beam's axis is called the *axial force*. The component V normal to the beam's axis is called the *shear force*, and the couple M is called the *bending moment*. The axial force, shear force, and bending moment on the free-body diagram of the other part of the beam are shown in Fig. 10.1d. Notice that they are equal in magnitude but opposite in direction to the internal forces and moment on the free-body diagram in Fig. 10.1c.

The directions of the axial force, shear force, and bending moment in Figs. 10.1c and 10.1d are the established definitions of the positive directions of these quantities. A positive axial force P subjects the beam to tension. A positive shear force V tends to rotate the axis of the beam clockwise (Fig. 10.2a). Bending moments are defined to be positive when they tend to bend the axis of the beam upward (Fig. 10.2b). In terms of the coordinate system we use, "upward" means in the direction of the positive y axis.

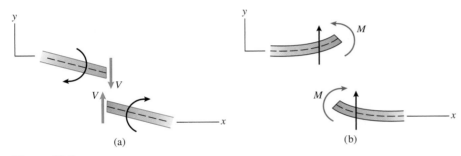

(a)   (b)

Figure 10.2
(a) Positive shear forces tend to rotate the axis of the beam clockwise.
(b) Positive bending moments tend to bend the axis of the beam upward.

Determining the internal forces and moment at a particular cross section of a beam typically involves three steps:

1. Determine the external forces and moments—Draw the free-body diagram of the beam, and determine the reactions at its supports. If the beam is a member of a structure, you must analyze the structure.

2. Draw the free-body diagram of part of the beam—Cut the beam at the point at which you want to determine the internal forces and moment, and draw the free-body diagram of one of the resulting parts. You can choose the part with the simplest free-body diagram. If your cut divides a distributed load, don't represent the distributed load by an equivalent force until after you have obtained your free-body diagram.

3. Apply the equilibrium equations—Use the equilibrium equations to determine P, V, and M.

Study Questions

1. What are the axial force, shear force, and bending moment?
2. How is the positive direction of the shear force defined?
3. How is the positive direction of the bending moment defined?

Example 10.1

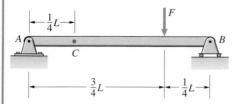

Figure 10.3

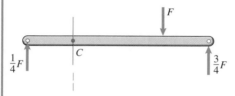

(a) The free-body diagram of the beam and a plane through point C.

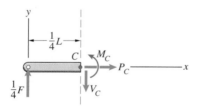

(b) The free-body diagram of the part of the beam to the left of the plane through point C.

Determining the Internal Forces and Moment

For the beam in Fig. 10.3, determine the internal forces and moment at C.

Solution

Determine the External Forces and Moments   We begin by drawing the free-body diagram of the beam and determining the reactions at its supports; the results are shown in Fig. (a).

Draw the Free-Body Diagram of Part of the Beam   We cut the beam at C (Fig. a) and draw the free-body diagram of the left part, including the internal forces and moment in their defined positive directions (Fig. b).

Apply the Equilibrium Equations   From the equilibrium equations

$$\Sigma F_x = P_C = 0,$$

$$\Sigma F_y = \frac{1}{4}F - V_C = 0,$$

$$\Sigma M_{(\text{point } C)} = M_C - \left(\frac{1}{4}L\right)\left(\frac{1}{4}F\right) = 0,$$

we obtain $P_C = 0$, $V_C = \frac{1}{4}F$, and $M_C = \frac{1}{16}LF$.

Discussion

We should check our results with the free-body diagram of the other part of the beam (Fig. c). The equilibrium equations are

$$\Sigma F_x = -P_C = 0,$$

$$\Sigma F_y = V_C - F + \frac{3}{4}F = 0,$$

$$\Sigma M_{(\text{point } C)} = -M_C - \left(\frac{1}{2}L\right)F + \left(\frac{3}{4}L\right)\left(\frac{3}{4}F\right) = 0,$$

confirming that $P_C = 0$, $V_C = \frac{1}{4}F$, and $M_C = \frac{1}{16}LF$.

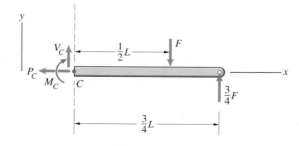

(c) The free-body diagram of the part of the beam to the right of the plane through point C.

Example 10.2

Determining the Internal Forces and Moment

For the beam in Fig. 10.4, determine the internal forces and moment (a) at B; (b) at C.

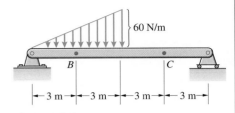

Figure 10.4

Solution

Determine the External Forces and Moments   We draw the free-body diagram of the beam and represent the distributed load by an equivalent force in Fig. a. The equilibrium equations are

$$\Sigma F_x = A_x = 0,$$

$$\Sigma F_y = A_y + D - 180 = 0,$$

$$\Sigma M_{(\text{point } A)} = 12D - (4)(180) = 0.$$

Solving them, we obtain $A_x = 0$, $A_y = 120\ N$, and $D = 60\ N$.

Draw the Free-Body Diagram of Part of the Beam   We cut the beam at B, obtaining the free-body diagram in Fig. b. Because point B is at the mid-point of the triangular distributed load, the value of the distributed load at B is 30 N/m. By representing the distributed load in Fig. b by an equivalent force, we obtain the free-body diagram in Fig. c. From the equilibrium equations

$$\Sigma F_x = P_B = 0,$$

$$\Sigma F_y = 120 - 45 - V_B = 0,$$

$$\Sigma M_{(\text{point } B)} = M_B + (1)(45) - (3)(120) = 0,$$

we obtain $P_B = 0$, $V_B = 75\ N$, and $M_B = 315$ N-m.

   To determine the internal forces and moment at C, we obtain the simplest free-body diagram by isolating the part of the beam to the right of C (Fig. d). From the equilibrium equations

$$\Sigma F_x = -P_C = 0,$$

$$\Sigma F_y = V_C + 60 = 0,$$

$$\Sigma M_{(\text{point } C)} = -M_C + (3)(60) = 0,$$

we obtain $P_C = 0$, $V_C = -60\ N$, and $M_C = 180$N-m.

Discussion

If you attempt to determine the internal forces and moment at B by cutting the freebody diagram in Fig. a at B, you do *not* obtain correct results. (You can confirm that the resulting free-body diagram of the part of the beam to the left of B gives $P_B = 0$, $V_B = 120\ N$, and $M_B = 360$ N-m.) The reason is that you do not properly account for the effect of the distributed load on your free-body diagram. You must wait until *after* you have obtained the free-body diagram of part of the beam before representing distributed loads by equivalent forces.

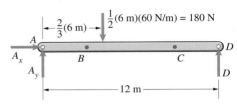

(a) Free-body diagram of the entire beam with the distributed load represented by an equivalent force.

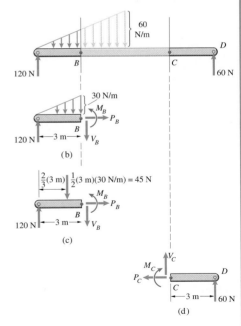

(b), (c) Free-body diagram of the part of the beam to the left of B.
(d) Free-body diagram of the part of the beam to the right of C.

10.2 Shear Force and Bending Moment Diagrams

To design a beam, an engineer must know the internal forces and moments throughout its length. Of special concern are the maximum and minimum values of the shear force and bending moment and where they occur. In this section we show how the values of P, V, and M can be determined as functions of x and introduce shear force and bending moment diagrams.

Consider a simply supported beam loaded by a force (Fig. 10.5a). Instead of cutting the beam at a specific cross section to determine the internal forces and moment, we cut it at an arbitrary position x between the left end of the beam and the load F (Fig. 10.5b). Applying the equilibrium equations to this free-body diagram, we obtain

$$\left. \begin{array}{l} P = 0 \\[4pt] V = \dfrac{1}{3}F \\[6pt] M = \dfrac{1}{3}Fx \end{array} \right\} \quad 0 < x < \dfrac{2}{3}L.$$

To determine the internal forces and moment for values of x greater than $\frac{2}{3}L$, we obtain a free-body diagram by cutting the beam at an arbitrary posi-

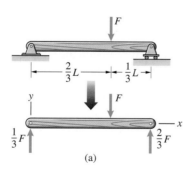

(a)

Figure 10.5

(a) A beam loaded by a force F and its free-body diagram.

(b) Cutting the beam at an arbitrary position x to the left of F.

(c) Cutting the beam at an arbitrary position x to the right of F.

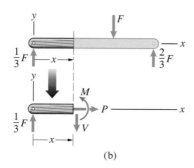

(b)

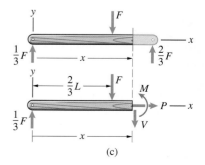

(c)

tion x between the load F and the right end of the beam (Fig. 10.5c). The results are

$$
\left.
\begin{array}{l}
P = 0 \\[4pt]
V = -\dfrac{2}{3}F \\[8pt]
M = \dfrac{2}{3}F(L - x)
\end{array}
\right\} \quad \dfrac{2}{3}L < x < L.
$$

The *shear force and bending moment diagrams* are simply the graphs of V and M, respectively, as functions of x (Fig. 10.6). They permit you to see the changes in the shear force and bending moment that occur along the beam's length as well as their maximum and minimum values. (By *maximum* we mean the least upper bound of the shear force or bending moment, and by *minimum* we mean the greatest lower bound.)

Thus you can determine the distributions of the internal forces and moment in a beam by considering a plane at an arbitrary distance x from the end of the beam and solving for P, V, and M as functions of x. Depending on the complexity of the loading, you may have to draw several free-body diagrams to determine the distributions over the entire length of the beam. The resulting equations for V and M allow you to draw the shear force and bending moment diagrams.

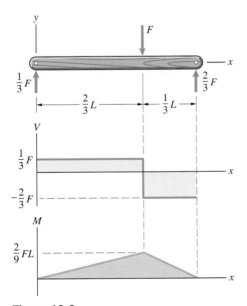

Figure 10.6
The shear force and bending moment diagrams indicating the maximum and minimum values of V and M.

Example 10.3

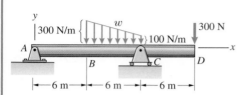

Figure 10.7

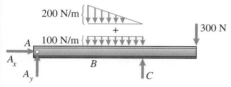

(a) Free-body diagram of the entire beam.

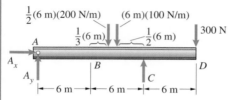

(b) Representing the distributed loads by equivalent forces.

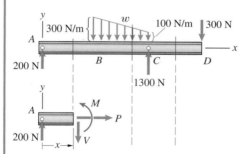

(c) Free-body diagram for $0 < x < 6$ m.

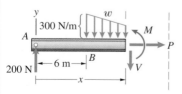

(d) Free-body diagram for $6 < x < 12$ m.

Shear Force and Bending Moment Diagrams

For the beam in Fig. 10.7, (a) draw the shear force and bending moment diagrams; (b) determine the locations and values of the maximum and minimum shear forces and bending moments.

Strategy

To determine the internal forces and moment as functions of x for the entire beam, we must use three free-body diagrams: one for the range $0 < x < 6$ m, one for $6 < x < 12$ m, and one for $12 < x < 18$ m.

Solution

(a) We begin by drawing the free-body diagram of the beam, treating the distributed load as the sum of uniform and triangular distributed loads (Fig. a). We then represent these distributed loads by equivalent forces (Fig. b). From the equilibrium equations

$$\Sigma F_x = A_x = 0,$$

$$\Sigma F_y = A_y + C - 600 - 600 - 300 = 0,$$

$$\Sigma M_{(\text{point } A)} = 12C - (8)(600) - (9)(600) - (18)(300) = 0,$$

we obtain the reactions $A_x = 0$, $A_y = 200$ N, and $C = 1300$ N.

We draw the free-body diagram for the range $0 < x < 6$ m in Fig. c. From the equilibrium equations

$$\Sigma F_x = P = 0,$$

$$\Sigma F_y = 200 - V = 0,$$

$$\Sigma M_{(\text{right end})} = M - 200x = 0,$$

we obtain

$$\left.\begin{array}{c} P = 0 \\ V = 200 \text{ N} \\ M = 200x \text{ N-m} \end{array}\right\} \quad 0 < x < 6 \text{ m}.$$

We draw the free-body diagram for the range $6 < x < 12$ m in Fig. d. To obtain the equilibrium equations, we determine w, as a function of x and integrate to determine the force and moment exerted by the distributed load. We can express w, in the form $w = cx + d$, where c and d are constants. Using the conditions $w = 300$N/m at $x = 6$ m and $w = 100$ N/m at $x = 12$ m, we obtain the equation $w = -(100/3)x + 500$N/m. The downward force on the free body in Fig. d due to the distributed load is

$$F = \int_L w \, dx = \int_6^x \left(-\frac{100}{3}x + 500\right) dx = -\frac{50}{3}x^2 + 500x - 2400 \text{ N}.$$

The clockwise moment about the origin (point A) due to the distributed load is

$$\int_L xw \, dx = \int_6^x \left(-\frac{100}{3}x^2 + 500x\right) dx = -\frac{100}{9}x^3 + 250x^2 - 6600 \text{ N-m.}$$

The equilibrium equations are

$$\Sigma F_x = P = 0,$$

$$\Sigma F_y = 200 - V + \frac{50}{3}x^2 - 500x + 2400 = 0,$$

$$\Sigma M_{(\text{point } A)} = M - Vx + \frac{100}{9}x^3 - 250x^2 + 6600 = 0.$$

Solving them, we obtain

$$\left.\begin{array}{l} P = 0 \\[2mm] V = \dfrac{50}{3}x^2 - 500x + 2600 \text{ N} \\[3mm] M = \dfrac{50}{9}x^3 - 250x^2 + 2600x - 6600 \text{ N-m} \end{array}\right\} \quad 6 < x < 12 \text{ m.}$$

For the range $12 < x < 18$ m, we obtain a very simple free-body diagram by using the part of the beam on the right of the cut (Fig. e). From the equilibrium equations

$$\Sigma F_x = -P = 0,$$

$$\Sigma F_y = V - 300 = 0,$$

$$\Sigma M_{(\text{left end})} = -M - 300(18 - x) = 0,$$

we obtain

$$\left.\begin{array}{l} P = 0 \\ V = 300 \text{ N} \\ M = 300x - 5400 \text{ N-m} \end{array}\right\} \quad 12 < x < 18 \text{ m.}$$

The shear force and bending moment diagrams obtained by plotting the equations for V and M for the three ranges of x are shown in Fig. 10.8.
(b) From the shear force diagram, the minimum shear force is -1000 N at $x = 12$ m, and its maximum value is 300 N over the range $12 < x < 18$ m. The minimum bending moment is -1800 N-m at $x = 12$ m. The bending moment has its maximum value in the range $6 < x < 12$ m. It occurs where $dM/dx = 0$. Using the equation for M as a function of x in the range $6 < x < 12$ m, we obtain

$$\frac{dM}{dx} = \frac{150}{9}x^2 - 500x + 2600 = 0.$$

The applicable root is $x = 6.69$ m. Substituting it into the equation for M, we determine that the value of the maximum bending moment is 1270 N-m.

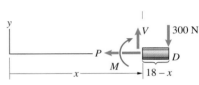

(e) Free-body diagram for $12 < x < 18$ m.

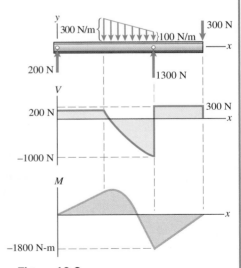

Figure 10.8
The shear force and bending moment diagrams.

10.3 Relations Between Distributed Load, Shear Force, and Bending Moment

The shear force and bending moment in a beam subjected to a distributed load are governed by simple differential equations. In this section we derive these equations and show that they provide an interesting and enlightening way to obtain shear force and bending moment diagrams. These equations are also useful for determining the deflections of beams.

Suppose that a portion of a beam is subjected to a distributed load w (Fig. 10.9a). In Fig. 10.9b, we obtain a free-body diagram by cutting the beam at x and at $x + \Delta x$. The terms ΔP, ΔV, and ΔM are the changes in the axial force, shear force, and bending moment, respectively, from x to $x + \Delta x$. From this free-body diagram we obtain the equilibrium equations

$$\Sigma F_x = P + \Delta P - P = 0,$$

$$\Sigma F_y = V - V - \Delta V - w\Delta x - O(\Delta x^2) = 0,$$

$$\Sigma M_{(\text{point } Q)} = M + \Delta M - M - (V + \Delta V)\Delta x - wO(\Delta x^2) = 0,$$

where the notation $O(\Delta x^2)$ means a term of order Δx^2. Dividing these equations by Δx and taking the limit as $\Delta x \to 0$, we obtain

$$\frac{dP}{dx} = 0, \tag{10.1}$$

$$\frac{dV}{dx} = -w, \tag{10.2}$$

$$\frac{dM}{dx} = V. \tag{10.3}$$

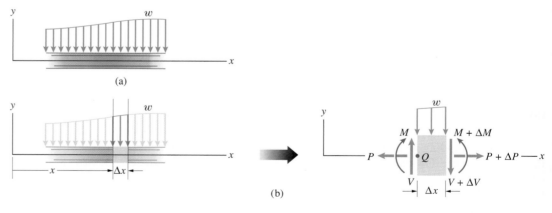

Figure 10.9
(a) A portion of a beam subjected to a distributed force w.
(b) Obtaining the free-body diagram of an element of the beam.

Equation (10.1) simply states that the axial force does not depend on x in a portion of a beam subjected only to a lateral distributed load. But notice that you can integrate Eq. (10.2) to determine V as a function of x if you know w, and then you can integrate Eq. (10.3) to determine M as a function of x.

We derived Eqs. (10.2) and (10.3) for a portion of beam subjected only to a distributed load. When you use these equations to determine shear force and bending moment diagrams, you must also account for the effects of forces and couples. Let's determine what happens to the shear force and bending moment diagrams where a beam is subjected to a force F in the positive y direction (Fig. 10.10a). By cutting the beam just to the left and just to the right of the force, we obtain the free-body diagram in Fig. 10.10b, where the subscripts $-$ and $+$ denote values to the left and right of the force, respectively. Equilibrium requires that

$$V_+ - V_- = F, \qquad (10.4)$$
$$M_+ - M_- = 0. \qquad (10.5)$$

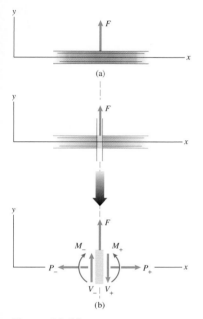

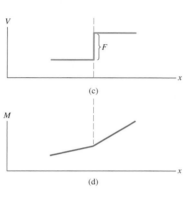

Figure 10.10
(a) A portion of a beam subjected to a distributed force F in the positive y direction.
(b) Obtaining a free-body diagram by cutting the beam to the left and right of F.
(c) The shear force diagram undergoes a positive jump of magnitude F.
(d) The bending moment diagram is continuous.

The shear force diagram undergoes a jump discontinuity of magnitude F (Fig. 10.10c), but the bending moment diagram is continuous (Fig. 10.10d). The jump in the shear force is positive if the force is in the positive y direction.

Now we consider what happens to the shear force and bending moment diagrams when a beam is subjected to a counterclockwise couple C (Fig. 10.11 a). Cutting the beam just to the left and just to the right of the couple (Fig. 10.11b), we determine that

$$V_+ - V_- = 0, \qquad (10.6)$$
$$M_+ - M_- = -C. \qquad (10.7)$$

The shear force diagram is continuous (Fig. 10.11c), but the bending moment diagram undergoes a jump discontinuity of magnitude C (Fig. 10.11d) where a beam is subjected to a couple. The jump in the bending moment is *negative* if the couple is in the counterclockwise direction.

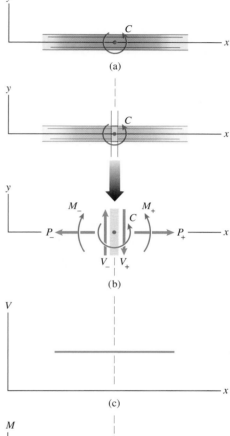

Figure 10.11
(a) A portion of a beam subjected to a counterclockwise couple C.
(b) Obtaining a free-body diagram by cutting the beam to the left and right of C.
(c) The shear force diagram is continuous.
(d) The bending moment diagram undergoes a *negative* jump of magnitude C.

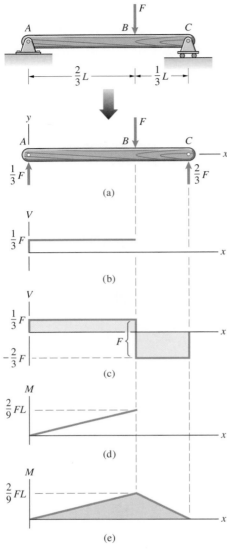

Figure 10.12
(a) A beam loaded by a force F.
(b) The shear force diagram from A to B.
(c) The complete shear force diagram.
(d) The bending moment diagram from A to B.
(e) The complete bending moment diagram.

We illustrate the application of these results by considering the simply supported beam in Fig. 10.12a. To determine its shear force diagram, we begin at $x = 0$, where the upward reaction at A results in a positive value of V of magnitude $\frac{1}{3}F$. Since there is no load between A and B, Eq. (10.2) states that $dV/dx = 0$. The shear force remains constant between A and B (Fig. 10.12b). At B, the downward load F causes a negative jump in V of magnitude F. There is no load between B and C, so the shear force remains constant between B and C (Fig. 10.12c).

Now that we have completed the shear force diagram, we begin again at $x = 0$ to determine the bending moment diagram. There is no couple at $x = 0$, so the bending moment is zero there. Between A and B, $V = \frac{1}{3}F$. Integrating Eq. (10.3) from $x = 0$ to an arbitrary value of x between A and B,

$$\int_0^M dM = \int_0^x V\,dx = \int_0^x \frac{1}{3}F\,dx,$$

we determine M as a function of x from A to B:

$$M = \frac{1}{3}Fx, \qquad 0 < x < \frac{2}{3}L.$$

The bending moment diagram from A to B is shown in Fig. 10.12d. The value of the bending moment at B is $M_B = \frac{2}{9}FL$.

Between B and C, $V = -\frac{2}{3}F$. Integrating Eq. (10.3) from $x = \frac{2}{3}L$ to an arbitrary value of x between B and C,

$$\int_{M_B}^M dM = \int_{2L/3}^x V\,dx = \int_{2L/3}^x -\frac{2}{3}F\,dx,$$

we obtain M as a function of x from B to C:

$$M = M_B - \frac{2}{3}F\left(x - \frac{2}{3}L\right) = \frac{2}{3}F(L - x), \qquad \frac{2}{3}L < x < L.$$

The completed bending moment diagram is shown in Fig. 10.12e. Compare the shear and bending moment diagrams in Figs. 10.12c and 10.12e with those in Fig. 10.6, which we obtained by cutting the beam and solving equilibrium equations.

We see that Eqs. (10.2)–(10.7) can be used to obtain the shear force and bending moment diagrams:

1. Shear force diagram—For segments of the beam that are unloaded or are subjected to a distributed load, you can integrate Eq. (10.2) to determine V as a function of x. In addition, you must use Eq. (10.4) to determine the effects of forces on V.

2. Bending moment diagram—Once you have determined V as a function of x, integrate Eq. (10.3) to determine M as a function of x. Use Eq. (10.7) to determine the effects of couples on M.

Study Questions

1. For a portion of a beam that is subjected only to a distributed load w, how are the shear force and bending moment distributions determined from Eqs. (10.2) and (10.3)?

2. What effect does a force F have on the shear force and bending moment distributions?

3. What effect does a couple C have on the shear force and bending moment distributions?

Example 10.4

Applying Eqs. (10.2)–(10.7)

Determine the shear force and bending moment diagrams for the beam in Fig. 10.13.

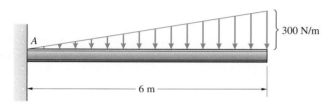

Figure 10.13

Solution

We must first determine the reactions at A. The results are shown on the free-body diagram of the beam in Fig. a. The equation describing the distributed load as a function of x is $w = (x/6)300 = 50x$ N/m.

Shear Force Diagram   The upward force at A causes a positive value of V of 900-N magnitude, so that $V_A = 900$ N. Integrating Eq. (10.2) from $x = 0$ to an arbitrary value of x,

$$\int_{V_A}^{V} dV = \int_{0}^{x} -w\, dx = \int_{0}^{x} -50x\, dx,$$

we obtain V as a function of x:

$$V = V_A - 25x^2 = 900 - 25x^2.$$

The shear force diagram is shown in Fig. b.

Bending Moment Diagram   The counterclockwise couple at A causes a *negative* value of M of 3600 N-m magnitude, so that $M_A = -3600$ N-m. Integrating Eq. (10.3) from $x = 0$ to an arbitrary value of x,

$$\int_{M_A}^{M} dM = \int_{0}^{x} V\, dx = \int_{0}^{x} (900 - 25x^2)\, dx,$$

we obtain

$$M = M_A + 900x - \frac{25}{3}x^3 = -3600 + 900x - \frac{25}{3}x^3.$$

The bending moment diagram is shown in Fig. c.

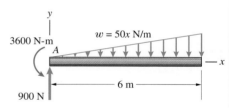

(a) Free-body diagram of the beam.

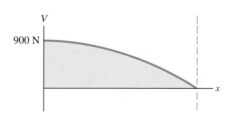

(b) Shear force diagram.

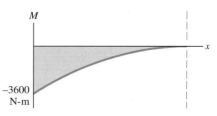

(c) Bending moment diagram.

Example 10.5

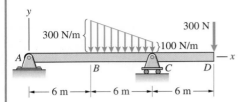

Figure 10.14

Applying Eqs. (10.2)–(10.7)

Determine the shear force and bending moment diagrams for the beam in Fig. 10.14.

Solution

The first step, determining the reactions at the supports, was carried out for this beam and loading in Example 10.3. The results are shown in Fig. a.

Shear Force Diagram   *From A to B*. There is no load between A and B, so the shear force increases by 200 N at A and then remains constant from A to B:

$$V = 200 \text{ N}, \qquad 0 < x < 6 \text{ m}.$$

From B to C. We can express the distributed load w between B and C in the form $w = cx + d$, where c and d are constants. Using the conditions $w = 300$ N/m at $x = 6$ m and $w = 100$ N/m at $x = 12$ m, we obtain the equation $w = -(100/3)x + 500$ N/m. Integrating Eq. (10.2) from $x = 6$ m to an arbitrary value of x between B and C,

$$\int_{V_B}^{V} dV = \int_{6}^{x} -w \, dx = \int_{6}^{x} \left(\frac{100}{3} x - 500 \right) dx,$$

we obtain an equation for V between B and C:

$$V = \frac{50}{3} x^2 - 500x + 2600 \text{ N}, \qquad 6 < x < 12 \text{ m}.$$

At $x = 12$ m, $V = -1000$ N.

   From C to D. At C, V undergoes a positive jump of 1300-N magnitude, so that its value becomes $-1000 + 1300 = 300$ N. There is no loading between C and D, so V remains constant from C to D:

$$V = 300 \text{ N}, \qquad 12 < x < 18 \text{ m}.$$

The shear force diagram is shown in Fig. b.

Bending Moment Diagram   *From A to B*. There is no couple at $x = 0$, so the bending moment is zero there. Integrating Eq. (10.3) from $x = 0$ to an arbitrary value of x between A and B,

$$\int_{0}^{M} dM = \int_{0}^{x} V \, dx = \int_{0}^{x} 200 \, dx,$$

we obtain

$$M = 200x \text{ N-m}, \qquad 0 < x < 6 \text{ m}.$$

At $x = 6$ m, $M_B = 1200$ N-m.

   From B to C. Integrating from $x = 6$ m to an arbitrary value of x between B and C,

$$\int_{M_B}^{M} dM = \int_{6}^{x} V \, dx = \int_{6}^{x} \left(\frac{50}{3} x^2 - 500x + 2600 \right) dx,$$

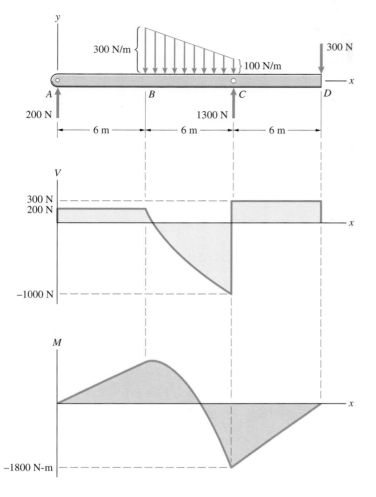

(a) Free-body diagram of the beam.

(b) Shear force diagram.

(c) Bending moment diagram.

we obtain

$$M = \frac{50}{9} x^3 - 250x^2 + 2600x - 6600 \text{ N-m}, \qquad 6 < x < 12 \text{ m.}$$

At $x = 12$ m, $M_C = -1800$ N-m.

From C to D. Integrating from $x = 12$ m to an arbitrary value of x between C and D,

$$\int_{M_C}^{M} dM = \int_{12}^{x} V \, dx = \int_{12}^{x} 300 \, dx,$$

we obtain

$$M = 300x - 5400 \text{ N-m}, \qquad 12 < x < 18 \text{ m.}$$

The bending moment diagram is shown in Fig. c.

Figure 10.15
The use of cables to suspend the roof of this sports stadium provides spectators with a view unencumbered by supporting columns.

Cables

Because of their unique combination of strength, lightness, and flexibility, ropes and cables are often used to support loads and transmit forces in structures, machines, and vehicles. The great suspension bridges are supported by enormous steel cables. Architectural engineers use cables to create aesthetic structures with open interior spaces (Fig. 10.15). In the following sections we determine the tensions in cables subjected to distributed and discrete loads.

10.4 Loads Distributed Uniformly Along Straight Lines

The main cable of a suspension bridge is the classic example of a cable subjected to a load uniformly distributed along a straight line (Fig. 10.16). The weight of the bridge is (approximately) uniformly distributed horizontally. The load, transmitted to the main cable by the large number of vertical cables, can be modeled as a distributed load. In this section we determine the shape and the variation in the tension of a cable loaded in this way.

Figure 10.16
(a) Main cable of a suspension bridge.
(b) The load is distributed horizontally.

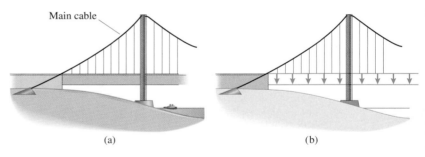

Main cable

(a)   (b)

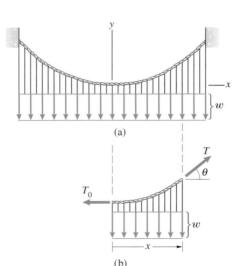

Figure 10.17
(a) A cable subjected to a load uniformly distributed along a horizontal line.
(b) Free-body diagram of the cable between $x = 0$ and an arbitrary position x.

Consider a suspended cable subjected to a load distributed uniformly along a horizontal line (Fig. 10.17a). We neglect the weight of the cable. The origin of the coordinate system is located at the cable's lowest point. Let the function $y(x)$ be the curve described by the cable in the x–y plane. Our objective is to determine the curve $y(x)$ and the tension in the cable.

Shape of the Cable

We obtain a free-body diagram by cutting the cable at its lowest point and at an arbitrary position x (Fig. 10.17b). The term T_0 is the tension in the cable at its lowest point, and T is the tension at x. The downward force exerted by the distributed load is wx. From this free-body diagram we obtain the equilibrium equations

$$T \cos \theta = T_0,$$

$$T \sin \theta = wx. \tag{10.8}$$

We eliminate the tension T by dividing the second equation by the first one, obtaining

$$\tan\theta = \frac{w}{T_0}x = ax,$$

where

$$a = \frac{w}{T_0}.$$

The slope of the cable at x is $dy/dx = \tan\theta$, so we obtain a differential equation governing the curve described by the cable:

$$\frac{dy}{dx} = ax. \tag{10.9}$$

We have chosen the coordinate system so that $y = 0$ at $x = 0$. Integrating Eq. (10.9),

$$\int_0^y dy = \int_0^x ax\,dx,$$

we find that the curve described by the cable is the parabola

$$y = \frac{1}{2}ax^2. \tag{10.10}$$

Tension of the Cable

To determine the distribution of the tension in the cable, we square both sides of Eqs. (10.8) and then sum them, obtaining

$$T = T_0\sqrt{1 + a^2x^2}. \tag{10.11}$$

The tension is a minimum at the lowest point of the cable and increases monotonically with distance from the lowest point.

Length of the Cable

In some applications it is useful to have an expression for the length of the cable in terms of x. We can write the relation $ds^2 = dx^2 + dy^2$, where ds is an element of length of the cable (Fig. 10.18), in the form

$$ds = \sqrt{1 + \left(\frac{dy}{dx}\right)^2}\,dx.$$

Substituting Eq. (10.9) into this expression and integrating, we obtain an equation for the length s of the cable in the horizontal interval from 0 to x:

$$s = \frac{1}{2}\left\{x\sqrt{1 + a^2x^2} + \frac{1}{a}\ln\left[ax + \sqrt{1 + a^2x^2}\right]\right\}. \tag{10.12}$$

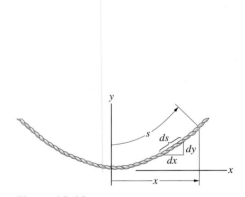

Figure 10.18
The length s of the cable in the horizontal interval from 0 to x.

Study Questions

1. If a cable is subjected to a load that is uniformly distributed along a straight line and its weight is negligible, what mathematical curve describes its shape?
2. Equation (10.10) describes the shape of a cable loaded as described in Question 1. Where must the origin of the x–y coordinate system be located?

Example 10.6

Cable with a Horizontally Distributed Load

The horizontal distance between the supporting towers of the Golden Gate Bridge in San Francisco, California, is 1280 m (Fig. 10.19). The tops of the towers are 160 m above the lowest point of the main supporting cables. Obtain the equation for the curve described by the cables.

Figure 10.19

Strategy

We know the coordinates of the cables' attachment points relative to their lowest points. By substituting the coordinates into Eq. (10.10), we can determine a. Once a is known, Eq. (10.10) describes the shapes of the cables.

Solution

The coordinates of the top of the right supporting tower relative to the lowest point of the support cables are $x_R = 640$ m, $y_R = 160$ m (Fig. a). By substituting these values into Eq. (10.10),

$$160 = \frac{1}{2} a(640)^2,$$

we obtain

$$a = 7.81 \times 10^{-4} \text{ m}^{-1}.$$

The curve described by the supporting cables is

$$y = \frac{1}{2} ax^2 = \left(3.91 \times 10^{-4}\right)x^2.$$

Fig. a compares this parabola with a photograph of the supporting cables.

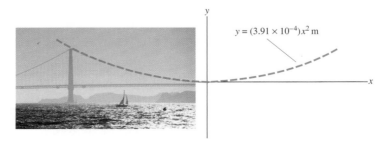

$$y = (3.91 \times 10^{-4})x^2 \text{ m}$$

(a) The theoretical curve superimposed on a photograph of the supporting cable.

Example 10.7

Maximum Tension in a Cable

The cable in Fig. 10.20 supports a distributed load of 100 lb/ft. What is the maximum tension in the cable?

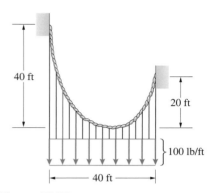

Figure 10.20

Strategy

We are given the vertical coordinate of each attachment point, but we are told only the total horizontal span. However, the coordinates of each attachment point relative to a coordinate system with its origin at the lowest point of the cable must satisfy Eq. (10.10). This permits us to determine the horizontal coordinates of the attachment points. Once we know them, we can use Eq. (10.10) to determine $a = w/T_0$, which tells us the tension at the lowest point, and then use Eq. (10.11) to obtain the maximum tension.

Solution

We introduce a coordinate system with its origin at the lowest point of the cable, denoting the coordinates of the left and right attachment points by (x_L, y_L) and (x_R, y_R) respectively (Fig. a). Equation (10.10) must be satisfied for both of these points:

$$y_L = 40 \text{ ft} = \frac{1}{2} ax_L^2,$$

$$y_R = 20 \text{ ft} = \frac{1}{2} ax_R^2. \qquad (10.13)$$

We don't know a, but we can eliminate it by dividing the first equation by the second one, obtaining

$$\frac{x_L^2}{x_R^2} = 2.$$

We also know that

$$x_R - x_L = 40 \text{ ft}.$$

(The reason for the minus sign is that x_L is negative.) We therefore have two equations we can solve for x_L and x_R; the results are $x_L = -23.4$ ft and $x_R = 16.6$ ft.

We can now use either of Eqs. (10.13) to determine a. We obtain $a = 0.146 \text{ ft}^{-1}$, so the tension T_0 at the lowest point of the cable is

$$T_0 = \frac{w}{a} = \frac{100}{0.146} = 686 \text{ lb}.$$

From Eq. (10.11), we know that the maximum tension in the cable occurs at the maximum horizontal distance from its lowest point, which in this example is the left attachment point. The maximum tension is therefore

$$T_{\max} = T_0\sqrt{1 + a^2 x_L^2} = 686\sqrt{1 + (0.146)^2(-23.4)^2} = 2440 \text{ lb}.$$

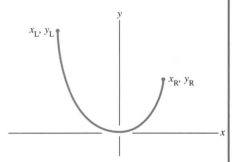

(a) A coordinate system with its origin at the lowest point and the coordinates of the left and right attachment points.

10.5 Loads Distributed Uniformly Along Cables

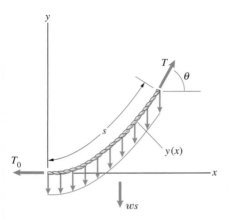

Figure 10.21
A cable subjected to a load distributed uniformly along its length.

A cable's own weight subjects it to a load that is distributed uniformly along its length. If a cable is subjected to equal, parallel forces spaced uniformly along its length, the load on the cable can often be modeled as a load distributed uniformly along its length. In this section we show how to determine both the cable's resulting shape and the variation in its tension.

Suppose that a cable is acted on by a distributed load that subjects each element ds of its length to a force $w\,ds$, where w is constant. In Fig. 10.21 we show the free-body diagram obtained by cutting the cable at its lowest point and at a point a distance s along its length. The terms T_0 and T are the tensions at the lowest point and at s, respectively. The distributed load exerts a downward force ws. The origin of the coordinate system is located at the lowest point of the cable. Let the function $y(x)$ be the curve described by the cable in the x–y plane. Our objective is to determine $y(x)$ and the tension T.

Shape of the Cable

From the free-body diagram in Fig. 10.21 we obtain the equilibrium equations

$$T \sin \theta = ws, \tag{10.14}$$

$$T \cos \theta = T_0. \tag{10.15}$$

Dividing the first equation by the second one, we obtain

$$\tan \theta = \frac{w}{T_0} s = as, \tag{10.16}$$

where

$$a = \frac{w}{T_0}.$$

The slope of the cable $dy/dx = \tan \theta$, so

$$\frac{dy}{dx} = as.$$

The derivative of this equation with respect to x is

$$\frac{d}{dx}\left(\frac{dy}{dx}\right) = a\frac{ds}{dx}. \tag{10.17}$$

By using the relation

$$ds^2 = dx^2 + dy^2,$$

we can write the derivative of s with respect to x as

$$\frac{ds}{dx} = \sqrt{1 + \left(\frac{dy}{dx}\right)^2} = \sqrt{1 + \sigma^2}, \tag{10.18}$$

where σ is the slope

$$\sigma = \frac{dy}{dx} = \tan\theta.$$

With Eq. (10.18), we can write Eq. (10.17) as

$$\frac{d\sigma}{\sqrt{1 + \sigma^2}} = a\,dx.$$

The slope $\sigma = 0$ at $x = 0$. Integrating this equation,

$$\int_0^\sigma \frac{d\sigma}{\sqrt{1 + \sigma^2}} = \int_0^x a\,dx,$$

we obtain the slope as a function of x:

$$\sigma = \frac{dy}{dx} = \frac{1}{2}\left(e^{ax} - e^{-ax}\right) = \sinh ax. \tag{10.19}$$

Then integrating this equation with respect to x, yields the curve described by the cable, which is called a *catenary*:

$$y = \frac{1}{2a}\left(e^{ax} + e^{-ax} - 2\right) = \frac{1}{a}\left(\cosh ax - 1\right). \tag{10.20}$$

Tension of the Cable

Using Eq. (10.15) and the relation $dx = \cos\theta\,ds$, we obtain

$$T = \frac{T_0}{\cos\theta} = T_0\frac{ds}{dx}.$$

Substituting Eq. (10.18) into this expression and using Eq. (10.19) yields the tension in the cable as a function of x:

$$T = T_0\sqrt{1 + \frac{1}{4}\left(e^{ax} - e^{-ax}\right)^2} = T_0\cosh ax. \tag{10.21}$$

Length of the Cable

From Eq. (10.16), the length s is

$$s = \frac{1}{a}\tan\theta = \frac{\sigma}{a}.$$

Substituting Eq. (10.19) into this equation, we obtain an expression for the length s of the cable in the horizontal interval from its lowest point to x:

$$s = \frac{1}{2a}\left(e^{ax} - e^{-ax}\right) = \frac{\sinh ax}{a}. \tag{10.22}$$

Example 10.8

Cable Loaded by Its Own Weight

The mass per unit length of the cable in Fig. 10.22 is 1 kg/m. The tension at its lowest point is 50 N. Determine the distance h and the maximum tension in the cable.

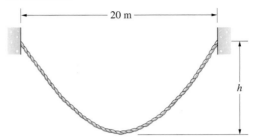

Figure 10.22

Strategy

The cable is subjected to a load $w = (9.81 \text{ m/s}^2)(1 \text{ kg/m}) = 9.81$ N/m distributed uniformly along its length. Since we know w and T_0, we can determine $a = w/T_0$. Then we can determine h from Eq. (10.20). Since the maximum tension occurs at the greatest distance from the lowest point of the cable, we can determine it by letting $x = 10$ m in Eq. (10.21).

Solution

The parameter a is

$$a = \frac{w}{T_0} = \frac{9.81}{50} = 0.196 \text{ m}^{-1}.$$

In terms of a coordinate system with its origin at the lowest point of the cable (Fig. a), the coordinates of the right attachment point are $x = 10$ m, $y = h$. From Eq. (10.20),

$$h = \frac{1}{a}(\cosh ax - 1) = \frac{1}{0.196}\left\{\cosh\left[(0.196)(10)\right] - 1\right\} = 13.4 \text{ m}.$$

From Eq. (10.21), the maximum tension is

$$T_{\max} = T_0\sqrt{1 + \sinh^2 ax} = 50\sqrt{1 + \sinh^2\left[(0.196)(10)\right]} = 181 \text{ N}.$$

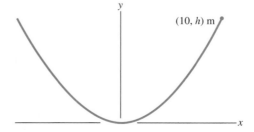

(a) A coordinate system with its origin at the lowest point of the cable.

10.6 Discrete Loads

Our first applications of equilibrium in Chapter 3 involved determining the tensions in cables supporting suspended objects. In this section we consider the case of an arbitrary number N of objects suspended from a cable (Fig. 10.23a). We assume that the weight of the cable can be neglected in comparison to the suspended weights and that the cable is sufficiently flexible that we can approximate its shape by a series of straight segments.

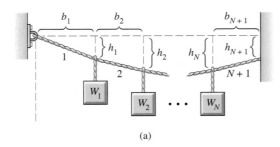

(a)

Figure 10.23
(a) N weights suspended from a cable.
(b) The first free-body diagram.
(c) The second free-body diagram.

Determining the Configuration and Tensions

Suppose that the horizontal distances $b_1, b_2, \ldots, b_{N+1}$ are known and that the vertical distance h_{N+1} specifying the position of the cable's right attachment point is known. We have two objectives: (1) to determine the configuration (shape) of the cable by solving for the vertical distances $h_1, h_2, \ldots, h_N$ specifying the positions of the attachment points of the weights and (2) to determine the tensions in the segments $1, 2, \ldots, N + 1$ of the cable.

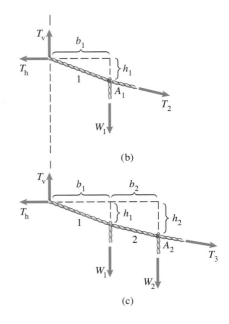

We begin by drawing a free-body diagram, cutting the cable at its left attachment point and just to the right of the weight W_1 (Fig. 10.23b). We resolve the tension in the cable at the left attachment point into its horizontal and vertical components T_h and T_v. Summing moments about the attachment point A_1, we obtain the equation

$$\Sigma M_{(\text{point } A_1)} = h_1 T_h - b_1 T_v = 0.$$

Our next step is to obtain a free-body diagram by cutting the cable at its left attachment point and just to the right of the weight W_2 (Fig. 10.23c). Summing moments about A_2, we obtain

$$\Sigma M_{(\text{point } A_2)} = h_2 T_h - (b_1 + b_2) T_v + b_2 W_1 = 0.$$

Proceeding in this way, cutting the cable just to the right of each of the N weights, we obtain N equations. We can also draw a free-body diagram by cutting the cable at its left and right attachment points and sum moments about the right attachment point. In this way, we obtain $N + 1$ equations in terms of $N + 2$ unknowns: the two components of the tension T_h and T_v and the vertical positions of the attachment points $h_1, h_2 \ldots, h_N$. If the vertical position of just one attachment point is also specified, we can solve the system of equations for the vertical positions of the other attachment points, determining the configuration of the cable.

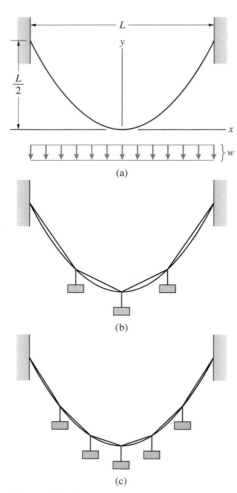

Figure 10.24
(a) A cable subjected to a continuous load.
(b) A cable with three discrete loads.
(c) A cable with five discrete loads.

Once we know the configuration of the cable and the force T_h, we can determine the tension in any segment by cutting the cable at the left attachment point and within the segment and summing forces in the horizontal direction.

Comments on Continuous and Discrete Models

By comparing cables subjected to distributed and discrete loads, we can make some observations about how continuous and discrete systems are modeled in engineering. Consider a cable subjected to a horizontally distributed load w (Fig. 10.24a). The total force exerted on it is wL. Since the cable passes through the point $x = L/2$, $y = L/2$, we find from Eq. (10.10) that $a = 4/L$, so the equation for the curve described by the cable is $y = (2/L)x^2$.

In Fig. 10.24b, we compare the shape of the cable with the distributed load to that of a cable of negligible weight subjected to three discrete loads $W = wL/3$ with equal horizontal spacing. (We chose the dimensions of the cable with discrete loads so that the heights of the two cables would be equal at their midpoints.) In Fig. 10.24c, we compare the shape of the cable with the distributed load to that of a cable subjected to five discrete loads $W = wL/5$ with equal horizontal spacing. In Figs. 10.25a and 10.25b, we compare the tension in the cable subjected to the distributed load to those in the cables subjected to three and five discrete loads.

The shape and the tension in the cable with a distributed load are approximated by the shapes and tensions in the cables with discrete loads. Although the approximation of the tension is less impressive than the approximation of the shape, it is clear that the former can be improved by increasing the number of discrete loads.

This approach—approximating a continuous distribution by a discrete model—is very important in engineering. It is the starting point of the finite difference and finite element methods. The opposite approach—modeling discrete systems by continuous models—is also widely used, for example, when the forces exerted on a bridge by traffic are modeled as a distributed load.

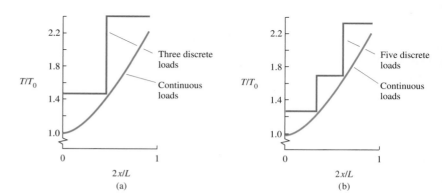

Figure 10.25
(a) The tension in a cable with a continuous load compared to the cable with three discrete loads.
(b) The tension in a cable with a continuous load compared to the cable with five discrete loads.

Example 10.9

Cable Subjected to Discrete Loads

Two masses $m_1 = 10$ kg and $m_2 = 20$ kg are suspended from the cable in Fig. 10.26.
(a) Determine the vertical distance h_2.
(b) Determine the tension in cable segment 2.

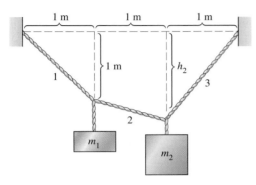

Figure 10.26

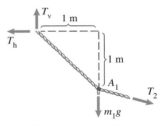

(a) First free-body diagram

Solution

(a) We begin by cutting the cable at the left attachment point and just to the right of the mass m_1, and resolve the tension at the left attachment point into horizontal and vertical components (Fig. a). Summing moments about A_1 yields

$$\Sigma M_{(\text{point } A_1)} = (1)T_h - (1)T_v = 0.$$

We then cut the cable just to the right of the mass m_2 (Fig. b) and sum moments about A_2:

$$\Sigma M_{(\text{point } A_2)} = h_2 T_h - (2)T_v + (1)m_1 g = 0.$$

The last step is to cut the cable at the right attachment point (Fig. c) and sum moments about A_3:

$$\Sigma M_{(\text{point } A_3)} = -(3)T_v + (2)m_1 g + (1)m_2 g = 0.$$

We have three equations in terms of the unknowns T_h, T_v, and h_2. Solving them yields $T_h = T_v = 131$ N and $h_2 = 1.25$ m.
(b) To determine the tension in segment 2, we use the free-body diagram in Fig. a. The angle between the force T_2 and the horizontal is arctan $[(h_2 - 1)/1] = 14.0°$. Summing forces in the horizontal direction,

$$T_2 \cos 14.0° - T_h = 0,$$

we obtain

$$T_2 = \frac{T_h}{\cos 14.0°} = 135 \text{ N}.$$

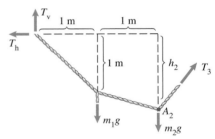

(b) Second free-body diagram.

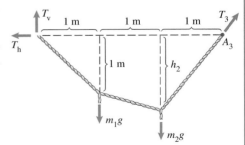

(c) Free-body diagram of the entire cable.

Computational Mechanics

The following example and problems are designed to be solved using a programmable calculator or computer.

Computational Example 10.10

As the first step in constructing a suspended pedestrian bridge, a cable is suspended across the span from attachment points of equal height (Fig. 10.27). The cable weighs 5 lb/ft and is 42 ft long. Determine the maximum tension in the cable and the vertical distance from the attachment points to the cable's lowest point.

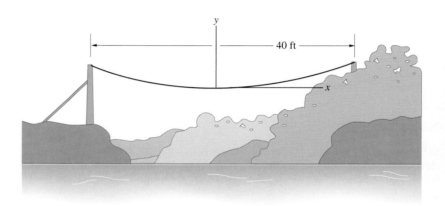

Figure 10.27

Strategy

Equation (10.22) gives the length s of the cable as a function of the horizontal distance x from the cable's lowest point and the parameter $a = w/T_0$. The term w is the weight per unit length, and T_0 is the tension in the cable at its lowest point. We know that the half-span of the cable is 20 ft, so we can draw a graph of s as a function of a and estimate the value of a for which $s = 21$ ft. Then we can determine the maximum tension from Eq. (10.21) and the vertical distance to the cable's lowest point from Eq. (10.20).

Solution

Setting $x = 20$ ft in Eq. (10.22),

$$s = \frac{\sinh 20a}{a},$$

we compute s as a function of a (Fig. 10.28). The length $s = 21$ ft when the parameter a is approximately 0.027. By examining the computed results near $a = 0.027$,

$a\ (\mathrm{ft}^{-1})$	$s\ (\mathrm{ft})$
0.0269	20.9789
0.0270	20.9863
0.0271	20.9937
0.0272	21.0012
0.0273	21.0086
0.0274	21.0162
0.0275	21.0237

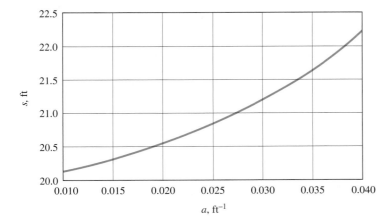

Figure 10.28
Graph of the length s as a function of the parameter a.

we see that s is approximately 21 ft when $a = 0.0272$ ft^{-1}. Therefore the tension at the cable's lowest point is

$$T_0 = \frac{w}{a} = \frac{5}{0.0272} = 184 \text{ lb},$$

and the maximum tension is

$$T_{\max} = T_0 \cosh ax = 184 \cosh\left[(0.0272)(20)\right] = 212 \text{ lb}.$$

From Eq. (10.20), the vertical distance from the cable's lowest point to the attachment points is

$$y_{\max} = \frac{1}{a}(\cosh ax - 1) = \frac{1}{0.0272}\left\{\cosh\left[(0.02272)(20)\right] - 1\right\} = 5.58 \text{ ft}.$$

Liquids and Gases

Wind forces on buildings and aerodynamic forces on cars and airplanes are examples of forces that are distributed over areas. The downward force exerted on the bed of a dump truck by a load of gravel is distributed over the area of the bed. The upward force that supports a building is distributed over the area of its foundation. Loads distributed over the roofs of buildings by accumulated snow can be hazardous. Many forces of concern in engineering are distributed over areas. In this section we analyze the most familiar example, the force exerted by the pressure of a gas or liquid.

10.7 Pressure and the Center of Pressure

A surface immersed in a gas or liquid is subjected to forces exerted by molecular impacts. If the gas or liquid is stationary, the load can be described by a function p, the *pressure*, defined such that the normal force exerted on a differential element dA of the surface is $p\,dA$ (Figs. 10.29a and 10.29b). (Notice the parallel between the pressure and a load w distributed along a line, which is defined such that the force on a differential element dx of the line is $w\,dx$.)

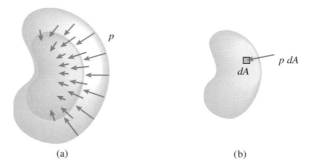

Figure 10.29
(a) The pressure on an area.
(b) The force on an element dA is $p\,dA$.

(a)   (b)

The dimensions of p are (force)/(area). In U.S. Customary units, pressure can be expressed in pounds per square foot or pounds per square inch (psi). In SI units, pressure can be expressed in newtons per square meter, which are called pascals (Pa).

In some applications, it is convenient to use the *gage pressure*

$$p_{\mathrm{g}} = p - p_{\mathrm{atm}}, \tag{10.23}$$

where p_{atm} is the pressure of the atmosphere. Atmospheric pressure varies with location and climatic conditions. Its value at sea level is approximately 1×10^5 Pa in SI units and 14.7 psi or 2120 lb/ft^2 in U.S. Customary units.

If the distributed force due to pressure on a surface is represented by an equivalent force, the point at which the line of action of the force intersects the surface is called the *center of pressure*. Let's consider a plane

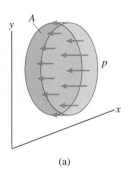 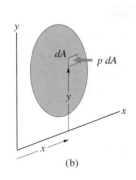 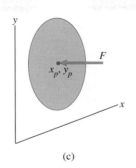

(a)   (b)   (c)

Figure 10.30
(a) A plane area subjected to pressure.
(b) The force on a differential element dA.
(c) The total force acting at the center of pressure.

area A subjected to a pressure p and introduce a coordinate system such that the area lies in the x–y plane (Fig. 10.30a). The normal force on each differential element of area dA is $p\, dA$ (Fig. 10.30b), so the total normal force on A is

$$F = \int_A p\, dA \tag{10.24}$$

Now we will determine the coordinates (x_p, y_p) of the center of pressure (Fig. 10.30c). Equating the moment of F about the origin to the total moment due to the pressure about the origin,

$$(x_p \mathbf{i} + y_p \mathbf{j}) \times (-F\mathbf{k}) = \int_A (x\mathbf{i} + y\mathbf{j}) \times (-p\, dA\mathbf{k}),$$

and using Eq. (10.24), we obtain

$$x_p = \frac{\displaystyle\int_A xp\, dA}{\displaystyle\int_A p\, dA}, \qquad y_p = \frac{\displaystyle\int_A yp\, dA}{\displaystyle\int_A p\, dA}. \tag{10.25}$$

These equations determine the position of the center of pressure when the pressure p is known. If the pressure p is uniform, the total normal force is $F = pA$ and Eqs. (10.25) indicate that the center of pressure is the centroid of A.

In Chapter 7 we showed that if you calculate the "area" defined by a load distributed along a line and place the resulting force at its centroid, the force is equivalent to the distributed load. A similar result holds for a pressure distributed on a plane area. The term $p\, dA$ in Eq. (10.24) is equal to a differential element dV of the "volume" between the surface defined by the pressure distribution and the area A (Fig. 10.31a). The total force exerted by the pressure is therefore equal to this "volume":

$$F = \int_V dV = V.$$

Substituting $p\, dA = dV$ into Eqs. (10.25), we obtain

$$x_p = \frac{\displaystyle\int_V x\, dV}{\displaystyle\int_V dV}, \qquad y_p = \frac{\displaystyle\int_V y\, dV}{\displaystyle\int_V dV}.$$

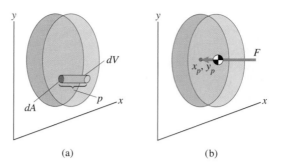

Figure 10.31
(a) The differential element $dV = p\, dA$.
(b) The line of action of F passes through the centroid of V.

(a)   (b)

The center of pressure coincides with the x and y coordinates of the centroid of the "volume" (Fig. 10.31b).

Study Questions

1. What is the definition of the pressure p?
2. What is the gage pressure?
3. What is the center of pressure? How can the "volume" defined by the pressure distribution be used to determine the location of the center of pressure?

10.8 Pressure in a Stationary Liquid

Designers of pressure vessels and piping, ships, dams, and other submerged structures must be concerned with forces and moments exerted by water pressure. If you swim toward the bottom of a swimming pool, you can feel the pressure on your ears increase—the pressure in a liquid at rest increases with depth. We can determine the dependence of pressure on depth by using a simple free-body diagram.

Introducing a coordinate system with its origin at the surface of the liquid and the positive x axis downward (Fig. 10.32a), we draw a free-body diagram of a cylinder of liquid that extends from the surface to a depth x (Fig. 10.32b). The top of the cylinder is subjected to the pressure at the surface, which we call p_0. The sides and bottom of the cylinder are subjected to

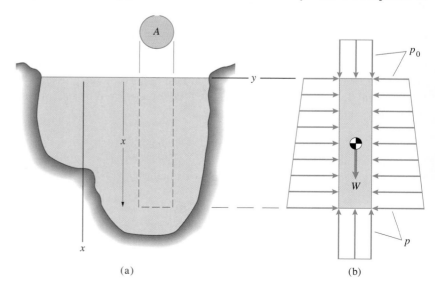

Figure 10.32
(a) A cylindrical volume that extends to a depth x in a body of stationary liquid.
(b) Free-body diagram of the cylinder.

(a)   (b)

pressure by the surrounding liquid, which increases from p_0 at the surface to a value p at the depth x. The volume of the cylinder is Ax, where A is its cross-sectional area. Therefore its weight is $W = \gamma Ax$, where γ is the weight density of the liquid. (Recall that the weight and mass densities are related by $\gamma = \rho g$.) Since the liquid is stationary, the cylinder is in equilibrium. From the equilibrium equation

$$\Sigma F_x = p_0 A - pA + \gamma Ax = 0,$$

we obtain a simple expression for the pressure p the liquid at depth x:

$$p = p_0 + \gamma x. \tag{10.26}$$

Thus the pressure increases linearly with depth, and the derivation we have used illustrates why: The pressure at a given depth literally holds up the liquid above that depth. If the surface of the liquid is open to the atmosphere, $p_0 = p_{atm}$ and we can write Eq. (10.26) in terms of the gage pressure $p_g = p - p_{atm}$ as

$$p_g = \gamma x. \tag{10.27}$$

In SI units, the mass density of water at sea level conditions is $\rho = 1000 \text{ kg/m}^3$, so its weight density is approximately $\gamma = \rho g = 9.81 \text{ kN/m}^3$. In U.S. Customary units, the weight density of water is approximately 62.4 lb/ft^3.

We have seen that the force and moment due to the pressure on a submerged plane area can be determined in two ways:

1. Integration—Integrate Eq. (10.26) or Eq. (10.27).
2. Volume analogy—Determine the total force by calculating the "volume" between the surface defined by the pressure distribution and the area A. The center of pressure coincides with the x and y coordinates of the centroid of the volume.

Example 10.11

Pressure Force and Center of Pressure

An engineer making preliminary design studies for a canal lock needs to determine the total pressure force on a submerged rectangular plate (Fig. 10.33) and the location of the center of pressure. The top of the plate is 6 m below the surface. Atmospheric pressure is $p_{atm} = 1 \times 10^5$ Pa, and the weight density of the water is $\gamma = 9.81 \text{ kN/m}^3$.

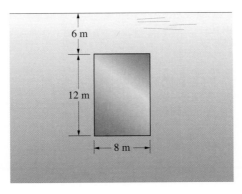

Figure 10.33

Strategy

We will determine the pressure force on a differential element of area of the plate in the form of a horizontal strip and integrate to determine the total force and moment exerted by the pressure.

Solution

In terms of a coordinate system with its origin at the surface and the positive x axis downward (Fig. a), the pressure of the water is $p = p_{atm} + \gamma x$. The horizontal strip $dA = 8\ dx$. Therefore the total force exerted on the face of the plate by the pressure is

$$F = \int_A p\ dA = \int_6^{18} (p_{atm} + \gamma x)(8\ dx) = 96p_{atm} + 1150\gamma$$

$$= (96)(1 \times 10^5) + (1150)(9810) = 20.9\ \text{MN}.$$

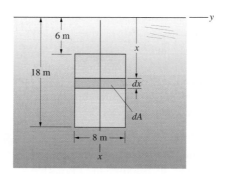

(a) An element of area in the form of a horizontal strip.

The moment about the y axis due to the pressure on the plate is

$$M = \int_A xp\ dA = \int_6^{18} x(p_{atm} + \gamma x)(8\ dx) = 262\ \text{MN-m}.$$

The force F acting at the center of pressure (Fig. b) exert a moment about the y axis equal to M:

$$x_p F = M.$$

Therefore the location of the center of pressure is

$$x_p = \frac{M}{F} = \frac{262\ \text{MN-m}}{20.9\ \text{MN}} = 12.5\ \text{m}.$$

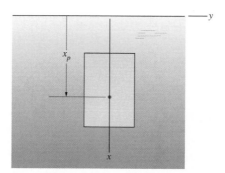

(b) The center of pressure.

Example 10.12

Gate Loaded by a Pressure Distribution

The gate AB in Fig. 10.34 has water of 2-ft depth on the right side. The width of the gate (the dimension into the page) is 3 ft, and its weight is 100 lb. The weight density of the water is $\gamma = 62.4 \text{ lb/ft}^3$. Determine the reactions on the gate at the supports A and B.

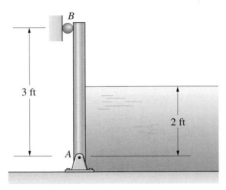

Figure 10.34

Strategy

The left face of the gate and the right face above the level of the water are exposed to atmospheric pressure. From Eqs. (10.23) and (10.26), the pressure in the water is the sum of atmospheric pressure and the gage pressure $p_g = \gamma x$, where x is measured downward from the surface of the water. The effects of atmospheric pressure cancel (Fig. a), so we need to consider only the forces and moments exerted on the gate by the gage pressure. We will determine them by integrating and also by calculating the "volume" of the pressure distribution.

Solution

Integration The face of the gate is shown in Fig. b. In terms of the differential element of area dA, the force exerted on the gate by the gage pressure is

$$F = \int_A p_g \, dA = \int_0^2 (\gamma x) 3 \, dx = 374 \text{ lb},$$

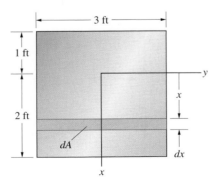

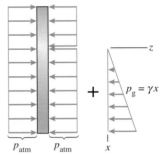

(a) The pressures acting on the faces of the gate.

(b) The face of the gate and the differential element dA.

and the moment about the y axis is

$$M = \int_A x p_g \, dA = \int_0^2 x(\gamma x)3 \, dx = 499 \text{ ft-lb}.$$

The position of the center of pressure is

$$x_p = \frac{M}{F} = \frac{499 \text{ ft-lb}}{374 \text{ lb}} = 1.33 \text{ ft}.$$

Volume Analogy   The gage pressure at the bottom of the gate is $p_g = (2 \text{ ft})\gamma$ (Fig. c), so the "volume" of the pressure distribution is

$$F = \frac{1}{2}[2 \text{ ft}][(2 \text{ ft})(62.4 \text{ lb/ft}^3)][3 \text{ ft}] = 374 \text{ lb}.$$

The x coordinate of the centroid of the triangular distribution, which is the center of pressure, is $\frac{2}{3}(2) = 1.33$ ft.

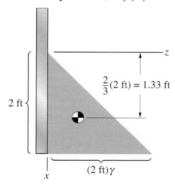

(c) Determining the "volume" of the pressure distribution and its centroid.

Determining the Reactions   We draw the free-body diagram of the gate in Fig. d. From the equilibrium equations.

$$\Sigma F_x = A_x + 100 = 0,$$

$$\Sigma F_z = A_z + B - 374 = 0,$$

$$\Sigma M_{(y \text{ axis})} = (1)B - (2)A_z + (1.33)(374) = 0,$$

we obtain $A_x = -100$ lb, $A_z = 291.2$ lb, and $B = 83.2$ lb.

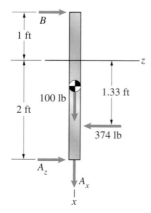

(d) Free-body diagram of the gate.

Example 10.13

Determination of a Pressure Force

The container in Fig. 10.35 is filled with a liquid with weight density γ. Determine the force exerted by the pressure of the liquid on the cylindrical wall AB.

Figure 10.35

Strategy

The pressure of the liquid on the cylindrical wall varies with depth (Fig. a). It is the force exerted by this pressure distribution we want to determine. We could determine it by integrating over the cylindrical surface, but we can avoid that by drawing a free-body diagram of the quarter-cylinder of liquid to the right of A.

(a) The pressure of the liquid on the wall AB.

Solution

We draw the free-body diagram of the quarter-cylinder of liquid in Fig. b. The pressure distribution on the cylindrical surface of the liquid is the same one that acts on the cylindrical wall. If we denote the force exerted on the liquid by this pressure distribution by $\mathbf{F}_p$, the force exerted by the liquid on the cylindrical wall is $-\mathbf{F}_p$.

The other forces parallel to the x–y plane that act on the quarter-cylinder of liquid are its weight, atmospheric pressure at the upper surface, and the pressure distribution of the liquid on the left side. The volume of liquid is $\left(\frac{1}{4}\pi R^2\right)b$, so the force exerted on the free-body diagram by the weight of the liquid is $\frac{1}{4}\gamma\pi R^2 b\mathbf{i}$. The force exerted on the upper surface by atmospheric pressure is $Rbp_{\text{atm}}\mathbf{i}$.

We can integrate to determine the force exerted by the pressure on the left side of the free-body diagram. Its magnitude is

$$\int_A p\,dA = \int_0^R (p_{\text{atm}} + \gamma x)b\,dx = Rb\left(p_{\text{atm}} + \frac{1}{2}\gamma R\right).$$

From the equilibrium equation

$$\Sigma\mathbf{F} = \frac{1}{4}\gamma\pi R^2 b\mathbf{i} + Rbp_{\text{atm}}\mathbf{i} + Rb\left(p_{\text{atm}} + \frac{1}{2}\gamma R\right)\mathbf{j} + \mathbf{F}_p = \mathbf{0},$$

we obtain the force exerted on the wall AB by the pressure of the liquid:

$$-\mathbf{F}_p = Rb\left(p_{\text{atm}} + \frac{\pi}{4}\gamma R\right)\mathbf{i} + Rb\left(p_{\text{atm}} + \frac{1}{2}\gamma R\right)\mathbf{j}.$$

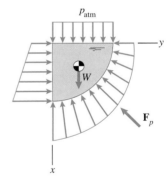

(b) Free-body diagram of the liquid to the right of A.

Chapter Summary

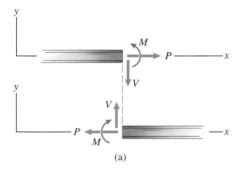

(a)

Beams

The internal forces and moment in a beam are expressed as the *axial force P*, *shear force V*, and *bending moment M*. Their positive directions are defined in Fig. a.

By cutting a beam at an arbitrary position x, the axial load P, shear force V, and bending moment M can be determined as functions of x. Depending on the loading and supports of the beam, it may be necessary to draw several free-body diagrams to determine the distributions for the entire beam. The graphs of V and M as functions of x are the *shear force and bending moment diagrams*.

The distributed load, shear force, and bending moment in a portion of a beam subjected only to a distributed load satisfy the relations

$$\frac{dV}{dx} = -w, \qquad \text{Eq. (10.2)}$$

$$\frac{dM}{dx} = V. \qquad \text{Eq. (10.3)}$$

For segments of a beam that are unloaded or are subjected to a distributed load, these equations can be integrated to determine V and M as functions of x. To obtain the complete shear force and bending moment diagrams, forces and couples must also be accounted for.

Cables

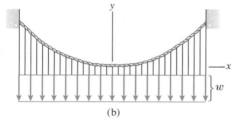

(b)

Loads Distributed Uniformly Along a Straight Line   If a suspended cable is subjected to a horizontally distributed load w (Fig. b), the curve described by the cable is the parabola

$$y = \frac{1}{2} ax^2, \qquad \text{Eq. (10.10)}$$

where $a = w/T_0$ and T_0 is the tension in the cable at $x = 0$. The tension in the cable at a position x is

$$T = T_0\sqrt{1 + a^2x^2}, \qquad \text{Eq. (10.11)}$$

and the length of the cable in the horizontal interval from 0 to x is

$$s = \frac{1}{2}\left\{ x\sqrt{1 + a^2x^2} + \frac{1}{a}\ln\left[ax + \sqrt{1 + a^2x^2} \right] \right\}. \qquad \text{Eq. (10.12)}$$

Loads Distributed Uniformly Along a Cable   If a suspended cable is subjected to a load w distributed along its length, the curve described by the cable is the catenary

$$y = \frac{1}{2a}\left(e^{ax} + e^{-ax} - 2 \right) = \frac{1}{a}\left(\cosh ax - 1 \right), \qquad \text{Eq. (10.20)}$$

where $a = w/T_0$ and T_0 is the tension in the cable at $x = 0$. The tension in the cable at a position x is

$$T = T_0\sqrt{1 + \frac{1}{4}(e^{ax} - e^{-ax})^2} = T_0\cosh ax, \qquad \text{Eq. (10.21)}$$

and the length of the cable in the horizontal interval from 0 to x is

$$s = \frac{1}{2a}(e^{ax} - e^{-ax}) = \frac{\sinh ax}{a}. \qquad \text{Eq. (10.22)}$$

Discrete Loads If N known weights are suspended from a cable and the positions of the attachment points of the cable, the horizontal positions of the attachment points of the weights, and the vertical position of the attachment point of one of the weights are known, the configuration of the cable and the tension in each of its segments can be determined.

Liquids and Gases

The pressure p on a surface is defined so that the normal force exerted on an element dA of the surface is $p\,dA$. The total normal force exerted by pressure on a plane area A is

$$F = \int_A p\,dA. \qquad \text{Eq. (10.24)}$$

The *center of pressure* is the point on A at which F must be placed to be equivalent to the pressure on A. The coordinates of the center of pressure are

$$x_p = \frac{\displaystyle\int_A xp\,dA}{\displaystyle\int_A p\,dA}, \qquad y_p = \frac{\displaystyle\int_A yp\,dA}{\displaystyle\int_A p\,dA}. \qquad \text{Eq. (10.25)}$$

The pressure in a stationary liquid is

$$p = p_0 + \gamma x, \qquad \text{Eq. (10.26)}$$

where p_0 is the pressure at the surface, γ is the weight density of the liquid, and x is the depth. If the surface of the liquid is open to the atmosphere, $p_0 = p_{atm}$, the atmospheric pressure.

Review Problems

10.1 Determine the internal forces and moment at B (a) if $x = 250$ mm; (b) if $x = 750$ mm.

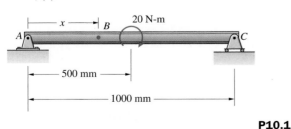

P10.1

10.2 Determine the internal forces and moment (a) at B; (b) at C.

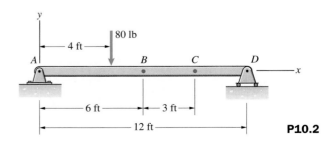

P10.2

10.3 (a) Determine the maximum bending moment in the beam and the value of x where it occurs.
(b) Show that the equations for V and M as functions of x satisfy the equation $V = dM/dx$.

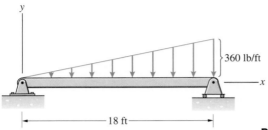

P10.3

10.4 Draw the shear and bending moment diagrams for the beam in Problem 10.3.

10.5 Determine the shear force and bending moment diagrams for the beam.

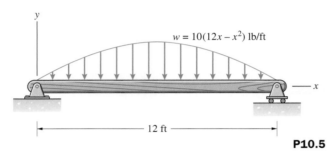

P10.5

10.6 Draw the shear force and bending moment diagrams for beam ABC.

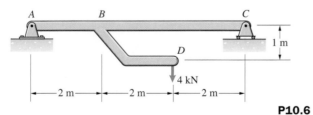

P10.6

10.7 Draw the shear force and bending moment diagrams for beam ABC.

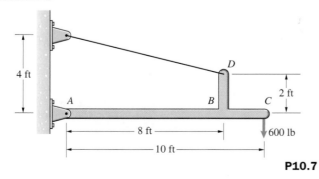

P10.7

10.8 Determine the internal forces and moments at A.

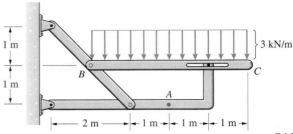

P10.8

10.9 Draw the shear force and bending moment diagrams of beam BC in Problem 10.8.

10.10 Determine the internal forces and moment at B (a) if $x = 250$ mm; (b) if $x = 750$ mm.

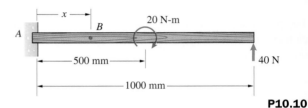

P10.10

10.11 Draw the shear force and bending moment diagrams for the beam in Problem 10.10.

10.12 The homogeneous beam weighs 1000 lb. What are the internal forces and bending moment at its midpoint?

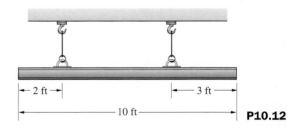

P10.12

10.13 Draw the shear force and bending moment diagrams for the beam in Problem 10.12.

10.14 At A the main cable of the suspension bridge is horizontal and its tension is 1×10^8 lb.
(a) Determine the distributed load acting on the cable.
(b) What is the tension at B?

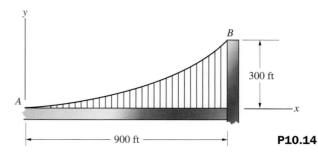

P10.14

10.15 The power line has a mass of 1.4 kg/m. If the line will safely support a tension of 5 kN, determine whether it will safely support an ice accumulation of 4 kg/m.

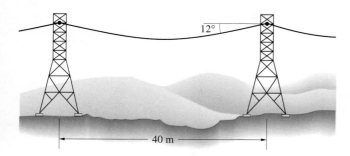

40 m

P10.15

10.16 Consider a plane, vertical area A below the surface of a liquid. Let p_0 be the pressure at the surface.

(a) Show that the force exerted by pressure on the area is $F = \bar{p}A$, where $\bar{p} = p_0 + \gamma\bar{x}$ is the pressure of the liquid at the centroid of the area.

(b) Show that the x coordinate of the center of pressure is

$$x_p = \bar{x} + \frac{\gamma I_{y'}}{\bar{p}A},$$

where $I_{y'}$ is the moment of inertia of the area about the y' axis through its centroid.

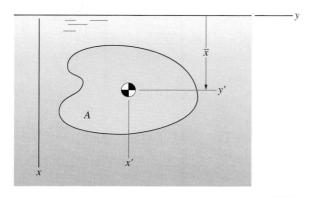

P10.16

10.17 A circular plate of 1-m radius is below the surface of a stationary pool of water. Atmospheric pressure is $p_{atm} = 1 \times 10^5$ Pa, and the mass density of the water is $\rho = 1000$ kg/m³. Determine (a) the force exerted on a face of the plate by the pressure of the water; (b) the x coordinate of the center of pressure. (See Problem 10.16.)

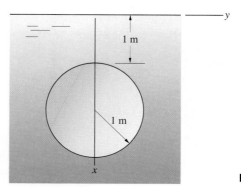

P10.17

10.18 The gate has water of 2-m depth on one side. The width of the gate (the dimension into the page) is 4 m, and its mass is 160 kg. The mass density of the water is $\rho = 1000$ kg/m³, and atmospheric pressure is $p_{atm} = 1 \times 10^5$ Pa. Determine the reactions on the gate at A and B. (The support at B exerts only a horizontal reaction on the gate.)

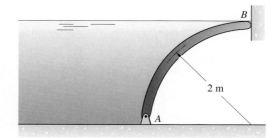

P10.18

10.19 The dam has water of depth 4 ft on one side. The width of the dam (the dimension into the page) is 8 ft. The weight density of the water is $\gamma = 62.4$ lb/ft³, and atmospheric pressure is $p_{atm} = 2120$ lb/ft². If you neglect the weight of the dam, what are the reactions at A and B?

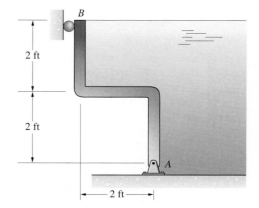

P10.19

The force necessary to hold the extensible platform in equilibrium can be determined by subjecting the platform to a hypothetical motion and applying the concept of virtual work.

Virtual Work and Potential Energy

When a spring is stretched, the work performed is stored in the spring as potential energy. Raising an extensible platform increases its gravitational potential energy. In this chapter we define work and potential energy and introduce a general and powerful result called the principle of virtual work.

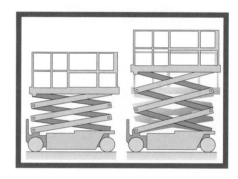

11.1 Virtual Work

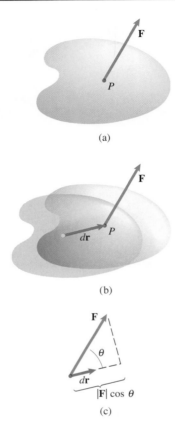

Figure 11.1
(a) A force **F** acting on an object.
(b) A displacement $d\mathbf{r}$ of P.
(c) The work $dU = (|\mathbf{F}| \cos\theta) |d\mathbf{r}|$.

The principle of virtual work is a statement about work done by forces and couples when an object or structure is subjected to various hypothetical motions. Before we can introduce this principle, we must define work.

Work

Consider a force acting on an object at a point P (Fig. 11.1a). Suppose that the object undergoes an infinitesimal motion, so that P has a differential displacement $d\mathbf{r}$ (Fig. 11.1b). The *work* dU done by **F** as a result of the displacement $d\mathbf{r}$ is defined to be

$$dU = \mathbf{F} \cdot d\mathbf{r}. \tag{11.1}$$

From the definition of the dot product, $dU = (|\mathbf{F}| \cos\theta) |d\mathbf{r}|$, where θ is the angle between **F** and $d\mathbf{r}$ (Fig. 11.1c). The work is equal to the product of the component of **F** in the direction of $d\mathbf{r}$ and the magnitude of $d\mathbf{r}$. Notice that if the component of **F** parallel to $d\mathbf{r}$ points in the direction opposite to $d\mathbf{r}$, the work is negative. Also notice that if **F** is perpendicular to $d\mathbf{r}$, the work is zero. The dimensions of work are (force) × (length).

Now consider a couple acting on an object (Fig. 11.2a). The moment due to the couple is $M = Fh$ in the counterclockwise direction. If the object rotates through an infinitesimal counterclockwise angle $d\alpha$ (Fig. 11.2b), the points of application of the forces are displaced through differential distances $\frac{1}{2}h\,d\alpha$. Consequently, the total work done is $dU = F(\frac{1}{2}h\,d\alpha) + F(\frac{1}{2}h\,d\alpha) = M\,d\alpha$.

We see that when an object acted on by a couple M is rotated through an angle $d\alpha$ in the same direction as the couple (Fig. 11.2c), the resulting work is

$$dU = M\,d\alpha. \tag{11.2}$$

If the direction of the couple is opposite to the direction of $d\alpha$, the work is negative.

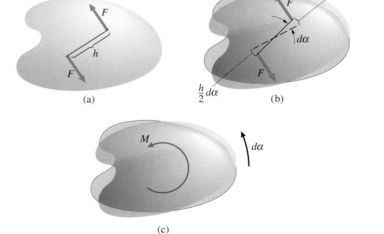

Figure 11.2
(a) A couple acting on an object.
(b) An infinitesimal rotation of the object.
(c) An object acted on by a couple M rotating through an angle $d\alpha$.

Principle of Virtual Work

Now that we have defined the work done by forces and couples, we can introduce the principle of virtual work. Before stating it, we first discuss an example to give you context for understanding the principle.

The homogeneous bar in Fig. 11.3a is supported by the wall and by the pin support at A and is loaded by a couple M. The free-body diagram of the bar is shown in Fig. 11.3b. The equilibrium equations are

$$\Sigma F_x = A_x - N = 0, \tag{11.3}$$

$$\Sigma F_y = A_y - W = 0, \tag{11.4}$$

$$\Sigma M_{(\text{point } A)} = NL \sin \alpha - W \frac{1}{2} L \cos \alpha - M = 0. \tag{11.5}$$

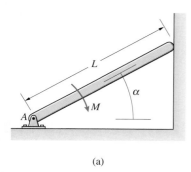
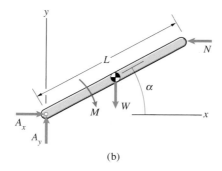

(a)   (b)

Figure 11.3
(a) A bar subjected to a couple M.
(b) Free-body diagram of the bar.

We can solve these three equations for the reactions A_x, A_y, and N. However, we have a different objective.

Let's ask the following question: If the bar is acted on by the forces and couple in Fig. 11.3b and we subject it to a hypothetical infinitesimal translation in the x direction, as shown in Fig. 11.4, what work is done? The hypothetical displacement δx is called a *virtual displacement* of the bar, and the resulting work δU is called the *virtual work*. The pin support and the wall prevent the bar from actually moving in the x direction: the virtual displacement is a theoretical artifice. Our objective is to calculate the resulting virtual work:

$$\delta U = A_x \delta x + (-N)\delta x = (A_x - N)\delta x. \tag{11.6}$$

The forces A_y and W do no work because they are perpendicular to the displacements of their points of application. The couple M also does no work, because the bar does not rotate. Comparing this equation with Eq. (11.3), we find that the virtual work equals zero.

Next we give the bar a virtual translation in the y direction (Fig. 11.5). The resulting virtual work is

$$\delta U = A_y \delta y + (-W)\delta y = (A_y - W)\delta y. \tag{11.7}$$

From Eq. (11.4), the virtual work again equals zero.

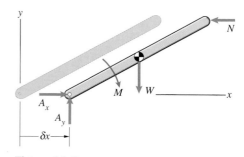

Figure 11.4
A virtual displacement δx.

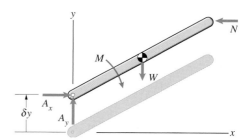

Figure 11.5
A virtual displacement δy.

Finally, we give the bar a virtual rotation while holding point A fixed (Fig. 11.6a). The forces A_x and A_y do no work because their point of application does not move. The work done by the couple M is $-M\ \delta\alpha$, because its direction is opposite to that of the rotation. The displacements of the points of application of the forces N and W are shown in Fig. 11.6b, and the components of the forces in the direction of the displacements are shown in Fig. 11.6c. The work done by N is $(N\sin\alpha)(L\ \delta\alpha)$, and the work done by W is $(-W\cos\alpha)(\frac{1}{2}L\ \delta\alpha)$. The total work is

$$\delta U = (N\sin\alpha)(L\ \delta\alpha) + (-W\cos\alpha)\left(\frac{1}{2}L\ \delta\alpha\right) - M\ \delta\alpha$$

$$= \left(NL\sin\alpha - W\frac{1}{2}L\cos\alpha - M\right)\delta\alpha. \tag{11.8}$$

From Eq. (11.5), the virtual work resulting from the virtual rotation is also zero.

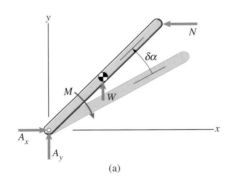
(a)

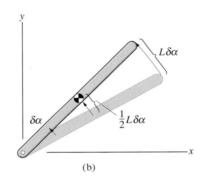
(b)

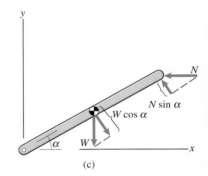
(c)

Figure 11.6
(a) A virtual rotation $\delta\alpha$.
(b) Displacements of the points of application of N and W.
(c) Components of N and W in the direction of the displacements.

We have shown that for three virtual motions of the bar, the virtual work is zero. These results are examples of a form of the principle of virtual work: *If an object is in equilibrium, the virtual work done by the external forces and couples acting on it is zero for any virtual translation or rotation:*

$$\delta U = 0. \tag{11.9}$$

As our example illustrates, this principle can be used to derive the equilibrium equations for an object. Subjecting the bar to virtual translations δx and δy and a virtual rotation $\delta\alpha$ results in Eqs. (11.6)–(11.8). Because the virtual work must be zero in each case, we obtain Eqs. (11.3)–(11.5). But there is no advantage to this approach compared to simply drawing the free-body diagram of the object and writing the equations of equilibrium in the usual way. The advantages of the principle of virtual work become evident when we consider structures.

Application to Structures

The principle of virtual work stated in the preceding section applies to each member of a structure. By subjecting certain types of structures in equilibrium to virtual motions and calculating the total virtual work, we can determine
unknown reactions at their supports as well as internal forces in their members. The procedure involves finding virtual motions that result in virtual work being done both by known loads and by unknown forces and couples.

Suppose that we want to determine the axial load in member BD of the truss in Fig. 11.7a. The other members of the truss are subjected to the 4-kN load and the forces exerted on them by member BD (Fig. 11.7b). If we give the structure a virtual rotation $\delta\alpha$ as shown in Fig. 113c, virtual work is done by the force T_{BD} acting at B and by the 4-kN load at C. Furthermore, the virtual work done by these two forces is the total virtual work done on the members of the structure, because the virtual work done by the internal forces they exert on each other cancels out. For example, consider joint C (Fig. 11.7d). The force T_{BC} is the axial load in member BC. The virtual work done at C on member BC is $T_{BC}(1.4 \text{ m}) \, \delta\alpha$, and the work done at C on member CD is $(4 \text{ kN} - T_{BC})(1.4 \text{ m}) \, \delta\alpha$. When we add up the virtual work done on the members to obtain the total virtual work on the structure, the virtual work due to the internal force T_{BC} cancels out. (If the members exerted an internal *couple* on each other at C—for example, as a result of friction in the pin support—the virtual work would not cancel out.) Therefore we can ignore internal forces in calculating the total virtual work on the structure:

$$\delta U = (T_{BD} \cos\theta)(1.4 \text{ m}) \, \delta\alpha + (4 \text{ kN})(1.4 \text{ m}) \, \delta\alpha = 0.$$

The angle $\theta = \arctan(1.4/1) = 54.5°$. Solving this equation, we obtain $T_{BD} = -6.88$ kN.

We have seen that using virtual work to determine reactions on members of structures involves two steps:

1. Choose a virtual motion—Identify a virtual motion that results in virtual work being done by known loads and by an unknown force or couple you want to determine.

2. Determine the virtual work—Calculate the total virtual work resulting from the virtual motion to obtain an equation for the unknown force or couple.

Study Questions

1. What work is done by a force **F** when its point of application undergoes a displacement $d\mathbf{r}$?
2. What work is done by a couple M when the object on which it acts rotates through an angle $d\alpha$ in the same direction as the couple?
3. What does the principle of virtual work say about the work done when an object in equilibrium is subjected to a virtual translation or rotation?

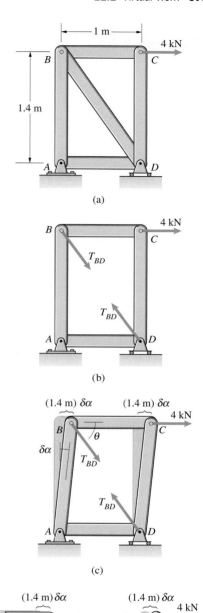

Figure 11.7
(a) A truss with a 4-kN load.
(b) Forces exerted by member BD.
(c) A virtual motion of the structure.
(d) Calculating the virtual work on members BC and CD at the joint C.

Example 11.1

Applying Virtual Work to a Structure

For the structure in Fig. 11.8, use the principle of virtual work to determine the horizontal reaction at C.

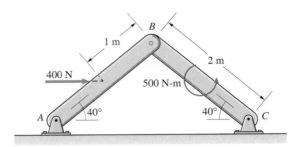

Figure 11.8

Solution

Choose a Virtual Motion We draw the free-body diagram of the structure in Fig. a. Our objective is to determine C_x. If we hold point A fixed and subject bar AB to a virtual rotation $\delta\alpha$ while requiring point C to move horizontally (Fig. b), virtual work is done only by the external loads on the structure and by C_x. The reactions A_x and A_y do no work because A does not move, and the reaction C_y does no work because it is perpendicular to the virtual displacement of point C.

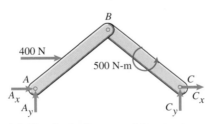
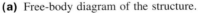
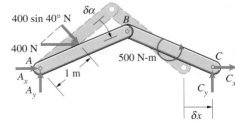

(a) Free-body diagram of the structure.

(b) A virtual displacement in which A remains fixed and C moves horizontally.

Determine the Virtual Work The virtual work done by the 400-N force is $(400 \sin 40°\ \text{N})(1\ \text{m})\ \delta\alpha$. The bar BC undergoes a virtual rotation $\delta\alpha$ in the counterclockwise direction, so the work done by the couple is $(500\ \text{N-m})\ \delta\alpha$. In terms of the virtual displacement δx of point C, the work done by the reaction C_x is $C_x\delta x$. The total virtual work is

$$\delta U = (400 \sin 40°)(1)\delta\alpha + 500\delta\alpha + C_x\delta x = 0.$$

To obtain C_x from this equation we must determine the relationship between $\delta\alpha$ and δx. From the geometry of the structure (Fig. c), the relationship between the angle α and the distance x from A to C is

$$x = 2(2\cos\alpha).$$

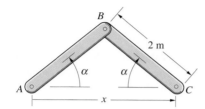

(c) The geometry for determining the relation between $\delta\alpha$ and δx.

The derivative of this equation with respect to α is

$$\frac{dx}{d\alpha} = -4\sin\alpha.$$

Therefore an infinitesimal change in x is related to an infinitesimal change in α by

$$dx = -4\sin\alpha\, d\alpha.$$

Because the virtual rotation $\delta\alpha$ in Fig. b is a decrease in α, we conclude that δx is related to $\delta\alpha$ by

$$\delta x = 4\sin 40°\, \delta\alpha.$$

Substituting this expression into our equation for the virtual work gives

$$\delta U = \left[400\sin 40° + 500 + C_x(4\sin 40°)\right]\delta\alpha = 0.$$

Solving, we obtain $C_x = -294$ N.

Discussion

Notice that we ignored the internal forces the members exert on each other at B. The virtual work done by these internal forces cancels out. To obtain the solution, we needed to determine the relationship between the virtual displacements δx and $\delta\alpha$. Determining the geometrical relationships between virtual displacements is often the most challenging aspect of applying the principle of virtual work.

Example 11.2

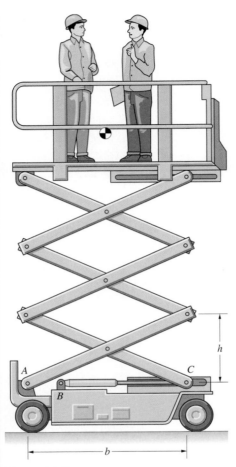

Figure 11.9

Applying Virtual Work to a Machine

The extensible platform in Fig. 11.9 is raised and lowered by the hydraulic cylinder BC. The total weight of the platform and men is W. The weights of the beams supporting the platform can be neglected. What axial force must the hydraulic cylinder exert to hold the platform in equilibrium in the position shown?

Strategy

We can use a virtual motion that coincides with the actual motion of the platform and beams when the length of the hydraulic cylinder changes. By calculating the virtual work done by the hydraulic cylinder and by the weight of the men and platform, we can determine the force exerted by the hydraulic cylinder.

Solution

Choose a Virtual Motion We draw the free-body diagram of the platform and beams in Fig. a. Our objective is to determine the force F exerted by the hydraulic cylinder. If we hold point A fixed and subject point C to a horizontal virtual displacement δx, the only external forces that do virtual work are F and the weight W. (The reaction due to the roller support at C is perpendicular to the virtual displacement.)

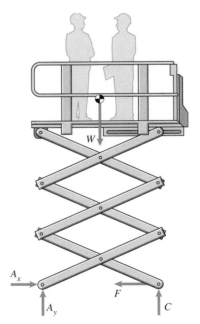

(a) Free-body diagram of the platform and supporting beams.

Determine the Virtual Work The virtual work done by the force F as point C undergoes a virtual displacement δx to the right (Fig. b) is $-F\,\delta x$. To determine the virtual work done by the weight W, we must determine the vertical displacement of point D in Fig. b when point C moves to the right a distance δx. The dimensions b and h are related by

$$b^2 + h^2 = L^2,$$

where L is the length of the beam AD. Taking the derivative of this equation with respect to b, we obtain

$$2b + 2h\frac{dh}{db} = 0,$$

which we can solve for dh in terms of db:

$$dh = -\frac{b}{h}\, db.$$

Thus when b increases an amount δx, the dimension h *decreases* an amount $(b/h)\,\delta x$. Because there are three pairs of beams, the platform moves downward a distance $(3b/h)\,\delta x$, and the virtual work done by the weight is $(3b/h)W\,\delta x$. The total virtual work is

$$\delta U = \left[-F + \left(\frac{3b}{h}\right)W\right]\delta x = 0,$$

and we obtain $F = (3b/h)W$.

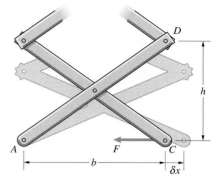

(b) A virtual displacement in which A remains fixed and C moves horizontally.

11.2 Potential Energy

The work of a force $\mathbf{F}$ due to a differential displacement of its point of application is

$$dU = \mathbf{F} \cdot d\mathbf{r}.$$

If a function of position V exists such that for any $d\mathbf{r}$,

$$dU = \mathbf{F} \cdot d\mathbf{r} = -dV, \tag{11.10}$$

the function V is called the *potential energy* associated with the force $\mathbf{F}$, and $\mathbf{F}$ is said to be *conservative*. (The negative sign in this equation is in keeping with the interpretation of V as "potential" energy. Positive work results from a decrease in V.) If the forces that do work on a system are conservative, we will show that you can use the total potential energy of the system to determine its equilibrium positions.

Examples of Conservative Forces

Weights of objects and the forces exerted by linear springs are conservative. In the following sections we derive the potential energies associated with these forces.

Weight in terms of a coordinate system with its y axis upward, the force exerted by the weight of an object is $\mathbf{F} = -W\mathbf{j}$ (Fig. 11.10a). If we give the object an arbitrary displacement $d\mathbf{r} = dx\,\mathbf{i} + dy\,\mathbf{j} + dz\,\mathbf{k}$ (Fig. 11.10b), the work done by its weight is

$$dU = \mathbf{F} \cdot d\mathbf{r} = (-W\mathbf{j}) \cdot (dx\,\mathbf{i} + dy\,\mathbf{j} + dz\,\mathbf{k}) = -W\,dy.$$

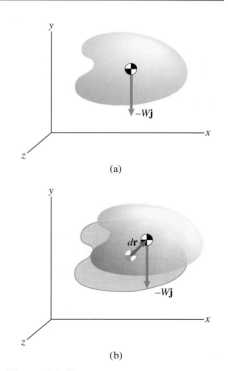

Figure 11.10
(a) Force exerted by the weight of an object.
(b) A differential displacement.

We seek a potential energy V such that

$$dU = -W \, dy = -dV, \tag{11.11}$$

or

$$\frac{dV}{dy} = W.$$

If we neglect the variation in the weight with height and integrate, we obtain

$$V = Wy + C.$$

The constant C is arbitrary, since this function satisfies Eq. (11.11) for any value of C, and we will let $C = 0$. The position of the origin of the coordinate system can also be chosen arbitrarily. Thus the potential energy associated with the weight of an object is

$$V = Wy, \tag{11.12}$$

where y is the height of the object above some chosen reference level, or *datum*.

Springs   Consider a linear spring connecting an object to a fixed support (Fig. 11.11a). In terms of the stretch $S = r - r_0$, where r is the length of the spring and r_0 is its unstretched length, the force exerted on the object is kS (Fig. 11.11b). If the point at which the spring is attached to the object undergoes a differential displacement $d\mathbf{r}$ (Fig. 11.11c), the work done by the force on the object is

$$dU = -kS \, dS,$$

where dS is the increase in the stretch of the spring resulting from the displacement (Fig. 11.11d). We seek a potential energy V such that

$$dU = -kS \, dS = -dV, \tag{11.13}$$

or

$$\frac{dV}{dS} = kS.$$

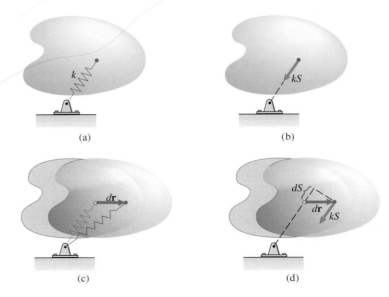

(a)

(b)

(c)

(d)

Figure 11.11

(a) A spring connected to an object.
(b) The force exerted on the object.
(c) A differential displacement of the object.
(d) The work done by the force is $dU = -kS \, dS$.

Integrating this equation and letting the integration constant be zero, we obtain the potential energy associated with the force exerted by a linear spring:

$$V = \frac{1}{2} kS^2. \tag{11.14}$$

Notice that V is positive if the spring is either stretched (S is positive) or compressed (S is negative). Potential energy (the potential to do work) is stored in a spring by either stretching or compressing it.

Principle of Virtual Work for Conservative Forces

Because the work done by a conservative force is expressed in terms of its potential energy through Eq. (11.10), we can give an alternative statement of the principle of virtual work when an object is subjected to conservative forces: *Suppose that an object is in equilibrium. If the forces that do work as the result of a virtual translation or rotation are conservative, the change in the total potential energy is zero:*

$$\delta V = 0. \tag{11.15}$$

We emphasize that it is not necessary that all of the forces acting on the object be conservative for this result to hold; it is necessary only that the forces that do work be conservative. This principle also applies to a system of interconnected objects if the external forces that do work are conservative and the internal forces at the connections between objects either do no work or are conservative. Such a system is called a *conservative system*.

If the position of a system can be specified by a single coordinate q, the system is said to have one *degree of freedom*. The total potential energy of a conservative, one-degree-of-freedom system can be expressed in terms of q, and we can write Eq. (11.15) as

$$\delta V = \frac{dV}{dq} \delta q = 0.$$

Thus when the object or system is in equilibrium, the derivative of its total potential energy with respect to q is zero:

$$\frac{dV}{dq} = 0. \tag{11.16}$$

We can use this equation to determine the values of q at which the system is in equilibrium.

Stability of Equilibrium

Suppose that a homogeneous bar of weight W and length L is pinned at one end. In terms of the angle α shown in Fig. 11.12a, the height of the center of mass relative to the pinned end is $-\frac{1}{2}L\cos\alpha$. Choosing the level of the pin support as the datum, we can therefore express the potential energy associated with the weight of the bar as

$$V = -\frac{1}{2} WL\cos\alpha.$$

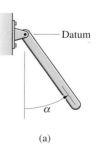

Datum

α

(a)

Figure 11.12
(a) A bar suspended from one end.

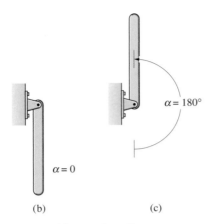

Figure 11.12 (continued)
(b) The equilibrium position $\alpha = 0$.
(c) The equilibrium position $\alpha = 180°$.

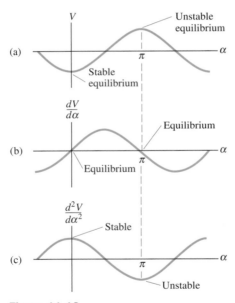

Figure 11.13
Graphs of V, $dV/d\alpha$, and $d^2V/d\alpha^2$.

When the bar is in equilibrium,

$$\frac{dV}{d\alpha} = \frac{1}{2}WL \sin\alpha = 0.$$

This condition is satisfied when $\alpha = 0$ (Fig. 11.12b) and also when $\alpha = 180°$ (Fig. 11.12c).

There is a fundamental difference between the two equilibrium positions of the bar. In the position shown in Fig. 11.12b, if we displace the bar slightly from its equilibrium position and release it, the bar will remain near the equilibrium position. We say that this equilibrium position is *stable*. When the bar is in the position shown in Fig. 11.12c, if we displace it slightly and release it, the bar will move away from the equilibrium position. This equilibrium position is *unstable*.

The graph of the bar's potential energy V as a function of α is shown in Fig. 11.13a. The potential energy is a minimum at the stable equilibrium position $\alpha = 0$ and a maximum at the unstable equilibrium position $\alpha = 180°$. The derivative of V (Fig. 11.13b) equals zero at both equilibrium positions. The second derivative of V (Fig. 11.13c) is positive at the stable equilibrium position $\alpha = 0$ and negative at the unstable equilibrium position $\alpha = 180°$.

If a conservative, one-degree-of-freedom system is in equilibrium and the second derivative of V evaluated at the equilibrium position is positive, the equilibrium position is stable. If the second derivative of V is negative, it is unstable (Fig. 11.14).

$$\frac{dV}{dq} = 0, \qquad \frac{d^2V}{dq^2} > 0: \qquad \textbf{Stable equilibrium}$$

$$\frac{dV}{dq} = 0, \qquad \frac{d^2V}{dq^2} < 0: \qquad \textbf{Unstable equilibrium}$$

Proving these results requires analyzing the motion of the system near an equilibrium position.

Using potential energy to analyze the equilibrium of one-degree-of-freedom systems typically involves three steps:

1. Determine the potential energy—Express the total potential energy in terms of a single coordinate that specifies the position of the system.

2. Find the equilibrium positions—By calculating the first derivative of the potential energy, determine the equilibrium position or positions.

3. Examine the stability—Use the sign of the second derivative of the potential energy to determine whether the equilibrium positions are stable.

Figure 11.14
Graphs of the potential energy V as a function of the coordinate q that exhibit stable and unstable equilibrium positions.

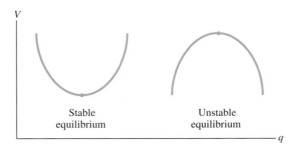

Study Questions

1. What is the definition of the potential energy of a conservative force?
2. If an object in equilibrium is subjected only to conservative forces, what do you know about the total potential energy when the object undergoes a virtual translation or rotation?
3. What does it mean when an equilibrium position of an object is said to be stable or unstable? How can you distinguish whether an equilibrium position of a conservative, one-degree-of-freedom system is stable or unstable?

Example 11.3

Stability of a Conservative System

In Fig. 11.15 a crate of weight W is suspended from the ceiling by a wire modeled as a linear spring with constant k. The coordinate x measures the position of the centered mass of the crate relative to its position when the wire is unstretched. Find the equilibrium position of the crate, and determine whether it is stable or unstable.

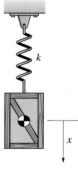

Strategy

The forces acting on the crate—its weight and the force exerted by the spring—are conservative. Therefore the system is conservative, and we can use the potential energy to determine both the equilibrium position and whether the equilibrium position is stable.

Figure 11.15

Solution

Determine the Potential Energy We can use $x = 0$ as the datum for the potential energy associated with the weight. Because the coordinate x is positive downward, the potential energy is $-Wx$. The stretch of the spring equals x, so the potential energy associated with the force of the spring is $\frac{1}{2} kx^2$. The total potential energy is

$$V = \frac{1}{2} kx^2 - Wx.$$

Find the Equilibrium Positions When the crate is in equilibrium,

$$\frac{dV}{dx} = kx - W = 0.$$

The equilibrium position is $x = W/k$.

Examine the Stability The second derivative of the potential energy is

$$\frac{d^2V}{dx^2} = k.$$

The equilibrium position is stable.

Example 11.4

Figure 11.16

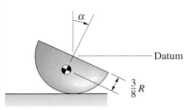

(a) The hemisphere rotated through an angle α.

Stability of an Equilibrium Position

The homogeneous hemisphere in Fig. 11.16 is at rest on the plane surface. Show that it is in equilibrium in the position shown. Is the equilibrium position stable?

Strategy

To determine whether the hemisphere is in equilibrium and whether its equilibrium is stable, we must introduce a coordinate that specifies its orientation and express its potential energy in terms of that coordinate. We can use as the coordinate the angle of rotation of the hemisphere relative to the position shown.

Solution

Determine the Potential Energy Suppose that the hemisphere is rotated through an angle α relative to its original position (Fig. a). Using the datum shown, the potential energy associated with the weight W of the hemisphere is

$$V = -\frac{3}{8} RW \cos \alpha.$$

Find the Equilibrium Positions When the hemisphere is in equilibrium,

$$\frac{dV}{d\alpha} = \frac{3}{8} RW \sin \alpha = 0,$$

which confirms that $\alpha = 0$ is an equilibrium position.

Examine the Stability The second derivative of the potential energy is

$$\frac{d^2V}{d\alpha^2} = \frac{3}{8} RW \cos \alpha.$$

This expression is positive at $\alpha = 0$, so the equilibrium position is stable.

Discussion

Notice that we ignored the normal force exerted on the hemisphere by the plane surface. This force does no work and so does not affect the potential energy.

Example 11.5

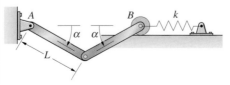

Figure 11.17

Stability of an Equilibrium Position

The pinned bars in Fig. 11.17 are held in place by the linear spring. Each bar has weight W and length L. The spring is unstretched when $\alpha = 0$, and the bars are in equilibrium when $\alpha = 60°$. Determine the spring constant k, and determine whether the equilibrium position is stable or unstable.

Strategy

The only forces that do work on the bars are their weights and the force exerted by the spring. By expressing the total potential energy in terms of α and using Eq. (11.16), we will obtain an equation we can solve for the spring constant k.

Solution

Determine the Potential Energy   If we use the datum shown in Fig. a, the potential energy associated with the weights of the two bars is

$$W\left(-\frac{1}{2}L\sin\alpha\right) + W\left(-\frac{1}{2}L\sin\alpha\right) = -WL\sin\alpha.$$

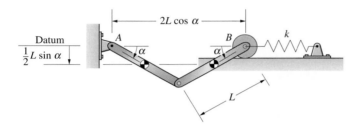

(a) Determining the total potential energy.

The spring is unstretched when $\alpha = 0$ and the distance between points A and B is $2L\cos\alpha$ (Fig. a), so the stretch of the spring is $2L - 2L\cos\alpha$. Therefore the potential energy associated with the spring is $\frac{1}{2}k(2L - 2L\cos\alpha)^2$, and the total potential energy is

$$V = -WL\sin\alpha + 2kL^2(1 - \cos\alpha)^2.$$

When the system is in equilibrium,

$$\frac{dV}{d\alpha} = -WL\cos\alpha + 4kL^2\sin\alpha(1 - \cos\alpha) = 0.$$

Because the system is in equilibrium when $\alpha = 60°$, we can solve this equation for the spring constant in terms of W and L:

$$k = \frac{W\cos\alpha}{4L\sin\alpha(1 - \cos\alpha)} = \frac{W\cos 60°}{4L\sin 60°(1 - \cos 60°)} = \frac{0.289W}{L}.$$

Examine the Stability   The second derivative of the potential energy is

$$\frac{d^2V}{d\alpha^2} = WL\sin\alpha + 4kL^2(\cos\alpha - \cos^2\alpha + \sin^2\alpha)$$

$$= WL\sin 60° + 4kL^2(\cos 60° - \cos^2 60° + \sin^2 60°)$$

$$= 0.866WL + 4kL^2.$$

This is a positive number, so the equilibrium position is stable.

Computational Mechanics

The following example and problems are designed for the use of a programmable calculator or computer.

Computational Example 11.6

The two bars in Fig. 11.18 are held in place by the linear spring. Each bar has weight W and length L. The spring is unstretched when $\alpha = 0$. If $W = kL$, what is the value of α for which the bars are in equilibrium? Is the equilibrium position stable?

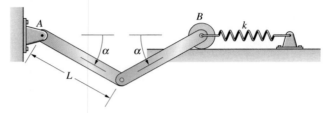

Figure 11.18

Strategy

By obtaining a graph of the derivative of the total potential energy as a function of α, we can estimate the value of α corresponding to equilibrium and determine whether the equilibrium position is stable.

Solution

We derived the total potential energy of the system and determined its derivative with respect to α in Example 11.5, obtaining

$$\frac{dV}{d\alpha} = -WL\cos\alpha + 4kL^2\sin\alpha(1 - \cos\alpha).$$

Substituting $W = kL$, we obtain

$$\frac{dV}{d\alpha} = kL^2\big[-\cos\alpha + 4\sin\alpha(1 - \cos\alpha)\big].$$

From the graph of this function (Fig. 11.19), we estimate that the system is in equilibrium when $\alpha = 43°$.

The slope of $dV/d\alpha$, which is the second derivative of V, is positive at $\alpha = 43°$. The equilibrium position is therefore stable.

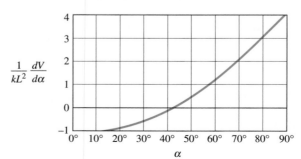

Figure 11.19
Graph of the derivative of V.

Chapter Summary

Work

The *work* done by a force $\mathbf{F}$ as a result of a displacement $d\mathbf{r}$ of its point of application is defined by

$$dU = \mathbf{F} \cdot d\mathbf{r}. \qquad \text{Eq. (11.1)}$$

The work done by a counterclockwise couple M due to a counterclockwise rotation $d\alpha$ is

$$dU = M \, d\alpha. \qquad \text{Eq. (11.2)}$$

Principle of Virtual Work

If an object is in equilibrium, the *virtual work* done by the external forces and couples acting on it is zero for any *virtual translation* or *rotation*:

$$\delta U = 0. \qquad \text{Eq. (11.9)}$$

Potential Energy

If a function of position V exists such that for any displacement $d\mathbf{r}$, the work done by a force $\mathbf{F}$ is

$$dU = \mathbf{F} \cdot d\mathbf{r} = -dV, \qquad \text{Eq. (11.10)}$$

V is called the *potential energy* associated with the force, and $\mathbf{F}$ is said to be *conservative*.

The potential energy associated with the weight W of an object is

$$V = Wy, \qquad \text{Eq. (11.12)}$$

where y is the height of the center of mass above some reference level, or *datum*.

The potential energy associated with the force exerted by a linear spring is

$$V = \frac{1}{2} kS^2, \qquad \text{Eq. (11.14)}$$

where k is the spring constant and S is the stretch of the spring.

Principle of Virtual Work for Conservative Forces

An object or a system of interconnected objects is *conservative* if the external forces and couples that do work are conservative and internal forces at the connections between objects either do no work or are conservative. The change in the total potential energy resulting from any virtual motion of a conservative object or system is zero:

$$\delta V = 0. \qquad \text{Eq. (11.15)}$$

If the position of an object or a system can be specified by a single coordinate q, it is said to have one *degree of freedom*. When a conservative, one-degree-of-freedom object or system is in equilibrium,

$$\frac{dV}{dq} = 0. \qquad \text{Eq. (11.16)}$$

If the second derivative of V is positive, the equilibrium position is stable, and if the second derivative of V is negative, it is unstable.

Review Problems

11.1 (a) Determine the couple exerted on the beam at A.
(b) Determine the vertical force exerted on the beam at A.

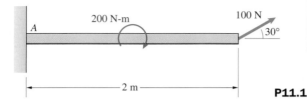

P11.1

11.2 The structure is subjected to a 20 kN-m couple. Determine the horizontal reaction at C.

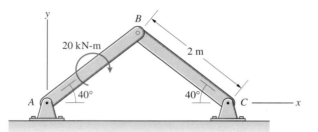

P11.2

11.3 The "rack and pinion" mechanism is used to exert a vertical force on a sample at A for a stamping operation. If a force $F = 30$ lb is exerted on the handle, use the principle of virtual work to determine the force exerted on the sample.

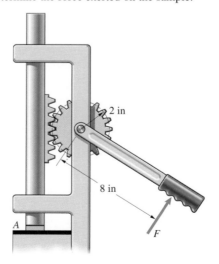

P11.3

11.4 If you were assigned to calculate the force exerted on the bolt by the pliers when the grips are subjected to forces F as shown in Fig. a, you could carefully measure the dimensions, draw free-body diagrams, and use the equilibrium equations. But another approach would be to measure the change in the distance between the jaws when the distance between the handles is changed by a small amount. If your measurements indicate that the distance d in Fig. b decreases by 1 mm when D is decreased 8 mm, what is the approximate value of the force exerted on the bolt by each jaw when the forces F are applied?

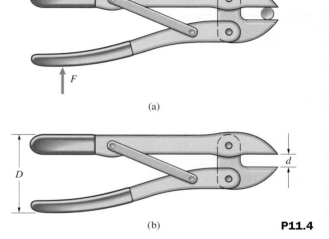

(a)

(b)

P11.4

11.5 The system is in equilibrium. The total weight of the suspended load and assembly A is 300 lb.

(a) By using equilibrium, determine the force F.
(b) Using the result of (a) and the principle of virtual work, determine the distance the suspended load rises if the cable is pulled downward 1 ft at B.

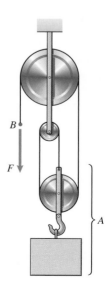

P11.5

11.6 The system is in equilibrium.

(a) By drawing free-body diagrams and using equilibrium equations, determine the couple M.
(b) Using the result of (a) and the principle of virtual work, determine the angle through which pulley B rotates if pulley A rotates through an angle α.

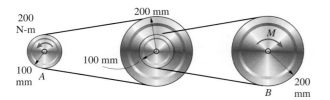

P11.6

11.7 The mechanism is in equilibrium. Neglect friction between the horizontal bar and the collar. Determine M in terms of F, α, and L.

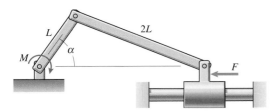

P11.7

11.8 In an injection casting machine, a couple M applied to arm AB exerts a force on the injection piston at C. Given that the horizontal component of the force exerted at C is 4 kN, use the principle of virtual work to determine M.

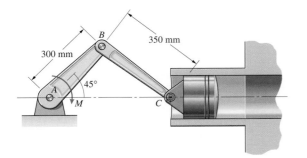

P11.8

11.9 Show that if bar AB is subjected to a clockwise virtual rotation $\delta\alpha$, bar CD undergoes a counterclockwise virtual rotation $(b/a)\,\delta\alpha$.

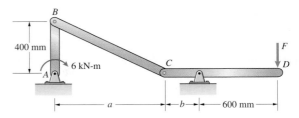

P11.9

11.10 The system in Problem 11.9 is in equilibrium, $a = 800$ mm, and $b = 400$ mm. Use the principle of virtual work to determine the force F.

11.11 Show that if bar AB is subjected to a clockwise virtual rotation $\delta\alpha$, bar CD undergoes a clockwise virtual rotation $\left[ad/(ac + bc - bd)\right]\delta\alpha$.

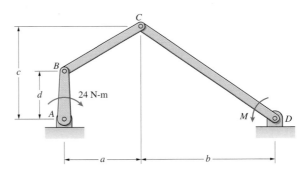

P11.11

11.12 The system in Problem 11.11 is in equilibrium, $a = 300$ mm, $b = 350$ mm, $c = 350$ mm, and $d = 200$ mm. Use the principle of virtual work to determine the couple M.

11.13 The mass of the bar is 10 kg, and it is 1 m in length. Neglect the masses of the two collars. The spring is unstretched when the bar is vertical ($\alpha = 0$), and the spring constant is $k = 100$ N/m. Determine the values of α at which the bar is in equilibrium.

P11.13

11.14 Determine whether the equilibrium positions of the bar in Problem 11.13 are stable or unstable.

11.15 The spring is unstretched when $\alpha = 90°$. Determine the value of α in the range $0 < \alpha < 90°$ for which the system is in equilibrium.

11.16 Determine whether the equilibrium position found in Problem 11.15 is stable or unstable.

11.17 The hydraulic cylinder C exerts a horizontal force at A, raising the weight W. Determine the magnitude of the force the hydraulic cylinder must exert to support the weight in terms of W and α.

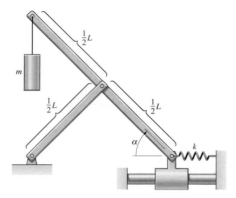

P11.15

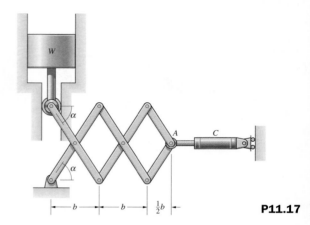

P11.17

A.1 Algebra

Quadratic Equations

The solutions of the quadratic equation

$$ax^2 + bx + c = 0$$

are

$$x = \frac{-b \pm \sqrt{b^2 - 4ac}}{2a}.$$

Natural Logarithms

The nautural logarithm of a positive real number x is denoted by $\ln x$. It is defined to be the number such that

$$e^{\ln x} = x,$$

where $e = 2.7182 \ldots$ is the base of natural logarithms.

Logarithms have the following properties:

$$\ln(xy) = \ln x + \ln y,$$

$$\ln(x/y) = \ln x - \ln y,$$

$$\ln y^x = x \ln y.$$

A.2 Trigonometry

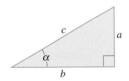

The trigonometric functions for a right triangle are

$$\sin\alpha = \frac{1}{\csc\alpha} = \frac{a}{c}, \qquad \cos\alpha = \frac{1}{\sec\alpha} = \frac{b}{c}, \qquad \tan\alpha = \frac{1}{\cot\alpha} = \frac{a}{b}.$$

The sine and cosine satisfy the relation

$$\sin^2\alpha + \cos^2\alpha = 1,$$

and the sine and cosine of the sum and difference of two angles satisfy

$$\sin(\alpha + \beta) = \sin\alpha\cos\beta + \cos\alpha\sin\beta,$$

$$\sin(\alpha - \beta) = \sin\alpha\cos\beta - \cos\alpha\sin\beta,$$

$$\cos(\alpha + \beta) = \cos\alpha\cos\beta - \sin\alpha\sin\beta,$$

$$\cos(\alpha - \beta) = \cos\alpha\cos\beta + \sin\alpha\sin\beta.$$

The **law of cosines** for an arbitrary triangle is

$$c^2 = a^2 + b^2 - 2ab\cos\alpha_c,$$

and the **law of sines** is

$$\frac{\sin\alpha_a}{a} = \frac{\sin\alpha_b}{b} = \frac{\sin\alpha_c}{c}.$$

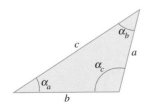

A.3 Derivatives

$$\frac{d}{dx}x^n = nx^{n-1}$$

$$\frac{d}{dx}\sin x = \cos x$$

$$\frac{d}{dx}\sinh x = \cosh x$$

$$\frac{d}{dx}e^x = e^x$$

$$\frac{d}{dx}\cos x = -\sin x$$

$$\frac{d}{dx}\cosh x = \sinh x$$

$$\frac{d}{dx}\ln x = \frac{1}{x}$$

$$\frac{d}{dx}\tan x = \frac{1}{\cos^2 x}$$

$$\frac{d}{dx}\tanh x = \frac{1}{\cosh^2 x}$$

A.4 Integrals

$$\int x^n \, dx = \frac{x^{n+1}}{n+1} \qquad (n \neq -1)$$

$$\int x^{-1} \, dx = \ln x$$

$$\int (a + bx)^{1/2} \, dx = \frac{2}{3b}(a + bx)^{3/2}$$

$$\int x(a + bx)^{1/2} \, dx = -\frac{2(2a - 3bx)(a + bx)^{3/2}}{15b^2}$$

$$\int (1 + a^2x^2)^{1/2} \, dx = \frac{1}{2}\left\{ x(1 + a^2x^2)^{1/2} + \frac{1}{a}\ln\left[x + \left(\frac{1}{a^2} + x^2 \right)^{1/2} \right] \right\}$$

$$\int x(1 + a^2x^2)^{1/2} \, dx = \frac{a}{3}\left(\frac{1}{a^2} + x^2 \right)^{3/2}$$

$$\int x^2(1 + a^2x^2)^{1/2} \, dx = \frac{1}{4}ax\left(\frac{1}{a^2} + x^2 \right)^{3/2} - \frac{1}{8a^2}x(1 + a^2x^2)^{1/2}$$
$$- \frac{1}{8a^3}\ln\left[x + \left(\frac{1}{a^2} + x^2 \right)^{1/2} \right]$$

$$\int (1 - a^2x^2)^{1/2} \, dx = \frac{1}{2}\left[x(1 - a^2x^2)^{1/2} + \frac{1}{a}\arcsin ax \right]$$

$$\int x(1 - a^2x^2)^{1/2} \, dx = -\frac{a}{3}\left(\frac{1}{a^2} - x^2 \right)^{3/2}$$

$$\int x^2(a^2 - x^2)^{1/2} \, dx = -\frac{1}{4}x(a^2 - x^2)^{3/2}$$
$$+ \frac{1}{8}a^2\left[x(a^2 - x^2)^{1/2} + a^2\arcsin \frac{x}{a} \right]$$

$$\int \frac{dx}{(1 + a^2x^2)^{1/2}} = \frac{1}{a}\ln\left[x + \left(\frac{1}{a^2} + x^2 \right)^{1/2} \right]$$

$$\int \frac{dx}{(1 - a^2x^2)^{1/2}} = \frac{1}{a}\arcsin ax, \quad \text{or} \quad -\frac{1}{a}\arccos ax$$

$$\int \sin x \, dx = -\cos x$$

$$\int \cos x \, dx = \sin x$$

$$\int \sin^2 x \, dx = -\frac{1}{2}\sin x \cos x + \frac{1}{2}x$$

$$\int \cos^2 x \, dx = \frac{1}{2}\sin x \cos x + \frac{1}{2}x$$

$$\int \sin^3 x \, dx = -\frac{1}{3}\cos x(\sin^2 x + 2)$$

$$\int \cos^3 x \, dx = \frac{1}{3}\sin x(\cos^2 x + 2)$$

$$\int \cos^4 x \, dx = \frac{3}{8}x + \frac{1}{4}\sin 2x + \frac{1}{32}\sin 4x$$

$$\int \sin^n x \cos x \, dx = \frac{(\sin x)^{n+1}}{n + 1} \qquad (n \neq -1)$$

$$\int \sinh x \, dx = \cosh x$$

$$\int \cosh x \, dx = \sinh x$$

$$\int \tanh x \, dx = \ln \cosh x$$

$$\int e^{ax} \, dx = \frac{e^{ax}}{a}$$

$$\int xe^{ax} \, dx = \frac{e^{ax}}{a^2}(ax - 1)$$

A.5 Taylor Series

The Taylor series of a function $f(x)$ is

$$f(a + x) = f(a) + f'(a)x + \frac{1}{2!} f''(a)x^2 + \frac{1}{3!} f'''(a)x^3 + \cdots,$$

where the primes indicate derivatives.

Some useful Taylor series are

$$e^x = 1 + x + \frac{x^2}{2!} + \frac{x^3}{3!} + \cdots,$$

$$\sin(a + x) = \sin a + (\cos a)x - \frac{1}{2}(\sin a)x^2 - \frac{1}{6}(\cos a)x^3 + \cdots,$$

$$\cos(a + x) = \cos a - (\sin a)x - \frac{1}{2}(\cos a)x^2 + \frac{1}{6}(\sin a)x^3 + \cdots,$$

$$\tan(a + x) = \tan a + \left(\frac{1}{\cos^2 a}\right)x + \left(\frac{\sin a}{\cos^3 a}\right)x^2$$

$$+ \left(\frac{\sin^2 a}{\cos^4 a} + \frac{1}{3\cos^2 a}\right)x^3 + \cdots.$$

A.6 Vector Analysis

Cartesian Coordinates

The gradient of a scalar field ψ is

$$\nabla\psi = \frac{\partial\psi}{\partial x}\mathbf{i} + \frac{\partial\psi}{\partial y}\mathbf{j} + \frac{\partial\psi}{\partial z}\mathbf{k}.$$

The divergence and curl of a vector field $\mathbf{v} = v_x\mathbf{i} + v_y\mathbf{j} + v_z\mathbf{k}$ are

$$\nabla \cdot \mathbf{v} = \frac{\partial v_x}{\partial x} + \frac{\partial v_y}{\partial y} + \frac{\partial v_z}{\partial z},$$

$$\nabla \times \mathbf{v} = \begin{vmatrix} \mathbf{i} & \mathbf{j} & \mathbf{k} \\ \dfrac{\partial}{\partial x} & \dfrac{\partial}{\partial y} & \dfrac{\partial}{\partial z} \\ v_x & v_y & v_z \end{vmatrix}.$$

Cylindrical Coordinates

The gradient of a scalar field ψ is

$$\nabla\psi = \frac{\partial\psi}{\partial r}\mathbf{e}_r + \frac{1}{r}\frac{\partial\psi}{\partial\theta}\mathbf{e}_\theta + \frac{\partial\psi}{\partial z}\mathbf{e}_z.$$

The divergence and curl of a vector field $\mathbf{v} = v_r\mathbf{e}_r + v_\theta\mathbf{e}_\theta + v_z\mathbf{e}_z$ are

$$\nabla \cdot \mathbf{v} = \frac{\partial v_r}{\partial r} + \frac{v_r}{r} + \frac{1}{r}\frac{\partial v_\theta}{\partial\theta} + \frac{\partial v_z}{\partial z},$$

$$\nabla \times \mathbf{v} = \frac{1}{r}\begin{vmatrix} \mathbf{e}_r & r\mathbf{e}_\theta & \mathbf{e}_z \\ \dfrac{\partial}{\partial r} & \dfrac{\partial}{\partial\theta} & \dfrac{\partial}{\partial z} \\ v_r & rv_\theta & v_z \end{vmatrix}.$$

Properties of Areas and Lines

B.1 Areas

The coordinates of the centroid of the area A are

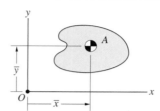

$$\bar{x} = \frac{\int_A x \, dA}{\int_A dA}, \qquad \bar{y} = \frac{\int_A y \, dA}{\int_A dA}.$$

The moment of inertia about the x axis I_x, the moment of inertia about the y axis I_y, and the product of inertia I_{xy} are

$$I_x = \int_A y^2 \, dA, \qquad I_y = \int_A x^2 \, dA, \qquad I_{xy} = \int_A xy \, dA.$$

The polar moment of inertia about O is

$$J_O = \int_A r^2 \, dA = \int_A (x^2 + y^2) \, dA = I_x + I_y.$$

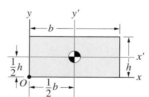

Rectangular area

Area $= bh$

$$I_x = \frac{1}{3} bh^3, \qquad I_y = \frac{1}{3} hb^3, \qquad I_{xy} = \frac{1}{4} b^2 h^2$$

$$I_{x'} = \frac{1}{12} bh^3, \qquad I_{y'} = \frac{1}{12} hb^3, \qquad I_{x'y'} = 0$$

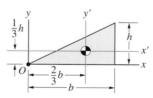

Triangular area

Area $= \frac{1}{2} bh$

$$I_x = \frac{1}{12} bh^3, \qquad I_y = \frac{1}{4} hb^3, \qquad I_{xy} = \frac{1}{8} b^2 h^2$$

$$I_{x'} = \frac{1}{36} bh^3, \qquad I_{y'} = \frac{1}{36} hb^3, \qquad I_{x'y'} = \frac{1}{72} b^2 h^2$$

Triangular area

$$\text{Area} = \frac{1}{2} bh \qquad I_x = \frac{1}{12} bh^3, \qquad I_{x'} = \frac{1}{36} bh^3$$

Circular area

$$\text{Area} = \pi R^2 \qquad I_{x'} = I_{y'} = \frac{1}{4} \pi R^4, \qquad I_{x'y'} = 0$$

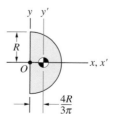

Semicircular area

$$\text{Area} = \frac{1}{2} \pi R^2 \qquad I_x = I_y = \frac{1}{8} \pi R^4, \qquad I_{xy} = 0$$

$$I_{x'} = \frac{1}{8} \pi R^4, \qquad I_{y'} = \left(\frac{\pi}{8} - \frac{8}{9\pi} \right) R^4, \qquad I_{x'y'} = 0$$

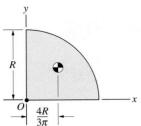

Quarter-circular area

$$\text{Area} = \frac{1}{4}\pi R^2 \qquad I_x = I_y = \frac{1}{16}\pi R^4, \qquad I_{xy} = \frac{1}{8}R^4$$

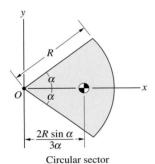

Circular sector

$$\text{Area} = \alpha R^2$$

$$I_x = \frac{1}{4}R^4\left(\alpha - \frac{1}{2}\sin 2\alpha\right), \qquad I_y = \frac{1}{4}R^4\left(\alpha + \frac{1}{2}\sin 2\alpha\right),$$

$$I_{xy} = 0$$

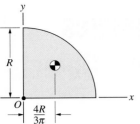

Quarter-elliptical area

$$\text{Area} = \frac{1}{4}\pi ab$$

$$I_x = \frac{1}{16}\pi ab^3, \qquad I_y = \frac{1}{16}\pi a^3 b, \qquad I_{xy} = \frac{1}{8}a^2 b^2$$

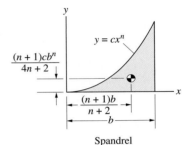

Spandrel

$$\text{Area} = \frac{cb^{n+1}}{n+1}$$

$$I_x = \frac{c^3 b^{3n+1}}{9n+3}, \qquad I_y = \frac{cb^{n+3}}{n+3}, \qquad I_{xy} = \frac{c^2 b^{2n+2}}{4n+4}$$

B.2 Lines

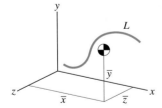

The coordinates of the centroid of the line L are

$$\bar{x} = \frac{\int_L x\,dL}{\int_L dL}, \qquad \bar{y} = \frac{\int_L y\,dL}{\int_L dL}, \qquad \bar{z} = \frac{\int_L z\,dL}{\int_L dL}.$$

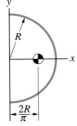

Semicircular arc

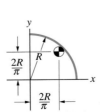

Quarter-circular arc

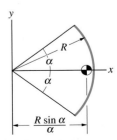

Circular arc

APPENDIX C

Properties of Volumes and Homogeneous Objects

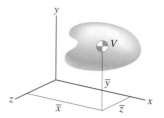

The coordinates of the centroid of the volume V are

$$\bar{x} = \frac{\displaystyle\int_V x\, dV}{\displaystyle\int_V dV}, \qquad \bar{y} = \frac{\displaystyle\int_V y\, dV}{\displaystyle\int_V dV}, \qquad \bar{z} = \frac{\displaystyle\int_V z\, dV}{\displaystyle\int_V dV}.$$

(The center of mass of a homogeneous object coincides with the centroid of its volume.)

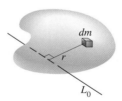

The mass moment of inertia of the object about the axis L_0 is

$$I_0 = \int_m r^2\, dm.$$

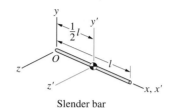

Slender bar

$$I_{(x\ \text{axis})} = 0, \qquad I_{(y\ \text{axis})} = I_{(z\ \text{axis})} = \frac{1}{3}\, ml^2$$

$$I_{(x'\ \text{axis})} = 0, \qquad I_{(y'\ \text{axis})} = I_{(z'\ \text{axis})} = \frac{1}{12}\, ml^2$$

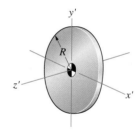

Thin circular plate

$$I_{(x'\ \text{axis})} = I_{(y'\ \text{axis})} = \frac{1}{4}\, mR^2, \qquad I_{(z'\ \text{axis})} = \frac{1}{2}\, mR^2$$

$$I_{(x\ \text{axis})} = \frac{1}{3}\, mh^2, \qquad I_{(y\ \text{axis})} = \frac{1}{3}\, mb^2, \qquad I_{(z\ \text{axis})} = \frac{1}{3}\, m(b^2 + h^2)$$

$$I_{(x'\ \text{axis})} = \frac{1}{12}\, mh^2, \qquad I_{(y'\ \text{axis})} = \frac{1}{12}\, mb^2, \qquad I_{(z'\ \text{axis})} = \frac{1}{12}\, m(b^2 + h^2)$$

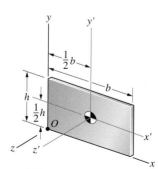

Thin rectangular plate

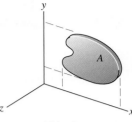

Thin plate

$$I_{(x\text{ axis})} = \frac{m}{A}\, I_x^A, \qquad I_{(y\text{ axis})} = \frac{m}{A}\, I_y^A, \qquad I_{(z\text{ axis})} = I_{(x\text{ axis})} + I_{(y\text{ axis})}$$

(The superscripts A denote moments of inertia of the plate's cross-sectional area A.)

Volume $= abc$

$$I_{(x'\text{ axis})} = \frac{1}{12}\, m(a^2 + b^2), \qquad I_{(y'\text{ axis})} = \frac{1}{12}\, m(a^2 + c^2),$$

$$I_{(z'\text{ axis})} = \frac{1}{12}\, m(b^2 + c^2),$$

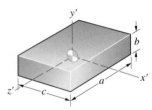

Rectangular prism

Volume $= \pi R^2 l$

$$I_{(x\text{ axis})} = I_{(y\text{ axis})} = m\left(\frac{1}{3} l^2 + \frac{1}{4} R^2\right), \qquad I_{(z\text{ axis})} = \frac{1}{2}\, mR^2$$

$$I_{(x'\text{ axis})} = I_{(y'\text{ axis})} = m\left(\frac{1}{12} l^2 + \frac{1}{4} R^2\right), \qquad I_{(z'\text{ axis})} = \frac{1}{2}\, mR^2$$

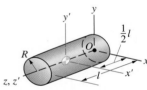

Circular cylinder

Volume $= \frac{1}{3}\, \pi R^2 h$

$$I_{(x\text{ axis})} = I_{(y\text{ axis})} = m\left(\frac{3}{5} h^2 + \frac{3}{20} R^2\right), \qquad I_{(z\text{ axis})} = \frac{3}{10}\, mR^2$$

$$I_{(x'\text{ axis})} = I_{(y'\text{ axis})} = m\left(\frac{3}{80} h^2 + \frac{3}{20} R^2\right), \qquad I_{(z'\text{ axis})} = \frac{3}{10}\, mR^2$$

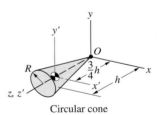

Circular cone

Volume $= \frac{4}{3}\, \pi R^3$

$$I_{(x'\text{ axis})} = I_{(y'\text{ axis})} = I_{(z'\text{ axis})} = \frac{2}{5}\, mR^2$$

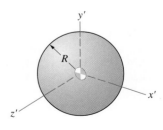

Sphere

Chapter 2

2.2 $|\mathbf{A}| = 1110$ lb, $\alpha = 29.7°$.

2.4 $|\mathbf{E}| = 313$ lb, $|\mathbf{F}| = 140$ lb.

2.6 $\mathbf{e}_{AB} = 0.625\mathbf{i} - 0.469\mathbf{j} - 0.625\mathbf{k}$.

2.8 $\mathbf{F}_p = 8.78\mathbf{i} - 6.59\mathbf{j} - 8.78\mathbf{k}$ (lb).

2.10 $\mathbf{r}_{BA} \times \mathbf{F} = -70\mathbf{i} + 40\mathbf{j} - 100\mathbf{k}$ (ft-lb).

2.12   (a), (b) $686\mathbf{i} - 486\mathbf{j} - 514\mathbf{k}$ (ft-lb).

2.14 $\mathbf{F}_A = 18.2\mathbf{i} + 19.9\mathbf{j} + 15.3\mathbf{k}$ (N),
  $\mathbf{F}_B = -7.76\mathbf{i} + 26.9\mathbf{j} + 13.4\mathbf{k}$ (N).

2.16 $\mathbf{F}_p = 1.29\mathbf{i} - 3.86\mathbf{j} + 2.57\mathbf{k}$ (kN),
  $\mathbf{F}_n = -1.29\mathbf{i} - 2.14\mathbf{j} - 2.57\mathbf{k}$ (kN).

2.18 $\mathbf{r}_{AG} \times \mathbf{W} = -16.4\mathbf{i} - 82.4\mathbf{k}$ (N-m).

2.20 $\mathbf{r}_{BC} \times \mathbf{T} = -12.0\mathbf{i} - 138.4\mathbf{j} - 117.4\mathbf{k}$ (N-m).

2.22   20.8 kN.

2.24 $34.9°$.

Chapter 3

3.2 $W = 25.0$ lb.

3.4   (a) 83.9 lb; (b) 230.5 lb.

3.6 $T = mg/26$.

3.8 $F = 162.0$ N.

3.10 $T_{AB} = 420$ N, $T_{AC} = 533$ N, $|\mathbf{F}_S| = 969$ N.

3.12 $T = mgL/(R + h)$.

3.14 $T_{AB} = 1.54$ lb, $T_{AC} = 1.85$ lb.

3.16   Normal force $= 12.15$ kN,
  friction force $= 4.03$ kN.

Chapter 4

4.2   (a) 160 N-m; (b) $160\mathbf{k}$ (N-m).

4.4   No. The moment is $mg \sin \alpha$ counterclockwise, where α is the clockwise angle.

4.6   (a) -76.2 N-m; (b) -66.3 N-m.

4.8 $|\mathbf{F}| = 224$ lb, $|\mathbf{M}| = 1600$ ft-lb.

4.10   671 lb.

4.12 $-228.1\mathbf{i} - 68.4\mathbf{k}$ (N-m).

4.14 $\mathbf{M}_{(x\,\text{axis})} = -153\mathbf{i}$ (ft-lb).

4.16 $\mathbf{M}_{CD} = -173\mathbf{i} + 1038\mathbf{k}$ (ft-lb).

4.18   (a) $T_{AB} = T_{CD} = 173$ lb;
  (b) $\mathbf{F} = 300\mathbf{j}$ (lb), intersects at $x = 4$ ft.

4.20 $\mathbf{F} = -20\mathbf{i} + 70\mathbf{j}$ (N), $M = 22$ N-m.

4.22 $\mathbf{F}' = -100\mathbf{i} + 40\mathbf{j} + 30\mathbf{k}$ (lb),
  $\mathbf{M} = -80\mathbf{i} + 200\mathbf{k}$ (in-lb).

4.24 $\mathbf{F} = 1166\mathbf{i} + 566\mathbf{j}$ (N), $y = 13.9$ m.

4.26 $\mathbf{F} = 190\mathbf{j}$ (N), $\mathbf{M} = -98\mathbf{i} + 184\mathbf{k}$ (N-m).

4.28 $\mathbf{F} = -0.364\mathbf{i} + 4.908\mathbf{j} + 1.090\mathbf{k}$ (kN),
  $\mathbf{M} = -0.131\mathbf{i} - 0.044\mathbf{j} + 1.112\mathbf{k}$ (kN-m).

Chapter 5

5.2 $A_x = -346.4$ N, $A_y = 47.6$ N, $B_y = 152.4$ N.

5.4   (a) There are four unknown reactions and three equilibrium equations; (b) $A_x = -50$ lb, $B_x = 50$ lb.

5.6   (b)   Force on nail $= 55$ lb, normal force $= 50.77$ lb, friction force $= 9.06$ lb.

5.8 $A = 500$ N, $B_x = 0$, $B_y = -800$ N.

5.10 $A = 727$ lb, $H_x = 225$ lb, $H_y = 113$ lb.

5.12 $\alpha = 0$ and $\alpha = 59.4°$.

5.14 $A_x = -32.0$ kN, $A_y = -61.7$ kN.

5.16   The force is 800 N upward; its line of action passes through the midpoint of the plate.

5.18 $m = 67.2$ kg.

5.20 $\alpha = 90°$, $T_{BC} = W/2$, $A = W/2$.

Chapter 6

6.2   (a) $B = 82.9$ N, $C_x = 40$ N, $C_y = -22.9$ N;
  (b) $AB : 82.9$ N (C); BC : zero; $AC : 46.1$ N (T).

6.4 $T_{AB} = 7.14$ kN (C), $T_{AC} = 5.71$ kN (T),
  $T_{BC} = 10$ kN (T).

6.6 $BC : 120$ kN (C); $BG : 42.4$ kN (T); $FG : 90$ kN (T).

6.8 $AB : 125$ lb (C); AC : zero; $BC : 188$ lb (T); $BD :$
  225 lb (C); $CD : 125$ lb (C); $CE : 225$ lb (T).

6.10 $T_{BD} = 13.3$ kN (T), $T_{CD} = 11.7$ kN (T),
  $T_{CE} = 28.3$ kN (C).

6.12 $AC : 480$ N (T); $CD : 240$ N (C); $CF : 300$ N (T).

6.14   Tension: member AC, 480 lb (T); Compression: member BD, 633 lb (C).

6.16 $CD : 11.42$ kN (C); $CJ : 4.17$ kN (C);
  $IJ : 12.00$ kN (T).

6.18   182 kg.

6.20 $A_x = -1.57$ kN, $A_y = 1.18$ kN, $B_x = 0$,
  $B_y = -2.35$ kN, $C_x = 1.57$ kN, $C_y = 1.18$ kN.

6.22   The force on the bolt is 972 N. The force at A is 576 N.

6.24 $F = 7.92$ kip; $BG : 19.35$ kip (T); $EF : 12.60$ kip (C).

6.26 $A_x = -52.33$ kN, $A_y = -43.09$ kN, $E_x = 0.81$ kN,
  $E_y = -14.86$ kN.

Chapter 7

7.2 $\bar{x} = 3/8$, $\bar{y} = 3/5$.

7.4 $\bar{x} = 87.3$ mm, $\bar{y} = 55.3$ mm.

7.6   917 N (T).

7.8 $T_B = T_C = 15.2$ kN.

7.10 $\bar{x} = 1.87$ m.

7.12 $A = 682$ in^2.

7.14 $\bar{x} = 110$ mm.

7.16 $\bar{x} = 1.70$ m.

7.18 $\bar{x} = 25.24$ mm, $\bar{y} = 8.02$ mm, $\bar{z} = 27.99$ mm.

7.20 (a) $\bar{x} = 1.511$ m. (b) $\bar{x} = 1.611$ m.

7.22 (a) $\bar{x} = 2$ ft, $\bar{y} = 2.33$ ft, $\bar{z} = 3.33$ ft.
   (b) $\bar{x} = 1.72$ ft, $\bar{y} = 2.39$ ft, $\bar{z} = 3.39$ ft.

Chapter 8

8.2 $I_y = \frac{1}{5}$, $k_y = \sqrt{\frac{3}{5}}$.

8.4 $J = \frac{26}{105}$, $k_0 = \sqrt{\frac{26}{35}}$.

8.6 $I_y = 12.8$, $k_y = 2.19$.

8.8 $I_{xy} = 2.13$.

8.10 $I_{x'} = 0.183$, $k_{x'} = 0.262$.

8.12 $I_y = 2.75 \times 10^7$ mm^4, $k_y = 43.7$ mm.

8.14 $I_x = 5.03 \times 10^7$ mm^4, $k_x = 59.1$ mm.

8.16 $I_y = 94.2$ ft^4, $k_y = 2.24$ ft.

8.18 $I_x = 396$ ft^4, $k_x = 3.63$ ft.

8.20 $\theta_p = 19.5°$, 20.3 m^4, 161 m^4.

8.22 $I_{y\,\text{axis}} = 0.0702$ kg-m^2.

8.24 $I_{z\,\text{axis}} = \frac{1}{10}mw^2$.

8.26 $I_{x\,\text{axis}} = 3.83$ slug-ft^2.

8.28 0.537 kg-m^2.

Chapter 9

9.2 (a) $f = 10.3$ lb.

9.4 $F = 290$ lb.

9.6 $\alpha = 65.7°$.

9.8 $\alpha = 24.2°$.

9.10 $b = (h/\mu_s - t)/2$.

9.12 $h = 5.82$ in.

9.14 286 lb.

9.16 1130 kg, torque = 2.67 kN-m.

9.18 $f = 2.63$ N.

9.20 $\mu_s = 0.272$.

9.22 $M = 1.13$ N-m.

9.24 $P = 43.5$ N.

9.26 146 lb.

9.28 (a) $W = 106$ lb; (b) $W = 273$ lb.

Chapter 10

10.2 (a) $P_B = 0$, $V_B = -26.7$ lb, $M_B = 160$ ft-lb;
   (b) $P_C = 0$, $V_C = -26.7$ lb, $M_C = 80$ ft-lb.

10.4

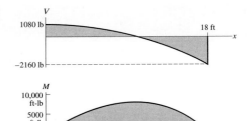

10.6

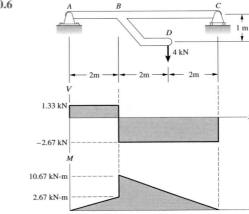

10.8 $P_A = 0$, $V_A = 8$ kN, $M_A = -8$ kN-m.

10.10 (a) $P_B = 0$, $V_B = -40$ N, $M_B = 10$ N-m;
   (b) $P_B = 0$, $V_B = -40$ N, $M_B = 10$ N-m.

10.12 $P = 0$, $V = -100$ lb, $M = -50$ ft-lb.

10.14 (a) $w = 74{,}100$ lb/ft; (b) 1.20×10^8 lb.

10.18 A : 44.2 kN to the left, 35.3 kN upward; B : 34.3 kN.

Chapter 11

11.2 $8F$.

11.4 $C_x = -7.78$ kN.

11.6 (a) $M = 800$ N-m; (b) $\alpha/4$.

11.8 $M = 1.50$ kN-m.

11.10 $F = 5$ kN.

11.12 $M = 63$ N-m.

11.14 $\alpha = 0$ is unstable and $\alpha = 59.4°$ is stable.

11.16 Unstable.

Index

C

T

ENGINEERING MECHANICS

DYNAMICS PRINCIPLES

To design and program an industrial robot, engineers must analyze its motion using the principles of dynamics. Dynamics is one of the sciences underlying the design of all machines.

Introduction

12

Engineers are responsible for the design, construction, and testing of the devices we use, from simple things such as chairs and pencil sharpeners to complicated ones such as dams, cars, airplanes, and spacecraft. They must have a deep understanding of the physics underlying these devices and must be familiar with the use of mathematical models to predict system behavior. Students of engineering begin to learn how to analyze and predict the behavior of physical systems by studying mechanics.

12.1 Engineering and Mechanics

How do engineers design complex systems and predict their characteristics before they are constructed? Engineers have always relied on their knowledge of previous designs, experiments, ingenuity, and creativity to develop new designs. Modern engineers add a powerful technique: They develop mathematical equations based on the physical characteristics of the devices they design. With these mathematical models, engineers predict the behavior of their designs, modify them, and test them prior to their actual construction. Aerospace engineers use mathematical models to predict the paths the space shuttle will follow in flight. Civil engineers use mathematical models to analyze the effects of loads on buildings and foundations.

At its most basic level, mechanics is the study of forces and their effects. Elementary mechanics is divided into *statics*, the study of objects in equilibrium, and *dynamics*, the study of objects in motion. The results obtained in elementary mechanics apply directly to many fields of engineering. Mechanical and civil engineers who design structures use the equilibrium equations derived in statics. Civil engineers who analyze the responses of buildings to earthquakes and aerospace engineers who determine the trajectories of satellites use the equations of motion derived in dynamics.

Mechanics was the first analytical science; consequently fundamental concepts, analytical methods, and analogies from mechanics are found in virtually every field of engineering. Students of chemical and electrical engineering gain a deeper appreciation for basic concepts in their fields such as equilibrium, energy, and stability by learning them in their original mechanical contexts. By studying mechanics, they retrace the historical development of these ideas.

12.2 Learning Mechanics

Mechanics consists of broad principles that govern the behavior of objects. In this book we describe these principles and provide examples that demonstrate some of their applications. Although it is essential that you practice working problems similar to these examples, and we include many problems of this kind, our objective is to help you understand the principles well enough to apply them to situations that are new to you. Each generation of engineers confronts new problems.

Problem Solving

In the study of mechanics you learn problem-solving procedures you will use in succeeding courses and throughout your career. Although different types of problems require different approaches, the following steps apply to many of them:

- Identify the information that is given and the information, or answer, you must determine. It's often helpful to restate the problem in your own words. When appropriate, make sure you understand the physical system or model involved.

- Develop a *strategy* for the problem. This means identifying the principles and equations that apply and deciding how you will use them to solve the problem. Whenever possible, draw diagrams to help visualize and solve the problem.

- Whenever you can, try to predict the answer. This will develop your intuition and will often help you recognize an incorrect answer.

- Solve the equations and, whenever possible, interpret your results and compare them with your prediction. This last step is a *reality check*. Is your answer reasonable?

Calculators and Computers

Most of the problems in this book are designed to lead to an algebraic expression with which to calculate the answer in terms of given quantities. A calculator with trigonometric and logarithmic functions is sufficient to determine the numerical value of such answers. The use of a programmable calculator or a computer with problem-solving software such as *Mathcad* or MATLAB is convenient, but be careful not to become too reliant on tools you will not have during tests.

Sections headed "Computational Mechanics" contain examples and problems that are suitable for solution with a programmable calculator or a computer.

Engineering Applications

Although the problems are designed primarily to help you learn mechanics, many of them illustrate uses of mechanics in engineering. Sections headed "Application to Engineering" describe how mechanics is applied in various fields of engineering.

We also include problems that emphasize two essential aspects of engineering:

- *Design.* Some problems ask you to choose values of parameters to satisfy stated design criteria.

- *Safety.* Some problems ask you to evaluate the safety of devices and choose values of parameters to satisfy stated safety requirements.

Subsequent Use of This Text

This book contains tables and information you will find useful in subsequent engineering courses and throughout your engineering career. In addition, you will often want to review fundamental engineering subjects, both during the remainder of your formal education and when you are a practicing engineer. The most efficient way to do so is by using the textbooks with which you are familiar. Your engineering textbooks will form the core of your professional library.

12.3 Fundamental Concepts

Some topics in mechanics will be familiar to you from everyday experience or from previous exposure to them in mathematics and physics courses. In this section we briefly review the foundations of elementary mechanics.

Numbers

Engineering measurements, calculations, and results are expressed in numbers. You need to know how we express numbers in the examples and problems and how to express the results of your own calculations.

Significant Digits   This term refers to the number of meaningful (that is, accurate) digits in a number, counting to the right starting with the first nonzero digit. The two numbers 7.630 and 0.007630 are each stated to four significant digits. If only the first four digits in the number 7,630,000 are known to be accurate, this can be indicated by writing the number in scientific notation as 7.630×10^6.

If a number is the result of a measurement, the significant digits it contains are limited by the accuracy of the measurement. If the result of a measurement is stated to be 2.43, this means that the actual value is believed to be closer to 2.43 than to 2.42 or 2.44.

Numbers may be rounded off to a certain number of significant digits. For example, we can express the value of π to three significant digits, 3.14, or we can express it to six significant digits, 3.14159. When you use a calculator or computer, the number of significant digits is limited by the number of digits the machine is designed to carry.

Use of Numbers in This Book   You should treat numbers given in problems as exact values and not be concerned about how many significant digits they contain. If a problem states that a quantity equals 32.2, you can assume its value is 32.200. ... We express intermediate results and answers in the examples and the answers to the problems to at least three significant digits. If you use a calculator, your results should be that accurate. Be sure to avoid round-off errors that occur if you round off intermediate results when making a series of calculations. Instead, carry through your calculations with as much accuracy as you can by retaining values in your calculator.

Space and Time

Space simply refers to the three-dimensional universe in which we live. Our daily experiences give us an intuitive notion of space and the locations, or positions, of points in space. The distance between two points in space is the length of the straight line joining them.

Measuring the distance between points in space requires a unit of length. We use both the International System of units, or SI units, and U.S. Customary units. In SI units, the unit of length is the meter (m). In U.S. Customary units, the unit of length is the foot (ft).

Time is, of course, familiar—our lives are measured by it. The daily cycles of light and darkness and the hours, minutes, and seconds measured by our clocks and watches give us an intuitive notion of time. Time is measured by the intervals between repeatable events, such as the swings of a clock pendulum or the vibrations of a quartz crystal in a watch. In both SI units and U.S. Customary units, the unit of time is the second (s). The minute (min), hour (hr), and day are also frequently used.

If the position of a point in space relative to some reference point changes with time, the rate of change of its position is called its *velocity*, and the rate of change of its velocity is called its *acceleration*. In SI units, the velocity is expressed in meters per second (m/s) and the acceleration is expressed in meters per second per second, or meters per second squared (m/s^2). In U.S. Customary units, the velocity is expressed in feet per second (ft/s) and the acceleration is expressed in feet per second squared (ft/s^2).

Newton's Laws

Elementary mechanics was established on a firm basis with the publication in 1687 of *Philosophiae naturalis principia mathematica*, by Isaac Newton. Although highly original, it built on fundamental concepts developed by many others during a long and difficult struggle toward understanding (Fig. 12.1).

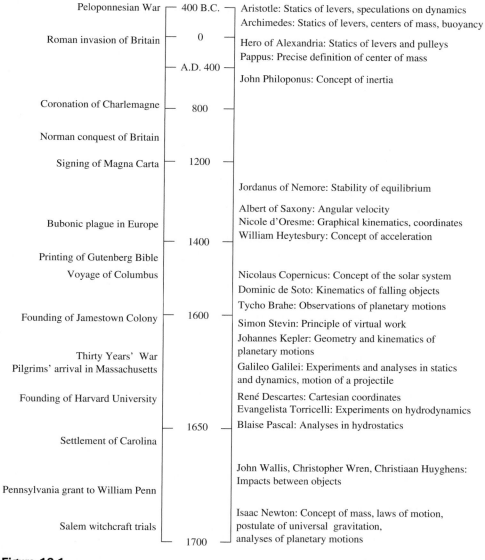

Figure 12.1

Chronology of developments in mechanics up to the publication of Newton's *Principia* in relation to other events in history.

Newton stated three "laws" of motion, which we express in modern terms:

1. *When the sum of the forces acting on a particle is zero, its velocity is constant. In particular, if the particle is initially stationary, it will remain stationary.*

2. *When the sum of the forces acting on a particle is not zero, the sum of the forces is equal to the rate of change of the linear momentum of the particle. If the mass is constant, the sum of the forces is equal to the product of the mass of the particle and its acceleration.*

3. *The forces exerted by two particles on each other are equal in magnitude and opposite in direction.*

Notice that we did not define force and mass before stating Newton's laws. The modern view is that these terms are defined by the second law. To demonstrate, suppose that we choose an arbitrary object and define it to have unit mass. Then we define a unit of force to be the force that gives our unit mass an acceleration of unit magnitude. In principle, we can then determine the mass of any object: We apply a unit force to it, measure the resulting acceleration, and use the second law to determine the mass. We can also determine the magnitude of any force: We apply it to our unit mass, measure the resulting acceleration, and use the second law to determine the force.

Thus Newton's second law gives precise meanings to the terms *mass* and *force*. In SI units, the unit of mass is the kilogram (kg). The unit of force is the newton (N), which is the force required to give a mass of one kilogram an acceleration of one meter per second squared. In U.S. Customary units, the unit of force is the pound (lb). The unit of mass is the slug, which is the amount of mass accelerated at one foot per second squared by a force of one pound.

Although the results we discuss in this book are applicable to many of the problems met in engineering practice, there are limits to the validity of Newton's laws. For example, they don't give accurate results if a problem involves velocities that are not small compared to the velocity of light $(3 \times 10^8 \text{ m/s})$. Einstein's special theory of relativity applies to such problems. Elementary mechanics also fails in problems involving dimensions that are not large compared to atomic dimensions. Quantum mechanics must be used to describe phenomena on the atomic scale.

Study Questions

1. What is the definition of the significant digits of a number?
2. What are the units of length, mass, and force in the SI system?

12.4 Units

The SI system of units has become nearly standard throughout the world. In the United States, U.S. Customary units are also used. In this section we summarize these two systems of units and explain how to convert units from one system to another.

International System of Units

In SI units, length is measured in meters (m) and mass in kilograms (kg). Time is measured in seconds (s), although other familiar measures such as minutes (min), hours (hr), and days are also used when convenient. Meters, kilograms, and seconds are called the *base units* of the SI system. Force is measured in newtons (N). Recall that these units are related by Newton's second law: One newton is the force required to give an object of one kilogram mass an acceleration of one meter per second squared:

$$1 \text{ N} = (1 \text{ kg})(1 \text{ m/s}^2) = 1 \text{ kg-m/s}^2.$$

Because the newton can be expressed in terms of the base units, it is called a *derived unit.*

To express quantities by numbers of convenient size, multiples of units are indicated by prefixes. The most common prefixes, their abbreviations, and the multiples they represent are shown in Table 12.1. For example, 1 km is 1 kilometer, which is 1000 m, and 1 Mg is 1 megagram, which is 10^6 g, or 1000 kg. We frequently use kilonewtons (kN).

Table 12.1 The common prefixes used in SI units and the multiples they represent.

Prefix	Abbreviation	Multiple
nano-	n	10^{-9}
micro-	μ	10^{-6}
milli-	m	10^{-3}
kilo-	k	10^3
mega-	M	10^6
giga-	G	10^9

U.S. Customary Units

In U.S. Customary units, length is measured in feet (ft) and force is measured in pounds (lb). Time is measured in seconds (s). These are the base units of the U.S. Customary system. In this system of units, mass is a derived unit. The unit of mass is the slug, which is the mass of material accelerated at one foot per second squared by a force of one pound. Newton's second law states that

$$1 \text{ lb} = (1 \text{ slug})(1 \text{ ft/s}^2).$$

From this expression we obtain

$$1 \text{ slug} = 1 \text{ lb-s}^2/\text{ft}.$$

We use other U.S. Customary units such as the mile (1 mi = 5280 ft) and the inch (1 ft = 12 in.). We also use the kilopound (kip), which is 1000 lb.

Angular Units

In both SI and U.S. Customary units, angles are normally expressed in radians (rad). We show the value of an angle θ in radians in Fig. 12.2. It is defined to be the ratio of the part of the circumference subtended by θ to the radius of the circle. Angles are also expressed in degrees. Since there are 360 degrees

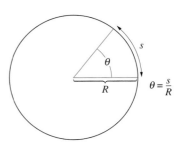

Figure 12.2
Definition of an angle in radians.

(360°) in a complete circle, and the complete circumference of the circle is $2\pi R$, 360° equals 2π rad.

Equations containing angles are nearly always derived under the assumption that angles are expressed in radians. Therefore when you want to substitute the value of an angle expressed in degrees into an equation, you should first convert it into radians. A notable exception to this rule is that many calculators are designed to accept angles expressed in either degrees or radians when you use them to evaluate functions such as $\sin\theta$.

Conversion of Units

Many situations arise in engineering practice that require you to convert values expressed in units of one kind into values in other units. If some data in a problem are given in terms of SI units and some are given in terms of U.S. Customary units, you must express all of the data in terms of one system of units. In problems expressed in terms of SI units, you will occasionally be given data in terms of units other than the base units of seconds, meters, kilograms, and newtons. You should convert these data into the base units before working the problem. Similarly, in problems involving U.S. Customary units, you should convert terms into the base units of seconds, feet, slugs, and pounds. After you gain some experience, you will recognize situations in which these rules can be relaxed, but for now the procedure we propose is the safest.

Converting units is straightforward, although you must do it with care. Suppose that we want to express 1 mi/hr in terms of ft/s. Since one mile equals 5280 ft and one hour equals 3600 seconds, we can treat the expressions

$$\left(\frac{5280 \text{ ft}}{1 \text{ mi}}\right) \quad \text{and} \quad \left(\frac{1 \text{ hr}}{3600 \text{ s}}\right)$$

as ratios whose values are 1. In this way we obtain

$$1 \text{ mi/hr} = 1 \text{ mi/hr} \times \left(\frac{5280 \text{ ft}}{1 \text{ mi}}\right) \times \left(\frac{1 \text{ hr}}{3600 \text{ s}}\right) = 1.47 \text{ ft/s}.$$

We give some useful unit conversions in Table 12.2.

Table 12.2 Unit conversions.

Time	1 minute	=	60 seconds
	1 hour	=	60 minutes
	1 day	=	24 hours
Length	1 foot	=	12 inches
	1 mile	=	5280 feet
	1 inch	=	25.4 millimeters
	1 foot	=	0.3048 meters
Angle	2π radians	=	360 degrees
Mass	1 slug	=	14.59 kilograms
Force	1 pound	=	4.448 newtons

Study Questions

1. What are the base units of the SI and U.S. Customary systems?
2. What is the definition of an angle in radians?

Example 12.1

Converting Units of Pressure

The pressure exerted at a point of the hull of the deep submersible in Fig. 12.3 is 3.00×10^6 Pa (pascals). A pascal is 1 newton per square meter. Determine the pressure in pounds per square foot.

Figure 12.3
Deep Submersible Vehicle.

Strategy

From Table 12.2, 1 pound = 4.448 newtons and 1 foot = 0.3048 meters. With these unit conversions we can calculate the pressure in pounds per square foot.

Solution

The pressure (to three significant digits) is

$$3.00 \times 10^6 \text{ N/m}^2 = 3.00 \times 10^6 \text{ N/m}^2 \times \left(\frac{1 \text{ lb}}{4.448 \text{ N}} \right) \times \left(\frac{0.3048 \text{ m}}{1 \text{ ft}} \right)^2$$

$$= 62{,}700 \text{ lb/ft}^2.$$

Discussion

From the table of unit conversions in the inside front cover, $1 \text{ Pa} = 0.0209 \text{ lb/ft}^2$. Therefore an alternative solution is

$$3.00 \times 10^6 \text{ N/m}^2 = 3.00 \times 10^6 \text{ N/m}^2 \times \left(\frac{0.0209 \text{ lb/ft}^2}{1 \text{ N/m}^2} \right)$$

$$= 62{,}700 \text{ lb/ft}^2.$$

Example 12.2

Determining Units from an Equation

Suppose that in Einstein's equation

$$E = mc^2,$$

the mass m is in kilograms and the velocity of light c is in meters per second.
(a) What are the SI units of E?
(b) If the value of E in SI units is 20, what is its value in U.S. Customary base units?

Strategy

(a) Since we know the units of the terms m and c, we can deduce the units of E from the given equation.
(b) We can use the unit conversions for mass and length from Table 12.2 to convert E from SI units to U.S. Customary units.

Solution

(a) From the equation for E,

$$E = (m \text{ kg})(c \text{ m/s})^2.$$

the SI units of E are kg-m^2/s^2.
(b) From Table 12.2, 1 slug $= 14.59$ kg and 1 ft $= 0.3048$ m. Therefore

$$1 \text{ kg-m}^2/\text{s}^2 = 1 \text{ kg-m}^2/\text{s}^2 \times \left(\frac{1 \text{ slug}}{14.59 \text{ kg}} \right) \times \left(\frac{1 \text{ ft}}{0.3048 \text{ m}} \right)^2$$

$$= 0.738 \text{ slug-ft}^2/\text{s}^2.$$

The value of E in U.S. Customary units is

$$E = (20)(0.738) = 14.8 \text{ slug-ft}^2/\text{s}^2.$$

Discussion

In part (a) we determined the units of E by using the fact that an equation must be dimensionally consistent. That is, the dimensions, or units, of each term must be the same.

12.5 Newtonian Gravitation

Newton postulated that the gravitational force between two particles of mass m_1 and m_2 that are separated by a distance r (Fig. 12.4) is

$$F = \frac{Gm_1m_2}{r^2},$$

(12.1)

Figure 12.4
The gravitational forces between two particles are equal in magnitude and directed along the line between them.

where G is called the universal gravitational constant. Based on this postulate, he calculated the gravitational force between a particle of mass m_1 and a homogeneous sphere of mass m_2 and found that it is also given by Eq. (12.1), with r denoting the distance from the particle to the center of the sphere. Although the earth is not a homogeneous sphere, we can use this result to approximate the weight of an object of mass m due to the gravitational attraction of the earth.

$$W = \frac{Gmm_E}{r^2},$$

(12.2)

where m_E is the mass of the earth and r is the distance from the center of the earth to the object. Notice that the weight of an object depends on its location relative to the center of the earth, whereas the mass of the object is a measure of the amount of matter it contains and doesn't depend on its position.

When an object's weight is the only force acting on it, the resulting acceleration is called the acceleration due to gravity. In this case, Newton's second law states that $W = ma$, and from Eq. (12.2) we see that the acceleration due to gravity is

$$a = \frac{Gm_E}{r^2}.$$

(12.3)

The *acceleration due to gravity at sea level* is denoted by g. Denoting the radius of the earth by R_E, we see from Eq. (12.3) that $Gm_E = gR_E^2$. Substituting this result into Eq. (12.3), we obtain an expression for the acceleration due to gravity at a distance r from the center of the earth in terms of the acceleration due to gravity at sea level:

$$a = g\frac{R_E^2}{r^2}.$$

(12.4)

Since the weight of the object $W = ma$, the weight of an object at a distance r from the center of the earth is

$$W = mg\frac{R_E^2}{r^2}.$$

(12.5)

At sea level ($r = R_E$), the weight of an object is given in terms of its mass by the simple relation

$$W = mg.$$

(12.6)

The value of g varies from location to location on the surface of the earth. The values we use in examples and problems are $g = 9.81$ m/s^2 in SI units and $g = 32.2$ ft/s^2 in U.S. Customary units.

Study Questions

1. Does the weight of an object depend on its location?
2. If you know an object's mass, how do you determine its weight at sea level?

Example 12.3

Determining an Object's Weight

In its final configuration, the International Space Station (Fig. 12.5) will have a mass of approximately 450,000 kg.

(a) What would be the weight of the ISS if it were at sea level?

(b) The orbit of the ISS is 354 km above the surface of the earth. The earth's radius is 6370 km. What is the weight of the ISS (the force exerted on it by gravity) when it is in orbit?

Figure 12.5
International Space Station.

Strategy

(a) The weight of an object at sea level is given by Eq. (12.6). Because the mass is given in kilograms, we will express g in SI units: $g = 9.81 \text{ m/s}^2$.
(b) The weight of an object at a distance r from the center of the earth is given by Eq. (12.5).

Solution

(a) The weight at sea level is

$$W = mg$$
$$= (450{,}000)(9.81)$$
$$= 4.41 \times 10^6 \text{ N.}$$

(b) The weight in orbit is

$$W = mg \frac{R_E^2}{r^2}$$

$$= (450{,}000)(9.81) \frac{(6{,}370{,}000)^2}{(6{,}370{,}000 + 354{,}000)^2}$$

$$= 3.96 \times 10^6 \text{ N.}$$

Discussion

Notice that the force exerted on the ISS by gravity when it is in orbit is approximately 90% of its weight at sea level.

The turning motorcycle has components of acceleration tangential and normal to its path.

Motion of a Point

I n this chapter, we begin the study of motion. We are not yet concerned with the properties of objects or the causes of their motions—our objective is to describe and analyze the motion of a point in space. After defining the position, velocity, and acceleration of a point, we consider the simplest example: motion along a straight line. We then show how motion of a point along an arbitrary path, or *trajectory*, is expressed and analyzed using various coordinate systems.

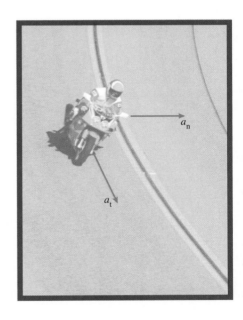

13.1 Position, Velocity, and Acceleration

If you observe people in a room, such as a group at a party, you perceive their positions relative to the room. For example, some people may be in the back of the room, some in the middle of the room, and so forth. The colloquial expression is that the room is your "frame of reference." To make this idea precise, we can introduce a cartesian coordinate system with its axes aligned with the walls of the room as in Fig. 13.1a and specify the position of a person (actually, the position of some point of the person, such as his or her center of mass) by specifying the components of the position vector **r** relative to the origin of the coordinate system. This coordinate system is a convenient reference frame for objects in the room. If you are sitting in an airplane, you perceive the positions of objects within the airplane relative to the airplane. In this case, the interior of the airplane is your frame of reference. To precisely specify the position of a person within the airplane, we can introduce a cartesian coordinate system that is fixed relative to the airplane and measure the position of the person's center of mass by specifying the components of the position vector **r** relative to the origin (Fig. 13.1b). A *reference frame* is simply a coordinate system that is suitable for specifying positions of points. You may be familiar only with cartesian coordinates. We discuss other examples in this chapter and continue our discussion of reference frames throughout the book.

We can describe the position of a point P relative to a given reference frame with origin O by the *position vector* **r** from O to P (Fig. 13.2a). Suppose that P is in motion relative to the chosen reference frame, so that **r** is a function of time t (Fig. 13.2b). We express this by the notation

$$\mathbf{r} = \mathbf{r}(t).$$

The *velocity* of P relative to the given reference frame at time t is defined by

$$\mathbf{v} = \frac{d\mathbf{r}}{dt} = \lim_{\Delta t \to 0} \frac{\mathbf{r}(t + \Delta t) - \mathbf{r}(t)}{\Delta t}, \tag{13.1}$$

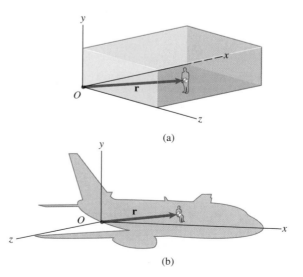

(a)

(b)

Figure 13.1
Convenient reference frames for specifying positions of objects:
(a) in a room;
(b) in an airplane.

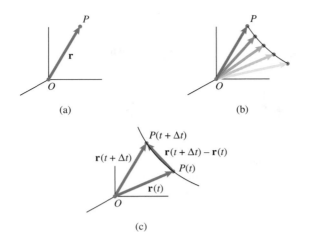

Figure 13.2
(a) The position vector **r** of P relative to O.
(b) Motion of P relative to the reference frame.
(c) Change in position of P from t to $t + \Delta t$.

where the vector $\mathbf{r}(t + \Delta t) - \mathbf{r}(t)$ is the change in position, or *displacement*, of P during the interval of time Δt (Fig. 13.2c). Thus, the velocity is the rate of change of the position of P.

The dimensions of a derivative are determined just as if it is a ratio, so the dimensions of **v** are (distance)/(time). The reference frame being used is often obvious, and we simply call **v** the velocity of P. However, remember that the position and velocity of a point can be specified only relative to some reference frame.

Notice in Eq. (13.1) that the derivative of a vector with respect to time is defined in exactly the same way as is the derivative of a scalar function. As a result, the derivative of a vector shares some of the properties of the derivative of a scalar function. We will use two of these properties: the derivative with respect to time, or time derivative, of the sum of two vector functions **u** and **w** is

$$\frac{d}{dt}(\mathbf{u} + \mathbf{w}) = \frac{d\mathbf{u}}{dt} + \frac{d\mathbf{w}}{dt},$$

and the time derivative of the product of a scalar function f and a vector function **u** is

$$\frac{d(f\mathbf{u})}{dt} = \frac{df}{dt}\mathbf{u} + f\frac{d\mathbf{u}}{dt}.$$

The *acceleration* of P relative to the given reference frame at time t is defined by

$$\mathbf{a} = \frac{d\mathbf{v}}{dt} = \lim_{\Delta t \to 0} \frac{\mathbf{v}(t + \Delta t) - \mathbf{v}(t)}{\Delta t}, \qquad (13.2)$$

where $\mathbf{v}(t + \Delta t) - \mathbf{v}(t)$ is the change in the velocity of P during the interval of time Δt (Fig. 13.3). The acceleration is the rate of change of the velocity of P at time t (the second time derivative of the displacement), and its dimensions are (distance)/(time)^2.

We have defined the velocity and acceleration of P relative to the origin O of the reference frame. We can show that *a point has the same velocity and acceleration relative to any fixed point in a given reference frame*. Let O' be

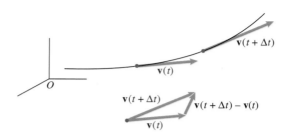

Figure 13.3
Change in the velocity of P from t to $t + \Delta t$.

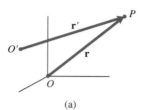

(a)

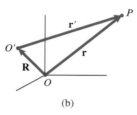

(b)

Figure 13.4
(a) Position vectors of P relative to O and O'.
(b) Position vector of O' relative to O.

an arbitrary fixed point, and let $\mathbf{r}'$ be the position vector from O' to P (Fig. 13.4a). The velocity of P relative to O' is $\mathbf{v}' = d\mathbf{r}'/dt$. The velocity of P relative to the origin O is $\mathbf{v} = d\mathbf{r}/dt$. We wish to show that $\mathbf{v}' = \mathbf{v}$. Let $\mathbf{R}$ be the vector from O to O' (Fig. 13.4b), so that

$$\mathbf{r}' = \mathbf{r} - \mathbf{R}.$$

Since the vector $\mathbf{R}$ is constant, the velocity of P relative to O' is

$$\mathbf{v}' = \frac{d\mathbf{r}'}{dt} = \frac{d\mathbf{r}}{dt} - \frac{d\mathbf{R}}{dt} = \frac{d\mathbf{r}}{dt} = \mathbf{v}.$$

The acceleration of P relative to O' is $\mathbf{a}' = d\mathbf{v}'/dt$, and the acceleration of P relative to O is $\mathbf{a} = d\mathbf{v}/dt$. Since $\mathbf{v}' = \mathbf{v}$, $\mathbf{a}' = \mathbf{a}$. Thus, the velocity and acceleration of a point P relative to a given reference frame do not depend on the location of the fixed reference point used to specify the position of P.

Study Questions

1.   What is a reference frame?
2.   What is the definition of the velocity of a point P relative to a given reference frame?
3.   Suppose that you know the velocity $\mathbf{v}$ and acceleration $\mathbf{a}$ of a point P relative to the origin O of a given reference frame. What do you know about the velocity and acceleration of P relative to an arbitrary fixed point O' of the reference frame?

13.2   Straight-Line Motion

We discuss this simple type of motion primarily so that you can gain experience and insight before proceeding to the general case of the motion of a point. But engineers must analyze straight-line motions in many practical situations, such as the motion of a vehicle on a straight road or track or the motion of a piston in an internal combustion engine.

Description of the Motion

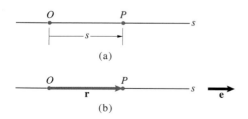

Figure 13.5
(a) The coordinate s from O to P.
(b) The unit vector $\mathbf{e}$ and position vector $\mathbf{r}$.

We can specify the position of a point P on a straight line relative to a reference point O by the coordinate s, measured along the line from O to P (Fig. 13.5a). In this case, the straight line is the reference frame we use to describe the position of P, and O is its origin. In Fig. 13.5a we define s to be positive to the right, so s is positive when P is to the right of O and negative when P is to the left of O. The *displacement* during an interval of time from t_0 to t is the change in the position, $s(t) - s(t_0)$.

By introducing a unit vector **e** parallel to the line and pointing in the positive s direction (Fig. 13.5b), we can write the position vector of P relative to O as

$$\mathbf{r} = s\mathbf{e}.$$

The velocity of P relative to O is

$$\mathbf{v} = \frac{d\mathbf{r}}{dt} = \frac{ds}{dt}\mathbf{e}.$$

We can write the velocity vector as $\mathbf{v} = v\mathbf{e}$, obtaining the scalar equation

$$v = \frac{ds}{dt}.$$

The velocity v of point P along the straight line is the rate of change of the position s of P. Notice that v is equal to the slope at time t of the line tangent to the graph of s as a function of time (Fig. 13.6).

The acceleration of P relative to O is

$$\mathbf{a} = \frac{d\mathbf{v}}{dt} = \frac{d}{dt}(v\mathbf{e}) = \frac{dv}{dt}\mathbf{e}.$$

Writing the acceleration vector as $\mathbf{a} = a\mathbf{e}$, we obtain the scalar equation

$$a = \frac{dv}{dt} = \frac{d^2s}{dt^2}.$$

The acceleration a is equal to the slope at time t of the line tangent to the graph of v as a function of time (Fig. 13.7).

By introducing the unit vector **e**, we have obtained scalar equations describing the motion of P. The position is specified by the coordinate s, and the velocity and acceleration are governed by the equations

$$v = \frac{ds}{dt} \tag{13.3}$$

and

$$a = \frac{dv}{dt}. \tag{13.4}$$

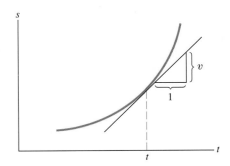

Figure 13.6
The slope of the straight line tangent to the graph of s versus t is the velocity at time t.

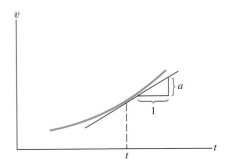

Figure 13.7
The slope of the straight line tangent to the graph of v versus t is the acceleration at time t.

Analysis of the Motion

In some situations, the position s of a point of an object is known as a function of time. Engineers use methods such as radar and laser-Doppler interferometry to measure positions as functions of time. In this case, we can obtain the velocity and acceleration as functions of time from Eqs. (13.3) and (13.4) by differentiation. For example, if the position of the truck in Fig. 13.8 during the interval of time from $t = 2$ s to $t = 4$ s is given by the equation

$$s = 6 + \frac{1}{3}t^3 \text{ m},$$

Figure 13.8
The coordinate s measures the position of the center of mass of the truck relative to a reference point.

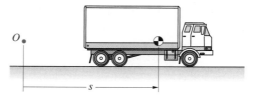

then the velocity and acceleration of the truck during that interval of time are

$$v = \frac{ds}{dt} = t^2 \text{ m/s}$$

and

$$a = \frac{dv}{dt} = 2t \text{ m/s}^2.$$

However, it is more common to know an object's acceleration than to know its position, because the acceleration of an object can be determined by Newton's second law when the forces acting on it are known. When the acceleration is known, we can determine the velocity and position from Eqs. (13.3) and (13.4) by integration. We discuss a number of important cases in the sections that follow.

Acceleration Specified as a Function of Time If the acceleration is a known function of time $a(t)$, we can integrate the relation

$$\frac{dv}{dt} = a(t) \tag{13.5}$$

with respect to time to determine the velocity as a function of time. We obtain

$$v = \int a(t)\,dt + A, \tag{13.6}$$

where A is an integration constant. Then we can integrate the relation

$$\frac{ds}{dt} = v \tag{13.7}$$

to determine the position as a function of time,

$$s = \int v\,dt + B, \tag{13.8}$$

where B is another integration constant. We would need additional information about the motion, such as the values of v and s at a given time, to determine the constants A and B.

Instead of using indefinite integrals, we can write Eq. (13.5) as

$$dv = a(t)\,dt$$

and integrate in terms of definite integrals:

$$\int_{v_0}^{v} dv = \int_{t_0}^{t} a(t)\,dt.$$

The lower limit v_0 is the velocity at time t_0, and the upper limit v is the velocity at an arbitrary time t. Evaluating the left integral, we obtain an expression for the velocity as a function of time:

$$v = v_0 + \int_{t_0}^{t} a(t)\,dt. \tag{13.9}$$

We can then write Eq. (13.7) as

$$ds = v\,dt$$

and integrate in terms of definite integrals to obtain

$$\int_{s_0}^{s} ds = \int_{t_0}^{t} v\,dt,$$

where the lower limit s_0 is the position at time t_0 and the upper limit s is the position at an arbitrary time t. Evaluating the left integral, we obtain the position as a function of time:

$$s = s_0 + \int_{t_0}^{t} v\,dt. \tag{13.10}$$

Although we have shown how to determine the velocity and position when the acceleration is known as a function of time, don't try to remember results such as Eqs. (13.9) and (13.10). As we will demonstrate in the examples, we recommend that straight-line motion problems be solved by using Eqs. (13.3) and (13.4).

We can make some useful observations from Eqs. (13.9) and (13.10):

- The area defined by the graph of the acceleration of P as a function of time from t_0 to t is equal to the change in the velocity from t_0 to t (Fig. 13.9a).

- The area defined by the graph of the velocity of P as a function of time from t_0 to t is equal to the change in position from t_0 to t (Fig. 13.9b).

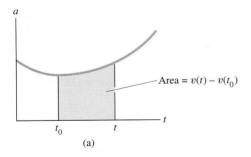

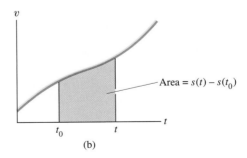

(a)　　　　　　　　　　(b)

Figure 13.9
Relations between areas defined by the graphs of the acceleration and velocity of P and changes in its velocity and position.

These relationships can often be used to obtain a qualitative understanding of an object's motion, and in some cases can even be used to determine the object's motion quantitatively.

Constant Acceleration　In some situations, the acceleration of an object is constant or nearly constant. For example, if a dense object such as a golf ball or rock is dropped and doesn't fall too far, the object's acceleration is approximately equal to the acceleration due to gravity at sea level.

Let the acceleration be a known constant a_0. From Eqs. (13.9) and (13.10), the velocity and position as functions of time are

$$v = v_0 + a_0(t - t_0) \tag{13.11}$$

and

$$s = s_0 + v_0(t - t_0) + \frac{1}{2} a_0(t - t_0)^2, \tag{13.12}$$

where s_0 and v_0 are the position and velocity, respectively, at time t_0. Notice that *if the acceleration is constant, the velocity is a linear function of time.*

We can use the *chain rule* to express the acceleration in terms of a derivative with respect to s:

$$a_0 = \frac{dv}{dt} = \frac{dv}{ds}\frac{ds}{dt} = \frac{dv}{ds} v.$$

Writing this expression as $v\,dv = a_0\,ds$ and integrating yields

$$\int_{v_0}^{v} v\,dv = \int_{s_0}^{s} a_0\,ds,$$

and we obtain an equation for the velocity as a function of position:

$$v^2 = v_0^2 + 2a_0(s - s_0). \tag{13.13}$$

Although Eqs. (13.11)–(13.13) can be useful *when the acceleration is constant*, they must not be used otherwise.

Study Questions

1. If you know the position s of a point P in straight-line motion as a function of time, how can you determine the velocity and acceleration of P as functions of time?
2. If you know the acceleration a of a point P in straight-line motion as a function of time, but have no other information, you can't determine the position and velocity of P as functions of time. Why not?
3. Suppose that you know the velocity v of a point P as a function of time. If you calculate the area defined by the graph of v from a time t_0 to a time t, what does that tell you?

Example 13.1

Straight-Line Motion with Constant Acceleration

Engineers testing a vehicle that will be dropped by parachute estimate that the vertical velocity of the vehicle when it reaches the ground will be 6 m/s. If they drop the vehicle from the test rig in Fig. 13.10, from what height h should they drop it to match the impact velocity of the parachute drop?

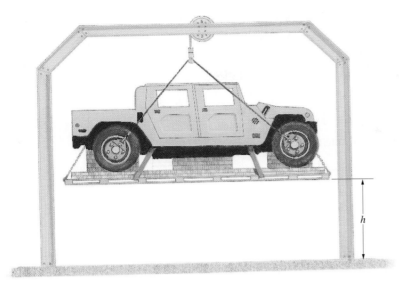

Figure 13.10

Strategy

If the only significant force acting on an object near the earth's surface is its weight, the acceleration of the object is approximately constant and equal to the acceleration due to gravity at sea level. Therefore, we can assume that the vehicle's acceleration during its short fall is $g = 9.81$ m/s^2. We will determine the height h in two ways:

- *First method*. We can integrate Eqs. (13.3) and (13.4) to obtain the vehicle's velocity and position as functions of time and then use them to determine the position of the vehicle when its velocity is 6 m/s.

- *Second method*. We can use Eq. (13.13), which relates the velocity to the position of the vehicle when the acceleration is constant.

Solution

Let $t = 0$ be the time at which the vehicle is dropped, and let s be the position of the bottom of the cushioning material beneath the vehicle relative to its position at $t = 0$ (Fig. a). The vehicle's acceleration is $a = 9.81$ m/s^2.

First Method   From Eq. (13.4),

$$\frac{dv}{dt} = a = 9.81.$$

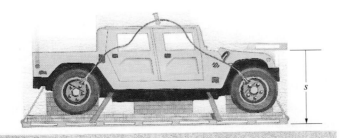

(a) The coordinate s measures the position of the bottom of the platform relative to its initial position.

Integrating, we obtain

$$v = 9.81t + A,$$

where A is an integration constant. Because the vehicle is at rest when it is released, $v = 0$ at $t = 0$. Therefore $A = 0$, and the vehicle's velocity as a function of time is

$$v = 9.81t \text{ m/s}.$$

We substitute this result into Eq. (13.3) to get

$$\frac{ds}{dt} = v = 9.81t$$

and integrate, obtaining

$$s = 4.91t^2 + B.$$

The position $s = 0$ when $t = 0$, so the integration constant $B = 0$, and the position as a function of time is

$$s = 4.91t^2.$$

From our equation for the velocity as a function of time, the time necessary for the vehicle to reach 6 m/s as it falls is

$$t = \frac{v}{9.81} = \frac{6}{9.81} = 0.612 \text{ s}.$$

Substituting this time into our equation for the position as a function of time yields the required height h:

$$h = 4.91t^2 = 4.91(0.612)^2 = 1.83 \text{ m}.$$

Second Method   Because the acceleration is constant, we can use Eq. (13.13) to determine the distance necessary for the velocity to increase to 6 m/s. We have

$$v^2 = v_0^2 + 2a_0(s - s_0):$$
$$(6)^2 = 0 + 2(9.81)(s - 0).$$

Solving for s, we obtain $h = 1.83$ m.

Example 13.2

Graphical Solution of Straight-Line Motion

The cheetah, *Acinonyx jubatus* (Fig. 13.11), can run as fast as 75 mi/hr. If you assume that the animal's acceleration is constant and that the cheetah reaches top speed in 4 s, what distance can it cover in 10 s?

Strategy

The acceleration has a constant value for the first 4 s and is then zero. We can determine the distance traveled during each of these "phases" of the motion and sum them to obtain the total distance covered. We do so both analytically and graphically.

Figure 13.11

Solution

The top speed in terms of feet per second is

$$75 \text{ mi/hr} = 75 \text{ mi/hr} \times \left(\frac{5280 \text{ ft}}{1 \text{ mi}}\right) \times \left(\frac{1 \text{ hr}}{3600 \text{ s}}\right) = 110 \text{ ft/s}.$$

First Method Let a_0 be the acceleration during the first 4 s. We integrate Eq. (13.4) to get

$$\int_0^v dv = \int_0^t a_0 \, dt,$$

$$[v]_0^v = [a_0 t]_0^t,$$

thereby obtaining the velocity as a function of time during the first 4 s:

$$v = a_0 t \text{ ft/s}.$$

When $t = 4$ s, $v = 110$ ft/s; so $a_0 = 110/4 = 27.5$ ft/s^2. Therefore, the velocity during the first 4 s is $v = 27.5t$ ft/s. Now we integrate Eq. (13.3),

$$\int_0^s ds = \int_0^t 27.5t \, dt,$$

$$[s]_0^s = 27.5\left[\frac{t^2}{2}\right]_0^t,$$

obtaining the position as a function of time during the first 4 s:

$$s = 13.75t^2 \text{ ft}$$

At $t = 4$ s, the position is $s = 13.75(4)^2 = 220$ ft.

From $t = 4$ to $t = 10$ s, the velocity is constant. The distance traveled is then

$$(110 \text{ ft/s})(6 \text{ s}) = 660 \text{ ft}.$$

The total distance the animal travels is $220 + 660 = 880$ ft, or 293 yd, in 10 s.

Second Method We draw a graph of the animal's velocity as a function of time in Fig. (a). The acceleration is constant during the first 4 s of motion, so the velocity is a linear function of time from $v = 0$ at $t = 0$ to $v = 110$ ft/s at $t = 4$ s. The velocity is constant during the last 6 s. The total distance covered is the sum of the areas during the two phases of motion:

$$\frac{1}{2}(4 \text{ s})(110 \text{ ft/s}) + (6 \text{ s})(110 \text{ ft/s}) = 220 \text{ ft} + 660 \text{ ft} = 880 \text{ ft}.$$

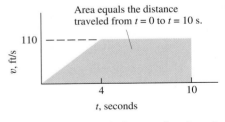

(a) The cheetah's velocity as a function of time.

Discussion

Notice that in the first method we used definite, rather than indefinite, integrals to determine the cheetah's velocity and position as functions of time. You should rework the example using indefinite integrals and compare your results with ours. Whether to use definite or indefinite integrals is primarily a matter of taste, but you need to be familiar with both procedures.

Example 13.3

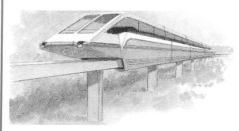

Figure 13.12

Acceleration That is a Function of Time

Suppose that the acceleration of the train in Fig. 13.12 during the interval of time from $t = 2$ s to $t = 4$ s is $a = 2t$ m/s^2, and at $t = 2$ s its velocity is $v = 180$ km/hr. What is the train's velocity at $t = 4$ s, and what is its displacement (change in position) from $t = 2$ s to $t = 4$ s?

Strategy

We can integrate Eqs. (13.3) and (13.4) to determine the train's velocity and position as functions of time.

Solution

The velocity at $t = 2$ s is

$$180 \text{ km/hr} \times \left(\frac{1000 \text{ m}}{1 \text{ km}} \right) \times \left(\frac{1 \text{ hr}}{3600 \text{ s}} \right) = 50 \text{ m/s}.$$

We write Eq. (13.4) as

$$dv = a \, dt = 2t \, dt$$

and integrate, introducing the condition $v = 50$ m/s at $t = 2$ s. We have

$$\int_{50}^{v} dv = \int_{2}^{t} 2t \, dt,$$

$$[v]_{50}^{v} = [t^2]_{2}^{t},$$

and it follows that

$$v - 50 = t^2 - 4.$$

Solving for v, we obtain

$$v = t^2 + 46 \text{ m/s}.$$

Now that we know the velocity as a function of time, we write Eq. (13.3) as

$$ds = v \, dt = (t^2 + 46) \, dt$$

and integrate, defining the position of the train at $t = 2$ s to be $s = 0$. We get

$$\int_{0}^{s} ds = \int_{2}^{t} (t^2 + 46) \, dt,$$

$$[s]_{0}^{s} = \left[\frac{t^3}{3} + 46t \right]_{2}^{t},$$

so that

$$s = \frac{t^3}{3} + 46t - \frac{2^3}{3} - 46(2).$$

The position as a function of time is

$$s = \frac{1}{3}t^3 + 46t - 94.7 \text{ m.}$$

Using our equations for the velocity and position, we find that the velocity at $t = 4$ s is

$$v = (4)^2 + 46 = 62 \text{ m/s,}$$

and the displacement from $t = 2$ s to $t = 4$ s is

$$\left[\frac{1}{3}(4)^3 + 46(4) - 94.7\right] - 0 = 111 \text{ m.}$$

Discussion

The acceleration in this example is not constant. You must not try to solve such problems by using equations that are valid only when the acceleration is constant. To convince yourself, try applying Eq. (13.11) to this example: Set $a_0 = 2t$ m/s^2, $t_0 = 2$ s, and $v_0 = 50$ m/s, and solve for the velocity at $t = 4$ s.

Acceleration Specified as a Function of Velocity Aerodynamic and hydrodynamic forces can cause an object's acceleration to depend on its velocity (Fig. 13.13). Suppose that the acceleration is a known function of velocity; that is,

$$\frac{dv}{dt} = a(v). \tag{13.14}$$

We cannot integrate this equation with respect to time to determine the velocity, because $a(v)$ is not known as a function of time. But we can *separate variables*, putting terms involving v on one side of the equation and terms involving t on the other side:

$$\frac{dv}{a(v)} = dt. \tag{13.15}$$

We can now integrate, obtaining

$$\int_{v_0}^{v} \frac{dv}{a(v)} = \int_{t_0}^{t} dt, \tag{13.16}$$

where v_0 is the velocity at time t_0. In principle, we can solve this equation for the velocity as a function of time and then integrate the relation

$$\frac{ds}{dt} = v$$

to determine the position as a function of time.

By using the chain rule, we can also determine the velocity as a function of the position. Writing the acceleration as

$$\frac{dv}{dt} = \frac{dv}{ds}\frac{ds}{dt} = \frac{dv}{ds}v$$

Figure 13.13
Aerodynamic and hydrodynamic forces depend on an object's velocity. As the bullet slows, the aerodynamic drag force resisting its motion decreases.

and substituting it into Eq. (13.14), we obtain

$$\frac{dv}{ds}v = a(v).$$

Separating variables yields

$$\frac{v\,dv}{a(v)} = ds.$$

Integrating,

$$\int_{v_0}^{v} \frac{v\,dv}{a(v)} = \int_{s_0}^{s} ds,$$

we can obtain a relation between the velocity and the position. (See Example 13.4.)

Acceleration Specified as a Function of Position Gravitational forces and forces exerted by springs can cause an object's acceleration to depend on its position. If the acceleration is a known function of position; that is,

$$\frac{dv}{dt} = a(s), \tag{13.17}$$

we cannot integrate with respect to time to determine the velocity, because $a(s)$ is not known as a function of time. Moreover, we cannot separate variables, because the equation contains three variables: v, t, and s. However, by using the chain rule

$$\frac{dv}{dt} = \frac{dv}{ds}\frac{ds}{dt} = \frac{dv}{ds}v,$$

we can write Eq. (13.17) as

$$\frac{dv}{ds}v = a(s).$$

Now we can separate variables:

$$v\,dv = a(s)\,ds. \tag{13.18}$$

Finally, we integrate:

$$\int_{v_0}^{v} v\,dv = \int_{s_0}^{s} a(s)\,ds. \tag{13.19}$$

In principle, we can solve this equation for the velocity as a function of the position:

$$v = \frac{ds}{dt} = v(s). \tag{13.20}$$

Then we can separate variables in this equation and integrate to determine the position as a function of time:

$$\int_{s_0}^{s} \frac{ds}{v(s)} = \int_{t_0}^{t} dt.$$

The procedures we have described when the acceleration is known as a function of velocity or position are summarized in Table 13.1.

Table 13.1 Determining the velocity when you know the acceleration as a function of velocity or position.

If you know $a = a(v)$,	separate variables:
	$$\frac{dv}{dt} = a(v);$$
	$$\frac{dv}{a(v)} = dt.$$
	Or apply the chain rule
	$$\frac{dv}{dt} = \frac{dv}{ds}\frac{ds}{dt} = \frac{dv}{ds}v = a(v),$$
	and then separate variables:
	$$\frac{v\,dv}{a(v)} = ds.$$
If you know $a = a(s)$,	apply the chain rule
	$$\frac{dv}{dt} = \frac{dv}{ds}\frac{ds}{dt} = \frac{dv}{ds}v = a(s),$$
	and then separate variables:
	$$v\,dv = a(s)\,ds.$$

Example 13.4

Acceleration That is a Function of Velocity

After deploying its drag parachute, the airplane in Fig. 13.14 has an acceleration $a = -0.004v^2$ m/s^2.
(a) Determine the time required for the velocity to decrease from 80 m/s to 10 m/s.
(b) What distance does the plane cover during that time?

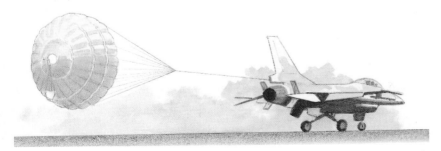

Figure 13.14

Strategy

In (b), we will use the chain rule to express the acceleration in terms of a derivative with respect to position and integrate to obtain a relation between the velocity and the position.

Solution

(a) The acceleration is

$$a = \frac{dv}{dt} = -0.004v^2.$$

We separate variables:

$$\frac{dv}{v^2} = -0.004 \, dt.$$

Then we integrate, defining $t = 0$ to be the time at which $v = 80$ m/s. We have

$$\int_{80}^{v} \frac{dv}{v^2} = \int_{0}^{t} - 0.004 \, dt,$$

or

$$\left[\frac{-1}{v}\right]_{80}^{v} = -0.004[t]_0^t,$$

yielding

$$\left(\frac{1}{v} - \frac{1}{80}\right) = 0.004t.$$

Solving for t, we obtain

$$t = 250\left(\frac{1}{v} - \frac{1}{80}\right).$$

From this equation, we find that the time required for the plane to slow to $v = 10$ m/s is 21.9 s. We show the velocity of the airplane as a function of time in Fig. 13.15.

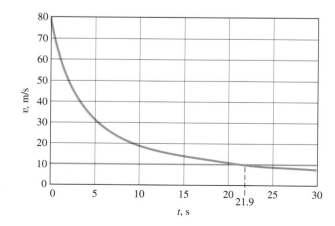

Figure 13.15
Graph of the airplane's velocity as a function of time.

(b) We write the acceleration as

$$a = \frac{dv}{dt} = \frac{dv}{ds}\frac{ds}{dt} = \frac{dv}{ds}v = -0.004v^2,$$

Separating variables yields

$$\frac{dv}{v} = -0.004\ ds.$$

We then integrate, defining $s = 0$ to be the position at which $v = 80$ m/s.

$$\int_{80}^{v}\frac{dv}{v} = \int_{0}^{s} - 0.004\ ds,$$

or

$$[\ln v]_{80}^{v} = -0.004[s]_{0}^{s}.$$

Evaluating each side yields

$$\ln v - \ln 80 = -\ln\left(\frac{80}{v}\right) = -0.004s.$$

Solving for s, we obtain

$$s = 250\ln\left(\frac{80}{v}\right).$$

The distance required for the plane to slow to $v = 10$ m/s is 520 m.

Discussion

Notice that our results predict that the time elapsed and the distance traveled continue to increase without bound as the airplane's velocity decreases. The reason is that the modeling is incomplete: The equation for the acceleration includes only aerodynamic drag and does not account for other forces, such as friction in the airplane's wheels.

Example 13.5

Gravitational (Position-Dependent) Acceleration

In terms of the distance s from the center of the earth, the magnitude of the acceleration due to gravity is gR_E^2/s^2, where R_E is the radius of the earth. (See the discussion of gravity in Section 12.5.) If a spacecraft is a distance s_0 from the center of the earth (Fig. 13.16), what outward velocity v_0 must it be given to reach a specified distance h from the earth's center?

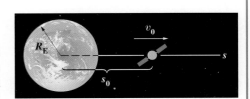

Figure 13.16

Solution

The acceleration due to gravity is *toward* the center of the earth:

$$a = -\frac{gR_E^2}{s^2}.$$

Applying the chain rule results in

$$a = \frac{dv}{dt} = \frac{dv}{ds}\frac{ds}{dt} = \frac{dv}{ds}v = -\frac{gR_E^2}{s^2}.$$

Separating variables, we obtain

$$v\,dv = -\frac{gR_E^2}{s^2}\,ds.$$

We integrate this equation using the initial condition ($v = v_0$ when $s = s_0$) as the lower limits and the final condition ($v = 0$ when $s = h$) as the upper limits. We have

$$\int_{v_0}^{0} v\,dv = -\int_{s_0}^{h} \frac{gR_E^2}{s^2}\,ds,$$

or

$$\left[\frac{v^2}{2}\right]_{v_0}^{0} = gR_E^2\left[\frac{1}{s}\right]_{s_0}^{h},$$

so that

$$-\frac{v_0^2}{2} = gR_E^2\left(\frac{1}{h} - \frac{1}{s_0}\right).$$

Solving for v_0, we obtain the initial velocity necessary for the spacecraft to reach a distance h:

$$v_0 = \sqrt{2gR_E^2\left(\frac{1}{s_0} - \frac{1}{h}\right)}.$$

Discussion

We can make an interesting and important observation from the result of this example. Notice that as the distance h increases, the necessary initial velocity v_0 approaches a finite limit. This limit,

$$v_{esc} = \lim_{h\to\infty} v_0 = \sqrt{\frac{2gR_E^2}{s_0}},$$

is called the *escape velocity*. In the absence of other effects, an object with this initial velocity will continue moving outward indefinitely. The existence of an escape velocity makes it feasible to send probes and persons to other planets. Once escape velocity is attained, it isn't necessary to expend additional fuel to keep going.

13.3 Curvilinear Motion

You have seen that the motion of a point along a straight line is described by the scalars s, v, and a. But if a point describes a *curvilinear* path relative to some reference frame, we must specify its motion in terms of its position, velocity, and acceleration vectors. Although the directions and magnitudes of these vectors do not depend on the particular coordinate system used to express them, we will show that the representations of these vectors are different in different coordinate systems. We can express many problems in terms of cartesian coordinates, but some situations, including the motions of satellites and rotating machines, can be expressed more naturally using other coordinate systems. In the sections that follow, we show how curvilinear motions of points are analyzed in terms of various coordinate systems.

Cartesian Coordinates

Let $\mathbf{r}$ be the position vector of a point P relative to the origin O of a cartesian reference frame (Fig. 13.17). The components of $\mathbf{r}$ are the x-, y-, and z-coordinates of P:

$$\mathbf{r} = x\mathbf{i} + y\mathbf{j} + z\mathbf{k}.$$

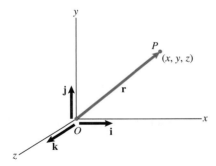

Figure 13.17
A cartesian coordinate system with origin O.

The velocity of P relative to the reference frame is

$$\mathbf{v} = \frac{d\mathbf{r}}{dt} = \frac{dx}{dt}\mathbf{i} + \frac{dy}{dt}\mathbf{j} + \frac{dz}{dt}\mathbf{k}. \tag{13.21}$$

Expressing the velocity in terms of scalar components yields

$$\mathbf{v} = v_x\mathbf{i} + v_y\mathbf{j} + v_z\mathbf{k}, \tag{13.22}$$

from which we obtain scalar equations relating the components of the velocity to the coordinates of P:

$$v_x = \frac{dx}{dt}, \quad v_y = \frac{dy}{dt}, \quad v_z = \frac{dz}{dt}. \tag{13.23}$$

The acceleration of P is

$$\mathbf{a} = \frac{d\mathbf{v}}{dt} = \frac{dv_x}{dt}\mathbf{i} + \frac{dv_y}{dt}\mathbf{j} + \frac{dv_z}{dt}\mathbf{k},$$

and by expressing the acceleration in terms of scalar components as

$$\mathbf{a} = a_x\mathbf{i} + a_y\mathbf{j} + a_z\mathbf{k}, \tag{13.24}$$

we obtain the scalar equations

$$a_x = \frac{dv_x}{dt}, \quad a_y = \frac{dv_y}{dt}, \quad a_z = \frac{dv_z}{dt}. \tag{13.25}$$

Equations (13.23) and (13.25) describe the motion of a point relative to a cartesian coordinate system. Notice that the equations describing the motion in each coordinate direction are identical in form to the equations that describe the motion of a point along a straight line. As a consequence, you often can analyze the motion in each coordinate direction using the methods you applied to straight-line motion.

The *projectile problem* is the classic example of this kind. If an object is thrown through the air and aerodynamic drag is negligible, the object accelerates downward with the acceleration due to gravity. In terms of a fixed cartesian coordinate system with its y-axis upward, the acceleration is given by $a_x = 0$, $a_y = -g$, and $a_z = 0$. Suppose that at $t = 0$ the projectile is located at the origin and has velocity v_0 in the xy-plane at an angle θ_0 above the horizontal (Fig. 13.18). At $t = 0$, $x = 0$ and $v_x = v_0 \cos\theta_0$. The acceleration in the x direction is zero; that is,

$$a_x = \frac{dv_x}{dt} = 0.$$

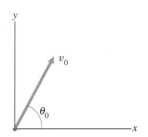

Figure 13.18
Initial conditions for a projectile problem.

Therefore v_x is constant and remains equal to its initial value:

$$v_x = \frac{dx}{dt} = v_0 \cos\theta_0. \tag{13.26}$$

(This result may seem unrealistic. The reason is that your intuition, based upon everyday experience, accounts for drag, whereas the analysis presented here does not.) Integrating Eq. (13.26) yields

$$\int_0^x dx = \int_0^t v_0 \cos\theta_0 \, dt,$$

whereupon we obtain the x-coordinate of the object as a function of time:

$$x = (v_0 \cos\theta_0)t. \tag{13.27}$$

Thus, we have determined the position and velocity of the projectile in the x direction as functions of time without considering the projectile's motion in the y or z direction.

At $t = 0$, $y = 0$ and $v_y = v_0 \sin\theta_0$. The acceleration in the y direction is

$$a_y = \frac{dv_y}{dt} = -g.$$

Integrating, we obtain

$$\int_{v_0 \sin \theta_0}^{v_y} dv_y = \int_0^t -g \, dt,$$

from which it follows that

$$v_y = \frac{dy}{dt} = v_0 \sin \theta_0 - gt. \tag{13.28}$$

Integrating this equation yields

$$\int_0^y dy = \int_0^t (v_0 \sin \theta_0 - gt) \, dt,$$

and we find that the y-coordinate as a function of time is

$$y = (v_0 \sin \theta_0)t - \frac{1}{2} gt^2. \tag{13.29}$$

Notice from this analysis that the same vertical velocity and position are obtained by throwing the projectile straight up with initial velocity $v_0 \sin \theta_0$ (Figs. 13.19a, b). The vertical motion is completely independent of the horizontal motion.

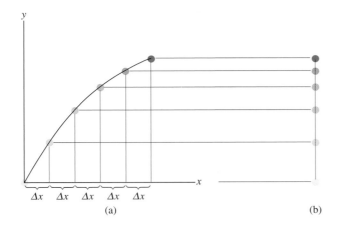

(a)   (b)

Figure 13.19
(a) Positions of the projectile at equal time intervals Δt. The distance $\Delta x = v_0 (\cos \theta_0) \Delta t$.
(b) Positions at equal time intervals Δt of a projectile given an initial vertical velocity equal to $v_0 \sin \theta_0$.

By solving Eq. (13.27) for t and substituting the result into Eq. (13.29), we obtain an equation describing the parabolic trajectory of the projectile:

$$y = (\tan \theta_0)x - \frac{g}{2v_0^2 \cos^2 \theta_0} x^2. \tag{13.30}$$

Study Questions

1. What are the components of the position vector $\mathbf{r}$ of a point P in terms of a cartesian reference frame?
2. In Eq. (13.21), we evaluated the time derivative of the position vector to obtain the velocity in terms of cartesian components. Why don't the time derivatives of the unit vectors appear in the result?
3. In the projectile problem, why is the horizontal component of the velocity constant?

Example 13.6

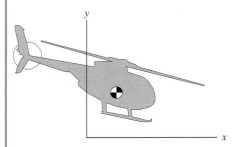

Figure 13.20

Analysis of Motion in Terms of Cartesian Components

During a test flight in which a helicopter starts from rest at $t = 0$ at the origin of the coordinate system shown in Fig. 13.20, accelerometers mounted on board the craft indicate that its components of acceleration from $t = 0$ to $t = 10$ s are closely approximated by

$$a_x = 0.6t \text{ m/s}^2,$$
$$a_y = 1.8 - 0.36t \text{ m/s}^2,$$

and

$$a_z = 0.$$

Determine the helicopter's velocity and position as functions of time.

Strategy

We can analyze the motion in each coordinate direction independently, integrating the acceleration to determine the velocity and then integrating the velocity to determine the position.

Solution

The velocity is zero at $t = 0$, and we assume that $x = y = z = 0$ at $t = 0$. The acceleration in the x direction is

$$a_x = \frac{dv_x}{dt} = 0.6t \text{ m/s}^2.$$

Integrating with respect to time yields

$$\int_0^{v_x} dv_x = \int_0^t 0.6t \, dt,$$

and we obtain the velocity component v_x as a function of time:

$$v_x = \frac{dx}{dt} = 0.3t^2 \text{ m/s}.$$

Integrating again results in

$$\int_0^x dx = \int_0^t 0.3t^2 \, dt,$$

and we obtain x as a function of time:

$$x = 0.1t^3 \text{ m}.$$

Now we analyze the motion in the y direction in the same way. The acceleration is

$$a_y = \frac{dv_y}{dt} = 1.8 - 0.36t \text{ m/s}^2.$$

Integrating, we have

$$\int_0^{v_y} dv_y = \int_0^t (1.8 - 0.36t) \, dt,$$

and we obtain the velocity:

$$v_y = \frac{dy}{dt} = 1.8t - 0.18t^2 \text{ m/s}.$$

Integrating again yields

$$\int_0^y dy = \int_0^t \left(1.8t - 0.18t^2\right) dt,$$

which determines the position:

$$y = 0.9t^2 - 0.06t^3 \text{ m}.$$

You can easily show that the z components of the velocity and position are $v_z = 0$ and $z = 0$. We show the position of the helicopter as a function of time in Fig. (a).

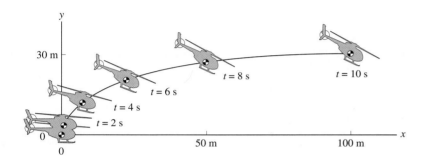

(a) Position of the helicopter at 2-s intervals.

Discussion

This example demonstrates how inertial navigation systems work. They contain accelerometers that measure the x, y, and z components of acceleration. (Gyroscopes maintain the alignments of the accelerometers.) By integrating the acceleration components twice with respect to time, the systems compute changes in the x-, y-, and z-coordinates of the airplane or ship.

Example 13.7

A Projectile Problem

The skier in Fig. 13.21 leaves the 20° surface at 10 m/s.
(a) Determine the distance d to the point where he lands.
(b) What are the magnitudes of his components of velocity parallel and perpendicular to the 45° surface just before he lands?

Strategy

(a) By neglecting aerodynamic drag and treating the skier as a projectile, we can determine his velocity and position as functions of time. Using the equation

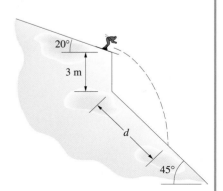

Figure 13.21

describing the straight surface on which he lands, we can relate his horizontal and vertical coordinates at impact and thereby obtain an equation for the time at which he lands. Knowing the time, we can determine his position and velocity.

(b) We can determine his velocity parallel and perpendicular to the 45° surface by using the result that the component of a vector **U** in the direction of a unit vector **e** is $(\mathbf{e} \cdot \mathbf{U})\mathbf{e}$.

Solution

(a) In Fig. a, we introduce a coordinate system with its origin where the skier leaves the surface. His components of velocity at that instant ($t = 0$) are

$$v_x = 10 \cos 20° = 9.40 \text{ m/s}$$

and

$$v_y = -10 \sin 20° = -3.42 \text{ m/s}.$$

The x component of acceleration is zero, so v_x is constant and the skier's x coordinate as a function of time is

$$x = 9.40t \text{ m}.$$

The y component of acceleration is

$$a_y = \frac{dv_y}{dt} = -9.81 \text{ m/s}^2.$$

Integrating to determine v_y as a function of time, we obtain

$$\int_{-3.42}^{v_y} dv_y = \int_0^t -9.81 \, dt,$$

from which it follows that

$$v_y = \frac{dy}{dt} = -3.42 - 9.81t \text{ m/s}.$$

We integrate this equation to determine the y coordinate as a function of time. We have

$$\int_0^y dy = \int_0^t (-3.42 - 9.81t) \, dt,$$

yielding

$$y = -3.42t - 4.905t^2 \text{ m}.$$

The slope of the surface on which the skier lands is -1, so the linear equation describing it is $y = (-1)x + A$, where A is a constant. At $x = 0$, the

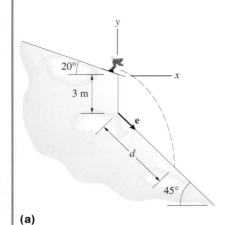

(a)

y coordinate of the surface is -3 m, so $A = -3$ m and the equation describing the $45°$ surface is

$$y = -x - 3 \text{ m.}$$

Substituting our equations for x and y as functions of time into this equation, we obtain an equation for the time at which the skier lands:

$$-3.42t - 4.905t^2 = -9.40t - 3.$$

Solving for t, we get $t = 1.60$ s. Therefore, his coordinates when he lands are

$$x = 9.40(1.60) = 15.0 \text{ m}$$

and

$$y = -3.42(1.60) - 4.905(1.60)^2 = -18.0 \text{ m,}$$

and the distance

$$d = \sqrt{(15.0)^2 + (18.0 - 3)^2} = 21.3 \text{ m.}$$

(b) The components of the skier's velocity just before he lands are

$$v_x = 9.40 \text{ m/s}$$

and

$$v_y = -3.42 - 9.81(1.60) = -19.1 \text{ m/s,}$$

and the magnitude of his velocity is $|\mathbf{v}| = \sqrt{(9.40)^2 + (-19.1)^2} = 21.3$ m/s. Let $\mathbf{e}$ be a unit vector that is parallel to the slope on which he lands (Fig. a):

$$\mathbf{e} = \cos 45° \, \mathbf{i} - \sin 45° \, \mathbf{j}.$$

The component of the velocity parallel to the surface is

$$(\mathbf{e} \cdot \mathbf{v})\mathbf{e} = \left[(\cos 45° \, \mathbf{i} - \sin 45° \, \mathbf{j}) \cdot (9.40\mathbf{i} - 19.1\mathbf{j})\right]\mathbf{e}$$

$$= 20.2\mathbf{e} \text{ (m/s).}$$

The magnitude of the skier's velocity parallel to the surface is 20.2 m/s. Therefore, the magnitude of his velocity perpendicular to the surface is

$$\sqrt{|\mathbf{v}|^2 - (20.2)^2} = 6.88 \text{ m/s.}$$

Angular Motion

We have seen that in some cases the curvilinear motion of a point can be analyzed by using cartesian coordinates. In the sections that follow, we describe problems that can be analyzed more simply in terms of other coordinate systems. To help you understand our discussion of these alternative coordinate systems, we introduce two preliminary topics in this section: the angular motion of a line in a plane and the time derivative of a unit vector rotating in a plane.

Angular Motion of a Line We can specify the angular position of a line L in a particular plane relative to a reference line L_0 in the plane by an angle θ (Fig. 13.22). The *angular velocity* of L relative to L_0 is defined by

$$\omega = \frac{d\theta}{dt}, \tag{13.31}$$

and the *angular acceleration* of L relative to L_0 is defined by

$$\alpha = \frac{d\omega}{dt} = \frac{d^2\theta}{dt^2}. \tag{13.32}$$

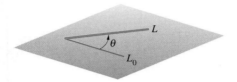

Figure 13.22
A line L and a reference line L_0 in a plane.

The dimensions of the angular position, angular velocity, and angular acceleration are rad, rad/s, and rad/s^2, respectively. Although these quantities are often expressed in terms of degrees or revolutions instead of radians, convert them into radians before using them in calculations.

Notice the analogy between Eqs. (13.31) and (13.32) and the equations relating the position, velocity, and acceleration of a point along a straight line (Table 13.2). In each case, the position is specified by a single scalar coordinate, which can be positive or negative. (In Fig. 13.22, the counterclockwise direction is positive.) Because the equations are identical in form, problems involving angular motion of a line can be analyzed by the same methods we applied to straight-line motion.

Table 13.2 The equations governing straight-line motion and the equations governing the angular motion of a line are identical in form.

Straight-Line Motion	Angular Motion
$v = \dfrac{ds}{dt}$	$\omega = \dfrac{d\theta}{dt}$
$a = \dfrac{dv}{dt} = \dfrac{d^2s}{dt^2}$	$\alpha = \dfrac{d\omega}{dt} = \dfrac{d^2\theta}{dt^2}$

Rotating Unit Vector The directions of the unit vectors $\mathbf{i}$, $\mathbf{j}$, and $\mathbf{k}$ relative to the cartesian reference frame are constant. However, in other coordinate systems, the unit vectors used to describe the motion of a point rotate as the point moves. To obtain expressions for the velocity and acceleration in such coordinate systems, we must know the time derivative of a rotating unit vector.

We can describe the angular motion of a unit vector **e** in a plane just as we described the angular motion of a line. The direction of **e** relative to a reference line L_0 is specified by the angle θ in Fig. 13.23a, and the rate of rotation of **e** relative to L_0 is specified by the angular velocity

$$\omega = \frac{d\theta}{dt}.$$

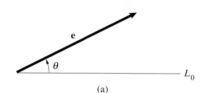

(a)

The time derivative of **e** is defined by

$$\frac{d\mathbf{e}}{dt} = \lim_{\Delta t \to 0} \frac{\mathbf{e}(t + \Delta t) - \mathbf{e}(t)}{\Delta t}.$$

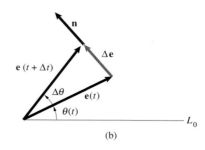

(b)

Figure 13.23b shows the vector **e** at time t and at time $t + \Delta t$. The change in **e** during this interval is $\Delta\mathbf{e} = \mathbf{e}(t + \Delta t) - \mathbf{e}(t)$, and the angle through which **e** rotates is $\Delta\theta = \theta(t + \Delta t) - \theta(t)$. The triangle in Fig. 13.23b is isosceles, so

$$|\Delta\mathbf{e}| = 2|\mathbf{e}| \sin(\Delta\theta/2) = 2 \sin(\Delta\theta/2).$$

To write the vector $\Delta\mathbf{e}$ in terms of this expression, we introduce a unit vector **n** that points in the direction of $\Delta\mathbf{e}$ (Fig. 13.23b):

$$\Delta\mathbf{e} = |\Delta\mathbf{e}|\mathbf{n} = 2 \sin(\Delta\theta/2)\mathbf{n}.$$

In terms of this expression, the time derivative of **e** is

$$\frac{d\mathbf{e}}{dt} = \lim_{\Delta t \to 0} \frac{\Delta\mathbf{e}}{\Delta t} = \lim_{\Delta t \to 0} \frac{2 \sin(\Delta\theta/2)\mathbf{n}}{\Delta t}.$$

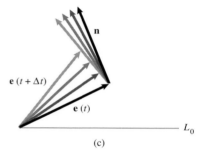

(c)

To evaluate the limit, we write it in the form

$$\frac{d\mathbf{e}}{dt} = \lim_{\Delta t \to 0} \frac{\sin(\Delta\theta/2)}{\Delta\theta/2} \frac{\Delta\theta}{\Delta t}\mathbf{n}.$$

In the limit as Δt approaches zero, $\sin(\Delta\theta/2)/(\Delta\theta/2) = 1$, $\Delta\theta/\Delta t = d\theta/dt$, and the unit vector **n** is perpendicular to $\mathbf{e}(t)$ (Fig. 13.23c). Therefore, the time derivative of **e** is

$$\frac{d\mathbf{e}}{dt} = \frac{d\theta}{dt}\mathbf{n} = \omega\mathbf{n}, \tag{13.33}$$

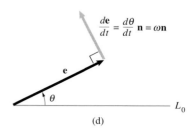

$$\frac{d\mathbf{e}}{dt} = \frac{d\theta}{dt}\mathbf{n} = \omega\mathbf{n}$$

(d)

where **n** is a unit vector that is perpendicular to **e** and points in the positive θ direction (Fig. 13.23d). In the sections that follow, we use this result in deriving expressions for the velocity and acceleration of a point in different coordinate systems.

Figure 13.23
(a) A unit vector **e** and reference line L_0.
(b) The change $\Delta\mathbf{e}$ in **e** from t to $t + \Delta t$.
(c) As Δt goes to zero, **n** becomes perpendicular to $\mathbf{e}(t)$.
(d) The time derivative of **e**.

Study Questions

1. What are the definitions of the angular velocity and angular acceleration of a line?
2. Is the time derivative of a rotating unit vector a scalar or a vector?
3. Suppose that a unit vector is rotating in a plane with angular velocity ω. What do you know about the time derivative of the vector?

Example 13.8

Analysis of Angular Motion

The rotor of a jet engine is rotating at 10,000 rpm (revolutions per minute) when the fuel is shut off. The ensuing angular acceleration is $\alpha = -0.02\omega$, where ω is the angular velocity in rad/s.
(a) How long does it take the rotor to slow to 1000 rpm?
(b) How many revolutions does the rotor turn while decelerating to 1000 rpm?

Strategy

To analyze the angular motion of the rotor, we define a line L that is fixed to the rotor and perpendicular to its axis (Fig. 13.24). Then we examine the motion of L relative to the reference line L_0. The angular position, velocity, and acceleration of L describe the angular motion of the rotor.

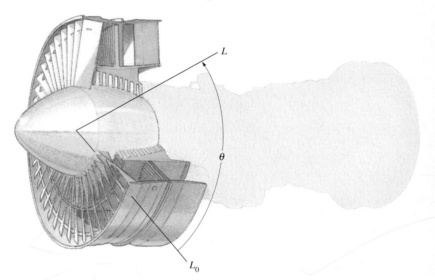

Figure 13.24
Introducing a line L and reference line L_0 to specify the angular position of the rotor.

Solution

The conversion from rpm to rad/s is

$$1 \text{ rpm} = 1 \text{ revolution/min} \times \left(\frac{2\pi \text{ rad}}{1 \text{ revolution}} \right) \times \left(\frac{1 \text{ min}}{60 \text{ s}} \right)$$

$$= \frac{\pi}{30} \text{ rad/s}.$$

(a) The angular acceleration is

$$\alpha = \frac{d\omega}{dt} = -0.02\omega.$$

We separate variables to get

$$\frac{d\omega}{\omega} = -0.02 \, dt.$$

Then we integrate, defining $t = 0$ to be the time at which the fuel is turned off:

$$\int_{10,000\pi/30}^{1000\pi/30} \frac{d\omega}{\omega} = \int_0^t -0.02 \, dt.$$

Evaluating the integrals and solving for t, we obtain

$$t = \left(\frac{1}{0.02}\right) \ln \left(\frac{10{,}000\pi/30}{1000\pi/30}\right) = 115 \text{ s}.$$

(b) Writing the angular acceleration as

$$\alpha = \frac{d\omega}{dt} = \frac{d\omega}{d\theta}\frac{d\theta}{dt} = \frac{d\omega}{d\theta}\,\omega = -0.02\omega,$$

we separate variables to obtain

$$d\omega = -0.02\,d\theta.$$

Next, we integrate, defining $\theta = 0$ to be the angular position at which the fuel is turned off:

$$\int_{10{,}000\pi/30}^{1000\pi/30} d\omega = \int_{0}^{\theta} -0.02\,d\theta.$$

Solving for θ yields

$$\theta = \left(\frac{1}{0.02}\right)\left[(10{,}000\pi/30) - (1000\pi/30)\right]$$

$$= 15{,}000\pi \text{ rad} = 7500 \text{ revolutions}.$$

Normal and Tangential Components

In this method of describing curvilinear motion, we specify the position of a point by a coordinate measured *along its path* and express the velocity and acceleration in terms of their components tangential and normal (perpendicular) to the path. Normal and tangential components are particularly useful when a point moves along a circular path. Furthermore, they give us unique insight into the character of the velocity and acceleration in curvilinear motion. We first discuss motion in a planar path because of its conceptual simplicity.

Planar Motion   Consider a point P moving along a plane, curvilinear path relative to some reference frame (Fig. 13.25a). The position vector $\mathbf{r}$ specifies

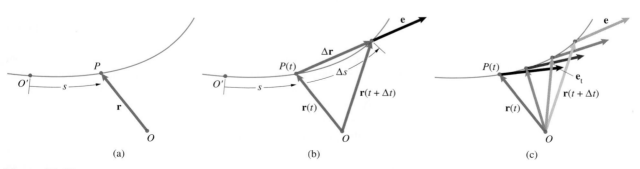

Figure 13.25
(a) The position of P along its path is specified by the coordinate s.
(b) Position of P at time t and at time $t + \Delta t$.
(c) The limit of $\mathbf{e}$ as $\Delta t \to 0$ is a unit vector tangent to the path.

the position of P relative to the reference point O, and the coordinate s measures P's position along the path relative to a point O' on the path. The velocity of P relative to O is

$$\mathbf{v} = \frac{d\mathbf{r}}{dt} = \lim_{\Delta t \to 0} \frac{\mathbf{r}(t + \Delta t) - \mathbf{r}(t)}{\Delta t} = \lim_{\Delta t \to 0} \frac{\Delta \mathbf{r}}{\Delta t}, \tag{13.34}$$

where $\Delta \mathbf{r} = \mathbf{r}(t + \Delta t) - \mathbf{r}(t)$ (Fig. 13.25b). We denote the distance traveled along the path from t to $t + \Delta t$ by Δs. By introducing a unit vector $\mathbf{e}$ defined to point in the direction of $\Delta \mathbf{r}$, we can write Eq. (13.34) as

$$\mathbf{v} = \lim_{\Delta t \to 0} \frac{\Delta s}{\Delta t} \mathbf{e}.$$

As Δt approaches zero, $\Delta s / \Delta t$ becomes ds/dt and $\mathbf{e}$ becomes a unit vector tangent to the path at the position of P at time t, which we denote by $\mathbf{e}_t$ (Fig. 13.25c):

$$\mathbf{v} = v\mathbf{e}_t = \frac{ds}{dt} \mathbf{e}_t. \tag{13.35}$$

The velocity of a point in curvilinear motion is a vector whose magnitude equals the rate of change of distance traveled along the path and whose direction is tangent to the path.

To determine the acceleration of P, we take the time derivative of Eq. (13.35):

$$\mathbf{a} = \frac{d\mathbf{v}}{dt} = \frac{dv}{dt} \mathbf{e}_t + v \frac{d\mathbf{e}_t}{dt}. \tag{13.36}$$

If the path is not a straight line, the unit vector $\mathbf{e}_t$ rotates as P moves. As a consequence, the time derivative of $\mathbf{e}_t$ is not zero. In the previous section, we derived an expression for the time derivative of a rotating unit vector in terms of the unit vector's angular velocity [Eq. (13.33)]. To use that result, we define the *path angle* θ specifying the direction of $\mathbf{e}_t$ relative to a reference line (Fig. 13.26). Then, from Eq. (13.33), the time derivative of $\mathbf{e}_t$ is

$$\frac{d\mathbf{e}_t}{dt} = \frac{d\theta}{dt} \mathbf{e}_n,$$

where $\mathbf{e}_n$ is a unit vector that is normal to $\mathbf{e}_t$ and points in the positive θ direction if $d\theta/dt$ is positive. Substituting this expression into Eq. (13.36), we obtain the acceleration of P:

$$\mathbf{a} = \frac{dv}{dt} \mathbf{e}_t + v \frac{d\theta}{dt} \mathbf{e}_n. \tag{13.37}$$

We can derive this result in another way that is less rigorous, but that gives additional insight into the meanings of the tangential and normal components of the acceleration. Figure 13.27a shows the velocity of P at times t and $t + \Delta t$. In Fig. 13.27b, you can see that the change in the velocity, $\mathbf{v}(t + \Delta t) - \mathbf{v}(t)$, consists of two components. The component Δv, which is tangent to the path at time t, is due to the change in the magnitude of the velocity. The component $v \Delta \theta$, which is perpendicular to the path at time t, is due to the change in the direction of the velocity vector. Thus, the change in the velocity is (approximately)

$$\mathbf{v}(t + \Delta t) - \mathbf{v}(t) = \Delta v \, \mathbf{e}_t + v\Delta \theta \, \mathbf{e}_n.$$

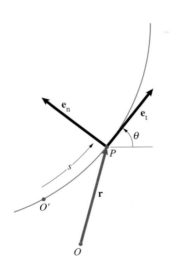

Figure 13.26
The path angle θ.

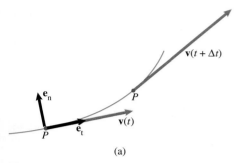

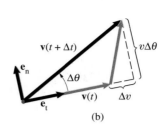

Figure 13.27
(a) Velocity of P at t and at $t + \Delta t$.
(b) The tangential and normal components of the change in the velocity.

To obtain the acceleration, we divide this expression by Δt and take the limit as $\Delta t \to 0$:

$$\mathbf{a} = \lim_{\Delta t \to 0} \frac{\Delta \mathbf{v}}{\Delta t} = \lim_{\Delta t \to 0} \left(\frac{\Delta v}{\Delta t} \mathbf{e}_t + v \frac{\Delta \theta}{\Delta t} \mathbf{e}_n \right)$$

$$= \frac{dv}{dt} \mathbf{e}_t + v \frac{d\theta}{dt} \mathbf{e}_n.$$

Thus, we again obtain Eq. (13.37). However, this derivation clearly points out that the tangential component of the acceleration arises from the rate of change of the magnitude of the velocity, whereas the normal component arises from the rate of change in the direction of the velocity vector. Notice that if the path is a straight line at time t, the normal component of the acceleration equals zero, because in that case $d\theta/dt$ is zero.

We can express the acceleration in another form that often is more convenient to use. Figure 13.28 shows the positions on the path reached by P at times t and $t + dt$. If the path is curved, straight lines extended from these points perpendicular to the path will intersect as shown. The distance ρ from the path to the point where these two lines intersect is called the *instantaneous radius of curvature* of the path. (If the path is circular, ρ is simply the radius of the path.) The angle $d\theta$ is the change in the path angle, and ds is the distance traveled from t to $t + \Delta t$. You can see from the figure that ρ is related to ds by

$$ds = \rho \, d\theta.$$

Dividing by dt, we obtain

$$\frac{ds}{dt} = v = \rho \frac{d\theta}{dt}.$$

Using this relation, we can write Eq. (13.37) as

$$\mathbf{a} = \frac{dv}{dt} \mathbf{e}_t + \frac{v^2}{\rho} \mathbf{e}_n.$$

For a given value of v, the normal component of the acceleration depends on the instantaneous radius of curvature. The greater the curvature of the path, the greater is the normal component of acceleration. When the acceleration is expressed in this way, the unit vector $\mathbf{e}_n$ must be defined to point toward the *concave* side of the path (Fig. 13.29).

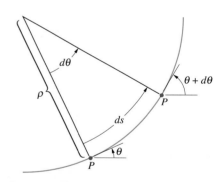

Figure 13.28
The instantaneous radius of curvature, ρ.

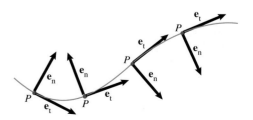

Figure 13.29
The unit vector normal to the path points toward the concave side.

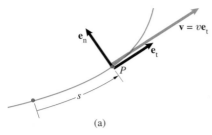

(a)

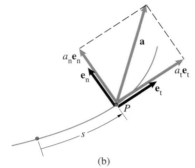

(b)

Figure 13.30
Normal and tangential components of the velocity (a) and acceleration (b).

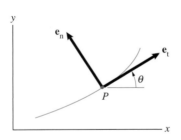

Figure 13.31
A point P moving in the xy-plane.

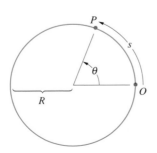

Figure 13.32
A point moving in a circular path.

Thus, the velocity and acceleration in terms of normal and tangential components are (Fig. 13.30)

$$\mathbf{v} = v\mathbf{e}_t = \frac{ds}{dt}\mathbf{e}_t \tag{13.38}$$

and

$$\mathbf{a} = a_t\mathbf{e}_t + a_n\mathbf{e}_n, \tag{13.39}$$

where

$$a_t = \frac{dv}{dt} \quad \text{and} \quad a_n = v\frac{d\theta}{dt} = \frac{v^2}{\rho}. \tag{13.40}$$

If the motion occurs in the xy-plane of a cartesian reference frame (Fig. 13.31) and θ is the angle between the x-axis and the unit vector $\mathbf{e}_t$, the unit vectors $\mathbf{e}_t$ and $\mathbf{e}_n$ are related to the cartesian unit vectors by

$$\mathbf{e}_t = \cos\theta\,\mathbf{i} + \sin\theta\,\mathbf{j}$$

and

$$\mathbf{e}_n = -\sin\theta\,\mathbf{i} + \cos\theta\,\mathbf{j}. \tag{13.41}$$

If the path in the xy-plane is described by a function $y = y(x)$, it can be shown that the instantaneous radius of curvature is given by

$$\rho = \frac{\left[1 + \left(\dfrac{dy}{dx}\right)^2\right]^{3/2}}{\left|\dfrac{d^2y}{dx^2}\right|}. \tag{13.42}$$

Circular Motion   If a point P moves in a plane circular path of radius R (Fig. 13.32), the distance s is related to the angle θ by

$$s = R\theta \quad \textbf{(circular path)}.$$

Using this relation, then, we can specify the position of P along the circular path by either s or θ. Taking the time derivative of the equation, we obtain a relation between $v = ds/dt$ and the angular velocity of the line from the center of the path to P:

$$v = R\frac{d\theta}{dt} = R\omega \quad \textbf{(circular path)}. \tag{13.43}$$

Taking another time derivative, we obtain a relation between the tangential component of the acceleration $a_t = dv/dt$ and the angular acceleration:

$$a_t = R\frac{d\omega}{dt} = R\alpha \quad \textbf{(circular path)}. \tag{13.44}$$

For this circular path, the instantaneous radius of curvature $\rho = R$, so the normal component of the acceleration is

$$a_n = \frac{v^2}{R} = R\omega^2 \quad \textbf{(circular path)}. \tag{13.45}$$

Because problems involving circular motion of a point are so common, these relations are worth remembering. But you must be careful to use them *only* when the path is circular.

Three-Dimensional Motion Although most applications of normal and tangential components involve the motion of a point in a plane, we briefly discuss three-dimensional motion for the insight it provides into the nature of the velocity and acceleration. If we consider the motion of a point along a three-dimensional path relative to some reference frame, the steps leading to Eq. (13.38) are unaltered. The velocity is

$$\mathbf{v} = v\mathbf{e}_t = \frac{ds}{dt}\mathbf{e}_t, \tag{13.46}$$

where $v = ds/dt$ is the rate of change of distance along the path and the unit vector $\mathbf{e}_t$ is tangent to the path and points in the direction of motion. We take the time derivative of this equation to obtain the acceleration:

$$\mathbf{a} = \frac{d\mathbf{v}}{dt} = \frac{dv}{dt}\mathbf{e}_t + v\frac{d\mathbf{e}_t}{dt}.$$

As the point moves along its three-dimensional path, the direction of the unit vector $\mathbf{e}_t$ changes. In the case of motion of a point in a plane, this unit vector rotates in the plane, but in three-dimensional motion, the picture is more complicated. Figure 13.33a shows the path seen from a viewpoint perpendicular to the plane containing the vector $\mathbf{e}_t$ at times t and $t + dt$. This plane is called the *osculating plane*. It can be thought of as the instantaneous plane of rotation of the unit vector $\mathbf{e}_t$, and its orientation will generally change as P moves along its path. Since $\mathbf{e}_t$ is rotating in the osculating plane at time t, its time derivative is

$$\frac{d\mathbf{e}_t}{dt} = \frac{d\theta}{dt}\mathbf{e}_n, \tag{13.47}$$

where $d\theta/dt$ is the angular velocity of $\mathbf{e}_t$ in the osculating plane and the unit vector $\mathbf{e}_n$ is defined as shown in Fig. 13.33b. The vector $\mathbf{e}_n$ is perpendicular to $\mathbf{e}_t$, parallel to the osculating plane, and directed toward the concave side of the path. Therefore the acceleration is

$$\mathbf{a} = \frac{dv}{dt}\mathbf{e}_t + v\frac{d\theta}{dt}\mathbf{e}_n. \tag{13.48}$$

In the same way as in the case of motion in a plane, we can also express the acceleration in terms of the instantaneous radius of curvature of the path (Fig. 13.33c):

$$\mathbf{a} = \frac{dv}{dt}\mathbf{e}_t + \frac{v^2}{\rho}\mathbf{e}_n. \tag{13.49}$$

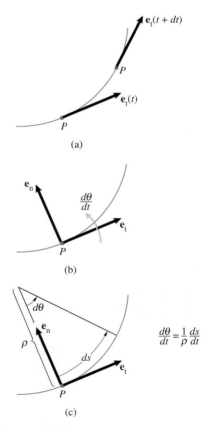

Figure 13.33
(a) Defining the osculating plane.
(b) Definition of the unit vector $\mathbf{e}_n$.
(c) The instantaneous radius of curvature.

We see that the expressions for the velocity and acceleration in normal and tangential components for three-dimensional motion are identical in form to the expressions for planar motion. The velocity is a vector whose magnitude equals the rate of change of distance traveled along the path and whose direction is tangent to the path. The acceleration has a component tangential to the path equal to the rate of change of the magnitude of the velocity and a component perpendicular to the path that depends on the magnitude of the velocity and the instantaneous radius of curvature of the path. In planar motion, the unit vector $\mathbf{e}_n$ is parallel to the plane of the motion. In three-dimensional motion, $\mathbf{e}_n$ is parallel to the osculating plane, whose orientation depends on the nature of the path. Notice from Eq. (13.47) that $\mathbf{e}_n$ can be expressed in terms of $\mathbf{e}_t$ by

$$\mathbf{e}_n = \frac{\dfrac{d\mathbf{e}_t}{dt}}{\left|\dfrac{d\mathbf{e}_t}{dt}\right|}.$$

As the final step necessary to establish a three-dimensional coordinate system, we introduce a third unit vector that is perpendicular to both $\mathbf{e}_t$ and $\mathbf{e}_n$ by the definition

$$\mathbf{e}_p = \mathbf{e}_t \times \mathbf{e}_n.$$

The unit vector $\mathbf{e}_p$ is perpendicular to the osculating plane (Fig. 13.34), which therefore defines its orientation.

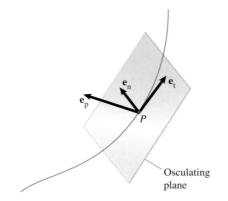

Figure 13.34
Defining the third unit vector $\mathbf{e}_p$.

Study Questions

1. What is the velocity vector of a point in terms of normal and tangential components?
2. If the magnitude of the velocity of a point moving in a curved path is constant, is the acceleration of the point zero?
3. What is the instantaneous radius of curvature?
4. If a point P moves in a circular path of radius R, what is the relation between the magnitude of the velocity of the point and the angular velocity of the line from the center of the path to P?

Example 13.9

Figure 13.35

Motion in Terms of Normal and Tangential Components

The motorcycle in Fig. 13.35 starts from rest at $t = 0$ on a circular track of 400-m radius. The tangential component of the cycle's acceleration is $a_t = 2 + 0.2t$ m/s^2. At $t = 10$ s, determine (a) the distance the motorcycle has moved along the track and (b) the magnitude of its acceleration.

Strategy

Let s be the distance along the track from the initial position O of the motor-cycle to its position at time t (Fig. a). Knowing the tangential acceleration as a function of time, we can integrate to determine v and s as functions of time.

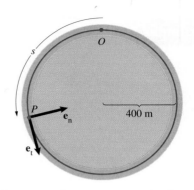

(a) The coordinate s measures the distance along the track.

Solution

(a) The tangential acceleration is

$$a_t = \frac{dv}{dt} = 2 + 0.2t \text{ m/s}^2.$$

Integrating

$$\int_0^v dv = \int_0^t (2 + 0.2t)\,dt,$$

we obtain v as a function of time:

$$v = \frac{ds}{dt} = 2t + 0.1t^2 \text{ m/s}.$$

Integrating

$$\int_0^s ds = \int_0^t (2t + 0.1t^2)\,dt,$$

we find that the coordinate s as a function of time is

$$s = t^2 + \frac{0.1}{3}t^3 \text{ m}.$$

At $t = 10$ s, the distance moved along the track is

$$s = (10)^2 + \frac{0.1}{3}(10)^3 = 133 \text{ m}.$$

(b) At $t = 10$ s, the tangential component of the acceleration is

$$a_t = 2 + 0.2(10) = 4 \text{ m/s}^2.$$

We must also determine the normal component of acceleration. The instantaneous radius of curvature of the path is the radius of the circular track, $\rho = 400$ m. The magnitude of the velocity at $t = 10$ s is

$$v = 2(10) + 0.1(10)^2 = 30 \text{ m/s}.$$

Therefore, the normal acceleration is

$$a_n = \frac{v^2}{\rho} = \frac{(30)^2}{400} = 2.25 \text{ m/s}^2.$$

The magnitude of the acceleration at $t = 10$ s is

$$|\mathbf{a}| = \sqrt{a_t^2 + a_n^2} = \sqrt{(4)^2 + (2.25)^2} = 4.59 \text{ m/s}^2.$$

Example 13.10

The Circular-Orbit Problem

A satellite is in a circular orbit of radius R around the earth. What is its velocity?

Strategy

The acceleration due to gravity at a distance R from the center of the earth is gR_E^2/R^2, where R_E is the radius of the earth. (See Eq. 12.4.) By using this expression together with the equation for the acceleration in terms of normal and tangential components, we can obtain an equation for the satellite's velocity.

Solution

In terms of normal and tangential components (Fig. a), the acceleration of the satellite is

$$\mathbf{a} = \frac{dv}{dt}\,\mathbf{e}_t + \frac{v^2}{R}\,\mathbf{e}_n.$$

This expression must equal the acceleration due to gravity toward the center of the earth:

$$\frac{dv}{dt}\,\mathbf{e}_t + \frac{v^2}{R}\,\mathbf{e}_n = \frac{gR_E^2}{R^2}\,\mathbf{e}_n.$$

Because there is no $\mathbf{e}_t$ component on the right side, we conclude that the magnitude of the satellite's velocity is constant:

$$\frac{dv}{dt} = 0.$$

Equating the $\mathbf{e}_n$ components and solving for v, we obtain

$$v = \sqrt{\frac{gR_E^2}{R}}.$$

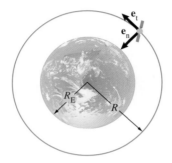

(a) Describing the satellite's motion in terms of normal and tangential components.

Discussion

In Example 13.5, we determined the escape velocity of an object traveling straight away from the earth in terms of its initial distance from the center of the earth. The escape velocity for an object a distance R from the center of the earth, $v_{esc} = \sqrt{2gR_E^2/R}$, is only $\sqrt{2}$ times the velocity of an object in a circular orbit of radius R. This explains why it was possible to begin launching probes to other planets not long after the first satellites were placed in orbit.

Example 13.11

Relating Cartesian Components to Normal and Tangential Components

During a flight in which a helicopter starts from rest at $t = 0$, the cartesian components of its acceleration are

$$a_x = 0.6t \text{ m/s}^2$$

and
$$a_y = 1.8 - 0.36t \ \text{m/s}^2.$$

What are the normal and tangential components of the helicopter's acceleration and the instantaneous radius of curvature of its path at $t = 4$ s?

Strategy

We can integrate the cartesian components of acceleration to determine the cartesian components of the velocity at $t = 4$ s, and then we can determine the components of the tangential unit vector $\mathbf{e}_t$ by dividing the velocity vector by its magnitude: $\mathbf{e}_t = \mathbf{v}/|\mathbf{v}|$. Next, we can determine the tangential component of the acceleration by evaluating the dot product of the acceleration vector with $\mathbf{e}_t$. Knowing the tangential component of the acceleration, we can then evaluate the normal component and determine the radius of curvature of the path from the relation $a_n = v^2/\rho$.

Solution

Integrating the components of acceleration with respect to time (see Example 13.6), we find that the cartesian components of the velocity are
$$v_x = 0.3t^2 \ \text{m/s}$$

and
$$v_y = 1.8t - 0.18t^2 \ \text{m/s}.$$

At $t = 4$ s, $v_x = 4.80$ m/s and $v_y = 4.32$ m/s. The tangential unit vector $\mathbf{e}_t$ at $t = 4$ s is (Fig. a)

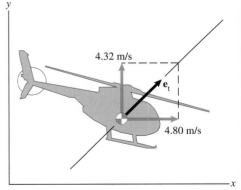

$$\mathbf{e}_t = \frac{\mathbf{v}}{|\mathbf{v}|} = \frac{4.80\mathbf{i} + 4.32\mathbf{j}}{\sqrt{(4.80)^2 + (4.32)^2}} = 0.743\mathbf{i} + 0.669\mathbf{j}.$$

The components of the acceleration at $t = 4$ s are
$$a_x = 0.6(4) = 2.4 \ \text{m/s}^2$$

and
$$a_y = 1.8 - 0.36(4) = 0.36 \ \text{m/s}^2,$$

so the tangential component of the acceleration at $t = 4$ s is

$$a_t = \mathbf{e}_t \cdot \mathbf{a}$$
$$= (0.743\mathbf{i} + 0.669\mathbf{j}) \cdot (2.4\mathbf{i} + 0.36\mathbf{j})$$
$$= 2.02 \ \text{m/s}^2$$

(a) Cartesian components of the velocity and the vector $\mathbf{e}_t$.

The magnitude of the acceleration is $\sqrt{(2.4)^2 + (0.36)^2} = 2.43$ m/s^2, so the magnitude of the normal component of the acceleration is
$$a_n = \sqrt{|\mathbf{a}|^2 - a_t^2} = \sqrt{(2.43)^2 - (2.02)^2} = 1.34 \ \text{m/s}^2.$$

The radius of curvature of the path is thus
$$\rho = \frac{|\mathbf{v}|^2}{a_n} = \frac{(4.80)^2 + (4.32)^2}{1.34} = 31.2 \ \text{m}.$$

Example 13.12

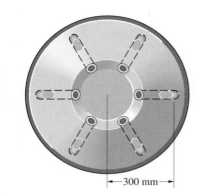

Figure 13.36

 Application to Engineering:

Centrifuge Design

The distance from the center of the medical centrifuge in Fig. 13.36 to its samples is 300 mm. When the centrifuge is turned on, its motor and control system give it an angular acceleration $\alpha = A - B\omega^2$. Choose the constants A and B so that the samples will be subjected to a maximum horizontal acceleration of 12,000 g's and the centrifuge will reach 90% of its maximum operating speed in 2 min.

Strategy

Since we know both the radius of the circular path in which the samples move and the horizontal acceleration to which they are to be subjected, we can solve for the operating angular velocity of the centrifuge. We will use the given angular acceleration to determine the centrifuge's angular velocity as a function of time in terms of the constants A and B. We can then use the operating angular velocity and the condition that the centrifuge reach 90% of the operating angular velocity in 2 min to determine the constants A and B.

Solution

From Eq. (13.45), the samples are subjected to a normal acceleration

$$a_n = R\omega^2.$$

Setting $a_n = (12,000)(9.81)$ m/s^2 and $R = 0.3$ m and solving for the angular velocity, we find that the desired maximum operating speed is $\omega_{max} = 626$ rad/s.

The angular acceleration is

$$\alpha = \frac{d\omega}{dt} = A - B\omega^2.$$

We separate variables to get

$$\frac{d\omega}{A - B\omega^2} = dt.$$

Then we integrate to determine ω as a function of time, assuming that the centrifuge starts from rest at $t = 0$:

$$\int_0^\omega \frac{d\omega}{A - B\omega^2} = \int_0^t dt.$$

Evaluating the integrals, we obtain

$$\frac{1}{2\sqrt{AB}} \ln\left(\frac{A + \sqrt{AB}\omega}{A - \sqrt{AB}\omega}\right) = t.$$

The solution of this equation for ω is

$$\omega = \sqrt{\frac{A}{B}}\left(\frac{e^{2\sqrt{AB}t} - 1}{e^{2\sqrt{AB}t} + 1}\right).$$

As t becomes large, ω approaches $\sqrt{A/B}$, so we have the condition that

$$\sqrt{\frac{A}{B}} = \omega_{max} = 626 \text{ rad/s}, \qquad (13.50)$$

and we can write the equation for ω as

$$\omega = \omega_{max}\left(\frac{e^{2\sqrt{AB}t} - 1}{e^{2\sqrt{AB}t} + 1}\right). \qquad (13.51)$$

We also have the condition that $\omega = 0.9\omega_{max}$ after 2 min. Setting $\omega = 0.9\omega_{max}$ and $t = 120$ s in Eq. (13.51) and solving for $\sqrt{AB}$, we obtain

$$\sqrt{AB} = \frac{\ln(19)}{240}.$$

We solve this equation together with Eq. (13.50), obtaining $A = 7.69 \text{ rad/s}^2$ and $B = 1.96 \times 10^{-5} \text{ rad}^{-1}$. Figure 13.37 shows the angular velocity of the centrifuge as a function of time.

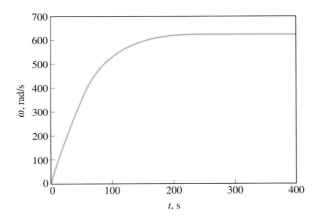

Figure 13.37
Graph of the angular velocity as a function of time.

$\mathscr{D}$esign Issues

A centrifuge is a device designed to rotate and thereby subject objects to large normal accelerations. The common spin dryer used to dry clothing is an example. Centrifuges are used extensively in biological and medical testing and research and also in the chemical processing industry to separate constituents of different densities. By using a centrifuge to separate the solid matter in blood (primarily red cells) from the liquid plasma, diagnostic tests can be carried out on the plasma, and the patient's *hematocrit*—the percentage of solid matter—can be determined. During the early days of manned space flight, there was concern that the large accelerations to which the astronauts would be subjected during their boost to orbit might injure them or affect their ability to carry out essential tasks. Large centrifuges consisting of rotating beams with "gondolas" at the end capable of carrying a person were built and used to study the effects of acceleration on human subjects. Centrifuges have also been designed to investigate the effects of large accelerations on mechanisms and to facilitate the modeling of geophysical phenomena such as slope stability and cratering. One such centrifuge is shown in Fig. 13.38. Consisting of a large beam that rotates samples in a horizontal circular path with a radius of 8 m, it can subject samples to 150 g's of acceleration.

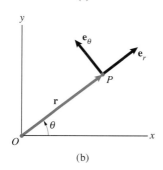

Centrifuges intended for commercial use, such as medical testing, must be designed so that samples can be installed easily, and they must have an electric motor and control system that permits specified accelerations to be achieved quickly and accurately. The bearing must be designed to sustain the very high angular velocities and to have a long lifetime. (The *ultracentrifuge*, used in biological research, reaches angular velocities of millions of revolutions per minute and is usually designed with an air bearing so that there is no solid contact between the spinning centrifuge and its support.) The structure of a centrifuge must be designed to support the large forces resulting from its rapid rotational motion, and the objects being accelerated must be adequately supported.

An essential consideration in the design of a centrifuge is the possibility of structural failure, which could result in projectiles moving outward at high velocities. A centrifuge must be designed with a surrounding structure strong enough that such projectiles would be safely contained and not cause additional damage or injury. This design requirement was achieved in the case of the large centrifuge in Fig. 13.38 by locating it underground.

Figure 13.38
A centrifuge designed to subject test articles to large accelerations. (Photograph courtesy of Sandia National Laboratories.)

Polar and Cylindrical Coordinates

Polar coordinates are often used to describe the curvilinear motion of a point. Circular motion, certain orbit problems, and, more generally, *central-force* problems, in which the acceleration of a point is directed toward a given point, can be expressed conveniently in polar coordinates.

Consider a point P in the xy-plane of a cartesian coordinate system. We can specify the position of P relative to the origin O either by its cartesian coordinates x,y or by its polar coordinates r,θ (Fig. 13.39a). To express vectors in terms of polar coordinates, we define a unit vector $\mathbf{e}_r$ that points in the direction of the radial line from the origin to P and a unit vector $\mathbf{e}_\theta$ that is perpendicular to $\mathbf{e}_r$ and points in the direction of increasing θ (Fig. 13.39b). In terms of these vectors, the position vector $\mathbf{r}$ from O to P is

$$\mathbf{r} = r\mathbf{e}_r. \tag{13.52}$$

(Notice that $\mathbf{r}$ has no component in the direction of $\mathbf{e}_\theta$.)

We can determine the velocity of P in terms of polar coordinates by taking the time derivative of Eq. (13.52):

$$\mathbf{v} = \frac{d\mathbf{r}}{dt} = \frac{dr}{dt}\mathbf{e}_r + r\frac{d\mathbf{e}_r}{dt}. \tag{13.53}$$

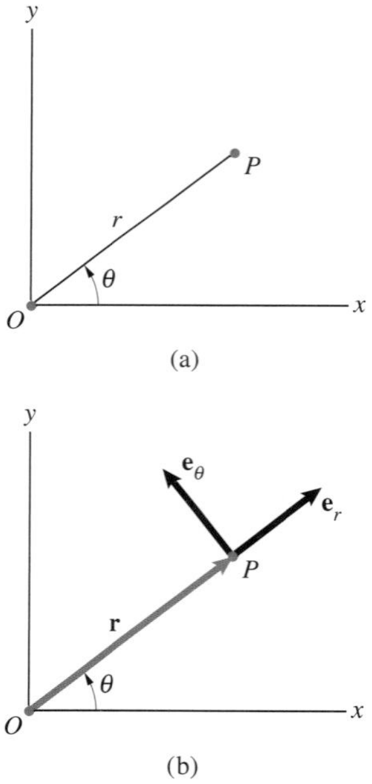

As P moves along a curvilinear path, the unit vector $\mathbf{e}_r$ rotates with angular velocity $\omega = d\theta/dt$. Therefore, from Eq. (13.33), we can express the time derivative of $\mathbf{e}_r$ in terms of $\mathbf{e}_\theta$ as

$$\frac{d\mathbf{e}_r}{dt} = \frac{d\theta}{dt}\mathbf{e}_\theta. \tag{13.54}$$

Substituting this result into Eq. (13.53), we obtain the velocity of P:

$$\mathbf{v} = \frac{dr}{dt}\mathbf{e}_r + r\frac{d\theta}{dt}\mathbf{e}_\theta = \frac{dr}{dt}\mathbf{e}_r + r\omega\mathbf{e}_\theta. \tag{13.55}$$

We can get this result in another way that is less rigorous, but more direct and intuitive. Figure 13.40 shows the position vector of P at times t and $t + \Delta t$. The change in the position vector, $\mathbf{r}(t + \Delta t) - \mathbf{r}(t)$, consists of two

Figure 13.39
(a) The polar coordinates of P.
(b) The unit vectors $\mathbf{e}_r$ and $\mathbf{e}_\theta$ and position vector $\mathbf{r}$.

components. The component Δr is due to the change in the radial position r and is in the $\mathbf{e}_r$ direction. The component $r\Delta\theta$ is due to the change in θ and is in the $\mathbf{e}_\theta$ direction. Thus, the change in the position of P is (approximately)

$$\mathbf{r}(t + \Delta t) - \mathbf{r}(t) = \Delta r\,\mathbf{e}_r + r\Delta\theta\,\mathbf{e}_\theta.$$

Dividing this expression by Δt and taking the limit as $\Delta t \to 0$, we obtain the velocity of P:

$$\mathbf{v} = \lim_{\Delta t \to 0}\left(\frac{\Delta r}{\Delta t}\,\mathbf{e}_r + r\,\frac{\Delta\theta}{\Delta t}\,\mathbf{e}_\theta\right)$$

$$= \frac{dr}{dt}\,\mathbf{e}_r + r\omega\,\mathbf{e}_\theta.$$

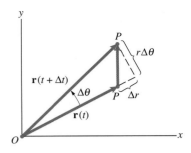

Figure 13.40
The position vector of P at t and $t + \Delta t$.

One component of the velocity is in the radial direction and is equal to the rate of change of the radial position r. The other component is normal, or *transverse*, to the radial direction and is proportional to the radial distance and to the rate of change of θ.

We obtain the acceleration of P by taking the time derivative of Eq. (13.55):

$$\mathbf{a} = \frac{d\mathbf{v}}{dt} = \frac{d^2r}{dt^2}\,\mathbf{e}_r + \frac{dr}{dt}\frac{d\mathbf{e}_r}{dt} + \frac{dr}{dt}\frac{d\theta}{dt}\,\mathbf{e}_\theta + r\frac{d^2\theta}{dt^2}\,\mathbf{e}_\theta + r\frac{d\theta}{dt}\frac{d\mathbf{e}_\theta}{dt}. \quad (13.56)$$

The time derivative of the unit vector $\mathbf{e}_r$ due to the rate of change of θ is given by Eq. (13.54). As P moves, $\mathbf{e}_\theta$ also rotates with angular velocity $d\theta/dt$ (Fig. 13.41). You can see from this figure that the time derivative of $\mathbf{e}_\theta$ is in the $-\mathbf{e}_r$ direction if $d\theta/dt$ is positive:

$$\frac{d\mathbf{e}_\theta}{dt} = -\frac{d\theta}{dt}\,\mathbf{e}_r.$$

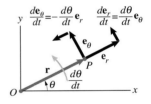

Figure 13.41
Time derivatives of $\mathbf{e}_r$ and $\mathbf{e}_\theta$.

Substituting this expression and Eq. (13.54) into Eq. (13.56), we obtain the acceleration of P:

$$\mathbf{a} = \left[\frac{d^2r}{dt^2} - r\left(\frac{d\theta}{dt}\right)^2\right]\mathbf{e}_r + \left[r\frac{d^2\theta}{dt^2} + 2\frac{dr}{dt}\frac{d\theta}{dt}\right]\mathbf{e}_\theta.$$

Thus, the velocity and acceleration are respectively (Fig. 13.42)

$$\mathbf{v} = v_r\mathbf{e}_r + v_\theta\mathbf{e}_\theta = \frac{dr}{dt}\,\mathbf{e}_r + r\omega\mathbf{e}_\theta \quad (13.57)$$

and

$$\mathbf{a} = a_r\mathbf{e}_r + a_\theta\mathbf{e}_\theta, \quad (13.58)$$

where

$$a_r = \frac{d^2r}{dt^2} - r\left(\frac{d\theta}{dt}\right)^2 = \frac{d^2r}{dt^2} - r\omega^2$$

and

$$a_\theta = r\frac{d^2\theta}{dt^2} + 2\frac{dr}{dt}\frac{d\theta}{dt} = r\alpha + 2\frac{dr}{dt}\omega. \quad (13.59)$$

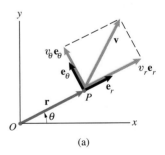

(a)

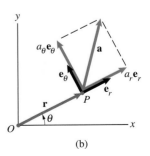

(b)

Figure 13.42
Radial and transverse components of the velocity (a) and acceleration (b).

The term $-r\omega^2$ in the radial component of the acceleration is called the *centripetal acceleration*, and the term $2(dr/dt)\omega$ in the transverse component is called the *Coriolis acceleration*.

The unit vectors $\mathbf{e}_r$ and $\mathbf{e}_\theta$ are related to the cartesian unit vectors by

$$\mathbf{e}_r = \cos\theta\mathbf{i} + \sin\theta\mathbf{j}$$

and

$$\mathbf{e}_\theta = -\sin\theta\mathbf{i} + \cos\theta\mathbf{j}. \tag{13.60}$$

Circular Motion Circular motion can be conveniently described using either radial and transverse or normal and tangential components. Let us compare these two methods of expressing the velocity and acceleration of a point P moving in a circular path of radius R (Fig. 13.43). Because the polar coordinate $r = R$ is constant, Eq. (13.57) for the velocity reduces to

$$\mathbf{v} = R\omega\mathbf{e}_\theta.$$

In terms of normal and tangential components, the velocity is

$$\mathbf{v} = v\mathbf{e}_t.$$

Notice in Fig. 13.43 that $\mathbf{e}_\theta = \mathbf{e}_t$. Comparing these two expressions for the velocity, we obtain the relation between the velocity and the angular velocity in circular motion:

$$v = R\omega.$$

From Eqs. (13.58) and (13.59), the acceleration for a circular path of radius R in terms of polar coordinates is

$$\mathbf{a} = -R\omega^2\mathbf{e}_r + R\alpha\mathbf{e}_\theta,$$

and the acceleration in terms of normal and tangential components is

$$\mathbf{a} = \frac{dv}{dt}\mathbf{e}_t + \frac{v^2}{R}\mathbf{e}_n.$$

The unit vector $\mathbf{e}_r = -\mathbf{e}_n$. Because of the relation $v = R\omega$, the normal components of acceleration are equal: $v^2/R = R\omega^2$. Equating the transverse and tangential components, we obtain the relation

$$\frac{dv}{dt} = a_t = R\alpha.$$

Cylindrical Coordinates Polar coordinates describe the motion of a point P in the xy-plane. We can describe three-dimensional motion by using *cylindrical coordinates* r, θ, and z (Fig. 13.44). The cylindrical coordinates r and θ are the polar coordinates of P, measured in the plane parallel to the xy-plane, and the definitions of the unit vectors $\mathbf{e}_r$ and $\mathbf{e}_\theta$ are unchanged. The position of P perpendicular to the xy-plane is measured by the coordinate z, and the unit vector $\mathbf{e}_z$ points in the positive z-axis direction.

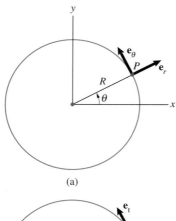

(a)

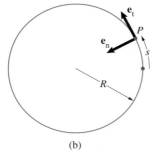

(b)

Figure 13.43
A point P moving in a circular path.
(a) Polar coordinates.
(b) Normal and tangential components.

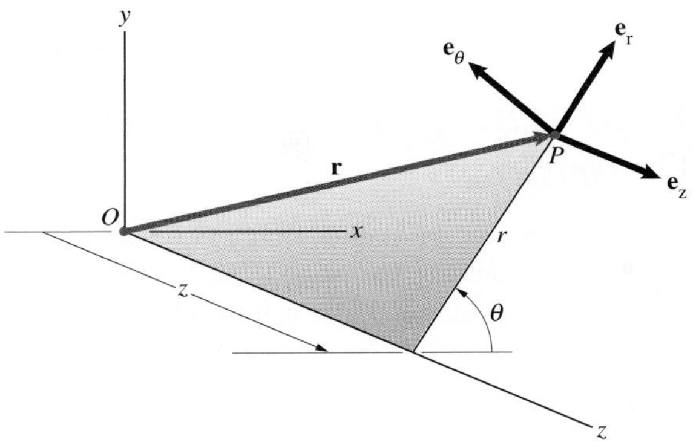

Figure 13.44
Cylindrical coordinates r, θ, and z of point P and the unit vectors $\mathbf{e}_r$, $\mathbf{e}_\theta$, and $\mathbf{e}_z$.

In terms of cylindrical coordinates, the position vector $\mathbf{r}$ is the sum of the expression for the position vector in polar coordinates and the z component:

$$\mathbf{r} = r\mathbf{e}_r + z\mathbf{e}_z. \tag{13.61}$$

(The polar coordinate r is not the magnitude of $\mathbf{r}$, except when P lies in the xy-plane.) By taking time derivatives, we obtain the velocity

$$\mathbf{v} = \frac{d\mathbf{r}}{dt} = v_r\mathbf{e}_r + v_\theta\mathbf{e}_\theta + v_z\mathbf{e}_z$$

$$= \frac{dr}{dt}\mathbf{e}_r + r\omega\mathbf{e}_\theta + \frac{dz}{dt}\mathbf{e}_z \tag{13.62}$$

and acceleration

$$\mathbf{a} = \frac{d\mathbf{v}}{dt} = a_r\mathbf{e}_r + a_\theta\mathbf{e}_\theta + a_z\mathbf{e}_z, \tag{13.63}$$

where

$$a_r = \frac{d^2r}{dt^2} - r\omega^2, \quad a_\theta = r\alpha + 2\frac{dr}{dt}\omega, \quad \text{and} \quad a_z = \frac{d^2z}{dt^2}. \tag{13.64}$$

Notice that Eqs. (13.62) and (13.63) reduce to the polar coordinate expressions for the velocity and acceleration, Eqs. (13.57) and (13.58), when P moves along a path in the xy-plane.

Study Questions

1. What are the definitions of the unit vectors $\mathbf{e}_r$ and $\mathbf{e}_\theta$?
2. What is the position vector of a point in terms of polar coordinates?
3. If a point moves in a circular path about the origin of a coordinate system, what is its Coriolis acceleration? What is its centripetal acceleration?

Example 13.13

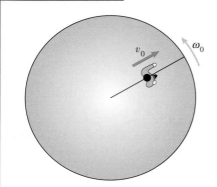

Figure 13.45

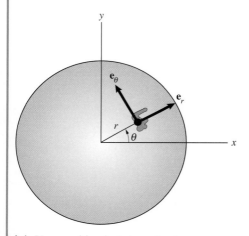

(a) Your position in terms of polar coordinates.

Expressing Motion in Terms of Polar Coordinates

Suppose that you are standing on a large disk (say, a merry-go-round) rotating with constant angular velocity ω_0 and you start walking at constant speed v_0 along a straight radial line painted on the disk (Fig. 13.45). What are your velocity and acceleration when you are a distance r from the center of the disk?

Strategy

We can describe your motion in terms of polar coordinates (Fig. a). By using the information given about your motion and the motion of the disk, we can evaluate the terms in the expressions for the velocity and acceleration in terms of polar coordinates.

Solution

The speed with which you walk along the radial line is the rate of change of r, $dr/dt = v_0$, and the angular velocity of the disk is the rate of change of θ, $\omega = \omega_0$. Your velocity is

$$\mathbf{v} = \frac{dr}{dt}\mathbf{e}_r + r\omega\mathbf{e}_\theta = v_0\mathbf{e}_r + r\omega_0\mathbf{e}_\theta.$$

Your velocity consists of two components: a radial component due to the speed at which you are walking and a transverse component due to the disk's rate of rotation. The transverse component increases as your distance from the center of the disk increases.

Your walking speed $v_0 = dr/dt$ is constant, so $d^2r/dt^2 = 0$. Also, the disk's angular velocity $\omega_0 = d\theta/dt$ is constant, so $d^2\theta/dt^2 = 0$. The radial component of your acceleration is

$$a_r = \frac{d^2r}{dt^2} - r\omega^2 = -r\omega_0^2,$$

and the transverse component is

$$a_\theta = r\alpha + 2\frac{dr}{dt}\omega = 2v_0\omega_0.$$

Discussion

If you have ever tried walking on a merry-go-round, you know that it is a difficult proposition. This example indicates why: Subjectively, you are walking along a straight line with constant velocity, but you are actually experiencing the centripetal acceleration a_r and the Coriolis acceleration a_θ due to the disk's rotation.

Example 13.14

Relating Polar Components to Cartesian Components

The robot arm in Fig. 13.46 is programmed so that point P traverses the path described by
$$r = 1 - 0.5 \cos 2\pi t \text{ m}$$
and
$$\theta = 0.5 - 0.2 \sin 2\pi t \text{ rad.}$$
At $t = 0.8$ s, determine (a) the velocity of P in terms of polar coordinates and (b) the cartesian components of the velocity of P.

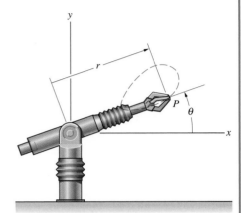

Strategy

(a) Since we are given r and θ as functions of time, we can calculate the derivatives in the expression for the velocity in terms of polar coordinates and obtain the velocity as a function of time.
(b) By determining the value of θ at $t = 0.8$ s, we can use trigonometry to determine the cartesian components in terms of the polar components.

Figure 13.46

Solution

(a) From Eq. (13.57), the velocity is
$$\mathbf{v} = \frac{dr}{dt}\mathbf{e}_r + r\frac{d\theta}{dt}\mathbf{e}_\theta$$
$$= (\pi \sin 2\pi t)\mathbf{e}_r + (1 - 0.5\cos 2\pi t)(-0.4\pi \cos 2\pi t)\mathbf{e}_\theta.$$

At $t = 0.8$ s,
$$\mathbf{v} = -2.99\mathbf{e}_r - 0.328\mathbf{e}_\theta \text{ (m/s)}.$$

(b) At $t = 0.8$ s, $\theta = 0.690$ rad $= 39.5°$ (Fig. a). The x component of the velocity of P is
$$v_x = v_r \cos 39.5° - v_\theta \sin 39.5°$$
$$= (-2.99) \cos 39.5° - (-0.328) \sin 39.5° = -2.09 \text{ m/s,}$$

and the y component is
$$v_y = v_r \sin 39.5° + v_\theta \cos 39.5°$$
$$= (-2.99) \sin 39.5° + (-0.328) \cos 39.5° = -2.16 \text{ m/s.}$$

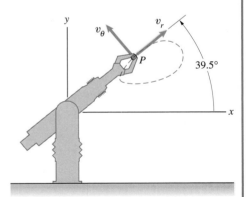

(a) Position at $t = 0.8$ s.

Discussion

When you determine components of a vector in terms of different coordinate systems, you should check them to make sure they give the same magnitude. In this example,
$$|\mathbf{v}| = \sqrt{(-2.99)^2 + (-0.328)^2} = \sqrt{(-2.09)^2 + (-2.16)^2} = 3.01 \text{ m/s.}$$

Remember that although the components of the velocity are different in the two coordinate systems, those components describe the same velocity vector.

Example 13.15

Velocity in Terms of Polar and Cartesian Components

In the cam–follower mechanism shown in Fig. 13.47, the slotted bar rotates with constant angular velocity $\omega = 4$ rad/s and the radial position of the follower is determined by the elliptic profile of the stationary cam. The path of the follower is described by the polar equation

$$r = \frac{0.15}{1 + 0.5 \cos \theta} \text{ m.}$$

Determine the velocity of the follower when $\theta = 45°$ in terms of (a) polar coordinates and (b) cartesian coordinates.

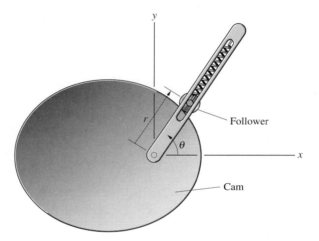

Figure 13.47

Strategy

By taking the time derivative of the polar equation for the profile of the cam, we can obtain a relation between the known angular velocity and the radial component of velocity that permits us to evaluate the velocity in terms of polar coordinates. Then, by using Eqs. (13.60), we can obtain the velocity in terms of cartesian coordinates.

Solution

(a) The polar equation for the cam profile is of the form $r = r(\theta)$. Taking its derivative with respect to time, we obtain

$$\frac{dr}{dt} = \frac{dr(\theta)}{d\theta} \frac{d\theta}{dt}$$

$$= \frac{d}{d\theta} \left(\frac{0.15}{1 + 0.5 \cos \theta} \right) \frac{d\theta}{dt}$$

$$= \left[\frac{0.075 \sin \theta}{(1 + 0.5 \cos \theta)^2} \right] \frac{d\theta}{dt}.$$

The velocity of the follower in polar coordinates is therefore

$$\mathbf{v} = \frac{dr}{dt}\mathbf{e}_r + r\frac{d\theta}{dt}\mathbf{e}_\theta$$

$$= \left[\frac{0.075\sin\theta}{(1+0.5\cos\theta)^2}\right]\frac{d\theta}{dt}\mathbf{e}_r + \left(\frac{0.15}{1+0.5\cos\theta}\right)\frac{d\theta}{dt}\mathbf{e}_\theta.$$

The angular velocity $\omega = d\theta/dt = 4$ rad/s, so we can evaluate the polar components of the velocity when $\theta = 45°$, obtaining

$$\mathbf{v} = 0.116\mathbf{e}_r + 0.443\mathbf{e}_\theta \text{ (m/s)}.$$

(b) Substituting Eqs. (13.60) with $\theta = 45°$ into the polar coordinate expression for the velocity, we obtain the velocity in terms of cartesian coordinates:

$$\mathbf{v} = 0.116\mathbf{e}_r + 0.443\mathbf{e}_\theta$$

$$= 0.116(\cos 45°\mathbf{i} + \sin 45°\mathbf{j}) + 0.443(-\sin 45°\mathbf{i} + \cos 45°\mathbf{j})$$

$$= -0.232\mathbf{i} + 0.395\mathbf{j} \text{ (m/s)}.$$

Computational Mechanics

The following examples and problems are designed to be worked with the use of a programmable calculator or computer.

Computational Example 13.16

With buoyancy accounted for, the downward acceleration of a steel ball falling in the container of liquid in Fig. 13.48 is $a = 0.9g - cv$, where c is a constant that is proportional to the viscosity of the liquid. To determine the viscosity, a rheologist releases the ball from rest at the surface of the liquid. If the ball requires 2 s to fall the 2 m to the bottom, what is the value of c?

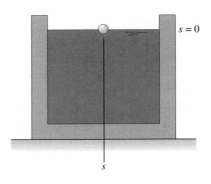

Figure 13.48

Strategy

We can obtain an equation for c by determining the distance the ball falls as a function of time.

Solution

We measure the ball's position s downward from the point of release and let $t = 0$ be the time of release. The acceleration is

$$a = \frac{dv}{dt} = 0.9g - cv.$$

Separating variables and integrating gives

$$\int_0^v \frac{dv}{0.9g - cv} = \int_0^t dt,$$

from which it follows that

$$\left[\frac{1}{(-c)} \ln(0.9g - cv) \right]_0^v = [t]_0^t,$$

or

$$\frac{1}{-c} \left[\ln(0.9g - cv) - \ln(0.9g) \right] = \frac{1}{(-c)} \ln\left(\frac{0.9g - cv}{0.9g} \right) = t.$$

Solving for v, we obtain

$$v = \frac{ds}{dt} = \frac{0.9g}{c} \left(1 - e^{-ct} \right).$$

Integrating this equation yields

$$\int_0^s ds = \int_0^t \frac{0.9g}{c} \left(1 - e^{-ct} \right) dt,$$

or

$$[s]_0^s = \frac{0.9g}{c} \left[t - \frac{e^{-ct}}{(-c)} \right]_0^t,$$

from which we obtain the distance the ball has fallen as a function of the time from its release:

$$s = \frac{0.9g}{c^2} \left(ct - 1 + e^{-ct} \right).$$

We know that $s = 2$ m when $t = 2$ s, so determining c requires solving the equation

$$f(c) = \frac{(0.9)(9.81)}{c^2} \left(2c - 1 + e^{-2c} \right) - 2 = 0.$$

We can't solve this transcendental equation in closed form to determine c. Problem-solving programs such as Maple, MathCad, and Mathematica, are designed to obtain roots of such equations. Another approach is to compute the value of $f(c)$ for a range of values of c and plot the results, as we have done in Fig. 13.49. From the graph, we estimate that $c = 8.3 \ s^{-1}$.

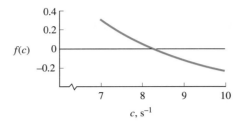

Figure 13.49
Graph of the function $f(c)$.

Computational Example 13.17

In an industrial process, the sprayer in Fig. 13.50 projects a stream of liquid onto the horizontal surface. An engineer wants to place the sprayer at a horizontal position x_e that maximizes its coverage of the horizontal surface. That is, she wants to maximize the distance $x_s - x_e$ reached by the spray. The angle θ_0 can be varied, but the spray velocity $v_0 = 3$ m/s is fixed. The dimensions $H = 0.2$ m, $L = 0.12$ m, and $h = 0.06$ m. If the droplets of the spray are treated as projectiles, what is the desired position x_e?

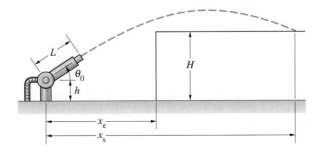

Figure 13.50

Strategy

We first observe that the trajectory that maximizes the coverage of the horizontal surface will be one that just clears the edge of the surface. For if a trajectory clears the edge by some horizontal distance, as in Fig. 13.50, the sprayer can be moved that distance to the right, increasing its coverage by that amount. Our procedure will be to choose a distance x_e, determine the angle θ_0 that causes the trajectory to just clear the edge, and calculate $x_s - x_e$. By doing this for a range of values of x_e, we can determine the horizontal placement that maximizes $x_s - x_e$.

Solution

Let $t = 0$ be the time at which a droplet leaves the nozzle. In terms of the coordinate system shown in Fig. (a), the x- and y-coordinates of the droplet are

$$x = L \cos\theta_0 + v_0 \cos\theta_0 t$$

and

$$y = h + L \sin\theta_0 + v_0 \sin\theta_0 t - \frac{1}{2} g t^2.$$

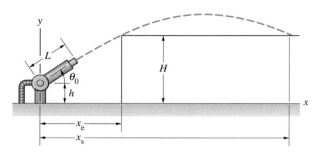

(a)

By setting $y = H$, we can solve for the times at which the droplet is at the height of the horizontal surface. The two resulting solutions for t are

$$t_1, t_2 = \frac{v_0 \sin\theta_0 \pm \sqrt{v_0^2 \sin^2\theta_0 - 2g(H - h - L \sin\theta_0)}}{g}.$$

We wish the smaller root t_1 to be the time at which the droplet passes the edge of the horizontal surface. Then the root t_2 will be the time at which the droplet lands on the surface. Setting $x = x_e$ when $t = t_1$, we obtain

$$x_e = L \cos\theta_0 + v_0 \cos\theta_0 t_1,$$

which is an equation we can use to determine θ_0 numerically for a given value of x_e. That is, we obtain the angle of the nozzle for which the spray just clears the edge of the surface. Once we know θ_0, we can determine x_s by substituting t_2 into the equation for x:

$$x_s = L \cos\theta_0 + v_0 \cos\theta_0 t_2.$$

Figure 13.51 shows the values of $x_s - x_e$ we obtained for a range of values of x_e. From the graph we estimate that a maximum coverage of $x_s - x_e = 0.82$ m is obtained at $x_e = 0.12$ m. The trajectory that gives the maximum coverage, obtained with a nozzle angle $\theta_0 = 50°$, is shown in Fig. 13.52.

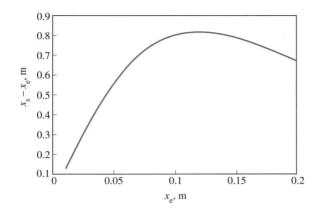

Figure 13.51
Graph of the extent of coverage as a function of the placement of the nozzle.

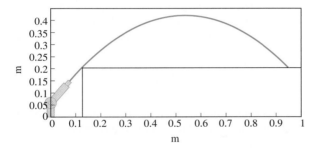

Figure 13.52
Placement of the nozzle and the trajectory that maximizes coverage of the surface.

Chapter Summary

In this chapter, we were concerned with the motion of a point relative to a given reference frame. We defined the position, velocity, and acceleration of the point and showed how to express them in terms of cartesian coordinates, normal and tangential components, and polar coordinates. We demonstrated how to determine the velocity and acceleration by differentiation when the position is known and how to determine the velocity and position by integration when the acceleration is known. In Chapter 14, we will use Newton's second law to determine the acceleration of an object when the forces acting on it are known. Once the acceleration is known, the methods we developed in the current chapter will be applied to obtain information about the velocity and position of the object.

The position of a point P relative to a given reference frame with origin O can be specified by the *position vector* $\mathbf{r}$ from O to P. The *velocity* of P relative to the reference frame is

$$\mathbf{v} = \frac{d\mathbf{r}}{dt},$$

Eq. (13.1)

and the *acceleration* of P relative to the reference frame is

$$\mathbf{a} = \frac{d\mathbf{v}}{dt}.$$ Eq. (13.2)

Straight-Line Motion

The position of a point P on a straight line relative to a reference point O is specified by a coordinate s measured along the line from O to P. The coordinate s and the velocity and acceleration of P along the line are related by

$$v = \frac{ds}{dt}$$ Eq. (13.3)

and

$$a = \frac{dv}{dt}.$$ Eq. (13.4)

If the acceleration is specified as a function of time, the velocity and position can be determined as functions of time by integration. If the acceleration is specified as a function of velocity, $dv/dt = a(v)$, the velocity can be determined as a function of time by separating variables and integrating:

$$\int_{v_0}^{v} \frac{dv}{a(v)} = \int_{t_0}^{t} dt.$$ Eq. (13.16)

If the acceleration is specified as a function of position, $dv/dt = a(s)$, the chain rule can be used to express the acceleration in terms of a derivative with respect to position:

$$\frac{dv}{dt} = \frac{dv}{ds}\frac{ds}{dt} = \frac{dv}{ds}v = a(s).$$

Separating variables, the velocity can be determined as a function of position:

$$\int_{v_0}^{v} v\, dv = \int_{s_0}^{s} a(s)\, ds.$$ Eq. (13.19)

Cartesian Coordinates

The position, velocity, and acceleration relative to the cartesian coordinate system in Fig. (a) are [Eqs. (13.21)–(13.25)]

$$\mathbf{r} = x\mathbf{i} + y\mathbf{j} + z\mathbf{k},$$

$$\mathbf{v} = v_x\mathbf{i} + v_y\mathbf{j} + v_z\mathbf{k} = \frac{dx}{dt}\mathbf{i} + \frac{dy}{dt}\mathbf{j} + \frac{dz}{dt}\mathbf{k},$$

and

$$\mathbf{a} = a_x\mathbf{i} + a_y\mathbf{j} + a_z\mathbf{k} = \frac{dv_x}{dt}\mathbf{i} + \frac{dv_y}{dt}\mathbf{j} + \frac{dv_z}{dt}\mathbf{k}.$$

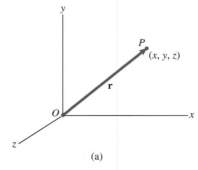

(a)

Figure (a)

The equations describing the motion in each coordinate direction are identical in form to the equations that describe the motion of a point along a straight line.

Angular Motion

The angular velocity ω and angular acceleration α of L relative to L_0 are (Fig. b)

$$\omega = \frac{d\theta}{dt} \qquad \text{Eq. (13.31)}$$

and

$$\alpha = \frac{d\omega}{dt} = \frac{d^2\theta}{dt^2}. \qquad \text{Eq. (13.32)}$$

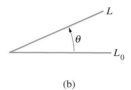

(b)

Figure (b)

Normal and Tangential Components

The velocity and acceleration are (Fig. c),

$$\mathbf{v} = v\mathbf{e}_t = \frac{ds}{dt}\,\mathbf{e}_t \qquad \text{Eq. (13.38)}$$

and

$$\mathbf{a} = a_t\mathbf{e}_t + a_n\mathbf{e}_n, \qquad \text{Eq. (13.39)}$$

where

$$a_t = \frac{dv}{dt} \quad \text{and} \quad a_n = v\frac{d\theta}{dt} = \frac{v^2}{\rho}. \qquad \text{Eq. (13.40)}$$

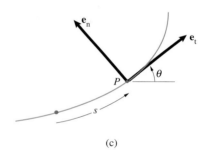

(c)

Figure (c)

The unit vector $\mathbf{e}_n$ points toward the concave side of the path. The term ρ is the instantaneous radius of curvature of the path.

Polar Coordinates

The position, velocity, and acceleration are (Fig. d),

$$\mathbf{r} = r\mathbf{e}_r, \qquad \text{Eq. (13.52)}$$

$$\mathbf{v} = v_r\mathbf{e}_r + v_\theta\mathbf{e}_\theta = \frac{dr}{dt}\,\mathbf{e}_r + r\omega\mathbf{e}_\theta, \qquad \text{Eq. (13.57)}$$

and

$$\mathbf{a} = a_r\mathbf{e}_r + a_\theta\mathbf{e}_\theta, \qquad \text{Eq. (13.58)}$$

where

$$a_r = \frac{d^2r}{dt^2} - r\left(\frac{d\theta}{dt}\right)^2 = \frac{d^2r}{dt^2} - r\omega^2$$

and

$$a_\theta = r\frac{d^2\theta}{dt^2} + 2\frac{dr}{dt}\frac{d\theta}{dt} = r\alpha + 2\frac{dr}{dt}\omega. \qquad \text{Eq. (13.59)}$$

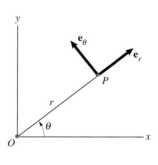

(d)

Figure (d)

Review Problems

13.1 The graph of the velocity v of a point as a function of time is a straight line. When $t = 2$ s, $v = 4$ ft/s, and when $t = 6$ s, $v = -12$ ft/s.

(a) What is the velocity of the point at $t = 0$?
(b) What is the acceleration of the point?

13.2 Suppose that you throw a ball straight up at 10 m/s and release it at 2 m above the ground.

(a) What maximum height above the ground does the ball reach?
(b) How long after you release it does the ball hit the ground?
(c) What is the magnitude of its velocity just before it hits the ground?

Ⓓ **13.3** Suppose that you must determine the duration of the yellow light at a highway intersection. Assume that cars will be approaching the intersection traveling as fast as 65 mi/hr, that drivers' reaction times are as long as 0.5 s, and that cars can safely achieve a deceleration of at least 0.4 g.

(a) How long must the light remain yellow to allow drivers to come to a stop safely before the light turns red?
(b) What is the minimum distance cars must be from the intersection when the light turns yellow to come to a stop safely at the intersection?

13.4 The acceleration of a point moving along a straight line is $a = 4t + 2$ m/s². When $t = 2$ s, the position of the point is $s = 36$ m, and when $t = 4$ s, its position is $s = 90$ m. What is the velocity of the point when $t = 4$ s?

13.5 A model rocket takes off straight up. Its acceleration during the 2 s its motor burns is 25 m/s². Neglect aerodynamic drag, and determine

(a) the maximum velocity of the rocket during the flight and
(b) the maximum altitude the rocket reaches.

13.6 In Problem 13.5, if the rocket's parachute fails to open, what is the total time of flight from takeoff until the rocket hits the ground?

13.7 The acceleration of a point moving along a straight line is $a = -cv^3$, where c is a constant. If the velocity of the point is v_0, what distance does the point move before its velocity decreases to $v_0/2$?

13.8 Water leaves the nozzle at 20° above the horizontal and strikes the wall at the point indicated. What is the velocity of the water as it leaves the nozzle?

Strategy: Determine the motion of the water by treating each particle of water as a projectile.

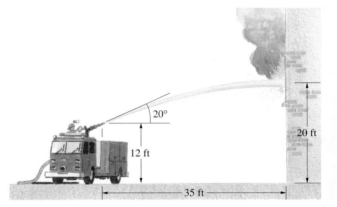

20°

20 ft

12 ft

35 ft

P13.8

13.9 In practice, the quarterback throws the football with velocity v_0 at 45° above the horizontal. At the same instant, the receiver standing 20 ft in front of him starts running straight downfield at 10 ft/s and catches the ball. Assume that the ball is thrown and caught at the same height above the ground. What is the velocity v_0?

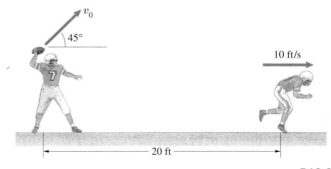

v_0

45°

10 ft/s

20 ft

P13.9

P13.5

13.10 The constant velocity $v = 2$ m/s. What are the magnitudes of the velocity and acceleration of point P when $x = 0.25$ m?

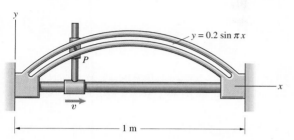

$y = 0.2 \sin \pi x$

1 m

P13.10

13.11 In Problem 13.10, what is the acceleration of point P in terms of normal and tangential components when $x = 0.25$ m? What is the instantaneous radius of curvature of the path?

13.12 In Problem 13.10, what is the acceleration of point P in terms of polar coordinates when $x = 0.25$ m?

13.13 A point P moves along the spiral path $r = (0.1)\theta$ ft, where θ is in radians. The angular position $\theta = 2t$ rad, where t is in seconds, and $r = 0$ at $t = 0$. Determine the magnitudes of the velocity and acceleration of P at $t = 1$ s.

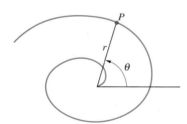

P13.13

13.14 In the cam–follower mechanism, the slotted bar rotates with constant angular velocity $\omega = 12$ rad/s, and the radial position of the follower A is determined by the profile of the stationary cam. The slotted bar is pinned a distance $h = 0.2$ m to the left of the center of the circular cam. The follower moves in a circular path 0.42 m in radius. Determine the velocity of the follower when $\theta = 40°$

(a) in terms of polar coordinates and
(b) in terms of cartesian coordinates.

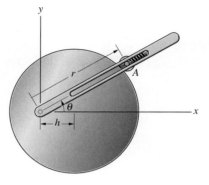

P13.14

13.15 In Problem 13.14, determine the acceleration of the follower when $\theta = 40°$

(a) in terms of polar coordinates and
(b) in terms of cartesian coordinates.

When the forces acting on the soccer ball are known, Newton's second law can be used to determine the acceleration of the ball's center of mass.

Force, Mass, and Acceleration

U ntil now, we have analyzed motions of objects without considering the forces that cause them. In this chapter, we relate cause and effect: By drawing the free-body diagram of an object to identify the forces acting on it, we can use Newton's second law to determine the acceleration of the object. Alternatively, when we know an object's acceleration, we can use Newton's second law to obtain information about the forces acting on the object.

14.1 Newton's Second Law

Newton stated that the total force on a particle is equal to the rate of change of its *linear momentum*, which is the product of its mass and velocity:

$$\mathbf{f} = \frac{d}{dt}(m\mathbf{v}).$$

If the particle's mass is constant, the total force equals the product of its mass and acceleration:

$$\mathbf{f} = m\frac{d\mathbf{v}}{dt} = m\mathbf{a}. \tag{14.1}$$

We pointed out in Chapter 12 that the second law gives precise meanings to the terms *force* and *mass*. Once a unit of mass is chosen, a unit of force is defined to be the force necessary to give one unit of mass an acceleration of unit magnitude. For example, the unit of force in SI units, the newton, is the force necessary to give a mass of one kilogram an acceleration of one meter per second squared. In principle, the second law then gives the value of any force and the mass of any object. By subjecting a one-kilogram mass to an arbitrary force and measuring the acceleration of the mass, we can solve the second law for the direction of the force and its magnitude in newtons. By subjecting an arbitrary mass to a one-newton force and again measuring the acceleration, we can solve the second law for the value of the mass in kilograms.

If the mass of a particle and the total force acting on it are known, Newton's second law determines its acceleration. In Chapter 13, we described how to determine the velocity, position, and trajectory of a point whose acceleration is known. Therefore, with the second law, a particle's motion can be determined when the total force acting on it is known, or the total force can be determined when the motion is known.

14.2 Equation of Motion for the Center of Mass

Newton's second law is postulated for a particle, or small element of matter, but an equation of precisely the same form describes the motion of the center of mass of an arbitrary object. We can show that the total external force on an arbitrary object is equal to the product of its mass and the acceleration of its center of mass.

To do so, we conceptually divide an arbitrary object into N particles. Let m_i be the mass of the ith particle, and let $\mathbf{r}_i$ be its position vector (Fig. 14.1a). The object's mass m is the sum of the masses of the particles; that is,

$$m = \sum_i m_i,$$

where the summation sign with subscript i means "the sum over i from 1 to N." The position of the object's center of mass is

$$\mathbf{r} = \frac{\sum_i m_i \mathbf{r}_i}{m}.$$

By taking two time derivatives of this expression, we obtain

$$\sum_i m_i \frac{d^2\mathbf{r}_i}{dt^2} = m\frac{d^2\mathbf{r}}{dt^2} = m\mathbf{a}, \tag{14.2}$$

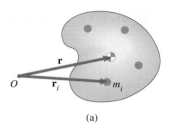

(a)

where $\mathbf{a}$ is the acceleration of the object's center of mass.

The ith particle of the object may be acted upon by forces exerted by the other particles of the object. Let $\mathbf{f}_{ij}$ be the force exerted on the ith particle by the jth particle. Then Newton's third law states that the ith particle exerts a force on the jth particle of equal magnitude and opposite direction: $\mathbf{f}_{ji} = -\mathbf{f}_{ij}$. If the external force on the ith particle (i.e., the total force exerted on the ith particle by objects other than the object we are considering) is denoted by $\mathbf{f}_i^E$, Newton's second law for the ith particle is (Fig. 14.1b)

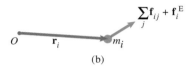

(b)

Figure 14.1
(a) Dividing an object into particles. The vector $\mathbf{r}_i$ is the position vector of the ith particle, and $\mathbf{r}$ is the position vector of the object's center of mass.
(b) Forces on the ith particle.

$$\sum_j \mathbf{f}_{ij} + \mathbf{f}_i^E = m_i \frac{d^2\mathbf{r}_i}{dt^2}.$$

We can write this equation for each particle of the object. Summing the resulting equations from $i = 1$ to N, we obtain

$$\sum_i \sum_j \mathbf{f}_{ij} + \sum_i \mathbf{f}_i^E = \sum_i m_i \frac{d^2\mathbf{r}_i}{dt^2}. \tag{14.3}$$

The first term on the left side, the sum of the internal forces on the object, is zero, due to Newton's third law:

$$\sum_i \sum_j \mathbf{f}_{ij} = \mathbf{f}_{12} + \mathbf{f}_{21} + \mathbf{f}_{13} + \mathbf{f}_{31} + \cdots = \mathbf{0}.$$

The second term on the left side of Eq. (14.3) is the sum of the external forces on the object. Denoting this sum by $\Sigma\mathbf{F}$ and using Eq. (14.2), we conclude that *the sum of the external forces equals the product of the total mass and the acceleration of the center of mass*:

$$\Sigma\mathbf{F} = m\mathbf{a}. \tag{14.4}$$

Because this equation is identical in form to Newton's postulate for a particle, for convenience we also refer to it as Newton's second law.

Notice that we made no assumptions restricting the nature of the object or its state of motion in obtaining Eq. (14.4). The sum of the external forces on any object or collection of objects, solid, liquid, or gas, equals the product of the total mass and the acceleration of the center of mass. For example, suppose that the space shuttle is in orbit and has fuel remaining in its tanks. If its engines are turned on, the fuel sloshes in a complicated manner, affecting the shuttle's motion due to internal forces between the fuel and the shuttle. Nevertheless, we can use Eq. (14.4) to determine the exact acceleration of the center of mass of the shuttle, including the fuel it contains, and thereby determine the velocity, position, and trajectory of the center of mass.

14.3 Inertial Reference Frames

When we discussed the motion of a point in Chapter 13, we specified the position, velocity, and acceleration of the point relative to an arbitrary reference frame. But Newton's second law cannot be expressed in terms of just any reference frame. Suppose that no force acts on a particle and that we measure the particle's motion relative to a particular reference frame and determine that its acceleration is zero. In terms of this reference frame, Newton's second law agrees with our observation. But if we then measure the particle's motion relative to a second reference frame that is accelerating or rotating with respect to the first one, we would find that the particle's acceleration is *not* zero. In terms of the second reference frame, Newton's second law, at least in the form given by Eq. (14.4), does not predict the correct result.

A well-known example is a person riding in an elevator. Suppose that you conduct an experiment in which you ride in an elevator while standing on a set of scales (Fig. 14.2a). The forces acting on you are your weight W and the force N exerted on you by the scales (Fig. 14.2b). You exert an equal and opposite force N on the scales, which is the force they measure. If the elevator is stationary, you observe that the scales read your weight, $N = W$. The sum of the forces on you is zero, and Newton's second law correctly states that your acceleration relative to the elevator is zero. If the elevator has an upward acceleration a (Fig. 14.2c), you know you will feel heavier, and indeed, you observe that the scales read a force greater than your weight, $N > W$. In terms of an earth-fixed reference frame, Newton's second law correctly relates the forces acting on you to your acceleration: $\Sigma F = N - W = ma$. But suppose that you use the elevator as your frame of reference. Then the sum of the forces acting on you is not zero, so Newton's second law states that you are accelerating relative to the elevator. But you are stationary relative to the elevator. Thus, expressed in terms of this accelerating reference frame, Newton's second law gives an erroneous result.

Newton stated that the second law should be expressed in terms of a reference frame at rest with respect to the "fixed stars." Even if the stars were fixed, that would not be practical advice, because virtually every convenient reference frame accelerates, rotates, or both. Newton's second law *can* be applied rigorously using reference frames that accelerate and rotate by properly accounting for the acceleration and rotation. We explain how to do this in Chapter 17, but for now, we need to give some indication of when Newton's second law can be applied.

Fortunately, in nearly all "down-to-earth" situations, Newton's second law can be expressed in the form given by Eq. (14.4) in terms of a reference frame that is fixed relative to the earth and obtain sufficiently accurate answers. For example, if you throw a piece of chalk across a room, you can use a reference frame that is fixed relative to the room to predict the chalk's motion. While the chalk is in motion, the earth rotates, and therefore the reference frame rotates. But *because the chalk's flight is brief*, the effect on your prediction is very small. (The earth rotates slowly—its angular velocity is one-half that of a clock's hour hand.) Equation (14.4) can usually be applied using a reference frame that translates (moves without rotating) at constant velocity relative to the earth. For example, if you and a friend play tennis on the deck of a cruise ship moving with constant velocity relative to the earth,

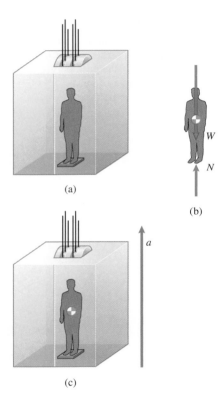

(a)

(b)

(c)

Figure 14.2
(a) Riding in an elevator while standing on scales.
(b) Your free-body diagram.
(c) Upward acceleration of the elevator.

you can apply Eq. (14.4) in terms of a reference frame fixed relative to the ship to analyze the ball's motion. But you cannot do so if the ship is turning or if it is changing its speed. In some situations that are not "down to earth," such as the motions of earth satellites and spacecraft near the earth, Eq. (14.4) can be applied using a nonrotating reference frame with its origin at the center of the earth.

A reference frame in which Eq. (14.4) can be applied is said to be *Newtonian*, or *inertial*. We discuss inertial reference frames in greater detail in Chapter 17. For now, assume that the examples and problems that follow are expressed in terms of inertial reference frames.

14.4 Applications

In the study of statics, you became familiar with different types of forces, including the weights of objects, the normal and friction forces exerted by contacting surfaces, and forces exerted by linear springs. By showing these forces acting on free-body diagrams, information was obtained about the systems of forces acting on objects in equilibrium. In this section, we show that the same types of forces are dealt with in dynamics. Furthermore, the techniques used to draw free-body diagrams in statics also apply to objects that are not in equilibrium.

By drawing the free-body diagram of an object, the external forces acting on it can be identified and Newton's second law used to determine the object's acceleration. Conversely, if the motion of an object is known, Newton's second law can be used to determine the total external force on the object. In particular, if an object's acceleration in a particular direction is known to be zero, the sum of the external forces in that direction must equal zero.

To apply Newton's second law in a particular situation, a coordinate system must be chosen. Often, the forces acting on an object can be resolved into components most conveniently in terms of a particular coordinate system, or the choice may be determined by the object's path. In the sections that follow, we show how to use different types of coordinate systems to analyze the motions of objects and the forces acting on them.

Cartesian Coordinates and Straight-Line Motion

If we express the sum of the forces acting on an object of mass m and the acceleration of its center of mass in terms of their components in a cartesian reference frame (Fig. 14.3), Newton's second law states that

$$\Sigma \mathbf{F} = m\mathbf{a},$$

or

$$\left(\Sigma F_x \mathbf{i} + \Sigma F_y \mathbf{j} + \Sigma F_z \mathbf{k} \right) = m \left(a_x \mathbf{i} + a_y \mathbf{j} + a_z \mathbf{k} \right).$$

Equating x, y, and z components, we obtain three scalar equations of motion:

$$\Sigma F_x = ma_x, \quad \Sigma F_y = ma_y, \quad \text{and} \quad \Sigma F_z = ma_z. \tag{14.5}$$

The total force in each coordinate direction equals the product of the mass and the component of the acceleration in that direction.

An important example is the projectile problem, in which an object is launched through the air and aerodynamic forces are neglected, so that the

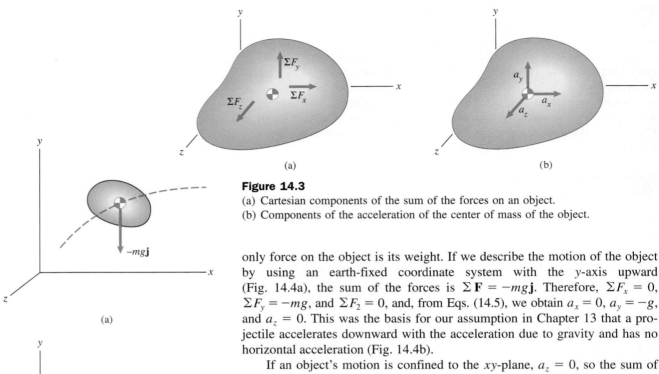

(a)

(b)

Figure 14.3
(a) Cartesian components of the sum of the forces on an object.
(b) Components of the acceleration of the center of mass of the object.

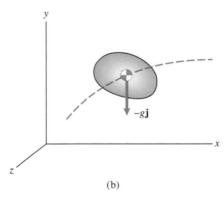

(a)

(b)

Figure 14.4
(a) Free-body diagram of a projectile.
(b) The resulting acceleration of the center of mass of the projectile.

only force on the object is its weight. If we describe the motion of the object by using an earth-fixed coordinate system with the y-axis upward (Fig. 14.4a), the sum of the forces is $\Sigma \mathbf{F} = -mg\mathbf{j}$. Therefore, $\Sigma F_x = 0$, $\Sigma F_y = -mg$, and $\Sigma F_z = 0$, and, from Eqs. (14.5), we obtain $a_x = 0$, $a_y = -g$, and $a_z = 0$. This was the basis for our assumption in Chapter 13 that a projectile accelerates downward with the acceleration due to gravity and has no horizontal acceleration (Fig. 14.4b).

If an object's motion is confined to the xy-plane, $a_z = 0$, so the sum of the forces in the z direction is zero. Thus, when the motion is confined to a fixed plane, the component of the total force normal to that plane equals zero. For straight-line motion along the x-axis (Fig. 14.5a), Eqs. (14.5) are

$$\Sigma F_x = ma_x, \qquad \Sigma F_y = 0, \quad \text{and} \quad \Sigma F_z = 0.$$

We see that in straight-line motion, the components of the total force perpendicular to the line equal zero, and the component of the total force tangent to the line equals the product of the mass and the acceleration along the line (Fig. 14.5b).

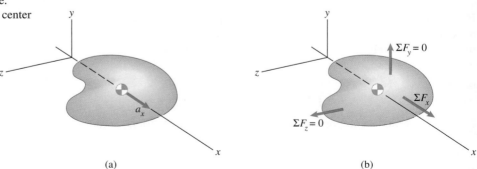

(a)

(b)

Figure 14.5
(a) Acceleration of an object in straight-line motion along the x-axis.
(b) The y and z components of the total force acting on the object equal zero.

Study Questions

1. Explain how Newton's second law gives precise meanings to the terms *force* and *mass*.
2. Newton's second law states that the total external force on an object equals the product of the object's mass and the acceleration of what point?
3. What is an inertial reference frame?
4. If you know that the sum of the external forces on an object is zero in a particular direction, what does that tell you?

Example 14.1

Application to Straight-Line Motion

The airplane in Fig. 14.6 touches down on the aircraft carrier with a horizontal velocity of 50 m/s relative to the carrier. The arresting gear exerts a horizontal force of magnitude $T_x = 10{,}000v$ newtons (N), where v is the plane's velocity in meters per second. The plane's mass is 6500 kg.

(a) What maximum horizontal force does the arresting gear exert on the plane?

(b) If other horizontal forces can be neglected, what distance does the plane travel before coming to rest?

Figure 14.6

Strategy

(a) Since the plane begins to decelerate when it contacts the arresting gear, the maximum force occurs at first contact when $v = 50$ m/s.

(b) The horizontal force exerted by the arresting gear equals the product of the plane's mass and its acceleration. Once we know the acceleration, we can integrate to determine the distance required for the plane to come to rest.

Solution

(a) We draw the free-body diagram of the airplane and introduce a coordinate system in Fig. (a). The forces T_x and T_y are the horizontal and vertical components of force exerted by the arresting gear, and N is the vertical force on the landing gear. The horizontal force on the plane is $\Sigma F_x = -T_x = -10{,}000v$ N. The magnitude of the maximum force is

$$10{,}000v = (10{,}000)(50) = 500{,}000 \text{ N,}$$

or 112,400 lb.

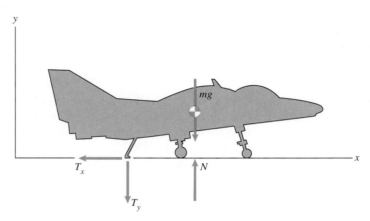

(a) Introducing a coordinate system with the x-axis parallel to the horizontal force.

(b) In terms of the plane's horizontal component of acceleration (Fig. b), we obtain the equation of motion

$$\Sigma F_x = ma_x,$$

or

$$-10{,}000v_x = ma_x.$$

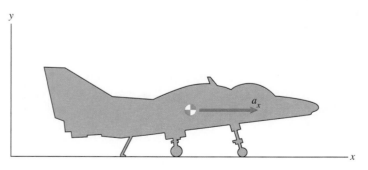

(b) The airplane's horizontal acceleration.

The airplane's acceleration' is a function of its velocity. We use the chain rule to express the acceleration in terms of a derivative with respect to x:

$$ma_x = m\frac{dv_x}{dt} = m\frac{dv_x}{dx}\frac{dx}{dt} = m\frac{dv_x}{dx}v_x = -10{,}000v_x.$$

Now we separate variables and integrate, defining $x = 0$ to be the position at which the plane contacts the arresting gear:

$$\int_{50}^{0} m \, dv_x = -\int_{0}^{x} 10{,}000 \, dx.$$

Evaluating the integrals and solving for x, we obtain

$$x = \frac{50m}{10,000} = \frac{(50)(6500)}{10,000} = 32.5 \text{ m}.$$

Discussion

As we demonstrate in this example, once you have used Newton's second law to determine an object's acceleration, you can apply the methods developed in Chapter 13 to determine the object's position and velocity.

Example 14.2

Connected Objects in Straight-Line Motion

The two crates in Fig. 14.7 are released from rest. Their masses are $m_A = 40$ kg and $m_B = 30$ kg, and the coefficients of friction between crate A and the inclined surface are $\mu_s = 0.2$ and $\mu_k = 0.15$. What is the acceleration of the crates?

Strategy

We must first determine whether A slips. We will assume that the crates remain stationary and see whether the force of friction necessary for equilibrium exceeds the maximum friction force. If slip occurs, we can determine the resulting acceleration by drawing free-body diagrams of the crates and applying Newton's second law to them individually.

Solution

We draw the free-body diagram of crate A and introduce a coordinate system in Fig. (a). If we assume that the crate does not slip, the following equilibrium equations apply:

$$\Sigma F_x = T + m_A g \sin 20° - f = 0;$$

$$\Sigma F_y = N - m_A g \cos 20° = 0.$$

In the first equation, the tension T equals the weight of crate B; therefore, the friction force necessary for equilibrium is

$$f = m_B g + m_A g \sin 20° = (30 + 40 \sin 20°)(9.81) = 429 \text{ N}.$$

The normal force $N = m_A g \cos 20°$, so the maximum friction force the surface will support is

$$f_{\max} = \mu_s N = (0.2)\big[(40)(9.81)\cos 20°\big] = 73.7 \text{ N}.$$

Crate A will therefore slip, and the friction force is $f = \mu_k N$. We show the crate's acceleration down the plane in Fig. (b). Its acceleration perpendicular to the plane is zero (i.e., $a_y = 0$). Applying Newton's second law yields

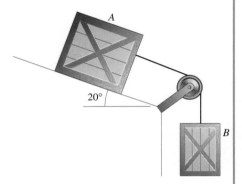

Figure 14.7

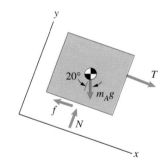

(a) Free-body diagram of crate A.

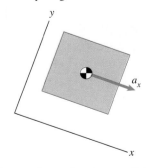

(b) The crate's acceleration parallel to the plane.

$$\Sigma F_x = T + m_A g \sin 20° - \mu_k N = m_A a_x$$

$$\Sigma F_y = N - m_A g \cos 20° = 0.$$

In this case, *we do not know* the tension T, because crate B is not in equilibrium. We show the free-body diagram of crate B and its vertical acceleration in Figs. (c) and (d). The equation of motion is

$$\Sigma F_x = m_B g - T = m_B a_x.$$

(In terms of the two coordinate systems we use, the two crates have the same acceleration a_x.) Thus, by applying Newton's second law to both crates, we have obtained three equations in terms of the unknowns T, N, and a_x. Solving for a_x, we obtain $a_x = 5.33$ m/s^2.

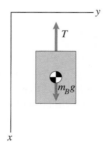

(c) Free-body diagram of crate B.

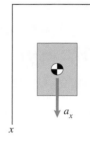

(d) Vertical acceleration of crate B.

Discussion

Notice that we assumed the tension in the cable to be the same on each side of the pulley (Fig. e). In fact, however, the tensions must be different, because a moment is necessary to cause angular acceleration of the pulley. For now, our only recourse is to assume that the pulley is light enough that the moment necessary to accelerate it is negligible. In Chapter 18, we include the analysis of the angular motion of the pulley in problems of this type and obtain more realistic solutions.

(e) The tension is assumed to be the same on both sides of the pulley.

Example 14.3

Newton's Second Law in Cartesian Coordinates

The sport utility vehicle in Fig. 14.8, which weighs 3000 lb with its driver, has left the ground after driving over a rise. At the instant shown, the vehicle is moving horizontally at 30 mi/hr and the bottoms of its tires are 24 in. above the (approximately) level ground. The earth-fixed coordinate system is placed with its origin 30 in. above the ground, at the height of the vehicle's center of mass when the tires first contact the ground. (Assume the vehicle remains horizontal.) Upon first contact, the vehicle's center of mass initially continues moving downward and then rebounds upward due to the flexure of the suspension system. While the tires are in contact with the ground, the force exerted on them by the road is

$$-R\mathbf{i} - N\mathbf{j} = -2400\mathbf{i} - 18{,}000y\mathbf{j} \text{ (lb)},$$

where y is the vertical position of the center of mass in feet. (The vertical component N is a function of y because of the flexure of the suspension.) When the vehicle hits the ground, what is the magnitude of the maximum acceleration to which it is subjected? If the 160-lb driver is subjected to the same acceleration, what maximum force is exerted on him by the vehicle?

Figure 14.8

Strategy

We will analyze the vehicle's motion in two phases: before and after its wheels contact the ground. If we neglect aerodynamic forces, the only force on the vehicle before its wheels contact the ground is its weight. As a result, it accelerates downward with the acceleration due to gravity and has no horizontal acceleration, so we can determine the vehicle's velocity when its wheels first touch the ground. We can then use Newton's second law to determine the vehicle's acceleration while its wheels are in contact with the ground. The maximum acceleration will occur when the suspension system has reached its maximum flexure, which means that the center of mass has reached its minimum height. By integrating the y component of the acceleration, we can determine the minimum height reached by the center of mass and evaluate the maximum acceleration.

Solution

Before the Wheels Contact the Ground The components of the vehicle's velocity at the instant shown in Fig. 14.8 are $v_x = (88/60)(30 \text{ mi/hr}) = 44 \text{ ft/s}$ and $v_y = 0$. Before the tires come into contact with the ground, the components of acceleration are $a_x = 0$ and $a_y = g$. The horizontal velocity therefore remains constant. We integrate to determine the y component of the velocity as a function of time. We have

$$a_y = \frac{dv_y}{dt} = g,$$

so that

$$\int_0^{v_y} dv_y = \int_0^t g \, dt,$$

which gives

$$v_y = \frac{dy}{dt} = gt.$$

The initial height of the center of mass is $y = -24$ in. $= -2$ ft. Integrating to determine the height as a function of time yields

$$\int_{-2}^y dy = \int_0^t gt \, dt,$$

or

$$y = -2 + \frac{1}{2} gt^2.$$

Setting $y = 0$, we determine the time at which the tires contact the ground:

$$t = \sqrt{\frac{4}{g}} = \sqrt{\frac{4}{32.2}} = 0.352 \text{ s}.$$

The vehicle's components of velocity when the tires contact the ground are (Fig. a)

$$v_x = 44 \text{ ft/s} \quad \text{and} \quad v_y = gt = (32.2)(0.352) = 11.3 \text{ ft/s}.$$

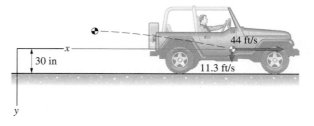

(a) Velocity of the center of mass when the tires contact the ground.

After the Wheels Contact the Ground In Figs. (b) and (c), we show the free-body diagram of the vehicle and its components of acceleration while its wheels are in contact with the ground. From the y component of Newton's second law,

$$\Sigma F_y = W - N = ma_y = \left(\frac{W}{g}\right) a_y,$$

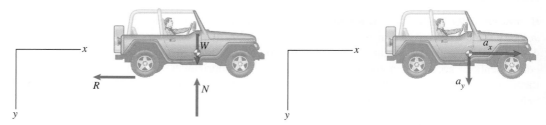

(b) Free-body diagram of the vehicle.

(c) Acceleration of the center of mass.

and we determine the vehicle's vertical acceleration:

$$a_y = \frac{dv_y}{dt} = g\left(1 - \frac{N}{W}\right)$$

$$= 32.2\left(1 - \frac{18{,}000y}{3000}\right)$$

$$= 32.2(1 - 6y).$$

Because the vehicle's vertical acceleration is a function of y, we use the chain rule to express the acceleration in terms of y instead of t:

$$\frac{dv_y}{dt} = \frac{dv_y}{dy}\frac{dy}{dt} = \frac{dv_y}{dy}v_y = 32.2(1 - 6y).$$

Separating variables and integrating yields

$$\int_{11.3}^{v_y} v_y\, dv_y = \int_0^y 32.2(1 - 6y)\, dy,$$

$$\left[\frac{v_y^2}{2}\right]_{11.3}^{v_y} = \left[32.2(y - 3y^2)\right]_0^y,$$

from which we obtain an equation for the vertical velocity as a function of y:

$$v_y = \sqrt{(11.3)^2 + 64.4(y - 3y^2)}.$$

To determine the maximum flexure of the suspension, we set $v_y = 0$ in this equation and solve for y, obtaining $y = 1$ ft. The maximum force exerted on the vehicle is therefore

$$\Sigma \mathbf{F} = -R\mathbf{i} + (W - N)\mathbf{j}$$

$$= -2400\mathbf{i} + \left[3000 - 18{,}000(1)\right]\mathbf{j}$$

$$= -2400\mathbf{i} - 15{,}000\mathbf{j}\ (\text{lb}).$$

The vehicle's maximum acceleration is

$$\mathbf{a} = \frac{1}{m}\Sigma \mathbf{F} = \frac{g}{W}\Sigma \mathbf{F} = \frac{32.2}{3000}(-2400\mathbf{i} - 15{,}000\mathbf{j})$$

$$= -25.8\mathbf{i} - 161.0\mathbf{j}\ \left(\text{ft/s}^2\right).$$

The magnitude of the maximum acceleration is

$$|\mathbf{a}| = \sqrt{(-25.8)^2 + (-161.0)^2} = 163\ \text{ft/s}^2,$$

which is 5 g's.

Now, let $\mathbf{F}_D$ be the force exerted by the vehicle on the 160-lb driver when the vehicle's acceleration is a maximum. Then newton's second law for the driver is

$$\Sigma \mathbf{F} = m\mathbf{a} = \left(\frac{W}{g}\right)\mathbf{a},$$

$$\mathbf{F}_D + 160\mathbf{j} = \left(\frac{160}{32.2}\right)(-25.8\mathbf{i} - 161.0\mathbf{j}).$$

Solving, we determine that the maximum force is

$$\mathbf{F}_D = -128\mathbf{i} - 960\mathbf{j} \text{ (lb)}.$$

Discussion

Engineers can use calculations of this kind in the design of vehicles and passenger restraint systems, but our modeling is incomplete in important ways. For example, we did not consider the vehicle's shock absorbers, which would affect the dynamics of the impact. To accurately model the vehicle's interaction with the ground, the dynamics of the vehicle's wheels and suspension would need to be included. Also, the vehicle's seat would help cushion the driver from the impact, decreasing the maximum force exerted on him.

Normal and Tangential Components

When an object moves in a curved path, we can resolve the sum of the forces acting it into normal and tangential components (Fig. 14.9a). We can also express the object's acceleration in terms of normal and tangential components (Fig. 14.9b) and write Newton's second law, $\Sigma \mathbf{F} = m\mathbf{a}$, in the form

$$\left(\Sigma F_t \mathbf{e}_t + \Sigma F_n \mathbf{e}_n\right) = m\left(a_t \mathbf{e}_t + a_n \mathbf{e}_n\right), \tag{14.6}$$

where

$$a_t = \frac{dv}{dt} \quad \text{and} \quad a_n = \frac{v^2}{\rho}.$$

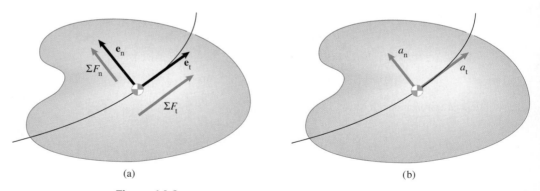

(a)   (b)

Figure 14.9
(a) Normal and tangential components of the sum of the forces on an object.
(b) Normal and tangential components of the acceleration of the center of mass of the object.

Equating the normal and tangential components in Eq. (14.6), we obtain two scalar equations of motion:

$$\Sigma F_t = ma_t = m\frac{dv}{dt}, \qquad \Sigma F_n = ma_n = m\frac{v^2}{\rho}. \tag{14.7}$$

The sum of the forces in the tangential direction equals the product of the mass and the rate of change of the magnitude of the velocity, and the sum of the forces in the normal direction equals the product of the mass and the normal component of acceleration. If the path of the object's center of mass lies in a plane, the acceleration of the center of mass perpendicular to the plane is zero, so the sum of the forces perpendicular to the plane is zero.

When an object moves in a circular path, normal and tangential components are usually the simplest choice for analyzing the motion of the object.

Example 14.4

Newton's Second Law in Normal and Tangential Components

Future space stations may be designed to rotate in order to provide simulated gravity for their inhabitants (Fig. 14.10). If the distance from the axis of rotation of the station to the occupied outer ring is $R = 100$ m, what rotation rate is necessary to simulate one-half of earth's gravity?

Figure 14.10

Strategy

By drawing the free-body diagram of a person and expressing Newton's second law in terms of normal and tangential components, we can relate the force exerted on the person by the floor to the angular velocity of the station. The person exerts an equal and opposite force on the floor, which is his effective weight.

Solution

We draw the free-body diagram of a person standing in the outer ring in Fig. (a), where N is the force exerted on him by the floor. Relative to a nonrotating reference frame with its origin at the center of the station, the person moves in a circular path of radius R. His normal and tangential

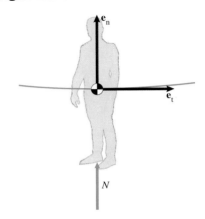

(a) Free-body diagram of a person standing in the occupied ring.

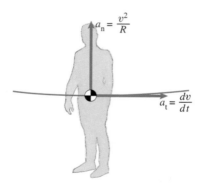

(b) The person's normal and tangential components of acceleration.

components of acceleration are shown in Fig. (b). Applying Eqs. (14.7), we obtain

$$\Sigma F_{\mathrm{t}} = 0 = m\frac{dv}{dt}$$

and

$$\Sigma F_{\mathrm{n}} = N = m\frac{v^2}{R}.$$

The first equation simply indicates that the magnitude of the person's velocity is constant. The second equation tells us the force N. The magnitude of his velocity is $v = R\omega$, where ω is the angular velocity of the station. If one-half of earth's gravity is simulated, $N = \frac{1}{2}mg$. Therefore

$$N = \frac{1}{2}mg = m\frac{(R\omega)^2}{R}.$$

Solving for ω, we obtain the necessary angular velocity of the station:

$$\omega = \sqrt{\frac{g}{2R}} = \sqrt{\frac{9.81}{(2)(100)}} = 0.221 \text{ rad/s}.$$

This is one revolution every 28.4 s.

Example 14.5

Forces on an Object in Circular Motion

The experimental magnetically levitated train in Fig. 14.11 is supported by magnetic repulsion forces exerted in a direction normal to the tracks. Motion of the train transverse to the tracks is prevented by lateral supports. The 20,000-kg train is traveling at 30 m/s on a circular segment of track of radius

Figure 14.11

$R = 150$ m, and the bank angle of the track is 40°. What force must the magnetic levitation system exert to support the train, and what total force is exerted by the lateral supports?

Strategy

We know the train's velocity and the radius of its circular path, so we can determine its normal component of acceleration. By expressing Newton's second law in terms of normal and tangential components, we can determine the components of force normal and transverse to the track.

Solution

Figure (a) shows the train viewed from above. The unit vector $\mathbf{e}_n$ is horizontal and points toward the center of the train's circular path, and $\mathbf{e}_t$ is tangential to the path. In Fig. (b) we draw the free-body diagram of the train seen from the front, where M is the magnetic force normal to the tracks and S is the transverse force. In Fig. (c) we show the train's acceleration, which is perpendicular to the circular path of the train and toward the center of the path. The sum of the forces in the vertical direction (perpendicular to the train's circular path) must equal zero:

$$M \cos 40° + S \sin 40° - mg = 0.$$

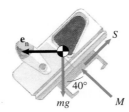

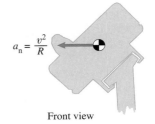

(a) The train's circular path viewed from above.

(b) Free-body diagram of the train.

(c) The train's acceleration.

The sum of the forces in the $\mathbf{e}_n$ direction equals the product of the mass and the normal component of the acceleration; that is,

$$\Sigma F_n = m \frac{v^2}{\rho},$$

or

$$M \sin 40° - S \cos 40° = m \frac{v^2}{R}.$$

Solving these two equations for M and S, we obtain $M = 227.4$ kN and $S = 34.2$ kN.

Example 14.6

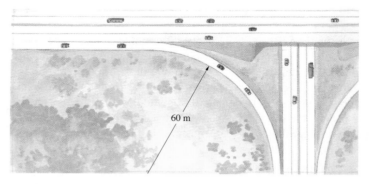

Application to Engineering:

Motor Vehicle Dynamics

A civil engineer's preliminary design for a freeway off-ramp is circular with radius $R = 60$ m (Fig. 14.12). If she assumes that the coefficient of static friction between tires and road is at least $\mu_s = 0.4$, what is the maximum speed at which vehicles can enter the ramp without losing traction?

Figure 14.12

Strategy

Since a vehicle on the off-ramp moves in a circular path, it has a normal component of acceleration that depends on its velocity. The necessary normal component of force is exerted by friction between the tires and the road, and the friction force cannot be greater than the product of μ_s and the normal force. By assuming that the friction force is equal to this value, we can determine the maximum velocity for which slipping will not occur.

Solution

We view the free-body diagram of a car on the off-ramp from above the car in Fig. (a) and from the front of the car in Fig. (b). In Fig. (c) we show the car's acceleration, which is perpendicular to the circular path of the car and toward the center of the path. The sum of the forces in the $\mathbf{e}_n$ direction equals the product of the mass and the normal component of the acceleration; that is,

$$\Sigma F_n = ma_n = m\frac{v^2}{R},$$

or

$$f = m\frac{v^2}{R}.$$

The required friction force increases as v increases. The maximum friction force the surfaces will support is $f_{max} = \mu_s N = \mu_s mg$. Therefore, the maximum velocity for which slipping does not occur is

$$v = \sqrt{\mu_s gR} = \sqrt{(0.4)(9.81)(60)} = 15.3 \text{ m/s},$$

or 55.2 km/hr (34.3 mi/hr).

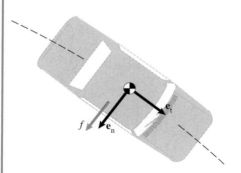

(a) Top view of the free-body diagram.

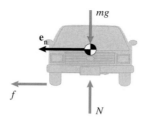

(b) Front view of the free-body diagram.

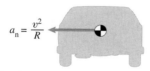

(c) The acceleration seen in the front view.

𝒟esign Issues

Automotive engineers, civil engineers who design highways, and engineers who study traffic accidents and their prevention must analyze and measure the motions of vehicles under different conditions. By using the methods discussed in this chapter, they can relate the forces acting on vehicles to their motions and study, for example, the factors influencing the distance necessary for a car to be brought to a stop in an emergency or the effects of banking and curvature on the velocity at which a car can safely be driven on a curved road (Fig. 14.13).

In this example, the analysis indicates that vehicles will lose traction if they enter the freeway off-ramp at speeds greater than 34.3 mi/hr. This result can be used as an indication of the speed limit that must be posted in order for vehicles to enter the ramp safely, or the off-ramp could be designed for a greater speed by increasing the radius of curvature of the road.

Figure 14.13
Tests of the capabilities of vehicles to negotiate curves influence the design of both vehicles and highways.

Polar and Cylindrical Coordinates

When an object moves in a planar curved path, we can describe the motion of the center of mass of the object in terms of polar coordinates. Resolving the sum of the forces parallel to the plane into radial and transverse components (Fig. 14.14a) and expressing the acceleration of the center of mass in terms of radial and transverse components (Fig. 14.14b), we can write Newton's second law, $\Sigma \mathbf{F} = m\mathbf{a}$, in the form

$$\left(\Sigma F_r \mathbf{e}_r + \Sigma F_\theta \mathbf{e}_\theta\right) = m\left(a_r \mathbf{e}_r + a_\theta \mathbf{e}_\theta\right), \tag{14.8}$$

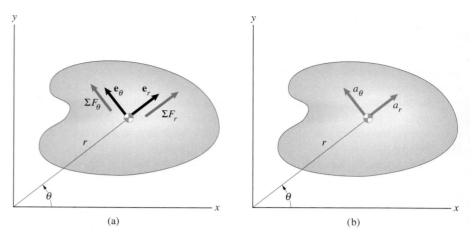

Figure 14.14
Radial and transverse components of the sum of the forces (a) and the acceleration of the center of mass (b).

where

$$a_r = \frac{d^2r}{dt^2} - r\left(\frac{d\theta}{dt}\right)^2 = \frac{d^2r}{dt^2} - r\omega^2$$

and

$$a_\theta = r\frac{d^2\theta}{dt^2} + 2\frac{dr}{dt}\frac{d\theta}{dt} = r\alpha + 2\frac{dr}{dt}\omega.$$

Equating the $\mathbf{e}_r$ and $\mathbf{e}_\theta$ components in Eq. (14.8), we obtain the scalar equations

$$\Sigma F_r = ma_r = m\left(\frac{d^2r}{dt^2} - r\omega^2\right) \tag{14.9}$$

and

$$\Sigma F_\theta = ma_\theta = m\left(r\alpha + 2\frac{dr}{dt}\omega\right). \tag{14.10}$$

The sum of the forces in the radial direction equals the product of the mass and the radial component of the acceleration, and the sum of the forces in the transverse direction equals the product of the mass and the transverse component of the acceleration. Since the object's acceleration perpendicular to the plane in which the motion takes place is zero, the sum of the forces perpendicular to the plane is zero.

We can describe the three-dimensional motion of an object using cylindrical coordinates, in which the position of the center of mass perpendicular to the x-y plane is measured by the coordinate z and the unit vector $\mathbf{e}_z$ points in the positive z direction. We resolve the sum of the forces into radial, transverse, and z components (Fig. 14.15a) and express the acceleration of the center of mass in terms of radial, transverse, and z components (Fig. 14.15b).

The three scalar equations of motion are the radial and transverse equations (14.9) and (14.10) and the equation of motion in the z direction,

$$\Sigma F_z = ma_z = m\frac{dv_z}{dt} = m\frac{d^2z}{dt^2}.$$

(14.11)

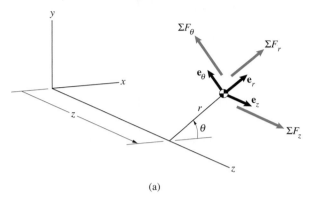

(a)

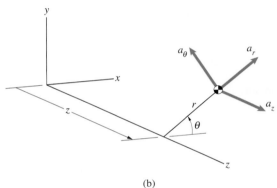

(b)

Figure 14.15
(a) Components of the sum of the forces on an object in cylindrical coordinates.
(b) Components of the acceleration of the center of mass.

Example 14.7

Newton's Second Law in Polar Coordinates

The smooth bar in Fig. 14.16 rotates *in the horizontal plane* with constant angular velocity ω_0. The unstretched length of the linear spring is r_0. The collar A has mass m and is released at $r = r_0$ with no radial velocity.
(a) Determine the radial velocity of the collar as a function of r.
(b) Determine the horizontal force exerted on the collar by the bar as a function of r.

Strategy

(a) The only force on the collar in the radial direction is the spring force, which we can express in polar coordinates in terms of r. By integrating Eq. (14.9), we can determine the radial velocity v_r as a function of r.
(b) Once $v_r = dr/dt$ is known in terms of r, we can use Eq. (14.10) to determine the transverse force exerted on the collar by the bar.

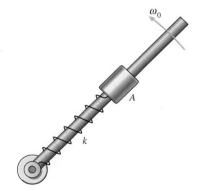

Figure 14.16

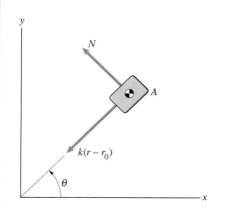

(a) Radial and transverse forces on A.

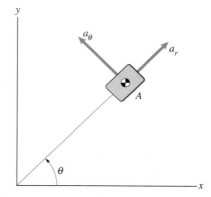

(b) Radial and transverse components of the acceleration of the center of mass.

Solution

(a) The spring exerts a radial force $k(r - r_0)$ in the negative r direction (Fig. a). Since the bar is smooth, it exerts no radial force on A, but may exert a transverse force N. Figure (b) shows the radial and transverse components of the collar's acceleration. Newton's second law in the radial direction is

$$\Sigma F_r = ma_r,$$

or

$$-k(r - r_0) = m\left(\frac{d^2r}{dt^2} - r\omega^2\right) = m\left(\frac{dv_r}{dt} - r\omega_0^2\right).$$

We solve this equation for the time derivative of v_r:

$$\frac{dv_r}{dt} = r\omega_0^2 - \frac{k}{m}(r - r_0).$$

Applying the chain rule,

$$\frac{dv_r}{dt} = \frac{dv_r}{dr}\frac{dr}{dt} = \frac{dv_r}{dr}v_r,$$

we obtain

$$v_r\, dv_r = \left[\left(\omega_0^2 - \frac{k}{m}\right)r + \frac{k}{m}r_0\right]dr.$$

Finally, integrating

$$\int_0^{v_r} v_r\, dv_r = \int_{r_0}^{r}\left[\left(\omega_0^2 - \frac{k}{m}\right)r + \frac{k}{m}r_0\right]dr$$

yields the radial velocity as a function of r:

$$v_r = \sqrt{\left(\omega_0^2 - \frac{k}{m}\right)(r^2 - r_0^2) + \frac{2k}{m}r_0(r - r_0)}.$$

(b) To determine the force N, we use Newton's second law in the transverse direction,

$$\Sigma F_\theta = ma_\theta,$$

or

$$N = m\left(r\alpha + 2\frac{dr}{dt}\omega\right) = 2m\omega_0 v_r.$$

Substituting our expression for v_r as a function of r, we obtain the transverse force exerted by the bar as a function of r:

$$N = 2m\omega_0\sqrt{\left(\omega_0^2 - \frac{k}{m}\right)(r^2 - r_0^2) + \frac{2k}{m}r_0(r - r_0)}.$$

14.5 Orbital Mechanics

It is appropriate to include a discussion of orbital mechanics in our chapter on applications of Newton's second law. Newton's analytical determination of the elliptical orbits of the planets, which had been deduced from observational data by Johannes Kepler, was a triumph for Newtonian mechanics and confirmation of the inverse-square relation for gravitational force.

We can use Newton's second law expressed in polar coordinates to determine the orbit of an earth satellite or a planet. Suppose that at $t = 0$ a satellite has an initial velocity v_0 at a distance r_0 from the center of the earth (Fig. 14.17a). We assume that the initial velocity is perpendicular to the line from the center of the earth to the satellite. The satellite's position during its subsequent motion is specified by its polar coordinates (r, θ), where θ is measured from the satellite's position at $t = 0$ (Fig. 14.17b). Our objective is to determine r as a function of θ.

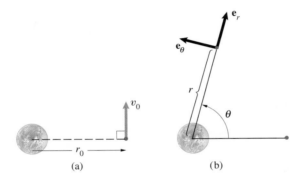

Figure 14.17
(a) Initial position and velocity of an earth satellite.
(b) Specifying the subsequent path in terms of polar coordinates.

Determination of the Orbit

If we model the earth as a homogeneous sphere, the force exerted on the satellite by gravity at a distance r from the center of the earth is mgR_E^2/r^2, where R_E is the earth's radius. (See Eq. 12.5.) From Eqs. (14.9) and (14.10), the equation of motion in the radial direction is

$$\Sigma F_r = ma_r,$$

or

$$-\frac{mgR_E^2}{r^2} = m\left[\frac{d^2r}{dt^2} - r\left(\frac{d\theta}{dt}\right)^2\right],$$

and the equation of motion in the transverse direction is

$$\Sigma F_\theta = ma_\theta,$$

or

$$0 = m\left(r\frac{d^2\theta}{dt^2} + 2\frac{dr}{dt}\frac{d\theta}{dt}\right).$$

We therefore obtain the two equations

$$\frac{d^2r}{dt^2} - r\left(\frac{d\theta}{dt}\right)^2 = -\frac{gR_E^2}{r^2} \qquad (14.12)$$

and

$$r \frac{d^2\theta}{dt^2} + 2 \frac{dr}{dt} \frac{d\theta}{dt} = 0. \tag{14.13}$$

We can write Eq. (14.13) in the form

$$\frac{1}{r} \frac{d}{dt} \left(r^2 \frac{d\theta}{dt} \right) = 0,$$

which indicates that

$$r^2 \frac{d\theta}{dt} = r v_\theta = \text{constant}. \tag{14.14}$$

At $t = 0$, the components of the velocity are $v_r = 0$ and $v_\theta = v_0$, and the radial position is $r = r_0$. We can therefore write the constant in Eq. (14.14) in terms of the initial conditions:

$$r^2 \frac{d\theta}{dt} = r v_\theta = r_0 v_0. \tag{14.15}$$

Using this equation to eliminate $d\theta/dt$ from Eq. (14.12), we obtain

$$\frac{d^2 r}{dt^2} - \frac{r_0^2 v_0^2}{r^3} = -\frac{g R_E^2}{r^2}. \tag{14.16}$$

We can solve this differential equation by introducing the change of variable

$$u = \frac{1}{r}. \tag{14.17}$$

In doing so, we also change the independent variable from t to θ, because we want to determine r as a function of the angle θ instead of time. To express Eq. (14.16) in terms of u, we must determine $d^2 r/dt^2$ in terms of u. Using the chain rule, we write the derivative of r with respect to time as

$$\frac{dr}{dt} = \frac{d}{dt} \left(\frac{1}{u} \right) = -\frac{1}{u^2} \frac{du}{dt} = -\frac{1}{u^2} \frac{du}{d\theta} \frac{d\theta}{dt}. \tag{14.18}$$

Notice from Eq. (14.15) that

$$\frac{d\theta}{dt} = \frac{r_0 v_0}{r^2} = r_0 v_0 u^2. \tag{14.19}$$

Substituting this expression into Eq. (14.18), we obtain

$$\frac{dr}{dt} = -r_0 v_0 \frac{du}{d\theta}. \tag{14.20}$$

We differentiate Eq. (14.20) with respect to time and apply the chain rule again:

$$\frac{d^2 r}{dt^2} = \frac{d}{dt} \left(-r_0 v_0 \frac{du}{d\theta} \right) = -r_0 v_0 \frac{d\theta}{dt} \frac{d}{d\theta} \left(\frac{du}{d\theta} \right) = -r_0 v_0 \frac{d\theta}{dt} \frac{d^2 u}{d\theta^2}.$$

Using Eq. (14.19) to eliminate $d\theta/dt$ from this expression, we obtain the second time derivative of r in terms of u:

$$\frac{d^2r}{dt^2} = -r_0^2 v_0^2 u^2 \frac{d^2u}{d\theta^2}.$$

Substituting this result into Eq. (14.16) yields a linear differential equation for u as a function of θ:

$$\frac{d^2u}{d\theta^2} + u = \frac{gR_E^2}{r_0^2 v_0^2}.$$

The general solution of this equation is

$$u = A\sin\theta + B\cos\theta + \frac{gR_E^2}{r_0^2 v_0^2}, \qquad (14.21)$$

where A and B are constants. We can use the initial conditions to determine A and B. When $\theta = 0$, $u = 1/r_0$. Also, when $\theta = 0$, the radial component of velocity, $v_r = dr/dt = 0$, so from Eq. (14.20), we see that $du/d\theta = 0$. From these two conditions, we obtain

$$A = 0 \quad\text{and}\quad B = \frac{1}{r_0} - \frac{gR_E^2}{r_0^2 v_0^2}.$$

Substituting these results into Eq. (14.21), we can write the resulting solution for $r = 1/u$ as

$$\frac{r}{r_0} = \frac{1 + \varepsilon}{1 + \varepsilon\cos\theta}, \qquad (14.22)$$

where

$$\varepsilon = \frac{r_0 v_0^2}{gR_E^2} - 1. \qquad (14.23)$$

Types of Orbits

The curve called a *conic section* (Fig. 14.18) has the property that the ratio of r to the perpendicular distance d to a straight line called the *directrix* is constant. This ratio, $r/d = r_0/d_0$, is called the *eccentricity* of the curve. From the figure, we see that

$$r\cos\theta + d = r_0 + d_0,$$

which we can write as

$$\frac{r}{r_0} = \frac{1 + (r_0/d_0)}{1 + (r_0/d_0)\cos\theta}.$$

Comparing this expression with Eq. (14.22), we see that the satellite's orbit describes a conic section with eccentricity ε. The value of the eccentricity determines the character of the orbit.

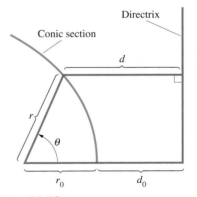

Figure 14.18
If the ratio r/d is constant, the curve describes a conic section.

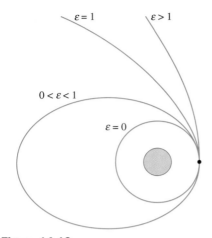

Figure 14.19
Orbits for different of eccentricity.

Circular Orbit If the initial velocity v_0 is chosen so that $\varepsilon = 0$, Eq. (14.22) reduces to $r = r_0$ and the orbit is circular (Fig. 14.19). Setting $\varepsilon = 0$ in Eq. (14.23) and solving for v_0, we obtain

$$v_0 = \sqrt{\frac{gR_E^2}{r_0}}, \tag{14.24}$$

which agrees with the velocity for a circular orbit we obtained by a different method in Example 13.10.

Elliptic Orbit If $0 < \varepsilon < 1$, orbit is an ellipse (Fig. 14.19). The maximum radius of the ellipse occurs when $\theta = 180°$. Setting θ equal to $180°$ in Eq. (14.22), we obtain an expression for the maximum radius of the ellipse in terms of the initial radius and ε:

$$r_{\max} = r_0\left(\frac{1 + \varepsilon}{1 - \varepsilon}\right). \tag{14.25}$$

Parabolic Orbit Notice from Eq. (14.25) that the maximum radius of the elliptic orbit increases without limit as $\varepsilon \to 1$. When $\varepsilon = 1$, the orbit is a parabola (Fig. 14.19). The corresponding velocity v_0 is the minimum initial velocity for which the radius r increases without limit, which is the escape velocity. Setting $\varepsilon = 1$ in Eq. (14.23) and solving for v_0, we obtain

$$v_0 = \sqrt{\frac{2gR_E^2}{r_0}}.$$

This is the same value for the escape velocity we obtained in Example 13.5 for the case of an object moving in a straight path directly away from the center of the earth.

Hyperbolic Orbit If $\varepsilon > 1$, the orbit is a hyperbola (Fig. 14.19).

The solution we have presented, based on the assumption that the earth is a homogeneous sphere, approximates the orbit of an earth satellite. Determining the orbit accurately requires taking into account the variations in the earth's gravitational field due to its actual mass distribution. Similarly, depending on the accuracy required, determining the orbit of a planet around the sun may require accounting for perturbations due to the gravitational attractions of the other planets.

Example 14.8

Orbit of an Earth Satellite

An earth satellite is in an elliptic orbit with a minimum radius of 6600 km and a maximum radius of 16,000 km. The earth's radius is 6370 km.
(a) Determine the satellite's velocity when it is at perigee (its minimum radius) and when it is at apogee (its maximum radius).
(b) Draw a graph of the orbit.

Strategy

We can regard the radius and velocity of the satellite at perigee as the initial conditions r_0 and v_0 used in obtaining Eq. (14.22). Since the maximum radius of the orbit is given, we can solve Eq. (14.25) for the eccentricity of the orbit and then use Eq. (14.23) to determine v_0. From Eq. (14.14), the product of r and the transverse component of the velocity is constant. From this condition, we can determine the velocity at apogee.

Solution

(a) The ratio of the radius at apogee to the radius at perigee is

$$\frac{r_{max}}{r_0} = \frac{1.60 \times 10^7 \text{ m}}{6.60 \times 10^6 \text{ m}} = 2.42.$$

Solving Eq. (14.25) for ε, we find that the eccentricity is

$$\varepsilon = \frac{\dfrac{r_{max}}{r_0} - 1}{\dfrac{r_{max}}{r_0} + 1} = \frac{2.42 - 1}{2.42 + 1} = 0.416.$$

From Eq. (14.23), the velocity at perigee is

$$v_0 = \sqrt{\frac{(\varepsilon + 1)gR_E^2}{r_0}} = \sqrt{\frac{(0.416 + 1)(9.81)(6.37 \times 10^6)^2}{6.60 \times 10^6}}$$

$$= 9240 \text{ m/s}.$$

At both perigee and apogee, the velocity has only a transverse component. From Eq. (14.14), the velocity at apogee, v_a, is related to the velocity v_0 by

$$r_0 v_0 = r_{max} v_a.$$

Therefore, the velocity at apogee is

$$v_a = \frac{v_0}{r_{max}/r_0} = \frac{9240}{2.42} = 3810 \text{ m/s}.$$

(b) By plotting Eq. (14.22) with $\varepsilon = 0.416$, we obtain the graph of the orbit (Fig. 14.20).

Figure 14.20
Orbit of an earth satellite with a perigee of 6600 km and an apogee of 16,000 km.

14.6 Numerical Solutions

So far in this chapter, we have described many situations in which we were able to determine the motion of an object by a simple procedure: After using Newton's second law to determine the acceleration, we integrated to obtain analytical, or *closed-form*, expressions for the object's velocity and position. These examples are very valuable—they demonstrate how to use free-body diagrams and express problems in different coordinate systems, and they develop intuitive understanding of forces and motions. But it would be misleading if we presented examples of this kind only, because most problems that must be dealt with in engineering cannot be solved in that way. The functions describing the forces, and therefore the acceleration, are often too complicated to integrate and obtain closed-form solutions. In other situations, the forces are not known in terms of functions, but instead are specified in terms of data, either as a continuous recording of force as a function of time (analog data) or as values of force measured at discrete times (digital data).

We can obtain approximate solutions to such problems by using numerical integration. Consider an object of mass m in straight-line motion along the x-axis (Fig. 14.21) and assume that the x component of the total force may depend on the time and on the position and velocity of the object:

$$\Sigma F_x = \Sigma F_x(t, x, v_x). \tag{14.26}$$

Suppose that at a particular time t_0 we know the position $x(t_0)$ and velocity $v_x(t_0)$ of the object. The acceleration of the object at t_0 is

$$\frac{dv_x}{dt}(t_0) = \frac{1}{m} \Sigma F_x(t_0, x(t_0), v_x(t_0)). \tag{14.27}$$

To determine the velocity at time $t_0 + \Delta t$, we express it as a Taylor series:

$$v_x(t_0 + \Delta t) = v_x(t_0) + \frac{dv_x}{dt}(t_0)\Delta t + \frac{1}{2}\frac{d^2v_x}{dt^2}(t_0)(\Delta t)^2 + \cdots.$$

By choosing a sufficiently small value of Δt, we can neglect terms in this equation that are of second and higher order in Δt and substitute Eq. (14.27) to obtain an approximation for the velocity at $t_0 + \Delta t$:

$$v_x(t_0 + \Delta t) = v_x(t_0) + \frac{1}{m} \Sigma F_x(t_0, x(t_0), v_x(t_0))\Delta t. \tag{14.28}$$

We approximate the position at $t_0 + \Delta t$ in the same way. Expressing it as a Taylor series, we have

$$x(t_0 + \Delta t) = x(t_0) + \frac{dx}{dt}(t_0)\Delta t + \frac{1}{2}\frac{d^2x}{dt^2}(t_0)(\Delta t)^2 + \cdots,$$

and neglecting higher-order terms in Δt, we obtain

$$x(t_0 + \Delta t) = x(t_0) + v_x(t_0)\Delta t. \tag{14.29}$$

Thus, if we know the position and velocity at a time t_0, we can approximate their values at $t_0 + \Delta t$ by using Eqs. (14.28) and (14.29). We can then repeat the procedure, using $x(t_0 + \Delta t)$ and $v_x(t_0 + \Delta t)$ as initial conditions to determine the approximate position and velocity at $t_0 + 2\Delta t$. By continuing in this way, we obtain approximate solutions for the position and velocity in terms of time. This procedure is easy to carry out using a calculator or a computer. It is called a *finite-difference method*, because it determines changes in the dependent variables over finite intervals of time. The particular finite-difference method we describe, due to Leonhard Euler (1707–1783), is called *forward differencing*: The value of the derivative of a function at t_0 is approximated by using its value at t_0 and its value forward in time at $t_0 + \Delta t$.

More elaborate finite-difference methods, based on retaining more terms in the Taylor series, produce smaller errors in each time step. For example, in the fourth-order Runge–Kutta method, terms through the fourth order in Δt are retained. Euler's method is adequate to introduce you to numerical solutions of problems in dynamics.

Figure 14.21
An object moving along the x axis.

Notice that Eq. (14.26) does not need to be a functional expression to carry out the process we have described. The values of the total force must be known at times $t_0, t_0 + \Delta t, \ldots$, and can be determined either from a function or from analog or digital data.

We can determine the velocity and position of an object in curvilinear motion by the same approach. Suppose that the object moves in the xy-plane and that the components of force may depend on the time and on the position and velocity of the object; that is,

$$\Sigma F_x = \Sigma F_x(t, x, y, v_x, v_y) \quad \text{and} \quad \Sigma F_y = \Sigma F_y(t, x, y, v_x, v_y).$$

If the position and velocity are known at a time t_0, we can use the same steps leading to Eqs. (14.28) and (14.29) to obtain approximate expressions for the components of position and velocity at $t_0 + \Delta t$, yielding

$$x(t_0 + \Delta t) = x(t_0) + v_x(t_0)\Delta t,$$

$$y(t_0 + \Delta t) = y(t_0) + v_y(t_0)\Delta t,$$

$$v_x(t_0 + \Delta t) = v_x(t_0) + \frac{1}{m} \Sigma F_x(t_0, x(t_0), y(t_0), v_x(t_0), v_y(t_0))\Delta t, \quad (14.30)$$

and

$$v_y(t_0 + \Delta t) = v_y(t_0) + \frac{1}{m} \Sigma F_y(t_0, x(t_0), y(t_0), v_x(t_0), v_y(t_0))\Delta t.$$

Computational Mechanics

The following example and problems are designed to be worked with the use of a programmable calculator or computer.

Computational Example 14.9

A 50-kg projectile is launched from $x = 0$, $y = 0$ with initial velocity $v_x = 100$ m/s, $v_y = 100$ m/s. (The y axis is positive upward.) The aerodynamic drag force on the projectile is of magnitude $C|\mathbf{v}|^2$, where C is a constant. Determine the trajectory of the projectile for $C = 0.005, 0.01$, and 0.02.

Solution

To apply Eqs. (14.30), we must determine the x and y components of the total force on the projectile. Let $\mathbf{D}$ be the drag force (Fig. 14.22). Because $\mathbf{v}/|\mathbf{v}|$ is a unit vector in the direction of $\mathbf{v}$, we can write

$$\mathbf{D} = -C|\mathbf{v}|^2 \frac{\mathbf{v}}{|\mathbf{v}|} = -C|\mathbf{v}|\mathbf{v}.$$

The external forces on the projectile are its weight and the drag, so we have

$$\Sigma \mathbf{F} = -mg\mathbf{j} - C|\mathbf{v}|\mathbf{v},$$

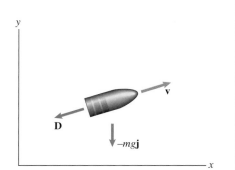

Figure 14.22
The forces on the projectile are its weight and the drag force $\mathbf{D}$.

and the components of the total force are

$$\Sigma F_x = -C\sqrt{v_x^2 + v_y^2}\, v_x, \qquad \Sigma F_y = -mg - C\sqrt{v_x^2 + v_y^2}\, v_y. \quad (14.31)$$

Consider the case in which $C = 0.005$, and let $\Delta t = 0.2$ s. At the initial time $t_0 = 0$, the position coordinates and the components of the velocity are $x(t_0) = 0$, $y(t_0) = 0$, $v_x(t_0) = 100$ m/s, and $v_y(t_0) = 100$ m/s. The x coordinate after the first time step is

$$x(t_0 + \Delta t) = x(t_0) + v_x(t_0)\Delta t,$$

or

$$x(0.2) = x(0) + v_x(0)\Delta t$$

$$= 0 + (100)(0.2)$$

$$= 20 \text{ m}.$$

The y coordinate after the first time step is

$$y(t_0 + \Delta t) = y(t_0) + v_y(t_0)\Delta t,$$

or

$$y(0.2) = y(0) + v_y(0)\Delta t$$

$$= 0 + (100)(0.2)$$

$$= 20 \text{ m}.$$

The x component of the velocity after the first time step is

$$v_x(t_0 + \Delta t) = v_x(t_0) + \frac{1}{m}\Sigma F_x(t_0, x(t_0), y(t_0), v_x(t_0), v_y(t_0))\Delta t,$$

or

$$v_x(0.2) = v_x(0) + \frac{1}{m}\left\{-C\sqrt{[v_x(0)]^2 + [v_y(0)]^2}\, v_x(0)\right\}\Delta t$$

$$= 100 + \frac{1}{50}\left[-0.005\sqrt{(100)^2 + (100)^2}\,(100)\right](0.2)$$

$$= 99.72 \text{ m/s},$$

and the y component of the velocity after the first time step is

$$v_y(t_0 + \Delta t) = v_y(t_0) + \frac{1}{m}\Sigma F_y(t_0, x(t_0), y(t_0), v_x(t_0), v_y(t_0))\Delta t,$$

or

$$v_y(0.2) = v_y(0) + \frac{1}{m}\left\{-mg - C\sqrt{[v_x(0)]^2 + [v_y(0)]^2}\, v_y(0)\right\}\Delta t$$

$$= 100 + \frac{1}{50}\left[-(50)(9.81) - 0.005\sqrt{(100)^2 + (100)^2}\,(100)\right](0.2)$$

$$= 97.76 \text{ m/s}.$$

Continuing in this way, we obtain the following results for the first five time steps:

Time, s	x, m	y, m	v_x, m/s	v_y, m/s
0.0	0.00	0.00	100.00	100.00
0.2	20.00	20.00	99.72	97.76
0.4	39.94	39.55	99.44	95.52
0.6	59.83	58.66	99.16	93.29
0.8	79.66	77.31	98.89	91.08
1.0	99.44	95.53	98.63	88.87

When there is no drag $(C = 0)$, we can obtain the closed-form solution for the trajectory. Figure 14.23 compares the closed-form solution with numerical solutions obtained using $\Delta t = 2.0$ s, 1.0 s, and 0.2 s. The numerical solutions approach the exact solution as Δt decreases.

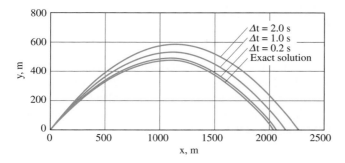

Figure 14.23
The closed-form solution for the trajectory when $C = 0$, compared with numerical solutions.

Figure 14.24 shows numerical solutions (obtained using $\Delta t = 0.01$ s) for the various values of C. As expected, the range of the projectile decreases as C increases. Also, notice that drag changes the shape of the trajectory: The angle at which the projectile descends is steeper than the angle at which it was launched.

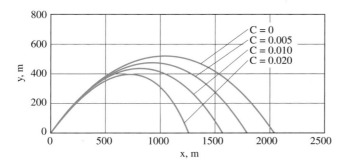

Figure 14.24
Trajectories for various values of C.

Discussion

The development of the first completely electronic digital computer, the ENIAC (Electronic Numerical Integrator and Computer), at the University of Pennsylvania between 1943 and 1945 was motivated in part by the need to calculate trajectories of projectiles. A room-sized machine with 18,000 vacuum tubes, the ENIAC had 20 bytes of RAM and 450 bytes of ROM.

Chapter Summary

We have used Newton's second law both to determine the acceleration of an object when the sum of the forces acting on it is known and to determine the sum of the forces when the acceleration is known. Once the acceleration of an object was known, we used the methods developed in Chapter 13 to obtain information about the object's velocity and position. In applying Newton's second law, we expressed it in terms of different coordinate systems. The choice of coordinate system was sometimes dictated by the nature of the forces acting on the object. When an object's path is known—especially when the object is constrained to move in a circle—normal and tangential components are often advantageous. In Chapter 15, we will use Newton's second law to derive a technique called the method of work and energy, which can greatly simplify the solution of particular types of problems in dynamics.

The total external force on an object is equal to the product of its mass and the acceleration of its center of mass relative to an inertial reference frame:

$$\Sigma \mathbf{F} = m\mathbf{a}. \qquad \text{Eq. (14.4)}$$

A reference frame is said to be inertial if it is one in which the second law can be applied in this form. A reference frame translating at constant velocity relative to an inertial reference frame is also inertial.

Expressing Newton's second law in terms of a coordinate system yields the following scalar equations of motion:

Cartesian Coordinates

$$\Sigma F_x = ma_x, \quad \Sigma F_y = ma_y, \quad \Sigma F_z = ma_z. \qquad \text{Eq. (14.5)}$$

Normal and Tangential Components

$$\Sigma F_t = m\frac{dv}{dt}, \quad \Sigma F_n = m\frac{v^2}{\rho}. \qquad \text{Eq. (14.7)}$$

Polar Coordinates

$$\Sigma F_r = m\left(\frac{d^2r}{dt^2} - r\omega^2\right), \qquad \text{Eq. (14.9)}$$

$$\Sigma F_\theta = m\left(r\alpha + 2\frac{dr}{dt}\omega\right). \qquad \text{Eq. (14.10)}$$

If the motion of an object is confined to a fixed plane, the component of the total force normal to the plane equals zero. In straight-line motion, the components of the total force perpendicular to the line equal zero and the component of the total force tangent to the line equals the product of the mass and the acceleration of the object along the line.

Review Problems

14.1 The total external force on an object as a function of time is $\Sigma\mathbf{F} = 10t\mathbf{i} + 6\mathbf{j}$ (N). At $t = 0$, the position vector of the object relative to an inertial reference frame is $\mathbf{r} = \mathbf{0}$ and the velocity is $\mathbf{v} = 2\mathbf{j}$ (m/s). At $t = 2$ s the magnitude of the position vector is measured and determined to be $|\mathbf{r}| = 7$ m. What is the object's mass?

14.2 The Acura NSX, which, together with its driver, weighs 3250 lb, can brake from 60 mi/hr to a stop in a distance of 112 ft. (a) If you assume that the vehicle's deceleration is constant, what are its deceleration and the magnitude of the horizontal force its tires exert on the road? (b) If the car's tires are at the limit of adhesion (i.e., slip is impending), and the normal force exerted on the car by the road equals the car's weight, what is the coefficient of friction μ_s? (This analysis neglects the effects of horizontal and vertical aerodynamic forces.)

14.3 Using the coefficient of friction obtained in Problem 14.1, determine the highest speed at which the NSX could drive on a flat, circular track of 600-ft radius without skidding.

14.4 A "cog" engine hauls three cars of sightseers to a mountaintop in Bavaria. The mass of each car, including its passengers, is 10,000 kg, and the friction forces exerted by the wheels of the cars are negligible. Determine the forces in the couplings 1, 2, and 3 if (a) the engine is moving at constant velocity and (b) the engine is accelerating up the mountain at 1.2 m/s^2.

P14.4

14.5 In a future mission, a spacecraft approaches the surface of an asteroid passing near the earth. Just before it touches down, the spacecraft is moving downward at a constant velocity relative to the surface of the asteroid and its downward thrust is 0.01 N. The computer decreases the downward thrust to 0.005 N, and an onboard

laser interferometer determines that the acceleration of the spacecraft relative to the surface becomes 5×10^{-6} m/s^2 downward. What is the gravitational acceleration of the asteroid near its surface?

P14.5

14.6 A car with a mass of 1470 kg, including its driver, is driven at 130 km/hr over a slight rise in the road. At the top of the rise, the driver applies the brakes. The coefficient of static friction between the tires and the road is $\mu_s = 0.9$, and the radius of curvature of the rise is 160 m. Determine the car's deceleration at the instant the brakes are applied, and compare it with the deceleration on a level road.

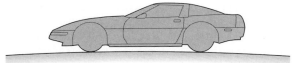

P14.6

14.7 The car drives at constant velocity up the straight segment of road on the left. If the car's tires continue to exert the same tangential force on the road after the car has gone over the crest of the hill and is on the straight segment of road on the right, what will be the car's acceleration?

P14.7

14.8 The aircraft carrier *Nimitz* weighs 91,000 tons. (A ton is 2000 lb.) Suppose that it is traveling at its top speed of approximately 30 knots (a knot is 6076 ft/hr) when its engines are shut down. If the water exerts a drag force of magnitude $20,000v$ lb, where v is the carrier's velocity in feet per second, what distance does the carrier move before coming to rest?

14.9 If $m_A = 10$ kg, $m_B = 40$ kg, and the coefficient of kinetic friction between all surfaces is $\mu_k = 0.11$, what is the acceleration of B down the inclined surface?

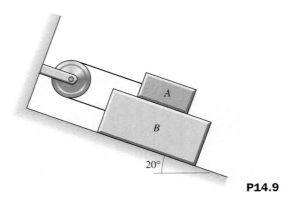

P14.9

14.10 In Problem 14.9, if A weighs 20 lb, B weighs 100 lb, and the coefficient of kinetic friction between all surfaces is $\mu_k = 0.15$, what is the tension in the cord as B slides down the inclined surface?

14.11 A gas gun is used to accelerate projectiles to high velocities for research on material properties. The projectile is held in place while gas is pumped into the tube to a high pressure p_0 on the left and the tube is evacuated on the right. The projectile is then released and is accelerated by the expanding gas. Assume that the pressure p of the gas is related to the volume V it occupies by $pV^\gamma = $ constant, where γ is a constant. If friction can be neglected, show that the velocity of the projectile at the position x is

$$v = \sqrt{\frac{2p_0 A x_0^\gamma}{m(\gamma - 1)} \left(\frac{1}{x_0^{\gamma-1}} - \frac{1}{x^{\gamma-1}} \right)},$$

where m is the mass of the projectile and A is the cross-sectional area of the tube.

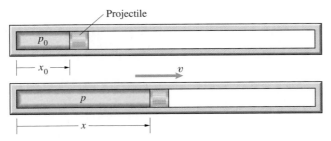

P14.11

14.12 The weights of the blocks are $W_A = 120$ lb and $W_B = 20$ lb, and the surfaces are smooth. Determine the acceleration of block A and the tension in the cord.

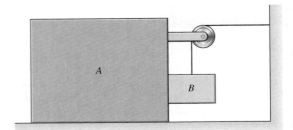

P14.12

14.13 The 100-Mg space shuttle is in orbit when its engines are turned on, exerting a thrust force $\mathbf{T} = 10\mathbf{i} - 20\mathbf{j} + 10\mathbf{k}$ (kN) for 2 s. Neglect the resulting change in mass of the shuttle. At the end of the 2-s burn, fuel is still sloshing back and forth in the shuttle's tanks. What is the change in the velocity of the center of mass of the shuttle (including the fuel it contains) due to the 2-s burn?

14.14 The water skier contacts the ramp with a velocity of 25 mi/hr parallel to the surface of the ramp. Neglecting friction and assuming that the tow rope exerts no force on him once he touches the ramp, estimate the horizontal length of the skier's jump from the end of the ramp.

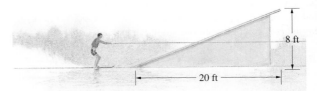

P14.14

14.15 Suppose you are designing a roller-coaster track that will take the cars through a vertical loop of 40-ft radius. If you decide that, for safety, the downward force exerted on a passenger by his or her seat at the top of the loop should be at least one-half the passenger's weight, what is the minimum safe velocity of the cars at the top of the loop?

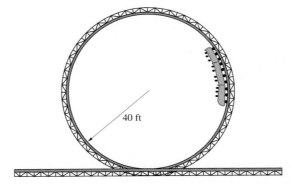

P14.15

14.16 As the smooth bar rotates *in the horizontal plane*, the string winds up on the fixed cylinder and draws the 1-kg collar A inward. The bar starts from rest at $t = 0$ in the position shown and rotates with constant angular acceleration. What is the tension in the string at $t = 1$ s?

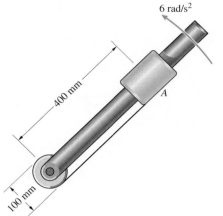

P14.16

14.17 In Problem 14.16, suppose that the coefficient of kinetic friction between the collar and the bar is $\mu_k = 0.2$. What is the tension in the string at $t = 1$ s?

14.18 If you want to design the cars of a train to tilt as the train goes around curves in order to achieve maximum passenger comfort, what is the relationship between the desired tilt angle α, the velocity v of the train, and the instantaneous radius of curvature, ρ, of the track?

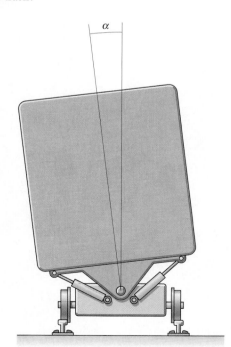

P14.18

14.19 To determine the coefficient of static friction between two materials, an engineer at the U.S. National Institute of Standards and Technology places a small sample of one material on a horizontal disk whose surface is made of the other material and then rotates the disk from rest with a constant angular acceleration of 0.4 rad/s². If she determines that the small sample slips on the disk after 9.903 s, what is the coefficient of friction?

P14.19

14.20 The 1-kg slider A is pushed along the curved bar by the slotted bar. The curved bar lies *in the horizontal plane*, and its profile is described by $r = 2(\theta/2\pi + 1)$ m, where θ is in radians. The angular position of the slotted bar is $\theta = 2t$ rad. Determine the radial and transverse components of the total external force exerted on the slider when $\theta = 120°$.

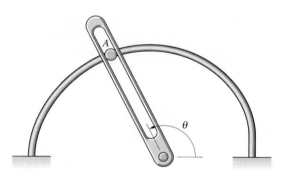

P14.20

14.21 In Problem 14.20, suppose that the curved bar lies *in the vertical plane*. Determine the radial and transverse components of the total force exerted on A by the curved and slotted bars at $t = 0.5$ s.

The ski jumper's kinetic energy is determined by the change in his gravitational potential energy and the work done on him by aerodynamic forces. In this chapter, we use energy methods to analyze motions of objects.

Energy Methods

The concepts of energy and conservation of energy originated in large part from the study of classical mechanics. A simple transformation of Newton's second law results in an equation that motivates the definitions of work, kinetic energy (energy due to an object's motion), and potential energy (energy due to an object's position). This equation can greatly simplify the solution of problems involving certain forces that depend on an object's position, including gravitational forces and forces exerted by springs.

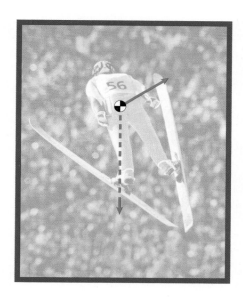

Work and Kinetic Energy

15.1 Principle of Work and Energy

We have used Newton's second law to relate the acceleration of an object's center of mass relative to an inertial reference frame to the mass of the object and the external forces acting on it. We will now show how this vector equation can be expressed in a scalar form that is extremely useful in particular circumstances. We begin with Newton's second law in the form

$$\Sigma \mathbf{F} = m \frac{d\mathbf{v}}{dt} \tag{15.1}$$

and take the dot product of both sides with the velocity:

$$\Sigma \mathbf{F} \cdot \mathbf{v} = m \frac{d\mathbf{v}}{dt} \cdot \mathbf{v}. \tag{15.2}$$

By writing the left and right sides of this equation as

$$\Sigma \mathbf{F} \cdot \mathbf{v} = \Sigma \mathbf{F} \cdot \frac{d\mathbf{r}}{dt}$$

and

$$m \frac{d\mathbf{v}}{dt} \cdot \mathbf{v} = \tfrac{1}{2} m \frac{d}{dt} (\mathbf{v} \cdot \mathbf{v}),$$

we can express Eq. (15.2) in the form

$$\Sigma \mathbf{F} \cdot d\mathbf{r} = \tfrac{1}{2} m \, d(v^2), \tag{15.3}$$

where $v^2 = \mathbf{v} \cdot \mathbf{v}$ is the square of the magnitude of the velocity. The term on the left of this equation is the *work*, expressed in terms of the total external force on the object and an infinitesimal displacement $d\mathbf{r}$ of its center of mass. We integrate Eq. (15.3), obtaining

$$\int_{\mathbf{r}_1}^{\mathbf{r}_2} \Sigma \mathbf{F} \cdot d\mathbf{r} = \tfrac{1}{2} m v_2^2 - \tfrac{1}{2} m v_1^2, \tag{15.4}$$

where v_1 and v_2 are the magnitudes of the velocity of the center of mass of the object when it is at the positions $\mathbf{r}_1$ and $\mathbf{r}_2$, respectively. The term $\tfrac{1}{2} m v^2$ is called the *kinetic energy* associated with the motion of the center of mass. Denoting the work done as the center of mass moves from position $\mathbf{r}_1$ to position $\mathbf{r}_2$ by

$$U_{12} = \int_{\mathbf{r}_1}^{\mathbf{r}_2} \Sigma \mathbf{F} \cdot d\mathbf{r}, \tag{15.5}$$

we obtain the *principle of work and energy*:

$$U_{12} = \tfrac{1}{2} m v_2^2 - \tfrac{1}{2} m v_1^2. \tag{15.6}$$

The work done on an object as it moves between two positions equals the change in its kinetic energy. The dimensions of work, and therefore the dimensions of kinetic energy, are (force) × (length). In U.S. Customary units, work is usually expressed in ft-lb. In SI units, work is usually expressed in N-m, or joules (J).

If the work done on an object as it moves between two positions can be evaluated, the principle of work and energy permits us to determine the change in the magnitude of the object's velocity. We can also apply this principle to a system of objects, equating the total work done by external forces to the change in the total kinetic energy of the system. But the principle must be applied with caution, because, as we demonstrate in Example 15.3, net work can be done on a system by internal forces.

Although the principle of work and energy relates changes in position to changes in velocity, we cannot use it to obtain other information about the motion of an object, such as the time it takes the object to move from one position to another. Furthermore, since the work is an integral with respect to position, we can usually evaluate it only when the forces doing work are known as functions of position. Despite these limitations, the principle is extremely useful for certain problems, because in some cases the work can be determined very easily.

15.2 Work and Power

In this section, we discuss how to determine the work done on an object. We also define the power transferred to or from an object by the forces acting on it and show how the power is calculated.

Evaluating the Work

Let's consider an object in curvilinear motion relative to an inertial reference frame (Fig. 15.1a) and specify its position by the coordinate s measured along its path from a reference point O. In terms of the tangential unit vector $\mathbf{e}_t$, the object's velocity is

$$\mathbf{v} = \frac{ds}{dt}\,\mathbf{e}_t.$$

Because $\mathbf{v} = d\mathbf{r}/dt$, we can multiply the velocity by dt to obtain an expression for the vector $d\mathbf{r}$ describing an infinitesimal displacement along the path (Fig. 15.1b):

$$d\mathbf{r} = \mathbf{v}\,dt = ds\,\mathbf{e}_t.$$

The work done by the external forces acting on the object as a result of the displacement $d\mathbf{r}$ is

$$\Sigma\mathbf{F} \cdot d\mathbf{r} = \left(\Sigma\mathbf{F} \cdot \mathbf{e}_t\right)ds = \Sigma F_t\,ds,$$

where ΣF_t is the tangential component of the total force. Therefore, as the object moves from a position s_1 to a position s_2 (Fig. 15.1c), the work is

$$U_{12} = \int_{s_1}^{s_2} \Sigma F_t\,ds. \tag{15.7}$$

The work is equal to the integral of the tangential component of the total force with respect to distance along the path. Thus, the work done is equal to the

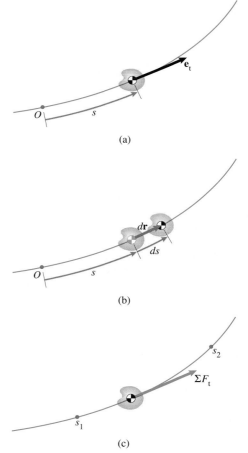

Figure 15.1
(a) The coordinate s and tangential unit vector.
(b) An infinitesimal displacement $d\mathbf{r}$.
(c) The work done from s_1 to s_2 is determined by the tangential component of the external forces.

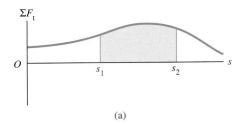

(a)

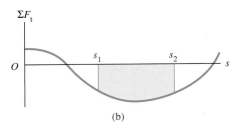

(b)

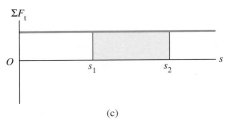

(c)

Figure 15.2
(a) The work equals the area defined by the graph of the tangential force as a function of the distance along the path.
(b) Negative work is done if the tangential force is opposite to the direction of the motion.
(c) The work done by a constant tangential force equals the product of the force and the distance.

area defined by the graph of the tangential force from s_1 to s_2 (Fig. 15.2a). Components of force perpendicular to the path do no work. Notice that if ΣF_t is opposite to the direction of motion over some part of the path, which means that the object is decelerating, the work is negative (Fig. 15.2b). If ΣF_t is constant between s_1 and s_2, the work is simply the product of the total tangential force and the displacement (Fig. 15.2c):

$$U_{12} = \Sigma F_t(s_2 - s_1). \qquad \textbf{Constant tangential force} \qquad (15.8)$$

Power

Power is the rate at which work is done. The work done by the external forces acting on an object during an infinitesimal displacement $d\mathbf{r}$ is

$$\Sigma \mathbf{F} \cdot d\mathbf{r}.$$

We obtain the power P by dividing this expression by the interval of time dt during which the displacement takes place:

$$P = \Sigma \mathbf{F} \cdot \mathbf{v}. \qquad (15.9)$$

This is the power transferred to or from the object, depending on whether P is positive or negative, respectively. In SI units, power is expressed in newton-meters per second, which is joules per second (J/s) or watts (W). In U.S. Customary units, power is expressed in foot-pounds per second or in the anachronistic horsepower (hp), which is 746 W, or approximately 550 ft-lb/s.

Notice from Eq. (15.3) that the power equals the rate of change of the kinetic energy of the object:

$$P = \frac{d}{dt}\left(\tfrac{1}{2}mv^2\right).$$

Transferring power to and from an object causes its kinetic energy to increase and decrease, respectively. Using the preceding relation, we can write the average with respect to time of the power during an interval of time from t_1 to t_2 as

$$P_{\text{av}} = \frac{1}{t_2 - t_1}\int_{t_1}^{t_2} P\,dt = \frac{1}{t_2 - t_1}\int_{v_1^2}^{v_2^2} \tfrac{1}{2}m\,d(v^2).$$

Performing the integration, we find that the average power transferred to or from an object during an interval of time is equal to the change in its kinetic energy, or the work done, divided by the interval of time:

$$P_{\text{av}} = \frac{\tfrac{1}{2}mv_2^2 - \tfrac{1}{2}mv_1^2}{t_2 - t_1} = \frac{U_{12}}{t_2 - t_1}. \qquad (15.10)$$

Study Questions

1. What is the definition of the kinetic energy associated with the motion of the center of mass of an object?
2. What is the principle of work and energy?
3. If the tangential component of the total force on an object is constant, what do you know about the work done as the object moves a given distance along its path?
4. If an object is subjected only to forces that are perpendicular to its path, what do you know about its kinetic energy?

Example 15.1

Work and Energy in Straight-Line Motion

The 180-kg container A in Fig. 15.3 starts from rest at position $s = 0$. The hydraulic cylinder exerts a horizontal force on the container that is given as a function of position by $F = 700 - 150s$ N. The coefficient of kinetic friction between the container and the floor is $\mu_k = 0.26$. What is the velocity of the container when it has reached the position $s = 1$ m?

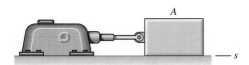

Figure 15.3

Strategy

We are asked to determine the change in the velocity of the container, given a change in its position, so we can apply the method of work and energy. We will determine the forces on the container in the direction tangent to its path and use Eq. (15.7) to evaluate the work.

Solution

Identify the Forces That Do Work We draw the free-body diagram of the container in Fig. (a). The forces tangent to the path of the container are the force exerted by the hydraulic cylinder and the friction force. The container's acceleration in the vertical direction is zero, so $N = (180)(9.81)$ N.

Apply Work and Energy Let v be the container's velocity (Fig. b). At the initial position $s_1 = 0$, the velocity is $v_1 = 0$. We can apply the principle of work and energy to determine the container's kinetic energy at the position $s_2 = 1$ m, using Eq. (15.7) to evaluate the work.

$$\int_{s_1}^{s_2} \Sigma F_t \, ds = \tfrac{1}{2} m v_2^2 - \tfrac{1}{2} m v_1^2:$$

$$\int_0^1 (F - \mu_k N) \, ds = \tfrac{1}{2} m v_2^2 - \tfrac{1}{2} m v_1^2,$$

$$\int_0^1 \left[(700 - 150s) - (0.26)(180)(9.81) \right] ds = \tfrac{1}{2}(180) v_2^2 - 0.$$

Evaluating the integral and solving for the velocity, we obtain $v_2 = 1.36$ m/s.

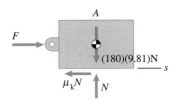

(a) Free-body diagram of the container.

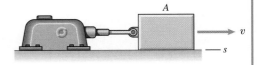

(b) Magnitude v of the container's velocity.

Example 15.2

Applying Work and Energy to a System

The two crates in Fig. 15.4 are released from rest. Their masses are $m_A = 40$ kg and $m_B = 30$ kg, and the kinetic coefficient of friction between crate A and the inclined surface is $\mu_k = 0.15$. What is the magnitude of the velocity of the crates when they have moved 400 mm?

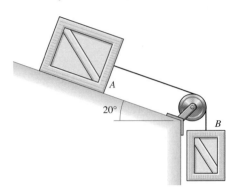

Figure 15.4

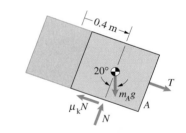

(a) Free-body diagram of A.

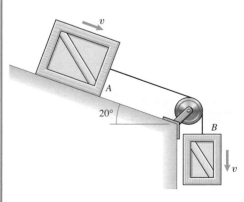

(b) The magnitude of the velocity of each crate is the same.

Strategy

We will determine the velocity in two ways.

First Method By drawing free-body diagrams of each of the crates and applying the principle of work and energy to them individually, we can obtain two equations in terms of the magnitude of the velocity and the tension in the cable.

Second Method We can draw a single free-body diagram of the two crates, the cable, and the pulley and apply the principle of work and energy to the entire system.

Solution

First Method We draw the free-body diagram of crate A in Fig. (a). The forces that do work as the crate moves down the plane are the forces tangential to its path: the tension T; the tangential component of the weight, $m_A g \sin 20°$; and the friction force $\mu_k N$. Because the acceleration of the crate normal to the surface is zero, $N = m_A g \cos 20°$. The magnitude v of the velocity at which A moves parallel to the surface equals the magnitude of the velocity at which B falls (Fig. b). Using Eq. (15.7) to determine the work, we equate the work done on A as it moves from $s_1 = 0$ to $s_2 = 0.4$ m to the change in the kinetic energy of A.

$$\int_{s_1}^{s_2} \Sigma F_t \, ds = \tfrac{1}{2} m v_2^2 - \tfrac{1}{2} m v_1^2 :$$

$$\int_0^{0.4} \left[T + m_A g \sin 20° - \mu_k (m_A g \cos 20°) \right] ds = \tfrac{1}{2} m_A v_2^2 - 0. \quad (15.11)$$

The forces that do work on crate B are its weight $m_B g$ and the tension T (Fig. c). The magnitude of B's velocity is the same as that of crate A. The work done on B equals the change in its kinetic energy.

$$\int_{s_1}^{s_2} \Sigma F_t \, ds = \tfrac{1}{2} m v_2^2 - \tfrac{1}{2} m v_1^2:$$

$$\int_0^{0.4} (m_B g - T) \, ds = \tfrac{1}{2} m_B v_2^2 - 0. \tag{15.12}$$

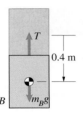

(c) Free-body diagram of B.

By summing Eqs. (15.11) and (15.12), we eliminate T, obtaining

$$\int_0^{0.4} (m_A g \sin 20° - \mu_k m_A g \cos 20° + m_B g) \, ds = \tfrac{1}{2}(m_A + m_B) v_2^2:$$

$$\left[40 \sin 20° - (0.15)(40) \cos 20° + 30 \right](9.81)(0.4) = \tfrac{1}{2}(40 + 30) v_2^2.$$

Solving for the velocity, we get $v_2 = 2.07$ m/s.

Second Method We draw the free-body diagram of the system consisting of the crates, cable, and pulley in Fig. (d). Notice that the cable tension does not appear in this diagram. The reactions at the pin support of the pulley do no work, because the support does not move. The total work done by external forces on the system as the boxes move 400 mm is equal to the change in the total kinetic energy of the system.

$$\int_0^{0.4} \left[m_A g \sin 20° - \mu_k (m_A g \cos 20°) \right] ds + \int_0^{0.4} m_B g \, ds$$

$$= \tfrac{1}{2} m_A v_2^2 + \tfrac{1}{2} m_B v_2^2 - 0:$$

$$\left[40 \sin 20° - (0.15)(40) \cos 20° + 30 \right](9.81)(0.4) = \tfrac{1}{2}(40 + 30) v_2^2.$$

This equation is identical to the one obtained by applying the principle of work and energy to the individual crates.

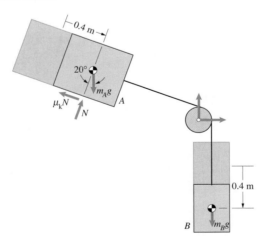
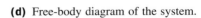

(d) Free-body diagram of the system.

Discussion

You will often find it simpler to apply the principle of work and energy to an entire system instead of its separate parts. However, as we demonstrate in the next example, you need to be aware that internal forces in a system can do net work.

Example 15.3

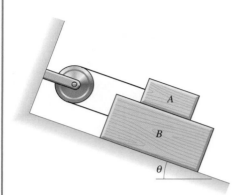

Figure 15.5

Net Work by Internal Forces

Crates A and B in Fig. 15.5 are released from rest. The coefficient of kinetic friction between A and B is μ_k, and friction between B and the inclined surface can be neglected. What is the velocity of the crates when they have moved a distance b?

Strategy

By applying the principle of work and energy to each crate, we can obtain two equations in terms of the tension in the cable and the velocity.

Solution

We draw the free-body diagrams of the crates in Figs. (a) and (b). The acceleration of A normal to the inclined surface is zero, so $N = m_A g \cos\theta$. The magnitudes of the velocities of A and B are equal (Fig. c). The work done on A equals the change in its kinetic energy.

$$U_{12} = \tfrac{1}{2} m_A v_2^2 - \tfrac{1}{2} m_A v_1^2:$$

$$\int_0^b \left(T - m_A g \sin\theta - \mu_k m_A g \cos\theta \right) ds = \tfrac{1}{2} m_A v_2^2. \qquad (15.13)$$

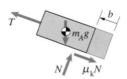

(a) Free-body diagram of A.

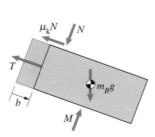

(b) Free-body diagram of B.

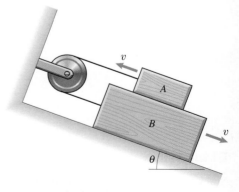

(c) The magnitude of the velocity of each crate is the same.

The work done on B equals the change in *its* kinetic energy.

$$U_{12} = \tfrac{1}{2}m_B v_2^2 - \tfrac{1}{2}m_B v_1^2:$$

$$\int_0^b \left(-T + m_B g \sin\theta - \mu_k m_A g \cos\theta\right) ds = \tfrac{1}{2}m_B v_2^2. \qquad (15.14)$$

Summing these equations to eliminate T and solving for v_2, we obtain

$$v_2 = \sqrt{2gb\left[(m_B - m_A)\sin\theta - 2\mu_k m_A \cos\theta\right]/(m_A + m_B)}.$$

Discussion

If we attempt to solve this example by applying the principle of work and energy to the system consisting of the crates, the cable, and the pulley (Fig. d), we obtain an incorrect result. Equating the work done by external forces to the change in the total kinetic energy of the system, we obtain

$$\int_0^b m_B g \sin\theta\, ds - \int_0^b m_A g \sin\theta\, ds = \tfrac{1}{2}m_A v_2^2 + \tfrac{1}{2}m_B v_2^2:$$

$$\left(m_B g \sin\theta\right)b - \left(m_A g \sin\theta\right)b = \tfrac{1}{2}m_A v_2^2 + \tfrac{1}{2}m_B v_2^2.$$

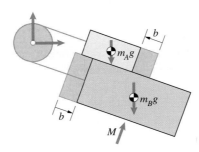

(d) Free-body diagram of the system.

But if we sum our work and energy equations for the individual crates—Eqs. (15.13) and (15.14)—we obtain the correct equation:

$$\underbrace{\left[(m_B g \sin\theta)b - (m_A g \sin\theta)b\right]}_{\substack{\text{Work done by}\\ \text{external forces}}} + \underbrace{\left[-(2\mu_k m_A g \cos\theta)b\right]}_{\substack{\text{Work done by}\\ \text{internal forces}}} = \tfrac{1}{2}m_A v_2^2 + \tfrac{1}{2}m_B v_2^2.$$

The internal frictional forces the crates exert on each other do net work on the system. We did not account for this work in applying the principle of work and energy to the free-body diagram of the system.

15.3 Work Done by Particular Forces

We have seen that if the tangential component of the total external force on an object is known as a function of distance along the object's path, the principle of work and energy can be used to relate a change in the position of the object to the change in its velocity. For certain types of forces, however, not only can we determine the work without knowing the tangential component of the force as a function of distance along the path, but we don't even need to know the path. Two important examples are weight and the force exerted by a spring.

Weight

To evaluate the work done by an object's weight, we orient a cartesian coordinate system with the y-axis upward and suppose that the object moves from position 1 with coordinates (x_1, y_1, z_1) to position 2 with coordinates (x_2, y_2, z_2) (Fig. 15.6a). The force exerted by the object's weight is $\mathbf{F} = -mg\mathbf{j}$. (Other forces may act on the object, but we are concerned only with the work done by its weight.) Because $\mathbf{v} = d\mathbf{r}/dt$, we can multiply the velocity, expressed in cartesian coordinates, by dt to obtain an expression for the vector $d\mathbf{r}$:

$$d\mathbf{r} = \left(\frac{dx}{dt}\mathbf{i} + \frac{dy}{dt}\mathbf{j} + \frac{dz}{dt}\mathbf{k} \right) dt = dx\,\mathbf{i} + dy\,\mathbf{j} + dz\,\mathbf{k}.$$

Taking the dot product of $\mathbf{F}$ and $d\mathbf{r}$ yields

$$\mathbf{F} \cdot d\mathbf{r} = (-mg\mathbf{j}) \cdot (dx\,\mathbf{i} + dy\,\mathbf{j} + dz\,\mathbf{k}) = -mg\,dy.$$

The work done as the object moves from position 1 to position 2 reduces to an integral with respect to y:

$$U_{12} = \int_{\mathbf{r}_1}^{\mathbf{r}_2} \mathbf{F} \cdot d\mathbf{r} = \int_{y_1}^{y_2} -mg\,dy.$$

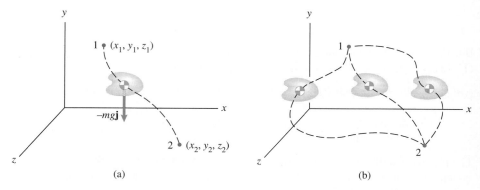

(a) (b)

Figure 15.6
(a) An object moving between two positions.
(b) The work done by the weight is the same for any path.

Evaluating the integral, we obtain the work done by the weight of an object as it moves between two positions:

$$U_{12} = -mg(y_2 - y_1).$$
(15.15)

The work is simply the product of the weight and the change in the object's height. The work done is negative if the height increases and positive if it decreases. Notice that *the work done is the same, no matter what path the object follows from position 1 to position 2* (Fig. 15.6b). Thus, we don't need to know the path to determine the work done by an object's weight—we only need to know the relative heights of the initial and final positions.

What work is done by an object's weight if we account for its variation with distance from the center of the earth? In terms of polar coordinates, we can write the weight of an object at a distance r from the center of the earth as (Fig. 15.7)

$$\mathbf{F} = -\frac{mgR_E^2}{r^2}\,\mathbf{e}_r.$$

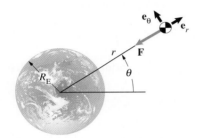

Figure 15.7
Expressing an object's weight in polar coordinates.

Using the expression for the velocity in polar coordinates, we obtain, for the vector $d\mathbf{r} = \mathbf{v}\,dt$,

$$d\mathbf{r} = \left(\frac{dr}{dt}\,\mathbf{e}_r + r\frac{d\theta}{dt}\,\mathbf{e}_\theta\right) dt = dr\,\mathbf{e}_r + r\,d\theta\,\mathbf{e}_\theta.$$
(15.16)

The dot product of $\mathbf{F}$ and $d\mathbf{r}$ is

$$\mathbf{F} \cdot d\mathbf{r} = \left(-\frac{mgR_E^2}{r^2}\,\mathbf{e}_r\right) \cdot \left(dr\,\mathbf{e}_r + r\,d\theta\,\mathbf{e}_\theta\right) = -\frac{mgR_E^2}{r^2}\,dr,$$

so the work reduces to an integral with respect to r:

$$U_{12} = \int_{\mathbf{r}_1}^{\mathbf{r}_2} \mathbf{F} \cdot d\mathbf{r} = \int_{r_1}^{r_2} -\frac{mgR_E^2}{r^2}\,dr.$$

Evaluating the integral, we obtain the work done by an object's weight, accounting for the variation of the weight with height:

$$U_{12} = mgR_E^2\left(\frac{1}{r_2} - \frac{1}{r_1}\right).$$
(15.17)

Again, the work is independent of the path from position 1 to position 2. To evaluate it, we only need to know the object's radial distance from the center of the earth at the two positions.

Springs

Suppose that a linear spring connects an object to a fixed support. In terms of polar coordinates (Fig. 15.8), the force exerted on the object is

$$\mathbf{F} = -k(r - r_0)\mathbf{e}_r,$$

where k is the spring constant and r_0 is the unstretched length of the spring. Using Eq. (15.16), we get the dot product of $\mathbf{F}$ and $d\mathbf{r}$:

$$\mathbf{F} \cdot d\mathbf{r} = \left[-k(r - r_0)\mathbf{e}_r\right] \cdot (dr\,\mathbf{e}_r + r\,d\theta\,\mathbf{e}_\theta) = -k(r - r_0)\,dr.$$

It is convenient to express the work done by a spring in terms of its *stretch*, defined by $S = r - r_0$. (Although the word stretch usually means an increase in length, we use the term more generally to denote the change in length of the spring. A negative stretch is a decrease in length.) In terms of this variable, $\mathbf{F} \cdot d\mathbf{r} = -kS\,dS$, and the work is

$$U_{12} = \int_{\mathbf{r}_1}^{\mathbf{r}_2} \mathbf{F} \cdot d\mathbf{r} = \int_{S_1}^{S_2} -kS\,dS.$$

The work done on an object by a spring attached to a fixed support is

$$U_{12} = -\tfrac{1}{2}k\left(S_2^2 - S_1^2\right), \tag{15.18}$$

where S_1 and S_2 are the values of the stretch at the initial and final positions. We don't need to know the object's path to determine the work done by the spring. Remember, however, that Eq. (15.18) applies only to a linear spring. In Fig. 15.9, we determine the work done in stretching a linear spring by calculating the area defined by the graph of the force as a function of S.

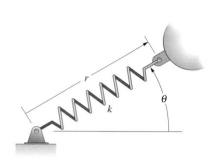

Figure 15.8
Expressing the force exerted by a linear spring in polar coordinates.

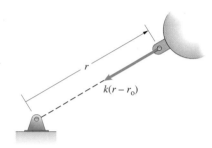

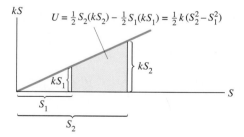

$$U = \tfrac{1}{2}S_2(kS_2) - \tfrac{1}{2}S_1(kS_1) = \tfrac{1}{2}k(S_2^2 - S_1^2)$$

Figure 15.9
Work done in stretching a linear spring from S_1 to S_2. (If $S_2 > S_1$, the work done *on* the spring is positive, so the work done *by* the spring is negative.)

Study Questions

1. If an object moves from a position 1 to a position 2, what do you need to know to calculate the work done by the object's weight?
2. If the height of an object increases, is the work done by its weight positive or negative?
3. What is the definition of the *stretch* of a spring?
4. If the stretch of a spring connecting an object to a fixed support changes from S_1 to S_2, how much work is done on the object by the spring?

Example 15.4

Work Done by Weight

At position 1, the skier in Fig. 15.10 is approaching his jump at 15 m/s. When he reaches the horizontal end of the ramp at position 2, 20 m below position 1, he jumps upward, achieving a vertical component of velocity of 3 m/s. (Disregard the small change in the vertical position of his center of mass due to his jumping motion.) Neglect aerodynamic drag and the frictional forces on his skis. (a) What is the magnitude of the skier's velocity as he leaves the ramp at position 2?
(b) At the highest point of his jump, position 3, what are the magnitude of his velocity and the height of his center of mass above position 2?

Figure 15.10

Strategy

(a) If we neglect aerodynamic and frictional forces, the only force doing work from position 1 to position 2 is the skier's weight. The normal force exerted on his skis by the ramp does no work because it is perpendicular to his path. We need to know only the change in the skier's height from position 1 to position 2 to determine the work done by his weight, so we can apply the principle of work and energy to determine his velocity at position 2 before he jumps.
(b) From the time he leaves the ramp at position 2 until he reaches position 3, the only force acting on the skier is his weight, so the horizontal component of his velocity is constant. This means that we know the magnitude of his velocity at position 3, because he is moving horizontally at that point. Therefore, we can apply the principle of work and energy to his motion from position 2 to position 3 to determine his height above position 2.

Solution

(a) We will use Eq. (15.15) to evaluate the work done by the skier's weight, measuring the height of his center of mass relative to position 2 (Fig. a). The principle of work and energy from position 1 to position 2 is

$$U_{12} = -mg(y_2 - y_1) = \tfrac{1}{2}mv_2^2 - \tfrac{1}{2}mv_1^2:$$
$$-m(9.81)(0 - 20) = \tfrac{1}{2}mv_2^2 - \tfrac{1}{2}m(15)^2.$$

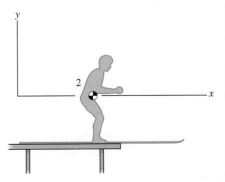

(a) The height of the skier's center of mass is measured relative to position 2.

Solving for v_2, we find that the skier's horizontal velocity at position 2 before he jumps upward is 24.8 m/s. After he jumps upward, the magnitude of his velocity at position 2 is $v_2' = \sqrt{(24.8)^2 + (3)^2} = 25.0$ m/s.

(b) The magnitude of the skier's velocity at position 3 is equal to the horizontal component of his velocity at position 2: $v_3 = v_2 = 24.8$ m/s. Applying work and energy to his motion from position 2 to position 3, we obtain

$$U_{23} = -mg(y_3 - y_2) = \tfrac{1}{2}mv_3^2 - \tfrac{1}{2}m(v_2')^2:$$
$$-m(9.81)(y_3 - 0) = \tfrac{1}{2}m(24.8)^2 - \tfrac{1}{2}m(25.0)^2,$$

from which it follows that $y_3 = 0.459$ m.

Discussion

Although we neglected aerodynamic effects, a ski jumper is actually subjected to substantial aerodynamic forces, both parallel to his path (drag) and perpendicular to it (lift).

Example 15.5

Work Done by Weight and Springs

In the forging device shown in Fig. 15.11, the 40-kg hammer is lifted to position 1 and released from rest. It falls and strikes a workpiece when it is in position 2. The spring constant $k = 1500$ N/m, and the tension in each spring is 150 N when the hammer is in position 2. Neglect friction.

(a) What is the velocity of the hammer just before it strikes the workpiece?

(b) Assuming that all the hammer's kinetic energy is transferred to the workpiece, what average power is transferred if the duration of the impact is 0.02 s?

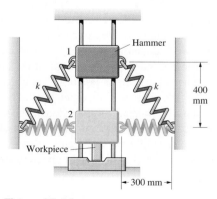

Figure 15.11

Strategy

Work is done on the hammer by its weight and by the forces exerted by the springs (Fig. a). We can apply the principle of work and energy to the motion of the hammer from position 1 to position 2 to determine its velocity at position 2.

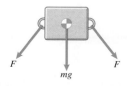

(a) Free-body diagram of the hammer.

Solution

(a) Let r_0 be the unstretched length of one of the springs. In position 2, the tension in the spring is 150 N and its length is 0.3 m. From the relation between the tension in a linear spring and its stretch, we obtain

$$150 = k(0.3 - r_0) = (1500)(0.3 - r_0),$$

from which it follows that $r_0 = 0.2$ m. The values of the stretch of each spring in positions 1 and 2 are $S_1 = \sqrt{(0.4)^2 + (0.3)^2} - 0.2 = 0.3$ m and $S_2 = 0.3 - 0.2 = 0.1$ m. From Eq. (15.18), the total work done on the hammer by the two springs from position 1 to position 2 is

$$U_{springs} = 2\left[-\tfrac{1}{2}k(S_2^2 - S_1^2)\right] = -(1500)\left[(0.1)^2 - (0.3)^2\right] = 120 \text{ N-m}.$$

The work done by the weight from position 1 to position 2 is positive and equal to the product of the weight and the change in height:

$$U_{weight} = mg(0.4 \text{ m}) = (40)(9.81)(0.4) = 157 \text{ N-m}.$$

From the principle of work and energy, we have

$$U_{springs} + U_{weight} = \tfrac{1}{2}mv_2^2 - \tfrac{1}{2}mv_1^2:$$

$$120 + 157 = \tfrac{1}{2}(40)v_2^2 - 0,$$

so that $v_2 = 3.72$ m/s.

(b) All the hammer's kinetic energy is transferred to the workpiece, so Eq. (15.10) indicates that the average power equals the kinetic energy of the hammer divided by the duration of the impact:

$$P_{av} = \frac{(1/2)(40 \text{ kg})(3.72 \text{ m/s})^2}{0.02 \text{ s}} = 13.8 \text{ kW (kilowatts)}.$$

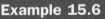

Example 15.6

Application to Engineering:

Automated Machining

The device in Fig. 15.12a machines a sample of material with the cutting tool at A. The hydraulic actuator attached at B pushes the toolholder to the right on its horizontal guide rail. The spring attached at C returns the toolholder and hydraulic actuator to their initial positions when the cut is completed. The position of the sample of material is changed, and the cycle is repeated. In the free-body diagram of the assembly consisting of the cutting tool and toolholder (Fig. 15.12b), A_x and A_y are the forces exerted on the cutting tool by the machined sample, B is the force exerted by the actuator, F is the force exerted by the spring, and N is the normal force exerted by the guide rail. The mass of the assembly is $m = 22$ kg, the unstretched length of the spring is 200 mm, and the spring constant is $k = 1800$ N/m.

(a) Suppose the assembly starts from rest in the position shown, the horizontal force on the cutting tool is $A_x = 1330$ N, and the force exerted by the actuator is $B = 1700$ N. Determine the velocity of the assembly when it has moved 0.05 m to the right.

(b) Determine the maximum power transferred from the actuator as the assembly moves from $x = 0$ to $x = 0.15$ m.

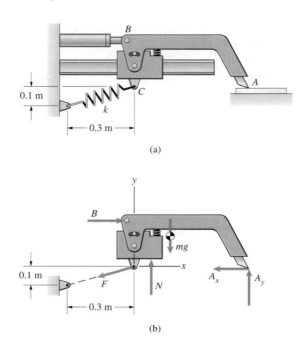

(a)

(b)

Figure 15.12

Strategy

(a) By determining the total work done by the spring and the horizontal forces A_x and B as a function of x, we can use work and energy to determine the velocity of the assembly as a function of x. (b) The power transferred equals the product of the force exerted by the actuator and the velocity. We will draw a graph of the velocity as a function of x to determine the maximum velocity.

Solution

(a) Let the initial position of the assembly be position 1, and let its position when it has moved a distance x to the right be position 2. The stretch of the spring in position 1 is.

$$S_1 = \sqrt{(0.1)^2 + (0.3)^2} - 0.2 \text{ m}.$$

and the stretch in position 2 is

$$S_2 = \sqrt{(0.1)^2 + (0.3 + x)^2} - 0.2 \text{ m}.$$

The work done from position 1 to position 2 by the horizontal forces A_x and B and the spring is

$$U_{12} = (B - A_x)x - \tfrac{1}{2}k(S_2^2 - S_1^2).$$

We apply the principle of work and energy:

$$U_{12} = \tfrac{1}{2}mv_2^2 - \tfrac{1}{2}mv_1^2;$$

$$(B - A_x)x - \tfrac{1}{2}k(S_2^2 - S_1^2) = \tfrac{1}{2}mv_2^2 - 0.$$

Setting $x = 0.05$ m, $A_x = 1330$ N, $B = 1700$ N, and $m = 22$ kg and solving for the velocity, we obtain $v_2 = 0.766$ m/s.

(b) In Fig. 15.13, we draw a graph of the velocity v_2 as a function of x for $0 \le x \le 0.15$ m. From the graph, we estimate that the maximum velocity occurs at $x = 0.1$ m. Solving for the velocity at this position, we obtain $v_2 = 0.884$ m/s. The power transferred from the actuator at the maximum velocity is

$$P = Bv_2 = (1700)(0.884) = 1500 \text{ N-m/s (watts)}.$$

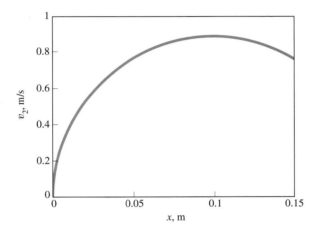

Figure 15.13
Velocity of the assembly as a function of x.

$\mathscr{D}$esign Issues

In developing a machine, the engineers must ensure that it will perform the tasks for which it is designed and do so reliably through a design lifetime that may be many years and millions of cycles of use. The machine should be designed for ease of use and maintenance and must be safe for its operators and other persons. It must also be economical. If it is an industrial machine, its original cost and the cost of its use and maintenance must be acceptable in comparison to the income derived through its use. The choices implied by these requirements are often mutually contradictory, and the design engineers must make successful compromises.

The use of dynamics to study the motions of machines is essential to designers, who analyze the behavior of the machines, predict their performance, and determine whether their structures will support the dynamic loads to which they will be subjected. Employing a simple context, we demonstrate in Example 15.6 that the principle of work and energy can be useful in analyzing a machine's motion. It can also provide information on energy requirements of machines. The work done by the hydraulic actuator in Example 15.6 provides an estimate of the energy that must be supplied (in the form of electrical energy to power the hydraulic pump) for each cycle of operation of the machine tool.

Figure 15.14

Example 15.7

Work Done by Earth's Gravity

A spacecraft at a distance $r_1 = 2R_E$ from the center of the earth has a velocity of magnitude $v_1 = \sqrt{2gR_E/3}$ relative to a nonrotating reference frame with its origin at the center of the earth (Fig. 15.14). Determine the magnitude of the spacecraft's velocity when it is at a distance $r_2 = 4R_E$ from the center of the earth.

Strategy

By applying Eq. (15.17) to determine the work done by the gravitational force on the spacecraft, we can use the principle of work and energy to determine the magnitude of the spacecraft's velocity.

Solution

From Eq. (15.17), the work done by gravity as the spacecraft moves from a distance r_1 from the center of the earth to a distance r_2 is

$$U_{12} = mgR_E^2\left(\frac{1}{r_2} - \frac{1}{r_1}\right).$$

Let v_2 be the magnitude of the velocity of the spacecraft when it is at a distance r_2 from the center of the earth. Applying the principle of work and energy yields

$$U_{12} = mgR_E^2\left(\frac{1}{r_2} - \frac{1}{r_1}\right) = \tfrac{1}{2}mv_2^2 - \tfrac{1}{2}mv_1^2.$$

We solve for v_2, obtaining

$$
\begin{aligned}
v_2 &= \sqrt{v_1^2 + 2gR_E^2\left(\frac{1}{r_2} - \frac{1}{r_1}\right)}\\[4pt]
&= \sqrt{\left(\frac{2gR_E}{3}\right) + 2gR_E^2\left(\frac{1}{4R_E} - \frac{1}{2R_E}\right)}\\[4pt]
&= \sqrt{\frac{gR_E}{6}}.
\end{aligned}
$$

The velocity $v_2 = v_1/2$.

Discussion

Notice that we did not need to specify the direction of the spacecraft's initial velocity to determine the magnitude of its velocity at a different distance from the center of the earth. This illustrates the power of the principle of work and energy, as well as one of its limitations: Even if we know the direction of the initial velocity, the principle of work and energy tells us only the *magnitude* of the velocity at a different distance.

Potential Energy

15.4   Conservation of Energy

The work done on an object by some forces can be expressed as the change of a function of the object's position called the potential energy. When all the forces that do work on a system have this property, we can state the principle of work and energy as a conservation law: The sum of the kinetic and potential energies is constant.

When we derived the principle of work and energy by integrating Newton's second law, we were able to evaluate the integral on one side of the equation, obtaining the change in the kinetic energy:

$$U_{12} = \int_{\mathbf{r}_1}^{\mathbf{r}_2} \Sigma \mathbf{F} \cdot d\mathbf{r} = \tfrac{1}{2}mv_2^2 - \tfrac{1}{2}mv_1^2. \tag{15.19}$$

Suppose we could determine a scalar function V of position such that

$$dV = -\Sigma \mathbf{F} \cdot d\mathbf{r}. \tag{15.20}$$

Then we could also evaluate the integral defining the work; that is,

$$U_{12} = \int_{\mathbf{r}_1}^{\mathbf{r}_2} \Sigma \mathbf{F} \cdot d\mathbf{r} = \int_{V_1}^{V_2} -dV = -(V_2 - V_1), \tag{15.21}$$

where V_1 and V_2 are the values of V at the positions $\mathbf{r}_1$ and $\mathbf{r}_2$, respectively. The principle of work and energy would then have the simple form

$$\tfrac{1}{2}mv_1^2 + V_1 = \tfrac{1}{2}mv_2^2 + V_2, \tag{15.22}$$

which means that the sum of the kinetic energy and the function V is constant:

$$\tfrac{1}{2}mv^2 + V = \text{constant}. \tag{15.23}$$

If the kinetic energy increases, V must decrease, and vice versa, as if V represents a reservoir of "potential" kinetic energy. For this reason, V is called the *potential energy.*

If a potential energy exists for a given force $\mathbf{F}$, meaning that a function V of position exists such that $dV = -\mathbf{F} \cdot d\mathbf{r}$, then $\mathbf{F}$ is said to be *conservative.* If all the forces that do work on a system are conservative, the total energy—the sum of the kinetic energy and the potential energies of the forces—is constant, or conserved. In that case, the system is said to be conservative, and we can use conservation of energy instead of the principle of work and energy to relate a change in position of the system to the change in its kinetic energy. The two approaches are equivalent, and we obtain the same quantitative information. But greater insight is gained by using conservation of energy, because the motion of the object or system can be interpreted in terms of transformations between potential and kinetic energies.

15.5 Conservative Forces

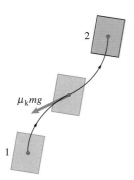

Figure 15.15
The book's path from position 1 to position 2.
The force of friction points opposite to the
direction of the motion.

We can apply conservation of energy only if the forces doing work on an object or system are conservative and we know (or can determine) their potential energies. In this section, we determine the potential energies of some conservative forces and use the results to demonstrate applications of conservation of energy. But before discussing forces that are conservative, we demonstrate with a simple example that frictional forces are not conservative.

The work done by a conservative force as an object moves from a position 1 to a position 2 is independent of the object's path. This result follows from Eq. (15.21), which states that the work depends only on the values of the potential energy at positions 1 and 2. That equation also implies that if the object moves along a closed path, returning to position 1, the work done by a conservative force is zero. Suppose that a book of mass m rests on a table and you push it horizontally so that it slides along a path of length L. The magnitude of the force of friction is $\mu_k mg$, and the direction of the force is opposite to that of the book's motion (Fig. 15.15). The work done is

$$U_{12} = \int_0^L -\mu_k mg \, ds = -\mu_k mgL.$$

The work is proportional to the length of the object's path and therefore is not independent of the path. Thus, frictional forces are not conservative.

Potential Energies of Particular Forces

The weight of an object and the force exerted by a spring attached to a fixed support are conservative forces. Using them as examples, we demonstrate how you can determine the potential energies of other conservative forces. We also use the potential energies of these forces in examples of the use of conservation of energy to analyze the motions of conservative systems.

Weight To determine the potential energy associated with an object's weight, we use a cartesian coordinate system with its y axis pointing upward (Fig. 15.16). The weight is $\mathbf{F} = -mg\mathbf{j}$, and its dot product with the vector $d\mathbf{r}$ is

$$\mathbf{F} \cdot d\mathbf{r} = (-mg\mathbf{j}) \cdot (dx\mathbf{i} + dy\mathbf{j} + dz\mathbf{k}) = -mg \, dy.$$

From Eq. (15.20), the potential energy V must satisfy the relation

$$dV = -\mathbf{F} \cdot d\mathbf{r} = mg \, dy, \tag{15.24}$$

which we can write as

$$\frac{dV}{dy} = mg.$$

Integrating this equation, we obtain

$$V = mgy + C,$$

where C, the constant of integration, is arbitrary: This expression satisfies Eq. (15.24) for any value of C. Another way of understanding why C is arbitrary is to notice in Eq. (15.22) that it is the difference in the potential energy

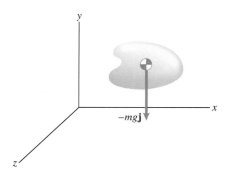

Figure 15.16
Weight of an object expressed in terms of a
coordinate system with the y-axis pointing
upward.

between two positions that determines the change in the kinetic energy. We will let $C = 0$ and write the potential energy of the weight of an object as

$$V = mgy \qquad (15.25)$$

The potential energy is the product of the object's weight and height. The height can be measured from any convenient reference level, or *datum*. Since the difference in potential energy determines the change in the kinetic energy, it is the difference in height that matters, not the level from which the height is measured.

The roller coaster (Fig. 15.17a) is a classic example of conservation of energy. If aerodynamic and frictional forces are neglected, the weight is the only force doing work, and the system is conservative. The potential energy of the roller coaster is proportional to the height of the track relative to a datum. In Fig. 15.17(b), we assume that the roller coaster started from rest at the datum level. The sum of the kinetic and potential energies is constant, so the kinetic energy "mirrors" the potential energy. At points of the track that have equal heights, the magnitudes of the velocities are equal.

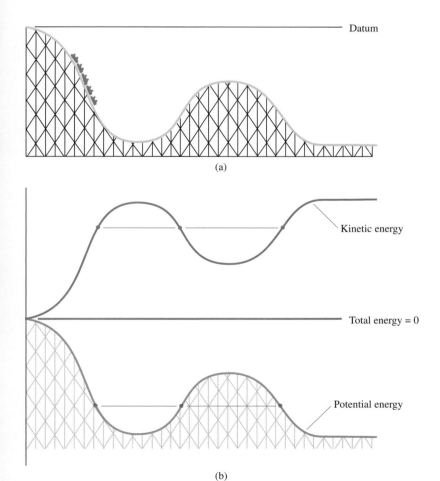

(a)

Datum

Kinetic energy

Total energy = 0

Potential energy

(b)

Figure 15.17
(a) Roller coaster and a reference level, or datum.
(b) The sum of the potential and kinetic energies is constant.

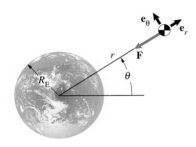

Figure 15.18
Expressing weight in terms of polar
coordinates.

To account for the variation of weight with distance from the center of
the earth, we can express the weight in polar coordinates as

$$\mathbf{F} = -\frac{mgR_E^2}{r^2}\mathbf{e}_r,$$

where r is the distance from the center of the earth (Fig. 15.18). From Eq.
(15.16), the vector $d\mathbf{r}$ in terms of polar coordinates is

$$d\mathbf{r} = dr\,\mathbf{e}_r + r\,d\theta\,\mathbf{e}_\theta. \tag{15.26}$$

The potential energy must satisfy

$$dV = -\mathbf{F} \cdot d\mathbf{r} = \frac{mgR_E^2}{r^2}\,dr,$$

or

$$\frac{dV}{dr} = \frac{mgR_E^2}{r^2}.$$

We integrate this equation and let the constant of integration be zero, obtaining the potential energy

$$V = -\frac{mgR_E^2}{r}. \tag{15.27}$$

Springs In terms of polar coordinates, the force exerted on an object by a
linear spring is

$$\mathbf{F} = -k(r - r_0)\mathbf{e}_r,$$

where r_0 is the unstretched length of the spring (Fig. 15.19). Using Eq.
(15.26), we see that the potential energy must satisfy

$$dV = -\mathbf{F} \cdot d\mathbf{r} = k(r - r_0)\,dr.$$

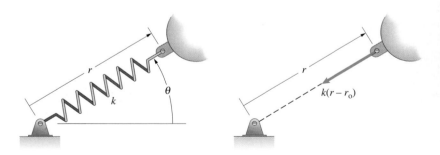

Figure 15.19
Expressing the force exerted by a linear
spring in polar coordinates.

Expressed in terms of the stretch of the spring $S = r - r_0$, this equation is
$dV = kS\,dS$, or

$$\frac{dV}{dS} = kS.$$

Integrating the foregoing equation, we obtain the potential energy of a linear spring:

$$V = \tfrac{1}{2}kS^2. \tag{15.28}$$

Using conservation of energy to relate changes in the positions of conservative systems to changes in their kinetic energies typically involves three steps:

1. *Determine whether the system is conservative.* Draw a free-body diagram to identify the forces that do work, and confirm that they are conservative.

2. *Determine the potential energy.* Evaluate the potential energies of the forces in terms of the position of the system.

3. *Apply conservation of energy.* Equate the sum of the kinetic and potential energies of the system at two positions to obtain an expression for the change in the kinetic energy.

Study Questions

1. What is the definition of a conservative force?
2. What condition is necessary for the total energy of a system to be conserved?
3. If you know the change in the total potential energy as an object subjected to conservative forces moves from a position 1 to a position 2, what can you infer about the work done on the object as it moves between the two positions? What can you infer about the change in the total energy?
4. If an object moves upward, does the potential energy associated with its weight increase or decrease? (If you don't know the answer, consider what happens when you throw an object upward.)

Example 15.8

Potential Energy of Weight and Springs

In Example 15.5, the 40-kg hammer is lifted into position 1 and released from rest. Its weight and the two springs $(k = 1500 \text{ N/m})$ accelerate the hammer downward to position 2, where it strikes a workpiece. Use conservation of energy to determine the hammer's velocity when it reaches position 2.

Solution

Determine whether the System Is Conservative From the free-body diagram of the hammer (Fig. a), we see that work is done only by its weight and the forces exerted by the springs. Therefore, the system is conservative.

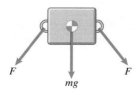

(a) Free-body diagram of the hammer.

Determine the Potential Energy The potential energy of each spring is $\frac{1}{2}kS^2$, where S is the stretch, so the total potential energy of the two springs is

$$V_{\text{springs}} = 2\left(\tfrac{1}{2}kS^2\right).$$

In Example 15.5, the stretches in positions 1 and 2 were determined to be $S_1 = 0.3$ m and $S_2 = 0.1$ m, respectively. The potential energy associated with the weight is

$$V_{\text{weight}} = mgy,$$

where y is the height relative to a convenient datum (Fig. b).

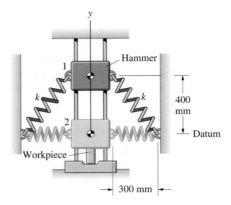

(b) Measuring the height of the hammer relative to position 2.

Apply Conservation of Energy The sums of the potential and kinetic energies at positions 1 and 2 must be equal:

$$2\left(\tfrac{1}{2}kS_1^2\right) + mgy_1 + \tfrac{1}{2}mv_1^2 = 2\left(\tfrac{1}{2}kS_2^2\right) + mgy_2 + \tfrac{1}{2}mv_2^2;$$

$$(1500)(0.3)^2 + (40)(9.81)(0.4) + 0 = (1500)(0.1)^2 + 0 + \tfrac{1}{2}(40)v_2^2.$$

Solving this equation, we obtain $v_2 = 3.72$ m/s.

Discussion

From the graphs of the total potential energy associated with the springs and the weight and the kinetic energy of the hammer as functions of y (Fig. 15.20), you can see the transformation of the potential energy into kinetic energy as the hammer falls. Notice that the total energy of the conservative system remains constant.

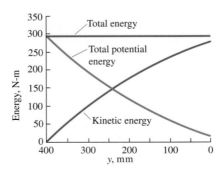

Figure 15.20
The potential and kinetic energies as functions of the y-coordinate of the hammer.

Example 15.9

Conservation of Energy of a System

The spring in Fig. 15.21 ($k = 300$ N/m) is connected to the floor and to the 90-kg collar A. Collar A is at rest, supported by the spring, when the 135-kg box B is released from rest in the position shown. What are the velocities of A and B when B has fallen 1 m?

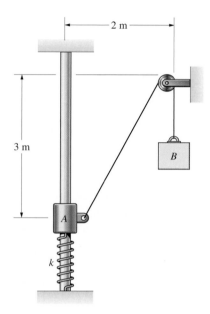

2 m

3 m

k

A

B

Figure 15.21

Solution

Determine whether the System Is Conservative We consider the collar A, box B, and pulley as a single system. From the free-body diagram of the system in Fig. (a), we see that work is done only by the weights of the collar and box and the spring force F. The system is therefore conservative.

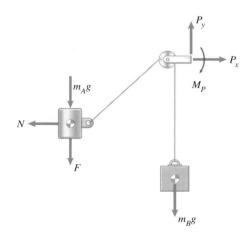

P_y

P_x

$m_A g$

M_P

N

F

$m_B g$

(a) Free-body diagram of the system.

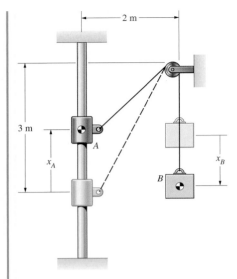

(b) Displacements of the collar and box.

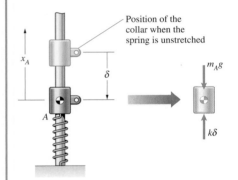

(c) Determining the initial compression of the spring.

Determine the Potential Energy   Using the initial position of collar A as its datum, the potential energy associated with the weight of A when it has risen a distance x_A (Fig. b) is $V_A = m_A g x_A$. Using the initial position of box B as its datum, the potential energy associated with its weight when it has fallen a distance x_B is $V_B = -m_B g x_B$. (The minus sign is necessary because x_B is positive downward.)

To determine the potential energy associated with the spring force, we must account for the fact that in the initial position the spring is compressed by the weight of collar A. The spring is initially compressed a distance δ such that $m_A g = k\delta$ (Fig. c). When the collar has moved upward a distance x_A, the stretch of the spring is $S = x_A - \delta = x_A - m_A g/k$, so its potential energy is

$$V_S = \tfrac{1}{2}kS^2 = \tfrac{1}{2}k\left(x_A - \frac{m_A g}{k}\right)^2.$$

The total potential energy of the system in terms of the displacements of the collar and box is

$$V = V_A + V_B + V_S$$

$$= m_A g x_A - m_B g x_B + \tfrac{1}{2}k\left(x_A - \frac{m_A g}{k}\right)^2.$$

Apply Conservation of Energy   The sum of the kinetic and potential energies of the system in its initial position and in the position shown in Fig. (b) must be equal. Denoting the total kinetic energy by T, we have

$$T_1 + V_1 = T_2 + V_2:$$

$$0 + \tfrac{1}{2}k\left(-\frac{m_A g}{k}\right)^2 = \tfrac{1}{2}m_A v_A^2 + \tfrac{1}{2}m_B v_B^2$$

$$+ m_A g x_A - m_B g x_B + \tfrac{1}{2}k\left(x_A - \frac{m_A g}{k}\right)^2. \quad (15.29)$$

We want to determine v_A and v_B when $x_B = 1$ m, but we have only one equation in terms of x_A, x_B, v_A, and v_B. To complete the solution, we must relate the displacement and velocity of the collar A to the displacement and velocity of the box B.

From Fig. b, the decrease in the length of the rope from A to the pulley as the collar rises must equal the distance the box falls:

$$\sqrt{3^2 + 2^2} - \sqrt{(3 - x_A)^2 + 2^2} = x_B.$$

Solving this equation for the value of x_A when $x_B = 1$ m, we obtain $x_A = 1.33$ m. By taking the derivative of this equation with respect to time, we also obtain a relation between v_A and v_B:

$$\left[\frac{3 - x_A}{\sqrt{(3 - x_A)^2 + 2^2}}\right]v_A = v_B.$$

Setting $x_A = 1.33$ m, we determine from the preceding equation that

$$0.641v_A = v_B.$$

We solve this equation together with Eq. (15.29) for the velocities of the collar and box when $x_A = 1.33$ m and $x_B = 1$ m, obtaining $v_A = 3.82$ m/s and $v_B = 2.45$ m/s.

Discussion

Why didn't we have to consider the forces exerted on the collar and box by the rope? The reason is that they are internal forces when the system considered is the collar, box, and pulley.

Example 15.10

Conservation of Energy of a Spacecraft

A spacecraft at a distance $r_0 = 2R_E$ from the center of the earth is moving outward with initial velocity $v_0 = \sqrt{2gR_E/3}$ (Fig. 15.22). Determine the velocity of the craft as a function of its distance from the center of the earth.

Figure 15.22

Solution

Determine whether the System Is Conservative   If (we assume that) work is done on the spacecraft by gravity alone, the system is conservative.

Determine the Potential Energy   The potential energy associated with the weight of the spacecraft is given in terms of its distance r from the center of the earth by Eq. (15.27):

$$V = -\frac{mgR_E^2}{r}.$$

Apply Conservation of Energy   Let v be the magnitude of the spacecraft's velocity at an arbitrary distance r. The sums of the potential and kinetic energies at r_0 and at r must be equal.

$$-\frac{mgR_E^2}{r_0} + \tfrac{1}{2}mv_0^2 = -\frac{mgR_E^2}{r} + \tfrac{1}{2}mv^2:$$

$$-\frac{mgR_E^2}{2R_E} + \tfrac{1}{2}m(\tfrac{2}{3}gR_E) = -\frac{mgR_E^2}{r} + \tfrac{1}{2}mv^2.$$

Solving for v, we find that the spacecraft's velocity as a function of r is

$$v = \sqrt{gR_E\left(\frac{2R_E}{r} - \frac{1}{3}\right)}.$$

Discussion

We show graphs of the kinetic energy, potential energy, and total energy as functions of r/R_E in Fig. 15.23. The kinetic energy decreases and the potential energy increases as the spacecraft moves outward until its velocity decreases to zero at $r = 6R_E$.

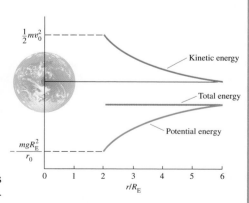

Figure 15.23
Energies as functions of the radial coordinate.

Example 15.11

Potential Energy of a Compressed Gas

The bore of the gas gun shown in Fig. 15.24 has cross-sectional area A. The bore is evacuated on the right of the projectile of mass m, and on the left it contains gas at pressure p. Let the value of the pressure when $s = s_0$ be p_0, and assume that the pressure of the gas is related to its volume V by $pV = $ constant.

Gas   Projectile

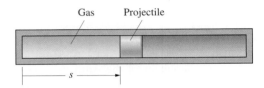

Figure 15.24

s

(a) Determine the potential energy associated with the force exerted on the projectile in terms of s.
(b) If the projectile starts from rest at $s = s_0$ and friction is negligible, what is the velocity the projectile as a function of s?

Strategy

(a) We can determine the force exerted on the piston by the pressure of the gas as a function of s and then use Eq. (15.20) to determine the potential energy.
(b) Knowing the potential energy, we can use conservation of energy to determine the velocity of the projectile as a function of s.

Solution

(a) The volume of the gas to the left of the projectile is $V = sA$. We know that pV is constant, so psA is constant. The value of the pressure when $s = s_0$ is p_0, so $psA = p_0 s_0 A$. The force F exerted on the projectile is the product of the pressure p and the cross-sectional area A of the projectile. Therefore,

$$F = pA = \frac{p_0 s_0 A}{s}.$$

To apply Eq. (15.20), we need to express the total force on the piston and an infinitesimal displacement $d\mathbf{r}$ of the piston as vectors. Let $\mathbf{e}_t$ be a unit vector that points to the right. Neglecting friction, we observe that the total force on the piston is

$$\Sigma \mathbf{F} = F\mathbf{e}_t = \frac{p_0 s_0 A}{s} \mathbf{e}_t,$$

and we can express $d\mathbf{r}$ as

$$d\mathbf{r} = ds\,\mathbf{e}_t.$$

Substituting these expressions into Eq. (15.20), we obtain

$$dV = -\Sigma\mathbf{F} \cdot d\mathbf{r} = -\left(\frac{p_0 s_0 A}{s}\mathbf{e}_{\mathrm{t}}\right) \cdot (ds\, \mathbf{e}_{\mathrm{t}}) = -\frac{p_0 s_0 A}{s}ds,$$

which we can write as

$$\frac{dV}{ds} = -\frac{p_0 s_0 A}{s}.$$

We integrate this equation, setting the constant of integration equal to zero, to obtain the potential energy:

$$V = -p_0 s_0 A \ln s.$$

(b) If the projectile starts from rest at $s = s_0$, we can use conservation of energy to determine its velocity v at an arbitrary position s.

$$\tfrac{1}{2}mv_1^2 + V_1 = \tfrac{1}{2}mv_2^2 + V_2:$$

$$0 - p_0 s_0 A \ln s_0 = \tfrac{1}{2}mv^2 - p_0 s_0 A \ln s.$$

Solving for the velocity, we obtain

$$v = \sqrt{\frac{2p_0 s_0 A}{m}\ln\left(\frac{s}{s_0}\right)}.$$

15.6 Relationships between Force and Potential Energy

Here we consider two questions: (1) Given a potential energy, how can we determine the corresponding force? (2) Given a force, how can we determine whether it is conservative? That is, how can we tell whether an associated potential energy exists?

The potential energy V of a force $\mathbf{F}$ is a function of position that satisfies the relation

$$dV = -\mathbf{F} \cdot d\mathbf{r}. \tag{15.30}$$

Let us express V in terms of a cartesian coordinate system:

$$V = V(x, y, z).$$

The differential of V is

$$dV = \frac{\partial V}{\partial x}dx + \frac{\partial V}{\partial y}dy + \frac{\partial V}{\partial z}dz. \tag{15.31}$$

Expressing $\mathbf{F}$ and $d\mathbf{r}$ in terms of cartesian components and taking their dot product yields

$$\mathbf{F} \cdot d\mathbf{r} = (F_x\mathbf{i} + F_y\mathbf{j} + F_z\mathbf{k}) \cdot (dx\mathbf{i} + dy\mathbf{j} + dz\mathbf{k})$$

$$= F_x\, dx + F_y\, dy + F_z\, dz.$$

Substituting this expression and Eq. (15.31) into Eq. (15.30), we obtain

$$\frac{\partial V}{\partial x}\, dx + \frac{\partial V}{\partial y}\, dy + \frac{\partial V}{\partial z}\, dz = -(F_x\, dx + F_y\, dy + F_z\, dz),$$

which implies that

$$F_x = -\frac{\partial V}{\partial x}, \qquad F_y = -\frac{\partial V}{\partial y}, \qquad \text{and} \quad F_z = -\frac{\partial V}{\partial z}. \qquad (15.32)$$

Given a potential energy V expressed in cartesian coordinates, we can use the foregoing relations to determine the corresponding force. The force

$$\mathbf{F} = -\left(\frac{\partial V}{\partial x}\mathbf{i} + \frac{\partial V}{\partial y}\mathbf{j} + \frac{\partial V}{\partial z}\mathbf{k}\right) = -\nabla V, \qquad (15.33)$$

where ∇V is the *gradient* of V. By using expressions for the gradient in terms of other coordinate systems, we can determine the force $\mathbf{F}$ when we know the potential energy in terms of those coordinate systems. For example, in terms of cylindrical coordinates,

$$\mathbf{F} = -\left(\frac{\partial V}{\partial r}\mathbf{e}_r + \frac{1}{r}\frac{\partial V}{\partial \theta}\mathbf{e}_\theta + \frac{\partial V}{\partial z}\mathbf{e}_z\right). \qquad (15.34)$$

If a force $\mathbf{F}$ is conservative, its *curl* $\nabla \times \mathbf{F}$ is zero. The expression for the curl of $\mathbf{F}$ in cartesian coordinates is

$$\nabla \times \mathbf{F} = \begin{vmatrix} \mathbf{i} & \mathbf{j} & \mathbf{k} \\ \dfrac{\partial}{\partial x} & \dfrac{\partial}{\partial y} & \dfrac{\partial}{\partial z} \\ F_x & F_y & F_z \end{vmatrix}. \qquad (15.35)$$

Substituting Eqs. (15.32) into this expression confirms that $\nabla \times \mathbf{F} = \mathbf{0}$ when $\mathbf{F}$ is conservative. The converse is also true: A force $\mathbf{F}$ is conservative if its curl is zero. We can use this condition to determine whether a given force is conservative. In terms of cylindrical coordinates, the curl of $\mathbf{F}$ is

$$\nabla \times \mathbf{F} = \frac{1}{r}\begin{vmatrix} \mathbf{e}_r & r\mathbf{e}_\theta & \mathbf{e}_z \\ \dfrac{\partial}{\partial r} & \dfrac{\partial}{\partial \theta} & \dfrac{\partial}{\partial z} \\ F_r & rF_\theta & F_z \end{vmatrix}. \qquad (15.36)$$

Example 15.12

Determining the Force from a Potential Energy

From Eq. (15.27), the potential energy associated with the weight of an object of mass m at a distance r from the center of the earth is (in polar coordinates)

$$V = -\frac{mgR_E^2}{r},$$

where R_E is the radius of the earth. Use this expression to determine the force exerted on the object by its weight.

Strategy

The force $\mathbf{F} = -\nabla V$. The potential energy is expressed in terms of polar coordinates, so we can use Eq. (15.34) to determine the force.

Solution

The partial derivatives of V with respect to r, θ, and z are

$$\frac{\partial V}{\partial r} = \frac{mgR_E^2}{r^2}, \quad \frac{\partial V}{\partial \theta} = 0, \quad \text{and} \quad \frac{\partial V}{\partial z} = 0.$$

From Eq. (15.34), the force is

$$\mathbf{F} = -\nabla V = -\frac{mgR_E^2}{r^2}\,\mathbf{e}_r.$$

Discussion

We already know that the force is conservative, because we know its potential energy, but we can use Eq. (15.36) to confirm that its curl is zero:

$$\nabla \times \mathbf{F} = \frac{1}{r}\begin{vmatrix} \mathbf{e}_r & r\mathbf{e}_\theta & \mathbf{e}_z \\ \dfrac{\partial}{\partial r} & \dfrac{\partial}{\partial \theta} & \dfrac{\partial}{\partial z} \\ -\dfrac{mgR_E^2}{r^2} & 0 & 0 \end{vmatrix} = \mathbf{0}.$$

Although we used cylindrical coordinates in determining $\mathbf{F}$ and in evaluating the cross product, the expression for V and our resulting expression for $\mathbf{F}$ are valid only if the object remains in the plane $z = 0$.

Computational Mechanics

The following example and problems are designed to be worked with the use of a programmable calculator or computer.

Computational Example 15.13

In the mechanical delay switch shown in Fig. 15.25, an electromagnet releases the 1-kg slider at position 1. Under the actions of gravity and the linear spring, the slider moves along the smooth bar from position 1 to position 2, closing the switch. The constant of the spring is $k = 40$ N/m, and the unstretched length of the spring is $r_0 = 50$ mm. The dimensions are $R = 200$ mm and $h = 100$ mm. What is the magnitude of the slider's maximum velocity, and where does it occur?

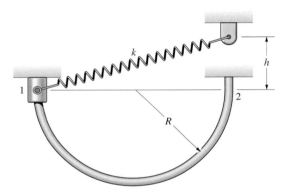

Figure 15.25

Strategy

We can use conservation of energy to obtain an equation relating the magnitude of the velocity of the slider to its position. By drawing a graph of the velocity as a function of the position, we can estimate the maximum velocity and the position where it occurs.

Solution

We can specify the slider's position by the angle θ through which it has moved relative to position 1 (Fig. a). In position 1, the stretch of the spring equals its length in position 1 minus its unstretched length:

$$S_1 = \sqrt{(2R)^2 + h^2} - r_0.$$

When the slider has moved through the angle θ, the stretch of the spring is

$$S = \sqrt{(R + R\cos\theta)^2 + (h + R\sin\theta)^2} - r_0.$$

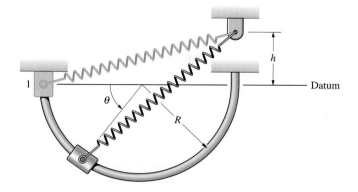

(a) The angle θ specifies the slider's position.

We express the potential energy of the slider's weight using the datum shown in Fig. (a). The sum of the potential and kinetic energies at position 1 must equal the sum of the potential and kinetic energies when the slider has moved through the angle θ.

$$\tfrac{1}{2}kS_1^2 + mgy_1 + \tfrac{1}{2}mv_1^2 = \tfrac{1}{2}kS^2 + mgy + \tfrac{1}{2}mv^2:$$

$$\tfrac{1}{2}k\left[\sqrt{(2R)^2 + h^2} - r_0\right]^2 + 0 + 0$$

$$= \tfrac{1}{2}k\left[\sqrt{(R + R\cos\theta)^2 + (h + R\sin\theta)^2} - r_0\right]^2$$

$$- mgR\sin\theta + \tfrac{1}{2}mv^2.$$

Solving for v yields

$$v = \Big\{(k/m)\left[\sqrt{(2R)^2 + h^2} - r_0\right]^2$$

$$- (k/m)\left[\sqrt{(R + R\cos\theta)^2 + (h + R\sin\theta)^2} - r_0\right]^2 + 2gR\sin\theta\Big\}^{1/2}.$$

Computing the values of this expression as a function of θ, we obtain the graph shown in Fig. 15.26. The graph indicates that the velocity is a maximum at approximately $\theta = 135°$, so we examine the computed results near 135°:

θ	v, m/s
132°	2.5393
133°	2.5397
134°	2.5399
135°	2.5398
136°	2.5394
137°	2.5389
138°	2.5380

We estimate that a maximum velocity of 2.54 m/s occurs at $\theta = 134°$.

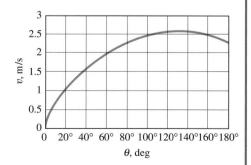

Figure 15.26
Magnitude of the velocity as a function of θ.

Chapter Summary

In Chapter 14, we used Newton's second law to determine an object's acceleration when the forces acting on it were known. Once the acceleration was known, it could be integrated to obtain information about the object's velocity and position. In this chapter, we have used Newton's second law to derive a new technique, the principle of work and energy, that relates the work done during a change in an object's position to the change in its kinetic energy. This principle is usually applicable only when the forces that do work are known as functions of the object's position, and the only information obtained is the change in the magnitude of the object's velocity. Nevertheless, the principle can be highly useful because the work done by certain forces, including gravitational forces and forces exerted by springs, is easy to determine and is independent of an object's path. This property motivated the definitions of the potential energy and conservative forces and resulted in the concept of conservation of energy. In Chapter 16, we will use Newton's second law to derive momentum principles, which are especially useful for applications involving collisions between objects.

Principle of Work and Energy

Defining the *work* done on an object as its center of mass moves from a position $\mathbf{r}_1$ to a position $\mathbf{r}_2$ by

$$U_{12} = \int_{\mathbf{r}_1}^{\mathbf{r}_2} \Sigma \mathbf{F} \cdot d\mathbf{r},$$

Eq. (15.5)

where $\Sigma\,\mathbf{F}$ is the sum of the forces acting on the object, *the principle of work and energy* states that the work equals the change in the kinetic energy:

$$U_{12} = \tfrac{1}{2}mv_2^2 - \tfrac{1}{2}mv_1^2$$

Eq. (15.6)

The total work done by external forces on a system of objects equals the change in the total kinetic energy of the system if no net work is done by internal forces.

Power

The *power* is the rate at which work is done. The power transferred to an object by the external forces acting on it is

$$P = \Sigma \mathbf{F} \cdot \mathbf{v}.$$

Eq. (15.9)

The power equals the rate of change of the object's kinetic energy. The average with respect to time of the power during an interval of time from t_1 to t_2 is equal to the change in kinetic energy of the object, or the work done, divided by the interval of time:

$$P_{av} = \frac{\tfrac{1}{2}mv_2^2 - \tfrac{1}{2}mv_1^2}{t_2 - t_1} = \frac{U_{12}}{t_2 - t_1}.$$

Eq. (15.10)

Evaluating the Work

Let s be the position of an object's center of mass along its path. The work done on the object from a position s_1 to a position s_2 is

$$U_{12} = \int_{s_1}^{s_2} \Sigma F_t\, ds,$$

Eq. (15.7)

where ΣF_t is the tangential component of the total external force on the object. Components of force perpendicular to the path do no work.

Weight   In terms of a coordinate system with the positive y-axis upward, the work done by an object's weight as the center of mass of the object moves from position 1 to position 2 is

$$U_{12} = -mg(y_2 - y_1).$$ Eq. (15.15)

The work is the product of the weight and the change in the height of the center of mass of the object. The work is negative if the height increases and positive if it decreases.

When the variation of an object's weight with distance r from the center of the earth is accounted for, the work done by the weight of the object is

$$U_{12} = mgR_E^2\left(\frac{1}{r_2} - \frac{1}{r_1}\right),$$ Eq. (15.17)

where R_E is the radius of the earth.

Springs   The work done on an object by a spring attached to a fixed support is

$$U_{12} = -\tfrac{1}{2}k(S_2^2 - S_1^2),$$ Eq. (15.18)

where S_1 and S_2 are the values of the stretch at the initial and final positions of the object.

Potential Energy

For a given force $\mathbf{F}$ acting on an object, if a function V of the object's position exists such that

$$dV = -\mathbf{F} \cdot d\mathbf{r},$$ Eq. (15.20)

then $\mathbf{F}$ is said to be *conservative* and V is called the *potential energy* associated with $\mathbf{F}$. The work done by $\mathbf{F}$ from a position 1 to a position 2 is

$$U_{12} = -(V_2 - V_1).$$ Eq. (15.21)

If all the forces that do work on an object are conservative, the total energy—the sum of the kinetic energy of the object and the potential energies of the forces acting on it—is conserved:

$$\tfrac{1}{2}mv^2 + V = \text{constant}.$$ Eq. (15.23)

Weight   In terms of a cartesian coordinate system with its y axis pointing upward, the potential energy of the weight of an object is

$$V = mgy.$$ Eq. (15.25)

The potential energy is the product of the object's weight and the height of its center of mass, measured from any convenient reference level, or *datum*.

When the variation of an object's weight with distance r from the center of the earth is accounted for, the potential energy of the weight is

$$V = -\frac{mgR_E^2}{r},$$ Eq. (15.27)

where R_E is the radius of the earth.

Springs   The potential energy of the force exerted on an object by a linear spring is

$$V = \tfrac{1}{2}kS^2,$$ (15.28)

where S is the stretch of the spring.

Relationships between Force and Potential Energy   A force **F** is related to its associated potential energy by

$$\mathbf{F} = -\left(\frac{\partial V}{\partial x}\mathbf{i} + \frac{\partial V}{\partial y}\mathbf{j} + \frac{\partial V}{\partial z}\mathbf{k}\right) = -\nabla V. \tag{15.33}$$

A force **F** is conservative if and only if its *curl* is zero:

$$\nabla \times \mathbf{F} = \begin{vmatrix} \mathbf{i} & \mathbf{j} & \mathbf{k} \\ \dfrac{\partial}{\partial x} & \dfrac{\partial}{\partial y} & \dfrac{\partial}{\partial z} \\ F_x & F_y & F_z \end{vmatrix} = \mathbf{0}.$$

Review Problems

15.1 A 10,000-kg airplane must reach a velocity of 60 m/s to take off. If the horizontal force exerted by the plane's engine is 60 kN and you neglect other horizontal forces, what length of runway is needed?

15.2 The driver of a 3000-lb car moving at 40 mi/hr applies an increasing force on the brake pedal. The magnitude of the resulting frictional force exerted on the car by the road is $f = 250 + 6s$ lb, where s is the car's horizontal position (in feet) relative to its position when the brakes were applied. Assuming that the car's tires do not slip; determine the distance required for the car to stop (a) by using Newton's second law and (b) by using the principle of work and energy.

15.3 Suppose that the car in Problem 15.2 is on wet pavement and the coefficients of friction between the tires and the road are $\mu_s = 0.4$ and $\mu_k = 0.35$. Determine the distance required for the car to stop.

15.4 An astronaut in a small rocket vehicle (combined mass = 450 kg) is hovering 100 m above the surface of the moon when he discovers that he is nearly out of fuel and can exert the thrust necessary to cause the vehicle to hover for only 5 more seconds. He quickly considers two strategies for getting to the surface: (a) Fall 20 m, turn on the thrust for 5 s, and then fall the rest of the way; (b) fall 40 m, turn on the thrust for 5 s, and then fall the rest of the way. Which strategy gives him the best chance of surviving? How much work is done by the engine's thrust in each case? $(g_{\text{moon}} = 1.62 \text{ m/s}^2.)$

15.5 The coefficients of friction between the 20-kg crate and the inclined surface are $\mu_s = 0.24$ and $\mu_k = 0.22$. If the crate starts from rest and the horizontal force $F = 200$ N, what is the magnitude of the velocity of the crate when it has moved 2 m?

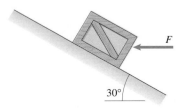

P15.5

15.6 In Problem 15.5, what is the magnitude of the velocity of the crate when it has moved 2 m if the horizontal force $F = 40$ N?

15.7 The Union Pacific Big Boy locomotive weighs 1.19 million lb, and the tractive effort (tangential force) of its drive wheels is 135,000 lb. If you neglect other tangential forces, what distance is required for the train to accelerate from zero to 60 mi/hr?

P15.7

15.8 In Problem 15.7, suppose that the acceleration of the locomotive as it accelerates from zero to 60 mi/hr is $(F_0/m)(1 - v/88)$, where $F_0 = 135,000$ lb, m is the mass of the locomotive, and v is its velocity in feet per second.

(a) How much work is done in accelerating the train to 60 mi/hr?
(b) Determine the locomotive's velocity as a function of time.

15.9 A car traveling 65 mi/hr hits the crash barrier. Determine the maximum deceleration to which the passengers are subjected if the car weighs (a) 2500 lb and (b) 5000 lb.

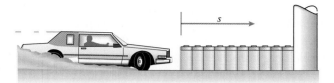

P15.9

15.10 In a preliminary design for a mail-sorting machine, parcels moving at 2 ft/s slide down a smooth ramp and are brought to rest by a linear spring. What should the spring constant be if you don't want a 10-lb parcel to be subjected to a maximum deceleration greater than 10 g's?

P15.10

15.11 When the 1-kg collar is in position 1, the tension in the spring is 50 N, and the unstretched length of the spring is 260 mm. If the collar is pulled to position 2 and released from rest, what is its velocity when it returns to position 1?

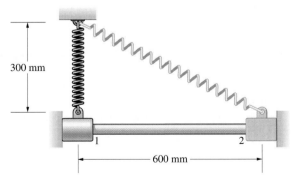

P15.11

15.12 In Problem 15.11, suppose that the tensions in the spring in positions 1 and 2 are 100 N and 400 N, respectively.

(a) What is the spring constant k?
(b) If the collar is given a velocity of 15 m/s at 1, what is its velocity when it reaches 2?

15.13 The 30-lb weight is released from rest with the two springs $(k_A = 30\text{ lb/ft}, k_B = 15\text{ lb/ft})$ unstretched.

(a) How far does the weight fall before rebounding?
(b) What maximum velocity does it attain?

P15.13

15.14 The piston and the load it supports are accelerated upward by the gas in the cylinder. The total weight of the piston and load is 1000 lb. The cylinder wall exerts a constant 50-lb frictional force on the piston as it rises. The net force exerted on the piston by pressure is $(p_2 - p_{atm})A$, where p is the pressure of the gas, $p_{atm} = 2117\text{ lb/ft}^2$ is atmospheric pressure, and $A = 1\text{ ft}^2$ is the cross-sectional area of the piston. Assume that the product of p and the volume of the cylinder is constant. When $s = 1$ ft, the piston is stationary and $p = 5000\text{ lb/ft}^2$. What is the velocity of the piston when $s = 2$ ft?

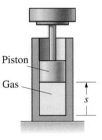

Piston

Gas

P15.14

15.15 When a 22,000-kg rocket's engine burns out at an altitude of 2 km, the velocity of the rocket is 3 km/s and it is traveling at an angle of 60° relative to the horizontal. Neglect the variation in the gravitational force with altitude.

(a) If you neglect aerodynamic forces, what is the magnitude of the velocity of the rocket when it reaches an altitude of 6 km?
(b) If the actual velocity of the rocket when it reaches an altitude of 6 km is 2.8 km/s, how much work is done by aerodynamic forces as the rocket moves from 2 km to 6 km altitude?

15.16 The 12-kg collar A is at rest in the position shown at $t = 0$ and is subjected to the tangential force $F = 24 - 12t^2$ N for 1.5 s. Neglecting friction, what maximum height h does the collar reach?

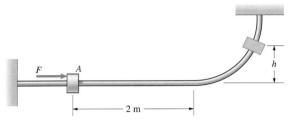

P15.16

15.17 Suppose that, in designing a loop for a roller coaster's track, you establish as a safety criterion that at the top of the loop the normal force exerted on a passenger by the roller coaster should equal 10 percent of the passenger's weight. (That is, the passenger's "effective weight" pressing him down into his seat is 10 percent of his actual weight.) The roller coaster is moving at 62 ft/s when it enters the loop. What is the necessary instantaneous radius of curvature ρ of the track at the top of the loop?

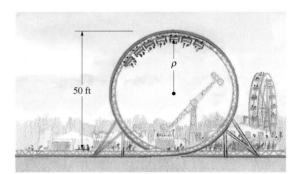

P15.17

15.18 A 180-lb student runs at 15 ft/s, grabs a rope, and swings out over a lake. He releases the rope when his velocity is zero.

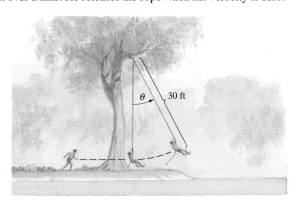

P15.18

(a) What is the angle θ when he releases the rope?
(b) What is the tension in the rope just before he releases it?
(c) What is the maximum tension in the rope?

15.19 If the student in Problem 15.18 releases the rope when $\theta = 25°$, what maximum height does he reach relative to his position when he grabs the rope?

15.20 A boy takes a running start and jumps on his sled at position 1. He leaves the ground at position 2 and lands in deep snow at a distance $b = 25$ ft. How fast was he going at position 1?

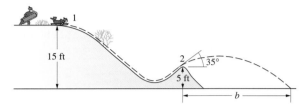

P15.20

15.21 In Problem 15.20, if the boy starts at position 1 going 15 ft/s, what distance b does he travel through the air?

15.22 The 1-kg collar A is attached to the linear spring $(k = 500$ N/m$)$ by a string. The collar starts from rest in the position shown, and the initial tension in the string is 100 N. What distance does the collar slide up the smooth bar?

P15.22

15.23 The masses $m_A = 40$ kg and $m_B = 60$ kg. The collar A slides on the smooth horizontal bar. The system is released from rest. Use conservation of energy to determine the velocity of the collar A when it has moved 0.5 m to the right.

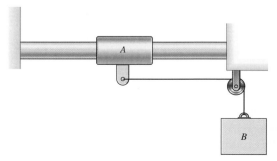

P15.23

15.24 The spring constant is $k = 850$ N/m, $m_A = 40$ kg, and $m_B = 60$ kg. The collar A slides on the smooth horizontal bar. The system is released from rest in the position shown with the spring unstretched. Use conservation of energy to determine the velocity of the collar A when it has moved 0.5 m to the right.

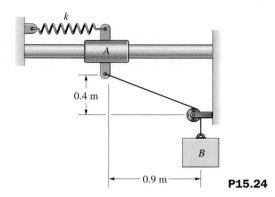

0.4 m

B

0.9 m

P15.24

15.25 The y-axis is vertical and the curved bar is smooth. If the magnitude of the velocity of the 4-lb slider is 6 ft/s at position 1, what is the magnitude of its velocity when it reaches position 2?

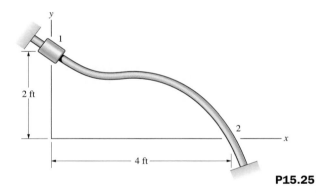

2 ft

4 ft

P15.25

15.26 In Problem 15.25, determine the magnitude of the velocity of the slider when it reaches position 2 if it is subjected to the additional force $\mathbf{F} = 3x\mathbf{i} - 2\mathbf{j}$ (lb) during its motion.

15.27 Suppose that an object of mass m is beneath the surface of the earth. In terms of a polar coordinate system with its origin at the earth's center, the gravitational force on the object is $-(mgr/R_E)\mathbf{e}_r$, where R_E is the radius of the earth. Show that the potential energy associated with the gravitational force is $V = mgr^2/2R_E$.

15.28 It has been pointed out that if tunnels could be drilled straight through the earth between points on the surface, trains could travel between those points using gravitational force for acceleration and deceleration. (The effects of friction and aerodynamic drag could be minimized by evacuating the tunnels and using magnetically levitated trains.) Suppose that such a train travels from the North Pole to a point on the equator. Determine the magnitude of the velocity of the train (a) when it arrives at the equator and (b) when it is halfway from the North Pole to the equator. The radius of the earth is $R_E = 3960$ mi.

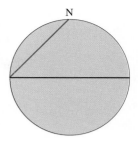

N

P15.28

15.29 In Problem 15.7, what is the maximum power transferred to the locomotive during its acceleration?

15.30 Just before it lifts off, a 10,500-kg airplane is traveling at 60 m/s. The total horizontal force exerted by the plane's engines is 189 kN, and the plane is accelerating at 15 m/s².

(a) How much power is being transferred to the plane by its engines?
(b) What is the total power being transferred to the plane?

P15.30

15.31 The "Paris Gun" used by Germany in World War I had a range of 120 km, a 37.5-m barrel, and a muzzle velocity of 1550 m/s and fired a 120-kg shell.

(a) If you assume the shell's acceleration to be constant, what maximum power was transferred to the shell as it traveled along the barrel?
(b) What average power was transferred to the shell?

P15.31

The total linear momentum of the test vehicles is approximately the same before and after their collision. In this chapter, we use methods based on linear and angular momentum to analyze motions of objects.

CHAPTER 16

Momentum Methods

ntegrating Newton's second law with respect to time yields a relation between the time integral of the forces acting on an object and the change in the object's linear momentum. With this result, called the principle of impulse and momentum, we can determine the change in an object's velocity when the external forces are known as functions of time, analyze impacts between objects, and evaluate forces exerted by continuous flows of mass.

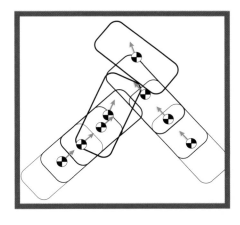

16.1 Principle of Impulse and Momentum

The principle of work and energy is a very useful tool in mechanics. We can derive another useful tool for the analysis of motion by integrating Newton's second law with respect to time. We express Newton's second law in the form

$$\Sigma \mathbf{F} = m \frac{d\mathbf{v}}{dt}.$$

Then we integrate with respect to time to obtain

$$\int_{t_1}^{t_2} \Sigma \mathbf{F}\, dt = m\mathbf{v}_2 - m\mathbf{v}_1, \tag{16.1}$$

where $\mathbf{v}_1$ and $\mathbf{v}_2$ are the velocities of the center of mass of the object at the times t_1 and t_2. The term on the left is called the *linear impulse*, and $m\mathbf{v}$ is the *linear momentum*. Equation (16.1) is called the *principle of impulse and momentum*: The impulse applied to an object during an interval of time is equal to the change in the object's linear momentum (Fig. 16.1). The dimensions of the linear impulse and linear momentum are (mass) × (length)/(time).

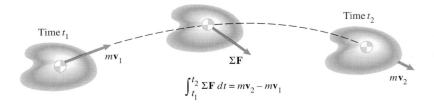

Figure 16.1
Principle of impulse and momentum.

The average with respect to time of the total force acting on an object from t_1 to t_2 is

$$\Sigma \mathbf{F}_{\text{av}} = \frac{1}{t_2 - t_1} \int_{t_1}^{t_2} \Sigma \mathbf{F}\, dt,$$

so we can write Eq. (16.1) as

$$(t_2 - t_1)\Sigma \mathbf{F}_{\text{av}} = m\mathbf{v}_2 - m\mathbf{v}_1. \tag{16.2}$$

With this equation, we can determine the average value of the total force acting on an object during a given interval of time if we know the change in the object's velocity.

A force of relatively large magnitude that acts over a small interval of time is called an *impulsive force* (Fig. 16.2). Determining the actual history of such a force is often impractical, but with Eq. (16.2) its average value can sometimes be determined. For example, a golf ball struck by a club is subjected to an impulsive force. By making high-speed motion pictures, the duration of the impact and the ball's velocity after the impact can be measured. Knowing the duration of the impact and the ball's change in linear momentum, we can determine the average force exerted by the club. (See Example 16.3.)

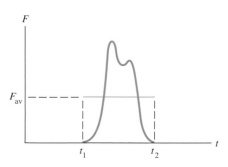

Figure 16.2
An impulsive force and its average value.

We can express Eqs. (16.1) and (16.2) in scalar forms that are often use-ful. The sum of the forces in the direction tangent to an object's path equals the product of the object's mass and the rate of change of its velocity along the path (see Eq. 14.7):

$$\Sigma F_t = ma_t = m\frac{dv}{dt}.$$

Integrating this equation with respect to time, we obtain

$$\int_{t_1}^{t_2} \Sigma F_t\, dt = mv_2 - mv_1, \tag{16.3}$$

where v_1 and v_2 are the velocities along the path at the times t_1 and t_2. The impulse applied to an object by the sum of the forces tangent to its path dur-ing an interval of time is equal to the change in the object's linear momentum along the path. In terms of the average with respect to time of the sum of the forces tangent to the path, or

$$\Sigma F_{t\,\text{av}} = \frac{1}{t_2 - t_1}\int_{t_1}^{t_2} \Sigma F_t\, dt,$$

we can write Eq. (16.3) as

$$\left(t_2 - t_1\right)\Sigma F_{t\,\text{av}} = mv_2 - mv_1. \tag{16.4}$$

This equation relates the average of the sum of the forces tangent to the path during an interval of time to the change in the velocity along the path.

Notice that Eq. (16.1) and the principle of work and energy, Eq. (15.6), are quite similar. They both relate an integral of the external forces to the change in an object's velocity. Equation (16.1) is a vector equation that determines the change in both the magnitude and direction of the velocity, whereas the principle of work and energy, a scalar equation, gives only the change in the magnitude of the velocity. But there is a greater difference between the two methods: In the case of impulse and momentum, there is no class of forces equivalent to the conservative forces that make the principle of work and energy so easy to apply.

When the external forces acting on an object are known as functions of time, the principle of impulse and momentum may be applied to determine the change in velocity of the object during an interval of time. Although this is an important result, it is not new: In Chapter 14, when we used Newton's second law to determine an object's acceleration and then integrated the acceleration with respect to time to determine the object's velocity, we were effectively applying the principle of impulse and momentum. In the rest of this chapter, we show that this principle can be extended to new and interest-ing applications.

Study Questions

1. What is the definition of the linear momentum?
2. What is the principle of impulse and momentum?
3. If you know the change in the velocity of an object during an interval of time, how can you determine the average of the total force exerted on the object during that interval?
4. How does the principle of impulse and momentum differ from the principle of work and energy?

Example 16.1

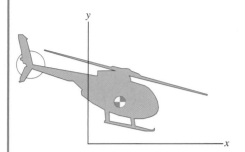

Figure 16.3

Applying Impulse and Momentum

A 1200-kg helicopter starts from rest at $t = 0$ (Fig. 16.3).
(a) The components of the total force on the helicopter from $t = 0$ to $t = 10$ s are given by

$$\Sigma F_x = 720t \text{ N},$$
$$\Sigma F_y = 2160 - 360t \text{ N},$$
$$\Sigma F_z = 0.$$

Determine the helicopter's velocity at $t = 10$ s.
(b) At $t = 20$ s, the helicopter's velocity is $36\mathbf{i} + 8\mathbf{j}$ (m/s). What is the average of the total force acting on the craft from $t = 10$ s to $t = 20$ s?

Strategy

(a) Since we know the helicopter's velocity at $t = 0$ and the components of the total force acting on the helicopter as functions of time, we can use the principle of impulse and momentum, Eq. (16.1), to determine the velocity at $t = 10$ s.
(b) Knowing the velocity at $t = 10$ s and at $t = 20$ s, we can determine the average of the total force from Eq. (16.2).

Solution

(a) Applying the principle of impulse and momentum from $t = 0$ to $t = 10$ s, we obtain

$$\int_{t_1}^{t_2} \Sigma \mathbf{F} \, dt = m\mathbf{v}_2 - m\mathbf{v}_1:$$

$$\int_0^{10} \left[720t\mathbf{i} + (2160 - 360t)\mathbf{j} \right] dt = (1200)\mathbf{v}_2 - (1200)(0),$$

$$36{,}000\mathbf{i} + 3600\mathbf{j} = 1200\mathbf{v}_2.$$

We find that the velocity at $t = 10$ s is $30\mathbf{i} + 3\mathbf{j}$ (m/s).
(b) To determine the average total force from $t = 10$ s to $t = 20$ s, we apply Eq. (16.2) to that interval of time:

$$(t_2 - t_1)\Sigma \mathbf{F}_{av} = m\mathbf{v}_2 - m\mathbf{v}_1:$$

$$(20 - 10)\Sigma \mathbf{F}_{av} = 1200(36\mathbf{i} + 8\mathbf{j}) - 1200(30\mathbf{i} + 3\mathbf{j}),$$

$$\Sigma \mathbf{F}_{av} = 720\mathbf{i} + 600\mathbf{j} \text{ (N)}.$$

Discussion

Although we did not know the total force acting on the helicopter during the interval from $t = 10$ s to $t = 20$ s, we were able to determine the average of the total force because we knew the velocities at the beginning and end of the interval.

Example 16.2

Impulse and Momentum Tangent to the Path

The motorcycle starts from rest at $t = 0$ and travels along a circular track with a 400-m radius. The tangential component of the total force on the motorcycle from $t = 0$ to $t = 30$ s is $\Sigma F_t = 200 - 6t$ N. The combined mass of the motorcycle and rider is 150 kg.
(a) What is the magnitude of the velocity at $t = 30$ s?
(b) What is the average of the tangential component of the total force from $t = 0$ to $t = 30$ s?
(c) What is the average of the normal component of the total force from $t = 0$ to $t = 30$ s?

Figure 16.4

Strategy

(a) Because the tangential component of the total force is given as a function of time, we can use Eq. (16.3) to determine the magnitude of the velocity as a function of time.
(b) We can simply calculate the average of the tangential component of the total force, or we can use Eq. (16.4).
(c) Since we will know the magnitude of the velocity as a function of time from part (a), we can determine the normal component of the total force as a function of time from Newton's second law.

Solution

(a) We use Eq. (16.3) to determine the magnitude v of the velocity as a function of time.

$$\int_{t_1}^{t_2} \Sigma F_t \, dt = mv_2 - mv_1:$$

$$\int_{0}^{t} (200 - 6t) \, dt = 150v - 0.$$

Evaluating the integral, we obtain

$$v = \frac{1}{150} (200t - 3t^2) \text{ m/s}.$$

At $t = 30$ s, the magnitude of the velocity is $v = 22$ m/s.
(b) We use Eq. (16.4) to determine the average of the tangential component of the total force from $t = 0$ to $t = 30$ s.

$$(t_2 - t_1)\Sigma F_{t\,\text{av}} = mv_2 - mv_1:$$

$$(30 - 0)\Sigma F_{t\,\text{av}} = (150)(22) - 0.$$

The result is $\Sigma F_{t\,\text{av}} = 110$ N.
(c) Using our expression for the magnitude of the velocity as a function of time, we apply Newton's second law in the normal direction to obtain the total normal force as a function of time:

$$\Sigma F_n = m \frac{v^2}{\rho}$$

$$= \frac{(150)(200t - 3t^2)^2}{(150)^2(400)} \text{ N.}$$

Therefore, the average of the normal component of the total force is

$$\Sigma F_{n\,av} = \frac{1}{t_2 - t_1} \int_{t_1}^{t_2} \Sigma F_n \, dt$$

$$= \frac{1}{30 - 0} \int_{0}^{30} \frac{(150)(200t - 3t^2)^2}{(150)^2(400)} \, dt$$

$$= 89.3 \text{ N.}$$

Example 16.3

Determining an Impulsive Force

A golf ball in flight is photographed at intervals of 0.001 s (Fig. 16.5). The 1.62-oz ball is 1.68 in. in diameter. If the club was in contact with the ball for 0.0006 s, estimate the average value of the impulsive force exerted by the club.

Figure 16.5

Strategy

By measuring the distance traveled by the ball in one of the 0.001-s intervals, we can estimate its velocity after being struck and then use Eq. (16.2) to determine the average total force on the ball.

Solution

By comparing the distance moved during one of the 0.001-s intervals with the known diameter of the ball, we estimate that the ball traveled 1.9 in. and that its direction is 21° above the horizontal (Fig. a). The magnitude of the ball's velocity is

$$\frac{(1.9/12) \text{ ft}}{0.001 \text{ s}} = 158 \text{ ft/s.}$$

The weight of the ball is $1.62/16 = 0.101$ lb, so its mass is $0.101/32.2 = 3.14 \times 10^{-3}$ slug. From Eq. (16.2), we obtain

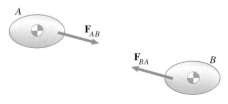

(a) Estimating the distance traveled during one 0.001-s interval.

$$(t_2 - t_1)\Sigma \mathbf{F}_{av} = m\mathbf{v}_2 - m\mathbf{v}_1:$$

$$(0.0006)\Sigma \mathbf{F}_{av} = (3.14 \times 10^{-3})(158)(\cos 21° \, \mathbf{i} + \sin 21° \, \mathbf{j}) - 0,$$

so that

$$\Sigma \mathbf{F}_{av} = 775\mathbf{i} + 297\mathbf{j} \text{ (lb)}.$$

Discussion

The average force during the time the club is in contact with the ball includes both the impulsive force exerted by the club and the ball's weight. In comparison with the large average impulsive force exerted by the club, the weight $(-0.101\mathbf{j}$ lb) is negligible.

16.2 Conservation of Linear Momentum

In this section, we consider the motions of several objects and show that if the effects of external forces can be neglected, the total linear momentum of the objects is conserved. (By *external forces*, we mean forces that are not exerted by the objects under consideration.) This result provides a powerful tool for analyzing interactions between objects, such as collisions, and also permits us to determine forces exerted on objects as a result of gaining or losing mass.

Consider the objects A and B in Fig. 16.6. $\mathbf{F}_{AB}$ is the force exerted on A by B, and $\mathbf{F}_{BA}$ is the force exerted on B by A. These forces could result from the two objects being in contact, for example, or could be exerted by a spring connecting them. As a consequence of Newton's third law, the two forces are equal and opposite, so that

$$\mathbf{F}_{AB} + \mathbf{F}_{BA} = \mathbf{0}. \tag{16.5}$$

Suppose that no external forces act on A and B or that the external forces are negligible in comparison with the forces that A and B exert on each other. Then we can apply the principle of impulse and momentum to each object for arbitrary times t_1 and t_2.

$$\int_{t_1}^{t_2} \mathbf{F}_{AB} \, dt = m_A \mathbf{v}_{A2} - m_A \mathbf{v}_{A1}:$$

$$\int_{t_1}^{t_2} \mathbf{F}_{BA} \, dt = m_B \mathbf{v}_{B2} - m_B \mathbf{v}_{B1}.$$

If we sum these equations, the terms on the left cancel, and we obtain

$$m_A \mathbf{v}_{A1} + m_B \mathbf{v}_{B1} = m_A \mathbf{v}_{A2} + m_B \mathbf{v}_{B2},$$

Figure 16.6
Two objects and the forces they exert on each other.

which means that the total linear momentum of A and B is conserved:

$$m_A \mathbf{v}_A + m_B \mathbf{v}_B = \text{constant.} \qquad (16.6)$$

We can show that the velocity of the combined center of mass of the objects A and B (that is, the center of mass of A and B regarded as a single object) is also constant. Let $\mathbf{r}_A$ and $\mathbf{r}_B$ be the position vectors of their individual centers of mass (Fig. 16.7). The position of the combined center of mass is

$$\mathbf{r} = \frac{m_A \mathbf{r}_A + m_B \mathbf{r}_B}{m_A + m_B}.$$

By taking the derivative of this equation with respect to time and using Eq. (16.6), we obtain

$$(m_A + m_B)\mathbf{v} = m_A \mathbf{v}_A + m_B \mathbf{v}_B = \text{constant,} \qquad (16.7)$$

where $\mathbf{v} = d\mathbf{r}/dt$ is the velocity of the combined center of mass. Although the goal will usually be to determine the individual motions of the objects, knowing that the velocity of the combined center of mass is constant can contribute to our understanding of a problem, and in some instances the motion of the combined center of mass may be the only information that can be obtained.

Even when significant external forces act on A and B, if the external forces are negligible in a particular direction, Eqs. (16.6) and (16.7) apply in that direction. These equations also apply to an arbitrary number of objects: If the external forces acting on any collection of objects are negligible, the total linear momentum of the objects is conserved, and the velocity of their center of mass is constant.

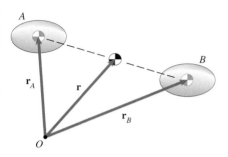

Figure 16.7
Position vector **r** of the center of mass of A and B.

Study Questions

1. What condition is necessary for the total linear momentum of two objects to be conserved?
2. If the only forces exerted on two objects are the forces they exert on each other, what can be inferred about the motion of their common center of mass?

Example 16.4

Conservation of Linear Momentum

A person of mass m_P stands at the center of a stationary barge of mass m_B (Fig. 16.8). Neglect horizontal forces exerted on the barge by the water.
(a) If the person starts running to the right with velocity v_P relative to the water, what is the resulting velocity of the barge relative to the water?
(b) If the person stops when he reaches the right-hand end of the barge, what are his position and the barge's position relative to their original positions?

Figure 16.8

Strategy

(a) The only horizontal forces exerted on the person and the barge are the forces they exert on each other. Therefore, their total linear momentum in

the horizontal direction is conserved, and we can use Eq. (16.6) to determine the barge's velocity while the person is running.
(b) The combined center of mass of the person and the barge is initially stationary, so it must remain stationary. Knowing the position of the combined center of mass, we can determine the positions of the person and barge when the person is at the right-hand end of the barge.

Solution

(a) Before the person starts running, the total linear momentum of the person and the barge in the horizontal direction is zero, so it must be zero after he starts running. Letting v_B be the value of the barge's velocity to the left while the person is running (Fig. a), we obtain

$$m_P v_P + m_B(-v_B) = 0,$$

so the velocity of the barge while he runs is

$$v_B = \left(\frac{m_P}{m_B}\right)v_P.$$

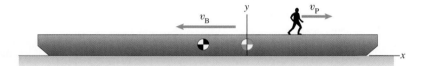

(a) Velocities of the person and barge.

(b) Let the origin of the coordinate system in Fig. (b) be the original horizontal position of the centers of mass of the barge and the person, and let x_B be the position of the barge's center of mass to the left of the origin. When the person has stopped at the right-hand end of the barge, the combined center of mass must still beat $x = 0$. That is,

$$\frac{x_P m_P + (-x_B)m_B}{m_P + m_B} = 0.$$

Solving this equation together with the relation $x_P + x_B = L/2$, we obtain

$$x_P = \frac{m_B L}{2(m_P + m_B)}, \qquad x_B = \frac{m_P L}{2(m_P + m_B)}.$$

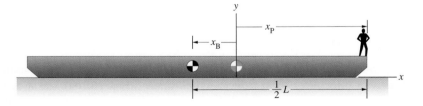

(b) Positions after the person has stopped running.

Discussion

This example is a well-known illustration of the power of momentum methods. Notice that we were able to determine the velocity of the barge and the final positions of the person and barge even though we did not know the complicated time dependence of the horizontal forces they exert on each other.

16.3 Impacts

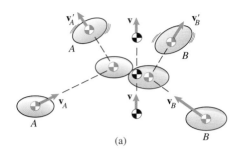

(a)

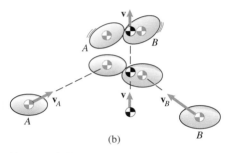

(b)

Figure 16.9
(a) Velocities of A and B before and after the impact, and the velocity $\mathbf{v}$ of their center of mass.
(b) A perfectly plastic impact.

In machines that perform stamping or forging operations, dies impact against workpieces. Mechanical printers create images by impacting metal elements against the paper and platen. Vehicles impact each other intentionally, as when railroad cars are rolled against each other to couple them, and unintentionally in accidents. Impacts occur in many situations of concern in engineering. In this section, we consider a basic question: If we know the velocities of two objects before they collide, how can we determine their velocities afterward? In other words, what is the effect of the impact on the motions of the objects?

If colliding objects are not subjected to external forces, their total linear momentum must be the same before and after the impact. Even when they are subjected to external forces, the force of the impact is often so large, and its duration so brief, that the effect of external forces on the motions of the objects during the impact is negligible. Suppose that objects A and B with velocities $\mathbf{v}_A$ and $\mathbf{v}_B$ collide, and let $\mathbf{v}'_A$ and $\mathbf{v}'_B$ be their velocities after the impact (Fig. 16.9a). If the effects of external forces are negligible, then the total linear momentum of the system composed of A and B is conserved:

$$m_A \mathbf{v}_A + m_B \mathbf{v}_B = m_A \mathbf{v}'_A + m_B \mathbf{v}'_B. \tag{16.8}$$

Furthermore, the velocity $\mathbf{v}$ of the center of mass of A and B is the same before and after the impact. Thus, from Eq. (16.7),

$$\mathbf{v} = \frac{m_A \mathbf{v}_A + m_B \mathbf{v}_B}{m_A + m_B}. \tag{16.9}$$

If A and B adhere and remain together after they collide, they are said to undergo a *perfectly plastic impact*. Equation (16.9) gives the velocity of the center of mass of the object they form after the impact (Fig. 16.9b). A remarkable feature of this result is that we determine the velocity following the impact *without considering the physical nature of the impact*.

If A and B do not adhere, linear momentum conservation alone does not provide enough equations to determine their velocities after the impact. We first consider the case in which they travel along the same straight line before and after they collide.

Direct Central Impacts

Suppose that the centers of mass of A and B travel along the same straight line with velocities v_A and v_B before their impact (Fig. 16.10a). Let R be the magnitude of the force A and B exert on each other during the impact (Fig. 16.10b). We assume that the contacting surfaces are oriented so that R is parallel to the line along which the two objects travel and is directed toward their centers of mass. This condition, called *direct central impact*, means that A and B continue to travel along the same straight line after their impact

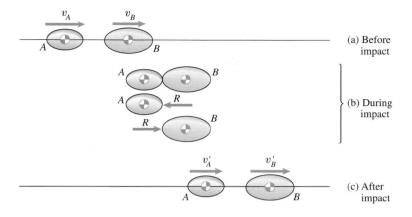

(a) Before impact

(b) During impact

(c) After impact

Figure 16.10
(a) Objects A and B traveling along the same straight line.
(b) During the impact, they exert a force R on each other.
(c) They travel along the same straight line after the central impact.

(Fig. 16.10c). If the effects of external forces during the impact are negligible, the total linear momentum of the objects is conserved:

$$m_A v_A + m_B v_B = m_A v_A' + m_B v_B'. \qquad (16.10)$$

However, we need another equation to determine the velocities v_A' and v_B'. To obtain it, we must consider the impact in more detail.

Let t_1 be the time at which A and B first come into contact (Fig. 16.11a). As a result of the impact, they will deform and their centers of mass will continue to approach each other. At a time t_C, their centers of mass will have reached their nearest proximity (Fig. 16.11b). At this time, the relative velocity of the two centers of mass is zero, so they have the same velocity. We denote it by v_C. The objects then begin to move apart and separate at a time t_2 (Fig. 16.11c). We apply the principle of impulse and momentum to A during the intervals of time from t_1 to the time of closest approach t_C and from t_C to t_2:

$$\int_{t_1}^{t_C} -R \, dt = m_A v_C - m_A v_A, \qquad (16.11)$$

$$\int_{t_C}^{t_2} -R \, dt = m_A v_A' - m_A v_C. \qquad (16.12)$$

Then we apply the principle to B for the same intervals of time:

$$\int_{t_1}^{t_C} R \, dt = m_B v_C - m_B v_B, \qquad (16.13)$$

$$\int_{t_C}^{t_2} R \, dt = m_B v_B' - m_B v_C. \qquad (16.14)$$

As a result of the impact, part of the objects' kinetic energy can be lost due to a variety of mechanisms, including permanent deformation and generation of heat and sound. As a consequence, the impulse they impart to each other during the "restitution" phase of the impact from t_C to t_2 is, in general,

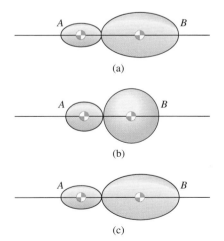

(a)

(b)

(c)

Figure 16.11
(a) First contact, $t = t_1$.
(b) Closest approach, $t = t_C$.
(c) End of contact, $t = t_2$.

smaller than the impulse they impart from t_1 to t_C. The ratio of these impulses is called the *coefficient of restitution*:

$$e = \frac{\displaystyle\int_{t_C}^{t_2} R\,dt}{\displaystyle\int_{t_1}^{t_C} R\,dt}. \tag{16.15}$$

The value of e depends on the properties of the objects as well as their velocities and orientations when they collide, and it can be determined only by experiment or by a detailed analysis of the deformations of the objects during the impact.

If we divide Eq. (16.12) by Eq. (16.11) and divide Eq. (16.14) by Eq. (16.13), we can express the resulting equations in the forms

$$\left(v_C - v_A\right)e = v_A' - v_C$$

and

$$\left(v_C - v_B\right)e = v_B' - v_C.$$

Subtracting the first equation from the second, we obtain

$$e = \frac{v_B' - v_A'}{v_A - v_B}. \tag{16.16}$$

Thus, the coefficient of restitution is related in a simple way to the relative velocities of the objects before and after the impact. If e is known, Eq. (16.16) can be used together with the equation of conservation of linear momentum, Eq. (16.10), to determine v_A' and v_B'.

If $e = 0$, Eq. (16.16) indicates that $v_B' = v_A'$. The objects remain together after the impact, and the impact is perfectly plastic. If $e = 1$, it can be shown that the total kinetic energy is the same before and after the impact:

$$\frac{1}{2}m_A v_A^2 + \frac{1}{2}m_B v_B^2 = \frac{1}{2}m_A\left(v_A'\right)^2 + \frac{1}{2}m_B\left(v_B'\right)^2 \quad \text{(when } e = 1\text{)}.$$

An impact in which kinetic energy is conserved is called *perfectly elastic*. Although this is sometimes a useful approximation, energy is lost in any impact in which material objects come into contact. If a collision can be heard, kinetic energy has been converted into sound. Permanent deformations of the colliding objects after the impact also represent losses of kinetic energy.

Oblique Central Impacts

We can extend the procedure used to analyze direct central impacts to the case in which the objects approach each other at an oblique angle. Suppose that A and B approach with arbitrary velocities $\mathbf{v}_A$ and $\mathbf{v}_B$ (Fig. 16.12) and that the forces they exert on each other during their impact are parallel to the x-axis and point toward their centers of mass. No forces are exerted on A and B in the y or z directions, so their velocities in those directions are unchanged by the impact:

$$\left(\mathbf{v}_A'\right)_y = \left(\mathbf{v}_A\right)_y, \qquad \left(\mathbf{v}_B'\right)_y = \left(\mathbf{v}_B\right)_y,$$
$$\left(\mathbf{v}_A'\right)_z = \left(\mathbf{v}_A\right)_z, \qquad \left(\mathbf{v}_B'\right)_z = \left(\mathbf{v}_B\right)_z. \tag{16.17}$$

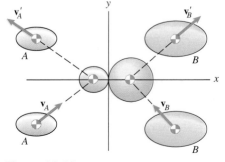

Figure 16.12
An oblique central impact.

In the x direction, linear momentum is conserved:

$$m_A(\mathbf{v}_A)_x + m_B(\mathbf{v}_B)_x = m_A(\mathbf{v}_A')_x + m_B(\mathbf{v}_B')_x. \tag{16.18}$$

By the same analysis we used to arrive at Eq. (16.16), the x components of velocity satisfy the relation

$$e = \frac{(\mathbf{v}_B')_x - (\mathbf{v}_A')_x}{(\mathbf{v}_A)_x - (\mathbf{v}_B)_x}. \tag{16.19}$$

We can analyze an oblique central impact in which an object A hits a stationary object B if friction is negligible. Suppose that B is constrained so that it cannot move relative to the inertial reference frame. For example, in Fig. 16.13, A strikes a wall B that is fixed relative to the earth. The y and z components of A's velocity are unchanged, because friction is neglected and the impact exerts no force in those directions. The x component of A's velocity after the impact is given by Eq. (16.19) with B's velocity equal to zero:

$$(\mathbf{v}_A')_x = -e(\mathbf{v}_A)_x.$$

In summary, the velocity of the common center of mass of two objects A and B after they collide can be determined from Eq. (16.9). In an oblique central impact, in terms of the coordinate system shown in Fig. 16.12, the y and z components of the velocities of the two objects are unchanged, and the x components of the velocities after the impact can be determined using Eqs. (16.18) and (16.19).

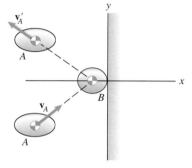

Figure 16.13
Impact with a stationary object.

Study Questions

1. If two objects collide, what can you infer about the motion of their common center of mass?
2. What are perfectly plastic and perfectly elastic impacts?
3. How is the coefficient of restitution defined?
4. If the coefficient of restitution is known for a direct central impact, what equations are used to determine the velocities of the objects after their collision?

Example 16.5

Analyzing an Impact

The two 10-lb weights in Fig. 16.14 slide on the smooth horizontal bar. Determine their velocities after they collide (a) if they are coated with Velcro and stick together and (b) if the coefficient of restitution is $e = 0.8$.

Figure 16.14

Strategy

(a) If the weights stick together, they have the same velocity after their collision. We can determine the velocity from conservation of linear momentum.
(b) Knowing the coefficient of restitution, we can determine the velocity of each weight after the collision by using conservation of linear momentum together with the definition of the coefficient of restitution, Eq. (16.16).

Solution

(a) The velocities before the collision are $v_A = 10$ ft/s and $v_B = -5$ ft/s. Let v be their common velocity after the collision. From Eq. (16.10), conservation of linear momentum requires that

$$m_A v_A + m_B v_B = m_A v_A' + m_B v_B':$$

$$\left(\frac{10}{32.2}\right)(10) + \left(\frac{10}{32.2}\right)(-5) = \left(\frac{10}{32.2} + \frac{10}{32.2}\right)v.$$

Solving, we obtain $v = 2.5$ ft/s. The connected weights move to the right at 2.5 ft/s after the collision.

(b) Conservation of linear momentum requires that

$$m_A v_A + m_B v_B = m_A v_A' + m_B v_B':$$

$$\left(\frac{10}{32.2}\right)(10) + \left(\frac{10}{32.2}\right)(-5) = \left(\frac{10}{32.2}\right)v_A' + \left(\frac{10}{32.2}\right)v_B'.$$

From Eq. (16.16),

$$e = \frac{v_B' - v_A'}{v_A - v_B}:$$

$$0.8 = \frac{v_B' - v_A'}{10 - (-5)}.$$

We now have two equations in v_A' and v_B'. Solving them, we obtain $v_A' = -3.5$ ft/s and $v_B' = 8.5$ ft/s. A moves to the left at 3.5 ft/s and B moves to the right at 8.5 ft/s after the collision.

Example 16.6

Applying Momentum Methods to Spacecraft Docking

The *Apollo* command-service module (CSM) attempts to dock with the *Soyuz* capsule July 15, 1975 (Fig. 16.15). Their masses are $m_A = 18$ Mg and $m_B = 6.6$ Mg. The *Soyuz* is stationary relative to the reference frame shown, and the CSM approaches with velocity $\mathbf{v}_A = 0.2\mathbf{i} + 0.03\mathbf{j} - 0.02\mathbf{k}$ (m/s).

(a) If the first attempt at docking is successful, what is the velocity of the center of mass of the combined vehicles afterward?

(b) If the first attempt is unsuccessful and the coefficient of restitution of the resulting impact is $e = 0.95$, what are the velocities of the two spacecraft after the impact?

Strategy

(a) If the docking is successful, the impact is perfectly plastic, and we can use Eq. (16.9) to determine the velocity of the center of mass of the combined object after the impact.

(b) By assuming an oblique central impact with the forces exerted by the docking collars parallel to the x-axis, we can use Eqs. (16.18) and (16.19) to determine the velocities of both spacecraft after the impact.

Figure 16.15
(A) Apollo command-service module (B) Soyuz capsule.

Solution

(a) From Eq. (16.9), the velocity of the center of mass of the combined vehicles is

$$\mathbf{v} = \frac{m_A \mathbf{v}_A + m_B \mathbf{v}_B}{m_A + m_B}$$

$$= \frac{(18)(0.2\mathbf{i} + 0.03\mathbf{j} - 0.02\mathbf{k}) + 0}{18 + 6.6}$$

$$= 0.146\mathbf{i} + 0.022\mathbf{j} - 0.015\mathbf{k} \ (\text{m/s}).$$

(b) The y and z components of the velocities of both spacecraft are unchanged. To determine the x components, we first use conservation of linear momentum, Eq. (16.18), to get

$$m_A(\mathbf{v}_A)_x + m_B(\mathbf{v}_B)_x = m_A(\mathbf{v}_A')_x + m_B(\mathbf{v}_B')_x:$$
$$(18)(0.2) + 0 = (18)(\mathbf{v}_A')_x + (6.6)(\mathbf{v}_B')_x.$$

We then use the coefficient of restitution, Eq. (16.19), to obtain

$$e = \frac{(\mathbf{v}_B')_x - (\mathbf{v}_A')_x}{(\mathbf{v}_A)_x - (\mathbf{v}_B)_x}:$$

and

$$0.95 = \frac{(\mathbf{v}_B')_x - (\mathbf{v}_A')_x}{0.2 - 0}.$$

Solving these two equations yields $(\mathbf{v}_A')_x = 0.095$ (m/s) and $(\mathbf{v}_B')_x = 0.285$ (m/s), so the velocities of the spacecraft after the impact are

$$\mathbf{v}_A' = 0.095\mathbf{i} + 0.03\mathbf{j} - 0.02\mathbf{k} \ (\text{m/s})$$

and

$$\mathbf{v}_B' = 0.285\mathbf{i} \ (\text{m/s}).$$

16.4 Angular Momentum

In this section we derive a result, analogous to the principle of impulse and momentum, that relates the integral of a moment with respect to time to the change in a quantity called the angular momentum.

Principle of Angular Impulse and Momentum

We describe the position of an object relative to an inertial reference frame with origin O by the position vector $\mathbf{r}$ from O to the object's center of mass (Fig. 16.16a). Recall that we obtained the very useful principle of work and energy by taking the dot product of Newton's second law with the velocity. Here we obtain another useful result by taking the cross product of Newton's second law with the position vector. This procedure gives us a relation between the moment of the external forces about O and the object's motion.

We take the cross product of Newton's second law with $\mathbf{r}$:

$$\mathbf{r} \times \Sigma\mathbf{F} = \mathbf{r} \times m\mathbf{a} = \mathbf{r} \times m\frac{d\mathbf{v}}{dt}. \tag{16.20}$$

Notice that the derivative of the quantity $\mathbf{r} \times m\mathbf{v}$ with respect to time is

$$\frac{d}{dt}(\mathbf{r} \times m\mathbf{v}) = \underbrace{\left(\frac{d\mathbf{r}}{dt} \times m\mathbf{v}\right)}_{=\,0} + \left(\mathbf{r} \times m\frac{d\mathbf{v}}{dt}\right).$$

(The first term on the right is zero because $d\mathbf{r}/dt = \mathbf{v}$ and the cross product of parallel vectors is zero.) Using this result, we can write Eq. (16.20) as

$$\mathbf{r} \times \Sigma\mathbf{F} = \frac{d\mathbf{H}_O}{dt}, \tag{16.21}$$

where the vector

$$\mathbf{H}_O = \mathbf{r} \times m\mathbf{v} \tag{16.22}$$

is called the *angular momentum* about O (Fig. 16.16b). If we interpret the angular momentum as the moment of the linear momentum of the object about point O, Eq. (16.21) states that the moment $\mathbf{r} \times \Sigma\mathbf{F}$ equals the rate of change of the moment of momentum about point O. If the moment is zero during an interval of time, $\mathbf{H}_O$ is constant.

Integrating Eq. (16.21) with respect to time, we obtain

$$\int_{t_1}^{t_2}(\mathbf{r} \times \Sigma\mathbf{F})\,dt = (\mathbf{H}_O)_2 - (\mathbf{H}_O)_1. \tag{16.23}$$

The integral on the left is called the *angular impulse*, and the equation itself is called the *principle of angular impulse and momentum*: The angular impulse applied to an object during an interval of time is equal to the change in the object's angular momentum. If we know the moment $\mathbf{r} \times \Sigma\mathbf{F}$ as a function of time, we can determine the change in the angular momentum. The dimensions of the angular impulse and angular momentum are (mass) $\times$ (length)^2/(time).

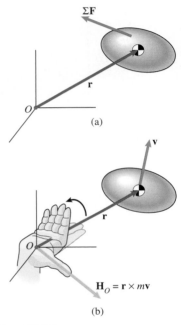

Figure 16.16
(a) The position vector and the total external force on an object.
(b) The angular momentum vector and the right-hand rule for determining its direction.

Central-Force Motion

If the total force acting on an object remains directed toward a point that is fixed relative to an inertial reference frame, the object is said to be in *central-force motion*. The fixed point is called the *center* of the motion. Orbits are the most familiar instances of central-force motion. For example, the gravitational force on an earth satellite remains directed toward the center of the earth.

If we place the reference point O at the center of the motion (Fig. 16.17a), the position vector $\mathbf{r}$ is parallel to the total force, so $\mathbf{r} \times \Sigma\mathbf{F}$ equals zero. Therefore, Eq. (16.23) indicates that in central-force motion, an object's angular momentum is conserved:

$$\mathbf{H}_O = \text{constant}. \qquad (16.24)$$

In plane central-force motion, we can express $\mathbf{r}$ and $\mathbf{v}$ in cylindrical coordinates (Fig. 16.17b):

$$\mathbf{r} = r\,\mathbf{e}_r, \quad \mathbf{v} = v_r\,\mathbf{e}_r + v_\theta\,\mathbf{e}_\theta.$$

Substituting these expressions into Eq. (16.22), we obtain the angular momentum:

$$\mathbf{H}_O = (r\,\mathbf{e}_r) \times m(v_r\,\mathbf{e}_r + v_\theta\,\mathbf{e}_\theta) = mrv_\theta\,\mathbf{e}_z.$$

From this expression, we see that in plane central-force motion, *the product of the radial distance from the center of the motion and the transverse component of the velocity is constant:*

$$rv_\theta = \text{constant}. \qquad (16.25)$$

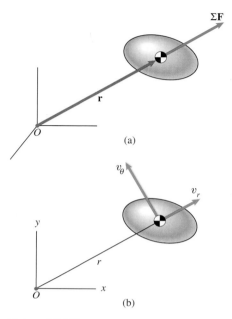

(a)

(b)

Figure 16.17
(a) Central-force motion.
(b) Expressing the position and velocity in cylindrical coordinates.

Study Questions

1. What is the definition of the angular momentum?
2. What is the principle of angular impulse and momentum?
3. What condition is necessary for angular momentum to be conserved?
4. In central-force motion, what can be inferred about the angular momentum?

Example 16.7

Applying Conservation of Angular Momentum

A disk of mass m attached to a string slides on a smooth horizontal table under the action of a constant transverse force F (Fig. 16.18). The string is drawn through a hole in the table at O at constant velocity v_0. At $t = 0$, $r = r_0$ and the transverse velocity of the disk is zero. What is the disk's velocity as a function of time?

Strategy

By expressing r as a function of time, we can determine the moment of the forces on the disk about O as a function of time. The angular momentum of the disk depends on its velocity, so we can apply the principle of angular impulse and momentum to obtain information about the velocity as a function of time.

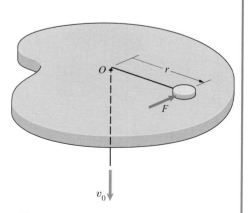

Figure 16.18

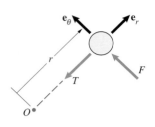

(a) Expressing the moment in terms of polar coordinates.

Solution

The radial position as a function of time is $r = r_0 - v_0 t$. In terms of polar coordinates (Fig. a), the moment about O of the forces on the disk is

$$\mathbf{r} \times \Sigma\mathbf{F} = r\,\mathbf{e}_r \times (-T\,\mathbf{e}_r + F\,\mathbf{e}_\theta) = F(r_0 - v_0 t)\mathbf{e}_z,$$

where T is the tension in the string. The angular momentum at time t is

$$\mathbf{H}_O = \mathbf{r} \times m\mathbf{v} = r\,\mathbf{e}_r \times m(v_r\,\mathbf{e}_r + v_\theta\,\mathbf{e}_\theta)$$

$$= m v_\theta (r_0 - v_0 t)\mathbf{e}_z.$$

Substituting these expressions into the principle of angular impulse and momentum, we get

$$\int_{t_1}^{t_2} (\mathbf{r} \times \Sigma\mathbf{F})\,dt = (\mathbf{H}_O)_2 - (\mathbf{H}_O)_1:$$

$$\int_0^t F(r_0 - v_0 t)\mathbf{e}_z\,dt = m v_\theta (r_0 - v_0 t)\mathbf{e}_z - \mathbf{0}.$$

Evaluating the integral, we obtain the transverse component of velocity as a function of time:

$$v_\theta = \frac{\left[r_0 t - (1/2)v_0 t^2\right]F}{(r_0 - v_0 t)m}.$$

The disk's velocity as a function of time is

$$\mathbf{v} = -v_0\,\mathbf{e}_r + \frac{\left[r_0 t - (1/2)v_0 t^2\right]F}{(r_0 - v_0 t)m}\,\mathbf{e}_\theta.$$

Example 16.8

Applying Momentum and Energy to a Satellite

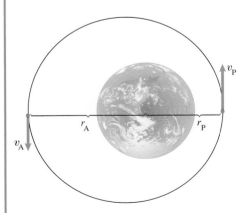

Figure 16.19

When an earth satellite is at perigee (the point at which it is nearest to the earth), the magnitude of its velocity is $v_P = 7000$ m/s and its distance from the center of the earth is $r_P = 10{,}000$ km (Fig. 16.19). What are the magnitude of the velocity v_A and the distance r_A of the satellite from the earth at apogee (the point at which it is farthest from the earth)? The radius of the earth is $R_E = 6370$ km.

Strategy

Because the satellite undergoes central-force motion about the center of the earth, the product of its distance from the center of the earth and the transverse component of its velocity is constant. This gives us one equation relating v_A and r_A. We can obtain a second equation relating v_A and r_A by using conservation of energy.

Solution

From Eq. (16.25), conservation of angular momentum requires that

$$r_A v_A = r_P v_P.$$

From Eq. (15.27), the potential energy of the satellite in terms of its distance from the center of the earth is

$$V = -\frac{mgR_E^2}{r}.$$

The sum of the kinetic and potential energies at apogee and perigee must be equal:

$$\frac{1}{2}mv_A^2 - \frac{mgR_E^2}{r_A} = \frac{1}{2}mv_P^2 - \frac{mgR_E^2}{r_P}.$$

Substituting $r_A = r_P v_P / v_A$ into this equation and rearranging terms, we obtain

$$(v_A - v_P)\left(v_A + v_P - \frac{2gR_E^2}{r_P v_P}\right) = 0.$$

This equation yields the trivial solution $v_A = v_P$ and also the solution for the velocity at apogee:

$$v_A = \frac{2gR_E^2}{r_P v_P} - v_P.$$

Substituting the values of g, R_E, r_P, and v_P, we obtain $v_A = 4370$ m/s and $r_A = 16{,}000$ km.

16.5 Mass Flows

In this section, we use conservation of linear momentum to determine the force exerted on an object as a result of emitting or absorbing a continuous flow of mass. The resulting equation applies to a variety of situations, including determining the thrust of a rocket and calculating the forces exerted on objects by flows of liquids or granular materials.

Suppose that an object of mass m and velocity $\mathbf{v}$ is subjected to no external forces (Fig. 16.20a) and emits an element of mass Δm_f with velocity $\mathbf{v}_f$ *relative to the object* (Fig. 16.20b). We denote the new velocity of the object by $\mathbf{v} + \Delta\mathbf{v}$. The linear momentum of the object before the element of mass is emitted equals the total linear momentum of the object and the element afterward:

$$m\mathbf{v} = (m - \Delta m_f)(\mathbf{v} + \Delta\mathbf{v}) + \Delta m_f(\mathbf{v} + \mathbf{v}_f).$$

Evaluating the products and simplifying, we obtain

$$m\Delta\mathbf{v} + \Delta m_f \mathbf{v}_f - \Delta m_f \Delta\mathbf{v} = 0. \qquad (16.26)$$

Now we assume that, instead of shedding a discrete element of mass, the object emits a continuous flow of mass and that Δm_f is the amount emitted in

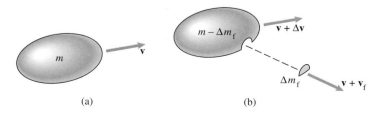

Figure 16.20
An object's mass and velocity (a) before and (b) after emitting an element of mass.

an interval of time Δt. We divide Eq. (16.26) by Δt and write the resulting equation as

$$m\frac{\Delta \mathbf{v}}{\Delta t} + \frac{\Delta m_\mathrm{f}}{\Delta t}\mathbf{v}_\mathrm{f} - \frac{\Delta m_\mathrm{f}}{\Delta t}\frac{\Delta \mathbf{v}}{\Delta t}\Delta t = 0.$$

Taking the limit of this equation as $\Delta t \to 0$, we obtain

$$-\frac{dm_\mathrm{f}}{dt}\mathbf{v}_\mathrm{f} = m\mathbf{a},$$

where $\mathbf{a}$ is the acceleration of the object's center of mass and the term dm_f/dt is the *mass flow rate*—the rate at which mass flows from the object. Comparing this equation with Newton's second law, we conclude that a flow of mass *from* an object exerts a force

$$\mathbf{F}_\mathrm{f} = -\frac{dm_\mathrm{f}}{dt}\mathbf{v}_\mathrm{f} \tag{16.27}$$

on the object. The force is proportional to the mass flow rate and to the magnitude of the relative velocity of the flow, and its direction is *opposite* to the direction of the relative velocity. Conversely, a flow of mass *to* an object exerts a force in the *same* direction as the relative velocity.

Example 16.9

Force Resulting from a Mass Flow

The rocket sled in Fig. 16.21 is slowed by a water brake after its motor has burned out. There is a trough of water between the tracks where the sled is brought to rest. A tube from the sled extends into the water with its open end pointing forward, so that water enters the tube in the direction parallel to the x axis as the sled moves forward. The other end of the tube points upward, so that the water flows out in the direction parallel to the y axis. If v is the sled's velocity, the water enters the tube with velocity v relative to the sled and flows out with the same velocity. The mass flow rate of water through the tube is $\rho v A$, where $\rho = 1000 \ \mathrm{kg/m^3}$ is the mass density of the water and $A = 0.01 \ \mathrm{m^2}$ is the cross-sectional area of the tube. At an instant when $v = 300 \ \mathrm{m/s}$, what forces are exerted on the sled by the flows of water entering and leaving it?

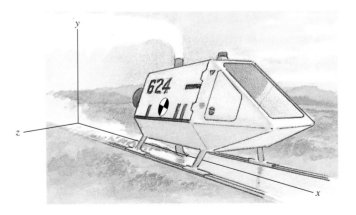

Figure 16.21

Strategy

We can use Eq. (16.27) to determine the forces exerted by the flows of water entering and leaving the sled.

Solution

Relative to the sled, the velocity vector of the water entering it is $\mathbf{v}_f = -v\mathbf{i}$. Because water is entering the sled, the force exerted is in the direction of the relative velocity:

$$\mathbf{F}_f = \frac{dm_f}{dt} \mathbf{v}_f$$

$$= \rho v A(-v\mathbf{i})$$

$$= (1000)(300)(0.01)(-300\mathbf{i})$$

$$= -900{,}000\mathbf{i} \text{ (N)}.$$

The water leaves the sled with relative velocity $\mathbf{v}_f = v\mathbf{j}$. The force on the sled is opposite to the direction of the relative velocity:

$$\mathbf{F}_f = -\frac{dm_f}{dt} \mathbf{v}_f$$

$$= -\rho v A(v\mathbf{j})$$

$$= -(1000)(300)(0.01)(300\mathbf{j})$$

$$= -900{,}000\mathbf{j} \text{ (N)}.$$

Example 16.10

Thrust of a Rocket

The classic example of a force created by a mass flow is the rocket. The one in Fig. 16.22 has a uniform, constant exhaust velocity v_f parallel to the x axis.
(a) What force is exerted on the rocket by the mass flow of its exhaust?
(b) If the force determined in (a) is the only force acting on the rocket, and it starts from rest with an initial mass m_0, determine the rocket's velocity as a function of its mass m.

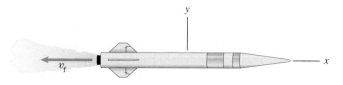

Figure 16.22

Strategy

(a) Equation (16.27) gives the force exerted on the rocket in terms of the exhaust velocity and the mass flow rate of fuel.

(b) We can use Newton's second law to obtain an equation for the rocket's velocity as a function of its mass.

Solution

(a) In terms of the coordinate system in Fig. 16.22, the velocity vector of the exhaust is $\mathbf{v}_f = -v_f \mathbf{i}$. From Eq. (16.27), the force exerted on the rocket is

$$\mathbf{F}_f = -\frac{dm_f}{dt}\mathbf{v}_f = \frac{dm_f}{dt}v_f \mathbf{i},$$

where dm_f/dt is the mass flow rate of the rocket's fuel. The force exerted on the rocket by its exhaust is toward the right, opposite to the direction of the flow of the exhaust.

(b) Newton's second law applied to the rocket is

$$\Sigma F_x = \frac{dm_f}{dt}v_f = m\frac{dv_x}{dt},$$

where m is the rocket's mass. The mass flow rate of fuel is the rate at which the rocket's mass is being consumed. Therefore, the rate of change of the mass of the rocket is

$$\frac{dm}{dt} = -\frac{dm_f}{dt}.$$

Using this expression, we can write Newton's second law as

$$dv_x = -v_f\frac{dm}{m}.$$

Because the exhaust velocity v_f is constant, we can integrate the preceding equation to determine the velocity of the rocket as a function of its mass:

$$\int_0^{v_x} dv_x = -v_f \int_{m_0}^{m} \frac{dm}{m}.$$

The result is

$$v_x = v_f \ln\left(\frac{m_0}{m}\right).$$

Discussion

The velocity attained by the rocket is dependent on the exhaust velocity and the amount of mass expended. Thus, a rocket can gain more velocity by expending more of its mass. However, notice that increasing the ratio m_0/m from 10 to 100 increases the velocity attained by only a factor of two. In contrast, increasing the exhaust velocity results in a proportional increase in the rocket's velocity. Rocket engineers use fuels such as liquid oxygen and liquid hydrogen because they produce a large exhaust velocity. This objective has also led to research on rocket engines that use electromagnetic fields to accelerate charged particles of fuel to large velocities.

Example 16.11

Force Resulting from a Mass Flow

A horizontal stream of water with velocity v_0 and mass flow rate dm_f/dt hits a plate that deflects the water in the horizontal plane through an angle θ (Fig. 16.23). Assume that the magnitude of the velocity of the water when it leaves the plate is approximately equal to v_0. What force is exerted on the plate by the water?

Strategy

We can determine the force exerted on the plate by treating the part of the stream in contact with the plate as an object with mass flows entering and leaving it.

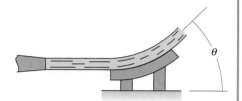

Figure 16.23

Solution

In Fig. (a), we draw the free-body diagram of the part of the stream in contact with the plate. Streams of mass with velocity v_0 enter and leave this "object," and $\mathbf{F}_P$ is the force exerted on the stream by the plate. We wish to determine the force $-\mathbf{F}_P$ exerted on the plate by the stream. First we consider the departing stream of water. The mass flow rate of water leaving the free-body diagram must be equal to the mass flow rate entering. In terms of the coordinate system shown, the velocity of the departing stream is

$$\mathbf{v}_f = v_0 \cos\theta\, \mathbf{i} + v_0 \sin\theta\, \mathbf{j}.$$

Let $\mathbf{F}_D$ be the force exerted on the object by the departing stream. From Eq. (16.27),

$$\mathbf{F}_D = -\frac{dm_f}{dt}\mathbf{v}_f = -\frac{dm_f}{dt}(v_0\cos\theta\,\mathbf{i} + v_0\sin\theta\,\mathbf{j}).$$

The velocity of the entering stream is $\mathbf{v}_f = v_0\,\mathbf{i}$. Since this flow is entering the object rather than leaving it, the resulting force $\mathbf{F}_E$ is in the same direction as the relative velocity:

$$\mathbf{F}_E = \frac{dm_f}{dt}\mathbf{v}_f = \frac{dm_f}{dt}v_0\,\mathbf{i}.$$

The sum of the forces on the free-body diagram must equal zero:

$$\mathbf{F}_D + \mathbf{F}_E + \mathbf{F}_P = \mathbf{0}.$$

Hence, the force exerted on the plate by the water is (Fig. b)

$$-\mathbf{F}_P = \mathbf{F}_D + \mathbf{F}_E = \frac{dm_f}{dt}v_0\big[(1 - \cos\theta)\mathbf{i} - \sin\theta\,\mathbf{j}\big].$$

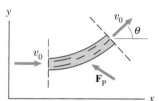

(a) Free-body diagram of the stream.

(b) Force exerted on the plate.

Discussion

This simple example affords an insight into how turbine blades and airplane wings can create forces by deflecting streams of liquid or gas (Fig. c).

(c) Pattern of moving fluid around an airplane wing.

Example 16.12

Application to Engineering:

Jet Engines

In a turbojet engine (Fig. 16.24), a mass flow rate dm_c/dt of inlet air enters the compressor with velocity v_i. The air is mixed with fuel and ignited in the combustion chamber. The mixture then flows through the turbine, which powers the compressor. The exhaust, with a mass flow rate equal to that of the air plus the mass flow rate of the fuel $(dm_c/dt + dm_f/dt)$, exits at a high exhaust velocity v_e, exerting a large force on the engine. Suppose that $dm_c/dt = 13.5$ kg/s and $dm_f/dt = 0.130$ kg/s. The inlet air velocity is $v_i = 120$ m/s and the exhaust velocity is $v_e = 480$ m/s. What is the engine's thrust?

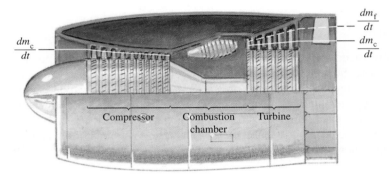

Figure 16.24

Strategy

We can determine the engine's thrust by using Eq. (16.27). We must include both the force exerted by the engine's exhaust and the force exerted by the mass flow of air entering the compressor to determine the net thrust.

Solution

The engine's exhaust exerts a force to the left equal to the product of the mass flow rate of the fuel–air mixture and the exhaust velocity. The inlet air exerts a force to the right equal to the product of the mass flow rate of the inlet air and the inlet velocity. The engine's thrust (the net force to the left) is

$$T = \left(\frac{dm_c}{dt} + \frac{dm_f}{dt} \right) v_e - \frac{dm_c}{dt} v_i$$

$$= (13.5 + 0.130)(480) - (13.5)(120)$$

$$= 4920 \text{ N.}$$

$\mathcal{D}$esign Issues

The jet engine was developed in Europe in the years just prior to World War II. Although the turbojet engine in Fig. 16.24 was a successful design that dominated both military and commercial aviation for many years, it has the drawback that it consumes a relatively large amount of fuel. During the last 30 years, the fan-jet engine, shown in Fig. 16.25, has become the most commonly used design, particularly for commercial airplanes. Part of its thrust is provided by air that is accelerated by the fan. The ratio of the mass flow rate of air entering the fan, dm_b/dt, to the mass flow rate of air entering the compressor, dm_c/dt, is called the *bypass ratio*.

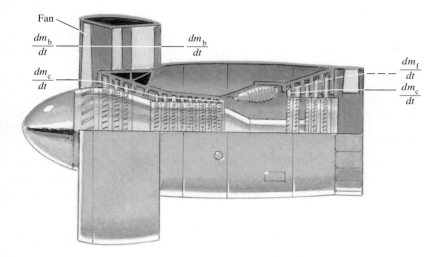

Figure 16.25
A fan-jet engine. Part of the entering mass flow of air is accelerated by the fan and does not enter the compressor.

The force exerted by a jet engine's exhaust equals the product of the mass flow rate and the exhaust velocity. In the fan-jet engine, the air passing through the fan is not heated by the combustion of fuel and therefore has a higher density than the exhaust of the turbojet engine. As a result, the fan-jet engine can provide a given thrust with a lower average exhaust velocity. Since the work that must be expended to create the thrust depends on the kinetic energy of the exhaust, the fan-jet engine creates thrust more efficiently.

Chapter Summary

Principle of Impulse and Momentum

The linear impulse applied to an object during an interval of time is equal to the change in the object's linear momentum:

$$\int_{t_1}^{t_2} \Sigma \mathbf{F}\, dt = m\mathbf{v}_2 - m\mathbf{v}_1. \qquad \text{Eq. (16.1)}$$

This result can also be expressed in terms of the average of the total force with respect to time:

$$(t_2 - t_1)\Sigma \mathbf{F}_{av} = m\mathbf{v}_2 - m\mathbf{v}_1. \qquad \text{Eq. (16.2)}$$

The principle of impulse and momentum can also be expressed in terms of the tangential component of the total force and the linear momentum along the object's path:

$$\int_{t_1}^{t_2} \Sigma F_t\, dt = mv_2 - mv_1. \qquad \text{Eq. (16.3)}$$

The average of the tangential component of the total force with respect to time is related to the change in the linear momentum along the path by

$$(t_2 - t_1)\Sigma F_{t\,\text{av}} = mv_2 - mv_1.$$ Eq. (16.4)

Conservation of Linear Momentum

If objects A and B are not subjected to external forces other than the forces they exert on each other (or if the effects of other external forces are negligible), their total linear momentum is conserved:

$$m_A \mathbf{v}_A + m_B \mathbf{v}_B = \text{constant}.$$ Eq. (16.6)

Also, the velocity of the common center of mass of A and B is constant.

Impacts

If colliding objects are not subjected to external forces, their total linear momentum must be the same before and after the impact. Even when they are subjected to external forces, the force of the impact is often so large, and its duration so brief, that the effect of the external forces on their motions during the impact is negligible.

If objects A and B adhere and remain together after they collide, they are said to undergo a *perfectly plastic impact*. The velocity of their common center of mass before and after the impact is given by

$$\mathbf{v} = \frac{m_A \mathbf{v}_A + m_B \mathbf{v}_B}{m_A + m_B}.$$ Eq. (16.9)

Central Impacts

In a *direct central impact* (Fig. a), linear momentum is conserved, or

$$m_A v_A + m_B v_B = m_A v'_A + m_B v'_B,$$ Eq. (16.10)

and the velocities are related by the *coefficient of restitution*:

$$e = \frac{v'_B - v'_A}{v_A - v_B}.$$ Eq. (16.16)

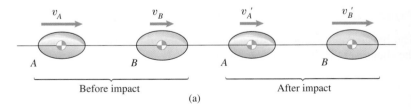

Before impact   After impact
(a)

If $e = 0$, the impact is perfectly plastic. If $e = 1$, the total kinetic energy is conserved and the impact is called *perfectly elastic*.

In an *oblique central impact* (Fig. b), the components of velocity in the y and z directions are unchanged by the impact:

$$(\mathbf{v}'_A)_y = (\mathbf{v}_A)_y, \qquad (\mathbf{v}'_B)_y = (\mathbf{v}_B)_y,$$

$$(\mathbf{v}'_A)_z = (\mathbf{v}_A)_z, \qquad (\mathbf{v}'_B)_z = (\mathbf{v}_B)_z.$$ Eq. (16.17)

In the x direction, linear momentum is conserved, or

$$m_A(\mathbf{v}_A)_x + m_B(\mathbf{v}_B)_x = m_A(\mathbf{v}'_A)_x + m_B(\mathbf{v}'_B)_x,$$ Eq. (16.18)

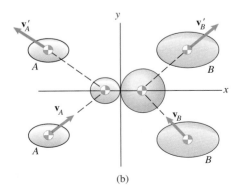

(b)

and the velocity components are related by the coefficient of restitution:

$$e = \frac{(v'_B)_x - (v'_A)_x}{(v_A)_x - (v_B)_x}.$$
Eq. (16.19)

Principle of Angular Impulse and Momentum

For a fixed point O, the angular impulse applied to an object during an interval of time is equal to the change in the object's angular momentum about O; that is,

$$\int_{t_1}^{t_2} (\mathbf{r} \times \Sigma\mathbf{F}) \, dt = (\mathbf{H}_O)_2 - (\mathbf{H}_O)_1,$$
Eq. (16.23)

where the angular momentum is

$$\mathbf{H}_O = \mathbf{r} \times m\mathbf{v}$$
Eq. (16.22)

Central-Force Motion

If the total force acting on an object remains directed toward a fixed point, the object is said to be in *central-force motion*, and its angular momentum about the fixed point is conserved:

$$\mathbf{H}_O = \text{constant.}$$
Eq. (16.24)

In plane central-force motion, the product of the radial distance and the transverse component of the velocity is constant:

$$rv_\theta = \text{constant.}$$
Eq. (16.25)

Mass Flows

A flow of mass *from* an object with velocity $\mathbf{v}_f$ *relative to the object* exerts a force

$$\mathbf{F}_f = -\frac{dm_f}{dt}\mathbf{v}_f$$
Eq. (16.27)

on the object, where dm_f/dt is the *mass flow rate*. The direction of the force is opposite to the direction of the relative velocity. A flow of mass *to* an object exerts a force in the same direction as the relative velocity.

Review Problems

16.1 The total external force on a 10-kg object is constant and equal to $90\mathbf{i} - 60\mathbf{j} + 20\mathbf{k}$ (N). At $t = 2$ s, the object's velocity is $-8\mathbf{i} + 6\mathbf{j}$ (m/s).
(a) What impulse is applied to the object from $t = 2$ s to $t = 4$ s?
(b) What is the object's velocity at $t = 4$ s?

16.2 The total external force on an object is $\mathbf{F} = 10t\mathbf{i} + 60\mathbf{j}$ (lb). At $t = 0$, the object's velocity is $\mathbf{v} = 20\mathbf{j}$ (ft/s). At $t = 12$ s, the x component of its velocity is 48 ft/s.
(a) What impulse is applied to the object from $t = 0$ to $t = 6$ s?
(b) What is the object's velocity at $t = 6$ s?

16.3 An aircraft arresting system is used to stop airplanes whose braking systems fail. The system stops a 47.5-Mg airplane moving at 80 m/s in 9.15 s.

(a) What impulse is applied to the airplane during the 9.15 s?
(b) What is the average deceleration to which the passengers are subjected?

P16.3

16.4 The 1895 Austrian 150-mm howitzer had a 1.94-m-long barrel, possessed a muzzle velocity of 300 m/s, and fired a 38-kg shell. If the shell took 0.013 s to travel the length of the barrel, what average force was exerted on the shell?

16.5 An athlete throws a shot weighing 16 lb. When he releases it, the shot is 7 ft above the ground and its components of velocity are $v_x = 31$ ft/s and $v_y = 26$ ft/s.

(a) Suppose the athlete accelerates the shot from rest in 0.8 s, and assume as a first approximation that the force **F** he exerts on the shot is constant. Use the principle of impulse and momentum to determine the x and y components of **F**.
(b) What is the horizontal distance from the point where he releases the shot to the point where it strikes the ground?

P16.5

16.6 The 6000-lb pickup truck A moving at 40 ft/s collides with the 4000-lb car B moving at 30 ft/s.

(a) What is the magnitude of the velocity of their common center of mass after the impact?
(b) Treat the collision as a perfectly plastic impact. How much kinetic energy is lost?

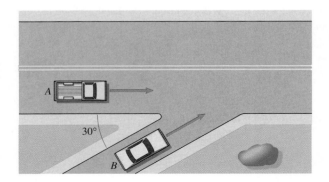

P16.6

16.7 Two hockey players $(m_A = 80$ kg, $m_B = 90$ kg$)$ converging on the puck at $x = 0$, $y = 0$ become entangled and fall. Before the collision, $v_A = 9\mathbf{i} + 4\mathbf{j}$ (m/s) and $v_B = -3\mathbf{i} + 6\mathbf{j}$ (m/s). If the coefficient of kinetic friction between the players and the ice is $\mu_k = 0.1$, what is their approximate position when they stop sliding?

P16.7

16.8 The cannon weighed 400 lb, fired a cannonball weighing 10 lb, and had a muzzle velocity of 200 ft/s. For the 10° elevation angle shown, determine (a) the velocity of the cannon after it was fired and (b) the distance the cannonball traveled. (Neglect drag.)

P16.8

16.9 A 1-kg ball moving horizontally at 12 m/s strikes a 10-kg block. The coefficient of restitution of the impact is $e = 0.6$, and the coefficient of kinetic friction between the block and the inclined surface is $\mu_k = 0.4$. What distance does the block slide before stopping?

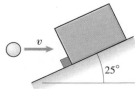

P16.9

⚙ **16.10** A Peace Corps volunteer designs the simple device shown for drilling water wells in remote areas. A 70-kg "hammer," such as a section of log or a steel drum partially filled with concrete, is hoisted to $h = 1$ m and allowed to drop onto a protective cap on the section of pipe being pushed into the ground. The combined mass of the cap and section of pipe is 20 kg. Assume that the coefficient of restitution is nearly zero.
(a) What is the velocity of the cap and pipe immediately after the impact?
(b) If the pipe moves 30 mm downward when the hammer is dropped, what resistive force was exerted on the pipe by the ground? (Assume that the resistive force is constant during the motion of the pipe.)

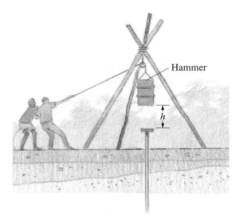

Hammer

h

P16.10

16.11 A tugboat (mass $= 40$ Mg) and a barge (mass $= 160$ Mg) are stationary with a slack hawser connecting them. The tugboat accelerates to 2 knots (1 knot $= 1852$ m/hr) before the hawser becomes taut. Determine the velocities of the tugboat and the barge just after the hawser becomes taut (a) if the "impact" is perfectly plastic ($e = 0$) and (b) if the "impact" is perfectly elastic ($e = 1$). Neglect the forces exerted by the water and the tugboat's engines.

P16.11

16.12 In Problem 16.11, determine the magnitude of the impulsive force exerted on the tugboat in the two cases if the duration of the "impact" is 4 s. Neglect the forces exerted by the water and the tugboat's engines during this period.

16.13 The balls are of equal mass m. Balls B and C are connected by an unstretched linear spring and are stationary. Ball A moves toward ball B with velocity v_A. The impact of A with B is perfectly elastic ($e = 1$). Neglect external forces.

(a) What is the velocity of the common center of mass of balls B and C immediately after the impact?
(b) What is the velocity of the common center of mass of B and C at time t after the impact?

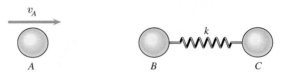

v_A

A B k C

P16.13

16.14 In Problem 16.13, what is the maximum compressive force in the spring as a result of the impact?

16.15 Suppose you interpret Problem 16.13 as an impact between the ball A and an "object" D consisting of the connected balls B and C.

(a) What is the coefficient of restitution of the impact between A and D?
(b) If you consider the total energy after the impact to be the sum of the kinetic energies, $\frac{1}{2}m(v_A')^2 + \frac{1}{2}(2m)(v_D')^2$, where v_D' is the velocity of the center of mass of D after the impact, how much energy is "lost" as a result of the impact?
(c) How much energy is actually lost as a result of the impact? (This problem is an interesting model for one of the mechanisms of energy loss in impacts between objects. The energy "loss" calculated in part (b) is transformed into "internal energy"—the vibrational motions of B and C relative to their common center of mass.)

16.16 A small object starts from rest at A and slides down the smooth ramp. The coefficient of restitution of the impact of the object with the floor is $e = 0.8$. At what height above the floor does the object hit the wall?

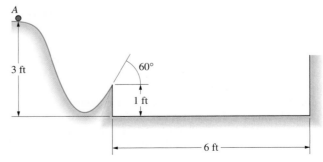

A

3 ft

60°

1 ft

6 ft

P16.16

16.17 The cue gives the cue ball A a velocity of magnitude 3 m/s. The angle $\beta = 0$ and the coefficient of restitution of the impact of the cue ball and the eight ball B is $e = 1$. If the magnitude of the eight ball's velocity after the impact is 0.9 m/s, what was the coefficient of restitution of the cue ball's impact with the cushion? (The balls are of equal mass.)

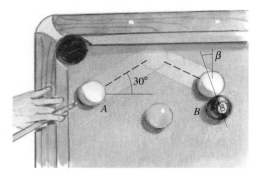

P16.17

16.18 What is the solution to Problem 16.17 if the angle $\beta = 10°$?

16.19 What is the solution to Problem 16.17 if the angle $\beta = 15°$ and the coefficient of restitution of the impact between the two balls is $e = 0.9$?

16.20 A ball is given a horizontal velocity of 3 m/s at 2 m above the smooth floor. Determine the distance D between the ball's first and second bounces if the coefficient of restitution is $e = 0.6$.

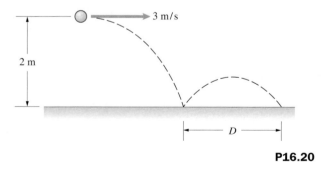

P16.20

16.21 A basketball dropped on the floor from a height of 4 ft rebounds to a height of 3 ft. In the layup shot shown, the magnitude of the ball's velocity is 5 ft/s, and the angles between its velocity vector and the positive coordinate axes are $\theta_x = 42°$, $\theta_y = 68°$, and $\theta_z = 124°$ just before it hits the backboard. What are the magnitude of its velocity and the angles between its velocity vector and the positive coordinate axes just after the ball hits the backboard?

P16.21

16.22 In Problem 16.21, the basketball's diameter is 9.5 in., the coordinates of the center of the basket rim are $x = 0$, $y = 0$, $z = 12$ in., and the backboard lies in the x-y plane. Determine the x and y coordinates of the point where the ball must hit the backboard so that the center of the ball passes through the center of the basket rim.

16.23 A satellite at $r_0 = 10,000$ mi from the center of the earth is given an initial velocity $v_0 = 20,000$ ft/s in the direction shown. Determine the magnitude of the transverse component of the satellite's velocity when $r = 20,000$ mi. (The radius of the earth is 3960 mi.)

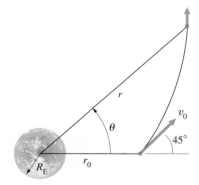

P16.23

16.24 In Problem 16.23, determine the magnitudes of the radial and transverse components of the satellite's velocity when $r = 15,000$ mi.

16.25 The snow is 2 ft deep and weighs 20 lb/ft³, the snowplow is 8 ft wide, and the truck travels at 5 mi/hr. What force does the snow exert on the truck?

P16.25

P16.27

16.26 An empty 55-lb drum, 3 ft in diameter, stands on a set of scales. Water begins pouring into the drum at 1200 lb/min from 8 ft above the bottom of the drum. The weight density of water is approximately 62.4 lb/ft³. What do the scales read 40 s after the water starts pouring?

16.28 The ski boat in Problem 16.27 weighs 2800 lb. The mass flow rate of water through the engine is 2.5 slugs/s, and the craft starts from rest at $t = 0$. Determine the boat's velocity at (a) $t = 20$ s and (b) $t = 60$ s.

16.29 A crate of mass m slides across a smooth floor pulling a chain from a stationary pile. The mass per unit length of the chain is ρ_L. If the velocity of the crate is v_0 when $s = 0$, what is its velocity as a function of s?

8 ft

P16.26

P16.29

16.27 The ski boat's jet propulsive system draws water in at A and expels it at B at 80 ft/s relative to the boat. Assume that the water drawn in enters with no horizontal velocity relative to the surrounding water. The maximum mass flow rate of water through the engine is 2.5 slugs/s. Hydrodynamic drag exerts a force on the boat of magnitude $1.5v$ lb, where v is the boat's velocity in feet per second. Neglecting aerodynamic drag, what is the ski boat's maximum velocity?

Project By making measurements, determine the coefficient of restitution of a tennis ball bouncing on a rigid surface. Try to determine whether your result is independent of the velocity with which the ball strikes the surface. Write a brief report describing your procedure and commenting on possible sources of error.

To design the excavator, the engineer needed to analyze the translational and rotational motions of its members.

Planar Kinematics of Rigid Bodies

I f the external forces acting on an object are known, Newton's second law can be used to determine the motion of the object's center of mass without considering any angular motion of the object about its center of mass. In many situations, however, angular motion must be considered. The rotational motions of some objects are central to their functions, as in the cases of gears, wheels, generators, and turbines. In this chapter, we analyze motions of objects, including their rotational motions.

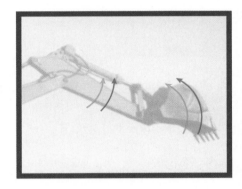

17.1 Rigid Bodies and Types of Motion

If a brick is thrown (Fig. 17.1a), we can determine the motion of its center of mass without having to be concerned about its rotational motion. The only significant force is the weight of the brick, and Newton's second law determines the acceleration of its center of mass. But suppose that the brick is standing on the floor, it is tipped over (Fig. 17.1b), and we want to determine the motion of its center of mass as it falls. In this case, the brick is subjected to its weight and also to a force exerted by the floor. We cannot determine either the force exerted by the floor or the motion of the brick's center of mass without also considering its rotational motion.

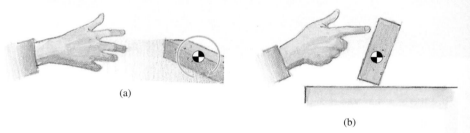

(a)

(b)

Figure 17.1
(a) A thrown brick—its rotation doesn't affect the motion of its center of mass.
(b) A tipped brick—its rotation and the motion of its center of mass are interrelated.

Before we can analyze such motions, we must consider how to describe them. A brick is an example of an object whose motion can be described by treating the object as a rigid body. A *rigid body* is an idealized model of an object that does not deform, or change shape. The precise definition is that the distance between every pair of points of a rigid body remains constant. Although any object does deform as it moves, if its deformation is sufficiently small, we can approximate its motion by modeling it as a rigid body. For example, in normal use a twirler's baton (Fig. 17.2a) can be modeled as a rigid body, but a fly-casting rod (Fig. 17.2b) cannot.

(a)

(b)

Figure 17.2
(a) A baton can be modeled as a rigid body.
(b) Under normal use, a fishing rod is too flexible to be modeled as a rigid body.

Describing the motion of a rigid body requires a reference frame (coordinate system) relative to which the motions of the points of the rigid body and its angular motion are measured. In many situations, it is convenient to use a reference frame that is fixed with respect to the earth. For example, we would use such an *earth-fixed* reference frame to describe the motion of the center of mass and the angular motion of the brick in Fig. 17.1. In the paragraphs that follow, we discuss some types of rigid-body motions relative to a given reference frame that occur frequently in applications.

Translation   If a rigid body in motion relative to a given reference frame does not rotate, it is said to be in *translation* (Fig. 17.3a). For example, the child's swing in Fig. 17.3b is designed so that the horizontal bar to which the seats are attached is in translation. Although each point of the horizontal bar moves in a circular path, the bar does not rotate. It remains horizontal, making it easier for the child to ride safely. Every point of a rigid body in translation has the same velocity and acceleration, so we describe the motion of the rigid body completely if we describe the motion of a single point.

(a)

(b)

Figure 17.3
(a) An object in translation does not rotate.
(b) The translating part of the swing on which the child sits remains level.

Rotation about a Fixed Axis   After translation, the simplest type of rigid-body motion is rotation about an axis that is fixed relative to a given reference frame (Fig. 17.4a). Each point of the rigid body on the axis is stationary, and each point not on the axis moves in a circular path about the axis as the rigid body rotates. The rotor of an electric motor (Fig. 17.4b) is an example of an object rotating about a fixed axis. The motion of a ship's propeller relative to the ship is rotation about a fixed axis. We discuss this type of motion in more detail in the next section.

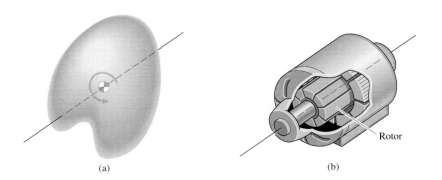

Rotor

(a)

(b)

Figure 17.4
(a) A rigid body rotating about a fixed axis.
(b) Relative to the frame of an electric motor, the rotor rotates about a fixed axis.

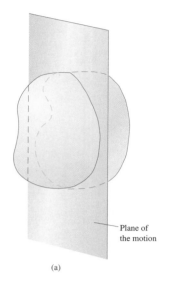

Plane of the motion

(a)

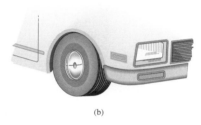

(b)

Figure 17.5
(a) A rigid body intersected by a fixed plane.
(b) A wheel undergoing planar motion.

Planar Motion Consider a plane that is fixed relative to a given reference frame and a rigid body intersected by the plane (Fig. 17.5a). If the rigid body undergoes a motion in which the points intersected by the plane remain in the plane, the body is said to be in *two-dimensional*, or *planar*, motion. We refer to the fixed plane as the plane of the motion. Rotation of a rigid body about a fixed axis is a special case of planar motion. As another example, when a car moves in a straight path, its wheels are in planar motion (Fig. 17.5b).

The components of an internal combustion engine illustrate these types of motion (Fig. 17.6). Relative to a reference frame that is fixed with respect to the engine, the pistons translate within the cylinders. The connecting rods are in general planar motion, and the crankshaft rotates about a fixed axis.

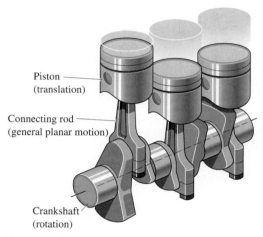

Piston (translation)

Connecting rod (general planar motion)

Crankshaft (rotation)

Figure 17.6
Translation, rotation about a fixed axis, and planar motion in an automobile engine.

We begin our analysis of rigid-body motion in the next section with a discussion of rotation about a fixed axis. By using normal and tangential components, we express the velocity and acceleration of a point of the rigid body in terms of the rigid body's angular velocity and angular acceleration. In the succeeding sections, we consider general planar motion and obtain expressions relating the relative velocity and acceleration of points of a rigid body to its angular velocity and angular acceleration. With these relations, we analyze particular examples of planar motion, such as rolling, and also analyze motions of connected rigid bodies.

17.2 Rotation about a Fixed Axis

By considering the rotation of an object about an axis that is fixed relative to a given reference frame, we can introduce some of the concepts of rigid-body motion in a familiar context. In this type of motion, each point of the rigid body moves in a circular path around the fixed axis, so we can analyze the motions of points using results developed in Chapter 13.

In Fig. 17.7, we show a rigid body rotating about a fixed axis and introduce two lines perpendicular to the axis. The reference line is fixed, and the body-fixed line rotates with the rigid body. The angle θ between the reference line and the body-fixed line describes the position, or orientation, of the rigid body about the fixed axis. The rigid body's *angular velocity*, or rate of rotation, and its *angular acceleration* are

$$\omega = \frac{d\theta}{dt}, \qquad \alpha = \frac{d\omega}{dt} = \frac{d^2\theta}{dt^2}. \qquad (17.1)$$

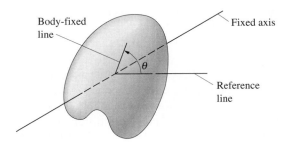

Body-fixed line

Fixed axis

θ

Reference line

Figure 17.7
Specifying the orientation of an object rotating about a fixed axis.

Each point of the object not on the fixed axis moves in a circular path about the axis. Using our knowledge of the motion of a point in a circular path, we can relate the velocity and acceleration of a point to the object's angular velocity and angular acceleration. In Fig. 17.8, we view the object in the direction parallel to the fixed axis. The velocity of a point at a distance r from the fixed axis is tangent to the point's circular path (Fig. 17.8a) and is given in terms of the angular velocity of the object by

$$v = r\omega. \qquad (17.2)$$

A point has components of acceleration tangential and normal to its circular path (Fig. 17.8b). In terms of the angular velocity and angular acceleration of the object, the components of acceleration are

$$a_t = r\alpha, \qquad a_n = \frac{v^2}{r} = r\omega^2. \qquad (17.3)$$

With these relations, we can analyze problems involving objects rotating about fixed axes. For example, suppose that we know the angular velocity ω_A

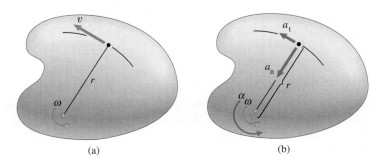

(a)   (b)

Figure 17.8
(a) Velocity and (b) acceleration of a point of a rigid body rotating about a fixed axis.

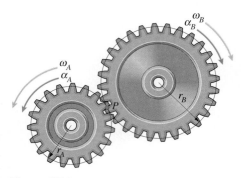

Figure 17.9
Relating the angular velocities and angular accelerations of meshing gears.

and angular acceleration α_A of the gear in Fig. 17.9 relative to a particular reference frame, and we want to determine ω_B and α_B. The velocities of the gears must be equal at P, because there is no relative motion between them in the tangential direction at P. Therefore, $r_A\omega_A = r_B\omega_B$, and we find that the angular velocity of gear B is

$$\omega_B = \left(\frac{r_A}{r_B}\right)\omega_A.$$

By taking the derivative of this equation with respect to time, we determine the angular acceleration of gear B:

$$\alpha_B = \left(\frac{r_A}{r_B}\right)\alpha_A.$$

From this result, we see that the tangential components of the accelerations of the gears at P are equal: $r_A\alpha_A = r_B\alpha_B$. However, the normal components of the accelerations of the gears at P are different in direction and, if the gears have different radii, are different in magnitude as well. The normal component of the acceleration of gear A at P points toward the center of gear A, and its magnitude is $r_A\omega_A^2$. The normal component of the acceleration of gear B at P points toward the center of gear B, and its magnitude is $r_B\omega_B^2 = (r_A/r_B)(r_A\omega_A^2)$.

Study Questions

1. What is the definition of a rigid body?
2. What is meant by two-dimensional, or planar, motion of a rigid body?
3. If a rigid body rotates about a fixed axis, what can you infer about the path of each point not on the axis?
4. If a rigid body is rotating about a fixed axis with angular velocity ω and angular acceleration α, what can you infer about the velocity and acceleration of a point at a distance r from the axis?

Example 17.1

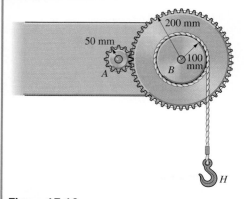

Figure 17.10

Objects Rotating about Fixed Axes

Gear A of the winch in Fig. 17.10 turns gear B, raising the hook H. If gear A starts from rest at $t = 0$ and its clockwise angular acceleration is $\alpha_A = 0.2t$ rad/s^2, what vertical distance has the hook H risen and what is its velocity at $t = 10$ s?

Strategy

By equating the tangential components of acceleration of gears A and B at their point of contact, we can determine the angular acceleration of gear B. Then we can integrate twice to obtain the angular velocity of gear B and the angle through which it has turned at $t = 10$ s.

Solution

The tangential acceleration of the point of contact of the two gears (Fig. a) is

$$a_t = (0.05 \text{ m})(0.2t \text{ rad/s}^2) = (0.2 \text{ m})(\alpha_B).$$

Therefore, the angular acceleration of gear B is

$$\alpha_B = \frac{d\omega_B}{dt} = \frac{(0.05 \text{ m})(0.2t \text{ rad/s}^2)}{(0.2 \text{ m})} = 0.05t \text{ rad/s}^2.$$

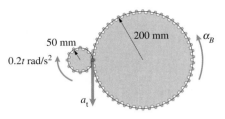

(a) The tangential accelerations of the gears are equal at their point of contact.

Integrating this equation yields

$$\int_0^{\omega_B} d\omega_B = \int_0^t 0.05t \, dt,$$

and we obtain the angular velocity of gear B:

$$\omega_B = \frac{d\theta_B}{dt} = 0.025t^2 \text{ rad/s}.$$

Integrating again, we obtain the angle through which gear B has turned:

$$\theta_B = 0.00833t^3 \text{ rad}.$$

At $t = 10$ s, $\theta_B = 8.33$ rad. The amount of cable wound around the drum, which is the distance the hook H has risen, is the product of θ_B and the radius of the drum: $(8.33 \text{ rad})(0.1 \text{ m}) = 0.833$ m.

At $t = 10$ s, $\omega_B = 2.5$ rad/s. The velocity of a point on the rim, which equals the velocity of the hook H (Fig. b), is

$$v_H = (0.1 \text{ m})(2.5 \text{ rad/s}) = 0.25 \text{ m/s}.$$

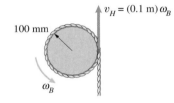

(b) Determining the hook's velocity.

17.3 General Motions: Velocities

Each point of a rigid body in translation undergoes the same motion. Each point of a rigid body rotating about a fixed axis undergoes circular motion about the axis. To analyze more complicated motions that combine translation and rotation, we must develop equations that relate the relative motions of points of a rigid body to its angular motion.

Relative Velocities

In Fig. 17.11(a), we view a rigid body from a perspective perpendicular to the plane of its motion. Points A and B are points of the rigid body contained in the latter plane, and O is the origin of a given reference frame. The position of A relative to B, $\mathbf{r}_{A/B}$, is related to the position of A and B relative to O by

$$\mathbf{r}_A = \mathbf{r}_B + \mathbf{r}_{A/B}.$$

Taking the derivative of this equation with respect to time, we obtain

$$\mathbf{v}_A = \mathbf{v}_B + \mathbf{v}_{A/B}, \tag{17.4}$$

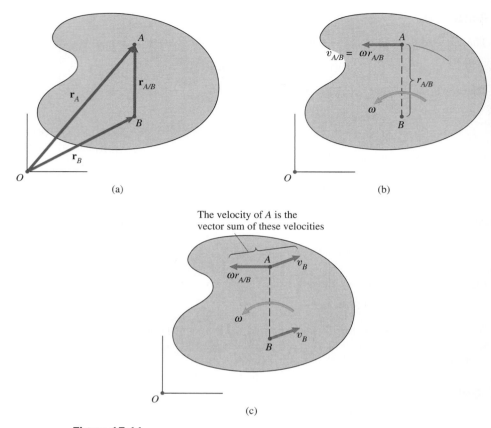

Figure 17.11
(a) A rigid body in planar motion.
(b) The velocity of A relative to B.
(c) The velocity of A is the sum of its velocity relative to B and the velocity of B.

where $\mathbf{v}_A$ and $\mathbf{v}_B$ are the velocities of A and B relative to the given reference frame and $\mathbf{v}_{A/B} = d\mathbf{r}_{A/B}/dt$ is the velocity of A relative to B. (*When we simply speak of the velocity of a point, we will mean its velocity relative to the given reference frame.*)

We can show that $\mathbf{v}_{A/B}$ is related in a simple way to the rigid body's angular velocity. Since A and B are points of the rigid body, the distance between them, $r_{A/B} = |\mathbf{r}_{A/B}|$, is constant. This means that, relative to B, A moves in a circular path as the rigid body rotates. The velocity of A relative to B is therefore tangent to the circular path and equal to the product of $r_{A/B}$ and the angular velocity ω of the rigid body (Fig. 17.11b). From Eq. (17.4), the velocity of A is the sum of the velocity of B and the velocity of A relative to B (Fig. 17.11c). This result can be used to relate velocities of points of a rigid body in planar motion when the angular velocity of the body is known.

For example, in Fig. 17.12(a), we show a circular disk of radius R rolling with counterclockwise angular velocity ω on a stationary plane surface. Saying that the surface is stationary means that we are describing the motion of the disk in terms of a reference frame that is fixed with respect to the surface. By *rolling*, we mean that the velocity of the disk relative to the surface is zero at the point of contact C. The velocity of the center B of the disk

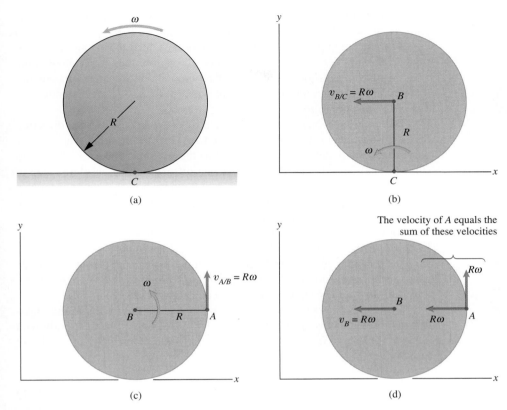

Figure 17.12
(a) A disk rolling with angular velocity ω.
(b) The velocity of the center B relative to C.
(c) The velocity of A relative to B.
(d) The velocity of A equals the sum of the velocity of B and the velocity of A relative to B.

relative to C is illustrated in Fig. 17.12b. Since $\mathbf{v}_C = \mathbf{0}$, the velocity of B in terms of the fixed coordinate system shown is

$$\mathbf{v}_B = \mathbf{v}_C + \mathbf{v}_{B/C} = -R\omega\mathbf{i}.$$

This result is worth remembering: *The magnitude of the velocity of the center of a round object rolling on a stationary plane surface equals the product of the radius and the magnitude of the angular velocity.*

We can determine the velocity of any other point of the disk in the same way. Figure 17.12(c) shows the velocity of a point A relative to point B. The velocity of A is the sum of the velocity of B and the velocity of A relative to B (Fig. 17.12d):

$$\mathbf{v}_A = \mathbf{v}_B + \mathbf{v}_{A/B} = -R\omega\mathbf{i} + R\omega\mathbf{j}.$$

The Angular Velocity Vector

We can express the rate of rotation of a rigid body as a vector. *Euler's theorem* states that a rigid body constrained to rotate about a fixed point B can move between any two positions by a single rotation about some axis

Instantaneous axis of rotation

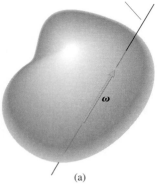

(a)

Direction of rotation

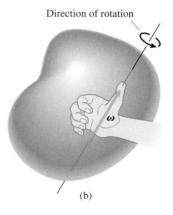

(b)

Figure 17.13
(a) An angular velocity vector.
(b) Right-hand rule for the direction of the vector.

through B. Suppose that we choose an arbitrary point B of a rigid body that is undergoing an arbitrary motion at a time t. Euler's theorem allows us to express the rigid body's change in position relative to B during an interval of time from t to $t + dt$ as a single rotation through an angle $d\theta$ about some axis. At time t, the rigid body's rate of rotation about the axis is its angular velocity $\omega = d\theta/dt$, and the axis about which the body rotates is called the *instantaneous axis of rotation*.

The *angular velocity vector*, denoted by $\boldsymbol{\omega}$, specifies both the direction of the instantaneous axis of rotation and the angular velocity. The angular velocity vector is defined to be parallel to the instantaneous axis of rotation (Fig. 17.13a), and its magnitude is the rate of rotation, the absolute value of ω. Its direction is related to the direction of the rigid body's rotation through a right-hand rule: Pointing the thumb of the right hand in the direction of $\boldsymbol{\omega}$, the fingers curl around $\boldsymbol{\omega}$ in the direction of the rotation (Fig. 17.13b).

For example, the axis of rotation of the rolling disk in Fig. 17.12 is parallel to the z axis, so the angular velocity vector of the disk is parallel to the z axis and its magnitude is ω. Curling the fingers of the right hand around the z axis in the direction of the rotation, the thumb points in the positive z direction (Fig. 17.14). The angular velocity vector of the disk is $\boldsymbol{\omega} = \omega\mathbf{k}$.

The angular velocity vector allows us to express the results of the previous section in a convenient form. Let A and B be points of a rigid body with angular velocity $\boldsymbol{\omega}$ (Fig. 17.15a). We can show that the velocity of A relative to B is

$$\mathbf{v}_{A/B} = \frac{d\mathbf{r}_{A/B}}{dt} = \boldsymbol{\omega} \times \mathbf{r}_{A/B}. \tag{17.5}$$

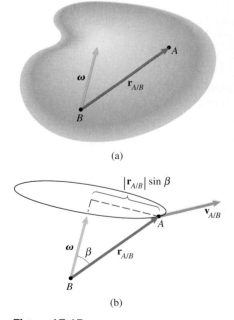

(a)

(b)

Figure 17.15
(a) Points A and B of a rotating rigid body.
(b) A is moving in a circular path relative to B.

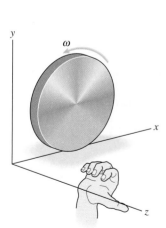

Figure 17.14
Determining the direction of the angular velocity vector of a rolling disk.

At the present instant, relative to B, point A is moving in a circular path of radius $|\mathbf{r}_{A/B}| \sin \beta$, where β is the angle between the vectors $\mathbf{r}_{A/B}$ and $\boldsymbol{\omega}$ (Fig. 17.15b). The magnitude of the velocity of A relative to B is equal to the product of the radius of the circular path and the angular velocity of the rigid body; that is, $|\mathbf{v}_{A/B}| = (|\mathbf{r}_{A/B}| \sin \beta)|\boldsymbol{\omega}|$. The right-hand side of this equation is the magnitude of the cross product of $\mathbf{r}_{A/B}$ and $\boldsymbol{\omega}$. In addition, $\mathbf{v}_{A/B}$ is perpendicular both to $\boldsymbol{\omega}$ and $\mathbf{r}_{A/B}$. But is $\mathbf{v}_{A/B}$ equal to $\boldsymbol{\omega} \times \mathbf{r}_{A/B}$ or $\mathbf{r}_{A/B} \times \boldsymbol{\omega}$? Notice in Fig. 17.15(b) that, pointing the fingers of the right hand in the direction of $\boldsymbol{\omega}$ and closing them toward $\mathbf{r}_{A/B}$, the thumb points in the direction of the velocity of A relative to B, so $\mathbf{v}_{A/B} = \boldsymbol{\omega} \times \mathbf{r}_{A/B}$. Substituting Eq. (17.5) into Eq. (17.4), we obtain an equation for the relation between the velocities of two points of a rigid body in terms of its angular velocity:

$$\mathbf{v}_A = \mathbf{v}_B + \underbrace{\boldsymbol{\omega} \times \mathbf{r}_{A/B}}_{\mathbf{v}_{A/B}}. \tag{17.6}$$

If the angular velocity vector and the velocity of one point of a rigid body are known, Eq. (17.6) can be used to determine the velocity of any other point of the rigid body. Returning to the example of a disk of radius R rolling with angular velocity ω (Fig. 17.16), let us use Eq. (17.6) to determine the velocity of point A. The velocity of the center of the disk is given in terms of the angular velocity by $\mathbf{v}_B = -R\omega\mathbf{i}$, the disk's angular velocity vector is $\boldsymbol{\omega} = \omega\mathbf{k}$, and the position vector of A relative to the center is $\mathbf{r}_{A/B} = R\mathbf{i}$. The velocity of A is

$$\mathbf{v}_A = \mathbf{v}_B + \boldsymbol{\omega} \times \mathbf{r}_{A/B} = -R\omega\mathbf{i} + (\omega\mathbf{k}) \times (R\mathbf{i})$$
$$= -R\omega\mathbf{i} + R\omega\mathbf{j}.$$

Compare this result with the velocity of A shown in Fig. 17.12(d).

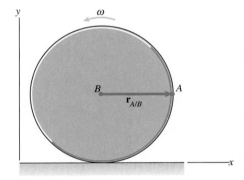

Figure 17.16
A rolling disk and the position vector of A relative to B.

Study Questions

1. What is meant when an object is said to be rolling on a surface?
2. How is the angular velocity vector of a rigid body defined?
3. If you know the direction of the angular velocity vector of a rigid body, how do you determine the direction of rotation of the body?
4. If you know the angular velocity vector and the velocity of one point of a rigid body, how can you determine the velocity of a different point of the rigid body?

Example 17.2

Determining Velocities and Angular Velocities

Bar AB in Fig. 17.17 rotates with a clockwise angular velocity of 10 rad/s. Determine the angular velocity of bar BC and the velocity of point C.

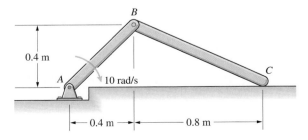

Figure 17.17

Strategy

Bar AB rotates about the fixed point A with a known angular velocity, so we can determine the velocity of B. Then, by expressing the horizontal velocity of C in terms of the velocity of B and the angular velocity of bar BC, we can obtain two equations in terms of the velocity of C and the angular velocity of bar BC.

Solution

Because the angular velocity of bar AB is given and point A is stationary, we can use Eq. (17.6) to determine the velocity of point B. From Fig. (a), the position vector of B relative to A is $\mathbf{r}_{B/A} = 0.4\mathbf{i} + 0.4\mathbf{j}$ (m). The angular velocity vector of bar AB is $\boldsymbol{\omega}_{AB} = -10\mathbf{k}$ (rad/s), so the velocity of B is

$$\mathbf{v}_B = \mathbf{v}_A + \boldsymbol{\omega} \times \mathbf{r}_{B/A}$$

$$= \mathbf{0} + \begin{vmatrix} \mathbf{i} & \mathbf{j} & \mathbf{k} \\ 0 & 0 & -10 \\ 0.4 & 0.4 & 0 \end{vmatrix}$$

$$= 4\mathbf{i} - 4\mathbf{j} \ (\text{m/s}).$$

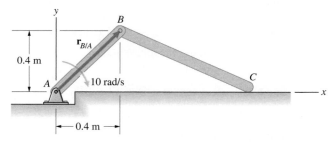

(a) The position vector of B relative to A.

We now use Eq. (17.6) to express the velocity of point C in terms of the velocity of point B. Let ω_{BC} be the counterclockwise angular velocity of bar BC (Fig. b), so that the angular velocity vector of bar BC is $\boldsymbol{\omega}_{BC} = \omega_{BC}\mathbf{k}$. Because point C is moving horizontally (Fig. b), we can write its velocity as $\mathbf{v}_C = v_C\mathbf{i}$. Therefore,

$$\mathbf{v}_C = \mathbf{v}_B + \boldsymbol{\omega}_{BC} \times \mathbf{r}_{C/B}:$$

$$v_C\mathbf{i} = 4\mathbf{i} - 4\mathbf{j} + \begin{vmatrix} \mathbf{i} & \mathbf{j} & \mathbf{k} \\ 0 & 0 & \omega_{BC} \\ 0.8 & -0.4 & 0 \end{vmatrix}$$

$$= (4 + 0.4\omega_{BC})\mathbf{i} - (4 - 0.8\omega_{BC})\mathbf{j}.$$

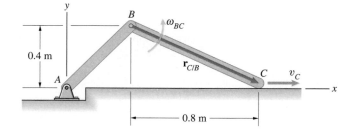

(b) The angular velocity of bar BC, the velocity of C, and the position vector of C relative to B.

Equating the **i** and **j** components in this equation, we obtain

$$v_C = 4 + 0.4\omega_{BC},$$

$$0 = 4 - 0.8\omega_{BC}.$$

Solving, we get $v_C = 6$ m/s and $\omega_{BC} = 5$ rad/s. The angular velocity of bar BC is $\boldsymbol{\omega}_{BC} = 5\mathbf{k}$ (rad/s), and the velocity of point C is $\mathbf{v}_C = 6\mathbf{i}$ (m/s).

Discussion

By expressing the velocity of point C as $\mathbf{v}_C = v_C\mathbf{i}$, we introduced the constraint imposed on the motion of bar BC by the horizontal surface. This critical step in the solution states that the vertical component of the velocity of point C is zero.

Example 17.3

Analysis of a Linkage

Bar AB in Fig. 17.18 rotates with a clockwise angular velocity of 10 rad/s. What is the vertical velocity v_R of the rack of the rack-and-pinion gear?

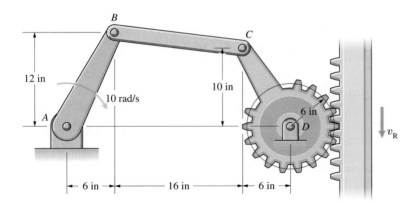

Figure 17.18

Strategy

To determine the velocity of the rack, we must determine the angular velocity of the member CD. Since we know the angular velocity of bar AB, we can apply Eq. (17.6) to points A and B to determine the velocity of point B. Then we can apply Eq. (17.6) to points C and D to obtain an equation for $\mathbf{v}_C$ in terms of the angular velocity of the member CD. We can also apply Eq. (17.6) to points B and C to obtain an equation for $\mathbf{v}_C$ in terms of the angular velocity of bar BC. By equating the two expressions for $\mathbf{v}_C$, we will obtain a vector equation in two unknowns: the angular velocities of bars BC and CD.

Solution

We first apply Eq. (17.6) to points A and B (Fig. a). In terms of the coordinate system shown, the position vector of B relative to A is $\mathbf{r}_{B/A} = 0.5\mathbf{i} + \mathbf{j}$ (ft), and the angular velocity vector of bar AB is $\boldsymbol{\omega}_{AB} = -10\mathbf{k}$ (rad/s). The velocity of B is

$$\mathbf{v}_B = \mathbf{v}_A + \boldsymbol{\omega}_{AB} \times \mathbf{r}_{B/A} = \mathbf{0} + \begin{vmatrix} \mathbf{i} & \mathbf{j} & \mathbf{k} \\ 0 & 0 & -10 \\ 0.5 & 1 & 0 \end{vmatrix}$$

$$= 10\mathbf{i} - 5\mathbf{j} \text{ (ft/s)}.$$

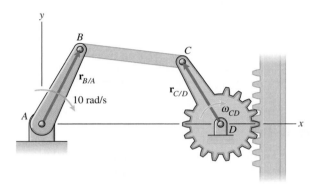

(a) Determining the velocities of points B and C.

We now apply Eq. (17.6) to points C and D. Let ω_{CD} be the unknown angular velocity of member CD (Fig. a). The position vector of C relative to D is $\mathbf{r}_{C/D} = -0.500\mathbf{i} + 0.833\mathbf{j}$ (ft), and the angular velocity vector of member CD is $\boldsymbol{\omega}_{CD} = -\omega_{CD}\mathbf{k}$. The velocity of C is

$$\mathbf{v}_C = \mathbf{v}_D + \boldsymbol{\omega}_{CD} \times \mathbf{r}_{C/D} = \mathbf{0} + \begin{vmatrix} \mathbf{i} & \mathbf{j} & \mathbf{k} \\ 0 & 0 & -\omega_{CD} \\ -0.500 & 0.833 & 0 \end{vmatrix}$$

$$= 0.833\omega_{CD}\mathbf{i} + 0.500\omega_{CD}\mathbf{j}.$$

Now we apply Eq. (17.6) to points B and C (Fig. b). We denote the unknown angular velocity of bar BC by ω_{BC}. The position vector of C relative to B is $\mathbf{r}_{C/B} = 1.333\mathbf{i} - 0.167\mathbf{j}$ (ft), and the angular velocity vector of bar BC is

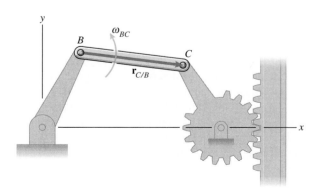

(b) Expressing the velocity of point C in terms of the velocity of point B.

$\boldsymbol{\omega}_{BC} = \omega_{BC}\mathbf{k}$. Expressing the velocity of C in terms of the velocity of B, we obtain

$$\mathbf{v}_C = \mathbf{v}_B + \boldsymbol{\omega}_{BC} \times \mathbf{r}_{C/B} = \mathbf{v}_B + \begin{vmatrix} \mathbf{i} & \mathbf{j} & \mathbf{k} \\ 0 & 0 & \omega_{BC} \\ 1.333 & -0.167 & 0 \end{vmatrix}$$

$$= \mathbf{v}_B + 0.167\omega_{BC}\mathbf{i} + 1.333\omega_{BC}\mathbf{j}.$$

Substituting our expressions for $\mathbf{v}_B$ and $\mathbf{v}_C$ into this equation, we get

$$0.833\omega_{CD}\mathbf{i} + 0.500\omega_{CD}\mathbf{j} = 10\mathbf{i} - 5\mathbf{j} + 0.167\omega_{BC}\mathbf{i} + 1.333\omega_{BC}\mathbf{j}.$$

Equating the $\mathbf{i}$ and $\mathbf{j}$ components yields two equations in terms of ω_{BC} and ω_{CD}:

$$0.833\omega_{CD} = 10 + 0.167\omega_{BC},$$
$$0.500\omega_{CD} = -5 + 1.333\omega_{BC}.$$

Solving these equations, we obtain $\omega_{BC} = 8.92$ rad/s and $\omega_{CD} = 13.78$ rad/s.

The vertical velocity of the rack is equal to the velocity of the gear where it contacts the rack:

$$v_R = (0.5\text{ ft})\omega_{CD} = (0.5)(13.78) = 6.89\text{ ft/s}.$$

Instantaneous Centers

A point of a rigid body whose velocity is zero at a given instant is called an *instantaneous center*. "Instantaneous" means that the point may have zero velocity *only* at the instant under consideration, although we also refer to a fixed point, such as a point of a fixed axis about which a rigid body rotates, as an instantaneous center.

When we know the location of an instantaneous center of a rigid body in two-dimensional motion and we know its angular velocity, the velocities of other points are easy to determine. For example, suppose that point C in Fig. 17.19a is the instantaneous center of a rigid body in plane motion with angular velocity ω. Relative to C, a point A moves in a circular path. The velocity of A relative to C is tangent to the circular path and equal to the product of the distance from C to A and the angular velocity. But since C is stationary at the present instant, the velocity of A relative to C is the velocity of A. At this instant, every point of the rigid body rotates about C (Fig. 17.19b).

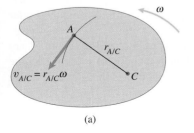

(a)

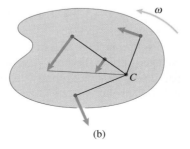

(b)

Figure 17.19
(a) An instantaneous center C and a different point A.
(b) Every point is rotating about the instantaneous center.

The instantaneous center of a rigid body in planar motion can often be located by a simple procedure. Suppose that the directions of the motions of two points A and B are known and are not parallel (Fig. 17.20a). If we draw lines through A and B perpendicular to their directions of motion, then the point C where the lines intersect is the instantaneous center. To show that this is true, let us express the velocity of C in terms of the velocity of A (Fig. 17.20b):

$$\mathbf{v}_C = \mathbf{v}_A + \boldsymbol{\omega} \times \mathbf{r}_{C/A}.$$

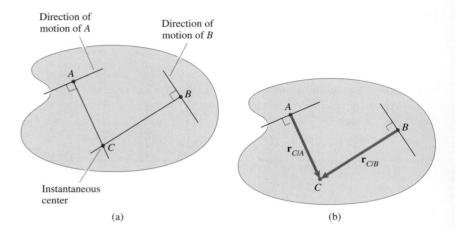

Figure 17.20
(a) Locating the instantaneous center in planar motion.
(b) Proving that $\mathbf{v}_C = \mathbf{0}$.

The vector $\boldsymbol{\omega} \times \mathbf{r}_{C/A}$ is perpendicular to $\mathbf{r}_{C/A}$, so the preceding equation indicates that $\mathbf{v}_C$ is parallel to the direction of motion of A. We also express the velocity of C in terms of the velocity of B:

$$\mathbf{v}_C = \mathbf{v}_B + \boldsymbol{\omega} \times \mathbf{r}_{C/B}.$$

The vector $\boldsymbol{\omega} \times \mathbf{r}_{C/B}$ is perpendicular to $\mathbf{r}_{C/B}$, so the foregoing equation indicates that $\mathbf{v}_C$ is parallel to the direction of motion of B. We have shown that the component of $\mathbf{v}_C$ perpendicular to the direction of motion of A is zero and that the component of $\mathbf{v}_C$ perpendicular to the direction of motion of B is zero, so $\mathbf{v}_C = \mathbf{0}$.

The instantaneous center may not be a point of the rigid body (Fig. 17.21a). This simply means that at the instant in question, the rigid body is rotating about an external point. It is helpful to imagine extending the rigid body so that it includes the instantaneous center (Fig. 17.21b). The velocity of point C of the extended body would be zero at the instant under consideration.

Notice in Fig. 17.21a that if the directions of motion of A and B are changed so that the lines perpendicular to their directions of motion become parallel, C moves to infinity. In that case, the rigid body is in pure translation, with an angular velocity of zero.

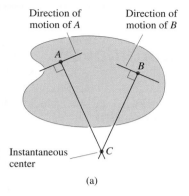

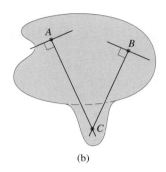

Figure 17.21
(a) An instantaneous center external to the rigid body.
(b) A hypothetical extended body. Point C would be stationary.

(a)

(b)

Instantaneous center

Direction of motion of A

Direction of motion of B

Returning once again to our example of a disk of radius R rolling with angular velocity ω (Fig. 17.22a), the point C in contact with the floor is stationary at the instant shown—it is the instantaneous center of the disk. Therefore, the velocity of any other point is perpendicular to the line from C to the point, and its magnitude equals the product of ω and the distance from C to the point. In terms of the coordinate system given in Fig. 17.22b, the velocity of point A is

$$\mathbf{v}_A = -\sqrt{2}R\omega \cos 45°\mathbf{i} + \sqrt{2}R\omega \sin 45°\mathbf{j}$$
$$= -R\omega\mathbf{i} + R\omega\mathbf{j}.$$

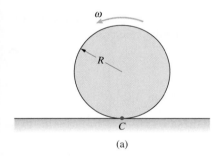

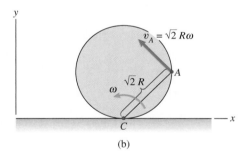

(a)

(b)

Figure 17.22
(a) Point C is the instantaneous center of the rolling disk.
(b) Determining the velocity of point A.

Study Questions

1. If a point of a rigid body is an instantaneous center, what is its velocity?
2. If you know both the angular velocity of a rigid body in planar motion and the position of the body's instantaneous center, what can you infer about the velocity of a different point of the rigid body?
3. If you know the directions of motion of two points of a rigid body in planar motion and these directions are not parallel, how can you determine the position of the instantaneous center of the body?
4. Where is the instantaneous center of a rolling object?

Example 17.4

Linkage Analysis by Instantaneous Centers

Bar AB in Fig. 17.23 rotates with a counterclockwise angular velocity of 10 rad/s. What are the angular velocities of bars BC and CD?

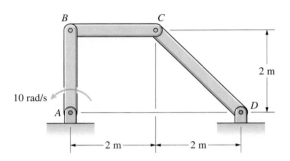

Figure 17.23

Strategy

Because bars AB and CD rotate about fixed axes, we know the directions of motion of points B and C and so can locate the instantaneous center of bar BC. Beginning with bar AB (because we know its angular velocity), we can use the instantaneous centers of the bars to determine both the velocities of the points where they are connected and their angular velocities.

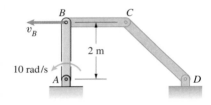

(a) Determining v_B.

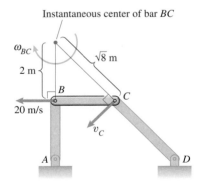

(b) Determining ω_{BC} and v_C.

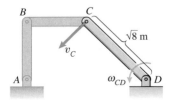

(c) Determining ω_{CD}.

Solution

The velocity of B due to the rotation of bar AB about A (Fig. a) is

$$v_B = (2 \text{ m})(10 \text{ rad/s}) = 20 \text{ m/s}.$$

Drawing lines perpendicular to the directions of motion of B and C, we locate the instantaneous center of bar BC (Fig. b). The velocity of B is equal to the product of its distance from the instantaneous center of bar BC and the angular velocity ω_{BC}:

$$v_B = 20 \text{ m/s} = (2 \text{ m})\omega_{BC}.$$

Hence, $\omega_{BC} = 10$ rad/s. (Notice that bar BC rotates in the clockwise direction.) Using the instantaneous center of bar BC and its angular velocity ω_{BC}, we can determine the velocity of point C:

$$v_C = (\sqrt{8} \text{ m})\omega_{BC} = 10\sqrt{8} \text{ m/s}.$$

Our last step is to use the velocity of point C to determine the angular velocity of bar CD about point D (Fig. c). We have

$$v_C = 10\sqrt{8} \text{ m/s} = (\sqrt{8} \text{ m})\omega_{CD},$$

so that $\omega_{CD} = 10$ rad/s counterclockwise.

Discussion

In this example, the use of instantaneous centers greatly simplified determining the angular velocities of bars BC and CD in comparison to our previous approach. However, notice that the lengths and positions of the bars made it very easy for us to locate the instantaneous center of bar BC. If the geometry is too complicated, the use of instantaneous centers can be impractical.

17.4 General Motions: Accelerations

In Chapter 18, we will be concerned with determining the motion of a rigid body when we know the external forces and couples acting on it. The governing equations are expressed in terms of the acceleration of the center of mass of the rigid body and its angular acceleration. To solve such problems, we need the relationship between the accelerations of points of a rigid body and its angular acceleration. In this section, we extend the methods we have used to analyze velocities of points of rigid bodies to accelerations.

Consider points A and B of a rigid body in planar motion relative to a given reference frame (Fig. 17.24a). Their velocities are related by

$$\mathbf{v}_A = \mathbf{v}_B + \mathbf{v}_{A/B}.$$

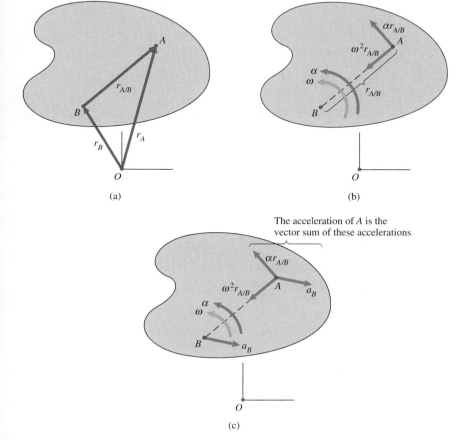

(a)

(b)

The acceleration of A is the vector sum of these accelerations

(c)

Figure 17.24
(a) Points A and B of a rigid body in planar motion and the position vector of A relative to B.
(b) Components of the acceleration of A relative to B.
(c) The acceleration of A.

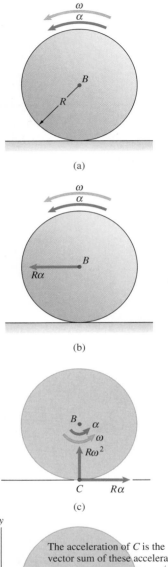

(a)

(b)

(c)

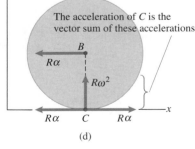

The acceleration of C is the vector sum of these accelerations

(d)

Figure 17.25
(a) A disk rolling with angular velocity ω and angular acceleration α.
(b) Acceleration of the center B.
(c) Components of the acceleration of C relative to B.
(d) The acceleration of C.

Taking the derivative of this equation with respect to time, we obtain

$$\mathbf{a}_A = \mathbf{a}_B + \mathbf{a}_{A/B},$$

where $\mathbf{a}_A$ and $\mathbf{a}_B$ are the accelerations of A and B relative to the reference frame and $\mathbf{a}_{A/B}$ is the acceleration of A relative to B. (*When we simply speak of the acceleration of a point, we will mean its acceleration relative to the given reference frame.*) Because A moves in a circular path relative to B as the rigid body rotates, $\mathbf{a}_{A/B}$ has normal and tangential components (Fig. 17.24b). The tangential component equals the product of the distance $r_{A/B} = |\mathbf{r}_{A/B}|$ and the angular acceleration α of the rigid body. The normal component points toward the center of the circular path, and its magnitude is $|\mathbf{v}_{A/B}|^2/r_{A/B} = \omega^2 r_{A/B}$. The acceleration of A equals the sum of the acceleration of B and the acceleration of A relative to B (Fig. 17.24c).

For example, let us consider a circular disk of radius R rolling on a stationary plane surface. The disk has counterclockwise angular velocity ω and counterclockwise angular acceleration α (Fig. 17.25a). The disk's center B is moving in a straight line with velocity $R\omega$, toward the left if ω is positive. Therefore, the acceleration of B is $d/dt(R\omega) = R\alpha$ and is toward the left if α is positive (Fig. 17.25b). In other words, *the magnitude of the acceleration of the center of a round object rolling on a stationary plane surface is the product of the radius and the angular acceleration.*

Now that we know the acceleration of the disk's center, let us determine the acceleration of the point C that is in contact with the surface. Relative to B, C moves in a circular path of radius R with angular velocity ω and angular acceleration α. The tangential and normal components of the acceleration of C relative to B are shown in Fig. 17.25c. The acceleration of C is the sum of the acceleration of B and the acceleration of C relative to B (Fig. 17.25d). In terms of the coordinate system shown,

$$\mathbf{a}_C = \mathbf{a}_B + \mathbf{a}_{C/B} = -R\alpha\mathbf{i} + R\alpha\mathbf{i} + R\omega^2\mathbf{j}$$
$$= R\omega^2\mathbf{j}.$$

The acceleration of point C parallel to the surface is zero, but C does have an acceleration normal to the surface.

Expressing the acceleration of a point A relative to a point B in terms of A's circular path about B as we have done is useful for visualizing and understanding the relative acceleration. However, just as we did in the case of the relative velocity, we can obtain $\mathbf{a}_{A/B}$ in a form more convenient for applications by using the angular velocity vector $\boldsymbol{\omega}$. The velocity of A relative to B is given in terms of $\boldsymbol{\omega}$ by Eq. (17.5):

$$\mathbf{v}_{A/B} = \boldsymbol{\omega} \times \mathbf{r}_{A/B}.$$

Taking the derivative of this equation with respect to time, we obtain

$$\mathbf{a}_{A/B} = \frac{d\boldsymbol{\omega}}{dt} \times \mathbf{r}_{A/B} + \boldsymbol{\omega} \times \mathbf{v}_{A/B}$$

$$= \frac{d\boldsymbol{\omega}}{dt} \times \mathbf{r}_{A/B} + \boldsymbol{\omega} \times (\boldsymbol{\omega} \times \mathbf{r}_{A/B}).$$

We next define the *angular acceleration vector* $\boldsymbol{\alpha}$ to be the rate of change of the angular velocity vector:

$$\boldsymbol{\alpha} = \frac{d\boldsymbol{\omega}}{dt}. \tag{17.7}$$

Then the acceleration of A relative to B is

$$\mathbf{a}_{A/B} = \boldsymbol{\alpha} \times \mathbf{r}_{A/B} + \boldsymbol{\omega} \times (\boldsymbol{\omega} \times \mathbf{r}_{A/B}).$$

Using this expression, we can write equations relating the velocities and accelerations of two points of a rigid body in terms of its angular velocity and angular acceleration:

$$\mathbf{v}_A = \mathbf{v}_B + \boldsymbol{\omega} \times \mathbf{r}_{A/B}, \tag{17.8}$$

$$\mathbf{a}_A = \mathbf{a}_B + \boldsymbol{\alpha} \times \mathbf{r}_{A/B} + \boldsymbol{\omega} \times (\boldsymbol{\omega} \times \mathbf{r}_{A/B}). \tag{17.9}$$

In the case of planar motion, the term $\boldsymbol{\alpha} \times \mathbf{r}_{A/B}$ in Eq. (17.9) is the tangential component of the acceleration of A relative to B, and $\boldsymbol{\omega} \times (\boldsymbol{\omega} \times \mathbf{r}_{A/B})$ is the normal component (Fig. 17.26). Therefore, for planar motion, we can write Eq. (17.9) in the simpler form

$$\mathbf{a}_A = \mathbf{a}_B + \boldsymbol{\alpha} \times \mathbf{r}_{A/B} - \omega^2 \mathbf{r}_{A/B}. \qquad \text{planar motion} \tag{17.10}$$

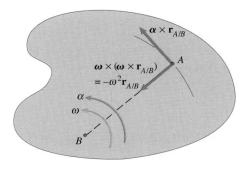

Figure 17.26
Vector components of the acceleration of A relative to B in planar motion.

To use Eq. (17.9) or Eq. (17.10) to determine accelerations of points and angular accelerations of a system of rigid bodies, it is usually necessary first to determine the angular velocities of the rigid bodies, because the angular velocity appears in Eqs. (17.9) and (17.10). But this necessary initial step provides a guide for completing the solution: Once a sequence of steps using Eq. (17.8) has been found for determining the velocities and angular velocities, the same sequence of steps using Eq. (17.9) or Eq. (17.10) will determine the accelerations and angular accelerations. (See Example 17.6.)

Study Questions

1. If a rigid body is in planar motion with angular velocity ω and angular acceleration α, what can you infer about the acceleration of a point A of the rigid body relative to a point B?
2. What is the definition of the angular acceleration vector?
3. If you know the angular velocity and angular acceleration vectors and the acceleration of one point of a rigid body, how can you determine the acceleration of a different point?

Example 17.5

Acceleration of a Point

The rolling disk in Fig. 17.27 has counterclockwise angular velocity ω and counterclockwise angular acceleration α. What is the acceleration of point A?

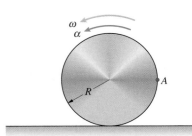

Figure 17.27

Strategy

We know that the magnitude of the acceleration of the center of the disk is the product of the radius and the angular acceleration. Therefore, we can express the acceleration of A as the sum of the acceleration of the center of the disk and the acceleration of A relative to the center. We will do so first by using vector diagrams as shown in Fig. 17.24(c) and then by using Eq. (17.10).

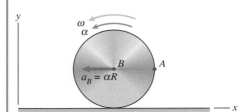

(a) Acceleration of the center of the disk.

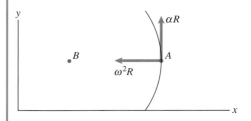

(b) Components of the acceleration of A relative to B.

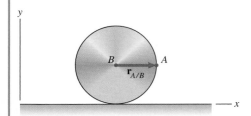

(c) Position of A relative to B.

Solution

First Method   In terms of the coordinate system in Fig. (a), the acceleration of the center B is $\mathbf{a}_B = -\alpha R \mathbf{i}$. A's motion in a circular path of radius R relative to B results in the tangential and normal components of relative acceleration shown in Fig. (b):

$$\mathbf{a}_{A/B} = -\omega^2 R \mathbf{i} + \alpha R \mathbf{j}.$$

Therefore, the acceleration of A is

$$\mathbf{a}_A = \mathbf{a}_B + \mathbf{a}_{A/B} = -\alpha R \mathbf{i} - \omega^2 R \mathbf{i} + \alpha R \mathbf{j}$$

$$= \left(-\alpha R - \omega^2 R\right)\mathbf{i} + \alpha R \mathbf{j}.$$

Second Method   The angular acceleration vector of the disk is $\boldsymbol{\alpha} = \alpha \mathbf{k}$, and the position of A relative to B is $\mathbf{r}_{A/B} = R\mathbf{i}$ (Fig. c). From Eq. (17.10), the acceleration of A is

$$\mathbf{a}_A = \mathbf{a}_B + \boldsymbol{\alpha} \times \mathbf{r}_{A/B} - \omega^2 \mathbf{r}_{A/B}$$

$$= -\alpha R \mathbf{i} + (\alpha \mathbf{k}) \times (R\mathbf{i}) - \omega^2 (R\mathbf{i})$$

$$= \left(-\alpha R - \omega^2 R\right)\mathbf{i} + \alpha R \mathbf{j}.$$

Example 17.6

Angular Accelerations of Members of a Linkage

Bar AB in Fig. 17.28 has a counterclockwise angular velocity of 10 rad/s and a clockwise angular acceleration of 300 rad/s². What are the angular accelerations of bars BC and CD?

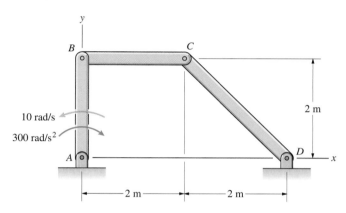

Figure 17.28

Strategy

Since we know the angular velocity of bar AB, we can determine the velocity of point B. Then we can apply Eq. (17.8) to points C and D to obtain an equation for $\mathbf{v}_C$ in terms of the angular velocity of bar CD. We can also apply Eq. (17.8) to points B and C to obtain an equation for $\mathbf{v}_C$ in terms of the angular velocity of bar BC. By equating the two expressions for $\mathbf{v}_C$, we will obtain a vector equation in two unknowns: the angular velocities of bars BC and CD. Then, by following the same sequence of steps, but using Eq. (17.10), we can obtain the angular accelerations of bars BC and CD.

Solution

The velocity of B is (Fig. a)

$$\mathbf{v}_B = \mathbf{v}_A + \boldsymbol{\omega}_{AB} \times \mathbf{r}_{B/A}$$
$$= \mathbf{0} + (10\mathbf{k}) \times (2\mathbf{j})$$
$$= -20\mathbf{i} \ (\text{m/s}).$$

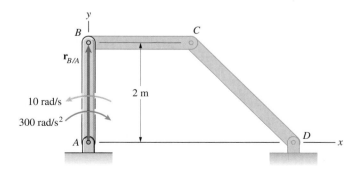

(a) Determining the motion of B.

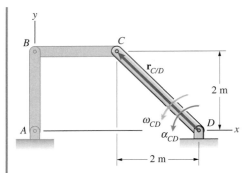

(b) Determining the motion of C in terms of the angular motion of bar CD.

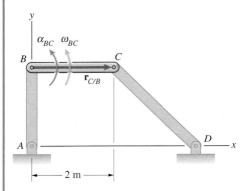

(c) Determining the motion of C in terms of the angular motion of bar BC.

Let ω_{CD} be the unknown angular velocity of bar CD (Fig. b). The velocity of C in terms of the velocity of D is

$$\mathbf{v}_C = \mathbf{v}_D + \boldsymbol{\omega}_{CD} \times \mathbf{r}_{C/D}$$

$$= \mathbf{0} + \begin{vmatrix} \mathbf{i} & \mathbf{j} & \mathbf{k} \\ 0 & 0 & \omega_{CD} \\ -2 & 2 & 0 \end{vmatrix}$$

$$= -2\omega_{CD}\mathbf{i} - 2\omega_{CD}\mathbf{j}.$$

Denoting the angular velocity of bar BC by ω_{BC} (Fig. c), we obtain the velocity of C in terms of the velocity of B:

$$\mathbf{v}_C = \mathbf{v}_B + \boldsymbol{\omega}_{BC} \times \mathbf{r}_{C/B}$$

$$= -20\mathbf{i} + (\omega_{BC}\mathbf{k}) \times (2\mathbf{i})$$

$$= -20\mathbf{i} + 2\omega_{BC}\mathbf{j}.$$

Equating our two expressions for $\mathbf{v}_C$ yields

$$-2\omega_{CD}\mathbf{i} - 2\omega_{CD}\mathbf{j} = -20\mathbf{i} + 2\omega_{BC}\mathbf{j}.$$

Equating the $\mathbf{i}$ and $\mathbf{j}$ components, we obtain $\omega_{CD} = 10 \text{ rad/s}$ and $\omega_{BC} = -10 \text{ rad/s}$.

We can use the same sequence of steps to determine the angular accelerations. The acceleration of B is (Fig. a)

$$\mathbf{a}_B = \mathbf{a}_A + \boldsymbol{\alpha}_{AB} \times \mathbf{r}_{B/A} - \omega_{AB}^2 \mathbf{r}_{B/A}$$

$$= \mathbf{0} + (-300\mathbf{k}) \times (2\mathbf{j}) - (10)^2(2\mathbf{j})$$

$$= 600\mathbf{i} - 200\mathbf{j} \, (\text{m/s}^2).$$

The acceleration of C in terms of the acceleration of D is (Fig. b)

$$\mathbf{a}_C = \mathbf{a}_D + \boldsymbol{\alpha}_{CD} \times \mathbf{r}_{C/D} - \omega_{CD}^2 \mathbf{r}_{C/D}$$

$$= \mathbf{0} + \begin{vmatrix} \mathbf{i} & \mathbf{j} & \mathbf{k} \\ 0 & 0 & \alpha_{CD} \\ -2 & 2 & 0 \end{vmatrix} - (10)^2(-2\mathbf{i} + 2\mathbf{j})$$

$$= (200 - 2\alpha_{CD})\mathbf{i} - (200 + 2\alpha_{CD})\mathbf{j}.$$

The acceleration of C in terms of the acceleration of B is (Fig. c)

$$\mathbf{a}_C = \mathbf{a}_B + \boldsymbol{\alpha}_{BC} \times \mathbf{r}_{C/B} - \omega_{BC}^2 \mathbf{r}_{C/B}$$

$$= 600\mathbf{i} - 200\mathbf{j} + (\alpha_{BC}\mathbf{k}) \times (2\mathbf{i}) - (-10)^2(2\mathbf{i})$$

$$= 400\mathbf{i} - (200 - 2\alpha_{BC})\mathbf{j}.$$

Equating the expressions for $\mathbf{a}_C$, we obtain

$$(200 - 2\alpha_{CD})\mathbf{i} - (200 + 2\alpha_{CD})\mathbf{j} = 400\mathbf{i} - (200 - 2\alpha_{BC})\mathbf{j}.$$

Equating $\mathbf{i}$ and $\mathbf{j}$ components, we obtain the angular accelerations $\alpha_{BC} = 100 \text{ rad/s}^2$ and $\alpha_{CD} = -100 \text{ rad/s}^2$.

17.5 Sliding Contacts

Figure 17.29
Linkage with a sliding contact.

In this section, we consider a type of problem that is superficially similar to those we have discussed previously in this chapter, but that actually requires a different method of solution. For example, in Fig. 17.29, pin A of the bar connected at C slides in a slot in the bar connected at B. Suppose that we know the angular velocity of the bar connected at B, and we want to determine the angular velocity of bar AC. We cannot use the equation $\mathbf{v}_A = \mathbf{v}_B + \boldsymbol{\omega} \times \mathbf{r}_{A/B}$ to express the velocity of pin A in terms of the angular velocity of the bar fixed at B, because we derived that equation under the assumption that A and B are points of the same rigid body. Here pin A moves relative to the bar connected at B as it slides along the slot. This is an example of a *sliding contact* between rigid bodies. To solve this type of problem, we must rederive Eqs. (17.8), (17.9), and (17.10) without making the assumption that A is a point of the rigid body.

To describe the motion of a point that moves relative to a given rigid body, it is convenient to use a reference frame that moves with the rigid body. We say that such a reference frame is *body-fixed*. In Fig. 17.30, we introduce a body-fixed reference frame xyz with its origin at a point B of the rigid body, in addition to the *primary reference frame* with origin O. (The primary reference frame is the reference frame relative to which we are describing the motion of the rigid body.) We do not assume A to be a point of the rigid body. The position of A relative to O is

$$\mathbf{r}_A = \mathbf{r}_B + \underbrace{x\mathbf{i} + y\mathbf{j} + z\mathbf{k}}_{\mathbf{r}_{A/B}},$$

where x, y, and z are the coordinates of A in terms of the body-fixed reference frame. Our next step is to take the derivative of this expression with respect to time in order to obtain an equation for the velocity of A. In doing

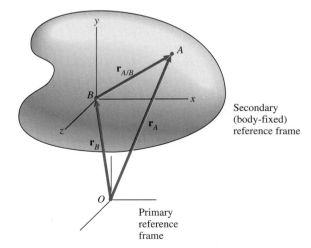

Secondary
(body-fixed)
reference frame

Primary
reference
frame

Figure 17.30
A point B of a rigid body, a body-fixed secondary reference frame, and an arbitrary point A.

so, we recognize that the unit vectors **i**, **j**, and **k** are not constant, because they rotate with the body-fixed reference frame:

$$\mathbf{v}_A = \mathbf{v}_B + \frac{dx}{dt}\mathbf{i} + x\frac{d\mathbf{i}}{dt} + \frac{dy}{dt}\mathbf{j} + y\frac{d\mathbf{j}}{dt} + \frac{dz}{dt}\mathbf{k} + z\frac{d\mathbf{k}}{dt}.$$

Now, what are the derivatives of the unit vectors? In Section 17.3, we showed that if $\mathbf{r}_{P/B}$ is the position of a point P of a rigid body relative to another point B of the same rigid body, $d\mathbf{r}_{P/B}/dt = \mathbf{v}_{P/B} = \boldsymbol{\omega} \times \mathbf{r}_{P/B}$. Since we can regard the unit vector **i** as the position vector of a point P of the rigid body (Fig. 17.31), its derivative is $d\mathbf{i}/dt = \boldsymbol{\omega} \times \mathbf{i}$. Applying the same argument to the unit vectors **j** and **k**, we obtain

$$\frac{d\mathbf{i}}{dt} = \boldsymbol{\omega} \times \mathbf{i}, \qquad \frac{d\mathbf{j}}{dt} = \boldsymbol{\omega} \times \mathbf{j}, \qquad \frac{d\mathbf{k}}{dt} = \boldsymbol{\omega} \times \mathbf{k}.$$

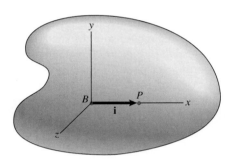

Figure 17.31
Interpreting **i** as the position vector of a point P relative to B.

Using these expressions, we can write the velocity of point A as

$$\mathbf{v}_A = \mathbf{v}_B + \underbrace{\mathbf{v}_{A\,\text{rel}} + \boldsymbol{\omega} \times \mathbf{r}_{A/B}}_{\mathbf{v}_{A/B}}, \tag{17.11}$$

where

$$\mathbf{v}_{A\,\text{rel}} = \frac{dx}{dt}\mathbf{i} + \frac{dy}{dt}\mathbf{j} + \frac{dz}{dt}\mathbf{k} \tag{17.12}$$

is the velocity of A relative to the body-fixed reference frame. That is, $\mathbf{v}_A$ is the velocity of A relative to the primary reference frame, and $\mathbf{v}_{A\,\text{rel}}$ is the velocity of A relative to the rigid body.

Equation (17.11) expresses the velocity of a point A as the sum of three terms (Fig. 17.32): the velocity of a point B of the rigid body, the velocity $\boldsymbol{\omega} \times \mathbf{r}_{A/B}$ of A relative to B due to the rotation of the rigid body, and the velocity $\mathbf{v}_{A\,\text{rel}}$ of A relative to the rigid body.

To obtain an equation for the acceleration of point A, we take the derivative of Eq. (17.11) with respect to time and use Eq. (17.12). The result is

$$\mathbf{a}_A = \mathbf{a}_B + \underbrace{\mathbf{a}_{A\,\text{rel}} + 2\boldsymbol{\omega} \times \mathbf{v}_{A\,\text{rel}} + \boldsymbol{\alpha} \times \mathbf{r}_{A/B} + \boldsymbol{\omega} \times (\boldsymbol{\omega} \times \mathbf{r}_{A/B})}_{\mathbf{a}_{A/B}}, \tag{17.13}$$

Figure 17.32
Expressing the velocity of A in terms of the velocity of a point B of the rigid body.

where

$$\mathbf{a}_{A\,\text{rel}} = \frac{d^2x}{dt^2}\mathbf{i} + \frac{d^2y}{dt^2}\mathbf{j} + \frac{d^2z}{dt^2}\mathbf{k} \tag{17.14}$$

is the acceleration of A relative to the body-fixed reference frame. That is, $\mathbf{a}_A$ is the acceleration of A relative to the primary reference frame, and $\mathbf{a}_{A\,\text{rel}}$ is the acceleration of A relative to the rigid body.

In the case of planar motion, we can express Eq. (17.13) in the simpler form

$$\mathbf{a}_A = \mathbf{a}_B + \underbrace{\mathbf{a}_{A\,\text{rel}} + 2\boldsymbol{\omega} \times \mathbf{v}_{A\,\text{rel}} + \boldsymbol{\alpha} \times \mathbf{r}_{A/B} - \omega^2\mathbf{r}_{A/B}}_{\mathbf{a}_{A/B}}. \tag{17.15}$$

In summary, $\mathbf{v}_A$ and $\mathbf{a}_A$ are the velocity and acceleration of point A relative to the primary reference frame—the reference frame relative to which the rigid body's motion is being described. The terms $\mathbf{v}_{A\,\text{rel}}$ and $\mathbf{a}_{A\,\text{rel}}$ are the velocity and acceleration of point A relative to the body-fixed reference frame. That is, they are the velocity and acceleration measured by an observer moving with the rigid body (Fig. 17.33). If A is a point of the rigid body, then $\mathbf{v}_{A\,\text{rel}}$ and $\mathbf{a}_{A\,\text{rel}}$ are zero, and Eqs. (17.11) and (17.13) are identical to Eqs. (17.8) and (17.9).

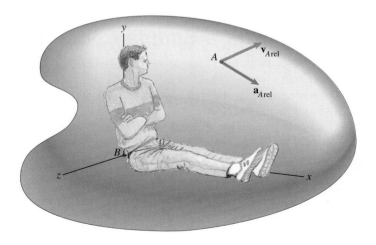

Figure 17.33
Imagine yourself to be stationary relative to the rigid body.

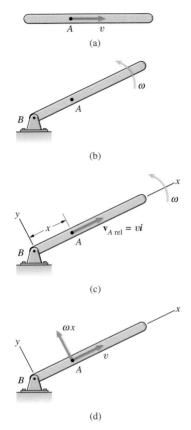

Figure 17.34
(a) A point moving along a bar.
(b) The bar is rotating.
(c) A body-fixed reference frame.
(d) Components of $\mathbf{v}_A$.

We can illustrate these concepts with a simple example. Figure 17.34(a) shows a point A moving with velocity v parallel to the axis of a bar. (Imagine that A is a bug walking along the bar.) Suppose that at the same time, the bar is rotating about a fixed point B with a constant angular velocity ω relative to an earth-fixed reference frame (Fig. 17.34b). We will use Eq. (17.11) to determine the velocity of A relative to the earth-fixed reference frame.

Let the coordinate system in Fig. 17.34(c) be fixed with respect to the bar, and let x be the present position of A. In terms of this body-fixed reference frame, the angular velocity vector of the bar (and the reference frame) relative to the primary earth-fixed reference frame is $\boldsymbol{\omega} = \omega\mathbf{k}$. Relative to the body-fixed reference frame, point A moves along the x axis with velocity v, so $\mathbf{v}_{A\,\text{rel}} = v\mathbf{i}$. From Eq. (17.11), the velocity of A relative to the earth-fixed reference frame is

$$\mathbf{v}_A = \mathbf{v}_B + \mathbf{v}_{A\,\text{rel}} + \boldsymbol{\omega} \times \mathbf{r}_{A/B}$$
$$= \mathbf{0} + v\mathbf{i} + (\omega\mathbf{k}) \times (x\mathbf{i})$$
$$= v\mathbf{i} + \omega x\mathbf{j}.$$

Relative to the earth-fixed reference frame, A has a component of velocity parallel to the bar and also a normal component due to the bar's rotation (Fig. 17.34d). Although $\mathbf{v}_A$ is the velocity of A relative to the earth-fixed reference frame, notice that it is expressed in components that are in terms of the body-fixed reference frame.

Study Questions

1. What is meant by a body-fixed reference frame?
2. How are the primary and secondary reference frames defined?
3. What is the definition of the term $\mathbf{v}_{A\,\text{rel}}$ in Eq. (17.11)?
4. What is the definition of the term $\mathbf{a}_{A\,\text{rel}}$ in Eq. (17.13)?

Example 17.7

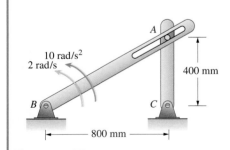

Figure 17.35

Linkage with a Sliding Contact

Bar AB in Fig. 17.35 has a counterclockwise angular velocity of 2 rad/s and a counterclockwise angular acceleration of 10 rad/s².
(a) Determine the angular velocity of bar AC and the velocity of pin A relative to the slot in bar AB.
(b) Determine the angular acceleration of bar AC and the acceleration of pin A relative to the slot in bar AB.

Strategy

By employing a secondary reference frame that is fixed with respect to the slotted bar, we can use Eq. (17.11) to express the velocity of pin A in terms of its velocity relative to the slot and the known angular velocity of bar AB. Pins A and C are both points of bar AC, so we can express $\mathbf{v}_A$ in terms of the angular velocity of bar AC in the usual way. By equating the resulting expressions for $\mathbf{v}_A$, we will obtain a vector equation in terms of the velocity of A relative to the slot and the angular velocity of bar AC. Then, by following the same sequence of steps, but this time using Eq. (17.15), we can obtain the acceleration of A relative to the slot and the angular acceleration of bar AC.

Solution

(a) Let the coordinate system in Fig. (a) be body fixed with respect to the slotted bar. Applying Eq. (17.11) to points A and B, we find that the velocity of A is

$$\mathbf{v}_A = \mathbf{v}_B + \mathbf{v}_{A\,\text{rel}} + \boldsymbol{\omega}_{AB} \times \mathbf{r}_{A/B}$$

$$= \mathbf{0} + \mathbf{v}_{A\,\text{rel}} + \begin{vmatrix} \mathbf{i} & \mathbf{j} & \mathbf{k} \\ 0 & 0 & 2 \\ 0.8 & 0.4 & 0 \end{vmatrix}.$$

The velocity of pin A *relative to the body-fixed coordinate system* is parallel to the slot (Fig. b). Therefore, we can express that velocity as

$$\mathbf{v}_{A\,\text{rel}} = v_{A\,\text{rel}} \cos\beta\,\mathbf{i} + v_{A\,\text{rel}} \sin\beta\,\mathbf{j},$$

where $\beta = \arctan(0.4/0.8)$. Substituting this expression into our equation for $\mathbf{v}_A$, we obtain

$$\mathbf{v}_A = \left(v_{A\,\text{rel}}\cos\beta - 0.8\right)\mathbf{i} + \left(v_{A\,\text{rel}}\sin\beta + 1.6\right)\mathbf{j}.$$

Let ω_{AC} be the angular velocity of bar AC (Fig. c). Expressing the velocity of A in terms of the velocity of C, we obtain

$$\mathbf{v}_A = \mathbf{v}_C + \boldsymbol{\omega}_{AC} \times \mathbf{r}_{A/C}$$

$$= \mathbf{0} + \left(\omega_{AC}\mathbf{k}\right) \times (0.4\mathbf{j})$$

$$= -0.4\omega_{AC}\mathbf{i}.$$

Notice that there is no relative velocity term in this equation, because A is a point of bar AC. Equating our two expressions for $\mathbf{v}_A$, we obtain

$$\left(v_{A\,\text{rel}}\cos\beta - 0.8\right)\mathbf{i} + \left(v_{A\,\text{rel}}\sin\beta + 1.6\right)\mathbf{j} = -0.4\omega_{AC}\mathbf{i}.$$

Equating $\mathbf{i}$ and $\mathbf{j}$ components yields the two equations

$$v_{A\,\text{rel}}\cos\beta - 0.8 = -0.4\omega_{AC}$$

and

$$v_{A\,\text{rel}}\sin\beta + 1.6 = 0.$$

Solving these equations, we obtain $v_{A\,\text{rel}} = -3.58$ m/s and $\omega_{AC} = 10$ rad/s. At this instant, pin A is moving relative to the slot at 3.58 m/s toward B. The vector

$$\mathbf{v}_{A\,\text{rel}} = -3.58(\cos\beta\,\mathbf{i} + \sin\beta\,\mathbf{j}) = -3.2\mathbf{i} - 1.6\mathbf{j}\ (\text{m/s}).$$

(b) Applying Eq. (17.15) to bar AB (Fig. b) yields the acceleration of A:

$$\mathbf{a}_A = \mathbf{a}_B + \mathbf{a}_{A\,\text{rel}} + 2\boldsymbol{\omega}_{AB} \times \mathbf{v}_{A\,\text{rel}} + \boldsymbol{\alpha}_{AB} \times \mathbf{r}_{A/B} - \omega_{AB}^2\mathbf{r}_{A/B}$$

$$= \mathbf{0} + \mathbf{a}_{A\,\text{rel}} + 2\begin{vmatrix} \mathbf{i} & \mathbf{j} & \mathbf{k} \\ 0 & 0 & 2 \\ -3.2 & -1.6 & 0 \end{vmatrix} + \begin{vmatrix} \mathbf{i} & \mathbf{j} & \mathbf{k} \\ 0 & 0 & 10 \\ 0.8 & 0.4 & 0 \end{vmatrix}$$

$$-(2)^2(0.8\mathbf{i} + 0.4\mathbf{j}).$$

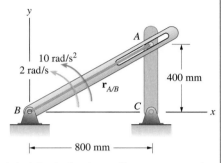

(a) A body-fixed coordinate system and the position vector of pin A relative to B.

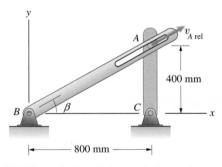

(b) The velocity of pin A relative to the body-fixed coordinate system.

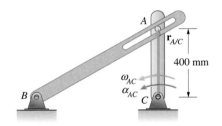

(c) The position vector of A relative to C.

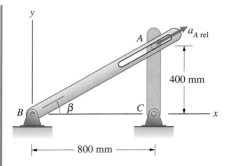

(d) The acceleration of pin A relative to the body-fixed coordinate system.

The acceleration of A relative to the body-fixed coordinate system is parallel to the slot (Fig. d), so we can write it in the same way we did $\mathbf{v}_{A\,\text{rel}}$:

$$\mathbf{a}_{A\,\text{rel}} = a_{A\,\text{rel}} \cos\beta\,\mathbf{i} + a_{A\,\text{rel}} \sin\beta\,\mathbf{j}.$$

Substituting this expression into our equation for $\mathbf{a}_A$ gives

$$\mathbf{a}_A = (a_{A\,\text{rel}} \cos\beta - 0.8)\mathbf{i} + (a_{A\,\text{rel}} \sin\beta - 6.4)\mathbf{j}.$$

Expressing the acceleration of A in terms of the acceleration of C (Fig. c), we obtain

$$
\begin{aligned}
\mathbf{a}_A &= \mathbf{a}_C + \boldsymbol{\alpha}_{AC} \times \mathbf{r}_{A/C} - \omega_{AC}^2 \mathbf{r}_{A/C} \\
&= \mathbf{0} + (\alpha_{AC}\mathbf{k}) \times (0.4\mathbf{j}) - (10)^2(0.4\mathbf{j}) \\
&= -0.4\alpha_{AC}\mathbf{i} - 40\mathbf{j}.
\end{aligned}
$$

Equating our two expressions for $\mathbf{a}_A$, we get

$$(a_{A\,\text{rel}} \cos\beta - 0.8)\mathbf{i} + (a_{A\,\text{rel}} \sin\beta - 6.4)\mathbf{j} = -0.4\alpha_{AC}\mathbf{i} - 40\mathbf{j}.$$

Equating $\mathbf{i}$ and $\mathbf{j}$ components yields the equations

$$a_{A\,\text{rel}} \cos\beta - 0.8 = -0.4\alpha_{AC}$$

and

$$a_{A\,\text{rel}} \sin\beta - 6.4 = -40.$$

Solving these equations, we obtain $a_{A\,\text{rel}} = -75.1 \text{ m/s}^2$ and $\alpha_{AC} = 170 \text{ rad/s}^2$. At the given instant, pin A is accelerating relative to the slot at 75.1 m/s² toward B.

Example 17.8

Bar Sliding Relative to a Support

The collar at B in Fig. 17.36 slides along the circular bar, causing pin B to move at constant speed v_0 in a circular path of radius R. Bar BC slides in the collar at A. At the instant shown, determine the angular velocity and angular acceleration of bar BC.

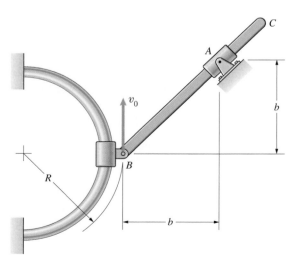

Figure 17.36

Strategy

We will use a secondary reference frame with its origin at B that is body fixed with respect to bar BC. By using Eqs. (17.11) and (17.15) to express the velocity and acceleration of the stationary pin A in terms of the velocity and acceleration of pin B, we can determine the angular velocity and angular acceleration of bar BC.

Solution

Angular Velocity Let the angular velocity and angular acceleration of bar BC, which are also the angular velocity and angular acceleration of the body-fixed coordinate system, be ω_{BC} and α_{BC} (Fig. a). The velocity of the stationary pin A is zero. From Eq. (17.11),

$$\mathbf{v}_A = \mathbf{0} = \mathbf{v}_B + \mathbf{v}_{A\,\text{rel}} + \boldsymbol{\omega} \times \mathbf{r}_{A/B}, \qquad (17.16)$$

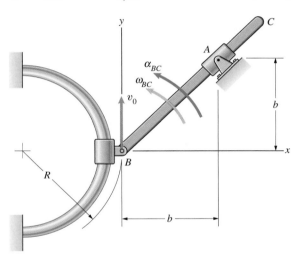

(a) A reference frame fixed with respect to bar BC.

where $\mathbf{v}_{A\,\text{rel}}$ is the velocity of A relative to the body-fixed coordinate system and $\boldsymbol{\omega} = \omega_{BC}\mathbf{k}$ is the angular velocity vector of the coordinate system. The velocity of pin B is $\mathbf{v}_B = v_0\mathbf{j}$. The velocity of the stationary pin A relative to the body-fixed coordinate system is parallel to the bar (Fig. b), so we can express it in the form

$$\mathbf{v}_{A\,\text{rel}} = v_{A\,\text{rel}} \cos 45°\mathbf{i} + v_{A\,\text{rel}} \sin 45°\mathbf{j}$$

and write Eq. (17.16) as

$$\mathbf{0} = v_0\mathbf{j} + v_{A\,\text{rel}} \cos 45°\mathbf{i} + v_{A\,\text{rel}} \sin 45°\mathbf{j}$$

$$+ \begin{vmatrix} \mathbf{i} & \mathbf{j} & \mathbf{k} \\ 0 & 0 & \omega_{BC} \\ b & b & 0 \end{vmatrix}.$$

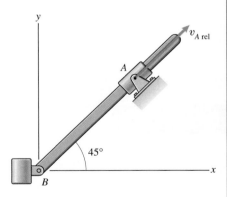

(b) Direction of the velocity of the fixed pin A relative to the body-fixed coordinate system.

From the $\mathbf{i}$ and $\mathbf{j}$ components of this equation, we obtain

$$v_{A\,\text{rel}} \cos 45° - b\omega_{BC} = 0,$$

$$v_0 + v_{A\,\text{rel}} \sin 45° + b\omega_{BC} = 0.$$

Solving these equations, we determine that the velocity of pin A relative to the body-fixed coordinate system is

$$\mathbf{v}_{A\,rel} = v_{A\,rel}\cos 45°\mathbf{i} + v_{A\,rel}\sin 45°\mathbf{j}$$

$$= -\frac{v_0}{2}\mathbf{i} - \frac{v_0}{2}\mathbf{j}$$

and the angular velocity of bar BC is

$$\omega_{BC} = -\frac{v_0}{2b}.$$

Angular Acceleration   The acceleration of pin A is zero. From Eq. (17.15),

$$\mathbf{a}_A = \mathbf{0} = \mathbf{a}_B + \mathbf{a}_{A\,rel} + 2\boldsymbol{\omega} \times \mathbf{v}_{A\,rel} + \boldsymbol{\alpha} \times \mathbf{r}_{A/B} - \omega^2\mathbf{r}_{A/B}. \qquad (17.17)$$

The acceleration of pin B is $\mathbf{a}_B = -(v_0^2/R)\mathbf{i}$. The acceleration of pin A relative to the body-fixed coordinate system is parallel to the bar (Fig. c). We can therefore express it as

$$\mathbf{a}_{A\,rel} = a_{A\,rel}\cos 45°\mathbf{i} + a_{A\,rel}\sin 45°\mathbf{j}$$

and write Eq. (17.17) as

$$\mathbf{0} = -\frac{v_0^2}{R}\mathbf{i} + a_{A\,rel}\cos 45°\mathbf{i} + a_{A\,rel}\sin 45°\mathbf{j}$$

$$+ 2\begin{vmatrix} \mathbf{i} & \mathbf{j} & \mathbf{k} \\ 0 & 0 & \omega_{BC} \\ -v_0/2 & -v_0/2 & 0 \end{vmatrix} + \begin{vmatrix} \mathbf{i} & \mathbf{j} & \mathbf{k} \\ 0 & 0 & \alpha_{BC} \\ b & b & 0 \end{vmatrix}$$

$$- \omega_{BC}^2(b\mathbf{i} + b\mathbf{j}).$$

From the $\mathbf{i}$ and $\mathbf{j}$ components of this equation, we obtain

$$-\frac{v_0^2}{R} + a_{A\,rel}\cos 45° + v_0\omega_{BC} - b\alpha_{BC} - b\omega_{BC}^2 = 0$$

and

$$a_{A\,rel}\sin 45° - v_0\omega_{BC} + b\alpha_{BC} - b\omega_{BC}^2 = 0.$$

Solving these equations, we determine that the angular acceleration of bar BC is

$$\alpha_{BC} = -\frac{v_0^2}{2b}\left(\frac{1}{R} + \frac{1}{b}\right).$$

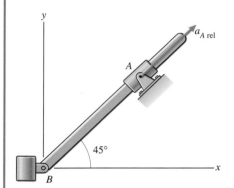

(c) Direction of the acceleration of the fixed pin A relative to the body-fixed coordinate system.

Example 17.9

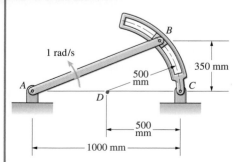

Figure 17.37

Analysis of a Sliding Contact

Bar AB in Fig. 17.37 rotates with a constant counterclockwise angular velocity of 1 rad/s. Block B slides in a circular slot in the curved bar BC. At the instant shown, the center of the circular slot is at D. Determine the angular velocity and angular acceleration of bar BC.

Strategy

Since we know the angular velocity of bar AB, we can determine the velocity of point B. Because B is not a point of bar BC, we must apply Eq. (17.11) to points B and C. By equating our expressions for $\mathbf{v}_B$, we can solve for the angular velocity of bar BC. Then, by following the same sequence of steps, but this time using Eq. (17.15), we can determine the angular acceleration of bar BC.

Solution

To determine the velocity of B, we express it in terms of the velocity of A and the angular velocity of bar AB: $\mathbf{v}_B = \mathbf{v}_A + \boldsymbol{\omega}_{AB} \times \mathbf{r}_{B/A}$. In terms of the coordinate system shown in Fig. (a), the position vector of B relative to A is

$$\mathbf{r}_{B/A} = (0.500 + 0.500 \cos\beta)\mathbf{i} + 0.350\mathbf{j} = 0.857\mathbf{i} + 0.350\mathbf{j} \text{ (m)},$$

where $\beta = \arcsin(350/500) = 44.4°$. Therefore, the velocity of B is

$$\mathbf{v}_B = \mathbf{v}_A + \boldsymbol{\omega}_{AB} \times \mathbf{r}_{B/A} = \mathbf{0} + \begin{vmatrix} \mathbf{i} & \mathbf{j} & \mathbf{k} \\ 0 & 0 & 1 \\ 0.857 & 0.350 & 0 \end{vmatrix} \quad (17.18)$$

$$= -0.350\mathbf{i} + 0.857\mathbf{j} \text{ (m/s)}.$$

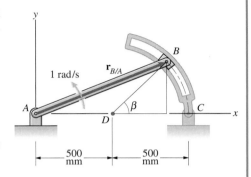

(a) Determining the velocity of point B.

To apply Eq. (17.11) to points B and C, we introduce a parallel secondary coordinate system that rotates with the curved bar (Fig. b). The velocity of B is

$$\mathbf{v}_B = \mathbf{v}_C + \mathbf{v}_{B\,\text{rel}} + \boldsymbol{\omega}_{BC} \times \mathbf{r}_{B/C}. \quad (17.19)$$

The position vector of B relative to C is

$$\mathbf{r}_{B/C} = -(0.500 - 0.500 \cos\beta)\mathbf{i} + 0.350\mathbf{j} = -0.143\mathbf{i} + 0.350\mathbf{j} \text{ (m)}.$$

Relative to the body-fixed coordinate system, point B moves in a circular path about point D (Fig. c). In terms of the angle β, the vector

$$\mathbf{v}_{B\,\text{rel}} = -v_{B\,\text{rel}} \sin\beta\mathbf{i} + v_{B\,\text{rel}} \cos\beta\mathbf{j}.$$

We substitute the preceding expressions for $\mathbf{r}_{B/C}$ and $\mathbf{v}_{B\,\text{rel}}$ into Eq. (17.19), obtaining

$$\mathbf{v}_B = -v_{B\,\text{rel}} \sin\beta\mathbf{i} + v_{B\,\text{rel}} \cos\beta\mathbf{j} + \begin{vmatrix} \mathbf{i} & \mathbf{j} & \mathbf{k} \\ 0 & 0 & \omega_{BC} \\ -0.143 & 0.350 & 0 \end{vmatrix}.$$

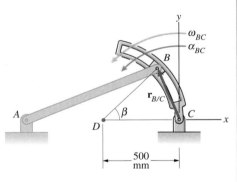

(b) A coordinate system fixed with respect to the curved bar.

Equating this expression for $\mathbf{v}_B$ to its value given in Eq. (17.18) yields the two equations

$$-v_{B\,\text{rel}} \sin\beta - 0.350\omega_{BC} = -0.350$$

and

$$v_{B\,\text{rel}} \cos\beta - 0.143\omega_{BC} = 0.857.$$

Solving these equations, we obtain $v_{B\,\text{rel}} = 1.0$ m/s and $\omega_{BC} = -1.0$ rad/s.

We follow the same sequence of steps to determine the angular acceleration of bar BC. The acceleration of point B is

$$\mathbf{a}_B = \mathbf{a}_A + \boldsymbol{\alpha}_{AB} \times \mathbf{r}_{B/A} - \omega_{AB}^2 \mathbf{r}_{B/A}$$

$$= \mathbf{0} + \mathbf{0} - (1)^2(0.857\mathbf{i} + 0.350\mathbf{j})$$

$$= -0.857\mathbf{i} - 0.350\mathbf{j} \text{ (m/s}^2). \quad (17.20)$$

Because the motion of point B relative to the body-fixed coordinate system is a circular path about point D, there is a tangential component of acceleration, which we denote a_{Bt}, and a normal component of acceleration $v_{B\,\text{rel}}^2/(0.5 \text{ m})$.

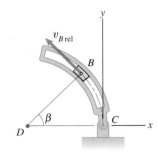

(c) The velocity of B relative to the body-fixed coordinate system.

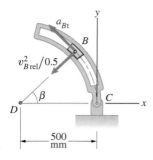

(d) Acceleration of B relative to the body-fixed coordinate system.

These components are shown in Fig. (d). In terms of the angle β, the vector

$$\mathbf{a}_{B\,\text{rel}} = -a_{Bt} \sin\beta\,\mathbf{i} + a_{Bt} \cos\beta\,\mathbf{j}$$
$$- \left(v_{B\,\text{rel}}^2/0.5\right) \cos\beta\,\mathbf{i} - \left(v_{B\,\text{rel}}^2/0.5\right) \sin\beta\,\mathbf{j}.$$

Applying Eq. (17.15) to points B and C, we obtain the acceleration of B:

$$\mathbf{a}_B = \mathbf{a}_C + \mathbf{a}_{B\,\text{rel}} + 2\boldsymbol{\omega}_{BC} \times \mathbf{v}_{B\,\text{rel}}$$
$$+ \boldsymbol{\alpha}_{BC} \times \mathbf{r}_{B/C} - \omega_{BC}^2 \mathbf{r}_{B/C}$$
$$= \mathbf{0} - a_{Bt} \sin\beta\,\mathbf{i} + a_{Bt} \cos\beta\,\mathbf{j}$$
$$- \left[(1)^2/0.5\right] \cos\beta\,\mathbf{i} - \left[(1)^2/0.5\right] \sin\beta\,\mathbf{j}$$
$$+ 2 \begin{vmatrix} \mathbf{i} & \mathbf{j} & \mathbf{k} \\ 0 & 0 & -1 \\ -(1)\sin\beta & (1)\cos\beta & 0 \end{vmatrix}$$
$$+ \begin{vmatrix} \mathbf{i} & \mathbf{j} & \mathbf{k} \\ 0 & 0 & \alpha_{BC} \\ -0.143 & 0.350 & 0 \end{vmatrix} - (-1)^2(-0.143\mathbf{i} + 0.350\mathbf{j}).$$

Equating this expression for $\mathbf{a}_B$ to its value given in Eq. (17.20) yields the two equations

$$-a_{Bt} \sin\beta - 0.350\alpha_{BC} + 0.143 = -0.857,$$
$$a_{Bt} \cos\beta - 0.143\alpha_{BC} - 0.350 = -0.350.$$

Solving these equations, we obtain $a_{Bt} = 0.408 \text{ m/s}^2$ and $\alpha_{BC} = 2.040 \text{ rad/s}^2$.

17.6 Moving Reference Frames

In this section, we revisit the subjects of Chapters 13 and 14—the motion of a point and Newton's second law. In many situations, it is convenient to describe the motion of a point by using a secondary reference frame that moves relative to some primary reference frame. For example, to measure the motion of a point relative to a moving vehicle, we would choose a secondary reference frame that is fixed with respect to the vehicle. Here we show how the velocity and acceleration of a point relative to a primary reference frame are related to their values relative to a moving secondary reference frame. We also discuss how to apply Newton's second law using moving reference frames. In Chapter 14, we mentioned the example of playing tennis on the deck of a cruise ship. If the ship translates with constant velocity, we can use the equation $\Sigma\mathbf{F} = m\mathbf{a}$ expressed in terms of a reference frame fixed with respect to the ship to analyze the ball's motion. We cannot do so if the ship is turning or changing its speed. However, we can apply the second law using reference frames that accelerate and rotate relative to an inertial reference frame by properly accounting for the acceleration and rotation. In what follows, we explain how this is done.

Motion of a Point Relative to a Moving Reference Frame

Equations (17.11) and (17.13) give the velocity and acceleration of an arbitrary point A relative to a point B of a rigid body in terms of a body-fixed secondary reference frame:

$$\mathbf{v}_A = \mathbf{v}_B + \mathbf{v}_{A\,\text{rel}} + \boldsymbol{\omega} \times \mathbf{r}_{A/B}, \tag{17.21}$$

$$\mathbf{a}_A = \mathbf{a}_B + \mathbf{a}_{A\,\text{rel}} + 2\boldsymbol{\omega} \times \mathbf{v}_{A\,\text{rel}} + \boldsymbol{\alpha} \times \mathbf{r}_{A/B} + \boldsymbol{\omega} \times (\boldsymbol{\omega} \times \mathbf{r}_{A/B}). \tag{17.22}$$

But these equations do not require us to assume that the secondary reference frame is connected to some rigid body. They apply to any reference frame having a moving origin B and rotating with angular velocity $\boldsymbol{\omega}$ and angular acceleration $\boldsymbol{\alpha}$ relative to a primary reference frame (Fig. 17.38). The terms $\mathbf{v}_A$ and $\mathbf{a}_A$ are the velocity and acceleration of A relative to the primary reference frame. The terms $\mathbf{v}_{A\,\text{rel}}$ and $\mathbf{a}_{A\,\text{rel}}$ are the velocity and acceleration of A relative to the secondary reference frame. That is, they are the velocity and acceleration measured by an observer moving with the secondary reference frame (Fig. 17.39).

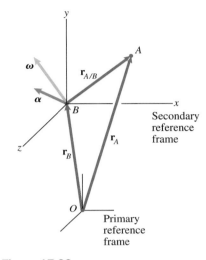

Figure 17.38
A secondary reference frame with origin B and an arbitrary point A.

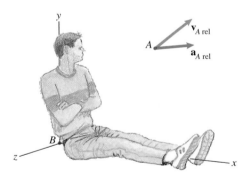

Figure 17.39
Imagine yourself to be stationary relative to the secondary reference frame.

When the velocity and acceleration of a point A relative to a moving secondary reference frame are known, we can use Eqs. (17.21) and (17.22) to determine the velocity and acceleration of A relative to the primary reference frame. There will also be situations in which the velocity and acceleration of A relative to the primary reference frame will be known and we will want to use Eqs. (17.21) and (17.22) to determine the velocity and acceleration of A relative to a moving secondary reference frame.

Example 17.10

A Rotating Secondary Reference Frame

The merry-go-round in Fig. 17.40 rotates with constant counterclockwise angular velocity ω. Suppose that you are in the center at B and observe the motion of a second person A, using a coordinate system that rotates with the merry-go-round. Consider two cases.

Case 1   Person A is not on the merry-go-round, but stands on the ground next to it. At the instant shown, what are his velocity and acceleration relative to your coordinate system?

Case 2   Person A is on the edge of the merry-go-round and moves with it. What are his velocity and acceleration relative to the earth?

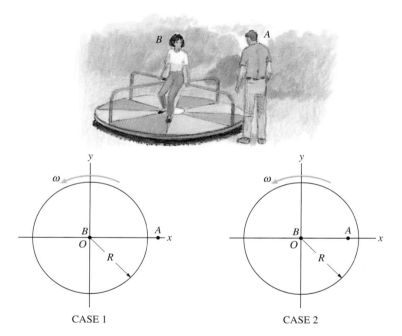

CASE 1   CASE 2

Figure 17.40

Strategy

This simple example clarifies the distinction between the terms $\mathbf{v}_A$, $\mathbf{a}_A$ and the terms $\mathbf{v}_{A\,\text{rel}}$, $\mathbf{a}_{A\,\text{rel}}$ in Eqs. (17.21) and (17.22). We choose the coordinate system that rotates with the merry-go-round as the secondary reference frame and let the primary reference frame be fixed with respect to the earth. In case 1, A's velocity and acceleration relative to the earth, $\mathbf{v}_A$ and $\mathbf{a}_A$, are known: He is standing still. We can use Eqs. (17.21) and (17.22) to determine $\mathbf{v}_{A\,\text{rel}}$ and $\mathbf{a}_{A\,\text{rel}}$, which are his velocity and acceleration relative to your rotating coordinate system. In case 2, $\mathbf{v}_{A\,\text{rel}}$ and $\mathbf{a}_{A\,\text{rel}}$ are known: A is stationary relative to your coordinate system. We can use Eqs. (17.21) and (17.22) to determine $\mathbf{v}_A$ and $\mathbf{a}_A$.

Solution

Case 1 A is standing on the ground, so his velocity relative to the earth is $\mathbf{v}_A = \mathbf{0}$. The angular velocity vector of your coordinate system is $\boldsymbol{\omega} = \omega\mathbf{k}$, and at the instant shown $\mathbf{r}_{A/B} = R\mathbf{i}$. From Eq. (17.21),

$$\mathbf{v}_A = \mathbf{v}_B + \mathbf{v}_{A\,\text{rel}} + \boldsymbol{\omega} \times \mathbf{r}_{A/B}:$$

$$\mathbf{0} = \mathbf{0} + \mathbf{v}_{A\,\text{rel}} + (\omega\mathbf{k}) \times (R\mathbf{i}).$$

We find that $\mathbf{v}_{A\,\text{rel}} = -\omega R\mathbf{j}$. Although A is stationary relative to the earth, $\mathbf{v}_{A\,\text{rel}}$ is not zero. What does this term represent? As you sit at the center of the merry-go-round, you see A moving around you in a circular path. *Relative to your rotating coordinate system*, A moves in a circular path of radius R in the clockwise direction with a velocity of constant magnitude ωR. At the instant shown, A's velocity relative to your coordinate system is $-\omega R\mathbf{j}$.

You know that a point moving in a circular path of radius R with velocity v has a normal component of acceleration equal to v^2/R. Relative to your coordinate system, person A moves in a circular path of radius R with velocity ωR. Therefore, *relative to your coordinate system*, A has a normal component of acceleration $(\omega R)^2/R = \omega^2 R$. At the instant shown, the normal acceleration points in the negative x direction. We conclude that A's acceleration relative to your coordinate system is $\mathbf{a}_{A\,\text{rel}} = -\omega^2 R\mathbf{i}$.

We can confirm this result with Eq. (17.22). A's acceleration relative to the earth is $\mathbf{a}_A = \mathbf{0}$. The angular velocity vector of the coordinate system is constant, so $\boldsymbol{\alpha} = \mathbf{0}$. From Eq. (17.22),

$$\mathbf{a}_A = \mathbf{a}_B + \mathbf{a}_{A\,\text{rel}} + 2\boldsymbol{\omega} \times \mathbf{v}_{A\,\text{rel}} + \boldsymbol{\alpha} \times \mathbf{r}_{A/B} + \boldsymbol{\omega} \times (\boldsymbol{\omega} \times \mathbf{r}_{A/B}):$$

$$\mathbf{0} = \mathbf{0} + \mathbf{a}_{A\,\text{rel}} + 2(\omega\mathbf{k}) \times (-\omega R\mathbf{j}) + \mathbf{0} + (\omega\mathbf{k}) \times \left[(\omega\mathbf{k}) \times (R\mathbf{i})\right].$$

Solving this equation for $\mathbf{a}_{A\,\text{rel}}$, we obtain $\mathbf{a}_{A\,\text{rel}} = -\omega^2 R\mathbf{i}$. A's velocity and acceleration relative to your coordinate system are shown in Fig. (a).

Case 2   *Relative to your coordinate system*, A is stationary, so $\mathbf{v}_{A\,\text{rel}} = \mathbf{0}$ and $\mathbf{a}_{A\,\text{rel}} = \mathbf{0}$. From Eq. (17.21), A's velocity relative to the earth is

$$\mathbf{v}_A = \mathbf{v}_B + \mathbf{v}_{A\,\text{rel}} + \boldsymbol{\omega} \times \mathbf{r}_{A/B} = \mathbf{0} + \mathbf{0} + (\omega\mathbf{k}) \times (R\mathbf{i})$$

$$= \omega R\mathbf{j}.$$

In this case, A is moving in a circular path of radius R with a velocity of constant magnitude ωR relative to the earth.

From Eq. (17.22), A's acceleration relative to the earth is

$$\mathbf{a}_A = \mathbf{a}_B + \mathbf{a}_{A\,\text{rel}} + 2\boldsymbol{\omega} \times \mathbf{v}_{A\,\text{rel}} + \boldsymbol{\alpha} \times \mathbf{r}_{A/B} + \boldsymbol{\omega} \times (\boldsymbol{\omega} \times \mathbf{r}_{A/B})$$

$$= \mathbf{0} + \mathbf{0} + \mathbf{0} + \mathbf{0} + (\omega\mathbf{k}) \times \left[(\omega\mathbf{k}) \times (R\mathbf{i})\right]$$

$$= -\omega^2 R\mathbf{i}.$$

This is A's acceleration relative to the earth due to his circular motion. A's velocity and acceleration relative to the earth are shown in Fig. (b).

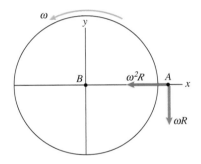

(a) The velocity and acceleration of A relative to the rotating coordinate system in case 1.

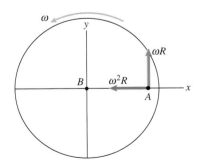

(b) The velocity and acceleration of A relative to the earth in case 2.

Example 17.11

A Reference Frame Fixed with Respect to a Ship

At the instant shown, the ship B in Fig. 17.41 is moving north at a constant speed of 15.0 m/s relative to the earth and is turning toward the west at a constant rate of 5.0° per second. Relative to the ship's body-fixed coordinate system, its radar indicates that the position, velocity, and acceleration of the helicopter A are

$$\mathbf{r}_{A/B} = 420.0\mathbf{i} + 236.2\mathbf{j} + 212.0\mathbf{k} \ (\text{m}),$$
$$\mathbf{v}_{A\,\text{rel}} = -53.5\mathbf{i} + 2.0\mathbf{j} + 6.6\mathbf{k} \ (\text{m/s}),$$

and

$$\mathbf{a}_{A\,\text{rel}} = 0.4\mathbf{i} - 0.2\mathbf{j} - 13.0\mathbf{k} \ (\text{m/s}^2).$$

What are the helicopter's velocity and acceleration relative to the earth?

Figure 17.41

Strategy

We are given the ship's velocity and enough information to determine its acceleration, angular velocity, and angular acceleration relative to the earth. We also know the position, velocity, and acceleration of the helicopter relative to the secondary body-fixed coordinate system. Therefore, we can use Eqs. (17.21) and (17.22) to determine the helicopter's velocity and acceleration relative to the earth.

Solution

In terms of the body-fixed coordinate system, the ship's velocity is $\mathbf{v}_B = 15.0\mathbf{i}$ (m/s). The ship's angular velocity due to its rate of turning is $\omega = (5.0/180)\pi = 0.0873$ rad/s. The ship is rotating about the y axis. Pointing the arc of the fingers of the right hand around the y axis in the direction of the ship's rotation, we find that the thumb points in the positive y direction, so the ship's angular velocity vector is $\boldsymbol{\omega} = 0.0873\mathbf{j}$ (rad/s). The helicopter's velocity relative to the earth is

$$\mathbf{v}_A = \mathbf{v}_B + \mathbf{v}_{A\,\text{rel}} + \boldsymbol{\omega} \times \mathbf{r}_{A/B}$$

$$= 15.0\mathbf{i} + (-53.5\mathbf{i} + 2.0\mathbf{j} + 6.6\mathbf{k}) + \begin{vmatrix} \mathbf{i} & \mathbf{j} & \mathbf{k} \\ 0 & 0.0873 & 0 \\ 420.0 & 236.2 & 212.0 \end{vmatrix}$$

$$= -20.0\mathbf{i} + 2.0\mathbf{j} - 30.1\mathbf{k} \ (\text{m/s}).$$

We can determine the ship's acceleration by expressing it in terms of normal and tangential components in the form given by Eq. (13.37) (Fig. a):

$$\mathbf{a}_B = \frac{dv}{dt}\mathbf{e}_t + v\frac{d\theta}{dt}\mathbf{e}_n = \mathbf{0} + (15)(0.0873)\mathbf{e}_n$$

$$= 1.31\mathbf{e}_n \ (\text{m/s}^2).$$

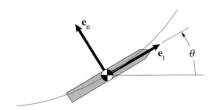

(a) Determining the ship's acceleration.

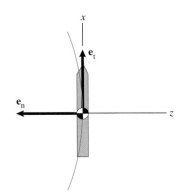

(b) Correspondence between the normal and tangential components and the body-fixed coordinate system.

The z axis is perpendicular to the ship's path and points toward the convex side of the path (Fig. b). Therefore, in terms of the body-fixed coordinate system, the ship's acceleration is $\mathbf{a}_B = -1.31\mathbf{k} \ (\text{m/s}^2)$. The ship's angular velocity vector is constant, so $\boldsymbol{\alpha} = \mathbf{0}$. The helicopter's acceleration relative to the earth is

$$\mathbf{a}_A = \mathbf{a}_B + \mathbf{a}_{A\,\text{rel}} + 2\boldsymbol{\omega} \times \mathbf{v}_{A\,\text{rel}} + \boldsymbol{\alpha} \times \mathbf{r}_{A/B}$$
$$+ \boldsymbol{\omega} \times (\boldsymbol{\omega} \times \mathbf{r}_{A/B})$$

$$= -1.31\mathbf{k} + (0.4\mathbf{i} - 0.2\mathbf{j} - 13.0\mathbf{k}) + 2\begin{vmatrix} \mathbf{i} & \mathbf{j} & \mathbf{k} \\ 0 & 0.0873 & 0 \\ -53.5 & 2.0 & 6.6 \end{vmatrix}$$

$$+ \mathbf{0} + (0.0873\mathbf{j}) \times \begin{vmatrix} \mathbf{i} & \mathbf{j} & \mathbf{k} \\ 0 & 0.0873 & 0 \\ 420.0 & 236.2 & 212.0 \end{vmatrix}$$

$$= -1.65\mathbf{i} - 0.20\mathbf{j} - 6.59\mathbf{k} \ (\text{m/s}^2).$$

Discussion

Notice the substantial differences between the helicopter's velocity and acceleration relative to the earth and the values the ship measures using its body-fixed coordinate system.

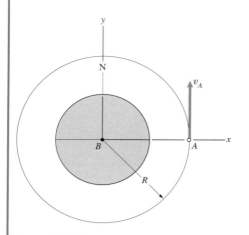

Figure 17.42

Example 17.12

An Earth-Fixed Reference Frame

The satellite A shown in Fig. 17.42 is in a circular polar orbit (an orbit that intersects the earth's axis of rotation). Relative to a nonrotating primary reference frame with its origin at the center of the earth, the satellite moves in a circular path of radius R with a velocity of constant magnitude v_A. At the present instant, the satellite is above the equator. The secondary earth-fixed reference frame shown is oriented with the y axis in the direction of the north pole and the x axis in the direction of the satellite. What are the satellite's velocity and acceleration relative to the earth-fixed reference frame? Let ω_E be the angular velocity of the earth.

Strategy

We are given enough information to determine the satellite's velocity and acceleration $\mathbf{v}_A$ and $\mathbf{a}_A$ relative to the nonrotating primary reference frame and the angular velocity vector $\boldsymbol{\omega}$ of the secondary reference frame. We can therefore use Eqs. (17.21) and (17.22) to determine the satellite's velocity and acceleration $\mathbf{v}_{A\,\text{rel}}$ and $\mathbf{a}_{A\,\text{rel}}$ relative to the earth-fixed reference frame.

Solution

At the present instant, the satellite's velocity and acceleration relative to a nonrotating primary reference frame with its origin at the center of the earth are $\mathbf{v}_A = v_A\mathbf{j}$ and $\mathbf{a}_A = -(v_A^2/R)\mathbf{i}$. The angular velocity vector of the earth points north (confirm this by using the right-hand rule), so the angular velocity of the earth-fixed reference frame is $\boldsymbol{\omega} = \omega_E\mathbf{j}$. From Eq. (17.21),

$$\mathbf{v}_A = \mathbf{v}_B + \mathbf{v}_{A\,\text{rel}} + \boldsymbol{\omega} \times \mathbf{r}_{A/B}:$$

$$v_A\mathbf{j} = \mathbf{0} + \mathbf{v}_{A\,\text{rel}} + \begin{vmatrix} \mathbf{i} & \mathbf{j} & \mathbf{k} \\ 0 & \omega_E & 0 \\ R & 0 & 0 \end{vmatrix}.$$

Solving for $\mathbf{v}_{A\,\text{rel}}$, we find that the satellite's velocity relative to the earth-fixed reference frame is

$$\mathbf{v}_{A\,\text{rel}} = v_A\mathbf{j} + R\omega_E\mathbf{k}.$$

The second term on the right side of this equation is the satellite's velocity toward the west relative to the rotating earth-fixed reference frame.

From Eq. (17.22),

$$\mathbf{a}_A = \mathbf{a}_B + \mathbf{a}_{A\,\text{rel}} + 2\boldsymbol{\omega} \times \mathbf{v}_{A\,\text{rel}} + \boldsymbol{\alpha} \times \mathbf{r}_{A/B} + \boldsymbol{\omega} \times (\boldsymbol{\omega} \times \mathbf{r}_{A/B}):$$

$$-\frac{v_A^2}{R}\mathbf{i} = \mathbf{0} + \mathbf{a}_{A\,\text{rel}} + 2\begin{vmatrix} \mathbf{i} & \mathbf{j} & \mathbf{k} \\ 0 & \omega_E & 0 \\ 0 & v_A & R\omega_E \end{vmatrix} + \mathbf{0} + \begin{vmatrix} \mathbf{i} & \mathbf{j} & \mathbf{k} \\ 0 & \omega_E & 0 \\ 0 & 0 & -R\omega_E \end{vmatrix}.$$

Solving for $\mathbf{a}_{A\,\text{rel}}$, we find the satellite's acceleration relative to the earth-fixed reference frame:

$$\mathbf{a}_{A\,\text{rel}} = -\left(\frac{v_A^2}{R} + \omega_E^2 R\right)\mathbf{i}.$$

Discussion

Instruments on earth observing the satellite's motion would measure its velocity and acceleration relative to the earth, $\mathbf{v}_{A\,rel}$ and $\mathbf{a}_{A\,rel}$. Equations (17.21) and (17.22) could then be used to calculate $\mathbf{v}_A$ and $\mathbf{a}_A$, the satellite's velocity and acceleration relative to a nonrotating reference frame.

Inertial Reference Frames

We say that a reference frame is *inertial* if it can be used to apply Newton's second law in the form $\Sigma\mathbf{F} = m\mathbf{a}$. Why can an earth-fixed reference frame be treated as inertial in many situations, even though it both accelerates and rotates? How can Newton's second law be applied using a reference frame that is fixed with respect to a ship or airplane? We are now in a position to answer these questions.

Earth-Centered, Nonrotating Reference Frame   We begin by showing why a nonrotating reference frame with its origin at the center of the earth can be assumed to be inertial for the purpose of describing motions of objects near the earth. Figure 17.43a shows a hypothetical nonaccelerating, nonrotating reference frame with origin O and a secondary nonrotating, *earth-centered reference frame*. The earth (and therefore the earth-centered reference frame) accelerates due to the gravitational attractions of the sun, moon, etc. We denote the earth's acceleration by the vector $\mathbf{g}_B$.

Suppose that we want to determine the motion of an object A of mass m (Fig. 17.43b). A is also subject to the gravitational attractions of the sun, moon, etc., and we denote the resulting gravitational acceleration by the vector $\mathbf{g}_A$. The vector $\Sigma\mathbf{F}$ is the sum of all other external forces acting on A, including the gravitational force exerted on it by the earth. The total external force acting on A is $\Sigma\mathbf{F} + m\mathbf{g}_A$. We can apply Newton's second law to A, using our hypothetical inertial reference frame:

$$\Sigma\mathbf{F} + m\mathbf{g}_A = m\mathbf{a}_A. \tag{17.23}$$

Here, $\mathbf{a}_A$ is the acceleration of A relative to O. Since the earth-centered reference frame does not rotate, we can use Eq. (17.22) to write $\mathbf{a}_A$ as

$$\mathbf{a}_A = \mathbf{a}_B + \mathbf{a}_{A\,rel},$$

where $\mathbf{a}_{A\,rel}$ is the acceleration of A relative to the earth-centered reference frame. Using this relation and our definition of the earth's acceleration $\mathbf{a}_B = \mathbf{g}_B$ in Eq. (17.23), we obtain

$$\Sigma\mathbf{F} = m\mathbf{a}_{A\,rel} + m(\mathbf{g}_B - \mathbf{g}_A). \tag{17.24}$$

If the object A is on or near the earth, its gravitational acceleration $\mathbf{g}_A$ due to the attraction of the sun, etc., is very nearly equal to the earth's gravitational acceleration $\mathbf{g}_B$. If we neglect the difference, Eq. (17.24) becomes

$$\Sigma\mathbf{F} = m\mathbf{a}_{A\,rel}. \tag{17.25}$$

Thus, we can apply Newton's second law using a nonrotating, earth-centered reference frame. Even though this reference frame accelerates, virtually the same gravitational acceleration acts on the object. Notice that this argument does not hold if the object is not near the earth.

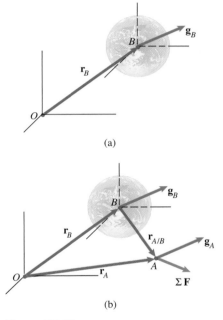

Figure 17.43
(a) An inertial reference frame and a nonrotating reference frame with its origin at the center of the earth.
(b) Determining the motion of an object A.

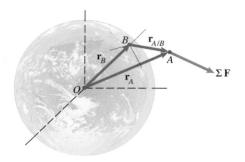

Figure 17.44
An earth-centered, nonrotating reference frame (origin O), an earth-fixed reference frame (origin B), and an object A.

Earth-Fixed Reference Frame   For many applications, the most convenient reference frame is a local, *earth-fixed reference frame*. Why can we usually assume that an earth-fixed reference frame is inertial? Figure 17.44 shows a nonrotating reference frame with its origin O at the center of the earth, and a secondary earth-fixed reference frame with its origin at a point B. Since we can assume that the earth-centered, nonrotating reference frame is inertial, we can write Newton's second law for an object A of mass m as

$$\Sigma \mathbf{F} = m\mathbf{a}_A, \tag{17.26}$$

where $\mathbf{a}_A$ is A's acceleration relative to O. The earth-fixed reference frame rotates with the angular velocity of the earth, which we denote by $\boldsymbol{\omega}_E$. We can use Eq. (17.22) to write Eq. (17.26) in the form

$$\Sigma \mathbf{F} = m\mathbf{a}_{A\,\text{rel}} + m[\mathbf{a}_B + 2\boldsymbol{\omega}_E \times \mathbf{v}_{A\,\text{rel}}$$
$$+ \boldsymbol{\omega}_E \times (\boldsymbol{\omega}_E \times \mathbf{r}_{A/B})], \tag{17.27}$$

where $\mathbf{a}_{A\,\text{rel}}$ is A's acceleration relative to the earth-fixed reference frame. If we can neglect the terms in brackets on the right side of Eq. (17.27), we can take the earth-fixed reference frame to be inertial. Let's consider each term. (Recall from the definition of the cross product that $|\mathbf{U} \times \mathbf{V}| = |\mathbf{U}||\mathbf{V}|\sin\theta$, where θ is the angle between the two vectors. Therefore, the magnitude of the cross product is bounded by the product of the magnitudes of the vectors.)

- The term $\boldsymbol{\omega}_E \times (\boldsymbol{\omega}_E \times \mathbf{r}_{A/B})$: The earth's angular velocity ω_E is approximately one revolution per day $= 7.27 \times 10^{-5}$ rad/s. Therefore, the magnitude of this term is bounded by $\omega_E^2|\mathbf{r}_{A/B}| = (5.29 \times 10^{-9})|\mathbf{r}_{A/B}|$. For example, if the distance $|\mathbf{r}_{A/B}|$ from the origin of the earth-fixed reference frame to the object A is 10,000 m, this term is no larger than 5.3×10^{-5} m/s^2.

- The term $\mathbf{a}_B$: This term is the acceleration of the origin B of the earth-fixed reference frame relative to the center of the earth. B moves in a circular path due to the earth's rotation. If B lies on the earth's surface, $\mathbf{a}_B$ is bounded by $\omega_E^2 R_E$, where R_E is the radius of the earth. Using the value $R_E = 6370$ km, we find that $\omega_E^2 R_E = 0.0337$ m/s^2. This value is too large to neglect for many purposes. However, under normal circumstances, $\mathbf{a}_B$ is accounted for as a part of the local value of the acceleration due to gravity.

- The term $2\boldsymbol{\omega}_E \times \mathbf{v}_{A\,\text{rel}}$: This term is called the *Coriolis acceleration*. Its magnitude is bounded by $2\omega_E|\mathbf{v}_{A\,\text{rel}}| = (1.45 \times 10^{-4})|\mathbf{v}_{A\,\text{rel}}|$. For example, if the magnitude of the velocity of A relative to the earth-fixed reference frame is 10 m/s, this term is no larger than 1.45×10^{-3} m/s^2.

We see that in most applications the terms in brackets in Eq. (17.27) can be neglected. However, in some cases this is not possible. The Coriolis acceleration becomes significant if an object's velocity relative to the earth is large, and even very small accelerations become significant if an object's motion must be predicted over a large period of time. In such cases, we can still use Eq. (17.27) to determine the motion, but must retain the significant terms. When this is done, the terms in brackets are usually moved to the left side:

$$\Sigma \mathbf{F} - m\mathbf{a}_B - 2m\boldsymbol{\omega}_E \times \mathbf{v}_{A\,\text{rel}} - m\boldsymbol{\omega}_E \times (\boldsymbol{\omega}_E \times \mathbf{r}_{A/B})$$
$$= m\mathbf{a}_{A\,\text{rel}}. \tag{17.28}$$

Written in this way, the equation has the usual form of Newton's second law, except that the left side contains additional "forces." (We use quotation marks because the quantities these terms represent are not forces, but arise from the motion of the earth-fixed reference frame.)

Coriolis Effects The term $-2m\boldsymbol{\omega}_E \times \mathbf{v}_{A\,\text{rel}}$ in Eq. (17.28) is called the *Coriolis force*. It explains a number of physical phenomena that exhibit different behaviors in the northern and southern hemispheres, such as the direction a liquid tends to rotate when going down a drain, the direction a vine tends to grow around a vertical shaft, and the direction of rotation of a storm. The earth's angular velocity vector $\boldsymbol{\omega}_E$ points north. When an object in the northern hemisphere that is moving tangent to the earth's surface travels north (Fig. 17.45a), the cross product $\boldsymbol{\omega}_E \times \mathbf{v}_{A\,\text{rel}}$ points west (Fig. 17.45b). Therefore, the Coriolis force points east; it causes an object moving north to turn to the right (Fig. 17.45c). If the object is moving south, the direction of $\mathbf{v}_{A\,\text{rel}}$ is reversed and the Coriolis force points west; its effect is to cause the object moving south to turn to the right (Fig. 17.45c). For example, in the northern hemisphere, winds converging on a center of low pressure tend to rotate about that center in the counterclockwise direction (Fig. 17.46a).

When an object in the southern hemisphere travels north (Fig. 17.45d), the cross product $\boldsymbol{\omega}_E \times \mathbf{v}_{A\,\text{rel}}$ points east (Fig. 17.45e). The Coriolis force points west and tends to cause the object to turn to the left (Fig. 17.45f). If the object is moving south, the Coriolis force points east and tends to cause the object to turn to the left (Fig. 17.45f). In the southern hemisphere, winds converging on a center of low pressure tend to rotate about that center in the clockwise direction (Fig. 17.46b).

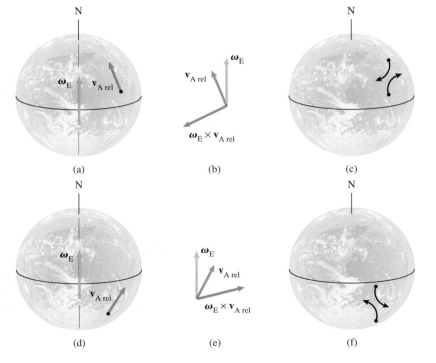

(a) (b) (c)

(d) (e) (f)

Figure 17.45
(a) An object in the northern hemisphere moving north.
(b) Cross product of the earth's angular velocity with the object's velocity.
(c) Effects of the Coriolis force in the northern hemisphere.
(d) An object in the southern hemisphere moving north.
(e) Cross product of the earth's angular velocity with the object's velocity.
(f) Effects of the Coriolis force in the southern hemisphere.

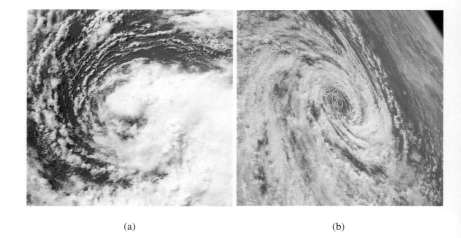

(a)   (b)

Figure 17.46
Storms in the (a) northern and (b) southern
hemispheres.

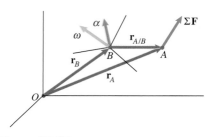

Figure 17.47
An inertial reference frame (origin O) and
a reference frame undergoing an arbitrary
motion (origin B).

Arbitrary Reference Frame How can we analyze an object's motion relative to a reference frame that undergoes an arbitrary motion, such as a reference frame fixed with respect to a moving vehicle? Suppose that the primary reference frame with its origin at O in Fig. 17.47 is inertial and the secondary reference frame with its origin at B undergoes an arbitrary motion with angular velocity $\boldsymbol{\omega}$ and angular acceleration $\boldsymbol{\alpha}$. We can write Newton's second law for an object A of mass m as

$$\Sigma \mathbf{F} = m\mathbf{a}_A, \tag{17.29}$$

where $\mathbf{a}_A$ is A's acceleration relative to O. We use Eq. (17.22) to write Eq. (17.29) in the form

$$\Sigma \mathbf{F} - m\big[\mathbf{a}_B + 2\boldsymbol{\omega} \times \mathbf{v}_{A\,\mathrm{rel}} + \boldsymbol{\alpha} \times \mathbf{r}_{A/B}$$
$$+ \boldsymbol{\omega} \times (\boldsymbol{\omega} \times \mathbf{r}_{A/B})\big] = m\mathbf{a}_{A\,\mathrm{rel}}, \tag{17.30}$$

where $\mathbf{a}_{A\,\mathrm{rel}}$ is A's acceleration relative to the secondary reference frame. This is Newton's second law expressed in terms of a secondary reference frame undergoing an arbitrary motion relative to an inertial primary reference frame. If the forces acting on A and the secondary reference frame's motion are known, Eq. (17.30) can be used to determine $\mathbf{a}_{A\,\mathrm{rel}}$.

Example 17.13

Inertial Reference Frames

Suppose that you and a friend play tennis on the deck of a cruise ship (Fig. 17.48), and you use the ship-fixed coordinate system with origin B to analyze the motion of the ball A. At the instant shown, the ball's position and velocity relative to the ship-fixed coordinate system are $\mathbf{r}_{A/B} = 5\mathbf{i} + 2\mathbf{j} + 12\mathbf{k}$ (m) and $\mathbf{v}_{A\,\mathrm{rel}} = \mathbf{i} - 2\mathbf{j} + 7\mathbf{k}$ (m/s). The mass of the ball is 0.056 kg. Ignore the aerodynamic force on the ball for this example. As a result of the

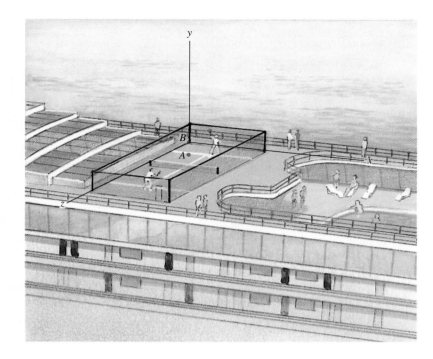

Figure 17.48

ship's motion, the acceleration of the origin B relative to the earth is $\mathbf{a}_B = 1.10\mathbf{i} + 0.07\mathbf{k}\ (\text{m/s}^2)$, and the angular velocity of the ship-fixed coordinate system is constant and equal to $\boldsymbol{\omega} = 0.1\mathbf{j}\ (\text{rad/s})$. Use Newton's second law to determine the ball's acceleration relative to the ship-fixed coordinate system, (a) assuming that the ship-fixed coordinate system is inertial and (b) not assuming that the ship-fixed coordinate system is inertial, but assuming that a local earth-fixed coordinate system is inertial.

Strategy

In part (a), we know the ball's mass and the external forces acting on it, so we can simply apply Newton's second law to determine the acceleration. In part (b), we can express Newton's second law in the form given by Eq. (17.30), which applies to a coordinate system undergoing an arbitrary motion relative to an inertial reference frame.

Solution

(a) If the ship-fixed coordinate system is assumed to be inertial, we can apply Newton's second law to the ball in the form $\Sigma\,\mathbf{F} = m\mathbf{a}_{A\,\text{rel}}$. The only external force on the ball is its weight,

$$-(0.056)(9.81)\mathbf{j} = 0.056\mathbf{a}_{A\,\text{rel}},$$

so we obtain $\mathbf{a}_{A\,\text{rel}} = -9.81\mathbf{j}\ (\text{m/s}^2)$.

(b) If we treat the earth-fixed reference frame as inertial, we can solve Eq. (17.30) for the ball's acceleration relative to the ship-fixed reference frame:

$$\mathbf{a}_{A\,rel} = \frac{1}{m} \Sigma \mathbf{F} - \mathbf{a}_B - 2\boldsymbol{\omega} \times \mathbf{v}_{A\,rel} - \boldsymbol{\alpha} \times \mathbf{r}_{A/B} - \boldsymbol{\omega} \times (\boldsymbol{\omega} \times \mathbf{r}_{A/B})$$

$$= \frac{1}{0.056}\left[-(0.056)(9.81)\mathbf{j}\right] - (1.10\mathbf{i} + 0.07\mathbf{k}) - 2\begin{bmatrix} \mathbf{i} & \mathbf{j} & \mathbf{k} \\ 0 & 0.1 & 0 \\ 1 & -2 & 7 \end{bmatrix}$$

$$- \mathbf{0} - (0.1\mathbf{j}) \times \begin{bmatrix} \mathbf{i} & \mathbf{j} & \mathbf{k} \\ 0 & 0.1 & 0 \\ 5 & 2 & 12 \end{bmatrix}$$

$$= -2.45\mathbf{i} - 9.81\mathbf{j} + 0.25\mathbf{k}\ (\mathrm{m/s^2}).$$

Discussion

This example demonstrates the care you must exercise in applying Newton's second law. When we treat the ship-fixed coordinate system as inertial, Newton's second law does not correctly predict the ball's acceleration relative to the coordinate system, because the effects of the motion of the coordinate system on the ball's relative motion are not accounted for.

Chapter Summary

In this chapter, we analyzed the motions of rigid bodies, showing how the velocities and accelerations of their points are related to their angular velocities and angular accelerations. In doing so, we did not consider the forces and couples causing the motions. In Chapter 18, we will use Newton's second law to determine the equations of motion for rigid bodies in planar motion. By drawing a free-body diagram of a rigid body, we will relate the acceleration of its center of mass and its angular acceleration to the forces and couples acting on it.

Types of Motion

If a rigid body in motion does not rotate relative to a given reference frame, it is said to be in *translation*. If the points of a rigid body intersected by a fixed plane remain in that plane, the rigid body is said to undergo *planar* motion. Rotation about a fixed axis is a special case of planar motion.

Relative Velocities and Accelerations

The *angular velocity vector* $\boldsymbol{\omega}$ of a rigid body is parallel to the axis of rotation of the body, and its magnitude $|\boldsymbol{\omega}|$ is the body's rate of rotation. If the thumb of the right hand points in the direction of $\boldsymbol{\omega}$, the fingers curl around $\boldsymbol{\omega}$ in the direction of the rotation. The *angular acceleration vector* $\boldsymbol{\alpha} = d\boldsymbol{\omega}/dt$ is the rate of change of the angular velocity vector of the body.

Consider a point B of a rigid body, a body-fixed reference frame with its origin at B, and an arbitrary point A (Fig. a). The velocities $\mathbf{v}_A$ and $\mathbf{v}_B$ of the points relative to the primary reference frame are related by

$$\mathbf{v}_A = \mathbf{v}_B + \mathbf{v}_{A\,rel} + \boldsymbol{\omega} \times \mathbf{r}_{A/B}, \qquad \text{Eq. (17.11)}$$

where $\mathbf{v}_{A\,rel}$ is the velocity of A relative to the body-fixed reference frame. If A is a point of the rigid body, $\mathbf{v}_{A\,rel}$ is zero.

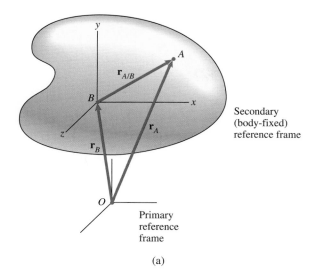

(a)

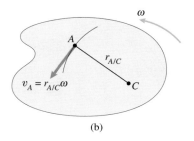

(b)

The accelerations $\mathbf{a}_A$ and $\mathbf{a}_B$ of the points relative to the primary reference frame are related by

$$\mathbf{a}_A = \mathbf{a}_B + \mathbf{a}_{A\,\text{rel}} + 2\boldsymbol{\omega} \times \mathbf{v}_{A\,\text{rel}} + \boldsymbol{\alpha} \times \mathbf{r}_{A/B}$$
$$+ \boldsymbol{\omega} \times (\boldsymbol{\omega} \times \mathbf{r}_{A/B}), \qquad \text{Eq. (17.13)}$$

where $\mathbf{a}_{A\,\text{rel}}$ is the acceleration of A relative to the body-fixed reference frame. In planar motion, the term $\boldsymbol{\omega} \times (\boldsymbol{\omega} \times \mathbf{r}_{A/B})$ can be written in the simpler form $-\omega^2 \mathbf{r}_{A/B}$. If A is a point of the rigid body, $\mathbf{a}_{A\,\text{rel}}$ is zero.

Instantaneous Centers

A point of a rigid body whose velocity is zero at a given instant (or a point exterior to the rigid body whose velocity would be zero if the rigid body were extended to include that point—see Fig. 17.21) is called an *instantaneous center*. Consider a rigid body in planar motion, and suppose that C is an instantaneous center. The velocity of a point A is perpendicular to the line from C to A, and its magnitude is the product of the distance from C to A and the angular velocity (Fig. b).

If the directions of motion of two points A and B of a rigid body in planar motion are not parallel, lines drawn through A and B perpendicular to their directions of motion intersect at the instantaneous center (Fig. c).

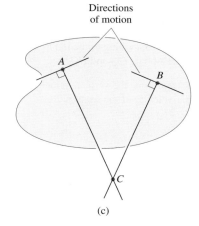

(c)

Moving Reference Frames

Consider a point A and a reference frame, with origin B, that rotates with angular velocity $\boldsymbol{\omega}$ and angular acceleration $\boldsymbol{\alpha}$ relative to a primary reference frame (Fig. d). The velocities of A and B relative to the primary reference frame are related by

$$\mathbf{v}_A = \mathbf{v}_B + \mathbf{v}_{A\,\text{rel}} + \boldsymbol{\omega} \times \mathbf{r}_{A/B}, \qquad \text{Eq. (17.21)}$$

where $\mathbf{v}_{A\,\text{rel}}$ is the velocity of A relative to the secondary reference frame. The accelerations of A and B relative to the primary reference frame are related by

$$\mathbf{a}_A = \mathbf{a}_B + \mathbf{a}_{A\,\text{rel}} + 2\boldsymbol{\omega} \times \mathbf{v}_{A\,\text{rel}}$$
$$+ \boldsymbol{\alpha} \times \mathbf{r}_{A/B} + \boldsymbol{\omega} \times (\boldsymbol{\omega} \times \mathbf{r}_{A/B}), \qquad \text{Eq. (17.22)}$$

where $\mathbf{a}_{A\,\text{rel}}$ is the acceleration of A relative to the secondary reference frame.

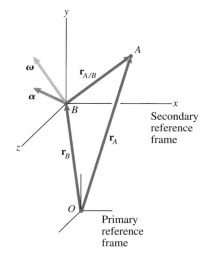

(d)

Review Problems

17.1 If $\theta = 60°$ and bar OQ rotates in the counterclockwise direction at 5 rad/s, what is the angular velocity of bar PQ?

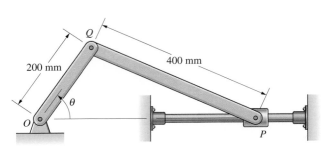

200 mm

400 mm

Q

θ

O

P

P17.1

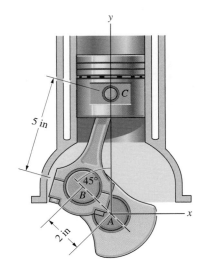

y

5 in

C

45°

B

A

2 in

x

P17.4

17.2 Consider the system shown in Problem 17.1. If $\theta = 55°$ and the sleeve P is moving to the left at 2 m/s, what are the angular velocities of bars OQ and PQ?

17.3 Determine the vertical velocity v_H of the hook and the angular velocity of the small pulley.

17.5 In Problem 17.4, if the piston is moving with velocity $\mathbf{v}_C = 20\mathbf{j}$ (ft/s), what are the angular velocities of the crankshaft AB and the connecting rod BC?

17.6 In Problem 17.4, if the piston is moving with velocity $\mathbf{v}_C = 20\mathbf{j}$ (ft/s), and its acceleration is zero, what are the angular accelerations of the crankshaft AB and the connecting rod BC?

17.7 Bar AB rotates at 6 rad/s in the counterclockwise direction. Use instantaneous centers to determine the angular velocity of bar BCD and the velocity of point D.

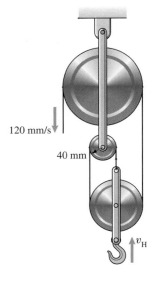

120 mm/s

40 mm

v_H

P17.3

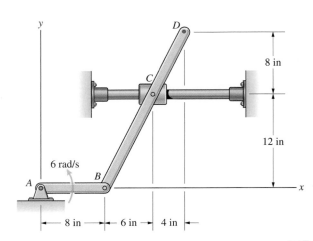

y

D

8 in

C

12 in

6 rad/s

B

A

x

8 in — 6 in — 4 in

17.4 If the crankshaft AB is turning in the counterclockwise direction at 2000 rpm, what is the velocity of the piston?

P17.7

17.8 In Problem 17.7, bar *AB* rotates with a constant angular velocity of 6 rad/s in the counterclockwise direction. Determine the acceleration of point *D*.

17.9 Point *C* is moving to the right at 20 in./s. What is the velocity of the midpoint *G* of bar *BC*?

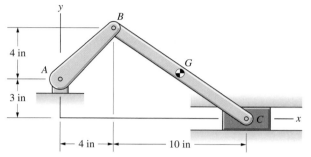

P17.9

17.10 In Problem 17.9, point *C* is moving to the right with a constant velocity of 20 in./s. What is the acceleration of the midpoint *G* of bar *BC*?

17.11 In Problem 17.9, if the velocity of point *C* is $\mathbf{v}_C = 1.0\mathbf{i}$ (in./s), what are the angular velocity vectors of arms *AB* and *BC*?

17.12 Points *B* and *C* are in the *x-y* plane. The angular velocity vectors of arms *AB* and *BC* are $\boldsymbol{\omega}_{AB} = -0.5\mathbf{k}$ (rad/s) and $\boldsymbol{\omega}_{BC} = -2.0\mathbf{k}$ (rad/s). Determine the velocity of point *C*.

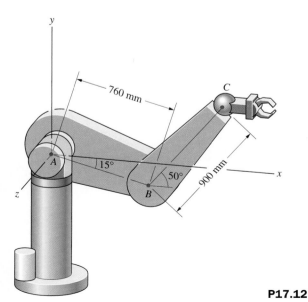

P17.12

⚇ **17.13** In Problem 17.12, if the velocity of point *C* is $\mathbf{v}_C = 1.0\mathbf{i}$ (m/s), what are the angular velocity vectors of arms *AB* and *BC*?

17.14 In Problem 17.12, if the angular velocity vectors of arms *AB* and *BC* are $\boldsymbol{\omega}_{AB} = -0.5\mathbf{k}$ (rad/s) and $\boldsymbol{\omega}_{BC} = 2.0\mathbf{k}$ (rad/s), and their angular acceleration vectors are $\boldsymbol{\alpha}_{AB} = 1.0\mathbf{k}$ (rad/s²) and $\boldsymbol{\alpha}_{BC} = 1.0\mathbf{k}$ (rad/s²), what is the acceleration of point *C*?

⚇ **17.15** In Problem 17.12, if the velocity of point *C* is $\mathbf{v}_C = 1.0\mathbf{i}$ (m/s) and $\mathbf{a}_C = \mathbf{0}$, what are the angular velocity and angular acceleration vectors of arm *BC*?

17.16 The crank *AB* has a constant clockwise angular velocity of 200 rpm. What are the velocity and acceleration of the piston *P*?

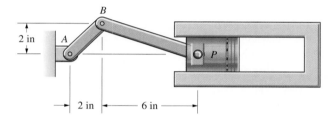

P17.16

17.17 Bar *AB* has a counterclockwise angular velocity of 10 rad/s and a clockwise angular acceleration of 20 rad/s². Determine the angular acceleration of bar *BC* and the acceleration of point *C*.

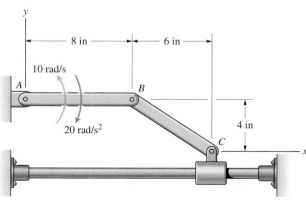

P17.17

17.18 The angular velocity of arm AC is 1 rad/s counterclockwise. What is the angular velocity of the scoop?

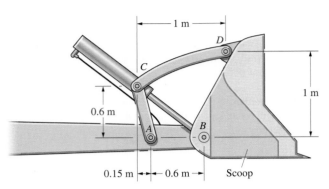

P17.18

17.19 The angular velocity of arm AC in Problem 17.18 is 2 rad/s counterclockwise, and its angular acceleration is 4 rad/s² clockwise. What is the angular acceleration of the scoop?

Ⓓ **17.20** If you want to program the robot so that, at the instant shown, the velocity of point D is $\mathbf{v}_D = 0.2\mathbf{i} + 0.8\mathbf{j}(\text{m/s})$ and the angular velocity of arm CD is 0.3 rad/s counterclockwise, what are the necessary angular velocities of arms AB and BC?

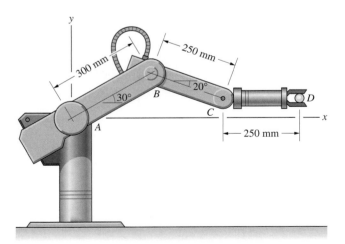

P17.20

17.21 The ring gear is stationary, and the sun gear rotates at 120 rpm in the counterclockwise direction. Determine the angular velocity of the planet gears and the magnitude of the velocity of their centerpoints.

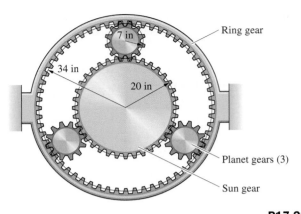

P17.21

17.22 Arm AB is rotating at 10 rad/s in the clockwise direction. Determine the angular velocity of arm BC and the velocity at which the arm slides relative to the sleeve at C.

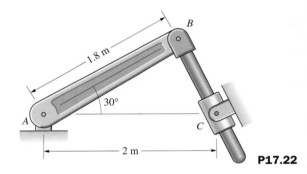

P17.22

17.23 In Problem 17.22, arm AB is rotating with an angular velocity of 10 rad/s and an angular acceleration of 20 rad/s², both in the clockwise direction. Determine the angular acceleration of arm BC.

17.24 Arm AB is rotating with a constant counterclockwise angular velocity of 10 rad/s. Determine the vertical velocity and acceleration of the rack R of the rack-and-pinion gear.

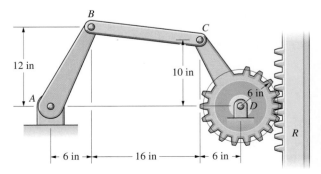

P17.24

17.25 In Problem 17.24, if the rack R of the rack-and-pinion gear is moving upward with a constant velocity of 10 ft/s, what are the angular velocity and angular acceleration of bar BC?

17.26 Bar AB has a constant counterclockwise angular velocity of 2 rad/s. The 1-kg collar C slides on the smooth horizontal bar. At the instant shown, what is the tension in the cable BC?

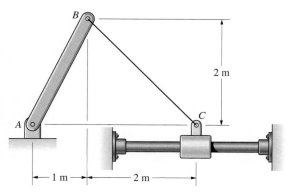

2 m

1 m 2 m

P17.26

🖉 **17.27** An athlete exercises his arm by raising the 8-kg mass m. The shoulder joint A is stationary. The distance AB is 300 mm, the distance BC is 400 mm, and the distance from C to the pulley is 340 mm. The angular velocities $\omega_{AB} = 1.5$ rad/s and $\omega_{BC} = 2$ rad/s are constant. What is the tension in the cable?

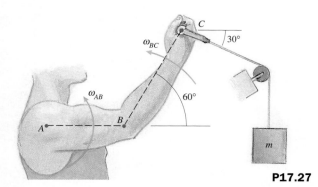

ω_{BC} C 30°

ω_{AB} 60°

A B m

P17.27

17.28 The secondary reference frame shown rotates with a constant angular velocity $\omega = 2\mathbf{k}$ (rad/s) relative to a primary reference frame. The point A moves outward along the x axis at a constant speed of 5 m/s.

(a) What are the velocity and acceleration of A relative to the secondary reference frame?
(b) If the origin B is stationary relative to the primary reference frame, what are the velocity and acceleration of A relative to the primary reference frame when A is at the position $x = 1$ m?

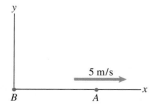

y

5 m/s

x

B A **P17.28**

17.29 The coordinate system shown is fixed relative to the ship B. The ship uses its radar to measure the position of a stationary buoy A and determines it to be $400\mathbf{i} + 200\mathbf{j}$ (m). The ship also measures the velocity of the buoy relative to its body-fixed coordinate system and determines it to be $2\mathbf{i} - 8\mathbf{j}$ (m/s). What are the ship's velocity and angular velocity relative to the earth? (Assume that the ship's velocity is in the direction of the y axis.)

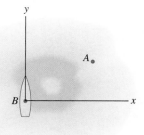

y

A

B x

P17.29

Electric motors exert torque on the wind-tunnel fan, accelerating it to its operating angular velocity.

Planar Dynamics of Rigid Bodies

18

n Chapter 17, we analyzed planar motions of rigid bodies without considering the forces and couples causing them. In this chapter, we derive planar equations of angular motion for a rigid body. By drawing the free-body diagram of an object and applying the equations of motion, we can determine both the acceleration of its center of mass and its angular acceleration in terms of the forces and couples to which it is subjected.

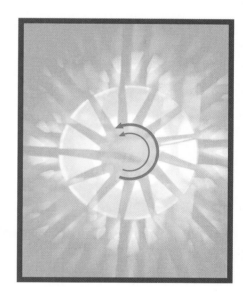

18.1 Preview of the Equations of Motion

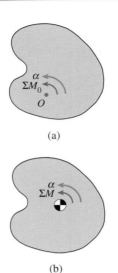

Figure 18.1
(a) A rigid body rotating about a fixed axis O.
(b) A rigid body in general planar motion.

The two-dimensional equations of angular motion for a rigid body are simple in form, but their derivations are rather involved. To provide help in following the derivations, we first summarize the equations in this section.

The equations of motion of a rigid body include Newton's second law,

$$\Sigma \mathbf{F} = m\mathbf{a},$$

which states that the sum of the external forces acting on the body equals the product of the mass of the body and the acceleration of its center of mass. The equations of motion are completed by an equation of angular motion. If the rigid body rotates about a fixed axis O (Fig. 18.1a), the sum of the moments about the axis due to external forces and couples acting on the body is related to its angular acceleration by

$$\Sigma M_0 = I_0 \alpha,$$

where I_0 is the moment of inertia of the rigid body about O. Just as an object's mass determines the acceleration resulting from the forces acting on it, its moment of inertia I_0 about a fixed axis determines the angular acceleration resulting from the sum of the moments about the axis.

In general planar motion (Fig. 18.1b), the sum of the moments about the center of mass of a rigid body is related to its angular acceleration by

$$\Sigma M = I\alpha,$$

where I is the moment of inertia of the rigid body about its center of mass. If the external forces and couples acting on a rigid body in planar motion are known, we can use these equations to determine the acceleration of the center of mass of the body and its angular acceleration.

18.2 Momentum Principles for a System of Particles

Our derivations of the equations of motion for rigid bodies are based on principles governing the motion of a system of particles. We summarize these general and important principles in this section.

Force–Linear-Momentum Principle

We begin by showing that the sum of the external forces on a system of particles equals the rate of change of the total linear momentum of the system. Let us consider a system of N particles. We denote the mass of the ith particle by m_i and denote its position vector relative to the origin O of an inertial reference frame by $\mathbf{r}_i$ (Fig. 18.2). Let $\mathbf{f}_{ij}$ be the force exerted on the ith particle by the jth particle, and let the external force on the ith particle (i.e., the total force exerted by objects other than the system of particles we are considering) be $\mathbf{f}_i^E$. Newton's second law states that the total force acting on the ith particle equals the product of its mass and the rate of change of its linear momentum; that is,

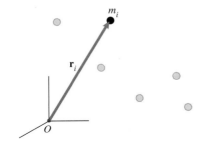

Figure 18.2
A system of particles. The vector $\mathbf{r}_i$ is the position vector of the ith particle.

$$\sum_j \mathbf{f}_{ij} + \mathbf{f}_i^E = \frac{d}{dt}(m_i \mathbf{v}_i), \tag{18.1}$$

where $\mathbf{v}_i = d\mathbf{r}_i/dt$ is the velocity of the ith particle. Writing this equation for each particle of the system and summing from $i = 1$ to N, we obtain

$$\sum_i \sum_j \mathbf{f}_{ij} + \sum_i \mathbf{f}_i^E = \frac{d}{dt} \sum_i m_i \mathbf{v}_i. \tag{18.2}$$

The first term on the left side of this equation is the sum of the internal forces on the system of particles. As a consequence of Newton's third law $\left(\mathbf{f}_{ij} + \mathbf{f}_{ji} = \mathbf{0}\right)$, that term equals zero:

$$\sum_i \sum_j \mathbf{f}_{ij} = \mathbf{f}_{12} + \mathbf{f}_{21} + \mathbf{f}_{13} + \mathbf{f}_{31} + \cdots = \mathbf{0}.$$

The second term on the left side of Eq. (18.2) is the sum of the external forces on the system. Denoting it by $\Sigma \mathbf{F}$, we conclude that the sum of the external forces on the system equals the rate of change of its total linear momentum:

$$\Sigma \mathbf{F} = \frac{d}{dt} \sum_i m_i \mathbf{v}_i. \tag{18.3}$$

Let m be the sum of the masses of the particles:

$$m = \sum_i m_i.$$

The position of the center of mass of the system is

$$\mathbf{r} = \frac{\sum_i m_i \mathbf{r}_i}{m}, \tag{18.4}$$

so the velocity of the center of mass is

$$\mathbf{v} = \frac{d\mathbf{r}}{dt} = \frac{\sum_i m_i \mathbf{v}_i}{m}.$$

By using this expression, we can write Eq. (18.3) as

$$\Sigma \mathbf{F} = \frac{d}{dt}(m\mathbf{v}).$$

The total external force on a system of particles thus equals the rate of change of the product of the total mass of the system and the velocity of its center of mass. Since any object or collection of objects, including a rigid body, can be modeled as a system of particles, this result is one of the most general and elegant in all of mechanics. Furthermore, if the total mass m is constant, we obtain

$$\Sigma \mathbf{F} = m\mathbf{a},$$

where $\mathbf{a} = d\mathbf{v}/dt$ is the acceleration of the center of mass. We see that the total external force equals the product of the total mass and the acceleration of the center of mass.

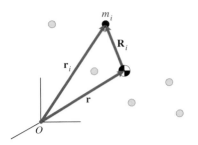

Figure 18.3
The vector $\mathbf{R}_i$ is the position vector of the ith particle relative to the center of mass.

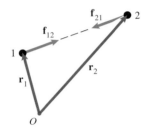

Figure 18.4
Particles 1 and 2 and the forces they exert on each other. If the forces act along the line between the particles, their total moment about O is zero.

Moment–Angular-Momentum Principles

We now obtain relations between the sum of the moments due to the forces acting on a system of particles and the rate of change of the total angular momentum of the system. In Fig. 18.3, $\mathbf{r}_i$ is the position vector of the ith particle of a system of particles, $\mathbf{r}$ is the position vector of the center of mass of the system, and $\mathbf{R}_i$ is the position vector of the ith particle relative to the center of mass. These vectors are related by

$$\mathbf{r}_i = \mathbf{r} + \mathbf{R}_i. \tag{18.5}$$

We take the cross product of Newton's second law for the ith particle, Eq. (18.1), with the position vector $\mathbf{r}_i$ and sum from $i = 1$ to N, writing the resulting equation in the form

$$\sum_i \sum_j \mathbf{r}_i \times \mathbf{f}_{ij} + \sum_i \mathbf{r}_i \times \mathbf{f}_i^E = \frac{d}{dt} \sum_i \mathbf{r}_i \times m_i \mathbf{v}_i. \tag{18.6}$$

The first term on the left side of this equation is the sum of the moments about O due to the forces exerted on the particles by the other particles of the system. This term vanishes if we assume that the mutual forces exerted by each pair of particles are not only equal and opposite, but directed along the straight line between the particles. For example, consider particles 1 and 2 in Fig. 18.4. If the forces the particles exert on each other are directed along the line between the particles, we can write the moment about O as

$$\mathbf{r}_1 \times \mathbf{f}_{12} + \mathbf{r}_1 \times \mathbf{f}_{21} = \mathbf{r}_1 \times \left(\mathbf{f}_{12} + \mathbf{f}_{21}\right) = \mathbf{0}. \tag{18.7}$$

The second term on the left side of Eq. (18.6) is the sum of the moments about O due to external forces, which we denote by $\Sigma \mathbf{M}_O$. We write Eq. (18.6) as

$$\Sigma \mathbf{M}_O = \frac{d\mathbf{H}_O}{dt}, \tag{18.8}$$

where

$$\mathbf{H}_O = \sum_i \mathbf{r}_i \times m_i \mathbf{v}_i \tag{18.9}$$

is the total angular momentum about O. The sum of the moments about O is equal to the rate of change of the total angular momentum about O. By using Eqs. (18.4) and (18.5), we can write Eq. (18.8) as

$$\Sigma \mathbf{M}_O = \frac{d}{dt} (\mathbf{r} \times m\mathbf{v} + \mathbf{H}), \tag{18.10}$$

where

$$\mathbf{H} = \sum_i \mathbf{R}_i \times m_i \frac{d\mathbf{R}_i}{dt} \tag{18.11}$$

is the total angular momentum of the system about the center of mass.

We also need to determine the relation between the sum of the moments about the center of mass, which we denote by $\Sigma \mathbf{M}$, and $\mathbf{H}$. We can obtain this relation by letting the fixed point O be coincident with the center of mass at the present instant. In that case, $\Sigma \mathbf{M}_O = \Sigma \mathbf{M}$ and $\mathbf{r} = \mathbf{0}$, and we see from Eq. (18.10) that

$$\Sigma \mathbf{M} = \frac{d\mathbf{H}}{dt}. \qquad (18.12)$$

The sum of the moments about the center of mass is equal to the rate of change of the total angular momentum about the center of mass. The angular momenta about point O and about the center of mass are related by (Fig. 18.5)

$$\mathbf{H}_O = \mathbf{H} + \mathbf{r} \times m\mathbf{v}. \qquad (18.13)$$

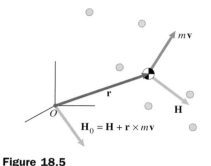

Figure 18.5
The angular momentum about O equals the sum of the angular momentum about the center of mass and the angular momentum about O due to the velocity of the center of mass.

18.3 Derivation of the Equations of Motion

We now derive the equations of motion for a rigid body in planar motion. We have shown that the total external force on any object equals the product of the mass of the object and the acceleration of its center of mass:

$$\Sigma \mathbf{F} = m\mathbf{a}.$$

This equation, which we refer to as Newton's second law, describes the motion of the center of mass of a rigid body. To derive the equations of angular motion, we consider first rotation about a fixed axis and then general planar motion.

Rotation about a Fixed Axis

Let O be a point that is stationary relative to an inertial reference frame, and let L_O be a nonrotating line through O. Suppose that a rigid body rotates about L_O. In terms of a coordinate system with its z axis aligned with L_O (Fig. 18.6a), we can express the rigid body's angular velocity vector as $\boldsymbol{\omega} = \omega \mathbf{k}$, and the velocity of the ith particle of the rigid body is

$$\frac{d\mathbf{r}_i}{dt} = \boldsymbol{\omega} \times \mathbf{r}_i = \omega \mathbf{k} \times \mathbf{r}_i.$$

Let $\Sigma M_O = \Sigma \mathbf{M}_O \cdot \mathbf{k}$ denote the sum of the moments about L_O. By taking the dot product of Eq. (18.8) with $\mathbf{k}$, we obtain

$$\Sigma M_O = \frac{dH_O}{dt}, \qquad (18.14)$$

where

$$H_O = \mathbf{H}_O \cdot \mathbf{k} = \sum_i \left[\mathbf{r}_i \times m_i(\omega \mathbf{k} \times \mathbf{r}_i) \right] \cdot \mathbf{k} \qquad (18.15)$$

is the angular momentum about L_O. Using the identity $\mathbf{U} \cdot (\mathbf{V} \times \mathbf{W}) = (\mathbf{U} \times \mathbf{V}) \cdot \mathbf{W}$, we can write Eq. (18.15) as

$$H_O = \sum_i m_i(\mathbf{k} \times \mathbf{r}_i) \cdot (\mathbf{k} \times \mathbf{r}_i)\omega = \sum_i m_i |\mathbf{k} \times \mathbf{r}_i|^2 \omega. \qquad (18.16)$$

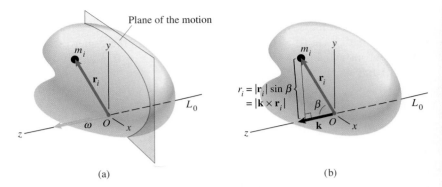

Figure 18.6

(a) A coordinate system with the z axis aligned with the axis of rotation, L_O.

(b) The magnitude of $\mathbf{k} \times \mathbf{r}_i$ is the perpendicular distance from the axis of rotation to m_i.

In Fig. 18.6b, we show that $|\mathbf{k} \times \mathbf{r}_i|$ is the perpendicular distance from L_O to the ith particle, which we denote by r_i. Using the definition of the moment of inertia of the rigid body about L_O,

$$I_O = \sum_i m_i r_i^2,$$

we can write Eq. (18.16) as

$$H_O = I_O \omega.$$

Substituting this expression into Eq. (18.14), we obtain the equation of angular motion for a rigid body rotating about a fixed axis. The sum of the moments about the fixed axis equals the product of the moment of inertia about the fixed axis and the angular acceleration:

$$\Sigma M_O = I_O \alpha. \tag{18.17}$$

General Planar Motion

Figure 18.7a shows the plane of the motion of a rigid body in general planar motion. Point O is a fixed point contained in the plane, L_O is the line through O that is perpendicular to the plane, and L is the line parallel to L_O that passes through the center of mass of the rigid body. In terms of the coordinate system shown, we can express the rigid body's angular velocity vector as $\boldsymbol{\omega} = \omega \mathbf{k}$, and the velocity of the ith particle of the rigid body relative to the center of mass is

$$\frac{d\mathbf{R}_i}{dt} = \boldsymbol{\omega} \times \mathbf{R}_i = \omega \mathbf{k} \times \mathbf{R}_i.$$

By taking the dot product of Eq. (18.10) with $\mathbf{k}$, we obtain

$$\Sigma M_O = \frac{d}{dt}\left[(\mathbf{r} \times m\mathbf{v}) \cdot \mathbf{k} + H\right], \tag{18.18}$$

where

$$H = \mathbf{H} \cdot \mathbf{k} = \sum_i \left[\mathbf{R}_i \times m_i(\omega \mathbf{k} \times \mathbf{R}_i)\right] \cdot \mathbf{k}$$

is the angular momentum about L. Using the same identity we applied to Eq. (18.15), we can write this equation for H as

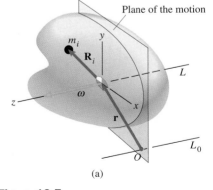

(a)

Figure 18.7

(a) A coordinate system with the z axis aligned with L.

$$H = \sum_i m_i(\mathbf{k} \times \mathbf{R}_i) \cdot (\mathbf{k} \times \mathbf{R}_i)\omega = \sum_i m_i |\mathbf{k} \times \mathbf{R}_i|^2 \omega.$$

The term $|\mathbf{k} \times \mathbf{R}_i| = r_i$ is the perpendicular distance from L to the ith particle (Fig. 18.7b). In terms of the moment of inertia of the rigid body about L,

$$I = \sum_i m_i r_i^2,$$

the rigid body's angular momentum about L is

$$H = I\omega.$$

Substituting this expression into Eq. (18.18), we obtain

$$\Sigma M_O = \frac{d}{dt}\left[(\mathbf{r} \times m\mathbf{v}) \cdot \mathbf{k} + I\omega\right] = (\mathbf{r} \times m\mathbf{a}) \cdot \mathbf{k} + I\alpha. \quad (18.19)$$

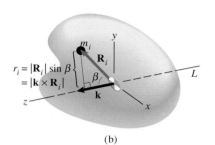

(b)

Figure 18.7 (continued)
(b) The magnitude of $\mathbf{k} \times \mathbf{R}_i$ is the perpendicular distance from L to m_i.

With this equation, we can obtain the relation between the sum of the moments about L, which we denote by ΣM, and the angular acceleration. If we let the fixed axis L_O be coincident with L at the present instant, then $\Sigma M_O = \Sigma M$ and $\mathbf{r} = \mathbf{0}$, and from Eq. (18.19) we obtain the equation of angular motion for a rigid body in general planar motion. The sum of the moments about L equals the product of the moment of inertia about the center of mass and the angular acceleration:

$$\Sigma M = I\alpha. \quad (18.20)$$

18.4 Applications

We have seen that the equations of motion for a rigid body in planar motion include Newton's second law,

$$\Sigma \mathbf{F} = m\mathbf{a}, \quad (18.21)$$

where $\mathbf{a}$ is the acceleration of the center of mass of the body, and an equation relating the moments due to forces and couples to the angular acceleration of the body. If the rigid body rotates about a fixed axis O, the total moment about O equals the product of the moment of inertia about O and the angular acceleration:

$$\Sigma M_O = I_O\alpha. \quad (18.22)$$

In *any* planar motion, the total moment about the center of mass equals the product of the moment of inertia about the center of mass and the angular acceleration:

$$\Sigma M = I\alpha. \quad (18.23)$$

Of course, this equation applies to the case of rotation about a fixed axis, but for that type of motion, it is often more convenient to use Eq. (18.22).

These equations can be used to obtain information about an object's motion, or to determine the values of unknown forces or couples acting on the object, or both. Doing so typically involves three steps:

1. *Draw the free-body diagram.* Isolate the object and identify the external forces and couples acting on it.

2. *Apply the equations of motion.* Write equations of motion suitable for the type of motion, choosing an appropriate coordinate system for applying Newton's second law. For example, if the center of mass moves in a circular path, it may be advantageous to use normal and tangential components.

3. *Determine kinematic relationships.* If necessary, supplement the equations of motion with relationships between the acceleration of the center of mass and the angular acceleration of the object.

As we show in the sections that follow, the best approach depends in part on the type of motion involved.

Translation

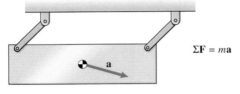

$$\Sigma \mathbf{F} = m\mathbf{a}$$

If a rigid body is in translation (Fig. 18.8), only Newton's second law is required to determine its motion. There is no rotational motion to determine. Nevertheless, it may be necessary to apply the angular equation of motion to determine unknown forces or couples. Since $\alpha = 0$, Eq. (18.23) states that the total moment *about the center of mass* equals zero:

$$\Sigma M = 0.$$

Figure 18.8
A rigid body in translation. There is no rotational motion to determine.

Example 18.1

Translating Object

The mass of the airplane in Fig. 18.9 is $m = 250$ Mg (megagrams), and the thrust of the engines during the plane's takeoff roll is $T = 700$ kN. Determine the acceleration of the airplane and the normal forces exerted on its wheels at A and B. Neglect the horizontal forces exerted on its wheels.

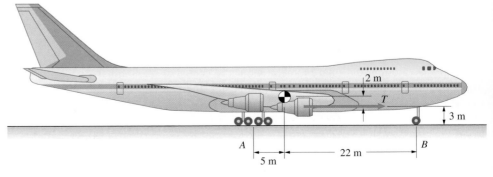

Figure 18.9

Strategy

The airplane is in translation during its takeoff roll, so the sum of the moments about its center of mass equals zero. Using this condition and Newton's second law, we can determine the acceleration of the plane and the normal forces exerted on its wheels.

Solution

Draw the Free-Body Diagram we draw the free-body diagram in Fig. (a), showing the airplane's weight and the normal forces A and B exerted on the wheels.

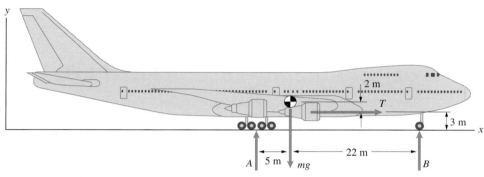

(a) Free-body diagram of the airplane.

Apply the Equations of Motion In terms of the coordinate system in Fig. (a), Newton's second law is

$$\Sigma F_x = T = ma_x,$$
$$\Sigma F_y = A + B - mg = 0.$$

From the first equation, the airplane's acceleration is

$$a_x = \frac{T}{m} = \frac{700{,}000 \text{ N}}{250{,}000 \text{ kg}} = 2.8 \text{ m/s}^2.$$

The equation of angular motion is

$$\Sigma M = (2)T + (22)B - (5)A = 0.$$

Solving this equation together with the second equation we obtained from Newton's second law yields $A = 2050$ kN and $B = 402$ kN.

Discussion

When an object is in equilibrium, the sum of the moments about any point due to the external forces and couples acting on it is zero. When a translating rigid body is not in equilibrium, you know only that the sum of the moments about the center of mass is zero. It would be instructive to try reworking this example by assuming that the sum of the moments about A or B is zero. You will not obtain the correct values for the normal forces exerted on the wheels.

Rotation about a Fixed Axis

In the case of rotation about a fixed axis (Fig. 18.10), only Eq. (18.22) is required to determine the rotational motion, although it may be necessary to use Newton's second law to determine unknown forces or couples.

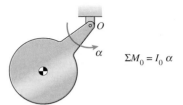

$$\Sigma M_O = I_O \, \alpha$$

Figure 18.10
A rigid body rotating about O. You need only the equation of angular motion about O to determine the angular acceleration of the body.

Example 18.2

Object Rotating about a Fixed Axis

The crate in Fig. 18.11 is pulled up the inclined surface by the winch. The mass of the crate is $m = 45$ kg and the dimension $b = 0.15$ m. The coefficient of kinetic friction between the crate and the surface is $\mu_k = 0.4$. The moment of inertia of the drum on which the cable is being wound, including the part of the cable already wound on the drum, is $I_A = 4$ kg-m^2. If the motor exerts a couple $M = 50$ N-m on the drum, what is the crate's acceleration?

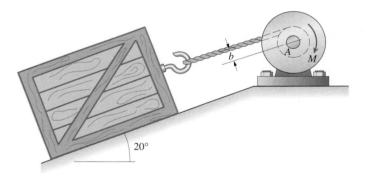

Figure 18.11

Strategy

We will draw separate free-body diagrams of the crate and drum and apply the equations of motion to them individually. The drum rotates about a fixed axis, so we can use the equation of angular motion about the axis to determine its angular acceleration. To complete the solution, we must determine the relationship between the crate's acceleration and the drum's angular acceleration.

Solution

Draw the Free-Body Diagrams   We draw the free-body diagrams in Fig. (a), showing the equal forces exerted on the crate and the drum by the cable.

Apply the Equations of Motion   We denote the crate's acceleration up the inclined surface by a_x and the *clockwise* angular acceleration of the drum by α (Fig. b). Newton's second law for the crate yields the equations

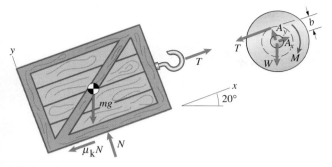

(a) Free-body diagrams of the crate and the drum.

(b) The crate's acceleration and the angular acceleration of the drum.

$$\Sigma F_x = T - mg \sin 20° - \mu_k N = ma_x,$$

$$\Sigma F_y = N - mg \cos 20° = 0.$$

The equation of angular motion for the drum is

$$\Sigma M_A = M - bT = I_A \alpha.$$

From the equations of motion, we have obtained three equations in four unknowns: T, N, a_x, and α. Eliminating T and N from these equations, we get one equation in terms of a_x and α:

$$M - bmg(\sin 20° + \mu_k \cos 20°) - bma_x = I_A \alpha. \qquad (18.24)$$

To complete the solution, we must determine the kinematic relationship between a_x and α.

Determine Kinematic Relationship   The tangential component of acceleration of the drum at the point where the cable begins winding onto it is equal to the crate's acceleration (Fig. b):

$$a_x = b\alpha.$$

Solving this equation together with Eq. (18.24), we find that the crate's acceleration is

$$
\begin{aligned}
a_x &= \frac{M - bmg(\sin 20° + \mu_k \cos 20°)}{bm + I_A/b} \\
&= \frac{50 - (0.15)(45)(9.81)(\sin 20° + 0.4 \cos 20°)}{(0.15)(45) + 4/0.15} \\
&= 0.0737 \text{ m/s}^2.
\end{aligned}
$$

Discussion

Notice that, for convenience, we defined the angular acceleration α to be positive in the clockwise direction so that a positive α would correspond to a positive a_x.

Example 18.3

Figure 18.12

Bar Rotating about a Fixed Axis

The slender bar of mass m in Fig. 18.12 is released from rest in the horizontal position shown. Determine the bar's angular acceleration and the force exerted on the bar by the support A at that instant.

Strategy

Since the bar rotates about a fixed point, we can use Eq. (18.22) to determine its angular acceleration. The advantage of using this equation instead of Eq. (18.23) is that the unknown reactions at A will not appear in the equation of angular motion. Once we know the angular acceleration, we can determine the acceleration of the center of mass and use Newton's second law to obtain the reactions at A.

Solution

Draw the Free-Body Diagram   In Fig. (a), we draw the free-body diagram of the bar, showing the reactions at the pin support.

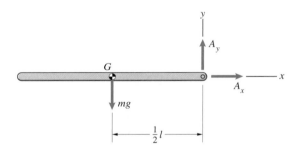

(a) Free-body diagram of the bar.

Apply the Equations of Motion   Let the acceleration of the center of mass G of the bar be $\mathbf{a}_G = a_x\mathbf{i} + a_y\mathbf{j}$, and let its counterclockwise angular acceleration be α (Fig. b). Newton's second law for the bar is

$$\Sigma F_x = A_x = ma_x,$$

$$\Sigma F_y = A_y - mg = ma_y.$$

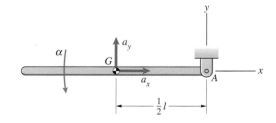

(b) The angular acceleration and components of the acceleration of the center of mass.

The equation of angular motion about the fixed point A is

$$\Sigma M_A = \left(\frac{1}{2}l\right)mg = I_A\alpha. \qquad (18.25)$$

The moment of inertia of a slender bar about its center of mass is $I = \frac{1}{12}ml^2$. (See Appendix C.) From the parallel-axis theorem, the moment of inertia of the bar about A is

$$I_A = I + d^2m = \frac{1}{12}ml^2 + \left(\frac{1}{2}l\right)^2 m = \frac{1}{3}ml^2.$$

Substituting this expression into Eq. (18.25), we obtain the angular acceleration:

$$\alpha = \frac{\frac{1}{2}mgl}{\frac{1}{3}ml^2} = \frac{3}{2}\frac{g}{l}.$$

Determine Kinematic Relationships To determine the reactions A_x and A_y, we need to determine the acceleration components a_x and a_y. We can do so by expressing the acceleration of G in terms of the acceleration of A:

$$\mathbf{a}_G = \mathbf{a}_A + \boldsymbol{\alpha} \times \mathbf{r}_{G/A} - \omega^2 \mathbf{r}_{G/A}.$$

At the instant the bar is released, its angular velocity $\omega = 0$. Also, $\mathbf{a}_A = \mathbf{0}$, so we obtain

$$\mathbf{a}_G = a_x\mathbf{i} + a_y\mathbf{j} = (\alpha\mathbf{k}) \times \left(-\frac{1}{2}l\mathbf{i}\right) = -\frac{1}{2}l\alpha\mathbf{j}.$$

Equating $\mathbf{i}$ and $\mathbf{j}$ components, we get

$$a_x = 0$$

and

$$a_y = -\frac{1}{2}l\alpha = -\frac{3}{4}g.$$

Substituting these acceleration components into Newton's second law, we find that the reactions at A at the instant the bar is released are

$$A_x = 0$$

and

$$A_y = mg + m\left(-\frac{3}{4}g\right) = \frac{1}{4}mg.$$

Discussion

We could have determined the acceleration of G in a less formal way. Since G describes a circular path about A, we know that the magnitude of the tangential component of acceleration equals the product of the radial distance from A to G and the angular acceleration. Because of the directions in which we define α and a_x to be positive, $a_y = -(\frac{1}{2}l)\alpha$. Also, the normal component of the acceleration of G equals the square of its velocity divided by the radius of its circular path. Since its velocity equals zero at the instant the bar is released, $a_x = 0$.

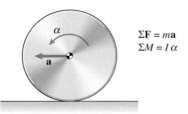

Figure 18.13
A rigid body in planar motion. You must apply both Newton's second law and the equation of angular motion about the center of mass.

General Planar Motion

If a rigid body undergoes general planar motion, both Newton's second law and the equation of angular motion are required to determine its motion. If the motion of the center of mass and the rotational motion are not independent—the rolling disk in Fig. 18.13 is an example—there will be more unknown quantities than equations of motion. In such cases, it is necessary to obtain additional equations by deriving kinematic relationships between the acceleration of the center of mass and the angular acceleration.

Example 18.4

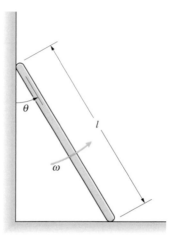

Figure 18.14

Bar in General Planar Motion

The slender bar of mass m in Fig. 18.14 slides on the smooth floor and wall and has counterclockwise angular velocity ω at the instant shown. What is the bar's angular acceleration?

Solution

Draw the Free-Body Diagram   We draw the free-body diagram in Fig. (a), showing the bar's weight and the normal forces exerted by the floor and wall.

Apply the Equations of Motion   Writing the acceleration of the center of mass G as $\mathbf{a}_G = a_x\mathbf{i} + a_y\mathbf{j}$ (Fig. b), we obtain from Newton's second law,

$$\Sigma F_x = P = ma_x$$

and

$$\Sigma F_y = N - mg = ma_y.$$

Let α be the bar's counterclockwise angular acceleration (Fig. b). The equation of angular motion is

$$\Sigma M = N\left(\frac{1}{2}l\sin\theta\right) - P\left(\frac{1}{2}l\cos\theta\right) = I\alpha,$$

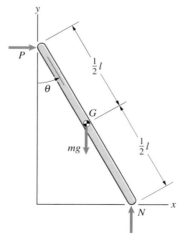

(a) Free-body diagram of the bar.

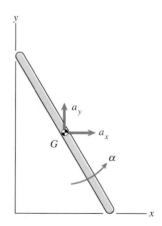

(b) Components of the acceleration of the center of mass and the angular acceleration.

where I is the moment of inertia of the bar about its center of mass. We have three equations of motion in terms of five unknowns: P, N, a_x, a_y, and α. To complete the solution, we must relate the acceleration of the center of mass of the bar to its angular acceleration.

Determine Kinematic Relationships   Although we don't know the accelerations of the endpoints A and B (Fig. c), we know that A moves horizontally and B moves vertically. We can use this information to obtain the needed relations between the acceleration of the center of mass and the angular acceleration. Expressing the acceleration of A as $\mathbf{a}_A = a_A\mathbf{i}$, we can write the acceleration of the center of mass as (Figs. b and c)

$$\mathbf{a}_G = \mathbf{a}_A + \boldsymbol{\alpha} \times \mathbf{r}_{G/A} - \omega^2\mathbf{r}_{G/A}:$$

$$a_x\mathbf{i} + a_y\mathbf{j} = a_A\mathbf{i} + \begin{vmatrix} \mathbf{i} & \mathbf{j} & \mathbf{k} \\ 0 & 0 & \alpha \\ -\dfrac{1}{2}l\sin\theta & \dfrac{1}{2}l\cos\theta & 0 \end{vmatrix} - \omega^2\left(-\dfrac{1}{2}l\sin\theta\,\mathbf{i} + \dfrac{1}{2}l\cos\theta\,\mathbf{j}\right).$$

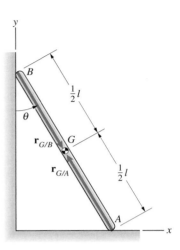

Taking advantage of the fact that $\mathbf{a}_A$ has no $\mathbf{j}$ component, we equate the $\mathbf{j}$ components in this equation, obtaining

$$a_y = -\frac{1}{2}l(\alpha\sin\theta + \omega^2\cos\theta).$$

(c) Position vectors of G relative to the endpoints A and B.

Now we express the acceleration of B as $\mathbf{a}_B = a_B\mathbf{j}$ and write the acceleration of the center of mass as

$$\mathbf{a}_G = \mathbf{a}_B + \boldsymbol{\alpha} \times \mathbf{r}_{G/B} - \omega^2\mathbf{r}_{G/B}:$$

$$a_x\mathbf{i} + a_y\mathbf{j} = a_B\mathbf{j} + \begin{vmatrix} \mathbf{i} & \mathbf{j} & \mathbf{k} \\ 0 & 0 & \alpha \\ \dfrac{1}{2}l\sin\theta & -\dfrac{1}{2}l\cos\theta & 0 \end{vmatrix} - \omega^2\left(\dfrac{1}{2}l\sin\theta\,\mathbf{i} - \dfrac{1}{2}l\cos\theta\,\mathbf{j}\right).$$

We equate the $\mathbf{i}$ components in this equation, obtaining

$$a_x = \frac{1}{2}l(\alpha\cos\theta - \omega^2\sin\theta).$$

With these two kinematic relationships, we have five equations in five unknowns. Solving them for the angular acceleration and using the relation $I = \frac{1}{12}ml^2$ for the bar's moment of inertia (Appendix C), we obtain

$$\alpha = \frac{3}{2}\frac{g}{l}\sin\theta.$$

Discussion

Notice that by expressing the acceleration of G in terms of the accelerations of the endpoints, we introduced into the solution the constraints imposed on the bar by the floor and wall: We required that point A move horizontally and that point B move vertically.

Example 18.5

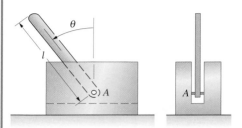

Figure 18.15

Connected Rigid Bodies

The slender bar in Fig. 18.15 has mass m and is pinned at A to a metal block of mass m_B that rests on a smooth, level surface. The system is released from rest in the position shown. What is the bar's angular acceleration at the instant of release?

Strategy

We must draw free-body diagrams of the bar and the block and apply the equations of motion to them individually. To complete the solution, we must also relate the acceleration of the bar's center of mass and its angular acceleration to the acceleration of the block.

Solution

Draw the Free-Body Diagrams   We draw the free-body diagrams of the bar and block in Fig. (a). Notice the opposite forces they exert on each other where they are pinned together.

Apply the Equations of Motion   Writing the acceleration of the center of mass of the bar as $\mathbf{a}_G = a_x\mathbf{i} + a_y\mathbf{j}$ (Fig. b), we have, from Newton's second law,

$$\Sigma F_x = A_x = ma_x$$

and

$$\Sigma F_y = A_y - mg = ma_y.$$

Letting α be the counterclockwise angular acceleration of the bar (Fig. b), its equation of angular motion is

$$\Sigma M = A_x\left(\frac{1}{2}l\cos\theta\right) + A_y\left(\frac{1}{2}l\sin\theta\right) = I\alpha.$$

We express the block's acceleration as $a_A\mathbf{i}$ (Fig. b) and write Newton's second law for the block:

$$\Sigma F_x = -A_x = m_B a_A,$$
$$\Sigma F_y = N - A_y - m_B g = 0.$$

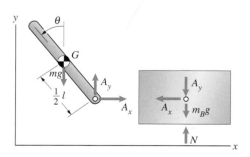

(a) Free-body diagrams of the bar and block.

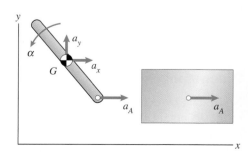

(b) Definitions of the accelerations.

Determine Kinematic Relationships To relate the bar's motion to that of the block, we express the acceleration of the bar's center of mass in terms of the acceleration of point A (Figs. b and c):

$$\mathbf{a}_G = \mathbf{a}_A + \boldsymbol{\alpha} \times \mathbf{r}_{G/A} - \omega^2 \mathbf{r}_{G/A},$$

$$a_x \mathbf{i} + a_y \mathbf{j} = a_A \mathbf{i} + \begin{vmatrix} \mathbf{i} & \mathbf{j} & \mathbf{k} \\ 0 & 0 & \alpha \\ -\dfrac{1}{2}l\sin\theta & \dfrac{1}{2}l\cos\theta & 0 \end{vmatrix} - \mathbf{0}.$$

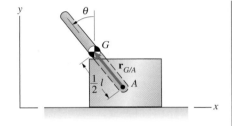

(c) Position vector of G relative to A.

Equating $\mathbf{i}$ and $\mathbf{j}$ components, we obtain

$$a_x = a_A - \frac{1}{2}l\alpha\cos\theta$$

and

$$a_y = -\frac{1}{2}l\alpha\sin\theta.$$

We have five equations of motion and two kinematic relations in terns of seven unknowns: A_x, A_y, N, a_x, a_y, α, and a_A. Solving them for the angular acceleration and using the relation $I = \frac{1}{12}ml^2$ for the bar's moment of inertia, we obtain

$$\alpha = \frac{\frac{3}{2}(g/l)\sin\theta}{1 - \frac{3}{4}\left[m/(m + m_B)\right]\cos^2\theta}.$$

Example 18.6

Wheel With an Offset Center of Mass

The drive wheel in Fig. 18.16 rolls on the horizontal track. The wheel is subjected to a downward force F_A by its axle A and a horizontal force F_C by the connecting rod. The mass of the wheel is m and the moment of inertia about its center of mass is I. The center of mass G is offset a distance b from the wheel's center. At the instant shown, the wheel has a counterclockwise angular velocity ω. What is the wheel's angular acceleration?

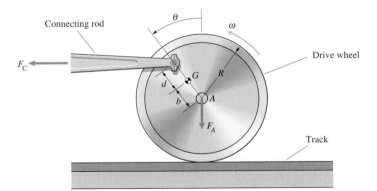

Figure 18.16

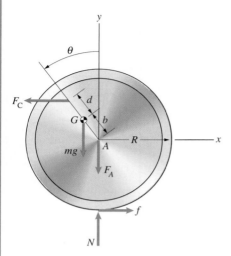

(a) Free-body diagram of the wheel.

Solution

Draw the Free-Body Diagram We draw the free-body diagram of the drive wheel in Fig. (a), showing its weight and the normal and frictional forces exerted by the track.

Apply the Equations of Motion Writing the acceleration of the center of mass G as $\mathbf{a}_G = a_x\mathbf{i} + a_y\mathbf{j}$ (Fig. b), we have, from Newton's second law,

$$\Sigma F_x = f - F_C = ma_x$$

and

$$\Sigma F_y = N - F_A - mg = ma_y.$$

Remember that we must express the equation of angular motion in terms of the sum of the moments about the center of mass, G, *not the center of the wheel*. The equation of angular motion is (Figs. a and b)

$$\Sigma M = F_C(d\cos\theta) - F_A(b\sin\theta) + N(b\sin\theta) + f(b\cos\theta + R)$$

$$= I\alpha.$$

We have three equation of motion in terms of five unknowns: N, f, a_x, a_y, and α. To complete the solution, we must relate the acceleration of the wheel's center of mass to its angular acceleration.

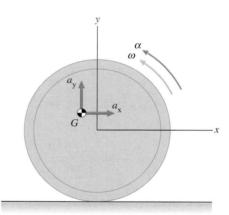

(b) Components of the acceleration of G, the angular velocity, and the angular acceleration.

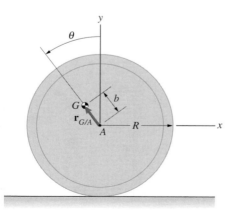

(c) Position vector of G relative to A.

Determine Kinematic Relationships The acceleration of the center A of the rolling wheel is $\mathbf{a}_A = -R\alpha\mathbf{i}$. By expressing the acceleration of the center of mass $\mathbf{a}_G$ in terms of $\mathbf{a}_A$ (Figs. b and c), we obtain relations between the components of $\mathbf{a}_G$ and α.

$$\mathbf{a}_G = \mathbf{a}_A + \boldsymbol{\alpha} \times \mathbf{r}_{G/A} - \omega^2\mathbf{r}_{G/A}:$$

$$a_x\mathbf{i} + a_y\mathbf{j} = -R\alpha\mathbf{i} + \begin{vmatrix} \mathbf{i} & \mathbf{j} & \mathbf{k} \\ 0 & 0 & \alpha \\ -b\sin\theta & b\cos\theta & 0 \end{vmatrix} - \omega^2(-b\sin\theta\mathbf{i} + b\cos\theta\mathbf{j}).$$

We equate the $\mathbf{i}$ and $\mathbf{j}$ components in this equation, obtaining

$$a_x = -R\alpha - b\alpha\cos\theta + b\omega^2\sin\theta$$

and

$$a_y = -b\alpha\sin\theta - b\omega^2\cos\theta.$$

With these two kinematic relationships, we have five equations in five unknowns. Solving them for the angular acceleration, we obtain

$$\alpha = \frac{F_C(R + b\cos\theta + d\cos\theta) + mgb\sin\theta + mbR\omega^2\sin\theta}{m(b^2 + 2bR\cos\theta + R^2) + I}.$$

Example 18.7

Application to Engineering:

Internal Forces and Moments in Beams

The slender bar of mass m in Fig. 18.17 starts from rest in the position shown and falls. When it has rotated through an angle θ, what is the maximum bending moment in the bar and where does it occur?

Figure 18.17

Strategy

The internal forces and moments in a beam subjected to two-dimensional loading are the axial force P, shear force V, and bending moment M (Fig. a). We must first use the equation of angular motion to determine the bar's angular acceleration. Then we can cut the bar at an arbitrary distance x from one end and apply the equations of motion to determine the bending moment as a function of x.

Solution

The moment of inertia of the bar about A is

$$I_A = I + d^2m = \frac{1}{12}ml^2 + \left(\frac{1}{2}l\right)^2 m = \frac{1}{3}ml^2.$$

When the bar has rotated through an angle θ (Fig. b), the total moment about A is $\Sigma M_A = mg\left(\frac{1}{2}l\sin\theta\right)$. Point A is fixed, so we can write the equation of angular motion as

$$\Sigma M_A = I_A\alpha:$$

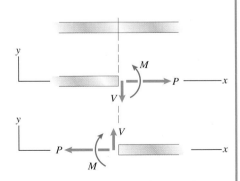

(a) The axial force, shear force, and bending moment in a beam.

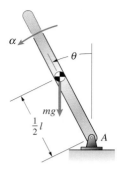

(b) Determining the moment about A.

$$\frac{1}{2} mgl \sin \theta = \frac{1}{3} ml^2 \alpha.$$

Solving for the angular acceleration, we obtain

$$\alpha = \frac{3}{2} \frac{g}{l} \sin \theta.$$

In Fig. (c), we introduce a coordinate system, cut the bar at a distance x from the top, and draw the free-body diagram of the top part. The center of mass is at the midpoint, and we determine the mass of the free body by multiplying the bar's mass by the ratio of the length of the free body to that of the bar. Applying Newton's second law in the y direction, we obtain

$$\Sigma F_y = -V - \frac{x}{l} mg \sin \theta = \frac{x}{l} ma_y.$$

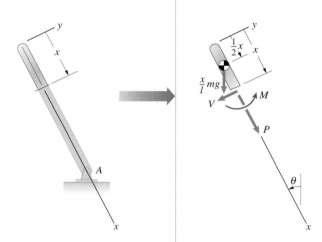

(c) Cutting the bar at an arbitrary distance x.

The moment of inertia of the free body about its center of mass is $\frac{1}{12}\left[(x/l)m\right]x^2$, so the equation of angular motion is

$$\Sigma M = I\alpha:$$

$$M - \left(\frac{1}{2}x\right)V = \frac{1}{12}\left(\frac{x}{l}m\right)x^2 \frac{3}{2}\frac{g}{l} \sin \theta.$$

The y component of the acceleration of the center of mass is equal to the product of its radial distance from A and the angular acceleration (Fig. d):

$$a_y = -\left(l - \frac{1}{2}x\right)\alpha = -\left(l - \frac{1}{2}x\right)\frac{3}{2}\frac{g}{l} \sin \theta.$$

Using this expression, we can solve the two equations of motion for V and M in terms of θ. The solution for M is

$$M = \frac{1}{4} mgl \sin \theta \left(\frac{x}{l}\right)^2\left(1 - \frac{x}{l}\right). \qquad (18.26)$$

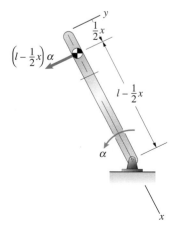

(d) Determining the acceleration of the center of mass of the free body.

The bending moment equals zero at both ends of the bar. Taking the derivative of this expression with respect to x and equating it to zero to determine where M is a maximum, we obtain $x = \frac{2}{3}l$. Substituting this value of x into Eq. (18.26), we obtain the maximum bending moment:

$$M_{\text{max}} = \frac{1}{27} mgl \sin \theta.$$

The distribution of M is shown in Fig. 18.18.

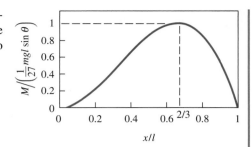

Figure 18.18
Distribution of the bending moment in a falling bar.

𝒟esign Issues

To design a member of a structure, engineers must consider both the external and internal forces and moments to which the member will be subjected. In the case of a beam, they must determine the distributions of the axial force P, shear force V, and bending moment M as the first step in determining whether the beam will support its design loads without failing. If they know the external loads and reactions, and if the beam is in equilibrium, they can apply the equilibrium equations to determine the internal forces and moment at a given cross section. But in many situations, a beam will not be in equilibrium. It could be a member of a structure, such as the internal frame of an airplane, that is accelerating, or it could be a connecting rod in an internal combustion engine. In such cases, the maximum internal forces and moments can far exceed the values predicted by a static analysis, and the procedure we describe in this example must be used.

The dynamic bending moment distribution we obtained in this example (Fig. 18.18) explains a phenomenon that has been observed during the demolition of masonry chimneys. An explosive charge at the base of the chimney causes it to fall, initially rotating as a rigid body about its base. As the chimney falls, it is observed to fracture near the location of the maximum bending moment (Fig. 18.19).

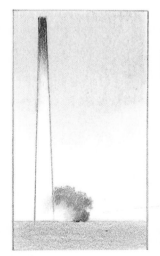  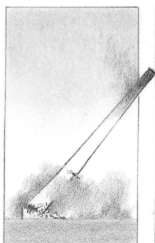 

Figure 18.19
A falling chimney fractures as it falls due to the bending moment to which it is subjected.

18.5 Numerical Solutions

When the forces and couples acting on a rigid body are known, the equations of motion can be used to determine the acceleration of the center of mass and the angular acceleration of the body. In some situations, we can then integrate to obtain closed-form expressions for the velocity and position of the center of mass and for the angular velocity and angular position as functions of time. But if the functions describing the accelerations are too complicated, or the forces and couples are known in terms of continuous or analog data instead of equations, a numerical method must be used to determine the velocities and positions as functions of time.

In Chapter 14, we described a simple finite-difference method for determining the position and velocity of the center of mass of an object as functions of time. We can determine the angular position and angular velocity in the same way. Let the angular acceleration of a rigid body be a function of time, the angular position, and angular velocity of the body:

$$\alpha = \alpha(t, \theta, \omega).$$

Suppose that at a particular time t_0, we know the angle $\theta(t_0)$ and angular velocity $\omega(t_0)$. The angular acceleration at t_0 is

$$\frac{d\omega}{dt}(t_0) = \alpha(t_0, \theta(t_0), \omega(t_0)). \qquad (18.27)$$

To determine the angular velocity at a time $t + \Delta t$, we express it as a Taylor series:

$$\omega(t_0 + \Delta t) = \omega(t_0) + \frac{d\omega}{dt}(t_0)\Delta t + \frac{1}{2}\frac{d^2\omega}{dt^2}(t_0)(\Delta t)^2 + \cdots.$$

By choosing a sufficiently small value of Δt, we can neglect terms of second and higher order in Δt and substitute Eq. (18.27) to obtain an approximation for the angular velocity at $t_0 + \Delta t$:

$$\omega(t_0 + \Delta t) = \omega(t_0) + \alpha(t_0, \theta(t_0), \omega(t_0))\Delta t. \qquad (18.28)$$

We approximate the angle at $t_0 + \Delta t$ in the same way. Expressing it as a Taylor series yields

$$\theta(t_0 + \Delta t) = \theta(t_0) + \frac{d\theta}{dt}(t_0)\Delta t + \frac{1}{2}\frac{d^2\theta}{dt^2}(t_0)(\Delta t)^2 + \cdots,$$

and neglecting higher order terms in Δt, we obtain

$$\theta(t_0 + \Delta t) = \theta(t_0) + \omega(t_0)\Delta t. \qquad (18.29)$$

With Eqs. (18.28) and (18.29), we can determine the approximate values of the angular velocity and position at $t_0 + \Delta t$. Using these values as initial conditions, we can repeat the procedure to determine the angular velocity and position at $t_0 + 2\Delta t$, and so forth.

Computational Mechanics

The following example and problems are designed to be worked with the use of a programmable calculator or computer.

Computational Example 18.8

The 18-kg ladder in Fig. 18.20 is released from rest in the position shown at $t = 0$. Neglecting friction, determine the angular position and angular velocity of the ladder as functions of time. Use time increments Δt of 0.1 s, 0.01 s, and 0.001 s.

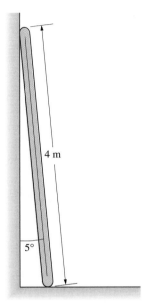

Strategy

The initial steps—drawing the free-body diagram of the ladder, applying the equations of motion, and determining the angular acceleration—are presented in Example 18.4. The ladder's angular acceleration is

$$\alpha = \frac{3g}{2l} \sin \theta,$$

where θ is the angle between the ladder and the wall and l is the length of the ladder. With this expression, we can use Eqs. (18.28) and (18.29) to approximate the ladder's angular position and angular velocity as functions of time.

Figure 18.20

Solution

The angular acceleration is

$$\alpha = \frac{(3)(9.81)}{(2)(4)} \sin \theta = 3.68 \sin \theta \ \text{rad/s}^2.$$

Let $\Delta t = 0.1$ s. At the initial time $t_0 = 0$, $\theta(t_0) = 5° = 0.0873$ rad and $\omega(t_0) = 0$. From Eq. (18.29), the angular position at time $t_0 + \Delta t = 0.1$ s is

$$\theta(t_0 + \Delta t) = \theta(t_0) + \omega(t_0)\Delta t:$$
$$\theta(0.1) = \theta(0) + \omega(0)\Delta t$$
$$= 0.0873 + (0)(0.1) = 0.0873 \text{ rad}.$$

From Eq. (18.28), the angular velocity is

$$\omega(t_0 + \Delta t) = \omega(t_0) + \alpha(t_0)\Delta t:$$
$$\omega(0.1) = 0 + \left[3.68 \sin (0.0873)\right](0.1) = 0.0321 \text{ rad/s}.$$

Using these values as the initial conditions for the next time step, we find that the angular position at $t = 0.2$ s is

$$\theta(0.2) = \theta(0.1) + \omega(0.1)\Delta t$$
$$= 0.0873 + (0.0321)(0.1) = 0.0905 \text{ rad},$$

and the angular velocity is

$$\omega(0.2) = \omega(0.1) + \alpha(0.1)\Delta t$$
$$= 0.0321 + \left[3.68 \sin (0.0873)\right](0.1) = 0.0641 \text{ rad/s}.$$

Continuing in this way, we obtain the following values for the first five time steps:

Time, s	θ, rad	ω, rad/s
0.0	0.0873	0.0000
0.1	0.0873	0.0321
0.2	0.0905	0.0641
0.3	0.0969	0.0974
0.4	0.1066	0.1329
0.5	0.1199	0.1721

Figures 18.21 and 18.22 show the numerical solutions for the angular position and angular velocity obtained using $\Delta t = 0.1$ s, $\Delta t = 0.01$ s, and $\Delta t = 0.001$ s Trials with smaller time intervals indicate that $\Delta t = 0.001$ s closely approximates the exact solution. We show the positions of the falling ladder at 0.2-s intervals in Fig. 18.23

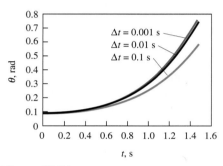

Figure 18.21
Numerical solutions for the ladder's angular position.

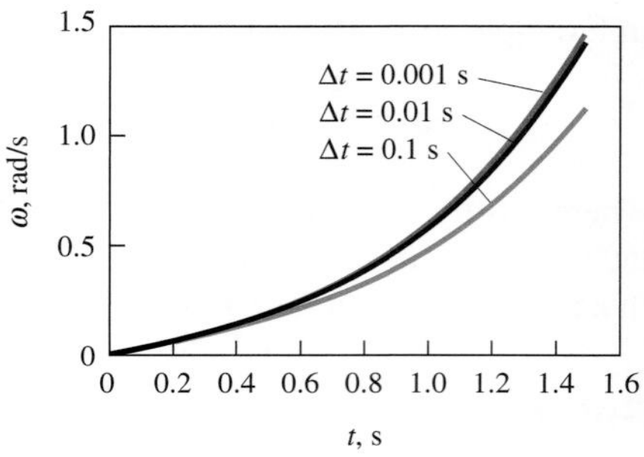

Figure 18.22
Numerical solutions for the ladder's angular velocity.

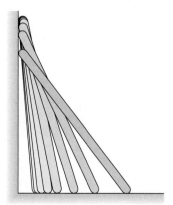

Figure 18.23
Position of the falling ladder at 0.2-s intervals from $t = 0$ to $t = 1.4$ s.

Discussion

By using the chain rule, we can write the ladder's angular acceleration as

$$\alpha = \frac{d\omega}{dt} = \frac{d\omega}{d\theta}\,\omega = \frac{3g}{2l}\sin\theta.$$

Separating variables, we can integrate to determine the angular velocity as a function of the angular position:

$$\int_0^\omega \omega \, d\omega = \int_{5°}^\theta \frac{3g}{2l} \sin \theta \, d\theta.$$

We obtain

$$\omega = \sqrt{\left(\frac{3g}{l}\right)(\cos 5° - \cos \theta)}.$$

This closed-form result is compared with the graph of our numerical solution (using $\Delta t = 0.001$ s) in Fig. 18.24. The curves are indistinguishable.

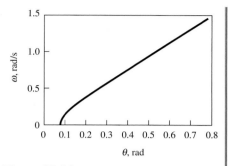

Figure 18.24
Analytical and numerical solutions for the ladder's angular velocity as a function of its angular position.

Appendix: Moments of Inertia

When a rigid body is subjected to forces and couples, the rotational motion that results depends not only on the mass of the body, but also on how its mass is *distributed*. Although the two objects in Fig. 18.25 have the same mass, the angular accelerations caused by the couple M are different. This difference is reflected in the equation of angular motion $M = I\alpha$ through the moment of inertia, I. The object in Fig. 18.25a has a smaller moment of inertia about the axis L, so its angular acceleration is greater.

In deriving the equations of motion of a rigid body in Sections 18.2 and 18.3, we modeled the body as a finite number of particles and expressed its moment of inertia about an axis L_O as

$$I_O = \sum_i m_i r_i^2,$$

where m_i is the mass of the ith particle and r_i is the perpendicular distance from L_O to the ith particle (Fig. 18.26a). To calculate moments of inertia of objects, it is often more convenient to model them as continuous distributions of mass and express the moment of inertia about L_O as

$$I_O = \int_m r^2 \, dm, \tag{18.30}$$

where r is the perpendicular distance from L_O to the differential element of mass dm (Fig. 18.26b). When the axis passes through the center of mass of the object, we denote the axis by L and the moment of inertia about L by I.

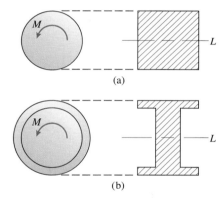

Figure 18.25
Objects of equal mass that have different moments of inertia about L.

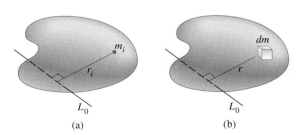

Figure 18.26
Determining the moment of inertia by modeling an object as (a) a finite number of particles and (b) a continuous distribution of mass.

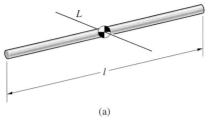

(a)

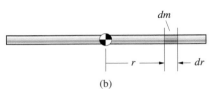

(b)

Figure 18.27
(a) A slender bar.
(b) A differential element of len;th dr.

Simple Objects

We begin by determining moments of inertia of some simple objects. Then, in the next section, we describe the parallel-axis theorem, which simplifies the task of determining moments of inertia of objects composed of combinations of simple parts.

Slender Bars   We will determine the moment of inertia of a straight slender bar about a perpendicular axis L through the center of mass of the bar (Fig. 18.27a). "Slender" means we assume that the bar's length is much greater than its width. Let the bar have length l, cross-sectional area A, and mass m. We assume that A is uniform along the length of the bar and that the material is homogeneous. Consider a differential element of the bar of length dr at a distance r from the center of mass (Fig. 18.27b). The element's mass is equal to the product of its volume and the mass density: $dm = \rho A\,dr$. Substituting this expression into Eq. (18.30), we obtain the moment of inertia of the bar about a perpendicular axis through its center of mass:

$$I = \int_m r^2\,dm = \int_{-l/2}^{l/2} \rho A r^2\,dr = \frac{1}{12}\rho A l^3.$$

The mass of the bar equals the product of the mass density and the volume of the bar ($m = \rho A l$), so we can express the moment of inertia as

$$I = \frac{1}{12}ml^2. \tag{18.31}$$

We have neglected the lateral dimensions of the bar in obtaining this result. That is, we treated the differential element of mass dm as if it were concentrated on the axis of the bar. As a consequence, Eq. (18.31) is an approximation for the moment of inertia of a bar. Later in this section, we will determine the moments of inertia for a bar of finite lateral dimension and show that Eq. (18.31) is a good approximation when the width of the bar is small in comparison to its length.

Thin Plates   Consider a homogeneous flat plate that has mass m and uniform thickness T. We will leave the shape of the cross-sectional area of the plate unspecified. Let a cartesian coordinate system be oriented so that the plate lies in the x-y plane (Fig. 18.28a). Our objective is to determine the moments of inertia of the plate about the x, y, and z axes.

We can obtain a differential element of volume of the plate by projecting an element of area dA through the thickness T of the plate (Fig. 18.28b). The resulting volume is $T\,dA$. The mass of this element of volume is equal to the product of the mass density and the volume: $dm = \rho T\,dA$. Substituting this expression into Eq. (18.30), we obtain the moment of inertia of the plate about the z axis in the form

$$I_{(z\,\text{axis})} = \int_m r^2\,dm = \rho T \int_A r^2\,dA,$$

where r is the distance from the z axis to dA. Since the mass of the plate is $m = \rho T A$, where A is the cross-sectional area of the plate, the product $\rho T = m/A$. The integral on the right is the polar moment of inertia, J_0, of the cross-sectional area of the plate. Therefore, we can write the moment of inertia of the plate about the z axis as

$$I_{(z\,\text{axis})} = \frac{m}{A}J_0. \tag{18.32}$$

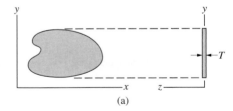

(a)

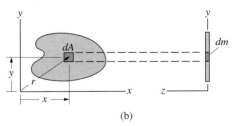

(b)

Figure 18.28
(a) A plate of arbitrary shape and uniform thickness T.
(b) An element of volume obtained by projecting an element of area dA through the plate.

From Fig. 18.28(b), we see that the perpendicular distance from the x axis to the element of area dA is the y coordinate of dA. Consequently, the moment of inertia of the plate about the x axis is

$$I_{(x \text{ axis})} = \int_m y^2 \, dm = \rho T \int_A y^2 \, dA = \frac{m}{A} I_x, \tag{18.33}$$

where I_x is the moment of inertia of the cross-sectional area of the plate about the x axis. The moment of inertia of the plate about the y axis is

$$I_{(y \text{ axis})} = \int_m x^2 \, dm = \rho T \int_A x^2 \, dA = \frac{m}{A} I_y, \tag{18.34}$$

where I_y is the moment of inertia of the cross-sectional area of the plate about the y axis.

Thus, we have expressed the moments of inertia of a thin homogeneous plate of uniform thickness in terms of the moments of inertia of the cross-sectional area of the plate. In fact, these results explain why the area integrals I_x, I_y, and J_O are called moments of inertia.

Since the sum of the area moments of inertia I_x and I_y is equal to the polar moment of inertia J_O, the moment of inertia of the thin plate about the z axis is equal to the sum of its moments of inertia about the x and y axes:

$$I_{(z \text{ axis})} = I_{(x \text{ axis})} + I_{(y \text{ axis})}. \tag{18.35}$$

Example 18.9

Moment of Inertia of an L-Shaped Bar

Two homogeneous slender bars, each of length l, mass m, and cross-sectional area A, are welded together to form an L-shaped object (Fig. 18.29). Determine the moment of inertia of the object about the axis L_O through point O. (The axis L_O is perpendicular to the two bars.)

Strategy

Using the same integration procedure we used for a single bar, we can determine the moment of inertia of each bar about L_O and sum the results.

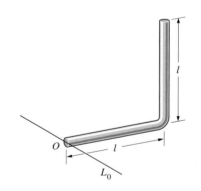

Figure 18.29

Solution

We orient a coordinate system with the z axis along L_O and the x axis colinear with bar 1 (Fig. a). The mass of the differential element of length dx of bar 1 is $dm = \rho A \, dx$. The moment of inertia of bar 1 about L_O is

$$(I_O)_1 = \int_m r^2 \, dm = \int_0^l \rho A x^2 \, dx = \frac{1}{3} \rho A l^3.$$

In terms of the mass of the bar, $m = \rho A l$, we can write this result as

$$(I_O)_1 = \frac{1}{3} m l^2.$$

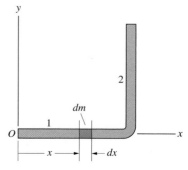

(a) Differential element of bar 1.

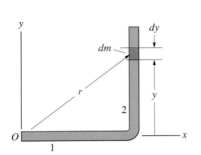

The mass of the element of length dy of bar 2 shown in Fig. (b) is $dm = \rho A\, dy$. From the figure, we see that the perpendicular distance from L_O to the element is $r = \sqrt{l^2 + y^2}$. Therefore, the moment of inertia of bar 2 about L_O is

$$(I_O)_2 = \int_m r^2 \, dm = \int_0^l \rho A(l^2 + y^2)\,dy = \frac{4}{3}\rho A l^3.$$

In terms of the mass of the bar, we obtain

$$(I_O)_2 = \frac{4}{3} ml^2.$$

The moment of inertia of the L-shaped object about L_O is

$$I_O = (I_O)_1 + (I_O)_2 = \frac{1}{3} ml^2 + \frac{4}{3} ml^2 = \frac{5}{3} ml^2.$$

(b) Differential element of bar 2.

Example 18.10

Moments of Inertia of a Triangular Plate

The thin, homogeneous plate in Fig. 18.30 is of uniform thickness and mass m. Determine its moments of inertia about the x, y and z axes.

Strategy

The moments of inertia about the x and y axes are given by Eqs. (18.33) and (18.34) in terms of the moments of inertia of the cross-sectional area of the plate. We can determine the moment of inertia of the plate about the z axis from Eq. (18.35).

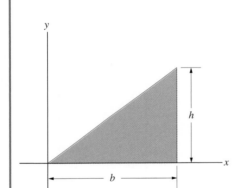

Figure 18.30

Solution

From Appendix B, the moments of inertia of the triangular area about the x and y axes are $I_x = \frac{1}{12}bh^3$ and $I_y = \frac{1}{4}hb^3$. Therefore, the moments of inertia of the plate about the x and y axes are

$$I_{(x\text{ axis})} = \frac{m}{A} I_x = \frac{m}{\frac{1}{2}bh}\left(\frac{1}{12}bh^3\right) = \frac{1}{6}mh^2$$

and

$$I_{(y\text{ axis})} = \frac{m}{A} I_y = \frac{m}{\frac{1}{2}bh}\left(\frac{1}{4}hb^3\right) = \frac{1}{2}mb^2.$$

The moment of inertia about the z axis is

$$I_{(z\text{ axis})} = I_{(x\text{ axis})} + I_{(y\text{ axis})} = m\left(\frac{1}{6}h^2 + \frac{1}{2}b^2\right).$$

Parallel-Axis Theorem

The parallel-axis theorem allows us to determine the moment of inertia of a composite object when we know the moments of inertia of its parts. Suppose that we know the moment of inertia I about an axis L through the center of mass of an object and we wish to determine its moment of inertia I_O about a parallel axis L_O (Fig. 18.31a). To determine I_O, we introduce parallel coordinate systems xyz and $x'y'z'$, with the z axis along L_O and the z' axis along L, as shown in Fig. 18.31b. (In this figure, the axes L_O and L are perpendicular to the page.) The origin O of the xyz coordinate system is contained in the x'-y' plane. The terms d_x and d_y are the coordinates of the center of mass relative to the xyz coordinate system.

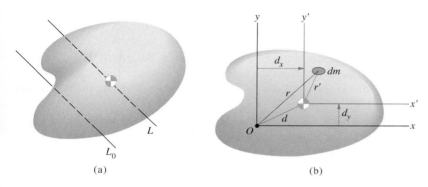

(a)   (b)

Figure 18.31
(a) An axis L through the center of mass of an object and a parallel axis L_0.
(b) The xyz and $x'y'z'$ coordinate systems.

The moment of inertia of the object about L_O is

$$I_O = \int_m r^2 \, dm = \int_m (x^2 + y^2) \, dm, \tag{18.36}$$

where r is the perpendicular distance from L_O to the differential element of mass dm and x, y are the coordinates of dm in the x-y plane. The x-y coordinates of dm are related to its x'-y' coordinates by

$$x = x' + d_x$$

and

$$y = y' + d_y.$$

Substituting these expressions into Eq. (18.36), we can write that equation as

$$I_O = \int_m \left[(x')^2 + (y')^2 \right] dm + 2d_x \int_m x' \, dm + 2d_y \int_m y' \, dm$$
$$+ \int_m (d_x^2 + d_y^2) \, dm. \tag{18.37}$$

Since $(x')^2 + (y')^2 = (r')^2$, where r' is the perpendicular distance from L to dm, the first integral on the right side of this equation is the moment of inertia I of the object about L. Recall that the x' and y' coordinates of the center of mass of the object relative to the $x'y'z'$ coordinate system are defined by

$$\bar{x}' = \frac{\displaystyle\int_m x' \, dm}{\displaystyle\int_m dm}, \qquad \bar{y}' = \frac{\displaystyle\int_m y' \, dm}{\displaystyle\int_m dm}.$$

Because the center of mass of the object is at the origin of the $x'y'z'$ system, $\bar{x}' = 0$ and $\bar{y}' = 0$. Therefore, the integrals in the second and third terms on the right side of Eq. (18.37) are equal to zero. From Fig. 18.31(b), we see that $d_x^2 + d_y^2 = d^2$, where d is the perpendicular distance between the axes L and L_O. Therefore, we obtain

$$I_O = I + d^2 m, \tag{18.38}$$

where m is the mass of the object. This is the *parallel-axis theorem*. If the moment of inertia of an object about a given axis is known, we can use the parallel-axis theorem to determine the moment of inertia of the object about any parallel axis. This makes it possible to determine moments of inertia of composite objects. Determining the moment of inertia of a composite object about a given axis L_O typically requires three steps:

1. *Choose the parts.* Try to divide the object into parts whose moments of inertia can easily be determined.
2. *Determine the moments of inertia of the parts.* Determine the moment of inertia of each part about the axis through its center of mass parallel to L_O. Then use the parallel-axis theorem to determine its moment of inertia about L_O.
3. *Sum the results.* Sum the moments of inertia of the parts (or subtract in the case of a hole or cutout) to obtain the moment of inertia of the composite object.

Example 18.11

Application of the Parallel-Axis Theorem

Two homogeneous, slender bars, each of length l and mass m, are welded together to form an L-shaped object (Fig. 18.32). Determine the moment of inertia of the object about the axis L_O through point O. (The axis L_O is perpendicular to the two bars.)

Figure 18.32

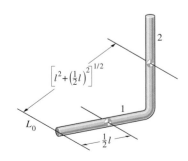

(a) The distances from L_O to parallel axes through the centers of mass of bars 1 and 2.

Solution

Choose the Parts The parts are the two bars, which we call bar 1 and bar 2 (Fig. a).

Determine the Moments of Inertia of the Parts From Appendix C, the moment of inertia of each bar about a perpendicular axis through its center of mass is $I = \frac{1}{12} ml^2$. The distance from L_O to the parallel axis through the center of mass of bar 1 is $\frac{1}{2} l$ (Fig. a). Therefore, the moment of inertia of bar 1 about L_O is

$$(I_O)_1 = I + d^2 m = \frac{1}{12} ml^2 + \left(\frac{1}{2} l\right)^2 m = \frac{1}{3} ml^2.$$

The distance from L_O to the parallel axis through the center of mass of bar 2 is $\left[l^2 + \left(\frac{1}{12} l\right)^2 \right]^{1/2}$. The moment of inertia of bar 2 about L_O is

$$(I_O)_2 = I + d^2 m = \frac{1}{12} ml^2 + \left[l^2 + \left(\frac{1}{2} l\right)^2 \right] m = \frac{4}{3} ml^2.$$

Sum the Results   The moment of inertia of the L-shaped object about L_O is

$$I_O = (I_O)_1 + (I_O)_2 = \frac{1}{3}ml^2 + \frac{4}{3}ml^2 = \frac{5}{3}ml^2.$$

Discussion

Compare this solution with Example 18.9, in which we used integration to determine the moment of inertia of the same object about L_O. We obtained the result much more easily with the parallel-axis theorem, but of course, we needed to know the moments of inertia of the bars about the axes through their centers of mass.

Example 18.12

Moments of Inertia of a Composite Object

The object in Fig. 18.33 consists of a slender 3-kg bar welded to a thin, circular 2-kg disk. Determine the moment of inertia of the object about the axis L through its center of mass. (The axis L is perpendicular to the bar and disk.)

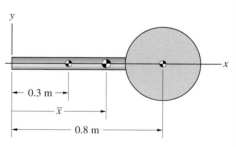

Figure 18.33

Strategy

We must locate the center of mass of the composite object and then apply the parallel-axis theorem. We can obtain the moments of inertia of the bar and disk from Appendix C.

Solution

Choose the Parts   The parts are the bar and the disk. Introducing the coordinate system in Fig. (a), we have, for the x coordinate of the center of mass of the composite object,

$$\bar{x} = \frac{\bar{x}_{(bar)}m_{(bar)} + \bar{x}_{(disk)}m_{(disk)}}{m_{(bar)} + m_{(disk)}} = \frac{(0.3)(3) + (0.6 + 0.2)(2)}{3 + 2} = 0.5 \text{ m}.$$

(a) The coordinate $\bar{x}$ of the center of mass of the object.

Determine the Moments of Inertia of the Parts   The distance from the center of mass of the bar to the center of mass of the composite object is 0.2 m (Fig. b). Therefore, the moment of inertia of the bar about L is

$$I_{(bar)} = \frac{1}{12}(3)(0.6)^2 + (0.2)^2(3) = 0.210 \text{ kg-m}^2.$$

The distance from the center of mass of the disk to the center of mass of the composite object is 0.3 m (Fig. c). The moment of inertia of the disk about L is

$$I_{(disk)} = \frac{1}{2}(2)(0.2)^2 + (0.3)^2(2) = 0.220 \text{ kg-m}^2.$$

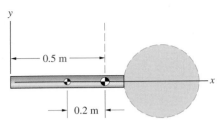

(b) Distance from L to the center of mass of the bar.

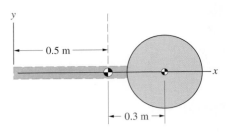

(c) Distance from L to the center of mass of the disk.

Sum the Results The moment of inertia of the composite object about L is

$$I = I_{(bar)} + I_{(disk)} = 0.430 \text{ kg-m}^2.$$

Example 18.13

Moments of Inertia of a Homogeneous Cylinder

The homogeneous cylinder in Fig. 18.34 has mass m, length l, and radius R. Determine the moments of inertia of the cylinder about the x, y, and z axes.

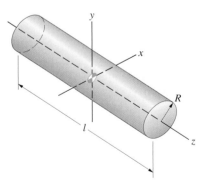

Figure 18.34

Strategy

We can determine the moments of inertia of the cylinder by an interesting application of the parallel-axis theorem. We use it to determine the moments of inertia about the x, y, and z axes of an infinitesimal element of the cylinder consisting of a disk of thickness dz. Then we integrate the results with respect to z to obtain the moments of inertia of the cylinder.

Solution

Consider an element of the cylinder of thickness dz at a distance z from the center of the cylinder (Fig. a). (You can imagine obtaining this element by "slicing" the cylinder perpendicular to its axis.) The mass of the element is equal to the product of the mass density and the volume of the element: $dm = \rho(\pi R^2 \, dz)$. We obtain the moments of inertia of the element by using the values for a thin circular plate given in Appendix C. The moment of inertia about the z axis is

$$dI_{(z \text{ axis})} = \frac{1}{2} \, dm \, R^2 = \frac{1}{2} \left(\rho \pi R^2 \, dz \right) R^2.$$

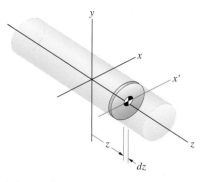

(a) A differential element of the cylinder in the form of a disk.

We integrate this result with respect to z from $-l/2$ to $l/2$, thereby summing the moments of inertia of the infinitesimal disk elements that make up the cylinder. The result is the moment of inertia of the cylinder about the z axis:

$$I_{(z \text{ axis})} = \int_{-l/2}^{l/2} \frac{1}{2} \rho \pi R^4 \, dz = \frac{1}{2} \rho \pi R^4 l.$$

We can write this result in terms of the mass of the cylinder, $m = \rho(\pi R^2 l)$, as

$$I_{(z \text{ axis})} = \frac{1}{2} m R^2.$$

The moment of inertia of the disk element about the x' axis is

$$dI_{(x' \text{ axis})} = \frac{1}{4} \, dm \, R^2 = \frac{1}{4} \left(\rho \pi R^2 \, dz \right) R^2.$$

We use this result and the parallel-axis theorem to determine the moment of inertia of the element about the x axis:

$$dI_{(x \text{ axis})} = dI_{(x' \text{ axis})} + z^2 \, dm = \frac{1}{4} \left(\rho \pi R^2 \, dz \right) R^2 + z^2 \left(\rho \pi R^2 \, dz \right).$$

Integrating this expression with respect to z from $-l/2$ to $l/2$, we obtain the moment of inertia of the cylinder about the x axis:

$$I_{(x \text{ axis})} = \int_{-l/2}^{l/2} \left(\frac{1}{4} \rho \pi R^4 + \rho \pi R^2 z^2 \right) dz = \frac{1}{4} \rho \pi R^4 l + \frac{1}{12} \rho \pi R^2 l^3.$$

In terms of the mass of the cylinder,

$$I_{(x \text{ axis})} = \frac{1}{4} m R^2 + \frac{1}{12} m l^2.$$

Due to the symmetry of the cylinder,

$$I_{(y \text{ axis})} = I_{(x \text{ axis})}.$$

Discussion

When the cylinder is very long in comparison to its width, $(l \gg R)$, the first term in the equation for $I_{(x \text{ axis})}$ can be neglected, and we obtain the moment of inertia of a slender bar about a perpendicular axis, Eq. (18.31). On the other hand, when the radius of the cylinder is much greater than its length, $(R \gg l)$, the second term in the equation for $I_{(x \text{ axis})}$ can be neglected, and we obtain the moment of inertia for a thin circular disk about an axis parallel to the disk. This indicates the sizes of the terms you neglect when you use the approximate expressions for the moments of inertia of a "slender" bar and a "thin" disk.

Chapter Summary

In this chapter, we derived the equations of planar motion for a rigid body. We used those equations together with the kinematics relationships developed in Chapter 17 to determine motions of rigid bodies resulting from the forces and couples acting on them. In Chapter 19, we will apply energy and momentum methods to the motions of rigid bodies and show that those methods can greatly simplify the solution of particular types of problems.

Moment–Angular-Momentum Relations

Let $\mathbf{r}_i$ be the position of the ith particle of a system of particles relative to a fixed point O, $\mathbf{r}$ the position of the center of mass of the system, and $\mathbf{R}_i$ the position of the ith particle relative to the center of mass. The sum of the moments due to external forces about O is equal to the rate of change of the total angular momentum about O:

$$\Sigma \mathbf{M}_O = \frac{d\mathbf{H}_O}{dt}. \qquad \text{Eq. (18.8)}$$

The angular momentum about O is

$$\mathbf{H}_O = \sum_i \mathbf{r}_i \times m_i \mathbf{v}_i, \qquad \text{Eq. (18.9)}$$

where m_i is the mass of the ith particle and $\mathbf{v}_i$ is its velocity. This relationship can also be written as

$$\Sigma \mathbf{M}_O = \frac{d}{dt} (\mathbf{r} \times m\mathbf{v} + \mathbf{H}), \qquad \text{Eq. (18.10)}$$

where m is the total mass of the system of particles, $\mathbf{v}$ is the velocity of the center of mass, and

$$\mathbf{H} = \sum_i \mathbf{R}_i \times m_i \frac{d\mathbf{R}_i}{dt} \qquad \text{Eq. (18.11)}$$

is the total angular momentum about the center of mass. The sum of the moments due to external forces about the center of mass is equal to the rate of change of the total angular momentum about the center of mass:

$$\Sigma \mathbf{M} = \frac{d\mathbf{H}}{dt}.$$
Eq. (18.12)

The angular momenta about point O and about the center of mass are related by

$$\mathbf{H}_O = \mathbf{H} + \mathbf{r} \times m\mathbf{v}.$$
Eq. (18.13)

Equations of Planar Motion

The equations of motion for a rigid body in planar motion include Newton's second law,

$$\Sigma \mathbf{F} = m\mathbf{a},$$
Eq. (18.21)

where $\mathbf{a}$ is the acceleration of the center of mass of the body. If the rigid body rotates about a fixed point O, the total moment about O equals the product of the moment of inertia about O and the angular acceleration (Fig. a):

$$\Sigma M_O = I_O \alpha.$$
Eq. (18.22)

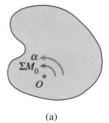

(a)

In any planar motion, the total moment about the center of mass equals the product of the moment of inertia about the center of mass and the angular acceleration (Fig. b):

$$\Sigma M = I\alpha.$$
Eq. (18.23)

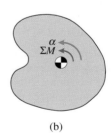

(b)

If a rigid body is in translation, Newton's second law is sufficient to determine its motion. Nevertheless, the angular equation of motion may be needed to determine unknown forces or couples. Since $\alpha = 0$, the total moment about the center of mass equals zero. In the case of rotation about a fixed axis, Eq. (18.22) is sufficient to determine the rotational motion, although Newton's second law may be needed to determine unknown forces or couples. If a rigid body undergoes general planar motion, both Newton's second law and the equation of angular motion are needed.

Moments of Inertia

The moment of inertia of an object about an axis L_O is

$$I_O = \int_m r^2 \, dm,$$
Eq. (18.30)

where r is the perpendicular distance from L_O to the differential element of mass dm (Fig. c).

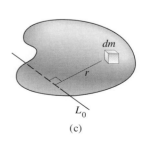

(c)

Let L be an axis through the center of mass of an object, and let L_O be a parallel axis. The moment of inertia, I_O, about L_O is given in terms of the moment of inertia, I, about L by the *parallel-axis theorem*

$$I_O = I + d^2 m, \qquad \text{Eq. (18.38)}$$

where m is the mass of the object and d is the distance between L and L_O.

Review Problems

18.1 The mass of each box is 20 kg. The moment of inertia of the pulley is 0.8 kg-m^2. The coefficient of kinetic friction between the boxes and the surface is $\mu_k = 0.1$. If the system starts from rest at $t = 0$, what distance have the boxes moved from their original position at $t = 1$ s?

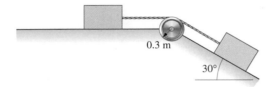

P18.1

18.2 The airplane is at the beginning of its takeoff run. Its weight is 1000 lb, and the initial thrust T exerted by its engine is 300 lb. Assume that the thrust is horizontal, and neglect the tangential forces exerted on the wheels.

(a) If the acceleration of the airplane remains constant, how long will it take to reach its takeoff speed of 80 mi/hr?
(b) Determine the normal force exerted on the forward landing gear at the beginning of the takeoff run.

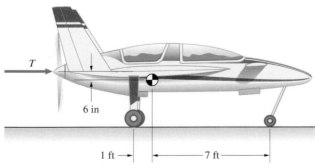

P18.2

18.3 The pulleys can turn freely on their pin supports. Their moments of inertia are $I_A = 0.002$ kg-m^2, $I_B = 0.036$ kg-m^2, and $I_C = 0.032$ kg-m^2. They are initially stationary, and at $t = 0$ a constant couple $M = 2$ N-m is applied to pulley A. What is the angular velocity of pulley C and how many revolutions has it turned at $t = 2$ s?

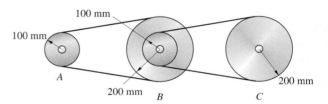

P18.3

18.4 A 2-kg box is subjected to a 40-N horizontal force. Neglect friction.

(a) If the box remains on the floor, what is its acceleration?
(b) Determine the range of values of c for which the box will remain on the floor when the force is applied.

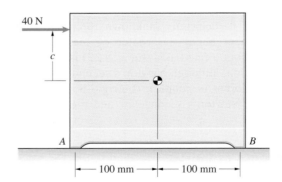

P18.4

18.5 The slender 2-slug bar AB is 3 ft long. It is pinned to the cart at A and leans against it at B.

(a) If the acceleration of the cart is $a = 20$ ft/s², what normal force is exerted on the bar by the cart at B?
(b) What is the largest acceleration a for which the bar will remain in contact with the surface at B?

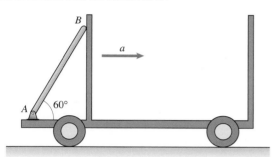

P18.5

18.6 To determine a 4.5-kg tire's moment of inertia, an engineer lets the tire roll down an inclined surface. If it takes the tire 3.5 s to start from rest and roll 3 m down the surface, what is the tire's moment of inertia about its center of mass?

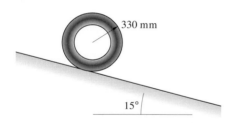

P18.6

18.7 Pulley A weighs 4 lb, $I_A = 0.060$ slug-ft², and $I_B = 0.014$ slug-ft². If the system is released from rest, what distance does the 16-lb weight fall in 0.5 s?

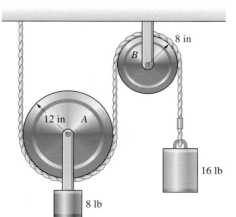

P18.7

18.8 Model the excavator's arm ABC as a single rigid body. Its mass is 1200 kg, and the moment of inertia *about its center of mass* is $I = 3600$ kg-m². If point A is stationary, the angular velocity of the arm is zero. and its angular acceleration is 1.0 rad/s² counterclockwise, what force does the vertical hydraulic cylinder exert on the arm at B?

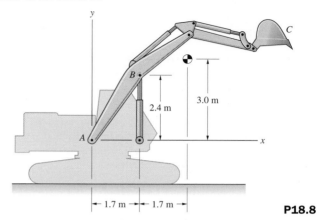

P18.8

18.9 In Problem 18.8, if the angular acceleration of arm ABC is 1.0 rad/s² counterclockwise and its angular velocity is 2.0 rad/s counterclockwise, what are the components of the force exerted on the arm at A?

18.10 To decrease the angle of elevation of the stationary 200-kg ladder, the gears that raised it are disengaged, and a fraction of a second later a second set of gears that lower it are engaged. At the instant the gears that raised the ladder are disengaged, what is the ladder's angular acceleration and what are the components of force exerted on the ladder by its support at O? The moment of inertia of the ladder about O is $I_O = 14{,}000$ kg-m², and the coordinates of its center of mass at the instant the gears are disengaged are $\bar{x} = 3$ m, $\bar{y} = 4$ m.

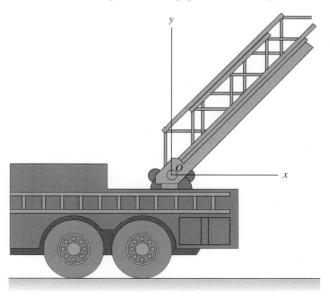

P18.10

18.11 The slender bars each weigh 4 lb and are 10 in. long. The homogeneous plate weighs 10 lb. If the system is released from rest in the position shown, what is the angular acceleration of the bars at that instant?

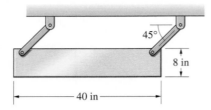

P18.11

18.12 A slender bar of mass m is released from rest in the position shown. The static and kinetic coefficients of friction at the floor and wall have the same value μ. If the bar slips, what is its angular acceleration at the instant of release?

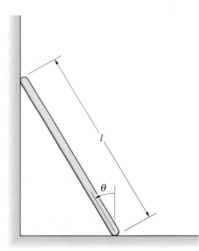

P18.12

18.13 Each of the go-cart's front wheels weighs 5 lb and has a moment of inertia of 0.01 slug-ft^2. The two rear wheels and rear axle form a single rigid body weighing 40 lb and having a moment of inertia of 0.1 slug-ft^2. The total weight of the go-cart and driver is 240 lb. (The location of the center of mass of the go-cart and driver, *not including* the front wheels or the rear wheels and rear axle, is shown.) If the engine exerts a torque of 12 ft-lb on the rear axle, what is the go-cart's acceleration?

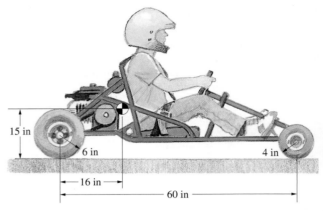

P18.13

18.14 Bar AB rotates with a constant angular velocity of 10 rad/s in the counterclockwise direction. The masses of the slender bars BC and CDE are 2 kg and 3.6 kg, respectively. The y axis points upward. Determine the components of the forces exerted on bar BC by the pins at B and C at the instant shown.

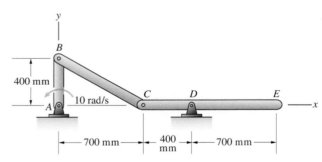

P18.14

18.15 At the instant shown, the arms of the robotic manipulator have constant counterclockwise angular velocities $\omega_{AB} = -0.5$ rad/s, $\omega_{BC} = 2$ rad/s, and $\omega_{CD} = 4$ rad/s. The mass of arm CD is 10 kg, and its center of mass is at its midpoint. At this instant, what force and couple are exerted on arm CD at C?

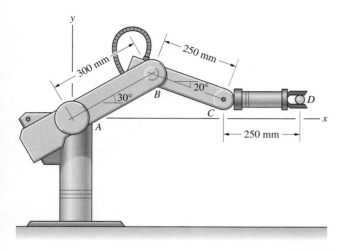

P18.15

be modeled as a slender bar. If a 1 kN-m counterclockwise couple is applied to the sun gear, what is the resulting angular acceleration of the bonded hub and planet gears?

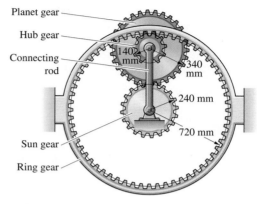

Planet gear
Hub gear
Connecting rod
140 mm
340 mm
240 mm
720 mm
Sun gear
Ring gear

P18.18

18.16 Each bar is 1 m in length and has a mass of 4 kg. The inclined surface is smooth. If the system is released from rest in the position shown, what are the angular accelerations of the bars at that instant?

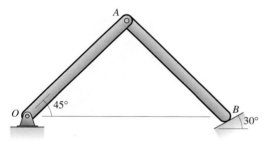

P18.16

18.17 At the instant the system in Problem 18.16 is released, what is the magnitude of the force exerted on bar OA by the support at O?

18.18 The fixed ring gear lies *in the horizontal plane*. The hub and planet gears are bonded together. The mass and moment of inertia of the combined hub and planet gears are $m_{HP} = 130$ kg and $I_{HP} = 130$ kg-m². The moment of inertia of the sun gear is $I_S = 60$ kg-m². The mass of the connecting rod is 5 kg, and it can

18.19 The system is stationary at the instant shown. The net force exerted on the piston by the exploding fuel–air mixture and friction is 5 kN to the left. A clockwise couple $M = 200$ N-m acts on the crank AB. The moment of inertia of the crank about A is 0.0003 kg-m². The mass of the connecting rod BC is 0.36 kg, and its center of mass is 40 mm from B on the line from B to C. The connecting rod's moment of inertia about its center of mass is 0.0004 kg-m². The mass of the piston is 4.6 kg. What is the piston's acceleration? (Neglect the gravitational forces on the crank and connecting rod.)

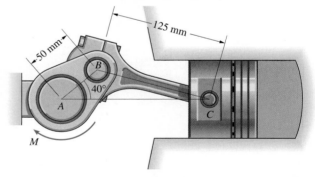

P18.19

18.20 If the crank AB in Problem 18.19 has a counterclockwise angular velocity of 2000 rpm at the instant shown, what is the piston's acceleration?

The wind exerts torque on the wind generators, performing work that is transformed into electrical energy. In this chapter, we use energy and momentum methods to analyze motions of rigid bodies.

Energy and Momentum in Rigid-Body Dynamics

I n Chapters 15 and 16 we demonstrated that energy and momentum methods are useful for solving particular types of problems in dynamics. If the forces on an object are known as functions of position, the principle of work and energy can be used to determine the change in the magnitude of the velocity of the object as it moves between two positions. If the forces are known as functions of time, the principle of impulse and momentum can be used to determine the change in the object's velocity during an interval of time. We now extend these methods to situations in which both the translational and rotational motions of objects must be considered.

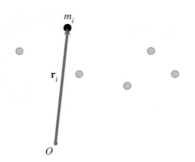

Figure 19.1
A system of particles. The vector $\mathbf{r}_i$ is the position vector of the ith particle.

19.1 Principle of Work and Energy

We will show that the work done on a rigid body by external forces and couples as it moves between two positions is equal to the change in its kinetic energy. To obtain this result, we adopt the same approach used in Chapter 18 to obtain the equations of motion for a rigid body. We derive the principle of work and energy for a system of particles and use it to deduce the principle for a rigid body.

Let m_i be the mass of the ith particle of a system of N particles. Let $\mathbf{r}_i$ be the position of the ith particle relative to a point O that is fixed with respect to an inertial reference frame (Fig. 19.1). We denote the sum of the kinetic energies of the particles by

$$T = \sum_i \tfrac{1}{2} m_i \mathbf{v}_i \cdot \mathbf{v}_i, \tag{19.1}$$

where $\mathbf{v}_i = d\mathbf{r}_i/dt$ is the velocity of the ith particle. Our objective is to relate the work done on the system of particles to the change in T. We begin with Newton's second law for the ith particle,

$$\sum_j \mathbf{f}_{ij} + \mathbf{f}_i^{\text{E}} = \frac{d}{dt}(m_i \mathbf{v}_i),$$

where $\mathbf{f}_{ij}$ is the force exerted on the ith particle by the jth particle and $\mathbf{f}_i^{\text{E}}$ is the external force on the ith particle. We take the dot product of this equation with $\mathbf{v}_i$ and sum from $i = 1$ to N:

$$\sum_i \sum_j \mathbf{f}_{ij} \cdot \mathbf{v}_i + \sum_i \mathbf{f}_i^{\text{E}} \cdot \mathbf{v}_i = \sum_i \mathbf{v}_i \cdot \frac{d}{dt}(m_i \mathbf{v}_i). \tag{19.2}$$

We can express the term on the right side of this equation as the rate of change of the total kinetic energy:

$$\sum_i \mathbf{v}_i \cdot \frac{d}{dt}(m_i \mathbf{v}_i) = \frac{d}{dt} \sum_i \frac{1}{2} m_i \mathbf{v}_i \cdot \mathbf{v}_i = \frac{dT}{dt}.$$

Multiplying Eq. (19.2) by dt yields

$$\sum_i \sum_j \mathbf{f}_{ij} \cdot d\mathbf{r}_i + \sum_i \mathbf{f}_i^{\text{E}} \cdot d\mathbf{r}_i = dT.$$

We integrate this equation, obtaining

$$\sum_i \sum_j \int_{(\mathbf{r}_i)_1}^{(\mathbf{r}_i)_2} \mathbf{f}_{ij} \cdot d\mathbf{r}_i + \sum_i \int_{(\mathbf{r}_i)_1}^{(\mathbf{r}_i)_2} \mathbf{f}_i^{\text{E}} \cdot d\mathbf{r}_i = T_2 - T_1. \tag{19.3}$$

The terms on the left side are the work done on the system by internal and external forces as the particles move from positions $(\mathbf{r}_i)_1$ to positions $(\mathbf{r}_i)_2$. We see that the work done by internal and external forces as a system of particles moves between two positions equals the change in the total kinetic energy of the system.

If the particles represent a rigid body, and we assume that the internal forces between each pair of particles are directed along the straight line between them, the work done by internal forces is zero. To show that this is true, we consider two particles of a rigid body designated 1 and 2 (Fig. 19.2).

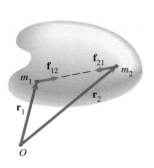

Figure 19.2
Particles 1 and 2 and the forces they exert on each other.

The sum of the forces the two particles exert on each other is zero $(\mathbf{f}_{12} + \mathbf{f}_{21} = \mathbf{0})$, so the rate at which the forces do work (the power) is

$$\mathbf{f}_{12} \cdot \mathbf{v}_1 + \mathbf{f}_{21} \cdot \mathbf{v}_2 = \mathbf{f}_{21} \cdot (\mathbf{v}_2 - \mathbf{v}_1).$$

We can show that $\mathbf{f}_{21}$ is perpendicular to $\mathbf{v}_2 - \mathbf{v}_1$, and therefore the rate at which work is done by the internal forces between these two particles is zero. Because the particles are points of a rigid body, we can express their relative velocity in terms of the rigid body's angular velocity $\boldsymbol{\omega}$ as

$$\mathbf{v}_2 - \mathbf{v}_1 = \boldsymbol{\omega} \times (\mathbf{r}_2 - \mathbf{r}_1). \tag{19.4}$$

This equation shows that the relative velocity $\mathbf{v}_2 - \mathbf{v}_1$ is perpendicular to $\mathbf{r}_2 - \mathbf{r}_1$, which is the position vector from particle 1 to particle 2. Since the force $\mathbf{f}_{21}$ is parallel to $\mathbf{r}_2 - \mathbf{r}_1$, it is perpendicular to $\mathbf{v}_2 - \mathbf{v}_1$. We can repeat this argument for each pair of particles of the rigid body, so the total rate at which work is done by internal forces is zero. This implies that the work done by internal forces as a rigid body moves between two positions is zero.

Therefore, in the case of a rigid body, the work done by internal forces in Eq. (19.3) vanishes. Denoting the work done by external forces by U_{12}, we obtain the principle of work and energy for a rigid body: *The work done by external forces and couples as a rigid body moves between two positions equals the change in the total kinetic energy of the body*:

$$U_{12} = T_2 - T_1. \tag{19.5}$$

We can also state this principle for a *system* of rigid bodies: *The work done by external and internal forces as a system of rigid bodies moves between two positions equals the change in the total kinetic energy of the system.*

19.2 Kinetic Energy

The kinetic energy of a rigid body can be expressed in terms of the velocity of the center of mass of the body and its angular velocity. We consider first general planar motion and then rotation about a fixed axis.

General Planar Motion

Let us model a rigid body as a system of particles, and let $\mathbf{R}_i$ be the position vector of the ith particle relative to the body's center of mass (Fig. 19.3). The position of the center of mass is

$$\mathbf{r} = \frac{\sum\limits_i m_i \mathbf{r}_i}{m}, \tag{19.6}$$

where m is the mass of the rigid body. The position of the ith particle relative to O is related to its position relative to the center of mass by

$$\mathbf{r}_i = \mathbf{r} + \mathbf{R}_i, \tag{19.7}$$

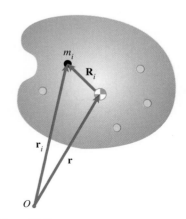

Figure 19.3
Representing a rigid body as a system of particles.

and the vectors $\mathbf{R}_i$ satisfy the relation

$$\sum_i m_i \mathbf{R}_i = \mathbf{0}. \tag{19.8}$$

The kinetic energy of the rigid body is the sum of the kinetic energies of its particles, given by Eq. (19.1):

$$T = \sum_i \tfrac{1}{2} m_i \mathbf{v}_i \cdot \mathbf{v}_i. \tag{19.9}$$

By taking the derivative of Eq. (19.7) with respect to time, we obtain

$$\mathbf{v}_i = \mathbf{v} + \frac{d\mathbf{R}_i}{dt},$$

where $\mathbf{v}$ is the velocity of the center of mass. Substituting this expression into Eq. (19.9) and using Eq. (19.8), we obtain the kinetic energy of the rigid body in the form

$$T = \tfrac{1}{2} m v^2 + \sum_i \frac{1}{2} m_i \frac{d\mathbf{R}_i}{dt} \cdot \frac{d\mathbf{R}_i}{dt}, \tag{19.10}$$

where v is the magnitude of the velocity of the center of mass.

Now, let L_0 be the axis through a fixed point O that is perpendicular to the plane of the motion, and let L be the parallel axis through the center of mass (Fig. 19.4a). Then, in terms of the coordinate system shown, we can express the angular velocity vector as $\boldsymbol{\omega} = \omega \mathbf{k}$. The velocity of the ith particle relative to the center of mass is $d\mathbf{R}_i/dt = \omega \mathbf{k} \times \mathbf{R}_i$, so we can write Eq. (19.10) as

$$T = \tfrac{1}{2} m v^2 + \tfrac{1}{2} \left[\sum_i m_i (\mathbf{k} \times \mathbf{R}_i) \cdot (\mathbf{k} \times \mathbf{R}_i) \right] \omega^2. \tag{19.11}$$

The magnitude of the vector $\mathbf{k} \times \mathbf{R}_i$ is the perpendicular distance r_i from L to the ith particle (Fig. 19.4b), so the term in brackets in Eq. (19.11) is the moment of inertia of the body about L:

$$\sum_i m_i (\mathbf{k} \times \mathbf{R}_i) \cdot (\mathbf{k} \times \mathbf{R}_i) = \sum_i m_i |\mathbf{k} \times \mathbf{R}_i|^2 = \sum_i m_i r_i^2 = I.$$

Thus, we obtain the kinetic energy of a rigid body in general planar motion in the form

$$T = \tfrac{1}{2} m v^2 + \tfrac{1}{2} I \omega^2. \tag{19.12}$$

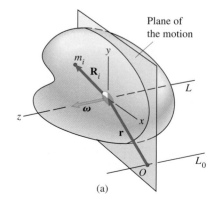

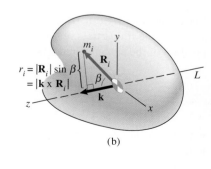

Figure 19.4
(a) A coordinate system with the z axis aligned with L.
(b) The magnitude of $\mathbf{k} \times \mathbf{R}_i$ is the perpendicular distance from L to m_i.

The kinetic energy consists of two terms: the *translational kinetic energy*, expressed in terms of the velocity of the center of mass, and the *rotational kinetic energy* (Fig. 19.5).

Fixed-Axis Rotation

An object rotating about a fixed axis is in general planar motion, and its kinetic energy is given by Eq. (19.12). But in this case there is another expression for the kinetic energy that is often convenient. Suppose that a rigid body rotates with angular velocity ω about a fixed axis O. In terms of the distance d from O to the center of mass of the body, the velocity of the center of mass is $v = \omega d$ (Fig. 19.6a). From Eq. (19.12), the kinetic energy is

$$T = \tfrac{1}{2}m(\omega d)^2 + \tfrac{1}{2}I\omega^2 = \tfrac{1}{2}(I + d^2m)\omega^2.$$

According to the parallel-axis theorem, the moment of inertia about O is $I_0 = I + d^2m$, so we obtain the kinetic energy of a rigid body rotating about a fixed axis O in the form (Fig. 19.6b)

$$T = \tfrac{1}{2}I_0\omega^2. \tag{19.13}$$

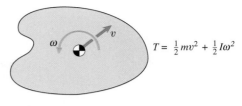

Figure 19.5
Kinetic energy in general planar motion.

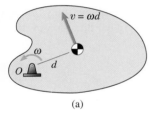
(a)

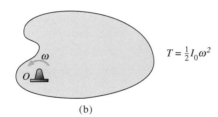
(b)

Figure 19.6
(a) Velocity of the center of mass.
(b) Kinetic energy of a rigid body rotating about a fixed axis.

Work and Potential Energy

The procedures for determining the work done by different types of forces and the expressions for the potential energies of forces discussed in Chapter 15 provide the essential tools for applying the principle of work and energy to a rigid body. The work done on a rigid body by a force $\mathbf{F}$ is given by

$$U_{12} = \int_{(\mathbf{r}_\mathrm{p})_1}^{(\mathbf{r}_\mathrm{p})_2} \mathbf{F} \cdot d\mathbf{r}_\mathrm{p}, \tag{19.14}$$

where $\mathbf{r}_\mathrm{p}$ is the position of the point of application of $\mathbf{F}$ (Fig. 19.7). If the point of application is stationary, or if its direction of motion is perpendicular to $\mathbf{F}$, no work is done.

A force $\mathbf{F}$ is conservative if a potential energy V exists such that

$$\mathbf{F} \cdot d\mathbf{r}_\mathrm{p} = -dV. \tag{19.15}$$

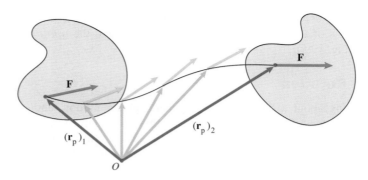

Figure 19.7
The work done by a force on a rigid body
is determined by the path of the point of
application of the force.

In terms of its potential energy, the work done by a conservative force $\mathbf{F}$ is

$$U_{12} = \int_{(\mathbf{r}_p)_1}^{(\mathbf{r}_p)_2} \mathbf{F} \cdot d\mathbf{r}_p = \int_{V_1}^{V_2} -dV = -(V_2 - V_1),$$

where V_1 and V_2 are the values of V at $(\mathbf{r}_p)_1$ and $(\mathbf{r}_p)_2$.

If a rigid body is subjected to a couple M (Fig. 19.8a), what work is done as the body moves between two positions? We can evaluate the work by representing the couple by forces (Fig. 19.8b) and determining the work done by the forces. If the rigid body rotates through an angle $d\theta$ in the direction of the couple (Fig. 19.8c), the work done by each force is $\left(\frac{1}{2} D\, d\theta\right)F$, so the total work is $DF\, d\theta = M\, d\theta$. Integrating this expression, we obtain the work done by a couple M as the rigid body rotates from θ_1 to θ_2 in the direction of M:

$$U_{12} = \int_{\theta_1}^{\theta_2} M\, d\theta. \tag{19.16}$$

Figure 19.8
(a) A rigid body subjected to a couple.
(b) An equivalent couple consisting of two
forces: $DF = M$.
(c) Determining the work done by the
forces.

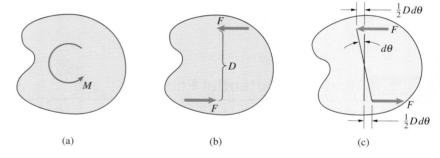

(a)   (b)   (c)

If M is constant, the work is simply the product of the couple and the angular displacement:

$$U_{12} = M(\theta_2 - \theta_1). \quad \text{(constant couple)}.$$

A couple M is conservative if a potential energy V exists such that

$$M\, d\theta = -dV. \tag{19.17}$$

We can express the work done by a conservative couple in terms of its potential energy:

$$U_{12} = \int_{\theta_1}^{\theta_2} M\, d\theta = \int_{V_1}^{V_2} -dV = -(V_2 - V_1).$$

For example, in Fig. 19.9, a torsional spring exerts a couple on a bar that is proportional to the bar's angle of rotation: $M = -k\theta$. From the relation

$$M \, d\theta = -k\theta \, d\theta = -dV,$$

we see that the potential energy must satisfy the equation

$$\frac{dV}{d\theta} = k\theta.$$

Integrating this equation, we find that the potential energy of the torsional spring is

$$V = \tfrac{1}{2} k\theta^2. \tag{19.18}$$

If *all* the forces and couples that do work on a rigid body are conservative, we can express the total work done as the body moves between two positions 1 and 2 in terms of the total potential energy of the forces and couples:

$$U_{12} = V_1 - V_2.$$

Combining this relation with the principle of work and energy, Eq. (19.5), we conclude that the sum of the kinetic energy and the total potential energy is constant—energy is conserved:

$$T + V = \text{constant.} \tag{19.19}$$

The results we have presented in Sections 19.1–19.3 can be used to relate changes in the translational and angular velocities of an object to a change in its position. This typically involves three steps:

1. *Identify the forces and couples that do work.* Use free-body diagrams to determine which external forces and couples do work.

2. *Apply the principle of work and energy or conservation of energy.* Either equate the total work done during a change in position to the change in the kinetic energy, or equate the sum of the kinetic and potential energies at two positions.

3. *Determine the kinematic relationships.* To complete the solution, it will often be necessary to obtain relations between velocities of points of rigid bodies and their angular velocities.

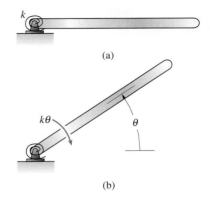

(a)

(b)

Figure 19.9
(a) A linear torsional spring connected to a bar.
(b) The spring exerts a couple of magnitude $k\theta$ in the direction opposite that of the bar's rotation.

Study Questions

1. What is the principle of work and energy for a rigid body?
2. What is the kinetic energy of a rigid body in general planar motion?
3. How do you determine the work done by a couple acting on a rigid body in planar motion?
4. If all of the forces and couples that do work on a rigid body are conservative, what can you infer about the sum of the kinetic energy of the rigid body and the total potential energy?

Example 19.1

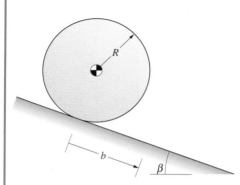

Figure 19.10

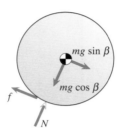

(a) Free-body diagram of the disk.

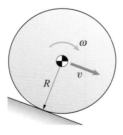

(b) Velocity of the center and the angular velocity when the disk has moved a distance b.

Applying Work and Energy to a Rolling Disk

A disk of mass m and moment of inertia I is released from rest on an inclined surface (Fig. 19.10). Assuming that the disk rolls, what is the velocity of its center when it has moved a distance b?

Strategy

We can determine the velocity by equating the total work done as the disk rolls a distance b to the change in its kinetic energy.

Solution

Identify the Forces and Couples That Do Work   We draw the free-body diagram of the disk in Fig. (a). The disk's weight does work as it rolls, but the normal force N and the friction force f do not. To explain why the friction force does no work, we can write the work done by a force $\mathbf{F}$ as

$$\int_{(\mathbf{r}_p)_1}^{(\mathbf{r}_p)_2} \mathbf{F} \cdot d\mathbf{r}_p = \int_{t_1}^{t_2} \mathbf{F} \cdot \frac{d\mathbf{r}_p}{dt}\, dt = \int_{t_1}^{t_2} \mathbf{F} \cdot \mathbf{v}_p\, dt,$$

where $\mathbf{v}_p$ is the velocity of the point of application of $\mathbf{F}$. Since the velocity of the point where f acts is zero as the disk rolls, the work done by f is zero.

Apply Work and Energy   We can determine the work done by the weight by multiplying the component of the weight in the direction of the motion of the center of the disk by the distance b:

$$U_{12} = (mg \sin \beta)b.$$

Letting v and ω be the velocity of the center and the angular velocity of the disk when it has moved a distance b (Fig. b), we equate the work to the change in the disk's kinetic energy:

$$mgb \sin \beta = \tfrac{1}{2}mv^2 + \tfrac{1}{2}I\omega^2 - 0. \tag{19.20}$$

Determine the Kinematic Relationship   The angular velocity ω of the rolling disk is related to the velocity v by $\omega = v/R$. Substituting this relation into Eq. (19.20) and solving for v, we obtain

$$v = \sqrt{\frac{2gb \sin \beta}{1 + I/mR^2}}.$$

Discussion

Suppose that the surface is smooth, so that the disk slides instead of rolling. In this case, the disk has no angular velocity, so Eq. (19.20) becomes

$$mgb \sin \beta = \tfrac{1}{2}mv^2 - 0,$$

and the velocity of the center of the disk is

$$v = \sqrt{2gb \sin \beta}.$$

The velocity is greater when the disk slides. You can see why by comparing the two expressions for the principle of work and energy. The work done by the disk's weight is the same in each case. When the disk rolls, part of the work increases the disk's translational kinetic energy, and part increases its rotational kinetic energy. When the disk slides, all of the work increases its translational kinetic energy.

Example 19.2

Applying Work and Energy to a Motorcycle

Each wheel of the motorcycle in Fig. 19.11 has mass $m_W = 9$ kg, radius $R = 330$ mm, and moment of inertia $I = 0.8$ kg-m^2. The combined mass of the rider and the motorcycle, not including the wheels, is $m_C = 142$ kg. The motorcycle starts from rest, and its engine exerts a constant couple $M = 140$ N-m on the rear wheel. Assume that the wheels do not slip. What horizontal distance b must the motorcycle travel to reach a velocity of 25 m/s?

Figure 19.11

Strategy

We can apply the principle of work and energy to the system consisting of the rider and the motorcycle, including its wheels, to determine the distance b.

Solution

Determining the distance b requires three steps.

Identify the Forces and Couples That Do Work   We draw the free-body diagram of the system in Fig. (a). The weights do no work because the motion is horizontal, and the forces exerted on the wheels by the road do no work because the velocity of their point of application is zero. (See Example 19.1.) Thus, no work is done by external forces and couples! However, work

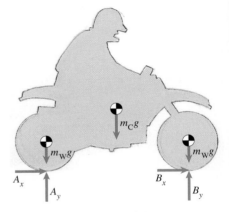

(a) Free-body diagram of the system.

is done by the couple M exerted on the rear wheel by the engine (Fig. b). Although this is an internal couple for the system we are considering—the wheel exerts an opposite couple on the body of the motorcycle—net work is done because the wheel rotates whereas the body does not.

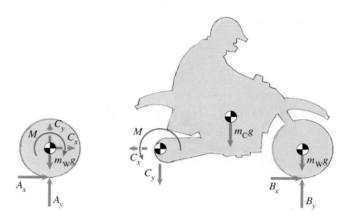

(b) Isolating the rear wheel.

Apply Work and Energy If the motorcycle moves a horizontal distance b, the wheels turn through an angle b/R rad, and the work done by the constant couple M is

$$U_{12} = M(\theta_2 - \theta_1) = M\left(\frac{b}{R}\right).$$

Let v be the motorcycle's velocity and ω the angular velocity of the wheels when the motorcycle has moved a distance b. The work equals the change in the total kinetic energy:

$$M\left(\frac{b}{R}\right) = \tfrac{1}{2}m_\text{C}v^2 + 2(\tfrac{1}{2}m_\text{W}v^2 + \tfrac{1}{2}I\omega^2) - 0. \tag{19.21}$$

Determine Kinematic Relationship The angular velocity of the rolling wheels is related to the velocity v by $\omega = v/R$. Substituting this relation into Eq. (19.21) and solving for b, we obtain

$$b = \left(\frac{1}{2}m_\text{C} + m_\text{W} + \frac{I}{R^2}\right)\frac{Rv^2}{M}$$

$$= \left[\frac{1}{2}(142) + (9) + \frac{(0.8)}{(0.33)^2}\right]\frac{(0.33)(25)^2}{(140)}$$

$$= 129 \text{ m}.$$

Discussion

Although we drew separate free-body diagrams of the motorcycle and its rear wheel to clarify the work done by the couple exerted by the engine, notice that we treated the motorcycle, including its wheels, as a single system in applying the principle of work and energy. By doing so, we did not need to consider the work done by the internal forces between the motorcycle's body and its wheels. When applying the principle of work and energy to a system of rigid bodies, you will usually find it simplest to express the principle for the system as a whole. This is in contrast to determining the motion of a system of rigid bodies by using the equations of motion, which usually requires that you draw free-body diagrams of each rigid body and apply the equations to them individually.

Example 19.3

Applying Conservation of Energy to a Linkage

The slender bars AB and BC of the linkage in Fig. 19.12 have mass m and length l, and the collar C has mass m_C. A torsional spring at A exerts a clockwise couple $k\theta$ on bar AB. The system is released from rest in the position $\theta = 0$ and allowed to fall. Neglecting friction, determine the angular velocity $\omega = d\theta/dt$ of bar AB as a function of θ.

Solution

Identify the Forces and Couples That Do Work   We draw the free-body diagram of the system in Fig. (a). The forces and couples that do work—the weights of the bars and collar and the couple exerted by the torsional spring—are conservative. We can use conservation of energy and the kinematic relationships between the angular velocities of the bars and the velocity of the collar to determine ω as a function of θ.

Apply Conservation of Energy   We denote the center of mass of bar BC by G and the angular velocity of bar BC by ω_{BC} (Fig. b). The moment of inertia

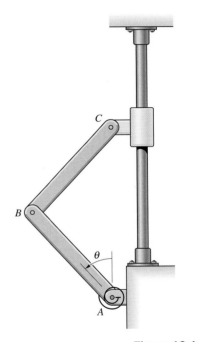

Figure 19.12

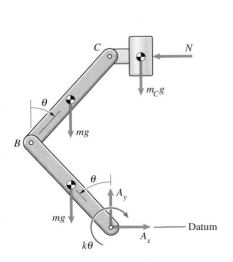

(a) Free-body diagram of the system.

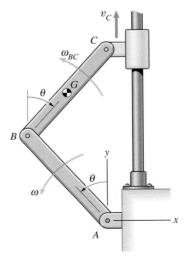

(b) Angular velocities of the bars and the velocity of the collar.

of each bar about its center of mass is $I = \frac{1}{12}ml^2$. Since bar AB rotates about the fixed point A, we can write its kinetic energy as

$$T_{\text{bar } AB} = \tfrac{1}{2}I_A\omega^2 = \tfrac{1}{2}\big[I + \big(\tfrac{1}{2}l\big)^2 m\big]\omega^2 = \tfrac{1}{6}ml^2\omega^2.$$

The kinetic energy of bar BC is

$$T_{\text{bar } BC} = \tfrac{1}{2}mv_G^2 + \tfrac{1}{2}I\omega_{BC}^2 = \tfrac{1}{2}mv_G^2 + \tfrac{1}{24}ml^2\omega_{BC}^2.$$

The kinetic energy of the collar C is

$$T_{\text{collar}} = \tfrac{1}{2}m_C v_C^2.$$

Using the datum in Fig. (a), we obtain the potential energies of the weights:

$$V_{\text{bar } AB} + V_{\text{bar } BC} + V_{\text{collar}} = mg\big(\tfrac{1}{2}l\cos\theta\big) + mg\big(\tfrac{3}{2}l\cos\theta\big) + m_C g(2l\cos\theta).$$

The potential energy of the torsional spring is given by Eq. (19.18):

$$V_{\text{spring}} = \tfrac{1}{2}k\theta^2.$$

We now have all the ingredients to apply conservation of energy. We equate the sum of the kinetic and potential energies at the position $\theta = 0$ to the sum of the kinetic and potential energies at an arbitrary value of θ:

$$T_1 + V_1 = T_2 + V_2:$$

$$0 + 2mgl + 2m_C gl = \tfrac{1}{6}ml^2\omega^2 + \tfrac{1}{2}mv_G^2 + \tfrac{1}{24}ml^2\omega_{BC}^2 + \tfrac{1}{2}m_C v_C^2$$
$$+ 2mgl\cos\theta + 2m_C gl\cos\theta + \tfrac{1}{2}k\theta^2.$$

To determine ω from this equation, we must express the velocities v_G, v_C, and ω_{BC} in terms of ω.

Determine Kinematic Relationships We can determine the velocity of point B in terms of ω and then express the velocity of point C in terms of the velocity of point B and the angular velocity ω_{BC}.

The velocity of B is

$$\mathbf{v}_B = \mathbf{v}_A + \boldsymbol{\omega}_{AB} \times \mathbf{r}_{B/A}$$

$$= \mathbf{0} + \begin{vmatrix} \mathbf{i} & \mathbf{j} & \mathbf{k} \\ 0 & 0 & \omega \\ -l\sin\theta & l\cos\theta & 0 \end{vmatrix}$$

$$= -l\omega\cos\theta\,\mathbf{i} - l\omega\sin\theta\,\mathbf{j}.$$

The velocity of C, expressed in terms of the velocity of B, is

$$v_C\mathbf{j} = \mathbf{v}_B + \boldsymbol{\omega}_{BC} \times \mathbf{r}_{C/B}$$

$$= -l\omega\cos\theta\,\mathbf{i} - l\omega\sin\theta\,\mathbf{j} + \begin{vmatrix} \mathbf{i} & \mathbf{j} & \mathbf{k} \\ 0 & 0 & \omega_{BC} \\ l\sin\theta & l\cos\theta & 0 \end{vmatrix}.$$

Equating $\mathbf{i}$ and $\mathbf{j}$ components, we obtain

$$\omega_{BC} = -\omega, \qquad v_C = -2l\omega\sin\theta.$$

(The minus signs indicate that the directions of the velocities are opposite to the directions we assumed in Fig. b.) Now that we know the angular velocity of bar BC in terms of ω, we can determine the velocity of its center of mass in terms of ω by expressing it in terms of $\mathbf{v}_B$:

$$\mathbf{v}_G = \mathbf{v}_B + \boldsymbol{\omega}_{BC} \times \mathbf{r}_{G/B}$$

$$= -l\omega\cos\theta\mathbf{i} - l\omega\sin\theta\mathbf{j} + \begin{vmatrix} \mathbf{i} & \mathbf{j} & \mathbf{k} \\ 0 & 0 & -\omega \\ \frac{1}{2}l\sin\theta & \frac{1}{2}l\cos\theta & 0 \end{vmatrix}$$

$$= -\tfrac{1}{2}l\omega\cos\theta\mathbf{i} - \tfrac{3}{2}l\omega\sin\theta\mathbf{j}.$$

Substituting these expressions for ω_{BC}, v_C, and $\mathbf{v}_G$ into our equation of conservation of energy and solving for ω, we obtain

$$\omega = \left[\frac{2gl(m + m_C)(1 - \cos\theta) - \frac{1}{2}k\theta^2}{\frac{1}{3}ml^2 + (m + 2m_C)l^2\sin^2\theta} \right]^{1/2}.$$

19.4 Power

The work done on a rigid body by a force $\mathbf{F}$ during an infinitesimal displacement $d\mathbf{r}_p$ of its point of application is

$$\mathbf{F} \cdot d\mathbf{r}_p.$$

We obtain the power P transmitted to the rigid body—the rate at which work is done on it—by dividing this expression by the interval of time dt during which the displacement takes place. We obtain

$$P = \mathbf{F} \cdot \mathbf{v}_p, \tag{19.22}$$

where $\mathbf{v}_p$ is the velocity of the point of application of $\mathbf{F}$.

Similarly, the work done on a rigid body in planar motion by a couple M during an infinitesimal rotation $d\theta$ in the direction of M is

$$M\,d\theta.$$

Dividing this expression by dt, we find that the power transmitted to the rigid body is the product of the couple and the angular velocity:

$$P = M\omega. \tag{19.23}$$

The total work done on a rigid body during an interval of time equals the change in kinetic energy of the body, so the total power transmitted equals the rate of change of the body's kinetic energy:

$$P = \frac{dT}{dt}.$$

The average with respect to time of the power during an interval of time from t_1 to t_2 is

$$P_{av} = \frac{1}{t_2 - t_1}\int_{t_1}^{t_2} P\,dt = \frac{1}{t_2 - t_1}\int_{T_1}^{T_2} dT = \frac{T_2 - T_1}{t_2 - t_1}.$$

This expression shows that we can determine the average power transferred to or from a rigid body during an interval of time by dividing the change in kinetic energy of the body, or the total work done, by the interval of time:

$$P_{av} = \frac{T_2 - T_1}{t_2 - t_1} = \frac{U_{12}}{t_2 - t_1}. \tag{19.24}$$

19.5 Principles of Impulse and Momentum

In this section, we review our discussion of the principle of linear impulse and momentum from Chapter 16 and then derive the principle of angular impulse and momentum for a rigid body. These principles relate time integrals of the forces and couples acting on a rigid body to changes in the velocity of its center of mass and its angular velocity.

Linear Momentum

Integrating Newton's second law with respect to time yields the principle of linear impulse and momentum for a rigid body:

$$\int_{t_1}^{t_2} \Sigma \mathbf{F}\, dt = m\mathbf{v}_2 - m\mathbf{v}_1. \tag{19.25}$$

Here, $\mathbf{v}_1$ and $\mathbf{v}_2$ are the velocities of the center of mass at the times t_1 and t_2 (Fig. 19.13). If the external forces acting on a rigid body are known as functions of time, this principle yields the change in the velocity of the center of mass of the body during an interval of time. In terms of the average of the total force from t_1 to t_2,

$$\Sigma \mathbf{F}_{\text{av}} = \frac{1}{t_2 - t_1} \int_{t_1}^{t_2} \Sigma \mathbf{F}\, dt,$$

we can write Eq. (19.25) as

$$(t_2 - t_1)\Sigma \mathbf{F}_{\text{av}} = m\mathbf{v}_2 - m\mathbf{v}_1. \tag{19.26}$$

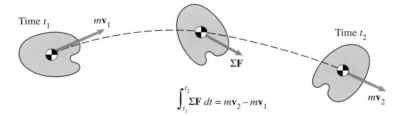

Figure 19.13
Principle of linear impulse and momentum.

This form of the principle of linear impulse and momentum is often useful when an object is subjected to impulsive forces.

If the only forces acting on two rigid bodies A and B are the forces they exert on each other, or if other forces are negligible, the total linear momentum of A and B is conserved:

$$m_A \mathbf{v}_A + m_B \mathbf{v}_B = \text{constant.} \tag{19.27}$$

Angular Momentum

When momentum principles are applied to rigid bodies, it is often necessary to determine both the velocities of their centers of mass and their angular velocities. For this task, linear momentum principles alone are not sufficient.

In this section, we derive angular momentum principles for a rigid body in planar motion.

Principles of Angular Impulse and Momentum   The total moment about the center of mass of a rigid body in planar motion equals the product of the moment of inertia of the body about its center of mass and the angular acceleration:

$$\Sigma M = I\alpha.$$

We can write this equation in the form

$$\Sigma M = \frac{dH}{dt}, \tag{19.28}$$

where

$$H = I\omega \tag{19.29}$$

is the rigid body's angular momentum about its center of mass. Integrating Eq. (19.28) with respect to time, we obtain one form of the principle of angular impulse and momentum:

$$\int_{t_1}^{t_2} \Sigma M \, dt = H_2 - H_1. \tag{19.30}$$

Here, H_1 and H_2 are the values of the angular momentum at the times t_1 and t_2. This equation says that angular impulse about the center of mass of the rigid body during the interval of time from t_1 to t_2 is equal to the change in the rigid body's angular momentum about its center of mass. If the total moment about the center of mass is known as a function of time, Eq. (19.30) can be used to determine the change in the angular velocity from t_1 to t_2.

We can derive another useful form of this principle: Let $\mathbf{r}$ be the position vector of the center of mass of the rigid body relative to a fixed point O (Fig. 19.14). In Chapter 18, we derived a relationship between the total moment about O due to external forces and couples and the rate of change of the rigid body's angular momentum about O:

$$\Sigma M_O = \frac{dH_O}{dt}, \tag{19.31}$$

where

$$H_O = (\mathbf{r} \times m\mathbf{v}) \cdot \mathbf{k} + I\omega. \tag{19.32}$$

Integrating Eq. (19.31) with respect to time, we obtain a second form of the principle of angular impulse and momentum:

$$\int_{t_1}^{t_2} \Sigma M_O \, dt = H_{O2} - H_{O1}. \tag{19.33}$$

The angular impulse about a fixed point O during the interval of time from t_1 to t_2 is equal to the change in the rigid body's angular momentum about O (Fig. 19.15).

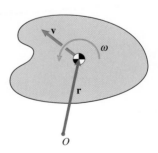

Figure 19.14
A rigid body in planar motion with velocity $\mathbf{v}$ and angular velocity ω.

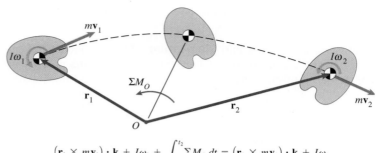

Figure 19.15
The impulse about O equals the change in the angular momentum about O.

$$(\mathbf{r}_1 \times m\mathbf{v}_1) \cdot \mathbf{k} + I\omega_1 + \int_{t_1}^{t_2} \Sigma M_O \, dt = (\mathbf{r}_2 \times m\mathbf{v}_2) \cdot \mathbf{k} + I\omega_2$$

The term $(\mathbf{r} \times m\mathbf{v}) \cdot \mathbf{k}$ in Eq. (19.32) is the rigid body's angular momentum about O due to the velocity of its center of mass. This term has the same form as the moment of a force, but with the linear momentum $m\mathbf{v}$ in place of the force. If we define ΣM_O and ω to be positive in the counterclockwise direction, the unit vector $\mathbf{k}$ points out of the page (Fig. 19.16) and $(\mathbf{r} \times m\mathbf{v}) \cdot \mathbf{k}$ is the counterclockwise "moment" of the linear momentum. The vector expression can be used to calculate this quantity, but it is often easier to use the fact that its magnitude is the product of the magnitude of the linear momentum and the perpendicular distance from point O to the line of action of the velocity. The "moment" is positive if it is counterclockwise (Fig. 19.17a) and negative if it is clockwise (Fig. 19.17b).

Impulsive Forces and Couples The average of the moment about the center of mass from t_1 to t_2 is

$$\Sigma M_{\mathrm{av}} = \frac{1}{t_2 - t_1} \int_{t_1}^{t_2} \Sigma M \, dt.$$

Using this equation, we can write Eq. (19.30) as

$$(t_2 - t_1)\Sigma M_{\mathrm{av}} = H_2 - H_1. \tag{19.34}$$

In the same way, we can express Eq. (19.33) in terms of the average moment about point O:

$$(t_2 - t_1)(\Sigma M_O)_{\mathrm{av}} = H_{O2} - H_{O1}. \tag{19.35}$$

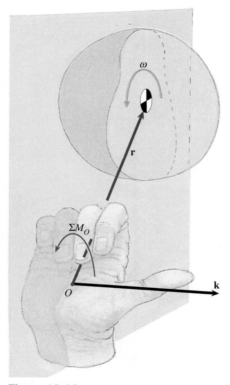

Figure 19.16
The direction of $\mathbf{k}$.

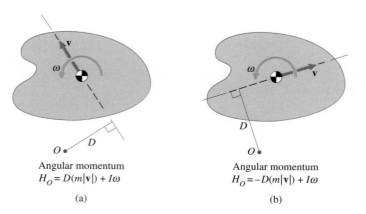

Figure 19.17
Determining the angular momentum about O by calculating the "moment" of the linear momentum.

Angular momentum
$H_O = D(m|\mathbf{v}|) + I\omega$

(a)

Angular momentum
$H_O = -D(m|\mathbf{v}|) + I\omega$

(b)

When the average value of the moment and its duration are known, we can use Eq. (19.34) or Eq. (19.35) to determine the change in the angular momentum. These equations are often useful when a rigid body is subjected to impulsive forces and couples.

Conservation of Angular Momentum We can use Eq. (19.33) to obtain an equation of conservation of total angular momentum for two rigid bodies. Let A and B be rigid bodies in two-dimensional motion in the same plane, and suppose that they are subjected only to the forces and couples they exert on each other or that other forces and couples are negligible. Let M_{OA} be the moment about a fixed point O due to the forces and couples acting on A, and let M_{OB} be the moment about O due to the forces and couples acting on B. Under the same assumption we made in deriving the equations of motion— the forces between each pair of particles are directed along the line between the particles—the moment $M_{OB} = -M_{OA}$. For example, in Fig. 19.18, A and B exert forces on each other by contact. The resulting moments about O are $M_{OA} = (\mathbf{r}_\mathrm{p} \times \mathbf{R}) \cdot \mathbf{k}$ and $M_{OB} = [\mathbf{r}_\mathrm{p} \times (-\mathbf{R})] \cdot \mathbf{k} = -M_{OA}$.

We apply Eq. (19.33) to A and B for arbitrary times t_1 and t_2, obtaining

$$\int_{t_1}^{t_2} M_{OA}\, dt = H_{OA2} - H_{OA1}$$

and

$$\int_{t_1}^{t_2} M_{OB}\, dt = H_{OB2} - H_{OB1}.$$

Summing these equations, the terms on the left cancel, and we obtain

$$H_{OA1} + H_{OB1} = H_{OA2} + H_{OB2}.$$

We see that the total angular momentum of A and B about O is conserved:

$$H_{OA} + H_{OB} = \text{constant}. \tag{19.36}$$

Notice that this result holds even when A and B are subjected to significant external forces and couples if the total moment about O due to the external forces and couples is zero. The point O can sometimes be chosen so that this condition is satisfied. The result also applies to an arbitrary number of rigid bodies: Their total angular momentum about O is conserved if the total moment about O due to external forces and couples is zero.

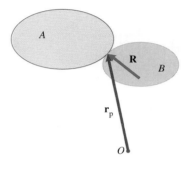

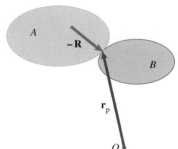

Figure 19.18
Rigid bodies A and B exerting forces on each other by contact.

Study Questions

1. If you know the total moment about the center of mass of a rigid body in planar motion during an interval of time, you can use a form of the principle of angular impulse and momentum to determine the change in the rigid body's angular velocity. Explain how.

2. Equation (19.32) is the angular momentum of a rigid body in planar motion about a fixed point O. You can evaluate the first term by calculating the "moment" of the linear momentum. How is this done?

3. If the total moment about a fixed point O due to the forces and couples acting on a rigid body is zero during an interval of time, what can you infer about the rigid body's angular momentum about O?

Example 19.4

Principle of Angular Impulse and Momentum

Disk A in Fig. 19.19 initially has a counterclockwise angular velocity ω_0, and disk B is stationary. At $t = 0$, the disks are moved into contact. As a result of friction at the point of contact, the angular velocity of A decreases and the angular velocity of B increases until there is no slip between them. What are their final angular velocities ω_A and ω_B? The disks are supported at their centers of mass, and their moments of inertia are I_A, I_B.

Strategy

Since the disks rotate about fixed axes through their centers of mass while they are in contact, we can apply the principle of angular impulse and momentum in the form given in Eq. (19.30) to each disk. When there is no longer any slip between the disks, their velocities are equal at their point of contact. With this kinematic relationship and the equations we obtain with the principle of angular impulse and momentum, we can determine the final angular velocities.

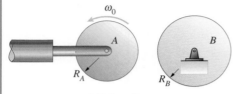

Initial position

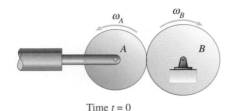

Time $t = 0$

Figure 19.19

Solution

We draw the free-body diagrams of the disks while slip occurs in Fig. (a), showing the normal and frictional forces they exert on each other. Let t_f be the time at which slip ceases. We apply Eq. (19.30) to disk A for the interval of time from $t = 0$ to $t = t_f$:

$$\int_{t_1}^{t_2} \Sigma M \, dt = H_2 - H_1 = I\omega_2 - I\omega_1:$$

$$\int_0^{t_f} -R_A f \, dt = I_A \omega_A - I_A \omega_0.$$

We also apply Eq. (19.30) to disk B:

$$\int_{t_1}^{t_2} \Sigma M \, dt = H_2 - H_1 = I\omega_2 - I\omega_1:$$

$$\int_0^{t_f} -R_B f \, dt = -I_B \omega_B - 0.$$

(Notice that because ω_B is clockwise, $\omega_2 = -\omega_B$. We divide the first equation by the second one and write the resulting equation as

$$\omega_A + \frac{R_A I_B}{R_B I_A} \omega_B = \omega_0.$$

When there is no slip, the velocities of the disks are equal at their point of contact:

$$R_A \omega_A = R_B \omega_B.$$

Solving these two equations, we obtain

$$\omega_A = \omega_0 \left[\frac{1}{1 + \dfrac{R_A^2 I_B}{R_B^2 I_A}} \right]$$

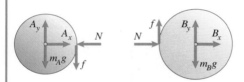

(a) Free-body diagrams of the disks.

and

$$\omega_B = \omega_0 \left[\frac{R_A/R_B}{1 + \frac{R_A^2 I_B}{R_B^2 I_A}} \right].$$

Notice that if the disks have the same radius and moment of inertia, $\omega_A = \frac{1}{2}\omega_0$ and $\omega_B = \frac{1}{2}\omega_0$.

Example 19.5

Impulsive Force on a Rigid Body

To help prevent injuries to passengers, engineers design a street light pole so that it shears off at ground level when struck by a vehicle (Fig. 19.20). From videotape of a test impact, the engineers estimate the angular velocity of the pole to be $\omega = 0.74$ rad/s and the horizontal velocity of its center of mass to be $v = 6.8$ m/s after the impact, and they estimate the duration of the impact to be $\Delta t = 0.01$ s. If the pole can be modeled as a 70-kg slender bar of length $l = 6$ m, the car strikes it at a height $h = 0.5$ m above the ground, and the couple exerted on the pole by its support during the impact is negligible, what average force was required to shear off the bolts supporting the pole?

Strategy

We will determine the average force by applying the principles of linear and angular impulse and momentum, expressed in terms of the average forces and moments exerted on the pole. We can apply the principle of angular impulse and momentum by using either Eq. (19.34) or Eq. (19.35). We will use Eq. (19.35) to demonstrate its use.

Figure 19.20

Solution

We draw the free-body diagram of the pole in Fig. (a), where F is the average force exerted by the car and S is the average shearing force exerted on the pole by the bolts. Let m be the mass of the pole, and let v and ω be the velocity of its center of mass and its angular velocity at the end of the impact (Fig. b). From Eq. (19.26), the principle of linear impulse and momentum expressed in terms of the average horizontal force is

$$(t_2 - t_1)(\Sigma F_x)_{\text{av}} = (mv_x)_2 - (mv_x)_1:$$

$$\Delta t(F - S) = mv - 0. \tag{19.37}$$

To apply the principle of angular impulse and momentum, we use Eq. (19.35), placing the fixed point O at the bottom of the pole (Figs. a and b). The pole's angular momentum about O at the end of the impact is

$$H_{O2} = [(\mathbf{r} \times m\mathbf{v}) \cdot \mathbf{k} + I\omega]_2 = [(\tfrac{1}{2}l\mathbf{j}) \times m(v\mathbf{i})] \cdot \mathbf{k} + I\omega = -\tfrac{1}{2}lmv + I\omega.$$

We can also obtain this result by calculating the "moment" of the linear momentum about O and adding the term $I\omega$. The magnitude of the "moment"

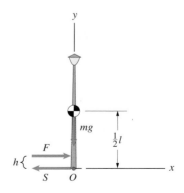

(a) Free-body diagram of the pole.

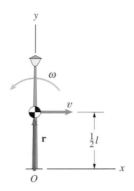

(b) Velocity and angular velocity at the end of the impact.

is the product of the magnitude of the linear momentum (mv) and the perpendicular distance from O to the line of action of the linear momentum $\left(\frac{1}{2}l\right)$, and it is negative because the "moment" is clockwise. (See Fig. 19.17.) From Eq. (19.35), we obtain

$$(t_2 - t_1)(\Sigma M_O)_{av} = H_{O2} - H_{O1}:$$

$$\Delta t(-hF) = -\tfrac{1}{2}lmv + I\omega - 0.$$

Solving this equation together with Eq. (19.37) for the average shear force S, we obtain

$$S = \frac{\left(\frac{1}{2}l - h\right)mv - I\omega}{h\Delta t}$$

$$= \frac{\left[\frac{1}{2}(6) - 0.5\right](70)(6.8) - \left[\frac{1}{12}(70)(6)^2\right](0.74)}{(0.5)(0.01)}$$

$$= 207{,}000 \text{ N}.$$

Example 19.6

Conservation of Angular Momentum

In a well-known demonstration of conservation of angular momentum, a person stands on a rotating platform holding a mass m in each hand (Fig. 19.21). The moment of inertia of the person and platform is $I_P = 0.4$ kg-m^2, the mass $m = 4$ kg, and the moment of inertia of each mass about the vertical axis through its center of mass is $I_M = 0.001$ kg-m^2. If the person's angular velocity with her arms extended to $r_1 = 0.6$ m is $\omega_1 = 1$ revolution per second, what is her angular velocity ω_2 when she pulls the masses inward to $r_2 = 0.2$ m? (You have observed skaters using this phenomenon to control their angular velocity in a spin by altering the positions of their arms.)

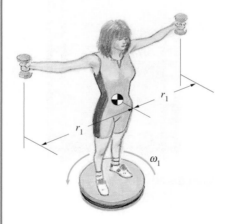

Strategy

If we neglect friction in the rotating platform, the total angular momentum of the person, platform, and masses about the vertical axis of rotation is conserved. We can use this condition to determine ω_2.

Solution

We begin by determining the angular momentum of one of the masses about the axis of rotation when the person's arms are extended. The magnitude of the velocity of this mass about the axis of rotation is $r_1\omega_1$, so the "moment" of its linear momentum about the axis of rotation is $r_1(mr_1\omega_1) = mr_1^2\omega_1$. The total angular momentum of the mass about the axis of rotation when the person's arms are extended is

$$(\mathbf{r} \times m\mathbf{v}) \cdot \mathbf{k} + I\omega = mr_1^2\omega_1 + I_M\omega_1.$$

Since the center of mass of the person and platform lie on the axis of rotation, their angular momentum about the axis of rotation is $I_P\omega_1$. Therefore, the com-

Figure 19.21

bined angular momentum of the person, platform, and masses when her arms are extended is

$$H_{O1} = I_P\omega_1 + 2(mr_1^2\omega_1 + I_M\omega_1) = \left[I_P + 2(I_M + mr_1^2)\right]\omega_1.$$

By replacing ω_1 by ω_2 and r_1 by r_2 in this expression, we obtain the combined angular momentum when the person has pulled the masses inward. Angular momentum is conserved, so

$$H_{O1} = H_{O2}:$$

$$\left[I_P + 2(I_M + mr_1^2)\right]\omega_1 = \left[I_P + 2(I_M + mr_2^2)\right]\omega_2,$$

$$\left\{0.4 + 2\left[0.001 + 4(0.6)^2\right]\right\}\omega_1 = \left\{0.4 + 2\left[0.001 + 4(0.2)^2\right]\right\}\omega_2.$$

Solving, we find that $\omega_2 = 4.55\omega_1 = 4.55$ revolutions per second.

Discussion

The parallel-axis theorem states that if the person holds the masses at a distance r from the axis of rotation, the moment of inertia of each mass about the axis of rotation is $I_M + mr^2$. The term $I_P + 2(I_M + mr^2)$ that appears in our equation of conservation of angular momentum is simply the total moment of inertia of the person, the platform, and the two masses about the axis of rotation.

19.6 Impacts

In Chapter 16, we analyzed impacts between objects with the aim of determining the velocities of their centers of mass after the collision. We now discuss how to determine the velocities of the centers of mass *and the angular velocities* of rigid bodies after they collide.

Conservation of Momentum

Suppose that two rigid bodies A and B, in two-dimensional motion in the same plane, collide. What do the principles of linear and angular momentum tell us about their motions after the collision?

Linear Momentum If other forces are negligible in comparison to the impact forces A and B exert on each other, then their total linear momentum is the same before and after the impact. But this result must be applied with care. For example, if one of the rigid bodies has a pin support (Fig. 19.22), the reactions exerted by the support cannot be neglected, and linear momentum is not conserved.

Angular Momentum If other forces and couples are negligible in comparison to the impact forces and couples that A and B exert on each other, their total angular momentum about *any* fixed point O is the same before and after

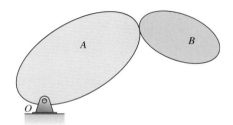

Figure 19.22
Rigid bodies A and B colliding. Because of the pin support, their total linear momentum is *not* conserved, but their total angular momentum about O is conserved.

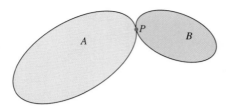

Figure 19.23
Rigid bodies A and B colliding at P. If forces are exerted only at P, the angular momentum of A about P and the angular momentum of B about P are each conserved.

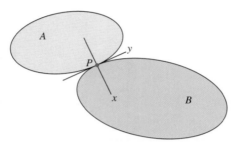

Figure 19.24
Rigid bodies A and B colliding at P. The x axis is perpendicular to the contacting surfaces.

the impact. [See Eq. (19.36).] If, in addition, A and B exert forces on each other only at their point of impact P, and exert no couples on each other, the angular momentum about P of *each* rigid body is the same before and after the impact (Fig. 19.23). This result follows from the principle of angular impulse and momentum, Eq. (19.33), because the impact forces on A and B exert no moment about P. If one of the rigid bodies has a pin support at a point O, as in Fig. 19.22, their total angular momentum about O is the same before and after the impact.

Coefficient of Restitution

If two rigid bodies adhere and move as a single rigid body after colliding, their velocities and angular velocity can be determined by using momentum conservation and kinematic relationships alone. These relationships are not sufficient if the objects do not adhere. But some impacts of the latter type can be analyzed by also using the concept of the coefficient of restitution.

Let P be the point of contact of rigid bodies A and B during an impact (Fig. 19.24), and let their velocities at P be $\mathbf{v}_{AP}$ and $\mathbf{v}_{BP}$ just before the impact and $\mathbf{v}'_{AP}$ and $\mathbf{v}'_{BP}$ just afterward. The x axis is perpendicular to the contacting surfaces at P. If the frictional forces resulting from the impact are negligible, we can show that the components of the velocities normal to the surfaces at P are related to the coefficient of restitution e by

$$e = \frac{(\mathbf{v}'_{BP})_x - (\mathbf{v}'_{AP})_x}{(\mathbf{v}_{AP})_x - (\mathbf{v}_{BP})_x}.$$

To derive this result, we must consider the effects of the impact on the individual objects. Let t_1 be the time at which they first come into contact. The objects are not actually rigid, but will deform as a result of the collision. At a time t_C, the maximum deformation will occur, and the objects will begin a "recovery" phase in which they tend to resume their original shapes. Let t_2 be the time at which they separate.

Our first step is to apply the principle of linear impulse and momentum to A and B for the intervals from t_1 to t_C and from t_C to t_2. Let R be the magnitude of the normal force exerted during the impact (Fig. 19.25). We denote the velocity of the center of mass of A at the times t_1, t_C, and t_2 by $\mathbf{v}_A$, $\mathbf{v}_{AC}$, and $\mathbf{v}'_A$, and denote the corresponding velocities of the center of mass of B by $\mathbf{v}_B$, $\mathbf{v}_{BC}$, and $\mathbf{v}'_B$. For A, we have

$$\int_{t_1}^{t_C} -R\,dt = m_A(\mathbf{v}_{AC})_x - m_A(\mathbf{v}_A)_x \qquad (19.38)$$

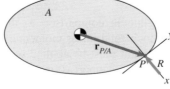
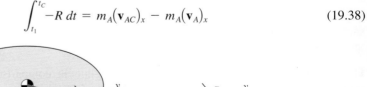
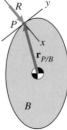

Figure 19.25
The normal force R resulting from the impact.

and

$$\int_{t_C}^{t_2} -R \, dt = m_A(\mathbf{v}_A')_x - m_A(\mathbf{v}_{AC})_x. \tag{19.39}$$

For B,

$$\int_{t_1}^{t_C} R \, dt = m_B(\mathbf{v}_{BC})_x - m_B(\mathbf{v}_B)_x \tag{19.40}$$

and

$$\int_{t_C}^{t_2} R \, dt = m_B(\mathbf{v}_B')_x - m_B(\mathbf{v}_{BC})_x. \tag{19.41}$$

The coefficient of restitution is the ratio of the linear impulse during the recovery phase to the linear impulse during the deformation phase:

$$e = \frac{\displaystyle\int_{t_C}^{t_2} R \, dt}{\displaystyle\int_{t_1}^{t_C} R \, dt}.$$

If we divide Eq. (19.39) by Eq. (19.38) and divide Eq. (19.41) by Eq. (19.40), the resulting equations can be written as

$$(\mathbf{v}_A')_x = -(\mathbf{v}_A)_x e + (\mathbf{v}_{AC})_x(1 + e)$$

and $\qquad (19.42)$

$$(\mathbf{v}_B')_x = -(\mathbf{v}_B)_x e + (\mathbf{v}_{BC})_x(1 + e).$$

We now apply the principle of angular impulse and momentum to A and B for the intervals of time from t_1 to t_C and from t_C to t_2. We denote the counterclockwise angular velocity of A at the times t_1, t_C, and t_2 by ω_A, ω_{AC}, and ω_A' and denote the corresponding angular velocities of B by ω_B, ω_{BC}, and ω_B'. We write the position vectors of P relative to the centers of mass of A and B as (Fig. 19.25)

$$\mathbf{r}_{P/A} = x_A \mathbf{i} + y_A \mathbf{j}$$

and

$$\mathbf{r}_{P/B} = x_B \mathbf{i} + y_B \mathbf{j}.$$

The moment about the center of mass of A due to the force exerted on A by the impact is $\mathbf{r}_{P/A} \times (-R\mathbf{i}) = y_A R\mathbf{k}$. From Eq. (19.30), we obtain

$$\int_{t_1}^{t_C} y_A R \, dt = I_A \omega_{AC} - I_A \omega_A \tag{19.43}$$

and

$$\int_{t_C}^{t_2} y_A R \, dt = I_A \omega_A' - I_A \omega_{AC}. \tag{19.44}$$

The corresponding equations for B are

$$\int_{t_1}^{t_C} -y_B R \, dt = I_B \omega_{BC} - I_B \omega_B \tag{19.45}$$

and

$$\int_{t_C}^{t_2} -y_B R \, dt = I_B \omega_B' - I_B \omega_{BC}. \tag{19.46}$$

Dividing Eq. (19.44) by Eq. (19.43) and dividing Eq. (19.46) by Eq. (19.45), we can write the resulting equations as

$$\omega_A' = -\omega_A e + \omega_{AC}(1 + e)$$

and

$$\omega_B' = -\omega_B e + \omega_{BC}(1 + e). \tag{19.47}$$

By expressing the velocity of the point of A at P in terms of the velocity of the center of mass of A and the angular velocity of A, and expressing the velocity of the point of B at P in terms of the velocity of the center of mass of B and the angular velocity of B, we obtain

$$\left(\mathbf{v}_{AP}\right)_x = \left(\mathbf{v}_A\right)_x - \omega_A y_A,$$

$$\left(\mathbf{v}_{AP}'\right)_x = \left(\mathbf{v}_A'\right)_x - \omega_A' y_A,$$

$$\left(\mathbf{v}_{BP}\right)_x = \left(\mathbf{v}_B\right)_x - \omega_B y_B, \tag{19.48}$$

and

$$\left(\mathbf{v}_{BP}'\right)_x = \left(\mathbf{v}_B'\right)_x - \omega_B' y_B.$$

At time t_C, the x components of the velocities of the two objects are equal at P, which yields the relation

$$\left(\mathbf{v}_{AC}\right)_x - \omega_{AC} y_A = \left(\mathbf{v}_{BC}\right)_x - \omega_{BC} y_B. \tag{19.49}$$

From Eqs. (19.48),

$$\frac{\left(\mathbf{v}_{BP}'\right)_x - \left(\mathbf{v}_{AP}'\right)_x}{\left(\mathbf{v}_{AP}\right)_x - \left(\mathbf{v}_{BP}\right)_x} = \frac{\left(\mathbf{v}_B'\right)_x - \omega_B' y_B - \left(\mathbf{v}_A'\right)_x + \omega_A' y_A}{\left(\mathbf{v}_A\right)_x - \omega_A y_A - \left(\mathbf{v}_B\right)_x + \omega_B y_B}.$$

Substituting Eqs. (19.42) and (19.47) into this equation and collecting terms yields

$$\frac{\left(\mathbf{v}_{BP}'\right)_x - \left(\mathbf{v}_{AP}'\right)_x}{\left(\mathbf{v}_{AP}\right)_x - \left(\mathbf{v}_{BP}\right)_x} = e - \left[\frac{\left(\mathbf{v}_{AC}\right)_x - \omega_{AC} y_A - \left(\mathbf{v}_{BC}\right)_x + \omega_{BC} y_B}{\left(\mathbf{v}_A\right)_x - \omega_A y_A - \left(\mathbf{v}_B\right)_x + \omega_B y_B}\right](e + 1).$$

From Eq. (19.49), the term in brackets vanishes, and we obtain the equation relating the normal components of the velocities at the point of contact to the coefficient of restitution:

$$e = \frac{\left(\mathbf{v}_{BP}'\right)_x - \left(\mathbf{v}_{AP}'\right)_x}{\left(\mathbf{v}_{AP}\right)_x - \left(\mathbf{v}_{BP}\right)_x}. \tag{19.50}$$

In arriving at this equation, we assumed that the contacting surfaces were smooth, so *the collision exerts no force on A or B in the direction tangential to their contacting surfaces.*

Although we derived Eq. (19.50) under the assumption that the motions of A and B are unconstrained, the relationship also holds if they are not—for example, if one of them is connected to a pin support.

In summary, we see that the approach used for analyzing collisions of rigid bodies in planar motion depends on the type of collision. If other forces are negligible in comparison to the impact forces, the total linear momentum is conserved. If other forces and couples are negligible in comparison to the impact forces and couples, the total angular momentum about any fixed point is conserved. If, in addition, forces are exerted only at the point of impact P, the angular momentum about P of each rigid body is conserved. If one of the rigid bodies has a pin support at a point O, the total angular momentum about O is conserved. If the impact is assumed to exert no forces on the colliding objects in the direction tangential to their surface of contact, the coefficient of restitution e relates the normal components of the velocities at the point of contact through Eq. (19.50).

Study Questions

1. When two rigid bodies collide, you cannot always assume that their total linear momentum is conserved. Explain why.
2. If P is the point of impact of two colliding rigid bodies and external forces and couples are negligible, is the total angular momentum of the rigid bodies about P conserved? Is the angular momentum of each rigid body about P conserved?
3. If two rigid bodies collide, one of which is free and the other of which is attached to a fixed pin support at a point O, is the total angular momentum about O conserved? Is the angular momentum of each rigid body about O conserved?
4. If P is the point of impact of two colliding rigid bodies, is the coefficient of restitution defined in terms of the velocities of their centers of mass or in terms of their velocities at P?

Example 19.7

Impact of a Sphere and a Suspended Bar

The homogeneous sphere in Fig. 19.26 is moving horizontally with velocity v_A and no angular velocity when it strikes the stationary slender bar. The sphere has mass m_A, and the bar has mass m_B and length l. The coefficient of restitution of the impact is e.
(a) What is the angular velocity of the bar after the impact?
(b) If the duration of the impact is Δt, what average horizontal force is exerted on the bar by the pin support C as a result of the impact?

Strategy

(a) From the definition of the coefficient of restitution, we can obtain an equation relating the horizontal velocity of the sphere and the velocity of the bar at the point of impact after the collision occurs. In addition, the total angular momentum of the sphere and bar about the pin C is conserved. With these two equations and kinematic relationships, we can determine the velocity of the sphere and the angular velocity of the bar after the impact.
(b) We can determine the average force exerted on the bar by the support by applying the principle of angular impulse and momentum to the bar.

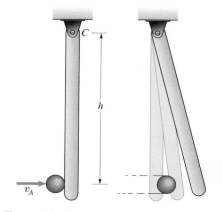

Figure 19.26

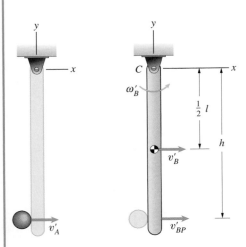

(a) Velocities of the sphere and bar after the impact.

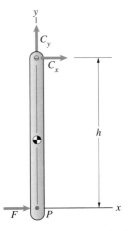

(b) Average forces exerted on the bar during the impact.

Solution

(a) In Fig. (a), we show the velocities just after the impact, where v'_{BP} is the bar's velocity at the point of impact. From the definition of the coefficient of restitution, Eq. (19.50), we obtain

$$e = \frac{v'_{BP} - v'_A}{v_A - 0}.$$

The equation of conservation of total angular momentum about C is

$$H_{CA} + H_{CB} = H'_{CA} + H'_{CB}:$$

$$(\mathbf{r}_A \times m_A \mathbf{v}_A) \cdot \mathbf{k} + I_A \omega_A + (\mathbf{r}_B \times m_B \mathbf{v}_B) \cdot \mathbf{k} + I_B \omega_B$$
$$= (\mathbf{r}_A \times m_A \mathbf{v}'_A) \cdot \mathbf{k} + I_A \omega'_A + (\mathbf{r}_B \times m_B \mathbf{v}'_B) \cdot \mathbf{k} + I_B \omega'_B,$$
$$(-h\mathbf{j} \times m_A v_A \mathbf{i}) \cdot \mathbf{k} = (-h\mathbf{j} \times m_A v'_A \mathbf{i}) \cdot \mathbf{k} + (-\tfrac{1}{2} l\mathbf{j} \times m_B v'_B \mathbf{i}) \cdot \mathbf{k} + I_B \omega'_B.$$

Carrying out the vector operations, we obtain

$$h m_A v_A = h m_A v'_A + \tfrac{1}{2} l m_B v'_B + I_B \omega'_B.$$

Notice in Fig. (a) that the velocities v'_B and v'_{BP} are related to the angular velocity of the bar ω'_B by

$$v'_B = \tfrac{1}{2} l \omega'_B, \qquad v'_{BP} = h \omega'_B.$$

We now have four equations in the four unknowns v'_A, v'_B, v'_{BP}, and ω'_B. Solving them for the angular velocity of the bar and using the relation $I_B = \tfrac{1}{12} m_B l^2$, we obtain

$$\omega'_B = \frac{(1+e) h m_A v_A}{h^2 m_A + \tfrac{1}{3} m_B l^2}.$$

(b) Let the forces on the free-body diagram of the bar in Fig. (b) represent the average forces exerted during the impact. We apply the principle of angular impulse and momentum, in the form given by Eq. (19.35), about the point of impact:

$$(t_2 - t_1)(\Sigma M_P)_{\text{av}} = H'_B - H_B:$$
$$(t_2 - t_1)(\Sigma M_P)_{\text{av}} = [(\mathbf{r}_B \times m_B \mathbf{v}'_B) \cdot \mathbf{k} + I_B \omega'_B] - [(\mathbf{r}_B \times m_B \mathbf{v}_B) \cdot \mathbf{k} + I_B \omega_B],$$
$$\Delta t(-hC_x) = [(h - \tfrac{1}{2} l)\mathbf{j} \times m_B v'_B \mathbf{i}] \cdot \mathbf{k} + I_B \omega'_B - 0.$$

Solving for C_x yields

$$C_x = \frac{(h - \tfrac{1}{2} l) m_B v'_B - I_B \omega'_B}{h \, \Delta t}.$$

Using our solution for ω'_B from part (a) and the relation $v'_B = \tfrac{1}{2} l \omega'_B$, we obtain the average horizontal force exerted by the support:

$$C_x = \frac{(1 + e)(\tfrac{1}{2} h - \tfrac{1}{3} l) l m_A m_B v_A}{(h^2 m_A + \tfrac{1}{3} m_B l^2) \Delta t}.$$

Discussion

Notice that the average horizontal force exerted on the bar by the support can be in either direction or can be zero, depending on where the impact occurs. The force is zero if $h = \tfrac{2}{3} l$.

Example 19.8

Impact with a Fixed Obstacle

The combined mass of the motorcycle and rider in Fig. 19.27 is $m = 170$ kg, and their combined moment of inertia about their center of mass is 22 kg-m^2. Following a jump, the motorcycle and rider are in the position shown just before the rear wheel contacts the ground. The velocity of their center of mass is of magnitude $|\mathbf{v}_G| = 8.8$ m/s, and their angular velocity is $\omega = 0.2$ rad/s. If the motorcycle and rider are modeled as a single rigid body and the coefficient of restitution of the impact is $e = 0.8$, what are the angular velocity ω' and velocity $\mathbf{v}'_G$ after the impact? Neglect the tangential component of force exerted on the motorcycle's wheel during the impact.

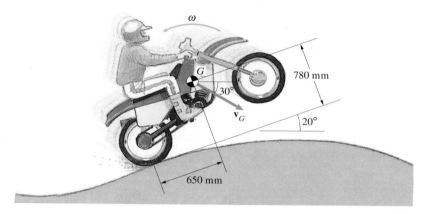

780 mm

30°

$\mathbf{v}_G$

20°

650 mm

Figure 19.27

Strategy

Since the tangential component of force on the motorcycle's wheel during the impact is neglected, the component of the velocity of the center of mass parallel to the ground can be taken to be unchanged by the impact. The coefficient of restitution relates the motorcycle's velocity normal to the ground at the point of impact before the impact to its value after the impact. Also, the force of the impact exerts no moment about the point of impact, so the motorcycle's angular momentum about that point is conserved. (We assume the impact to be so brief that the angular impulse due to the weight is negligible.) With these three relations, we can determine the two components of the velocity of the center of mass and the angular velocity after the impact.

Solution

In Fig. (a), we align a coordinate system parallel and perpendicular to the ground at the point P where the impact occurs. Let the components of the velocity of the center of mass before and after the impact be $\mathbf{v}_G = v_x\mathbf{i} + v_y\mathbf{j}$ and $\mathbf{v}'_G = v'_x\mathbf{i} + v'_y\mathbf{j}$, respectively. The x and y components of the velocity are

$$v_x = 8.8 \cos 50° = 5.66 \text{ m/s}$$

and

$$v_y = -8.8 \sin 50° = -6.74 \text{ m/s}.$$

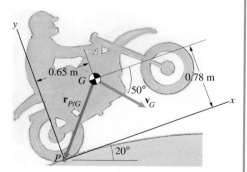

y

0.65 m

G

50°

0.78 m

$\mathbf{r}_{P/G}$

$\mathbf{v}_G$

x

P

20°

(a) Aligning the x axis of the coordinate system tangent to the ground at P.

Because the component of the impact force tangential to the ground is neglected, the x component of the velocity of the center of mass is unchanged:

$$v'_x = v_x = 5.66 \text{ m/s}.$$

We can express the y component of the wheel's velocity at P before the impact in terms of the velocity of the center of mass and the angular velocity (Fig. a) as

$$\mathbf{j} \cdot \mathbf{v}_P = \mathbf{j} \cdot (\mathbf{v}_G + \boldsymbol{\omega} \times \mathbf{r}_{P/G})$$

$$= \mathbf{j} \cdot \left\{ v_x \mathbf{i} + v_y \mathbf{j} + \begin{vmatrix} \mathbf{i} & \mathbf{j} & \mathbf{k} \\ 0 & 0 & \omega \\ -0.65 & -0.78 & 0 \end{vmatrix} \right\}$$

$$= v_y - 0.65\omega.$$

(Notice that this expression gives the y component of the velocity at P even though the wheel is spinning.) The y component of the wheel's velocity at P after the impact is

$$\mathbf{j} \cdot \mathbf{v}'_P = \mathbf{j} \cdot (\mathbf{v}'_G + \boldsymbol{\omega}' \times \mathbf{r}_{P/G})$$

$$= v'_y - 0.65\omega'.$$

The coefficient of restitution relates the y components of the wheel's velocity at P before and after the impact:

$$e = \frac{-(\mathbf{j} \cdot \mathbf{v}'_P)}{(\mathbf{j} \cdot \mathbf{v}_P)} = \frac{-(v'_y - 0.65\omega')}{(v_y - 0.65\omega)}. \qquad (19.51)$$

The force of the impact exerts no moment about P, so angular momentum about P is conserved:

$$H_P = H'_P:$$

$$\left[(\mathbf{r}_{G/P} \times m\mathbf{v}_G) \cdot \mathbf{k} + I\omega \right] = \left[(\mathbf{r}_{G/P} \times m\mathbf{v}'_G) \cdot \mathbf{k} + I\omega' \right],$$

$$\begin{vmatrix} \mathbf{i} & \mathbf{j} & \mathbf{k} \\ 0.65 & 0.78 & 0 \\ mv_x & mv_y & 0 \end{vmatrix} \cdot \mathbf{k} + I\omega = \begin{vmatrix} \mathbf{i} & \mathbf{j} & \mathbf{k} \\ 0.65 & 0.78 & 0 \\ mv'_x & mv'_y & 0 \end{vmatrix} \cdot \mathbf{k} + I\omega'.$$

Expanding the determinants and evaluating the dot products, we obtain

$$0.65mv_y - 0.78mv_x + I\omega = 0.65mv'_y - 0.78mv'_x + I\omega'. \qquad (19.52)$$

Since we have already determined v'_x, we can solve Eqs. (19.51) and (19.52) for v'_y and ω'. The results are

$$v'_y = -3.84 \text{ m/s}$$

and

$$\omega' = -14.4 \text{ rad/s}.$$

The velocity of the center of mass after the impact is $\mathbf{v}'_G = 5.66\mathbf{i} - 3.84\mathbf{j}$ m/s, and the angular velocity is 14.4 rad/s in the clockwise direction.

Example 19.9

Colliding Cars

An engineer simulates a collision between two 1600-kg cars by modeling them as rigid bodies (Fig. 19.28). The moment of inertia of each car about its center of mass is 960 kg-m^2. The engineer assumes the contacting surfaces at P to be smooth and parallel to the x axis and assumes the coefficient of restitution to be $e = 0.2$. What are the angular velocities of the cars and the velocities of their centers of mass after the collision?

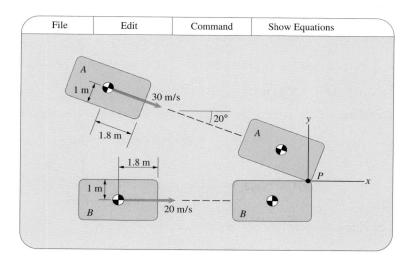

Figure 19.28

Strategy

Since the contacting surfaces are smooth, the x components of the velocities of the centers of mass are unchanged by the collision. The y components of the velocities must satisfy conservation of linear momentum, and the y components of the velocities at the point of impact before and after the impact are related by the coefficient of restitution. The force of the impact exerts no moment about P on either car, so the angular momentum of each car about P is conserved. From these conditions and kinematic relations between the velocities of the centers of mass and the velocities at P, we can determine the angular velocities and the velocities of the centers of mass after the impact.

Solution

The components of the velocities of the centers of mass before the impact are

$$\mathbf{v}_A = 30 \cos 20° \, \mathbf{i} - 30 \sin 20° \, \mathbf{j}$$
$$= 28.2\mathbf{i} - 10.3\mathbf{j} \ (\text{m/s})$$

and

$$\mathbf{v}_B = 20\mathbf{i} \ (\text{m/s}).$$

The x components of the velocities are unchanged by the impact:

$$v'_{Ax} = v_{Ax} = 28.2 \ \text{m/s}, \qquad v'_{Bx} = v_{Bx} = 20 \ \text{m/s}.$$

The y components of the velocities must satisfy conservation of linear momentum:

$$m_A v_{Ay} + m_B v_{By} = m_A v'_{Ay} + m_B v'_{By}. \tag{19.53}$$

Let the velocities of the two cars at P before the collision be $\mathbf{v}_{AP}$ and $\mathbf{v}_{BP}$. The coefficient of restitution $e = 0.2$ relates the y components of the velocities at P:

$$0.2 = \frac{v'_{BPy} - v'_{APy}}{v_{APy} - v_{BPy}} \tag{19.54}$$

We can express the velocities at P after the impact in terms of the velocities of the centers of mass and the angular velocities after the impact (Fig. a). The position of P relative to the center of mass of car A is

$$\mathbf{r}_{P/A} = \left[(1.8) \cos 20° - (1) \sin 20°\right]\mathbf{i} - \left[(1.8) \sin 20° + (1) \cos 20°\right]\mathbf{j}$$

$$= 1.35\mathbf{i} - 1.56\mathbf{j} \ (\text{m}).$$

Therefore, we can express the velocity of point P of car A after the impact as

$$\mathbf{v}'_{AP} = \mathbf{v}'_A + \boldsymbol{\omega}'_A \times \mathbf{r}_{P/A}:$$

$$v'_{APx}\mathbf{i} + v'_{APy}\mathbf{j} = v'_{Ax}\mathbf{i} + v'_{Ay}\mathbf{j} + \begin{vmatrix} \mathbf{i} & \mathbf{j} & \mathbf{k} \\ 0 & 0 & \omega'_A \\ 1.35 & -1.56 & 0 \end{vmatrix}.$$

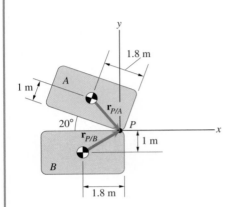

(a) Position vectors of P relative to the centers of mass.

Equating $\mathbf{i}$ and $\mathbf{j}$ components in this equation, we obtain

$$v'_{APx} = v'_{Ax} + 1.56\omega'_A$$

and $\hspace{8cm}$ (19.55)

$$v'_{APy} = v'_{Ay} + 1.35\omega'_A.$$

The position of P relative to the center of mass of car B is

$$\mathbf{r}_{P/B} = 1.8\mathbf{i} + \mathbf{j} \ (\text{m}).$$

We can express the velocity of point P of car B after the impact as

$$\mathbf{v}'_{BP} = \mathbf{v}'_B + \boldsymbol{\omega}'_B \times \mathbf{r}_{P/B}:$$

$$v'_{BPx}\mathbf{i} + v'_{BPy}\mathbf{j} = v'_{Bx}\mathbf{i} + v'_{By}\mathbf{j} + \begin{vmatrix} \mathbf{i} & \mathbf{j} & \mathbf{k} \\ 0 & 0 & \omega'_B \\ 1.8 & 1 & 0 \end{vmatrix}.$$

Equating $\mathbf{i}$ and $\mathbf{j}$ components yields

$$v'_{BPx} = v'_{Bx} - \omega'_B$$

and $\hspace{8cm}$ (19.56)

$$v'_{BPy} = v'_{By} + 1.8\omega'_B.$$

The angular momentum of car A about P is conserved:

$$H_{PA} = H'_{PA}:$$

$$\left[(\mathbf{r}_{A/P} \times m_A \mathbf{v}_A) \cdot \mathbf{k} + I_A \omega_A\right] = \left[(\mathbf{r}_{A/P} \times m_A \mathbf{v}'_A) \cdot \mathbf{k} + I_A \omega'_A\right],$$

$$\begin{vmatrix} \mathbf{i} & \mathbf{j} & \mathbf{k} \\ -1.35 & 1.56 & 0 \\ m_A v_{Ax} & m_A v_{Ay} & 0 \end{vmatrix} \cdot \mathbf{k} + 0 = \begin{vmatrix} \mathbf{i} & \mathbf{j} & \mathbf{k} \\ -1.35 & 1.56 & 0 \\ m_A v'_{Ax} & m_A v'_{Ay} & 0 \end{vmatrix} \cdot \mathbf{k} + I_A \omega'_A.$$

Expanding the determinants and evaluating the dot products, we obtain

$$-1.35 m_A v_{Ay} - 1.56 m_A v_{Ax}$$
$$= -1.35 m_A v'_{Ay} - 1.56 m_A v'_{Ax} + I_A \omega'_A. \tag{19.57}$$

The angular momentum of car B about P is also conserved,

$$H_{PB} = H'_{PB}:$$

$$\left[(\mathbf{r}_{B/P} \times m_B \mathbf{v}_B) \cdot \mathbf{k} + I_B \omega_B \right] = \left[(\mathbf{r}_{B/P} \times m_B \mathbf{v}'_B) \cdot \mathbf{k} + I_B \omega'_B \right],$$

$$\begin{vmatrix} \mathbf{i} & \mathbf{j} & \mathbf{k} \\ -1.8 & -1 & 0 \\ m_B v_{Bx} & 0 & 0 \end{vmatrix} \cdot \mathbf{k} + 0 = \begin{vmatrix} \mathbf{i} & \mathbf{j} & \mathbf{k} \\ -1.8 & -1 & 0 \\ m_B v'_{Bx} & m_B v'_{By} & 0 \end{vmatrix} \cdot \mathbf{k} + I_B \omega'_B.$$

From this equation, it follows that

$$m_B v_{Bx} = -1.8 m_B v'_{By} + m_B v'_{Bx} + I_B \omega'_B. \tag{19.58}$$

We can solve Eqs. (19.53)–(19.58) for v'_A, v'_{AP}, ω'_A, v'_B, v'_{BP}, and ω'_B. The results for the velocities of the centers of mass of the cars and their angular velocities are as follows:

$$\mathbf{v}'_A = 28.2\mathbf{i} - 9.08\mathbf{j} \text{ (m/s)}, \qquad \omega'_A = 2.65 \text{ rad/s},$$
$$\mathbf{v}'_B = 20.0\mathbf{i} - 1.18\mathbf{j} \text{ (m/s)}, \qquad \omega'_B = -3.54 \text{ rad/s}.$$

Chapter Summary

Work and Energy

The work done by external forces and couples as a rigid body moves between two positions is equal to the change in kinetic energy of the body:

$$U_{12} = T_2 - T_1. \qquad \text{Eq. (19.5)}$$

The work done on a system of rigid bodies by external and internal forces and couples equals the change in the total kinetic energy.

The kinetic energy of a rigid body in general planar motion is

$$T = \tfrac{1}{2} m v^2 + \tfrac{1}{2} I \omega^2, \qquad \text{Eq. (19.12)}$$

where v is the magnitude of the velocity of the center of mass of the body and I is the moment of inertia about the center of mass. If a rigid body rotates about a fixed axis O, its kinetic energy can also be expressed as

$$T = \tfrac{1}{2} I_0 \omega^2. \qquad \text{Eq. (19.13)}$$

The work done on a rigid body by a force $\mathbf{F}$ is

$$U_{12} = \int_{(\mathbf{r}_p)_1}^{(\mathbf{r}_p)_2} \mathbf{F} \cdot d\mathbf{r}_p, \qquad \text{Eq. (19.14)}$$

where $\mathbf{r}_p$ is the position of the point of application of $\mathbf{F}$. If the point of application is stationary, or if its direction of motion is perpendicular to $\mathbf{F}$, no work is done.

The work done by a couple M on a rigid body in planar motion as the body rotates from θ_1 to θ_2 in the direction of M is

$$U_{12} = \int_{\theta_1}^{\theta_2} M \, d\theta.$$

Eq. (19.16)

A couple M is conservative if a potential energy V exists such that

$$M \, d\theta = -dV.$$

Eq. (19.17)

The potential energy of a linear torsional spring that exerts a couple $k\theta$ in the direction opposite to angular displacement θ of the spring (Fig. a) is $\frac{1}{2}k\theta^2$.

If all the forces and couples that do work on a rigid body are conservative, the sum of the kinetic energy and the total potential energy of the body is constant:

$$T + V = \text{constant}.$$

Eq. (19.19)

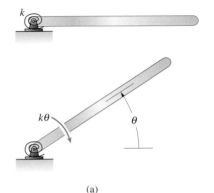

(a)

Power

The power transmitted to a rigid body by a force $\mathbf{F}$ is

$$P = \mathbf{F} \cdot \mathbf{v}_p,$$

Eq. (19.22)

where $\mathbf{v}_p$ is the velocity of the point of application of $\mathbf{F}$. The power transmitted to a rigid body in planar motion by a couple M is

$$P = M\omega.$$

Eq. (19.23)

The average power transferred to a rigid body during an interval of time is equal to the change in kinetic energy of the body, or the total work done during that time, divided by the interval of time:

$$P_{\text{av}} = \frac{T_2 - T_1}{t_2 - t_1} = \frac{U_{12}}{t_2 - t_1}.$$

Eq. (19.24)

Impulse and Momentum

The principle of linear impulse and momentum states that the linear impulse applied to a rigid body during an interval of time is equal to the change in linear momentum of the body during that time:

$$\int_{t_1}^{t_2} \Sigma \mathbf{F} \, dt = m\mathbf{v}_2 - m\mathbf{v}_1.$$

Eq. (19.25)

This result can also be expressed in terms of the average of the total force with respect to time:

$$(t_2 - t_1)\Sigma \mathbf{F}_{\text{av}} = m\mathbf{v}_2 - m\mathbf{v}_1.$$

Eq. (19.26)

If the only forces acting on two rigid bodies A and B are the forces they exert on each other, or if other forces are negligible, than the total linear momentum of the bodies is conserved:

$$m_A \mathbf{v}_A + m_B \mathbf{v}_B = \text{constant}.$$

Eq. (19.27)

The angular momentum about the center of mass of a rigid body in planar motion is

$$H = I\omega,$$

Eq. (19.29)

where I is the moment of inertia of the body about its center of mass. One form of the principle of angular impulse and momentum states that the angular impulse about the center of mass during the interval from t_1 to t_2 is equal to the change in the rigid body's angular momentum about its center of mass during that interval:

$$\int_{t_1}^{t_2} \Sigma M \, dt = H_2 - H_1.$$
Eq. (19.30)

The angular momentum of a rigid body about a fixed point O is

$$H_O = (\mathbf{r} \times m\mathbf{v}) \cdot \mathbf{k} + I\omega,$$
Eq. (19.32)

where $\mathbf{r}$ is the position of the center of mass relative to O and $\mathbf{v}$ is the velocity of the center of mass. A second form of the principle of angular impulse and momentum states that the angular impulse about O during the interval from t_1 to t_2 is equal to the change in the rigid body's angular momentum about O during that interval:

$$\int_{t_1}^{t_2} \Sigma M_O \, dt = H_{O2} - H_{O1}.$$
Eq. (19.33)

In terms of the averages of the total moments with respect to time, Eqs. (19.30) and (19.32) are

$$(t_2 - t_1)\Sigma M_{av} = H_2 - H_1.$$
Eq. (19.34)

and

$$(t_2 - t_1)(\Sigma M_O)_{av} = H_{O2} - H_{O1}.$$
Eq. (19.35)

If two rigid bodies A and B in planar motion are subjected only to internal forces and couples, or if the total moment due to external forces and couples about a fixed point O is zero, the total angular momentum of A and B about O is conserved:

$$H_{OA} + H_{OB} = \text{constant.}$$
Eq. (19.36)

Impacts

Suppose that two rigid bodies A and B, in two-dimensional motion in the same plane, collide. If other forces and couples are negligible in comparison to the impact forces and couples that A and B exert on each other, then the total linear momentum of A and B and their total angular momentum about any fixed point O are conserved. If, in addition, A and B exert only forces on each other at their point of impact, P, the angular momentum about P of *each* rigid body is conserved. If one of the rigid bodies has a pin support at a point O, the total angular momentum of both bodies about that point is conserved.

Let P be the point of impact (Fig. b). The normal components of the velocities at P are related to the coefficient of restitution e by

$$e = \frac{\left(\mathbf{v}'_{BP}\right)_x - \left(\mathbf{v}'_{AP}\right)_x}{\left(\mathbf{v}_{AP}\right)_x - \left(\mathbf{v}_{BP}\right)_x}.$$
Eq. (19.50)

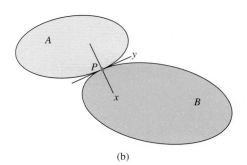

(b)

Review Problems

19.1 The moment of inertia of the pulley is 0.2 kg-m². The system is released from rest. Use the principle of work and energy to determine the velocity of the 10-kg cylinder when it has fallen 1 m.

150 mm

5 kg   10 kg

P19.1

19.2 Use momentum principles to determine the velocity of the 10-kg cylinder in Problem 19.1 1 s after the system is released from rest.

19.3 Arm *BC* has a mass of 12 kg, and the moment of inertia about its center of mass is 3 kg-m². Point *B* is stationary. Arm *BC* is initially aligned with the (horizontal) *x* axis with zero angular velocity, and a constant couple *M* applied at *B* causes the arm to rotate upward. When it is in the position shown, its counterclockwise angular velocity is 2 rad/s. Determine *M*.

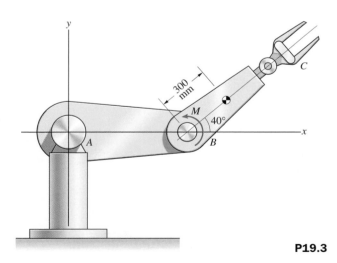

y

300 mm

M

40°

A   B

C

x

P19.3

19.4 The cart is stationary when a constant force *F* is applied to it. What will the velocity of the cart be when it has rolled a distance *b*? The mass of the body of the cart is m_c, and each of the four wheels has mass *m*, radius *R*, and moment of inertia, *I*.

F

P19.4

19.5 Each pulley has moment of inertia $I = 0.003$ kg-m², and the mass of the belt is 0.2 kg. If a constant couple $M = 4$ N-m is applied to the bottom pulley, what will its angular velocity be when it has turned 10 revolutions?

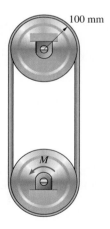

100 mm

M

P19.5

19.6 The ring gear is fixed. The mass and moment of inertia of the sun gear are $m_S = 22$ slugs and $I_S = 4400$ slug-ft². The mass and moment of inertia of each planet gear are $m_P = 2.7$ slugs and $I_P = 65$ slug-ft². A couple $M = 600$ ft-lb is applied to the sun gear. Use work and energy to determine the angular velocity of the sun gear after it has turned 100 revolutions.

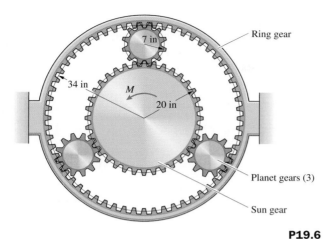

P19.6

moment of inertia of 0.1 slug-ft^2. The total weight of the rider and go-cart, including its wheels, is 240 lb. The go-cart starts from rest, its engine exerts a constant torque of 15 ft-lb on the rear axle, and its wheels do not slip. Neglecting friction and aerodynamic drag, how fast is the go-cart moving when it has traveled 50 ft?

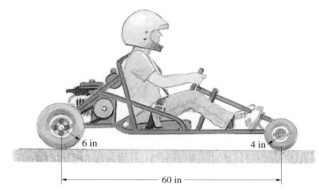

P19.9

19.7 The moments of inertia of gears A and B are $I_A = 0.014$ slug-ft^2 and $I_B = 0.100$ slug-ft^2. Gear A is connected to a torsional spring with constant $k = 0.2$ ft-lb/rad. If the spring is unstretched and the surface supporting the 5-lb weight is removed, what is the velocity of the weight when it has fallen 3 in.?

19.10 Determine the maximum power and the average power transmitted to the go-cart in Problem 19.9 by its engine.

19.11 The system starts from rest with the 4-kg slender bar horizontal. The mass of the suspended cylinder is 10 kg. What is the angular velocity of the bar when it is in the position shown?

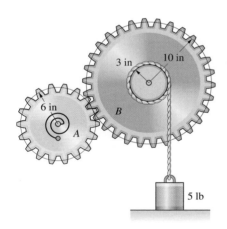

P19.7

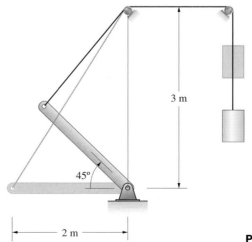

19.8 Consider the system in Problem 19.7.
(a) What maximum distance does the 5-lb weight fall when the supporting surface is removed?
(b) What maximum velocity does the weight achieve?

19.9 Each of the go-cart's front wheels weighs 5 lb and has a moment of inertia of 0.01 slug-ft^2. The two rear wheels and rear axle form a single rigid body weighing 40 lb and having a

P19.11

19.12 The 0.1-kg slender bar and 0.2-kg cylindrical disk are released from rest with the bar horizontal. The disk rolls on the curved surface. What is the angular velocity of the bar when it is vertical?

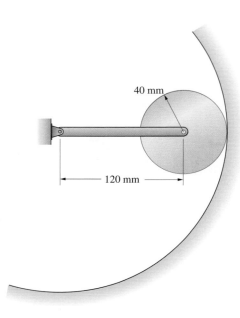

40 mm

120 mm

P19.12

19.13 A slender bar of mass m is released from rest in the vertical position and allowed to fall. Neglecting friction and assuming that it remains in contact with the floor and wall, determine the bar's angular velocity as a function of θ.

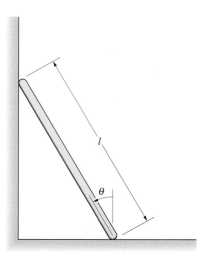

l

θ

P19.13

19.14 The 4-kg slender bar is pinned to 2-kg sliders at A and B. If friction is negligible and the system starts from rest in the position shown, what is the bar's angular velocity when the slider at A has fallen 0.5 m?

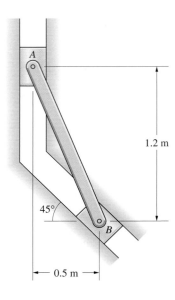

A

1.2 m

45°

B

0.5 m

P19.14

19.15 A homogeneous hemisphere of mass m is released from rest in the position shown. If it rolls on the horizontal surface, what is its angular velocity when its flat surface is horizontal?

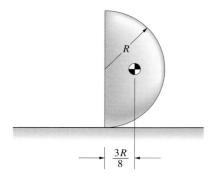

R

$\dfrac{3R}{8}$

P19.15

19.16 What normal force is exerted on the hemisphere in Problem 19.15 by the horizontal surface at the instant the flat surface of the hemisphere is horizontal?

19.17 The slender bar rotates freely *in the horizontal plane* about a vertical shaft at O. The bar weighs 20 lb and its length is 6 ft. The slider A weighs 2 lb. If the bar's angular velocity is $\omega = 10$ rad/s and the radial component of the velocity of A is zero when $r = 1$ ft, what is the angular velocity of the bar when $r = 4$ ft? (The moment of inertia of A about its center of mass is negligible; that is, treat A as a particle.)

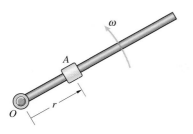

P19.17

19.20 The homogeneous cylindrical disk of mass m rolls on the horizontal surface with angular velocity ω. If the disk does not slip or leave the slanted surface when it comes into contact with it, what is the angular velocity ω' of the disk immediately afterward?

P19.20

$\mathscr{D}$ **19.18** A satellite is deployed with angular velocity $\omega = 1$ rad/s (Fig. a). Two internally stored antennas that span the diameter of the satellite are then extended, and the satellite's angular velocity decreases to ω' (Fig. b). By modeling the satellite as a 500-kg sphere of 1.2-m radius and each antenna as a 10-kg slender bar, determine ω'.

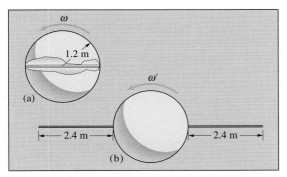

P19.18

19.21 The 10-lb slender bar falls from rest in the vertical position and hits the smooth projection at B. The coefficient of restitution of the impact is $e = 0.6$, the duration of the impact is 0.1 s, and $b = 1$ ft. Determine the average force exerted on the bar at B as a result of the impact.

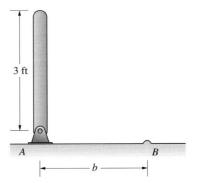

P19.21

$\mathscr{D}$ **19.19** An engineer decides to control the angular velocity of a satellite by deploying small masses attached to cables. If the angular velocity of the satellite in configuration (a) is 4 rpm, determine the distance d in configuration (b) that will cause the angular velocity to be 1 rpm. The moment of inertia of the satellite is $I = 500$ kg-m^2 and each mass is 2 kg. (Assume that the cables and masses rotate with the same angular velocity as the satellite. Neglect the masses of the cables and the moments of inertia of the masses about their centers of mass.)

19.22 In Problem 19.21, determine the value of b for which the average force exerted on the bar at A as a result of the impact is zero.

19.23 The 1-kg sphere A is moving at 2 m/s when it strikes the end of the 2-kg stationary slender bar B. If the velocity of the sphere after the impact is 0.8 m/s to the right, what is the coefficient of restitution?

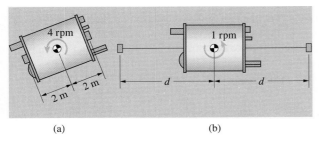

P19.19

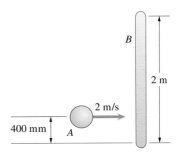

P19.23

19.24 The slender bar is released from rest in the position shown in Fig. (a) and falls a distance $h = 1$ ft. When the bar hits the floor, its tip is supported by a depression and remains on the floor (Fig. b). The length of the bar is 1 ft and its weight is 4 oz. What is angular velocity ω of the bar just after it hits the floor?

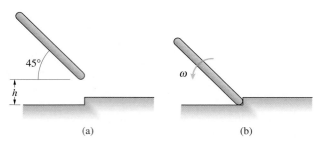

(a)   (b)

P19.24

19.25 The slender bar is released from rest with $\theta = 45°$ and falls a distance $h = 1$ m onto the smooth floor. The length of the bar is 1 m and its mass is 2 kg. If the coefficient of restitution of the impact is $e = 0.4$, what is the angular velocity of the bar just after it hits the floor?

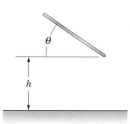

P19.25

19.26 In Problem 19.25, determine the angle θ for which the angular velocity of the bar just after it hits the floor is a maximum. What is the maximum angular velocity?

19.27 A nonrotating slender bar A moving with velocity v_0 strikes a stationary slender bar B. Each bar has mass m and length l. If the bars adhere when they collide, what is their angular velocity after the impact?

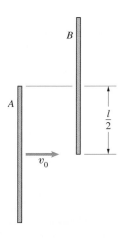

P19.27

19.28 An astronaut translates toward a nonrotating satellite at $1.0\mathbf{i}$ (m/s) relative to the satellite. Her mass is 136 kg, and the moment of inertia about the axis through her center of mass parallel to the z axis is 45 kg-m². The mass of the satellite is 450 kg and its moment of inertia about the z axis is 675 kg-m². At the instant the astronaut attaches to the satellite and begins moving with it, the position of her center of mass is $(-1.8, -0.9, 0)$ m. The axis of rotation of the satellite after she attaches is parallel to the z axis. What is their angular velocity?

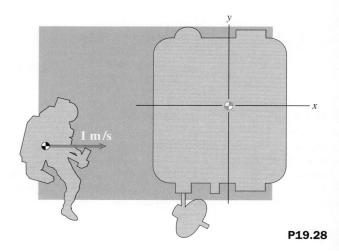

P19.28

19.29 In Problem 19.28, suppose that the design parameters of the satellite's control system require that the angular velocity of the satellite not exceed 0.02 rad/s. If the astronaut is moving parallel to the x axis and the position of her center of mass when she attaches is $(-1.8, -0.9, 0)$ m, what is the maximum relative velocity at which she should approach the satellite?

19.30 A 2800-lb car skidding on ice strikes a concrete abutment at 3 mi/hr. The car's moment of inertia about its center of mass is 1800 slug-ft^2. Assume that the impacting surfaces are smooth and parallel to the y axis and that the coefficient of restitution of the impact is $e = 0.8$. What are the angular velocity of the car and the velocity of its center of mass after the impact?

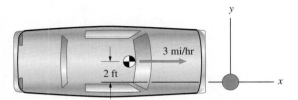

P19.30

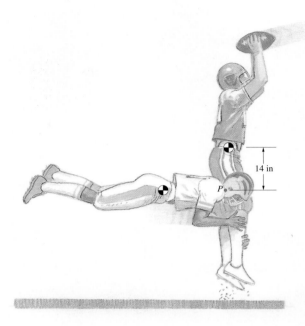

P19.31

19.31 A 170-lb wide receiver jumps vertically to receive a pass and is stationary at the instant he catches the ball. At the same instant, he is hit at P by a 180-lb linebacker moving horizontally at 15 ft/s. The wide receiver's moment of inertia about his center of mass is 7 slug-ft^2. If you model the players as rigid bodies and assume that the coefficient of restitution is $e = 0$, what is the wide receiver's angular velocity immediately after the impact?

Predicting the interesting behaviors of tops and gyroscopes requires a three-dimensional analysis. In this chapter, we discuss three-dimensional motions of rigid bodies.

Three-Dimensional Kinematics and Dynamics of Rigid Bodies

For many engineering applications of dynamics, such as the design of airplanes and other vehicles, we must consider three-dimensional motion. After explaining how three-dimensional motion of a rigid body is described, we derive the equations of motion and use them to analyze simple three-dimensional motions. Finally, we introduce the Euler angles used to specify the orientation of a rigid body in three dimensions and express the equations of angular motion in terms of those angles.

20.1 Kinematics

If a bicyclist rides in a straight path, the wheels undergo planar motion; but if the rider is turning, the motion of the wheels is three dimensional (Fig. 20.1a). Similarly, an airplane can remain in planar motion while it is in level flight or as it descends, climbs, or performs loops; but if it banks and turns, it is in three-dimensional motion (Fig. 20.1b). A spinning top may remain in planar motion for a brief period, rotating about a fixed vertical axis; but eventually, the top's axis begins to tilt and rotate. The top is then in three-dimensional motion and exhibits interesting, apparently gravity-defying behavior (Fig. 20.1c). In this section, we begin the analysis of such motions by discussing the kinematics of rigid bodies in three-dimensional motion.

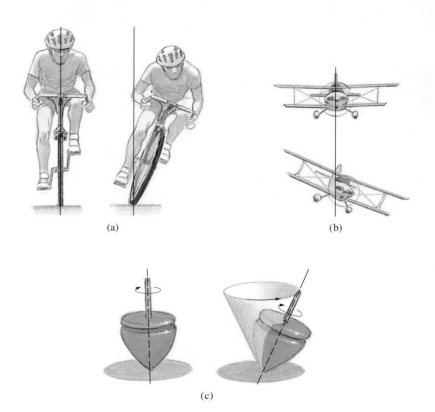

(a) (b)

(c)

Figure 20.1

Examples of planar and three-dimensional motions.

Velocities and Accelerations

We have already discussed some of the concepts needed to describe the three-dimensional motion of a rigid body relative to a given reference frame. In Chapter 17, we showed that Euler's theorem implies that a rigid body undergoing any motion other than translation has an instantaneous axis of rotation. The direction of this axis at a particular instant and the rate at which the rigid body rotates about the axis are specified by the angular velocity vector $\boldsymbol{\omega}$.

We have also shown that a rigid body's velocity is completely specified by its angular velocity and the velocity of a single point of the body. For the rigid body and reference frame in Fig. 20.2, suppose that we know the angular velocity $\boldsymbol{\omega}$ and the velocity $\mathbf{v}_B$ of a point B. Then the velocity of any other point A of the body is given by Eq. (17.8):

$$\mathbf{v}_A = \mathbf{v}_B + \boldsymbol{\omega} \times \mathbf{r}_{A/B}. \tag{20.1}$$

A rigid body's acceleration is completely specified by its angular acceleration vector $\boldsymbol{\alpha} = d\boldsymbol{\omega}/dt$, its angular velocity vector, and the acceleration of a single point of the body. If we know $\boldsymbol{\alpha}$, $\boldsymbol{\omega}$, and the acceleration $\mathbf{a}_B$ of the point B in Fig. 20.2, the acceleration of any other point A of the rigid body is given by Eq. (17.9):

$$\mathbf{a}_A = \mathbf{a}_B + \boldsymbol{\alpha} \times \mathbf{r}_{A/B} + \boldsymbol{\omega} \times (\boldsymbol{\omega} \times \mathbf{r}_{A/B}) \tag{20.2}$$

Moving Reference Frames

The velocities and accelerations in Eqs. (20.1) and (20.2) are measured relative to the reference frame shown in Fig. 20.2, which we refer to as the primary reference frame. In this chapter, we also use secondary reference frames that move relative to the primary reference frame. In some situations, the secondary reference frame will be fixed to the rigid body. In other situations, we will find it convenient to use a secondary reference frame that moves relative to the primary reference frame but is not fixed to the rigid body. The secondary reference frame and its motion will be chosen for convenience in describing the motion of a particular rigid body.

Let us consider (1) a secondary reference frame xyz whose rate of rotation relative to the primary reference frame is described by an angular velocity vector $\boldsymbol{\Omega}$ and (2) a rigid body whose angular velocity vector relative to the primary reference frame is $\boldsymbol{\omega}$ (Fig. 20.3). If the secondary reference frame is fixed to the rigid body, $\boldsymbol{\Omega} = \boldsymbol{\omega}$. If we express $\boldsymbol{\omega}$ in terms of its components in the secondary reference frame as

$$\boldsymbol{\omega} = \omega_x \mathbf{i} + \omega_y \mathbf{j} + \omega_z \mathbf{k},$$

then the rigid body's angular acceleration relative to the primary reference frame is

$$\boldsymbol{\alpha} = \frac{d\boldsymbol{\omega}}{dt} = \frac{d\omega_x}{dt}\mathbf{i} + \omega_x \frac{d\mathbf{i}}{dt} + \frac{d\omega_y}{dt}\mathbf{j} + \omega_y \frac{d\mathbf{j}}{dt} + \frac{d\omega_z}{dt}\mathbf{k} + \omega_z \frac{d\mathbf{k}}{dt}. \tag{20.3}$$

In terms of the angular velocity $\boldsymbol{\Omega}$ of the secondary reference frame, the derivatives of the unit vectors $\mathbf{i}$, $\mathbf{j}$, and $\mathbf{k}$ of the secondary reference frame with respect to time are (see Section 17.5)

$$\frac{d\mathbf{i}}{dt} = \boldsymbol{\Omega} \times \mathbf{i}, \qquad \frac{d\mathbf{j}}{dt} = \boldsymbol{\Omega} \times \mathbf{j}, \qquad \frac{d\mathbf{k}}{dt} = \boldsymbol{\Omega} \times \mathbf{k}.$$

Substituting these expressions into Eq. (20.3), we obtain the angular acceleration of the rigid body relative to the primary reference frame in the form

$$\boldsymbol{\alpha} = \frac{d\omega_x}{dt}\mathbf{i} + \frac{d\omega_y}{dt}\mathbf{j} + \frac{d\omega_z}{dt}\mathbf{k} + \boldsymbol{\Omega} \times \boldsymbol{\omega}. \tag{20.4}$$

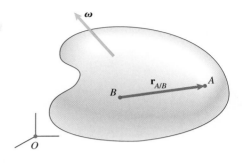

Figure 20.2
Points A and B of a rigid body. The velocity of A can be determined if the velocity of B and the rigid body's angular velocity $\boldsymbol{\omega}$ are known. The acceleration of A can be determined if the acceleration of B, the angular velocity, and the angular acceleration are known.

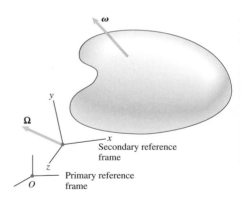

Figure 20.3
The primary and secondary reference frames. The vector $\boldsymbol{\Omega}$ is the angular velocity of the secondary reference frame relative to the primary reference frame. The vector $\boldsymbol{\omega}$ is the angular velocity of the rigid body relative to the primary reference frame.

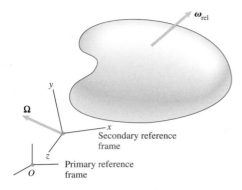

Figure 20.4
The vector $\boldsymbol{\omega}_{\text{rel}}$ is the rigid body's angular velocity relative to the secondary reference frame, and the vector $\boldsymbol{\Omega}$ is the angular velocity of the secondary reference frame relative to the primary reference frame. The rigid body's angular velocity relative to the primary reference frame is $\boldsymbol{\omega}_{\text{rel}} + \boldsymbol{\Omega}$.

When the secondary reference frame is not fixed to the rigid body, it is often convenient to express the body's angular velocity $\boldsymbol{\omega}$ as the sum of the angular velocity $\boldsymbol{\Omega}$ of the secondary reference frame and the angular velocity $\boldsymbol{\omega}_{\text{rel}}$ of the rigid body relative to the secondary reference frame (Fig. 20.4):

$$\boldsymbol{\omega} = \boldsymbol{\Omega} + \boldsymbol{\omega}_{\text{rel}}. \tag{20.5}$$

Study Questions

1. If you know the velocity of a point B of a rigid body relative to a given reference frame, what additional information do you need to determine the velocity of a second point A of the rigid body relative to the given reference frame?

2. If you know the acceleration of a point B of a rigid body relative to a given reference frame, what additional information do you need to determine the acceleration of a second point A of the rigid body relative to the given reference frame?

3. Suppose that the angular velocity of a secondary reference frame relative to a primary reference frame is $\boldsymbol{\Omega}$. If the angular velocity of a rigid body relative to the secondary reference frame is $\boldsymbol{\omega}_{\text{rel}}$, what is the rigid body's angular velocity relative to the primary reference frame?

Example 20.1

Use of a Secondary Reference Frame

The tire in Fig. 20.5 is rolling on the level surface. As the car turns, the midpoint B of the tire moves at 5 m/s in a circular path about the fixed point P (see the top view), and the tire remains perpendicular to the line from B to P.
(a) What is the tire's angular velocity $\boldsymbol{\omega}$ relative to an earth-fixed reference frame?
(b) Determine the velocity of point A, the rearmost point of the tire, at the instant shown.

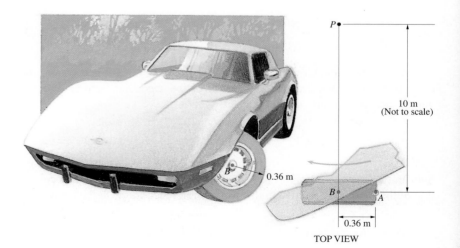

Figure 20.5

Strategy

(a) In Fig. (a) we introduce a secondary reference frame with its origin at B and its y axis along the line from B to P. We assume that the x axis *remains horizontal*. The motion of this reference frame is simple: It rotates about its z axis as the car turns. The motion of the tire relative to the secondary reference frame is also simple: It rotates about the y axis. By determining the angular velocity Ω of the secondary reference frame and the angular velocity ω_{rel} of the tire relative to the secondary reference frame, we can use Eq. (20.5) to determine the tire's angular velocity relative to an earth-fixed reference frame.

(b) Knowing the velocity of point B and the tire's angular velocity ω, we can use Eq. (20.1) to determine the velocity of point A.

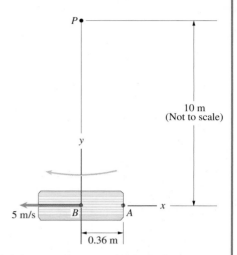

(a) A rotating secondary coordinate system. The y axis remains aligned with the line BP, and. the x axis remains horizontal.

Solution

(a) The magnitude of the angular velocity of the line PB about P is $(5 \text{ m/s})/(10 \text{ m}) = 0.5 \text{ rad/s}$. The secondary coordinate system rotates about its z axis in the clockwise direction as viewed in Fig. (a), so its angular velocity is

$$\Omega = -0.5\mathbf{k} \text{ (rad/s)}.$$

The center of the tire moves at 5 m/s, so the magnitude of the tire's angular velocity about the y axis is $(5 \text{ m/s})/(0.36 \text{ m}) = 13.9 \text{ rad/s}$. The tire's angular velocity relative to the secondary coordinate system is

$$\omega_{rel} = -13.9\mathbf{j} \text{ (rad/s)}.$$

Therefore, the tire's angular velocity relative to an earth-fixed reference frame is

$$\omega = \Omega + \omega_{rel} = -13.9\mathbf{j} - 0.5\mathbf{k} \text{ (rad/s)}.$$

(b) The position vector of point A relative to point B is $\mathbf{r}_{A/B} = 0.36\mathbf{i}$ (m). From Eq. (20.1), the velocity of point A is

$$\mathbf{v}_A = \mathbf{v}_B + \omega \times \mathbf{r}_{A/B}$$

$$= -5\mathbf{i} + \begin{vmatrix} \mathbf{i} & \mathbf{j} & \mathbf{k} \\ 0 & -13.9 & -0.5 \\ 0.36 & 0 & 0 \end{vmatrix}$$

$$= -5\mathbf{i} - 0.18\mathbf{j} + 5\mathbf{k} \text{ (m/s)}.$$

Discussion

Notice how using a secondary coordinate system rotating with the car simplified the determination of the tire's angular velocity. Although the tire's motion in space is quite complicated, its motion relative to the secondary coordinate system is simple.

Example 20.2

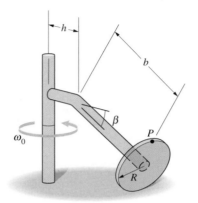

Figure 20.6

Angular Velocity and Acceleration of a Disk

The bent bar in Fig. 20.6 is rigidly attached to the vertical shaft, which rotates with constant angular velocity ω_0. The circular disk is pinned to the bent bar and rolls on the horizontal surface.
(a) Determine the disk's angular velocity ω_{disk} and angular acceleration α_{disk}.
(b) Determine the velocity of point P, which is the uppermost point of the circular disk, at the present instant.

Strategy

(a) In this example, the primary reference frame is fixed with respect to the surface on which the disk rolls. To simplify our analysis of the disk's angular motion, we will use a secondary coordinate system that is fixed with respect to the bent bar. By applying Eq. (20.1) to the bent bar, we will determine the velocity of the center of the disk. We will determine the disk's angular velocity by recognizing that the velocity of the point of the disk in contact with the horizontal surface is zero. We can then use Eq. (20.4) to determine the disk's angular acceleration.
(b) Knowing the velocity of the center of the disk and the disk's angular velocity, we can apply Eq. (20.1) to the disk to determine the velocity of point P.

Solution

(a) Let the coordinate system in Fig. (a) be fixed with respect to the bent bar. The x axis coincides with the horizontal part of the bar, and the y axis coincides with the vertical shaft. The angular velocity ω_{bar} of the bar and the angular velocity $\boldsymbol{\Omega}$ of the coordinate system are equal:

$$\omega_{\text{bar}} = \boldsymbol{\Omega} = \omega_0 \mathbf{j}.$$

Let point B be the stationary origin of the coordinate system, and let point A be the center of the disk (Fig. a). The position vector of A relative to B is

$$\mathbf{r}_{A/B} = (h + b \cos\beta)\mathbf{i} - b \sin\beta\mathbf{j}.$$

From Eq. (20.1), the velocity of point A is

$$\mathbf{v}_A = \mathbf{v}_B + \omega_{\text{bar}} \times \mathbf{r}_{A/B}$$

$$= \mathbf{0} + \begin{vmatrix} \mathbf{i} & \mathbf{j} & \mathbf{k} \\ 0 & \omega_0 & 0 \\ h + b \cos\beta & -b \sin\beta & 0 \end{vmatrix}$$

$$= -\omega_0(h + b \cos\beta)\mathbf{k}.$$

Because the coordinate system is fixed with respect to the bent bar, we can write the angular velocity of the disk relative to the coordinate system as (Fig. b)

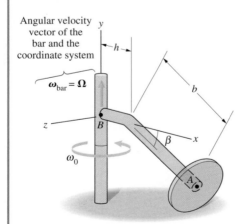

(a) A secondary coordinate system fixed to the bent bar.

$$\omega_{rel} = \omega_{rel} \cos\beta\mathbf{i} - \omega_{rel} \sin\beta\mathbf{j}.$$

Let point C in Fig. (b) be the point of the disk that is in contact with the surface. To determine ω_{rel}, we use the condition $\mathbf{v}_C = \mathbf{0}$. The position of C relative to A is

$$\mathbf{r}_{C/A} = -R\sin\beta\mathbf{i} - R\cos\beta\mathbf{j}.$$

Therefore,

$$\mathbf{v}_C = \mathbf{v}_A + \boldsymbol{\omega}_{disk} \times \mathbf{r}_{C/A}$$

$$= -\omega_0(h + b\cos\beta)\mathbf{k} + \begin{vmatrix} \mathbf{i} & \mathbf{j} & \mathbf{k} \\ \omega_{rel}\cos\beta & \omega_0 - \omega_{rel}\sin\beta & 0 \\ -R\sin\beta & -R\cos\beta & 0 \end{vmatrix} = \mathbf{0}.$$

Solving this equation for ω_{rel}, we obtain

$$\omega_{rel} = -\omega_0\left(\frac{h}{R} + \frac{b}{R}\cos\beta - \sin\beta\right).$$

The angular velocity of the coordinate system is $\boldsymbol{\Omega} = \omega_0\mathbf{j}$, so the disk's angular velocity is

$$\boldsymbol{\omega}_{disk} = \boldsymbol{\Omega} + \boldsymbol{\omega}_{rel} = \omega_{rel}\cos\beta\mathbf{i} + (\omega_0 - \omega_{rel}\sin\beta)\mathbf{j}.$$

Even though the components of $\boldsymbol{\omega}_{disk}$ are constants, we find from Eq. (20.4) that the disk's angular acceleration is not zero:

$$\boldsymbol{\alpha}_{disk} = \boldsymbol{\Omega} \times \boldsymbol{\omega}_{disk} = \begin{vmatrix} \mathbf{i} & \mathbf{j} & \mathbf{k} \\ 0 & \omega_0 & 0 \\ \omega_{rel}\cos\beta & \omega_0 - \omega_{rel}\sin\beta & 0 \end{vmatrix}$$

$$= -\omega_0\omega_{rel}\cos\beta\mathbf{k}.$$

(b) The position vector of point P relative to the center of the disk is
$$\mathbf{r}_{P/A} = R\sin\beta\mathbf{i} + R\cos\beta\mathbf{j}.$$
Using Eq. (20.1) and our result for the velocity $\mathbf{v}_A$ of the center of the disk, we determine the velocity of point P:

$$\mathbf{v}_P = \mathbf{v}_A + \boldsymbol{\omega}_{disk} \times \mathbf{r}_{P/A}$$

$$= -\omega_0(h + b\cos\beta)\mathbf{k} + \begin{vmatrix} \mathbf{i} & \mathbf{j} & \mathbf{k} \\ \omega_{rel}\cos\beta & \omega_0 - \omega_{rel}\sin\beta & 0 \\ R\sin\beta & R\cos\beta & 0 \end{vmatrix}$$

$$= \left[-\omega_0(h + b\cos\beta + R\sin\beta) + \omega_{rel}R\right]\mathbf{k}$$

$$= -2\omega_0(h + b\cos\beta)\mathbf{k}.$$

Discussion

The use of a secondary coordinate system rotating with the bent bar simplified our determination of the disk's angular velocity $\boldsymbol{\omega}_{disk}$ and angular acceleration $\boldsymbol{\alpha}_{disk}$. Although the disk's angular motion is complicated, its angular motion relative to the bar is simple.

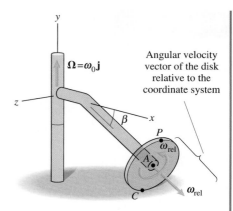

(b) Analyzing the motion of the disk.

Example 20.3

Angular Velocity and Acceleration of a Bar

The bar AB in Fig. 20.7 is connected by ball-and-socket joints to collars sliding on the fixed bars CD and EF. The collar at B has constant velocity $\mathbf{v}_B = 20\mathbf{i}$ (m/s). The angular velocity and angular acceleration of bar AB about its axis are zero. Determine the velocity and acceleration of the collar at A and the angular velocity and angular acceleration of bar AB.

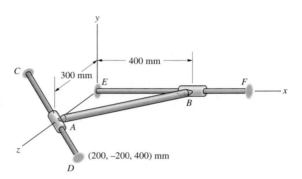

Figure 20.7

Strategy

Since we know the velocity of the collar at B, if we apply Eq. (20.1) to points A and B, we will obtain three scalar equations in four unknowns: the magnitude of the velocity of point A and the three components of the angular velocity $\boldsymbol{\omega}$ of the bar AB. Because the angular velocity of AB about its axis is zero, $\mathbf{r}_{A/B} \cdot \boldsymbol{\omega} = 0$, which supplies the required fourth equation. We can then determine the acceleration of the collar at A and the angular acceleration of bar AB using the same procedure, applying Eq. (20.2) to points A and B.

Solution

Velocity By dividing the position vector of point D relative to point A by its magnitude, we obtain a unit vector that points from A toward D:

$$\mathbf{e}_{AD} = \frac{0.2\mathbf{i} - 0.2\mathbf{j} + 0.1\mathbf{k}}{\sqrt{(0.2)^2 + (-0.2)^2 + (0.1)^2}} = 0.667\mathbf{i} - 0.667\mathbf{j} + 0.333\mathbf{k}.$$

We can write the velocity of the collar at A as

$$\mathbf{v}_A = v_A \mathbf{e}_{AD} = v_A(0.667\mathbf{i} - 0.667\mathbf{j} + 0.333\mathbf{k}).$$

We apply Eq. (20.1) to the bar AB:

$$\mathbf{v}_A = \mathbf{v}_B + \boldsymbol{\omega} \times \mathbf{r}_{A/B}:$$

$$v_A(0.667\mathbf{i} - 0.667\mathbf{j} + 0.333\mathbf{k}) = 20\mathbf{i} + \begin{vmatrix} \mathbf{i} & \mathbf{j} & \mathbf{k} \\ \omega_x & \omega_y & \omega_z \\ -0.4 & 0 & 0.3 \end{vmatrix}.$$

Equating **i**, **j**, and **k** components, we obtain

$$0.667v_A = 20 + 0.3\omega_y,$$

$$-0.667v_A = -0.4\omega_z - 0.3\omega_x,$$

and

$$0.333v_A = 0.4\omega_y.$$

Since the angular velocity of bar AB about its axis is zero, we know that

$$\mathbf{r}_{A/B} \cdot \boldsymbol{\omega} = -0.4\omega_x + 0.3\omega_z = 0.$$

We therefore have four equations in terms of v_A, ω_x, ω_y, and ω_z. Solving, we obtain

$$v_A = 48 \text{ m/s}$$

and

$$\boldsymbol{\omega} = 38.4\mathbf{i} + 40\mathbf{j} + 51.2\mathbf{k} \text{ (rad/s)}.$$

Acceleration   We can write the acceleration of the collar at A as

$$\mathbf{a}_A = a_A \mathbf{e}_{AD} = a_A(0.667\mathbf{i} - 0.667\mathbf{j} + 0.333\mathbf{k}).$$

We apply Eq. (20.2) to the bar AB:

$$\mathbf{a}_A = \mathbf{a}_B + \boldsymbol{\alpha} \times \mathbf{r}_{A/B} + \boldsymbol{\omega} \times (\boldsymbol{\omega} \times \mathbf{r}_{A/B}):$$

$$a_A(0.667\mathbf{i} - 0.667\mathbf{j} + 0.333\mathbf{k}) = \mathbf{0} + \begin{vmatrix} \mathbf{i} & \mathbf{j} & \mathbf{k} \\ \alpha_x & \alpha_y & \alpha_z \\ -0.4 & 0 & 0.3 \end{vmatrix}$$

$$+ \begin{vmatrix} \mathbf{i} & \mathbf{j} & \mathbf{k} \\ \omega_x & \omega_y & \omega_z \\ 0.3\omega_y & -0.4\omega_z - 0.3\omega_x & 0.4\omega_y \end{vmatrix}.$$

Equating the **i**, **j**, and **k** components yields

$$0.667a_A = 0.3\alpha_y + 0.4\omega_y^2 + 0.4\omega_z^2 + 0.3\omega_x\omega_z,$$
$$-0.667a_A = -0.4\alpha_z - 0.3\alpha_x + 0.3\omega_y\omega_z - 0.4\omega_x\omega_y,$$

and

$$0.333a_A = 0.4\alpha_y - 0.4\omega_x\omega_z - 0.3\omega_x^2 - 0.3\omega_y^2.$$

Because the angular acceleration of bar AB about its axis equals zero,

$$\mathbf{r}_{A/B} \cdot \boldsymbol{\alpha} = -0.4\alpha_x + 0.3\alpha_z = 0.$$

We know the components of the angular velocity, so we have four equations in terms of a_A, α_x, α_y, and α_z. Solving, we obtain

$$a_A = 8540 \text{ m/s}^2$$

and

$$\boldsymbol{\alpha} = 6840\mathbf{i} + 11{,}390\mathbf{j} + 9110\mathbf{k} \text{ (rad/s}^2).$$

20.2 Euler's Equations

The three-dimensional equations of motion for a rigid body are called *Euler's equations*. They include, in addition to equations of angular motion, the statement that the sum of the external forces equals the product of the mass of the rigid body and the acceleration of its center of mass:

$$\Sigma \mathbf{F} = m\mathbf{a}. \tag{20.6}$$

In deriving the equations of angular motion, we consider first the special case of rotation about a fixed point and then general three-dimensional motion.

Rotation about a Fixed Point

Let m_i be the mass of the ith particle of a rigid body, and let $\mathbf{r}_i$ be its position relative to a point O that is fixed with respect to an inertial primary reference frame (Fig. 20.8).

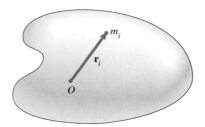

Figure 20.8
Mass and position of the ith particle of a
rigid body.

In Section 18.2, we showed that for an arbitrary system of particles, the sum of the moments about O equals the rate of change of the total angular momentum about O:

$$\Sigma \mathbf{M}_O = \frac{d\mathbf{H}_O}{dt}, \tag{20.7}$$

where the total angular momentum is

$$\mathbf{H}_O = \sum_i \mathbf{r}_i \times m_i \frac{d\mathbf{r}_i}{dt}.$$

If the rigid body rotates about O with angular velocity $\boldsymbol{\omega}$, the velocity of the ith particle is $d\mathbf{r}_i/dt = \boldsymbol{\omega} \times \mathbf{r}_i$, and the angular momentum is

$$\mathbf{H}_O = \sum_i \mathbf{r}_i \times m_i(\boldsymbol{\omega} \times \mathbf{r}_i). \tag{20.8}$$

In Fig. (20.9), we introduce a secondary coordinate system with its origin at O. We express the vectors $\boldsymbol{\omega}$ and $\mathbf{r}_i$ in terms of their components in this coordinate system as

$$\boldsymbol{\omega} = \omega_x \mathbf{i} + \omega_y \mathbf{j} + \omega_z \mathbf{k}$$

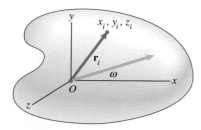

Figure 20.9
Secondary coordinate system with its origin
at O.

and

$$\mathbf{r}_i = x_i\mathbf{i} + y_i\mathbf{j} + z_i\mathbf{k},$$

where (x_i, y_i, z_i) are the coordinates of the ith particle. Substituting these expressions into Eq. (20.8) and evaluating the cross products, we can write the resulting components of the angular momentum vector in the forms

$$H_{Ox} = I_{xx}\omega_x - I_{xy}\omega_y - I_{xz}\omega_z,$$
$$H_{Oy} = -I_{yx}\omega_x + I_{yy}\omega_y - I_{yz}\omega_z,$$

and
(20.9)

$$H_{Oz} = -I_{zx}\omega_x - I_{zy}\omega_y + I_{zz}\omega_z.$$

The coefficients

$$I_{xx} = \sum_i m_i(y_i^2 + z_i^2), \qquad I_{yy} = \sum_i m_i(x_i^2 + z_i^2),$$

and
(20.10)

$$I_{zz} = \sum_i m_i(x_i^2 + y_i^2)$$

are called the *moments of inertia* about the x, y, and z axes, and the coefficients

$$I_{xy} = I_{yx} = \sum_i m_i x_i y_i, \qquad I_{yz} = I_{zy} = \sum_i m_i y_i z_i,$$

and
(20.11)

$$I_{xz} = I_{zx} = \sum_i m_i x_i z_i$$

are called the *products of inertia*. (The evaluation of the moments and products of inertia is discussed in the appendix to this chapter.)

To obtain the equations of angular motion, we must substitute the components of the angular momentum given by Eqs. (20.9) into Eq. (20.7). The secondary coordinate system in which these components are expressed is usually chosen to be body fixed, so it rotates relative to the primary reference frame with the angular velocity $\boldsymbol{\omega}$ of the rigid body. However, we have seen in the previous section that in some situations it is convenient to use a secondary coordinate system that rotates, but is not body fixed. We denote the secondary coordinate system's angular velocity vector by $\boldsymbol{\Omega}$, where $\boldsymbol{\Omega} = \boldsymbol{\omega}$ if the coordinate system is body fixed. Expressing the angular momentum vector in terms of its components as

$$\mathbf{H}_O = H_{Ox}\mathbf{i} + H_{Oy}\mathbf{j} + H_{Oz}\mathbf{k},$$

we obtain the derivative of $\mathbf{H}_O$ with respect to time:

$$\frac{d\mathbf{H}_O}{dt} = \frac{dH_{Ox}}{dt}\mathbf{i} + H_{Ox}\frac{d\mathbf{i}}{dt} + \frac{dH_{Oy}}{dt}\mathbf{j} + H_{Oy}\frac{d\mathbf{j}}{dt} + \frac{dH_{Oz}}{dt}\mathbf{k} + H_{Oz}\frac{d\mathbf{k}}{dt}.$$

Using this expression and writing the time derivatives of the unit vectors in terms of the angular velocity $\boldsymbol{\Omega}$ of the coordinate system,

$$\frac{d\mathbf{i}}{dt} = \boldsymbol{\Omega} \times \mathbf{i}, \qquad \frac{d\mathbf{j}}{dt} = \boldsymbol{\Omega} \times \mathbf{j}, \qquad \frac{d\mathbf{k}}{dt} = \boldsymbol{\Omega} \times \mathbf{k},$$

we can write Eq. (20.7) as

$$\Sigma \mathbf{M}_O = \frac{dH_{Ox}}{dt}\mathbf{i} + \frac{dH_{Oy}}{dt}\mathbf{j} + \frac{dH_{Oz}}{dt}\mathbf{k} + \boldsymbol{\Omega} \times \mathbf{H}_O.$$

Substituting the components of $\mathbf{H}_O$ from Eq. (20.9) into this equation, we obtain the equations of angular motion (see Problem 20.60):

$$
\begin{aligned}
\Sigma M_{Ox} = I_{xx}&\frac{d\omega_x}{dt} - I_{xy}\frac{d\omega_y}{dt} - I_{xz}\frac{d\omega_z}{dt} \\
&- \Omega_z(-I_{yx}\omega_x + I_{yy}\omega_y - I_{yz}\omega_z) \\
&+ \Omega_y(-I_{zx}\omega_x - I_{zy}\omega_y + I_{zz}\omega_z), \\[2mm]
\Sigma M_{Oy} = -I_{yx}&\frac{d\omega_x}{dt} + I_{yy}\frac{d\omega_y}{dt} - I_{yz}\frac{d\omega_z}{dt} \\
&+ \Omega_z(I_{xx}\omega_x - I_{xy}\omega_y - I_{xz}\omega_z) \\
&- \Omega_x(-I_{zx}\omega_x - I_{zy}\omega_y + I_{zz}\omega_z), \\[2mm]
\Sigma M_{Oz} = -I_{zx}&\frac{d\omega_x}{dt} - I_{zy}\frac{d\omega_y}{dt} + I_{zz}\frac{d\omega_z}{dt} \\
&- \Omega_y(I_{xx}\omega_x - I_{xy}\omega_y - I_{xz}\omega_z) \\
&+ \Omega_x(-I_{yx}\omega_x + I_{yy}\omega_y - I_{yz}\omega_z).
\end{aligned}
\tag{20.12}
$$

(In carrying out this final step, we have assumed that the moments and products of inertia are constants. This is so when the secondary reference frame is body fixed, but must be confirmed when it is not.) Equations (20.12) can be written as the matrix equation

$$
\begin{bmatrix} \Sigma M_{Ox} \\ \Sigma M_{Oy} \\ \Sigma M_{Oz} \end{bmatrix}
=
\begin{bmatrix} I_{xx} & -I_{xy} & -I_{xz} \\ -I_{yx} & I_{yy} & -I_{yz} \\ -I_{zx} & -I_{zy} & I_{zz} \end{bmatrix}
\begin{bmatrix} d\omega_x/dt \\ d\omega_y/dt \\ d\omega_z/dt \end{bmatrix}
$$
$$
+
\begin{bmatrix} 0 & -\Omega_z & \Omega_y \\ \Omega_z & 0 & -\Omega_x \\ -\Omega_y & \Omega_x & 0 \end{bmatrix}
\begin{bmatrix} I_{xx} & -I_{xy} & -I_{xz} \\ -I_{yx} & I_{yy} & -I_{yz} \\ -I_{zx} & -I_{zy} & I_{zz} \end{bmatrix}
\begin{bmatrix} \omega_x \\ \omega_y \\ \omega_z \end{bmatrix},
\tag{20.13}
$$

where

$$\begin{bmatrix} I_{xx} & -I_{xy} & -I_{xz} \\ -I_{yx} & I_{yy} & -I_{yz} \\ -I_{zx} & -I_{zy} & I_{zz} \end{bmatrix} = [I] \qquad (20.14)$$

is called the *inertia matrix* of the rigid body.

General Three-Dimensional Motion

Let $\mathbf{R}_i$ be the position of the *i*th particle of a rigid body relative to the center of mass of the body (Fig. 20.10). In Section 18.2, we showed that the sum of the moments about the center of mass equals the rate of change of the total angular momentum of the body relative to its center of mass; that is,

$$\Sigma \mathbf{M} = \frac{d\mathbf{H}}{dt}, \qquad (20.15)$$

where the total angular momentum is

$$\mathbf{H} = \sum_i \mathbf{R}_i \times m_i \frac{d\mathbf{R}_i}{dt}.$$

In terms of the rigid body's angular velocity $\boldsymbol{\omega}$, the velocity of the *i*th particle is $d\mathbf{R}_i/dt = \boldsymbol{\omega} \times \mathbf{R}_i$, and the angular momentum is

$$\mathbf{H} = \sum_i \mathbf{R}_i \times m_i(\boldsymbol{\omega} \times \mathbf{R}_i). \qquad (20.16)$$

We introduce a secondary coordinate system with its origin at the center of mass (Fig. 20.11) and express the vectors $\boldsymbol{\omega}$ and $\mathbf{R}_i$ in terms of their components in this coordinate system as

$$\boldsymbol{\omega} = \omega_x \mathbf{i} + \omega_y \mathbf{j} + \omega_z \mathbf{k}$$

and

$$\mathbf{R}_i = x_i \mathbf{i} + y_i \mathbf{j} + z_i \mathbf{k}.$$

Substituting these expressions into Eq. (20.16) and evaluating the cross products, we obtain the components of the angular momentum vector in the forms

$$H_x = I_{xx}\omega_x - I_{xy}\omega_y - I_{xz}\omega_z,$$
$$H_y = -I_{yx}\omega_x + I_{yy}\omega_y - I_{yz}\omega_z,$$

and $\qquad\qquad\qquad\qquad\qquad\qquad\qquad (20.17)$

$$H_z = -I_{zx}\omega_x - I_{zy}\omega_y + I_{zz}\omega_z,$$

where the expressions for the moments and products of inertia are again given by Eqs. (20.10) and (20.11). Denoting the coordinate system's angular velocity vector by $\boldsymbol{\Omega}$, and following the same steps we used to obtain Eqs. (20.12), we obtain the equations of angular motion,

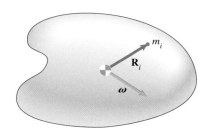

Figure 20.10
Position of the *i*th particle of a rigid body relative to the center of mass of the body.

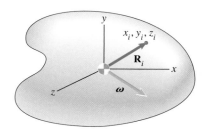

Figure 20.11
Coordinate system with its origin at the center of mass of the body.

$$\Sigma M_x = I_{xx}\frac{d\omega_x}{dt} - I_{xy}\frac{d\omega_y}{dt} - I_{xz}\frac{d\omega_z}{dt}$$
$$- \Omega_z(-I_{yx}\omega_x + I_{yy}\omega_y - I_{yz}\omega_z)$$
$$+ \Omega_y(-I_{zx}\omega_x - I_{zy}\omega_y + I_{zz}\omega_z),$$

$$\Sigma M_y = -I_{yx}\frac{d\omega_x}{dt} + I_{yy}\frac{d\omega_y}{dt} - I_{yz}\frac{d\omega_z}{dt}$$
$$+ \Omega_z(I_{xx}\omega_x - I_{xy}\omega_y - I_{xz}\omega_z)$$
$$- \Omega_x(-I_{zx}\omega_x - I_{zy}\omega_y + I_{zz}\omega_z),$$

$$\Sigma M_z = -I_{zx}\frac{d\omega_x}{dt} - I_{zy}\frac{d\omega_y}{dt} + I_{zz}\frac{d\omega_z}{dt}$$
$$- \Omega_y(I_{xx}\omega_x - I_{xy}\omega_y - I_{xz}\omega_z)$$
$$+ \Omega_x(-I_{yx}\omega_x + I_{yy}\omega_y - I_{yz}\omega_z),$$

$$(20.18)$$

which we can write as the matrix equation

$$
\begin{bmatrix} \Sigma M_x \\ \Sigma M_y \\ \Sigma M_z \end{bmatrix} = \begin{bmatrix} I_{xx} & -I_{xy} & -I_{xz} \\ -I_{yx} & I_{yy} & -I_{yz} \\ -I_{zx} & -I_{zy} & I_{zz} \end{bmatrix} \begin{bmatrix} d\omega_x/dt \\ d\omega_y/dt \\ d\omega_z/dt \end{bmatrix}
$$
$$
+ \begin{bmatrix} 0 & -\Omega_z & \Omega_y \\ \Omega_z & 0 & -\Omega_x \\ -\Omega_y & \Omega_x & 0 \end{bmatrix} \begin{bmatrix} I_{xx} & -I_{xy} & -I_{xz} \\ -I_{yx} & I_{yy} & -I_{yz} \\ -I_{zx} & -I_{zy} & I_{zz} \end{bmatrix} \begin{bmatrix} \omega_x \\ \omega_y \\ \omega_z \end{bmatrix}.
$$

$$(20.19)$$

We have thus obtained equations that are identical in form to the equations of angular motion for rotation about a fixed point. Equations (20.12) and (20.13) are expressed in terms of the total moment about a fixed point O about which the rigid body rotates, and the moments and products of inertia and the components of the vectors are expressed in terms of a coordinate system with its origin at O. Equations (20.18) and (20.19) are expressed in terms of the total moment about the center of mass of the body, and the moments and products of inertia and the components of the vectors are expressed in terms of a coordinate system with its origin at the center of mass.

Using the Euler equations to analyze three-dimensional motions of rigid bodies typically involves three steps:

1. *Choose a coordinate system.* If an object rotates about a fixed point O, it is usually preferable to use a secondary coordinate system with its origin at O and express the equations of angular motion in the forms given by Eqs. (20.12) or (20.13). Otherwise, it is necessary to use a coordinate system with its origin at the center of mass of the body and express the equations of angular motion in the forms given by Eqs. (20.18) or (20.19). In either case, it is usually preferable to choose a coordinate system that simplifies the determination of the moments and products of inertia.

2. *Draw the free-body diagram.* Isolate the object and identify the external forces and couples acting on it.

3. *Apply the equations of motion.* Use Newton's second law and the equations of angular motion to relate the forces and couples acting on the object to the acceleration of its center of mass and its angular acceleration.

Equations of Planar Motion

Here we demonstrate how the equations of angular motion for a rigid body in planar motion can be obtained from the three-dimensional equations. Consider a rigid body that rotates about a fixed axis L_O. We introduce a body-fixed secondary coordinate system with the z axis aligned with L_O, so that the rigid body's angular velocity vector is $\boldsymbol{\omega} = \omega_z \mathbf{k}$ (Fig. 20.12). Substituting $\Omega_x = \omega_x = 0$, $\Omega_y = \omega_y = 0$, and $\Omega_z = \omega_z$ into Eqs. (20.12), we find that the third equation reduces to $\Sigma M_{Oz} = I_{zz}(d\omega_z/dt)$. Introducing the simpler notation $\Sigma M_{Oz} = \Sigma M_O$, $I_{zz} = I_O$, and $\omega_z = \omega$, we obtain

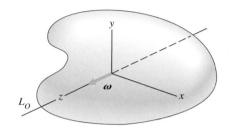

Figure 20.12

$$\Sigma M_O = I_O \frac{d\omega}{dt}.$$

This is the equation used in Chapter 18 to analyze the rotation of a rigid body about a fixed axis. [See Eq. (18.17).] The total moment about the fixed axis equals the product of the moment of inertia about the fixed axis and the angular acceleration.

For general planar motion, we introduce a body-fixed secondary coordinate system with its origin at the center of mass of the body and the z axis perpendicular to the plane of the motion (Fig. 20.13). The rigid body's angular velocity vector is $\boldsymbol{\omega} = \omega_z \mathbf{k}$. Substituting $\Omega_x = \omega_x = 0$, $\Omega_y = \omega_y = 0$, and $\Omega_z = \omega_z$ into Eqs. (20.18), the third equation reduces to $\Sigma M_z = I_{zz}(d\omega_z/dt)$. With the notation $\Sigma M_z = \Sigma M$, $I_{zz} = I$, and $\omega_z = \omega$, we obtain

$$\Sigma M = I \frac{d\omega}{dt}.$$

This is the equation of angular motion used in Chapter 18 to analyze the general planar motion of a rigid body. [See Eq. (18.20).] The total moment about the center of mass of the body equals the product of the moment of inertia about the center of mass and the angular acceleration. (The term I is the moment of inertia about the axis through the center of mass that is perpendicular to the plane of the motion.)

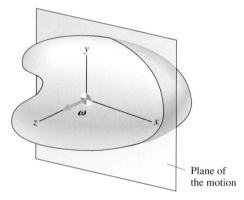

Plane of the motion

Figure 20.13

Study Questions

1. What are Euler's equations?
2. How are the moments and products of inertia of a rigid body defined?
3. Suppose that a rigid body rotates about a fixed point O with a known angular velocity. If you know the moments and products of inertia of the body in terms of a body-fixed secondary reference frame with its origin at O, how can you determine the rigid body's angular momentum about O?
4. What are the differences between Eqs. (20.12) and Eqs. (20.18)?

Example 20.4

Three-Dimensional Dynamics of a Plate

During an assembly process, the 4-kg rectangular plate in Fig. 20.14 is held at O by a robotic manipulator. Point O is stationary. At the instant shown, the plate is horizontal, its angular velocity is $\boldsymbol{\omega} = 4\mathbf{i} - 2\mathbf{j}$ (rad/s), and its angular acceleration is $\boldsymbol{\alpha} = -10\mathbf{i} + 6\mathbf{j}$ (rad/s²). Determine the couple exerted on the plate by the manipulator.

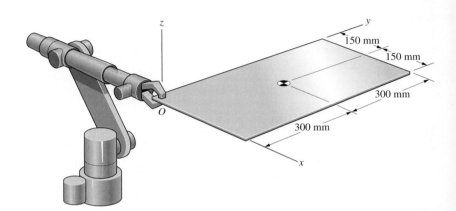

Figure 20.14

Strategy

The plate rotates about the fixed point O, so we can use Eq. (20.13) to determine the total moment exerted on the plate about O.

Solution

Draw the Free-Body Diagram We denote the force and couple exerted on the plate by the manipulator by $\mathbf{F}$ and $\mathbf{C}$ (Fig. a).

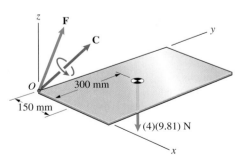

(a) Free-body diagram of the plate.

Apply the Equations of Motion   The total moment about O is the sum of the couple exerted by the manipulator and the moment about O due to the plate's weight:

$$\Sigma \mathbf{M}_O = \mathbf{C} + (0.15\mathbf{i} + 0.30\mathbf{j}) \times \left[-(4)(9.81)\mathbf{k}\right]$$

$$= \mathbf{C} - 11.77\mathbf{i} + 5.89\mathbf{j} \text{ (N-m)}. \tag{20.20}$$

To obtain the unknown couple $\mathbf{C}$, we can determine the total moment about O from Eq. (20.13).

We let the secondary coordinate system be body fixed, so its angular velocity $\boldsymbol{\Omega}$ equals the plate's angular velocity $\boldsymbol{\omega}$. We determine the plate's inertia matrix in Example 20.9, obtaining

$$[I] = \begin{bmatrix} 0.48 & -0.18 & 0 \\ -0.18 & 0.12 & 0 \\ 0 & 0 & 0.6 \end{bmatrix} \text{ kg-m}^2.$$

Therefore, the total moment about O exerted on the plate is

$$\begin{bmatrix} \Sigma M_{Ox} \\ \Sigma M_{Oy} \\ \Sigma M_{Oz} \end{bmatrix} = \begin{bmatrix} I_{xx} & -I_{xy} & -I_{xz} \\ -I_{yx} & I_{yy} & -I_{yz} \\ -I_{zx} & -I_{zy} & I_{zz} \end{bmatrix} \begin{bmatrix} d\omega_x/dt \\ d\omega_y/dt \\ d\omega_z/dt \end{bmatrix}$$

$$+ \begin{bmatrix} 0 & -\omega_z & \omega_y \\ \omega_z & 0 & -\omega_x \\ -\omega_y & \omega_x & 0 \end{bmatrix} \begin{bmatrix} I_{xx} & -I_{xy} & -I_{xz} \\ -I_{yx} & I_{yy} & -I_{yz} \\ -I_{zx} & -I_{zy} & I_{zz} \end{bmatrix} \begin{bmatrix} \omega_x \\ \omega_y \\ \omega_z \end{bmatrix}$$

$$= \begin{bmatrix} 0.48 & -0.18 & 0 \\ -0.18 & 0.12 & 0 \\ 0 & 0 & 0.6 \end{bmatrix} \begin{bmatrix} -10 \\ 6 \\ 0 \end{bmatrix}$$

$$+ \begin{bmatrix} 0 & 0 & -2 \\ 0 & 0 & -4 \\ 2 & 4 & 0 \end{bmatrix} \begin{bmatrix} 0.48 & -0.18 & 0 \\ -0.18 & 0.12 & 0 \\ 0 & 0 & 0.6 \end{bmatrix} \begin{bmatrix} 4 \\ -2 \\ 0 \end{bmatrix}$$

$$= \begin{bmatrix} -5.88 \\ 2.52 \\ 0.72 \end{bmatrix} \text{ N-m}.$$

We substitute this result into Eq. (20.20) to get

$$\Sigma \mathbf{M}_O = \mathbf{C} - 11.77\mathbf{i} + 5.89\mathbf{j} = -5.88\mathbf{i} + 2.52\mathbf{j} + 0.72\mathbf{k},$$

and then solve for the couple $\mathbf{C}$:

$$\mathbf{C} = 5.89\mathbf{i} - 3.37\mathbf{j} + 0.72\mathbf{k} \text{ (N-m)}$$

Example 20.5

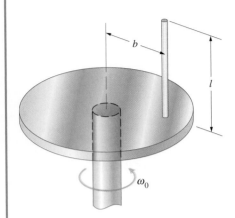

Figure 20.15

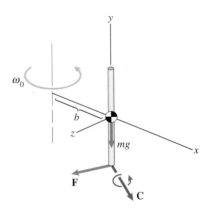

(a) Free-body diagram of the bar.

Three-Dimensional Dynamics of a Bar

A slender vertical bar of mass m is rigidly attached to a horizontal disk rotating with constant angular velocity ω_0 (Fig. 20.15). What force and couple are exerted on the bar by the disk?

Strategy

The external forces and couples on the bar are its weight and the force and couple exerted on it by the disk. The angular velocity and angular acceleration of the bar are given, and we can determine the acceleration of its center of mass, so we can use the Euler equations to determine the total force and couple exerted on the bar.

Solution

Choose a Coordinate System   In Fig. (a), we place the origin of a body-fixed coordinate system at the center of mass of the bar with the y axis vertical and the x axis in the radial direction. With this orientation, we will obtain simple expressions for the bar's angular velocity and the acceleration of its center of mass.

Draw the Free-Body Diagram   We draw the free-body diagram of the bar in Fig. (a), showing the force $\mathbf{F}$ and couple $\mathbf{C}$ exerted by the disk.

Apply the Equations of Motion   The acceleration of the center of mass of the bar due to its motion along its circular path is $\mathbf{a} = -\omega_0^2 b\mathbf{i}$. From Newton's second law,

$$\Sigma \mathbf{F} = \mathbf{F} - mg\mathbf{j} = m\left(-\omega_0^2 b\mathbf{i}\right),$$

we obtain the force exerted on the bar by the disk:

$$\mathbf{F} = -m\omega_0^2 b\mathbf{i} + mg\mathbf{j}.$$

The total moment about the center of mass is the sum of the couple $\mathbf{C}$ and the moment due to $\mathbf{F}$:

$$\Sigma \mathbf{M} = \mathbf{C} + \left(-\tfrac{1}{2}l\mathbf{j}\right) \times \left(-m\omega_0^2 b\mathbf{i} + mg\mathbf{j}\right)$$
$$= C_x\mathbf{i} + C_y\mathbf{j} + \left(C_z - \tfrac{1}{2}mlb\omega_0^2\right)\mathbf{k}.$$

The bar's inertia matrix in terms of the coordinate system in Fig. (a) is

$$[I] = \begin{bmatrix} \tfrac{1}{12}ml^2 & 0 & 0 \\ 0 & 0 & 0 \\ 0 & 0 & \tfrac{1}{12}ml^2 \end{bmatrix},$$

and its angular velocity, $\boldsymbol{\omega} = \omega_0\mathbf{j}$, is constant. The equation of angular motion, from Eq. (20.19), is

$$\begin{bmatrix} C_x \\ C_y \\ C_z - \frac{1}{2}mlb\omega_0^2 \end{bmatrix} = \begin{bmatrix} 0 & 0 & \omega_0 \\ 0 & 0 & 0 \\ -\omega_0 & 0 & 0 \end{bmatrix} \begin{bmatrix} \frac{1}{12}ml^2 & 0 & 0 \\ 0 & 0 & 0 \\ 0 & 0 & \frac{1}{12}ml^2 \end{bmatrix} \begin{bmatrix} 0 \\ \omega_0 \\ 0 \end{bmatrix}.$$

The right side of this equation equals zero, so the components of the couple exerted on the bar by the disk are $C_x = 0$, $C_y = 0$, and $C_z = \frac{1}{2}mlb\omega_0^2$.

Alternative Solution   The bar rotates about a fixed axis, so we can also determine the couple **C** by using Eq. (20.13). Let the fixed point O be the center of the disk (Fig. b), and let the body-fixed coordinate system be oriented with the x axis through the bottom of the bar. Then the total moment about O is

$$\Sigma \mathbf{M}_O = \mathbf{C} + (b\mathbf{i}) \times (-m\omega_0^2 b\mathbf{i} + mg\mathbf{j}) + (b\mathbf{i} + \tfrac{1}{2}l\mathbf{j}) \times (-mg\mathbf{j})$$

$$= \mathbf{C}.$$

Thus, the only moment about O is the couple exerted by the disk. Applying the parallel-axis theorems (Eqs. 20.42), we get for the bar's moments and products of inertia (Fig. c),

$$I_{xx} = I_{x'x'} + (d_y^2 + d_z^2)m = \tfrac{1}{12}ml^2 + (\tfrac{1}{2}l)^2 m = \tfrac{1}{3}ml^2,$$

$$I_{yy} = I_{y'y'} + (d_x^2 + d_z^2)m = mb^2,$$

$$I_{zz} = I_{z'z'} + (d_x^2 + d_y^2)m = \tfrac{1}{12}ml^2 + [b^2 + (\tfrac{1}{2}l)^2]m = \tfrac{1}{3}ml^2 + mb^2,$$

$$I_{xy} = I_{x'y'} + d_x d_y m = 0 + (b)(\tfrac{1}{2}l)m = \tfrac{1}{2}mbl,$$

$$I_{yz} = I_{y'z'} + d_y d_z m = 0,$$

and

$$I_{zx} = I_{z'x'} + d_z d_x m = 0.$$

Substituting these results into Eq. (20.13), we obtain

$$\begin{bmatrix} C_x \\ C_y \\ C_z \end{bmatrix} = \begin{bmatrix} 0 & 0 & \omega_0 \\ 0 & 0 & 0 \\ -\omega_0 & 0 & 0 \end{bmatrix} \begin{bmatrix} \frac{1}{3}ml^2 & -\frac{1}{2}mbl & 0 \\ -\frac{1}{2}mbl & mb^2 & 0 \\ 0 & 0 & \frac{1}{3}ml^2 + mb^2 \end{bmatrix} \begin{bmatrix} 0 \\ \omega_0 \\ 0 \end{bmatrix}$$

$$= \begin{bmatrix} 0 \\ 0 \\ \frac{1}{2}mlb\omega_0^2 \end{bmatrix}.$$

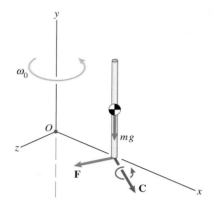

(b) Expressing the equation of angular motion in terms of the fixed point O.

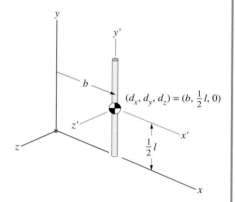

(c) Applying the parallel-axis theorem.

Discussion

If the bar were attached to the disk by a ball-and-socket support instead of a built-in support, the bar would rotate outward due to the disk's rotation. We have determined the couple the built-in support exerts on the bar that prevents it from rotating outward.

Example 20.6

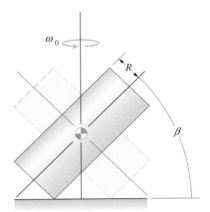

Figure 20.16

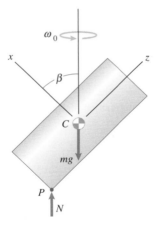

(a) Coordinate system with the z axis aligned with the cylinder axis and the y axis horizontal.

Three-Dimensional Dynamics of a Tilted Cylinder

The tilted homogeneous cylinder in Fig. 20.16 undergoes a steady motion in which one end rolls on the floor while the center of mass of the cylinder remains stationary. The angle β between the cylinder axis and the horizontal remains constant, and the cylinder axis rotates about the vertical axis with constant angular velocity ω_0. The cylinder has mass m, radius R, and length l. What is ω_0?

Strategy

By expressing the equations of angular motion in terms of ω_0, we can determine the value of ω_0 necessary for the equations to be satisfied. Therefore, our first task is to determine the cylinder's angular velocity $\boldsymbol{\omega}$ in terms of ω_0. We can simplify this task by using a secondary coordinate system that is not body-fixed.

Solution

Choose a Coordinate System We use a secondary coordinate system in which the z axis remains aligned with the cylinder axis and the y axis remains horizontal (Fig. a). The reason for this choice is that the angular velocity of the coordinate system is easy to describe—the coordinate system rotates about the vertical axis with the angular velocity ω_0—and the rotation of the cylinder relative to the coordinate system is also easy to describe. The angular velocity vector of the coordinate system is

$$\boldsymbol{\Omega} = \omega_0 \cos\beta\,\mathbf{i} + \omega_0 \sin\beta\,\mathbf{k}.$$

Relative to the coordinate system, the cylinder rotates about the z axis. Writing its angular velocity relative to the coordinate system as $\omega_{\text{rel}}\,\mathbf{k}$, we express the angular velocity vector of the cylinder as

$$\boldsymbol{\omega} = \boldsymbol{\Omega} + \omega_{\text{rel}}\,\mathbf{k} = \omega_0 \cos\beta\,\mathbf{i} + \left(\omega_0 \sin\beta + \omega_{\text{rel}}\right)\mathbf{k}.$$

We can determine ω_{rel} from the condition that the velocity of the point P in contact with the floor is zero. Expressing the velocity of P in terms of the velocity of the center of mass C, we obtain

$$\mathbf{v}_P = \mathbf{v}_C + \boldsymbol{\omega} \times \mathbf{r}_{P/C}:$$

$$\mathbf{0} = \mathbf{0} + \left[\omega_0 \cos\beta\,\mathbf{i} + \left(\omega_0 \sin\beta + \omega_{\text{rel}}\right)\mathbf{k}\right] \times \left[-R\mathbf{i} - \tfrac{1}{2}l\mathbf{k}\right]$$

$$= \left[\tfrac{1}{2}l\omega_0 \cos\beta - R\left(\omega_0 \sin\beta + \omega_{\text{rel}}\right)\right]\mathbf{j}.$$

Solving for ω_{rel} yields

$$\omega_{\text{rel}} = \left[\frac{1}{2}\left(\frac{l}{R}\right)\cos\beta - \sin\beta\right]\omega_0.$$

Therefore, the cylinder's angular velocity vector is

$$\boldsymbol{\omega} = \omega_0 \cos\beta\mathbf{i} + \frac{1}{2}\left(\frac{l}{R}\right)\omega_0 \cos\beta\mathbf{k}.$$

Draw the Free-Body Diagram We draw the free-body diagram of the cylinder in Fig. (a), showing the weight the of the cylinder and the normal force exerted by the floor. Because the cylinder's center of mass is stationary, the floor exerts no horizontal force on the cylinder, and the normal force is $N = mg$.

Apply the Equations of Motion The moment about the center of mass due to the normal force is

$$\Sigma\mathbf{M} = \left(mg\,R\sin\beta - \tfrac{1}{2}mgl\cos\beta\right)\mathbf{j}.$$

From Appendix C, the moments and products of inertia are

$$\begin{bmatrix} \frac{1}{4}mR^2 + \frac{1}{12}ml^2 & 0 & 0 \\ 0 & \frac{1}{4}mR^2 + \frac{1}{12}ml^2 & 0 \\ 0 & 0 & \frac{1}{2}mR^2 \end{bmatrix}.$$

Substituting our expressions for $\boldsymbol{\Omega}$, $\boldsymbol{\omega}$, $\Sigma\mathbf{M}$, and the moments and products of inertia into the equation of angular motion, Eq. (20.19), and evaluating the matrix products, we obtain the equation

$$mg\left(R\sin\beta - \tfrac{1}{2}l\cos\beta\right) = \left(\tfrac{1}{4}mR^2 + \tfrac{1}{12}ml^2\right)\omega_0^2\sin\beta\cos\beta$$

$$- \frac{1}{2}\left(\frac{1}{2}mR^2\right)\omega_0^2\left(\frac{l}{R}\right)\cos^2\beta.$$

We solve this equation for ω_0^2:

$$\omega_0^2 = \frac{g\left(R\sin\beta - \tfrac{1}{2}l\cos\beta\right)}{\left(\tfrac{1}{4}R^2 + \tfrac{1}{12}l^2\right)\sin\beta\cos\beta - \tfrac{1}{4}lR\cos^2\beta}. \qquad (20.21)$$

Discussion

If our solution yields a negative value for ω_0^2 for a given value of β, the assumed steady motion of the cylinder is not possible. For example, if the cylinder's diameter is equal to its length ($2R = l$), we can write Eq. (20.21) as

$$\frac{R\omega_0^2}{g} = \frac{\sin\beta - \cos\beta}{\frac{7}{12}\sin\beta\cos\beta - \frac{1}{2}\cos^2\beta}.$$

In Fig. 20.17, we show the graph of this equation as a function of β. For values of β from approximately 40° to 45°, there is no real solution for ω_0. Notice that at $\beta = 45°$, $\omega_0 = 0$, which means that the cylinder is stationary and balanced, with the center of mass directly above point P.

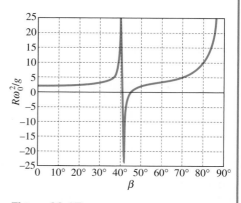

Figure 20.17
Graph of $R\omega_0^2/g$ as a function of β.

20.3 The Euler Angles

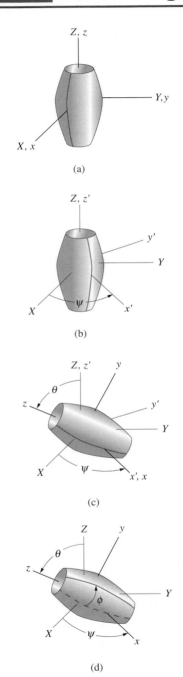

Figure 20.18
(a) The reference position.
(b) The rotation ψ about the Z axis.
(c) The rotation θ about the x' axis.
(d) The rotation ϕ of the object relative to the xyz system.

The equations of angular motion relate the total moment acting on a rigid body to its angular velocity and acceleration. If we know the total moment and the angular velocity, we can determine the angular acceleration. But how can we use the angular acceleration to determine the rigid body's angular position, or orientation, as a function of time? To explain how this is done, we must first show how to specify the orientation of a rigid body in three dimensions.

We have seen that describing the orientation of a rigid body in planar motion requires only the angle θ that specifies the body's rotation relative to some reference orientation. In three-dimensional motion, three angles are required. To understand why, consider a particular axis that is fixed relative to a rigid body. Two angles are necessary to specify the direction of the axis, and a third angle is needed to specify the rigid body's orientation about the axis. Although several systems of angles for describing the orientation of a rigid body are commonly used, the best-known system is the one called the *Euler angles*. In this section, we define these angles and express the equations of angular motion in terms of them.

Objects with an Axis of Symmetry

We first explain how the Euler angles are used to describe the orientation of an object with an axis of rotational symmetry, because this case results in simpler equations of angular motion.

Definitions We assume that an object has an axis of rotational symmetry, and we introduce two reference frames: a secondary coordinate system xyz, with its z axis coincident with the object's axis of symmetry, and an inertial primary coordinate system XYZ. We begin with the object in a reference position in which xyz and XYZ are superimposed on each other (Fig. 20.18a).

Our first step is to rotate the object and the xyz system together through an angle ψ about the Z axis (Fig. 20.18b). In this intermediate orientation, we denote the secondary coordinate system by $x'y'z'$. Next, we rotate the object and the xyz system together through an angle θ about the x' axis (Fig. 20.18c). Finally, we rotate the object relative to the xyz system through an angle ϕ about the object's axis of symmetry (Fig. 20.18d). Notice that the x axis remains in the XY plane.

The angles ψ and θ specify the orientation of the secondary xyz system relative to the primary XYZ system. The angle ψ is called the *precession angle*, and θ is called the *nutation angle*. The angle ϕ specifying the rotation of the rigid body relative to the xyz system is called the *spin angle*. These three angles specify the orientation of the rigid body relative to the primary coordinate system and are called the *Euler angles*. We can obtain any orientation of the object relative to the primary coordinate system by appropriate choices of the Euler angles: We choose ψ and θ to obtain the desired direction of the axis of symmetry and then choose ϕ to obtain the desired rotational position of the object about its axis of symmetry.

Equations of Angular Motion   To analyze an object's motion in terms of the Euler angles, we must express the equations of angular motion in terms of those angles. Figure 20.19a shows the rotation ψ from the reference orientation of the xyz system to its intermediate orientation $x'y'z'$. We represent the angular velocity of the coordinate system due to the rate of change of ψ by the angular velocity vector $\dot{\psi}$ pointing in the z' direction. (We use a dot to denote the derivative with respect to time.) Figure 20.19b shows the second rotation θ. We represent the angular velocity due to the rate of change of θ by the vector $\dot{\theta}$ pointing in the x direction. We also resolve the angular velocity vector $\dot{\psi}$ into components in the y and z directions. The components of the angular velocity of the xyz system relative to the primary coordinate system are

$$\Omega_x = \dot{\theta},$$

$$\Omega_y = \dot{\psi} \sin\theta,$$

and

$$\Omega_z = \dot{\psi} \cos\theta. \qquad (20.22)$$

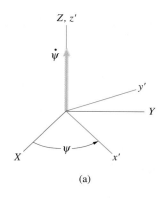

(a)

In Fig. 20.19c, we represent the angular velocity of the rigid body relative to the xyz system by the vector $\dot{\phi}$. Adding this angular velocity to the angular velocity of the xyz system, we obtain the components of the angular velocity of the rigid body relative to the XYZ system:

$$\omega_x = \dot{\theta},$$

$$\omega_y = \dot{\psi} \sin\theta,$$

and

$$\omega_z = \dot{\phi} + \dot{\psi} \cos\theta. \qquad (20.23)$$

Taking the derivatives of these equations with respect to time yields

$$\frac{d\omega_x}{dt} = \ddot{\theta},$$

$$\frac{d\omega_y}{dt} = \ddot{\psi} \sin\theta + \dot{\psi}\dot{\theta} \cos\theta,$$

and

$$\frac{d\omega_z}{dt} = \ddot{\phi} + \ddot{\psi} \cos\theta - \dot{\psi}\dot{\theta} \sin\theta. \qquad (20.24)$$

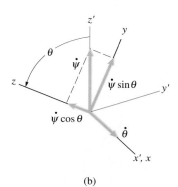

(b)

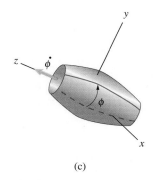

(c)

Figure 20.19
(a) The rotation ψ and the angular velocity $\dot{\psi}$.
(b) The rotation θ, the angular velocity $\dot{\theta}$, and the components of the angular velocity $\dot{\psi}$.
(c) The rotation ϕ and the angular velocity $\dot{\phi}$.

As a consequence of the object's rotational symmetry, the products of inertia I_{xy}, I_{xz}, and I_{yz} are zero and $I_{xx} = I_{yy}$. The inertia matrix is of the form

$$[I] = \begin{bmatrix} I_{xx} & 0 & 0 \\ 0 & I_{xx} & 0 \\ 0 & 0 & I_{zz} \end{bmatrix}. \qquad (20.25)$$

Substituting Eqs. (20.22)–(20.25) into Eqs. (20.18), we obtain the equations of angular motion in terms of the Euler angles:

$$\Sigma M_x = I_{xx}\ddot{\theta} + (I_{zz} - I_{xx})\dot{\psi}^2 \sin\theta \cos\theta + I_{zz}\dot{\phi}\dot{\psi} \sin\theta, \qquad (20.26)$$

$$\Sigma M_y = I_{xx}(\ddot{\psi} \sin\theta + 2\dot{\psi}\dot{\theta} \cos\theta) - I_{zz}(\dot{\phi}\dot{\theta} + \dot{\psi}\dot{\theta} \cos\theta), \qquad (20.27)$$

$$\Sigma M_z = I_{zz}(\ddot{\phi} + \ddot{\psi} \cos\theta - \dot{\psi}\dot{\theta} \sin\theta). \qquad (20.28)$$

To determine the Euler angles as functions of time when the total moment is known, these equations usually must be solved by numerical integration. However, we can obtain an important class of closed-form solutions by assuming a specific type of motion.

Steady Precession The motion called steady precession is commonly observed in tops and gyroscopes. The object's rate of spin $\dot{\phi}$ relative to the *xyz* coordinate system is assumed to be constant (Fig. 20.20). The nutation angle θ, the inclination of the *spin axis z* relative to the *Z* axis, is assumed to be constant, and the *precession rate* $\dot{\psi}$, the rate at which the *xyz* system rotates about the *Z* axis, is assumed to be constant. The last assumption explains the name given to this motion.

With these assumptions, Eqs. (20.26)–(20.28) reduce to

$$\Sigma M_x = (I_{zz} - I_{xx})\dot{\psi}^2 \sin\theta \cos\theta + I_{zz}\dot{\phi}\dot{\psi} \sin\theta, \qquad (20.29)$$

$$\Sigma M_y = 0, \qquad (20.30)$$

and

$$\Sigma M_z = 0. \qquad (20.31)$$

We discuss two examples: the steady precession of a spinning top and the steady precession of an axially symmetric object that is free of external moments.

Precession of a Top The peculiar behavior of a top (Fig. 20.21a) inspired some of the first analytical studies of three-dimensional motions of rigid bodies. When a top is set into motion, its spin axis may initially remain vertical, a motion called *sleeping*. As friction reduces the spin rate, the spin axis begins to lean over and rotate about the vertical axis. This phase of the top's motion approximates steady precession. (The top's spin rate continuously decreases due to friction, whereas in steady precession we assume the spin rate to be constant.)

To analyze the motion, we place the primary coordinate system *XYZ* with its origin at the point of the top and the *Z* axis upward. Then we align the *z* axis of the *xyz* system with the spin axis (Fig. 20.21b). We assume that the top's point rests in a small depression so that it remains at a fixed point on the floor. The precession angle ψ and nutation angle θ specify the orientation of the spin axis, and the spin rate of the top relative to the *xyz* system is $\dot{\phi}$.

The top's weight exerts a moment $\Sigma M_x = mgh \sin\theta$ about the origin, and the moments $\Sigma M_y = 0$ and $\Sigma M_z = 0$. Substituting $\Sigma M_x = mgh \sin\theta$ into Eq. (20.29), we obtain

$$mgh = (I_{zz} - I_{xx})\dot{\psi}^2 \cos\theta + I_{zz}\dot{\phi}\dot{\psi}, \qquad (20.32)$$

and Eqs. (20.30) and (20.31) are identically satisfied. Equation (20.32) relates the spin rate, nutation angle, and rate of precession. For example, if we know the spin rate $\dot{\phi}$ and nutation angle θ, we can solve for the top's precession rate $\dot{\psi}$.

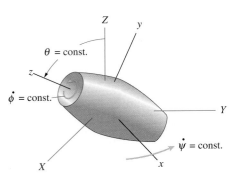

Figure 20.20
Steady procession.

(a)

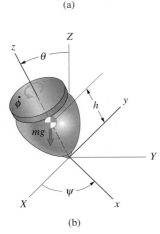

(b)

Figure 20.21
(a) A spinning top seems to defy gravity.
(b) The precession angle ψ and nutation angle θ specify the orientation of the spin axis.

Moment-Free Steady Precession   A spinning axisymmetric object that is free of external moments, such as an axisymmetric satellite in orbit, can exhibit a motion similar to the steady precessional motion of a top. This motion is observed when an American football is thrown in a "wobbly" spiral. To analyze it, we place the origin of the xyz system at the object's center of mass (Fig. 20.22a). Equation (20.29) then becomes

$$\left(I_{zz} - I_{xx}\right)\dot{\psi}\cos\theta + I_{zz}\dot{\phi} = 0, \tag{20.33}$$

and Eqs. (20.30) and (20.31) are identically satisfied. For a given value of the nutation angle, Eq. (20.33) relates the object's rates of precession and spin.

We can interpret Eq. (20.33) in a way that makes it possible to visualize the motion. We look for a point in the y-z plane at which the object's velocity relative to the center of mass is zero at the current instant. We want to find a point with coordinates $(0, y, z)$ such that

$$\boldsymbol{\omega} \times (y\mathbf{j} + z\mathbf{k}) = \left[(\dot{\psi}\sin\theta)\mathbf{j} + (\dot{\phi} + \dot{\psi}\cos\theta)\mathbf{k}\right] \times \left[y\mathbf{j} + z\mathbf{k}\right]$$

$$= \left[z\dot{\psi}\sin\theta - y(\dot{\phi} + \dot{\psi}\cos\theta)\right]\mathbf{i} = 0.$$

This equation is satisfied at points in the y-z plane such that

$$\frac{y}{z} = \frac{\dot{\psi}\sin\theta}{\dot{\phi} + \dot{\psi}\cos\theta}.$$

The preceding relation is satisfied by points on the straight line at an angle β relative to the z axis in Fig. 20.22b, where

$$\tan\beta = \frac{y}{z} = \frac{\dot{\psi}\sin\theta}{\dot{\phi} + \dot{\psi}\cos\theta}.$$

Solving Eq. (20.33) for $\dot{\phi}$ and substituting the result into this equation, we obtain

$$\tan\beta = \left(\frac{I_{zz}}{I_{xx}}\right)\tan\theta.$$

If $I_{xx} > I_{zz}$, the angle $\beta < \theta$. In Fig. 20.22c, we show an imaginary cone of half-angle β, called the *body cone*, whose axis is coincident with the z axis. The body cone is in contact with a fixed cone, called the *space cone*, whose axis is coincident with the Z axis. If the body cone rolls on the curved surface of the space cone as the z axis precesses about the Z axis (Fig. 20.22d), the points of the body cone lying on the straight line in Fig. 20.22b have zero velocity relative to the XYZ system. That means that the motion of the body cone is identical to the motion of the object. The object's motion can be visualized by visualizing the motion of the body cone as it rolls around the outer surface of the space cone. This motion is called *direct precession*.

If $I_{xx} < I_{zz}$, the angle $\beta > \theta$. In this case, we must visualize the *interior* surface of the body cone rolling on the fixed space cone (Fig. 20.22e). This motion is called *retrograde precession*.

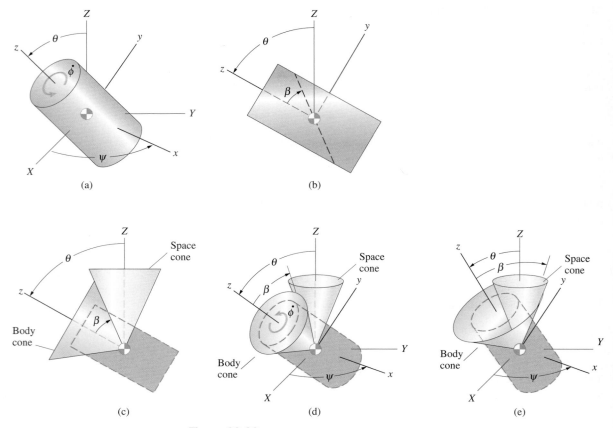

Figure 20.22
(a) An axisymmetric object.
(b) Points on the straight line at an angle β from the z axis are stationary relative to the XYZ coordinate system.
(c), (d) The body and space cones. The body cone rolls on the stationary space cone.
(e) When $\beta > \theta$, the interior surface of the body cone rolls on the stationary space cone.

Arbitrary Objects

In our analysis of axially symmetric objects, we let the object move relative to the secondary xyz coordinate system, rotating about the z axis. As a consequence, only two angles—the precession angle ψ and nutation angle θ—are needed to specify the orientation of the xyz coordinate system, and this simplifies the equations of angular motion. The object must be axially symmetric about the z axis, so that the moments and products of inertia will not vary as it rotates. In the case of an arbitrary object, the moments and products of inertia will be constants only if the xyz coordinate system is body fixed. This means that three angles are needed to specify the orientation of the coordinate system, and the resulting equations of angular motion are more complicated.

Definitions   We begin with a reference position in which the body-fixed xyz and primary XYZ coordinate systems are superimposed on each other (Fig. 20.23a). First, we rotate the xyz system through the precession angle ψ

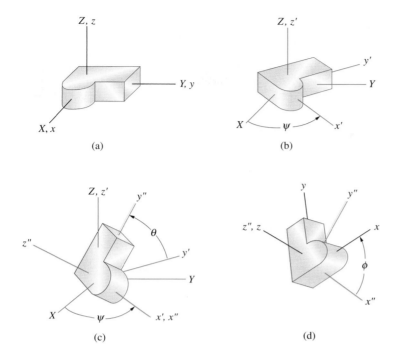

Figure 20.23
(a) The reference position.
(b) The rotation ψ about the Z axis.
(c) The rotation θ about the x' axis.
(d) The rotation ϕ about the z'' axis.

about the Z axis (Fig. 20.23b) and denote it by $x'y'z'$ in this intermediate orientation. Then we rotate the xyz system through the nutation angle θ about the x' axis (Fig. 20.23c), denoting it now by $x''y''z''$. We obtain the final orientation of the xyz system by rotating it through the angle ϕ about the z'' axis (Fig. 20.23d). Notice that we use one more rotation of the xyz system than in the case of an axially symmetric object.

We can obtain any orientation of the body-fixed coordinate system relative to the reference coordinate system by these three rotations. We choose ψ and θ to obtain the desired direction of the z axis and then choose ϕ to obtain the desired orientation of the x and y axes.

Just as in the case of an object with rotational symmetry, we must express the components of the rigid body's angular velocity in terms of the Euler angles to obtain the equations of angular motion. Figure 20.24a shows the rotation ψ from the reference orientation of the xyz system to the intermediate orientation $x'y'z'$. We represent the angular velocity of the body-fixed coordinate system due to the rate of change of ψ by the vector $\dot{\psi}$ pointing in the z' direction. Figure 20.24b shows the next rotation θ that takes the body-fixed coordinate system to the intermediate orientation $x''y''z''$. We represent the angular velocity due to the rate of change of θ by the vector $\dot{\theta}$ pointing in the x'' direction. In this figure, we also show the components of the angular velocity vector $\dot{\psi}$ in the y'' and z'' directions. Figure 20.24c shows the third rotation ϕ that takes the body-fixed coordinate system to its final orientation defined by the three Euler angles. We represent the angular velocity due to the rate of change of ϕ by the vector $\dot{\phi}$ pointing in the z direction.

To determine ω_x, ω_y, and ω_z in terms of the Euler angles, we need to determine the components of the angular velocities shown in Fig. 20.24c in the x, y, and z axis directions. The vectors $\dot{\phi}$ and $\dot{\psi}\cos\theta$ point in the z axis

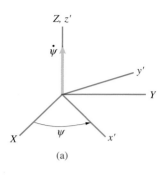

(a)

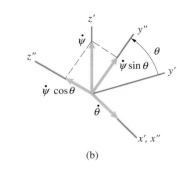

(b)

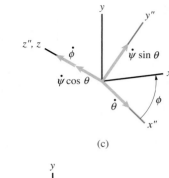

(c)

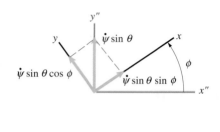

(d)

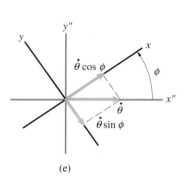

(e)

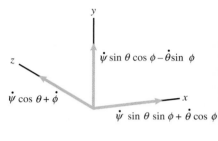

(f)

Figure 20.24
(a) The rotation ψ and the angular velocity $\dot{\psi}$.
(b) The rotation θ, the angular velocity $\dot{\theta}$, and the components of $\dot{\psi}$ in the $x''y''z''$ system.
(c) The rotation ϕ and the angular velocity $\dot{\phi}$.
(d), (e) The components of the angular velocities $\dot{\psi} \sin\theta$ and $\dot{\theta}$ in the xyz system.
(f) The angular velocities ω_x, ω_y, ω_z.

direction. In Figs. 20.24d and e, which are drawn with the z axis pointing out of the page, we determine the components of the vectors $\dot{\psi} \sin\theta$ and $\dot{\theta}$ in the x and y axis directions.

By summing the components of the angular velocities in the three coordinate directions (Fig. 20.24f), we obtain

$$\omega_x = \dot{\psi} \sin\theta \sin\phi + \dot{\theta} \cos\phi,$$

$$\omega_y = \dot{\psi} \sin\theta \cos\phi - \dot{\theta} \sin\phi,$$

and (20.34)

$$\omega_z = \dot{\psi} \cos\theta + \dot{\phi}.$$

The derivatives of these equations with respect to time are

$$\frac{d\omega_x}{dt} = \ddot{\psi} \sin\theta \sin\phi + \dot{\psi}\dot{\theta} \cos\theta \sin\phi + \dot{\psi}\dot{\phi} \sin\theta \cos\phi$$

$$+ \ddot{\theta} \cos\phi - \dot{\theta}\dot{\phi} \sin\phi,$$

$$\frac{d\omega_y}{dt} = \ddot{\psi} \sin\theta \cos\phi + \dot{\psi}\dot{\theta} \cos\theta \cos\phi - \dot{\psi}\dot{\phi} \sin\theta \sin\phi$$

$$- \ddot{\theta} \sin\phi - \dot{\theta}\dot{\phi} \cos\phi,$$

and   (20.35)

$$\frac{d\omega_z}{dt} = \ddot{\psi} \cos\theta - \dot{\psi}\dot{\theta} \sin\theta + \ddot{\phi}.$$

Equations of Angular Motion   With Eqs. (20.34) and (20.35), we can express the equations of angular motion in terms of the three Euler angles. To simplify the equations, *we assume that the body-fixed coordinate system xyz is a set of principal axes*. (See the appendix to this chapter.) Then the equations of angular motion, Eqs. (20.18), become

$$\Sigma M_x = I_{xx}\frac{d\omega_x}{dt} - \left(I_{yy} - I_{zz}\right)\omega_y\omega_z,$$

$$\Sigma M_y = I_{yy}\frac{d\omega_y}{dt} - \left(I_{zz} - I_{xx}\right)\omega_z\omega_x,$$

and

$$\Sigma M_z = I_{zz}\frac{d\omega_z}{dt} - \left(I_{xx} - I_{yy}\right)\omega_x\omega_y.$$

Substituting Eqs. (20.34) and (20.35) into these relations, we obtain the equations of angular motion in terms of Euler angles:

$$\begin{aligned}
\Sigma M_x &= I_{xx}\ddot{\psi} \sin\theta \sin\phi + I_{xx}\ddot{\theta} \cos\phi \\
&\quad + I_{xx}\left(\dot{\psi}\dot{\theta} \cos\theta \sin\phi + \dot{\psi}\dot{\phi} \sin\theta \cos\phi - \dot{\theta}\dot{\phi} \sin\phi\right) \\
&\quad - \left(I_{yy} - I_{zz}\right)\left(\dot{\psi} \sin\theta \cos\phi - \dot{\theta} \sin\phi\right)\left(\dot{\psi} \cos\theta + \dot{\phi}\right), \\
\Sigma M_y &= I_{yy}\ddot{\psi} \sin\theta \cos\phi - I_{yy}\ddot{\theta} \sin\phi \\
&\quad + I_{yy}\left(\dot{\psi}\dot{\theta} \cos\theta \cos\phi - \dot{\psi}\dot{\phi} \sin\theta \sin\phi - \dot{\theta}\dot{\phi} \cos\phi\right) \qquad \text{(20.36)} \\
&\quad - \left(I_{zz} - I_{xx}\right)\left(\dot{\psi} \cos\theta + \dot{\phi}\right)\left(\dot{\psi} \sin\theta \sin\phi + \dot{\theta} \cos\phi\right), \\
\Sigma M_z &= I_{zz}\ddot{\psi} \cos\theta + I_{zz}\ddot{\phi} - I_{zz}\dot{\psi}\dot{\theta} \sin\theta \\
&\quad - \left(I_{xx} - I_{yy}\right)\left(\dot{\psi} \sin\theta \sin\phi + \dot{\theta} \cos\phi\right)\left(\dot{\psi} \sin\theta \cos\phi - \dot{\theta} \sin\phi\right).
\end{aligned}$$

If the Euler angles and their first and second derivatives with respect to time are known, Eqs. (20.36) can be solved for the components of the total moment. Or if the total moment, the Euler angles, and the first derivatives of the Euler angles are known, Eqs. (20.36) can be solved for the second derivatives of the Euler angles. In this way, the Euler angles can be determined as function of time when the total moment is known as a function of time, but numerical integration is usually necessary.

Study Questions

1. How are the Euler angles defined?
2. What is steady precession?
3. How can the body and space cones be used to visualize moment-free steady precession?
4. If you know the Euler angles of a rigid body and their first derivatives with respect to time, how can you determine the rigid body's angular velocity?

Example 20.7

Steady Precession of a Disk

The thin circular disk of radius R and mass m in Fig. 20.25 rolls along a horizontal circular path of radius r. The angle θ between the disk's axis and the vertical remains constant. Determine the magnitude v of the velocity of the center of the disk as a function of the angle θ.

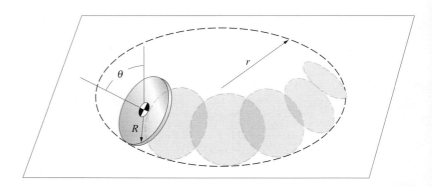

Figure 20.25

Strategy

We can obtain the velocity of the center of the disk by assuming that the disk is in steady precession and determining the conditions necessary for the equations of motion to be satisfied.

Solution

In Fig. (a), we align the z axis of the secondary coordinate system with the disk's spin axis and assume that the x axis remains parallel to the surface on which the disk rolls. The angle θ is the nutation angle. The center of mass moves in a circular path of radius $r_G = r - R\cos\theta$. Therefore, the precession rate—the rate at which the x axis rotates in the horizontal plane—is

$$\dot{\psi} = \frac{v}{r_G}.$$

From Eqs. (20.23), the components of the disk's angular velocity are

$$\omega_x = \dot{\theta} = 0,$$

$$\omega_y = \dot{\psi}\sin\theta = \frac{v}{r_G}\sin\theta,$$

and

$$\omega_z = \dot{\phi} + \dot{\psi}\cos\theta = \dot{\phi} + \frac{v}{r_G}\cos\theta,$$

where $\dot{\phi}$ is the spin rate. To determine $\dot{\phi}$, we use the condition that the velocity of the point of the disk that is in contact with the surface is zero. In

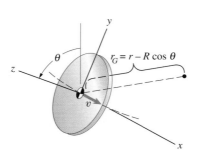

(a) Aligning the z axis with the spin axis. The x axis is horizontal.

$r_G = r - R\cos\theta$

terms of the velocity of the center of the disk, the velocity of the point of contact is

$$\mathbf{0} = v\mathbf{i} + \boldsymbol{\omega} \times (-R\mathbf{j}) = v\mathbf{i} + \begin{vmatrix} \mathbf{i} & \mathbf{j} & \mathbf{k} \\ 0 & \dfrac{v}{r_G}\sin\theta & \dot{\phi} + \dfrac{v}{r_G}\cos\theta \\ 0 & -R & 0 \end{vmatrix}.$$

Expanding the determinant and solving for $\dot{\phi}$, we obtain

$$\dot{\phi} = -\frac{v}{R} - \frac{v}{r_G}\cos\theta.$$

We draw the free-body diagram of the disk in Fig. (b). Because the center of mass moves in the horizontal plane, its acceleration in the vertical direction is zero. Therefore, the normal force $N = mg$. The acceleration of the center of mass in the direction perpendicular to its circular path is $a_\mathrm{n} = v^2/r_G$. Newton's second law in the direction perpendicular to the circular path is

$$T = m\frac{v^2}{r_G}.$$

Therefore, the components of the total moment about the center of mass are

$$\Sigma M_x = TR\sin\theta - NR\cos\theta = m\frac{v^2}{r_G}R\sin\theta - mgR\cos\theta,$$

$$\Sigma M_y = 0,$$

and

$$\Sigma M_z = 0.$$

Substituting our expressions for $\dot{\psi}$, $\dot{\phi}$, and ΣM_x into the equation of angular motion for steady precession [Eq. (20.29)] and solving for v, we obtain

$$v = \sqrt{\frac{\frac{2}{3}g\cot\theta(r - R\cos\theta)^2}{r - \frac{5}{6}R\cos\theta}}.$$

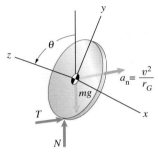

(b) Free-body diagram of the disk showing the normal acceleration of the center of mass.

Appendix: Moments and Products of Inertia

To use the equations of motion to predict the behavior of a rigid body in three dimensions, the moments and products of inertia of the body, given by Eqs. (20.10) and (20.11), must be known. In this appendix, we demonstrate how the moments and products can be evaluated for simple objects such as slender bars and thin plates. We derive the parallel-axis theorems, which make it possible to determine the moments and products of inertia of composite objects. We also introduce the concept of principal axes, which simplifies the equations of angular motion.

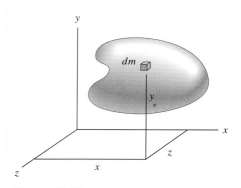

Figure 20.26
Determining the moments and products of inertia by modeling an object as a continuous distribution of mass.

Simple Objects

If we model a rigid body as a continuous distribution of mass, we can express its moments and products of inertia [Eqs. (20.10) and (20.11)] as

$$[I] = \begin{bmatrix} I_{xx} & -I_{xy} & -I_{xz} \\ -I_{yx} & I_{yy} & -I_{yz} \\ -I_{zx} & -I_{zy} & I_{zz} \end{bmatrix}$$

$$= \begin{bmatrix} \int_m (y^2 + z^2)\, dm & -\int_m xy\, dm & -\int_m xz\, dm \\ -\int_m yx\, dm & \int_m (x^2 + z^2)\, dm & -\int_m yz\, dm \\ -\int_m zx\, dm & -\int_m zy\, dm & \int_m (x^2 + y^2)\, dm \end{bmatrix}, \quad (20.37)$$

where x, y, and z are the coordinates of the differential element of mass dm (Fig. 20.26).

Slender Bars Let the origin of the coordinate system be at a slender bar's center of mass, with the x axis along the bar (Fig. 20.27a). The bar has length l, cross-sectional area A, and mass m. We assume that A is uniform along the length of the bar and that the material is homogeneous.

Consider a differential element of the bar of length dx at a distance x from the center of mass (Fig. 20.27b). The mass of the element is $dm = \rho A\, dx$, where ρ is the mass density. We neglect the lateral dimensions of the bar, assuming the coordinates of the differential element dm to be $(x, 0, 0)$. As a consequence of this approximation, the moment of inertia of the bar about the x axis is zero:

$$I_{xx} = \int_m (y^2 + z^2)\, dm = 0.$$

The moment of inertia about the y axis is

$$I_{yy} = \int_m (x^2 + z^2)\, dm = \int_{-l/2}^{l/2} \rho A x^2\, dx = \frac{1}{12} \rho A l^3.$$

Expressing this result in terms of the mass of the bar, $m = \rho A l$, we obtain

$$I_{yy} = \frac{1}{12} m l^2.$$

Figure 20.27
(a) A slender bar and a coordinate system with the x axis aligned with the bar.
(b) A differential element of mass of length dx.

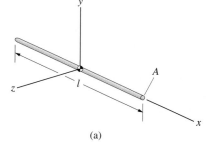

(a)

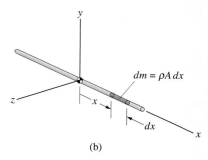

(b)

The moment of inertia about the z axis is equal to the moment of inertia about the y axis:

$$I_{zz} = \int_m (x^2 + y^2)\, dm = \frac{1}{12} ml^2.$$

Because the y and z coordinates of dm are zero, the products of inertia are zero, so the inertia matrix for the slender bar is

$$[I] = \begin{bmatrix} 0 & 0 & 0 \\ 0 & \frac{1}{12} ml^2 & 0 \\ 0 & 0 & \frac{1}{12} ml^2 \end{bmatrix}. \tag{20.38}$$

It is important to remember that the moments and products of inertia depend on the orientation of the coordinate system relative to the object. In terms of the alternative coordinate system shown in Fig. 20.28, the bar's inertia matrix is

$$[I] = \begin{bmatrix} \frac{1}{12} ml^2 & 0 & 0 \\ 0 & 0 & 0 \\ 0 & 0 & \frac{1}{12} ml^2 \end{bmatrix}.$$

Figure 20.28
Aligning the y axis with the bar.

Thin Plates Suppose that a homogeneous plate of uniform thickness T, area A, and unspecified shape lies in the x-y plane (Fig. 20.29a). We can express the moments of inertia of the plate in terms of the moments of inertia of its cross-sectional area.

By projecting an element of area dA through the thickness T of the plate (Fig. 20.29b), we obtain a differential element of mass $dm = \rho T\, dA$. We neglect the thickness of the plate in calculating the moments of inertia, so the coordinates of the element dm are $(x, y, 0)$. The plate's moment of inertia about the x axis is

$$I_{xx} = \int_m (y^2 + z^2)\, dm = \rho T \int_A y^2\, dA = \rho T I_x,$$

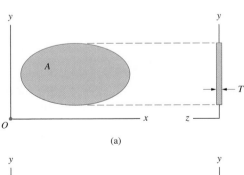

(a)

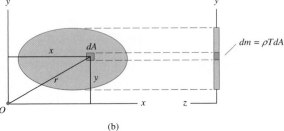

(b)

Figure 20.29
(a) A thin plate lying in the x-y plane.
(b) Obtaining a differential element of mass by projecting an element of area dA through the plate.

where I_x is the moment of inertia of the plate's cross-sectional area about the x axis. Since the mass of the plate is $m = \rho T A$, the product $\rho T = m/A$, and we obtain the moment of inertia in the form

$$I_{xx} = \frac{m}{A} I_x.$$

The moment of inertia about the y axis is

$$I_{yy} = \int_m (x^2 + z^2)\, dm = \rho T \int_A x^2\, dA = \frac{m}{A} I_y,$$

where I_y is the moment of inertia of the cross-sectional area about the y axis. The moment of inertia about the z axis is

$$I_{zz} = \int_m (x^2 + y^2)\, dm = \frac{m}{A} J_O,$$

where $J_O = I_x + I_y$ is the polar moment of inertia of the cross-sectional area. The product of inertia I_{xy} is

$$I_{xy} = \int_m xy\, dm = \frac{m}{A} I_{xy}^A,$$

where

$$I_{xy}^A = \int_A xy\, dA$$

is the product of inertia of the cross-sectional area. (We use a superscript A to distinguish the product of inertia of the plate's cross-sectional area from the product of inertia of its mass.) If the cross-sectional area A is symmetric about either the x axis or the y axis, $I_{xy}^A = 0$.

Because the z coordinate of dm is zero, the products of inertia I_{xz} and I_{yz} are zero. The inertia matrix for the thin plate is thus

$$[I] = \begin{bmatrix} \dfrac{m}{A} I_x & -\dfrac{m}{A} I_{xy}^A & 0 \\[2ex] -\dfrac{m}{A} I_{xy}^A & \dfrac{m}{A} I_y & 0 \\[2ex] 0 & 0 & \dfrac{m}{A} J_O \end{bmatrix}. \qquad (20.39)$$

If the moments of inertia and product of inertia of the plate's cross-sectional area are known, (Eq. 20.39) can be used to obtain the moments and products of inertia of the plate.

Parallel-Axis Theorems

Suppose that we know an object's inertia matrix $[I']$ in terms of a coordinate system $x'y'z'$ with its origin at the center of mass of the object, and we want to determine the inertia matrix $[I]$ in terms of a parallel coordinate system xyz (Fig. 20.30). Let (d_x, d_y, d_z) be the coordinates of the center of mass in the xyz coordinate system. The coordinates of a differential element of mass dm in the xyz system are given in terms of its coordinates in the $x'y'z'$ system by

$$x = x' + d_x, \qquad y = y' + d_y, \qquad z = z' + d_z. \qquad (20.40)$$

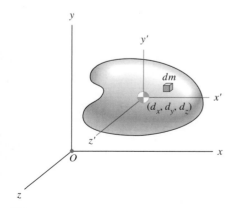

Figure 20.30
A coordinate system $x'y'z'$ with its origin at the center of mass and a parallel coordinate system xyz.

Substituting these expressions into the definition of I_{xx}, we obtain

$$I_{xx} = \int_m \left[(y')^2 + (z')^2 \right] dm + 2d_y \int_m y' \, dm$$

$$+ 2d_z \int_m z' \, dm + (d_y^2 + d_z^2) \int_m dm. \qquad (20.41)$$

The first integral on the right is the object's moment of inertia about the x' axis. We can show that the second and third integrals are zero by using the definitions of the object's center of mass, expressed in terms of the $x'y'z'$ coordinate system:

$$\bar{x}' = \frac{\displaystyle\int_m x' \, dm}{\displaystyle\int_m dm}, \qquad \bar{y}' = \frac{\displaystyle\int_m y' \, dm}{\displaystyle\int_m dm}, \qquad \bar{z}' = \frac{\displaystyle\int_m z' \, dm}{\displaystyle\int_m dm}.$$

The object's center of mass is at the origin of the $x'y'z'$ system, so $\bar{x}' = \bar{y}' = \bar{z}' = 0$. Therefore, the second and third integrals on the right of Eq. (20.41) are zero, and we obtain

$$I_{xx} = I_{x'x'} + (d_y^2 + d_z^2)m,$$

where m is the mass of the object. Substituting Eqs. (20.40) into the definition of I_{xy}, we obtain

$$I_{xy} = \int_m x'y' \, dm + d_x \int_m y' \, dm + d_y \int_m x' \, dm + d_x d_y \int_m dm$$

$$= I_{x'y'} + d_x d_y m.$$

Proceeding in this way for each of the moments and products of inertia, we obtain the *parallel-axis theorems*:

$$I_{xx} = I_{x'x'} + (d_y^2 + d_z^2)m,$$
$$I_{yy} = I_{y'y'} + (d_x^2 + d_z^2)m,$$
$$I_{zz} = I_{z'z'} + (d_x^2 + d_y^2)m,$$
$$I_{xy} = I_{x'y'} + d_x d_y m, \qquad (20.42)$$
$$I_{yz} = I_{y'z'} + d_y d_z m,$$
$$I_{zx} = I_{z'x'} + d_z d_x m.$$

If an object's inertia matrix is known in terms of a particular coordinate system, these theorems can be used to determine its inertia matrix in terms of any parallel coordinate system. They can also be used to determine the inertia matrices of composite objects.

Moment of Inertia about an Arbitrary Axis

If we know a rigid body's inertia matrix in terms of a given coordinate system with origin O, we can determine its moment of inertia about an arbitrary axis through O. Suppose that the rigid body rotates with angular velocity $\boldsymbol{\omega}$ about

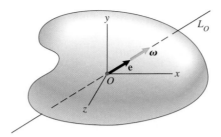

Figure 20.31
Rigid body rotating about L_O.

an arbitrary fixed axis L_O through O, and let **e** be a unit vector with the same direction as $\boldsymbol{\omega}$ (Fig. 20.31). Then, in terms of the moment of inertia I_O about L_O, the rigid body's angular momentum about L_O is

$$H_O = I_O|\boldsymbol{\omega}|.$$

We can express the angular velocity vector as

$$\boldsymbol{\omega} = |\boldsymbol{\omega}|(e_x\mathbf{i} + e_y\mathbf{j} + e_z\mathbf{k}),$$

so that $\omega_x = |\boldsymbol{\omega}|e_x$, $\omega_y = |\boldsymbol{\omega}|e_y$, and $\omega_z = |\boldsymbol{\omega}|e_z$. Using these expressions and Eqs. (20.9), we obtain the angular momentum about L_O:

$$\begin{aligned}
H_O = \mathbf{H}_O \cdot \mathbf{e} &= \left(I_{xx}|\boldsymbol{\omega}|e_x - I_{xy}|\boldsymbol{\omega}|e_y - I_{xz}|\boldsymbol{\omega}|e_z\right)e_x \\
&+ \left(-I_{yx}|\boldsymbol{\omega}|e_x + I_{yy}|\boldsymbol{\omega}|e_y - I_{yz}|\boldsymbol{\omega}|e_z\right)e_y \\
&+ \left(-I_{zx}|\boldsymbol{\omega}|e_x - I_{zy}|\boldsymbol{\omega}|e_y + I_{zz}|\boldsymbol{\omega}|e_z\right)e_z.
\end{aligned}$$

Equating our two expressions for H_O yields

$$I_O = I_{xx}e_x^2 + I_{yy}e_y^2 + I_{zz}e_z^2 - 2I_{xy}e_xe_y - 2I_{yz}e_ye_z - 2I_{zx}e_ze_x. \quad (20.43)$$

Notice that the moment of inertia about an arbitrary axis depends on the products of inertia, in addition to the moments of inertia about the coordinate axes. If an object's inertia matrix is known, Eq. (20.43) can be used to determine the object's moment of inertia about an axis through O whose direction is specified by the unit vector **e**.

Principal Axes

For *any* object and origin O, at least one coordinate system exists for which the products of inertia are zero:

$$[I] = \begin{bmatrix} I_{xx} & 0 & 0 \\ 0 & I_{yy} & 0 \\ 0 & 0 & I_{zz} \end{bmatrix} \quad (20.44)$$

These coordinate axes are called *principal axes*, and the moments of inertia about them are called the *principal moments of inertia*.

If the inertia matrix of a rigid body is known in terms of a coordinate system $x'y'z'$ and the products of inertia are zero, then $x'y'z'$ is a set of principal axes. Suppose that the products of inertia are not zero, and we want to find a set of principal axes xyz and the corresponding principal moments of

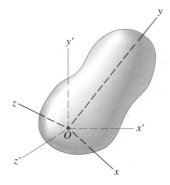

Figure 20.32
The $x'y'z'$ system with its origin at O and a set of principal axes xyz.

inertia (Fig. 20.32). It can be shown that the principal moments of inertia are roots of the cubic equation

$$I^3 - (I_{x'x'} + I_{y'y'} + I_{z'z'})I^2$$
$$+ (I_{x'x'}I_{y'y'} + I_{y'y'}I_{z'z'} + I_{z'z'}I_{x'x'} - I_{x'y'}^2 - I_{y'z'}^2 - I_{z'x'}^2)I \qquad (20.45)$$
$$- (I_{x'x'}I_{y'y'}I_{z'z'} - I_{x'x'}I_{y'z'}^2 - I_{y'y'}I_{x'z'}^2 - I_{z'z'}I_{x'y'}^2 - 2I_{x'y'}I_{y'z'}I_{z'x'}) = 0.$$

For each principal moment of inertia I, the vector $\mathbf{V}$ with components

$$V_{x'} = (I_{y'y'} - I)(I_{z'z'} - I) - I_{y'z'}^2,$$
$$V_{y'} = I_{x'y'}(I_{z'z'} - I) + I_{x'z'}I_{y'z'},$$

and $\qquad (20.46)$

$$V_{z'} = I_{x'z'}(I_{y'y'} - I) + I_{x'y'}I_{y'z'},$$

is parallel to the corresponding principal axis.

When the inertia matrix of an object is known in terms of a coordinate system with origin O, determining the associated principal moments of inertia and a set of principal axes involves two steps:

1. Determine the principal moments of inertia by obtaining the roots of Eq. (20.45).
2. If the three principal moments of inertia are distinct, substitute each one into Eqs. (20.46) to obtain the components of a vector parallel to the corresponding principal axis. The three principal axes can be denoted as x, y, and z arbitrarily, as long as the resulting coordinate system is right handed. If the three principal moments of inertia are equal, the moment of inertia about any axis through O has the same value, and any coordinate system with origin O is a set of principal axes. This is the case, for example, if the object is a homogeneous sphere with O at its center (Fig. 20.33a). If only two of the principal moments of inertia are equal, the third one can be substituted into Eqs. (20.46) to determine the associated principal axis. Then the moment of inertia about any axis through O that is perpendicular to the determined axis has the same value, so any coordinate system with origin O that has an axis coincident with the determined axis is a set of principal axes. This is the case when an object has an axis of rotational symmetry and O is on the axis (Fig. 20.33b).

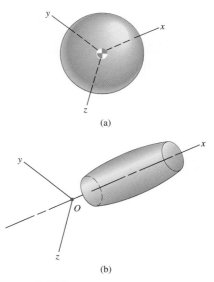

(a)

(b)

Figure 20.33
(a) A homogeneous sphere. Any coordinate system with its origin at the center is a set of principal axes.
(b) A rotationally symmetric object. The axis of symmetry is a principal axis, and any axis perpendicular to it is a principal axis.

Example 20.8

Parallel-Axis Theorems

The boom *AB* of the crane in Fig. 20.34 has a mass of 4800 kg, and the boom *BC* has a mass of 1600 kg and is perpendicular to *AB*. Modeling each boom as a slender bar and treating them as a single object, determine the moments and products of inertia of the object in terms of the coordinate system shown.

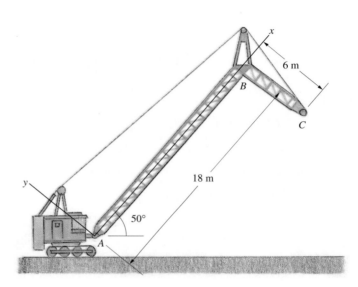

Figure 20.34

Strategy

We can apply the parallel-axis theorems to each boom to determine its moments and products of inertia in terms of the given coordinate system. The moments and products of inertia of the combined object are the sums of those for the two booms.

Solution

Boom *AB* In Fig. (a), we introduce a parallel coordinate system $x'y'z'$ with its origin at the center of mass of boom *AB*. In terms of the $x'y'z'$ system, the inertia matrix of boom *AB* is

$$[I'] = \begin{bmatrix} 0 & 0 & 0 \\ 0 & \frac{1}{12}ml^2 & 0 \\ 0 & 0 & \frac{1}{12}ml^2 \end{bmatrix} = \begin{bmatrix} 0 & 0 & 0 \\ 0 & \frac{1}{12}(4800)(18)^2 & 0 \\ 0 & 0 & \frac{1}{12}(4800)(18)^2 \end{bmatrix} \text{kg-m}^2.$$

The coordinates of the origin of the $x'y'z'$ system relative to the xyz system are $d_x = 9$ m, $d_y = 0$, $d_z = 0$. Applying the parallel-axis theorems, we obtain

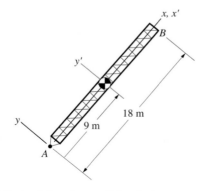

(a) Applying the parallel-axis theorems to boom *AB*.

$$I_{xx} = I_{x'x'} + \left(d_y^2 + d_z^2\right)m = 0,$$

$$I_{yy} = I_{y'y'} + \left(d_x^2 + d_z^2\right)m = \frac{1}{12}(4800)(18)^2 + (9)^2(4800)$$

$$= 518{,}400 \text{ kg -m}^2,$$

$$I_{zz} = I_{z'z'} + \left(d_x^2 + d_y^2\right)m = \frac{1}{12}(4800)(18)^2 + (9)^2(4800)$$

$$= 518{,}400 \text{ kg -m}^2,$$

$$I_{xy} = I_{x'y'} + d_x d_y m = 0,$$

$$I_{yz} = I_{y'z'} + d_y d_z m = 0,$$

and

$$I_{zx} = I_{z'x'} + d_z d_x m = 0.$$

Boom _BC_   In Fig. (b), we introduce a parallel coordinate system $x'y'z'$ with its origin at the center of mass of boom BC. In terms of the $x'y'z'$ system, the inertia matrix of boom BC is

$$[I'] = \begin{bmatrix} \frac{1}{12}ml^2 & 0 & 0 \\ 0 & 0 & 0 \\ 0 & 0 & \frac{1}{12}ml^2 \end{bmatrix} = \begin{bmatrix} \frac{1}{12}(1600)(6)^2 & 0 & 0 \\ 0 & 0 & 0 \\ 0 & 0 & \frac{1}{12}(1600)(6)^2 \end{bmatrix} \text{kg-m}^2.$$

The coordinates of the origin of the $x'y'z'$ system relative to the xyz system are $d_x = 18$ m, $d_y = -3$ m, $d_z = 0$. Applying the parallel-axis theorems, we obtain

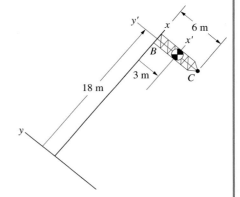

(b) Applying the parallel-axis theorems to boom _BC_.

$$I_{xx} = I_{x'x'} + \left(d_y^2 + d_z^2\right)m = \frac{1}{12}(1600)(6)^2 + (-3)^2(1600)$$

$$= 19{,}200 \text{ kg-m}^2,$$

$$I_{yy} = I_{y'y'} + \left(d_x^2 + d_z^2\right)m = 0 + (18)^2(1600) = 518{,}400 \text{ kg-m}^2,$$

$$I_{zz} = I_{z'z'} + \left(d_x^2 + d_y^2\right)m = \frac{1}{12}(1600)(6)^2 + \left[(18)^2 + (-3)^2\right](1600)$$

$$= 537{,}600 \text{ kg-m}^2,$$

$$I_{xy} = I_{x'y'} + d_x d_y m = 0 + (18)(-3)(1600) = -86{,}400 \text{ kg-m}^2,$$

$$I_{yz} = I_{y'z'} + d_y d_z m = 0,$$

and

$$I_{zx} = I_{z'x'} + d_z d_x m = 0.$$

Summing the results for the two booms, we obtain the inertia matrix for the single object:

$$[I] = \begin{bmatrix} 19{,}200 & -(-86{,}400) & 0 \\ -(-86{,}400) & 518{,}400 + 518{,}400 & 0 \\ 0 & 0 & 518{,}400 + 537{,}600 \end{bmatrix}$$

$$= \begin{bmatrix} 19{,}200 & 86{,}400 & 0 \\ 86{,}400 & 1{,}036{,}800 & 0 \\ 0 & 0 & 1{,}056{,}000 \end{bmatrix} \text{kg-m}^2.$$

Example 20.9

Inertia Matrix of a Plate

The 4-kg rectangular plate in Fig. 20.35 lies in the x-y plane of the body-fixed coordinate system.
(a) Determine the plate's moments and products of inertia.
(b) Determine the plate's moment of inertia about the diagonal axis L_O.
(c) If the plate is rotating about the fixed point O with angular velocity $\boldsymbol{\omega} = 4\mathbf{i} - 2\mathbf{j}$ (rad/s), what is the plate's angular momentum about O?

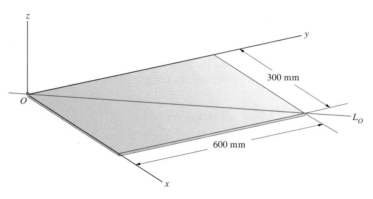

Figure 20.35

Strategy

(a) We can obtain the moments and products of inertia of the plate's rectangular area from Appendix B and use Eq. (20.39) to obtain the moments and products of inertia of the plate.
(b) Once we know the moments and products of inertia, we can use Eq. (20.43) to determine the moment of inertia about L_O.
(c) The angular momentum about O is given by Eqs. (20.9).

Solution

(a) From Appendix B, the moments of inertia of the plate's cross-sectional area are as follows (Fig. a):

$$I_x = \frac{1}{3} bh^3, \qquad I_y = \frac{1}{3} hb^3,$$

$$I_{xy}^A = \frac{1}{4} b^2 h^2, \qquad J_O = \frac{1}{3}\left(bh^3 + hb^3\right).$$

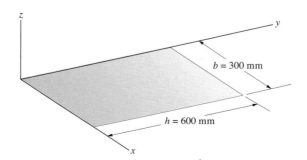

(a) Determining the moments of inertia of the plate's area.

Therefore, the moments and products of inertia of the plate are

$$I_{xx} = \frac{m}{A} I_x = \frac{(4)}{(0.3)(0.6)} \left(\frac{1}{3}\right)(0.3)(0.6)^3 = 0.48 \text{ kg-m}^2,$$

$$I_{yy} = \frac{m}{A} I_y = \frac{(4)}{(0.3)(0.6)} \left(\frac{1}{3}\right)(0.6)(0.3)^3 = 0.12 \text{ kg-m}^2,$$

$$I_{xy} = \frac{m}{A} I_{xy}^A = \frac{(4)}{(0.3)(0.6)} \left(\frac{1}{4}\right)(0.3)^2(0.6)^2 = 0.18 \text{ kg-m}^2,$$

$$I_{zz} = \frac{m}{A} J_O = \frac{(4)}{(0.3)(0.6)} \left(\frac{1}{3}\right)[(0.3)(0.6)^3 + (0.6)(0.3)^3] = 0.60 \text{ kg-m}^2,$$

and

$$I_{xz} = I_{yz} = 0.$$

(b) To apply Eq. (20.43), we must determine the components of a unit vector parallel to L_O:

$$\mathbf{e} = \frac{300\mathbf{i} + 600\mathbf{j}}{|300\mathbf{i} + 600\mathbf{j}|} = 0.447\mathbf{i} + 0.894\mathbf{j}.$$

The moment of inertia about L_O is

$$I_O = I_{xx}e_x^2 + I_{yy}e_y^2 + I_{zz}e_z^2 - 2I_{xy}e_xe_y - 2I_{yz}e_ye_z - 2I_{zx}e_ze_x$$

$$= (0.48)(0.447)^2 + (0.12)(0.894)^2 - 2(0.18)(0.447)(0.894)$$

$$= 0.048 \text{ kg-m}^2.$$

(c) The plate's angular momentum about O is

$$\begin{bmatrix} H_{Ox} \\ H_{Oy} \\ H_{Oz} \end{bmatrix} = \begin{bmatrix} I_{xx} & -I_{xy} & -I_{xz} \\ -I_{yx} & I_{yy} & -I_{yz} \\ -I_{zx} & -I_{zy} & I_{zz} \end{bmatrix} \begin{bmatrix} \omega_x \\ \omega_y \\ \omega_z \end{bmatrix}$$

$$= \begin{bmatrix} 0.48 & -0.18 & 0 \\ -0.18 & 0.12 & 0 \\ 0 & 0 & 0.6 \end{bmatrix} \begin{bmatrix} 4 \\ -2 \\ 0 \end{bmatrix}$$

$$= \begin{bmatrix} 2.28 \\ -0.96 \\ 0 \end{bmatrix} \text{ kg-m}^2/\text{s}.$$

Example 20.10

Principal Axes and Moments of Inertia

In terms of a coordinate system $x'y'z'$ with its origin at the center of mass, the inertia matrix of a rigid body is

$$[I'] = \begin{bmatrix} 4 & -2 & 1 \\ -2 & 2 & -1 \\ 1 & -1 & 3 \end{bmatrix} \text{kg-m}^2.$$

Determine the principal moments of inertia and the directions of a set of principal axes relative to the $x'y'z'$ system.

Solution

Substituting the moments and products of inertia into Eq. (20.45), we obtain the equation

$$I^3 - 9I^2 + 20I - 10 = 0.$$

We show the value of the left side of this equation as a function of I in Fig. 20.36. The three roots, which are the values of the principal moments of inertia in kg-m², are $I_1 = 0.708$, $I_2 = 2.397$, and $I_3 = 5.895$.

Substituting the principal moment of inertia $I_1 = 0.708$ kg-m² into Eqs. (20.46) and dividing the resulting vector **V** by its magnitude, we obtain a unit vector parallel to the corresponding principal axis:

$$\mathbf{e}_1 = 0.473\mathbf{i} + 0.864\mathbf{j} + 0.171\mathbf{k}.$$

Substituting $I_2 = 2.397$ kg-m² into Eqs. (20.46), we obtain the unit vector

$$\mathbf{e}_2 = -0.458\mathbf{i} + 0.076\mathbf{j} + 0.886\mathbf{k},$$

and substituting $I_3 = 5.895$ kg-m² into Eqs. (20.46), we obtain the unit vector

$$\mathbf{e}_3 = 0.753\mathbf{i} - 0.497\mathbf{j} + 0.432\mathbf{k}.$$

We have determined the principal moments of inertia and the components of unit vectors parallel to the corresponding principal axes. In Fig. 20.37, we show the principal axes, arbitrarily designating them so that $I_{xx} = 5.895$ kg-m², $I_{yy} = 0.708$ kg-m², and $I_{zz} = 2.397$ kg-m².

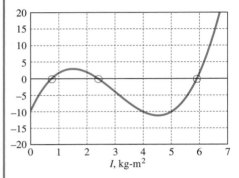

Figure 20.36
Graph of $I^3 - 9I^2 + 20I - 10$.

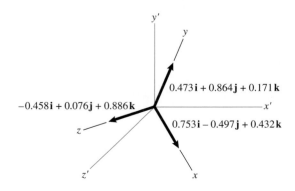

Figure 20.37
The principal axes. Our choice of which to call x, y, and z is arbitrary.

Chapter Summary

Kinematics

The velocity of a point A of a rigid body relative to a given reference frame is given in terms of the velocity of a point B of the rigid body and the rigid body's angular velocity by (Fig. a)

$$\mathbf{v}_A = \mathbf{v}_B + \boldsymbol{\omega} \times \mathbf{r}_{A/B}. \qquad \text{Eq. (20.1)}$$

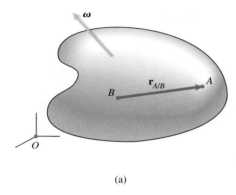

(a)

The acceleration of point A is given in terms of the acceleration of point B, the rigid body's angular acceleration, and its angular velocity by

$$\mathbf{a}_A = \mathbf{a}_B + \boldsymbol{\alpha} \times \mathbf{r}_{A/B} + \boldsymbol{\omega} \times (\boldsymbol{\omega} \times \mathbf{r}_{A/B}). \qquad \text{Eq. (20.2)}$$

Consider a secondary reference frame xyz with angular velocity $\boldsymbol{\Omega}$ relative to a primary reference frame, and a rigid body with angular velocity $\boldsymbol{\omega}$ relative to the primary reference frame (Fig. b). If the secondary reference frame is body fixed, $\boldsymbol{\Omega} = \boldsymbol{\omega}$. The rigid body's angular acceleration relative to the primary reference frame is

$$\boldsymbol{\alpha} = \frac{d\omega_x}{dt}\mathbf{i} + \frac{d\omega_y}{dt}\mathbf{j} + \frac{d\omega_z}{dt}\mathbf{k} + \boldsymbol{\Omega} \times \boldsymbol{\omega}. \qquad \text{Eq. (20.4)}$$

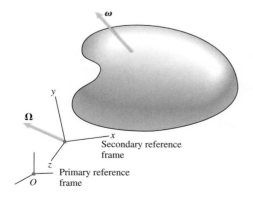

(b)

When the secondary reference frame is not body fixed, it is often convenient to express $\boldsymbol{\omega}$ as the sum of $\boldsymbol{\Omega}$ and the angular velocity $\boldsymbol{\omega}_{\text{rel}}$ of the rigid body relative to the secondary reference frame:

$$\boldsymbol{\omega} = \boldsymbol{\Omega} + \boldsymbol{\omega}_{\text{rel}}. \qquad \text{Eq. (20.5)}$$

Euler's Equations

The three-dimensional equations of motion for a rigid body are called *Euler's equations*. They include Newton's second law and equations of angular motion. For a rigid body rotating about a fixed point O (Fig. c), the equations of angular motion are expressed in terms of the components of the total moment about O as

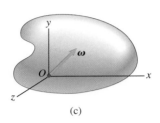

(c)

$$\Sigma M_{Ox} = I_{xx}\frac{d\omega_x}{dt} - I_{xy}\frac{d\omega_y}{dt} - I_{xz}\frac{d\omega_z}{dt}$$
$$- \Omega_z\left(-I_{yx}\omega_x + I_{yy}\omega_y - I_{yz}\omega_z\right)$$
$$+ \Omega_y\left(-I_{zx}\omega_x - I_{zy}\omega_y + I_{zz}\omega_z\right),$$

$$\Sigma M_{Oy} = -I_{yx}\frac{d\omega_x}{dt} + I_{yy}\frac{d\omega_y}{dt} - I_{yz}\frac{d\omega_z}{dt}$$
$$+ \Omega_z\left(I_{xx}\omega_x - I_{xy}\omega_y - I_{xz}\omega_z\right)$$
$$- \Omega_x\left(-I_{zx}\omega_x - I_{zy}\omega_y + I_{zz}\omega_z\right),$$

and

$$\Sigma M_{Oz} = -I_{zx}\frac{d\omega_x}{dt} - I_{zy}\frac{d\omega_y}{dt} + I_{zz}\frac{d\omega_z}{dt}$$
$$- \Omega_y\left(I_{xx}\omega_x - I_{xy}\omega_y - I_{xz}\omega_z\right)$$
$$+ \Omega_x\left(-I_{yx}\omega_x + I_{yy}\omega_y - I_{yz}\omega_z\right),$$

Eq. (20.12)

where $\boldsymbol{\omega}$ is the angular velocity of the rigid body and $\boldsymbol{\Omega}$ is the angular velocity of the chosen secondary reference frame with origin at O. If the secondary reference frame is body fixed, $\boldsymbol{\Omega} = \boldsymbol{\omega}$. In the case of general three-dimensional motion (Fig. d), the equations of angular motion [Eqs. (20.18)] are identical in form to those for motion about a fixed point, except that they are expressed in terms of the components of the total moment about the center of mass and the origin of the secondary reference frame is at the center of mass.

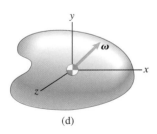

(d)

Euler Angles: Axisymmetric Objects

In the case of an object with an axis of rotational symmetry, the orientation of the xyz system relative to the reference XYZ system is specified by the *precession angle* ψ and the *nutation angle* θ (Fig. e). The rotation of the object relative to the xyz system is specified by the *spin angle* ϕ.

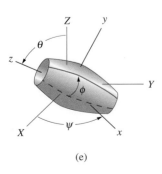

(e)

The components of the rigid body's angular velocity relative to the XYZ system are given by

$$\omega_x = \dot{\theta},$$

$$\omega_y = \dot{\psi} \sin\theta,$$

and   Eq. (20.23)

$$\omega_z = \dot{\phi} + \dot{\psi} \cos\theta.$$

The equations of angular motion expressed in terms of the Euler angles are

$$\Sigma M_x = I_{xx}\ddot{\theta} + (I_{zz} - I_{xx})\dot{\psi}^2 \sin\theta \cos\theta + I_{zz}\dot{\phi}\dot{\psi} \sin\theta, \qquad \text{Eq. (20.26)}$$

$$\Sigma M_y = I_{xx}(\ddot{\psi} \sin\theta + 2\dot{\psi}\dot{\theta} \cos\theta) - I_{zz}(\dot{\phi}\dot{\theta} + \dot{\psi}\dot{\theta} \cos\theta), \qquad \text{Eq. (20.27)}$$

$$\Sigma M_z = I_{zz}(\ddot{\phi} + \ddot{\psi} \cos\theta - \dot{\psi}\dot{\theta} \sin\theta). \qquad \text{Eq. (20.28)}$$

In the *steady precession* of an axisymmetric spinning object, the spin rate $\dot{\phi}$, the nutation angle θ, and the precession rate $\dot{\psi}$ are assumed to be constant. With these assumptions, the equations of angular motion reduce to

$$\Sigma M_x = (I_{zz} - I_{xx})\dot{\psi}^2 \sin\theta \cos\theta + I_{zz}\dot{\phi}\dot{\psi} \sin\theta, \qquad \text{Eq. (20.29)}$$

$$\Sigma M_y = 0, \qquad \text{Eq. (20.30)}$$

$$\Sigma M_z = 0. \qquad \text{Eq. (20.31)}$$

Moments and Products of Inertia

In terms of a given coordinate system xyz, the *inertia matrix* of an object is defined by

$$[I] = \begin{bmatrix} I_{xx} & -I_{xy} & -I_{xz} \\ -I_{yx} & I_{yy} & -I_{yz} \\ -I_{zx} & -I_{zy} & I_{zz} \end{bmatrix}$$

Eq. (20.37)

$$= \begin{bmatrix} \int_m (y^2 + z^2)\, dm & -\int_m xy\, dm & -\int_m xz\, dm \\ -\int_m yx\, dm & \int_m (x^2 + z^2)\, dm & -\int_m yz\, dm \\ -\int_m zx\, dm & -\int_m zy\, dm & \int_m (x^2 + y^2)\, dm \end{bmatrix},$$

where x, y, and z are the coordinates of the differential element of mass dm. The terms I_{xx}, I_{yy}, and I_{zz} are the *moments of inertia* about the x, y, and z axes, and I_{xy}, I_{yz}, and I_{zx} are the *products of inertia*.

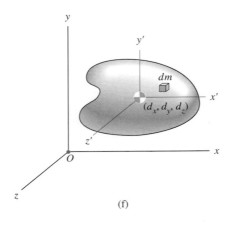

(f)

If $x'y'z'$ is a coordinate system with its origin at the center of mass of an object and xyz is a parallel system (Fig. f), the *parallel-axis theorems* state that

$$I_{xx} = I_{x'x'} + (d_y^2 + d_z^2)m,$$
$$I_{yy} = I_{y'y'} + (d_x^2 + d_z^2)m,$$
$$I_{zz} = I_{z'z'} + (d_x^2 + d_y^2)m,$$
$$I_{xy} = I_{x'y'} + d_x d_y m,$$
$$I_{yz} = I_{y'z'} + d_y d_z m,$$

Eq. (20.42)

and

$$I_{zx} = I_{z'x'} + d_z d_x m.$$

where (d_x, d_y, d_z) are the coordinates of the center of mass of the object in the xyz coordinate system.

The moment of inertia of the object about an axis through the origin O parallel to a unit vector $\mathbf{e}$ is given by

$$I_O = I_{xx}e_x^2 + I_{yy}e_y^2 + I_{zz}e_z^2 - 2I_{xy}e_xe_y - 2I_{yz}e_ye_z - 2I_{zx}e_ze_x.$$

Eq. (20.43)

Review Problems

20.1 Relative to a primary reference frame, a rigid body's angular velocity is $\boldsymbol{\omega} = 200\mathbf{i} + 900\mathbf{j} - 600\mathbf{k}$ (rad/s). The position of the center of mass of the body relative to the origin O of the reference frame is $\mathbf{r}_G = 6\mathbf{i} + 6\mathbf{j} + 2\mathbf{k}$ (m), and the velocity of the center of mass is $\mathbf{v}_G = 100\mathbf{i} + 80\mathbf{j} - 60\mathbf{k}$ (m/s). What is the velocity of a point A of the rigid body whose position relative to O is $\mathbf{r}_A = 5.8\mathbf{i} + 6.4\mathbf{j} + 1.6\mathbf{k}$ (m)?

Strategy: Use Eq. (20.1) to express $\mathbf{v}_A$ in terms of $\mathbf{v}_G$, $\boldsymbol{\omega}$, and $\mathbf{r}_{A/G}$.

20.2 The circular disk remains perpendicular to the horizontal shaft and rotates relative to it with angular velocity ω_d. The horizontal shaft is rigidly attached to a vertical shaft rotating with angular velocity ω_0.

(a) What is the disk's angular velocity vector $\boldsymbol{\omega}$?
(b) What is the velocity of point A of the disk?

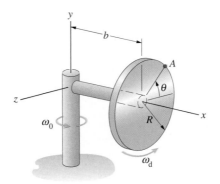

P20.2

20.3 If the angular velocities ω_d and ω_0 in Problem 20.2 are constant, what is the acceleration of point A of the disk?

20.4 The cone is connected by a ball-and-socket joint at its vertex to a 100-mm post. The radius of its base is 100 mm, and the base rolls on the floor. The velocity of the center of the base is $\mathbf{v}_C = 2\mathbf{k}$ (m/s).
(a) What is the cone's angular velocity vector $\boldsymbol{\omega}$?
(b) What is the velocity of point A?

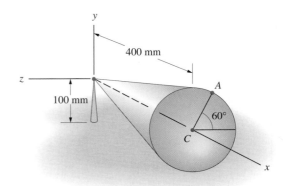

P20.4

20.5 The bar AB is connected by ball-and-socket joints to collars sliding on the fixed bars CD and EF. The bar EF is parallel to the y axis. At the instant shown, the collar at A has velocity $\mathbf{v}_A = 20\mathbf{k}$ (ft/s) and the angular velocity of bar AB about its axis is zero. Determine the velocity of the collar at B and the angular velocity of bar AB.

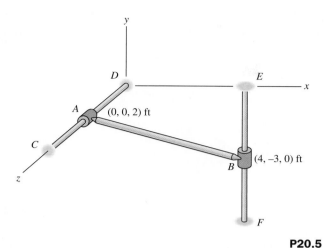

P20.5

20.6 The collar at *A* in Problem 20.5 has acceleration
$\mathbf{a}_A = -40\mathbf{k}\ (\text{ft/s}^2)$, and the angular acceleration of bar *AB* about
its axis is zero. Determine the acceleration of the collar at *B* and
the angular acceleration of bar *AB*.

𝒟 **20.7** The mechanism shown is a type of universal joint called a
yoke and spider. The axis *L* lies in the *x-z* plane. Determine the
angular velocity ω_L and the angular velocity vector $\boldsymbol{\omega}_S$ of the
cross-shaped "spider" in terms of the angular velocity ω_R at the
instant shown.

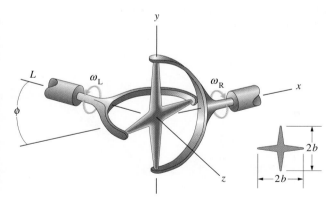

P20.7

20.8 The inertia matrix of a rigid body in terms of a body-fixed
coordinate system with its origin at the center of mass is

$$[I] = \begin{bmatrix} 4 & 1 & -1 \\ 1 & 2 & 0 \\ -1 & 0 & 6 \end{bmatrix} \text{kg-m}^2.$$

If the rigid body's angular velocity is $\boldsymbol{\omega} = 10\mathbf{i} - 5\mathbf{j} +
10\mathbf{k}\ (\text{rad/s})$, what is its angular momentum about its center
of mass?

20.9 What is the moment of inertia of the rigid body in Problem
20.8 about the axis that passes through the origin and the point
$(4, -4, 7)$ m?

Strategy: Determine the components of a unit vector
parallel to the axis, and use Eq. (20.43).

20.10 Determine the inertia matrix of the 0.6-slug thin plate in
terms of the coordinate system shown.

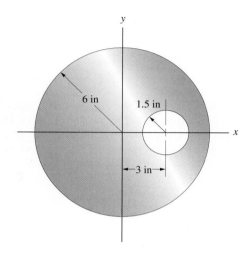

P20.10

20.11 At $t = 0$, the plate in Problem 20.10 has angular velocity
$\boldsymbol{\omega} = 10\mathbf{i} + 10\mathbf{j}\ (\text{rad/s})$ and is subjected to the force
$\mathbf{F} = -10\mathbf{k}\ (\text{lb})$ acting at the point $(0, 6, 0)$ in. No other forces or
couples act on the plate. What are the components of its angular
acceleration at that instant?

20.12 The inertia matrix of a rigid body in terms of a body-fixed
coordinate system with its origin at the center of mass is

$$[I] = \begin{bmatrix} 4 & 1 & -1 \\ 1 & 2 & 0 \\ -1 & 0 & 6 \end{bmatrix} \text{kg-m}^2.$$

If the rigid body's angular velocity is $\boldsymbol{\omega} = 10\mathbf{i} - 5\mathbf{j} +
10\mathbf{k}\ (\text{rad/s})$ and its angular acceleration is zero, what are the
components of the total moment about its center of mass?

20.13 If the total moment about the center of mass of the rigid
body in Problem 20.12 is zero, what are the components of its
angular acceleration?

20.14 The slender bar of length *l* and mass *m* is pinned to the
L-shaped bar at *O*. The L-shaped bar rotates about the vertical
axis with a constant angular velocity ω_0. Determine the value of
ω_0 necessary for the bar to remain at a constant angle β relative
to the vertical.

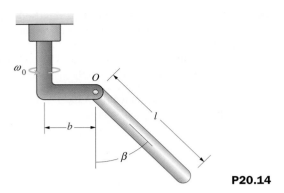

P20.14

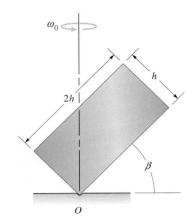

P20.16

20.15 A slender bar of length l and mass m is rigidly attached to the center of a thin circular disk of radius R and mass m. The composite object undergoes a motion in which the bar rotates in the horizontal plane with constant angular velocity ω_0 about the center of mass of the composite object and the disk rolls on the floor. Show that $\omega_0 = 2\sqrt{g/R}$.

20.17 In Problem 20.16, determine the range of values of the angle β for which the plate will remain in the steady motion described.

20.18 Arm BC has a mass of 12 kg, and its moments and products of inertia, in terms of the coordinate system shown, are $I_{xx} = 0.03$ kg-m^2, $I_{yy} = I_{zz} = 4$ kg-m^2, and $I_{xy} = I_{yz} = I_{xz} = 0$. At the instant shown, arm AB is rotating in the horizontal plane with a constant angular velocity of 1 rad/s in the counterclockwise direction viewed from above. Relative to arm AB, arm BC is rotating about the z axis with a constant angular velocity of 2 rad/s. Determine the force and couple exerted on arm BC at B.

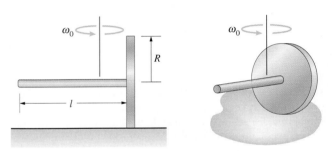

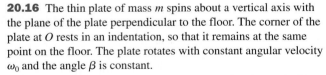

P20.15

20.16 The thin plate of mass m spins about a vertical axis with the plane of the plate perpendicular to the floor. The corner of the plate at O rests in an indentation, so that it remains at the same point on the floor. The plate rotates with constant angular velocity ω_0 and the angle β is constant.

(a) Show that the angular velocity ω_0 is related to the angle β by

$$\frac{h\omega_0^2}{g} = \frac{2\cos\beta - \sin\beta}{\sin^2\beta - 2\sin\beta\cos\beta - \cos^2\beta}.$$

(b) The equation you obtained in (a) indicates that $\omega_0 = 0$ when $2\cos\beta - \sin\beta = 0$. What is the interpretation of this result?

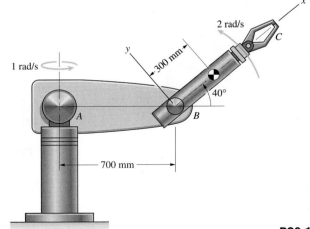

P20.18

20.19 Suppose that you throw a football in a wobbly spiral with a nutation angle of 25°. The football's moments of inertia are $I_{xx} = I_{yy} = 0.003$ slug-ft^2 and $I_{zz} = 0.001$ slug-ft^2. If the spin rate is $\dot{\phi} = 4$ revolutions per second, what is the magnitude of the precession rate (the rate at which the football wobbles)?

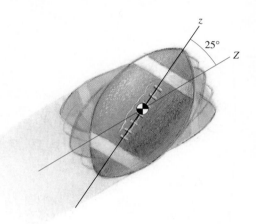

P20.19

20.20 Sketch the body and space cones for the motion of the football in Problem 20.19.

20.21 The mass of the homogeneous thin plate is 1 kg. For a coordinate system with its origin at O, determine the plate's principal moments of inertia and the directions of unit vectors parallel to the corresponding principal axes.

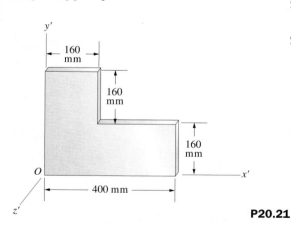

P20.21

20.22 The airplane's principal moments of inertia, in slug-ft^2, are $I_{xx} = 8000$, $I_{yy} = 48,000$, and $I_{zz} = 50,000$.
(a) The airplane begins in the reference position shown and maneuvers into the orientation $\psi = \theta = \phi = 45°$. Draw a sketch showing the plane's orientation relative to the XYZ system.
(b) If the airplane is in the orientation described in (a), the rates of change of the Euler angles are $\dot{\psi} = 0$, $\dot{\theta} = 0.2$ rad/s, and $\dot{\phi} = 0.2$ rad/s, and the second derivatives of the angles with respect to time are zero, what are the components of the total moment about the airplane's center of mass?

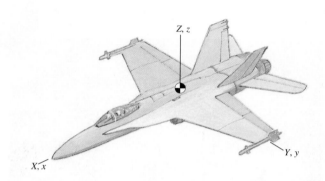

P20.22

20.23 What are the x, y, and z components of the angular acceleration of the airplane described in Problem 20.22?

20.24 If the orientation of the airplane in Problem 20.22 is $\psi = 45°$, $\theta = 60°$, and $\phi = 45°$, the rates of change of the Euler angles are $\dot{\psi} = 0$, $\dot{\theta} = 0.2$ rad/s, and $\dot{\phi} = 0.1$ rad/s, and the components of the total moment about the center of mass of the plane are $\Sigma M_x = 400$ ft-lb, $\Sigma M_y = 1200$ ft-lb, and $\Sigma M_z = 0$, what are the x, y, and z components of the airplane's angular acceleration?

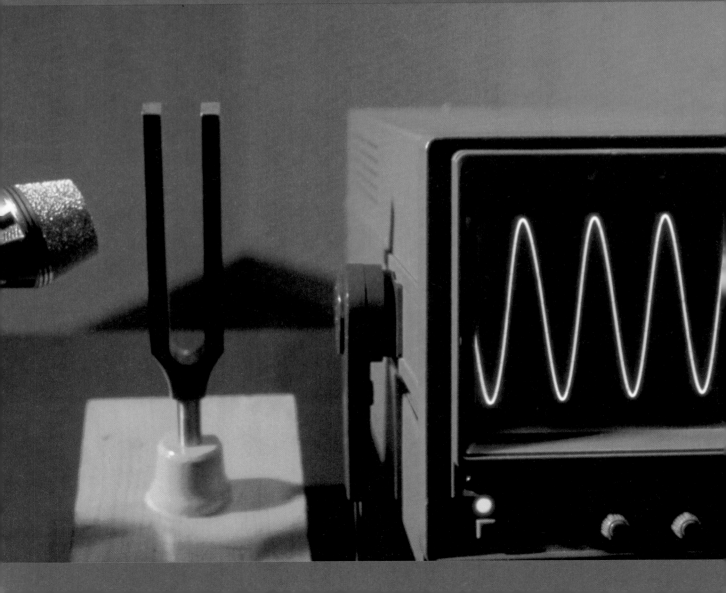

Vibrations of the tuning fork's prongs create sound waves that register on the microphone and are displayed by the oscilloscope. In this chapter, we analyze the vibrations of simple mechanical systems.

Vibrations

Vibrations have been of concern in engineering since the beginning of the industrial revolution. Beginning with the development of electro-mechanical devices capable of creating and measuring mechanical vibrations, engineering applications of vibrations have included the various areas of acoustics, from architectural acoustics to earthquake detection and analysis. We consider vibrating systems with one degree of freedom; that is, the position, or configuration, of a system can be specified by a single variable. The fundamental concepts we introduce, including amplitude, frequency, period, damping, and resonance, are also used in the analysis of systems with multiple degrees of freedom.

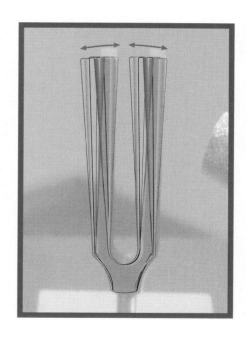

Conservative Systems

We begin by presenting different examples of one-degree-of-freedom systems subjected to conservative forces, demonstrating that their motions are described by the same differential equation. We then examine solutions of this equation and use them to describe the vibrations of one-degree-of-freedom conservative systems.

Examples

The *spring–mass oscillator* (Fig. 21.1a) is the simplest example of a one-degree-of-freedom vibrating system. A single coordinate x measuring the displacement of the mass relative to a reference point is sufficient to specify the position of the system. We draw the free-body diagram of the mass in Fig. 21.1b, neglecting friction and assuming that the spring is unstretched when $x = 0$. Applying Newton's second law, we can write the equation describing the horizontal motion of the mass as

$$\frac{d^2x}{dt^2} + \frac{k}{m}\, x = 0. \tag{21.1}$$

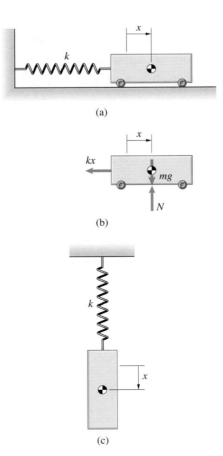

(a)

(b)

(c)

Figure 21.1

(a) The spring–mass oscillator has one degree of freedom.

(b) Free-body diagram of the mass.

(c) Suspending the mass.

We can obtain this equation by a different method that is very useful. The only force that does work on the mass, the force exerted by the spring, is conservative, which means that the sum of the kinetic and potential energies is constant:

$$\frac{1}{2} m \left(\frac{dx}{dt} \right)^2 + \frac{1}{2} kx^2 = \text{constant}.$$

Taking the derivative of this equation with respect to time, we can write the result as

$$\left(\frac{dx}{dt} \right) \left(\frac{d^2x}{dt^2} + \frac{k}{m} x \right) = 0,$$

again obtaining Eq. (21.1).

Suppose that the mass is suspended from the spring, as shown in Fig. 21.1c, and undergoes vertical motion. If the spring is unstretched when $x = 0$, it is easy to confirm that the equation of motion is

$$\frac{d^2x}{dt^2} + \frac{k}{m} x = g.$$

If the suspended mass is stationary, the magnitude of the force exerted by the spring must equal the weight ($kx = mg$), so the equilibrium position is $x = mg/k$. (Notice that we can also determine the equilibrium position by setting the acceleration equal to zero in the equation of motion.) Let us introduce a new variable $\tilde{x}$ that measures the position of the mass relative to its equilibrium position: $\tilde{x} = x - mg/k$. Writing the equation of motion in terms of this variable, we obtain

$$\frac{d^2\tilde{x}}{dt^2} + \frac{k}{m} \tilde{x} = 0,$$

which is identical to Eq. (21.1). The vertical motion of the mass in Fig. 21.1c relative to its equilibrium position is described by the same equation that describes the horizontal motion of the mass in Fig. 21.1a relative to its equilibrium position.

Now let's consider a different one-degree-of-freedom system. If we rotate the slender bar in Fig. 21.2a through some angle and release it, it will oscillate back and forth. (An object swinging from a fixed point is called a *pendulum*.) There is only one degree of freedom, since θ specifies the bar's position. Drawing the free-body diagram of the bar (Fig. 21.2b) and writing the equation of angular motion about A yields

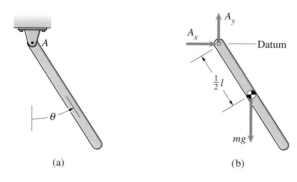

(a)   (b)

Figure 21.2
(a) A pendulum consisting of a slender bar.
(b) Free-body diagram of the bar.

$$\frac{d^2\theta}{dt^2} + \frac{3g}{2l}\sin\theta = 0. \qquad (21.2)$$

We can also obtain this equation by using conservation of energy. The bar's kinetic energy is $T = \frac{1}{2}I_A(d\theta/dt)^2$. If we place the datum at the level of point A (Fig. 21.2b), the potential energy associated with the bar's weight is $V = -mg(\frac{1}{2}l\cos\theta)$, so

$$T + V = \frac{1}{2}\left(\frac{1}{3}ml^2\right)\left(\frac{d\theta}{dt}\right)^2 - \frac{1}{2}mgl\cos\theta = \text{constant}.$$

Taking the derivative of this equation with respect to time and writing the result in the form

$$\left(\frac{d\theta}{dt}\right)\left(\frac{d^2\theta}{dt^2} + \frac{3g}{2l}\sin\theta\right) = 0,$$

we obtain Eq. (21.2). Note that Eq. (21.2) does not have the same form as Eq. (21.1). However, if we express $\sin\theta$ in terms of its Taylor series,

$$\sin\theta = \theta - \frac{1}{6}\theta^3 + \frac{1}{120}\theta^5 + \cdots,$$

and assume that θ remains small enough to approximate $\sin\theta$ by θ, then Eq. (21.2) becomes identical in form to Eq. (21.1):

$$\frac{d^2\theta}{dt^2} + \frac{3g}{2l}\theta = 0. \qquad (21.3)$$

Our analyses of the spring–mass oscillator and the pendulum resulted in equations of motion that are identical in form. To accomplish this in the case of the suspended spring–mass oscillator, we had to express the equation of motion in terms of displacement relative to the equilibrium position. In the case of the pendulum, we needed to assume that the motions were small. But within those restrictions, the form of equation we obtained describes the motions of many one-degree-of-freedom conservative systems.

Solutions

Let us consider the differential equation

$$\frac{d^2x}{dt^2} + \omega^2 x = 0, \qquad (21.4)$$

where ω is a constant. We have seen that with $\omega^2 = k/m$, this equation describes the motion of a spring–mass oscillator, and with $\omega^2 = 3g/2l$, it describes small motions of a suspended slender bar. Equation (21.4) is an *ordinary differential equation*, because it is expressed in terms of ordinary (not partial) derivatives of the dependent variable x with respect to the independent variable t. Also, Eq. (21.4) is *linear*, meaning that there are no nonlinear terms in x or its derivatives, and it is *homogeneous*, meaning that each term contains x or one of its derivatives. Finally, Eq. (21.4) has *constant coefficients*, meaning that the coefficients multiplying the dependent variable x or its derivative in each term do not depend on the independent variable t. The standard approach to solving a differential equation of this kind is to assume that the solution is of the form

$$x = Ce^{\lambda t}, \tag{21.5}$$

where C and λ are constants. Substituting this expression into Eq. (21.4) yields

$$Ce^{\lambda t}(\lambda^2 + \omega^2) = 0.$$

This equation is satisfied for any value of the constant C if $\lambda = i\omega$ or $\lambda = -i\omega$, where $i = \sqrt{-1}$, so there are two nontrivial solutions of the form of Eq. (21.5), which we write as

$$x = Ce^{i\omega t} + De^{-i\omega t}.$$

By using *Euler's identity* $e^{i\theta} = \cos\theta + i\sin\theta$, we can express this solution in the alternative form

$$x = A \sin\omega t + B \cos\omega t, \tag{21.6}$$

where A and B are arbitrary constants.

Although in practical applications Eq. (21.6) is usually the most convenient form of the solution of Eq. (21.4), we can describe the properties of the solution more easily by expressing it in the form

$$x = E \sin(\omega t - \phi), \tag{21.7}$$

where E and ϕ are constants. To show that this solution is equivalent to Eq. (21.6), we use the identity

$$E \sin(\omega t - \phi) = E(\sin\omega t \cos\phi - \cos\omega t \sin\phi)$$
$$= (E\cos\phi)\sin\omega t + (-E\sin\phi)\cos\omega t.$$

This expression is identical to Eq. (21.6) if the constants A and B are related to E and ϕ by

$$A = E\cos\phi \quad \text{and} \quad B = -E\sin\phi. \tag{21.8}$$

Equation (21.7) clearly demonstrates the oscillatory nature of the solution of Eq. (21.4). Called *simple harmonic motion*, it describes a sinusoidal function of ωt (Fig. 21.3). The positive constant E is called the *amplitude* of the vibration. By squaring Eqs. (21.8) and adding them, we obtain a relation between the amplitude and the constants A and B:

$$E = \sqrt{A^2 + B^2}. \tag{21.9}$$

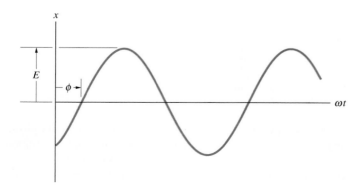

Figure 21.3
Graph of x as a function of ωt.

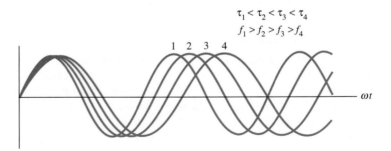

Figure 21.4
Correspondence of simple harmonic motion
with circular motion of a point.

Equation (21.7) can be interpreted in terms of the uniform motion of a
point along a circular path. We draw a circle whose radius equals the amplitude (Fig. 21.4) and assume that the line from O to P rotates in the counterclockwise direction with constant angular velocity ω. If we choose the
position of P at $t = 0$ as shown, the projection of the line OP onto the vertical axis is $E \sin(\omega t - \phi)$. Thus, there is a one-to-one correspondence between the circular motion of P and Eq. (21.7). Point P makes one complete
revolution, or *cycle*, during the time required for the angle ωt to increase by
2π radians. The time $\tau = 2\pi/\omega$ required for one cycle is called the *period*
of the vibration. Since τ is the time required for one cycle, its inverse
$f = 1/\tau$ is the number of cycles per unit time, or *frequency* of the vibration.
The frequency is usually expressed in cycles per second, or *Hertz* (Hz). The
effect of changing the period and frequency is illustrated in Fig. 21.5.

Figure 21.5
Effect of increasing the period (decreasing
the frequency) of simple harmonic motion.

We see that the period and frequency are given by

$$\tau = \frac{2\pi}{\omega}, \tag{21.10}$$

$$f = \frac{\omega}{2\pi}. \tag{21.11}$$

A system's period and frequency are determined by its physical properties,
and do not depend on the functional form in which its motion is expressed.
The frequency f is the number of revolutions the point P moves around the
circular path in Fig. 21.4 per unit time, so $\omega = 2\pi f$ is the number of radians

per unit time. Therefore, ω *is also a measure of the frequency* and is expressed in radians per second (rad/s).

Suppose that Eq. (21.7) describes the displacement of the spring–mass oscillator in Fig. 21.1(a), so that $\omega^2 = k/m$. Then the kinetic energy of the mass is

$$T = \frac{1}{2} m \left(\frac{dx}{dt} \right)^2 = \frac{1}{2} mE^2\omega^2 \cos^2(\omega t - \phi),$$

and the potential energy of the spring is

$$V = \frac{1}{2} kx^2 = \frac{1}{2} mE^2\omega^2 \sin^2(\omega t - \phi).$$

The sum of the kinetic and potential energies, $T + V = \frac{1}{2} mE^2\omega^2$, is constant (Fig. 21.6). As the system vibrates, its total energy oscillates between kinetic and potential energy. Notice that the total energy is proportional to the square of the amplitude and the square of the natural frequency.

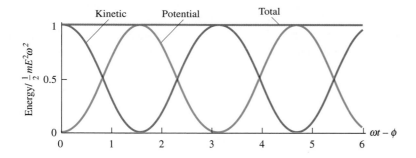

Figure 21.6
Kinetic, potential, and total energies of a spring–mass oscillator.

In summary, the motions of many one-degree-of-freedom conservative systems relative to an equilibrium position can be modeled by Eq. (21.4). To obtain the equation in that form, it may be necessary to linearize the system's equation of motion by assuming that the displacement from equilibrium is small, as we did in obtaining Eq. (21.3) from Eq. (21.2). Once the equation of motion has been expressed in the form of Eq. (21.4), the value of ω is known in terms of the physical parameters (spring constants and masses) of the system and can be used to determine the period and frequency of the system from Eqs. (21.10) and (21.11). The displacement of the system can be determined as a function of time from Eq. (21.6) or Eq. (21.7) if there is sufficient information to determine the arbitrary constants.

Study Questions

1. What is a one-degree-of-freedom system? Give examples.
2. How can you use conservation of energy to obtain the equation of motion for a conservative one-degree-of-freedom system?
3. If the motion of a one-degree-of-freedom system is described by Eq. (21.4), how can you determine the period and frequency of the system?
4. If you know the constants A and B in Eq. (21.6), how can you determine the amplitude of the vibrations the equation describes?

Example 21.1

R

k   m

Figure 21.7

(a) Free-body diagrams of the pulley and mass.

Vibration of a Conservative System

The pulley in Fig. 21.7 has radius R and moment of inertia I, and the cable does not slip relative to the pulley. The mass m is displaced downward a distance h from its equilibrium position and released from rest at $t = 0$.
(a) What is the frequency of the resulting vibrations?
(b) Determine the position of the mass relative to its equilibrium position as a function of time.

Strategy

A single coordinate specifying the vertical displacement of the mass specifies the positions of both the mass and pulley, so there is one degree of freedom. We can obtain the equation of motion of the system either by writing the individual equations of motion of the mass and pulley or by using conservation of energy.

Solution

Let x be the downward displacement of the mass relative to its position when the spring is unstretched. We draw the free-body diagrams of the pulley and mass in Fig. (a), where T_C is the tension in the cable and α is the angular acceleration of the pulley. Applying Newton's second law to the mass, we obtain

$$mg - T_C = m\frac{d^2x}{dt^2}. \tag{21.12}$$

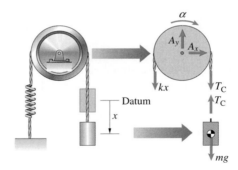

The equation of angular motion for the pulley is

$$T_C R - (kx)R = I\alpha.$$

The angular acceleration of the pulley and the acceleration of the mass are related by $\alpha = (d^2x/dt^2)/R$, so we can write the equation of angular motion as

$$T_C - kx = \left(\frac{I}{R^2}\right)\frac{d^2x}{dt^2}.$$

Summing this equation and Eq. (21.12), we obtain the equation of motion:

$$\left(m + \frac{I}{R^2}\right)\frac{d^2x}{dt^2} + kx = mg. \tag{21.13}$$

Alternative Method In terms of the velocity of the mass, the angular velocity of the pulley is $(dx/dt)/R$. Therefore, we can write the total kinetic energy of the mass and pulley as

$$T = \frac{1}{2} m \left(\frac{dx}{dt}\right)^2 + \frac{1}{2} I \left[\frac{1}{R}\left(\frac{dx}{dt}\right)\right]^2.$$

Placing the datum for the potential energy associated with the weight of the mass at $x = 0$, we obtain the total potential energy:

$$V = -mgx + \frac{1}{2} kx^2.$$

The sum of the kinetic and potential energies is constant:

$$T + V = \frac{1}{2}\left(m + \frac{I}{R^2}\right)\left(\frac{dx}{dt}\right)^2 - mgx + \frac{1}{2} kx^2 = \text{constant}.$$

Taking the derivative of this equation with respect to time, we again obtain Eq. (21.13).

By setting $d^2x/dt^2 = 0$ in Eq. (21.13), we see that the equilibrium position is $x = mg/k$. By expressing Eq. (21.13) in terms of a new variable $\tilde{x} = x - mg/k$ that measures the position of the mass relative to its equilibrium position, we obtain the equation of motion in the form of Eq. (21.4), namely,

$$\frac{d^2\tilde{x}}{dt^2} + \omega^2\tilde{x} = 0,$$

where

$$\omega^2 = \frac{k}{m + I/R^2}.$$

(a) The frequency of vibration of the system is

$$f = \frac{\omega}{2\pi} = \frac{1}{2\pi}\sqrt{\frac{k}{m + I/R^2}}.$$

(b) From Eq. (21.6), we can write the general solution for $\tilde{x}$ in the form

$$\tilde{x} = A \sin \omega t + B \cos \omega t.$$

When $t = 0$, $\tilde{x} = h$ and $d\tilde{x}/dt = 0$. The derivative of the general solution is

$$\frac{d\tilde{x}}{dt} = A\omega \cos \omega t - B\omega \sin \omega t.$$

The initial conditions yield the equations

$$h = B$$

and

$$0 = A\omega,$$

so the position of the mass relative to its equilibrium position as a function of time is

$$\tilde{x} = h \cos \omega t.$$

Example 21.2

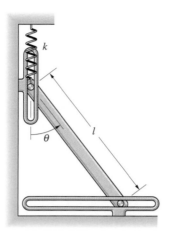

Figure 21.8

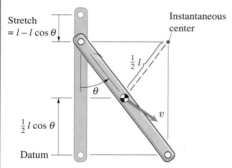

(a) Determining the velocity of the center of mass, the stretch of the spring, and the height of the center of mass above the datum.

Frequency of a System

The spring attached to the slender bar of mass m in Fig. 21.8 is unstretched when $\theta = 0$. Neglecting friction, determine the frequency of small vibrations of the bar relative to its equilibrium position.

Strategy

The angle θ specifies the bar's position, so there is one degree of freedom. We can express the kinetic and potential energies in terms of θ and its time derivative and then take the derivative of the total energy with respect to time to obtain the equation of motion.

Solution

The kinetic energy of the bar is

$$T = \frac{1}{2} mv^2 + \frac{1}{2} I \left(\frac{d\theta}{dt} \right)^2,$$

where v is the velocity of the center of mass and $I = \frac{1}{12} ml^2$. The distance from the bar's instantaneous center to its center of mass is $\frac{1}{2} l$ (Fig. a), so $v = \left(\frac{1}{2} l \right)(d\theta/dt)$, and the kinetic energy is

$$T = \frac{1}{2} m \left[\frac{1}{2} l \left(\frac{d\theta}{dt} \right) \right]^2 + \frac{1}{2} \left(\frac{1}{12} ml^2 \right) \left(\frac{d\theta}{dt} \right)^2 = \frac{1}{6} ml^2 \left(\frac{d\theta}{dt} \right)^2.$$

In terms of θ, the stretch of the spring is $l - l \cos \theta$. We place the datum for the potential energy associated with the weight at the bottom of the bar (Fig. a), so the total potential energy is

$$V = mg \left(\frac{1}{2} l \cos \theta \right) + \frac{1}{2} k(l - l \cos \theta)^2.$$

The sum of the kinetic and potential energies is constant:

$$T + V = \frac{1}{6} ml^2 \left(\frac{d\theta}{dt} \right)^2 + \frac{1}{2} mgl \cos \theta + \frac{1}{2} kl^2 (1 - \cos \theta)^2 = \text{constant}.$$

Taking the derivative of this equation with respect to time, we obtain the equation of motion:

$$\frac{1}{3} ml^2 \frac{d^2\theta}{dt^2} - \frac{1}{2} mgl \sin \theta + kl^2 (1 - \cos \theta) \sin \theta = 0. \tag{21.14}$$

To express this equation in the form of Eq. (21.4), we need to write it in terms of small vibrations relative to the equilibrium position. Let θ_e be the value of θ when the bar is in equilibrium. By setting $d^2\theta/dt^2 = 0$ in Eq. (21.14), we find that θ_e must satisfy the relation

$$\cos \theta_e = 1 - \frac{mg}{2kl}. \tag{21.15}$$

We define $\tilde{\theta} = \theta - \theta_e$, and expand $\sin \theta$ and $\cos \theta$ in Taylor series in terms of $\tilde{\theta}$:

$$\sin\theta = \sin\left(\theta_e + \widetilde{\theta}\right) = \sin\theta_e + \cos\theta_e\,\widetilde{\theta} + \cdots,$$

$$\cos\theta = \cos\left(\theta_e + \widetilde{\theta}\right) = \cos\theta_e - \sin\theta_e\,\widetilde{\theta} + \cdots.$$

Substituting these expressions into Eq. (21.14), neglecting terms in $\widetilde{\theta}$ of second and higher orders, and using Eq. (21.15), we obtain

$$\frac{d^2\widetilde{\theta}}{dt^2} + \omega^2\widetilde{\theta} = 0,$$

where

$$\omega^2 = \frac{3g}{l}\left(1 - \frac{mg}{4kl}\right).$$

From Eq. (21.11), the frequency of small vibrations of the bar is

$$f = \frac{\omega}{2\pi} = \frac{1}{2\pi}\sqrt{\frac{3g}{l}\left(1 - \frac{mg}{4kl}\right)}.$$

21.2 Damped Vibrations

If the mass of a spring–mass oscillator is displaced and released, it will not vibrate indefinitely. It will slow down and eventually stop as a result of frictional forces, or *damping mechanisms*, acting on the system. Damping mechanisms damp out, or *attenuate*, the vibration. In some cases, engineers intentionally include damping mechanisms in vibrating systems. For example, the shock absorbers in a car are designed to damp out vibrations of the suspension relative to the frame. In the previous section we neglected damping, so the solutions we obtained describe only motions of systems over periods of time brief enough that the effects of damping can be neglected. We now discuss a simple method for modeling damping in vibrating systems.

The spring–mass oscillator in Fig. 21.9a has a *damping element*. The schematic diagram for the damping element represents a piston moving in a cylinder of viscous fluid. The force required to lengthen or shorten a damping element is defined to be the product of a constant c, the *damping constant*, and the rate of change of the length of the element (Fig. 21.9b). Therefore, the equation of motion of the mass is

$$-c\frac{dx}{dt} - kx = m\frac{d^2x}{dt^2}.$$

By defining $\omega = \sqrt{k/m}$ and $d = c/2m$, we can write this equation in the form

$$\frac{d^2x}{dt^2} + 2d\frac{dx}{dt} + \omega^2 x = 0. \tag{21.16}$$

This equation describes the vibrations of many damped, one-degree-of-freedom systems. The form of its solution, and consequently, the character of the predicted behavior of the system the equation describes, depends on whether the constant d is less than, equal to, or greater than ω. We discuss these cases in the sections that follow.

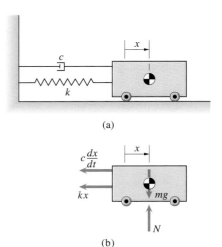

(a)

(b)

Figure 21.9
(a) Damped spring–mass oscillator.
(b) Free-body diagram of the mass.

Subcritical Damping

If $d < \omega$, the system is said to be *subcritically damped*. Assuming a solution of the form

$$x = Ce^{\lambda t} \tag{21.17}$$

and substituting it into Eq. (21.16), we obtain

$$\lambda^2 + 2d\lambda + \omega^2 = 0.$$

This quadratic equation yields two roots for the constant λ that we can write as

$$\lambda = -d \pm i\omega_d,$$

where

$$\omega_d = \sqrt{\omega^2 - d^2}. \tag{21.18}$$

Because we are assuming that $d < \omega$, the constant ω_d is a real number. The two roots for λ give us two solutions of the form of Eq. (21.17). The resulting general solution of Eq. (21.16) is

$$x = e^{-dt}\big(Ce^{i\omega_d t} + De^{-i\omega_d t}\big),$$

where C and D are constants. By using the Euler identity $e^{i\theta} = \cos\theta + i\sin\theta$, we can express this solution in the form

$$x = e^{-dt}\big(A \sin\omega_d t + B\cos\omega_d t\big), \tag{21.19}$$

where A and B are constants. Equation (21.19) is the product of an exponentially decaying function of time and an expression identical in form to the solution we obtained for an undamped system. The exponential function describes the expected effect of damping: The amplitude of the vibration attenuates with time. The coefficient d determines the rate at which the amplitude decreases.

Damping has an important effect in addition to causing attenuation. Because the oscillatory part of the solution is identical in form to Eq. (21.6), except that the term ω is replaced by ω_d, it follows from Eqs. (21.10) and (21.11) that the period and frequency of the damped system are

$$\tau_d = \frac{2\pi}{\omega_d}, \qquad f_d = \frac{\omega_d}{2\pi}. \tag{21.20}$$

From Eq. (21.18), we see that $\omega_d < \omega$, so *the period of the vibration is increased and its frequency is decreased as a result of subcritical damping.*

The rate of damping is often expressed in terms of the *logarithmic decrement* δ, which is the natural logarithm of the ratio of the amplitude at a time t to the amplitude at time $t + \tau_d$. Since the amplitude is proportional to e^{-dt}, we can obtain a simple relation between the logarithmic decrement, the coefficient d, and the period:

$$\delta = \ln\left[\frac{e^{-dt}}{e^{-d(t+\tau_d)}}\right] = d\tau_d.$$

Critical and Supercritical Damping

When $d \geq \omega$, the character of the solution of Eq. (21.16) is different from the case of subcritical damping. Suppose that $d > \omega$. When this is the case, the system is said to be *supercritically damped*. We again substitute a solution of the form

$$x = Ce^{\lambda t} \tag{21.21}$$

into Eq. (21.16), obtaining

$$\lambda^2 + 2d\lambda + \omega^2 = 0. \tag{21.22}$$

We can write the roots of this equation as

$$\lambda = -d \pm h,$$

where

$$h = \sqrt{d^2 - \omega^2}. \tag{21.23}$$

The resulting general solution of Eq. (21.16) is

$$x = Ce^{-(d-h)t} + De^{-(d+h)t}, \tag{21.24}$$

where C and D are constants.

When $d = \omega$, a system is said to be *critically damped*. Then the constant $h = 0$, so Eq. (21.22) has a repeated root $\lambda = -d$, and we obtain only one solution of the form (21.21). In this case, it can be shown that the general solution of Eq. (21.16) is

$$x = Ce^{-dt} + Dte^{-dt}, \tag{21.25}$$

where C and D are constants.

Equations (21.24) and (21.25) indicate that the motion of a system is not oscillatory when $d \geq \omega$. These equations are expressed in terms of exponential functions, and do not contain sines and cosines. The condition $d = \omega$ defines the minimum amount of damping necessary to avoid oscillatory behavior, which is why it is referred to as the critically damped case. Figure 21.10 shows the effect of increasing amounts of damping on the behavior of a vibrating system.

The concept of critical damping has important implications in the design of many systems. For example, it is desirable to introduce enough damping into a car's suspension so that its motion is not oscillatory, but too much damping would cause the suspension to be too "stiff."

In summary, the motions of many damped one-degree-of-freedom systems can be modeled by Eq. (21.16). Once the equation of motion has been expressed in that form, the values of d and ω are known in terms of the physical parameters of the system. Those values determine whether the damping is subcritical, critical, or supercritical, which indicates the form of the solution of Eq. (21.16). If the system is subcritically damped, the period and frequency of its vibrations are given by Eqs. (21.20).

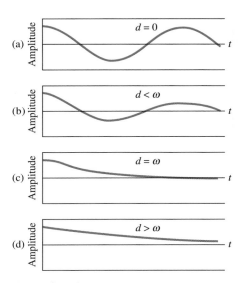

Figure 21.10
Amplitude history of a vibrating system that is (a) undamped; (b) subcritically damped; (c) critically damped; (d) supercritically damped.

Study Questions

Type of Damping		Solution
$d < \omega$:	Subcritical	Eq. (21.19)
$d = \omega$:	Critical	Eq. (21.25)
$d > \omega$:	Supercritical	Eq. (21.24)

1. What is a damping element? How is the damping constant defined?
2. How do you determine whether the damping of a system whose motion is described by Eq. (21.16) is subcritical, critical, or supercritical?
3. Why can you calculate the period and frequency for a system that is subcritically damped, but not for one that is critically or supercritically damped?
4. What is the logarithmic decrement?

Example 21.3

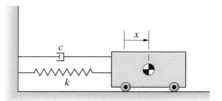

Figure 21.11

Damped Spring–Mass Oscillator

The damped spring–mass oscillator in Fig. 21.11 has mass $m = 2$ kg, spring constant $k = 8$ N/m, and damping constant $c = 1$ N-s/m. At $t = 0$, the mass is released from rest in the position $x = 0.1$ m. Determine its position as a function of time.

Solution

The constants $\omega = \sqrt{k/m} = 2$ rad/s and $d = c/2m = 0.25$ rad/s, so the damping is subcritical and the motion is described by Eq. (21.19). From Eq. (21.18),

$$\omega_d = \sqrt{\omega^2 - d^2} = 1.98 \text{ rad/s}.$$

From Eq. (21.19),

$$x = e^{-0.25t}(A \sin 1.98t + B \cos 1.98t),$$

and the velocity of the mass is

$$\frac{dx}{dt} = -0.25e^{-0.25t}(A \sin 1.98t + B \cos 1.98t)$$

$$+ e^{-0.25t}(1.98A \cos 1.98t - 1.98B \sin 1.98t).$$

From the conditions $x = 0.1$ m and $dx/dt = 0$ at $t = 0$, we obtain $A = 0.0126$ m and $B = 0.1$ m, so the position of the mass is

$$x = e^{-0.25t}(0.0126 \sin 1.98t + 0.1 \cos 1.98t) \text{ m}.$$

The graph of x for the first 10 s of motion in Fig. 21.12 clearly exhibits the attenuation of the amplitude.

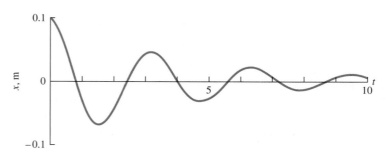

Figure 21.12
Position of the mass as a function of time.

Example 21.4

Motion of a Damped System

The mass of the stepped disk in Fig. 21.13 is $m = 20$ kg, its radius is $R = 0.3$ m, and its moment of inertia is $I = \frac{1}{2}mR^2$. The spring constant is $k = 60$ N/m and the damping constant is $c = 24$ N-s/m. Determine the position of the center of the disk as a function of time if the disk is released from rest with the spring unstretched.

Solution

Let x be the downward displacement of the center of the disk relative to its position when the spring is unstretched. From the position of the disk's instantaneous center (Fig. a), we can see that the rate at which the spring is stretched is $2(dx/dt)$ and the rate at which the damping element is lengthened is $3(dx/dt)$. When the center of the disk is displaced a distance x, the stretch of the spring is $2x$.

We draw the free-body diagram of the disk in Fig. (b), showing the forces exerted by the spring, the damping element, and the tension in the cable. Newton's second law is

$$mg - T - 2kx - 3c\frac{dx}{dt} = m\frac{d^2x}{dt^2},$$

and the equation of angular motion is

$$RT - R(2kx) - 2R\left(3c\frac{dx}{dt}\right) = \left(\frac{1}{2}mR^2\right)\alpha.$$

The angular acceleration is related to the acceleration of the center of the disk by $\alpha = (d^2x/dt^2)/R$. Eliminating T from Newton's second law and the equation of angular motion, we obtain the equation of motion:

$$\frac{3}{2}m\frac{d^2x}{dt^2} + 9c\frac{dx}{dt} + 4kx = mg.$$

By setting d^2x/dt^2 and dx/dt equal to zero in this equation, we find that the equilibrium position of the disk is $x = mg/4k$. Expressing the equation of motion in terms of the position of the center of the disk relative to its equilibrium position, $\tilde{x} = x - mg/4k$, we obtain

$$\frac{d^2\tilde{x}}{dt^2} + \left(\frac{6c}{m}\right)\frac{d\tilde{x}}{dt} + \left(\frac{8k}{3m}\right)\tilde{x} = 0.$$

This equation is identical in form to Eq. (21.16), where the constants are

$$d = \frac{6c}{2m} = \frac{(6)(24)}{(2)(20)} = 3.60 \text{ rad/s}$$

and

$$\omega = \sqrt{\frac{8k}{3m}} = \sqrt{\frac{(8)(60)}{(3)(20)}} = 2.83 \text{ rad/s}.$$

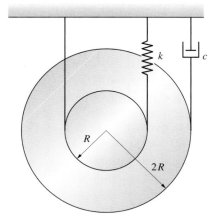

Figure 21.13

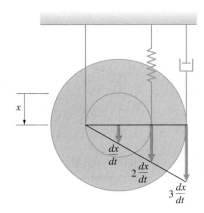

(a) Using the instantaneous center to determine the relationships between the velocities.

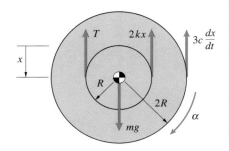

(b) Free-body diagram of the disk.

The damping is supercritical $(d > \omega)$ so the motion is described by Eq. (21.24) with $h = \sqrt{d^2 - \omega^2} = 2.23$ rad/s:

$$\tilde{x} = Ce^{-(d-h)t} + De^{-(d+h)t} = Ce^{-1.37t} + De^{-5.83t}.$$

The velocity is

$$\frac{d\tilde{x}}{dt} = -1.37Ce^{-1.37t} - 5.83De^{-5.83t}.$$

At $t = 0$, $\tilde{x} = -mg/4k = -0.818$ m and $d\tilde{x}/dt = 0$. From these conditions, we obtain $C = -1.069$ m and $D = 0.252$ m, so the position of the center of the disk relative to its equilibrium position is

$$\tilde{x} = -1.069e^{-1.37t} + 0.252e^{-5.83t} \text{ m}.$$

The graph of the position for the first 4 s of motion is shown in Fig. 21.14.

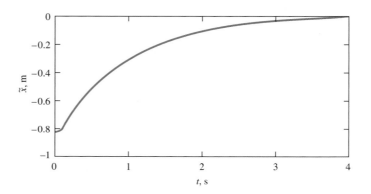

Figure 21.14
Position of the center of the disk as a function of time.

21.3 Forced Vibrations

The term *forced vibrations* means that external forces affect the vibrations of a system. Until now, we have discussed *free vibrations* of systems—vibrations unaffected by external forces. For example, during an earthquake, a building undergoes forced vibrations induced by oscillatory forces exerted on its foundations. After the earthquake subsides, the building vibrates freely until its motion damps out.

The damped spring–mass oscillator in Fig. 21.15a is subjected to a horizontal time-dependent force $F(t)$. From the free-body diagram of the mass (Fig. 21.15b), its equation of motion is

$$F(t) - kx - c\frac{dx}{dt} = m\frac{d^2x}{dt^2}.$$

Defining $d = c/2m$, $\omega^2 = k/m$, and $a(t) = F(t)/m$, we can write this equation in the form

$$\frac{d^2x}{dt^2} + 2d\frac{dx}{dt} + \omega^2x = a(t). \tag{21.26}$$

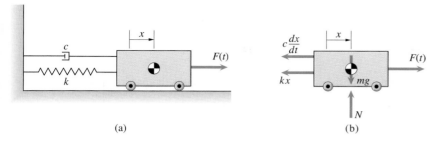

Figure 21.15
(a) A damped spring–mass oscillator subjected to a time-dependent force.
(b) Free-body diagram of the mass.

(a)

(b)

We call $a(t)$ the *forcing function*. Equation (21.26) describes the forced vibrations of many damped one-degree-of-freedom systems. It is nonhomogeneous, because the forcing function does not contain x or one of its derivatives. Its general solution consists of two parts—the homogeneous and particular solutions:

$$x = x_h + x_p.$$

The *homogeneous solution* x_h is the general solution of Eq. (21.26) with the right side set equal to zero. Therefore, the homogeneous solution is the general solution for free vibrations, which we described in Section 21.2. The *particular solution* x_p is a solution that satisfies Eq. (21.26). In the sections that follow, we discuss the particular solutions for two types of forcing functions that occur frequently in applications.

Oscillatory Forcing Function

Unbalanced wheels and shafts exert forces that oscillate at their frequency of rotation. When a car's wheels are out of balance, they exert oscillatory forces that cause vibrations passengers can feel. Engineers design electromechanical devices that transform oscillating currents into oscillating forces for use in testing vibrating systems. But the principal reason we are interested in this type of forcing function is that nearly any forcing function can be represented as a sum of oscillatory forcing functions with several different frequencies or with a continuous spectrum of frequencies.

By studying the motion of a vibrating system subjected to an oscillatory forcing function, we can determine the response of the system as a function of the frequency of the force. Suppose that the forcing function is an oscillatory function of the form

$$a(t) = a_0 \sin \omega_0 t + b_0 \cos \omega_0 t, \tag{21.27}$$

where a_0, b_0, and the frequency of the forcing function ω_0 are given constants. We can obtain the particular solution to Eq. (21.26) by seeking a solution of the form

$$x_p = A_p \sin \omega_0 t + B_p \cos \omega_0 t, \tag{21.28}$$

where A_p and B_p are constants we must determine. Substituting this expression and Eq. (21.27) into Eq. (21.26), we can write the resulting equation as

$$\left(-\omega_0^2 A_p - 2d\omega_0 B_p + \omega^2 A_p - a_0\right) \sin \omega_0 t$$
$$+ \left(-\omega_0^2 B_p + 2d\omega_0 A_p + \omega^2 B_p - b_0\right) \cos \omega_0 t = 0.$$

Equating the coefficients of $\sin \omega_0 t$ and $\cos \omega_0 t$ to zero and solving for A_p and B_p, we obtain

$$A_p = \frac{\left(\omega^2 - \omega_0^2\right)a_0 + 2d\omega_0 b_0}{\left(\omega^2 - \omega_0^2\right)^2 + 4d^2\omega_0^2}$$

and (21.29)

$$B_{\rm p} = \frac{-2d\omega_0 a_0 + (\omega^2 - \omega_0^2)b_0}{(\omega^2 - \omega_0^2)^2 + 4d^2\omega_0^2}.$$

Substituting these results into Eq. (21.28) yields the particular solution:

$$x_{\rm p} = \left[\frac{(\omega^2 - \omega_0^2)a_0 + 2d\omega_0 b_0}{(\omega^2 - \omega_0^2)^2 + 4d^2\omega_0^2}\right]\sin\omega_0 t$$

$$+ \left[\frac{-2d\omega_0 a_0 + (\omega^2 - \omega_0^2)b_0}{(\omega^2 - \omega_0^2)^2 + 4d^2\omega_0^2}\right]\cos\omega_0 t. \qquad (21.30)$$

The amplitude of the particular solution is

$$E_{\rm p} = \sqrt{A_{\rm p}^2 + B_{\rm p}^2} = \frac{\sqrt{a_0^2 + b_0^2}}{\sqrt{(\omega^2 - \omega_0^2)^2 + 4d^2\omega_0^2}}. \qquad (21.31)$$

In Section 21.2, we showed that the solution of the equation describing free vibration of a damped system attenuates with time. For this reason, the particular solution for the motion of a damped vibrating system subjected to an oscillatory external force is also called the *steady-state solution*. The motion approaches the steady-state solution with increasing time. (See Example 21.5.)

To illustrate the effects of damping and the frequency of the forcing function on the amplitude of the particular solution, in Fig. 21.16 we plot the nondimensional expression $\omega^2 E_{\rm p}/\sqrt{a_0^2 + b_0^2}$ as a function of ω_0/ω for several values of the parameter d/ω. When there is no damping $(d = 0)$, the amplitude of the particular solution approaches infinity as the frequency ω_0 of the forcing function approaches the frequency ω. When the damping is small, the amplitude of the particular solution approaches a finite maximum value at a value of ω_0 that is smaller than ω. The frequency at which the amplitude of the particular solution is a maximum is called the *resonant frequency*. (See Problem 21.67.)

The phenomenon of resonance is a familiar one in our everyday experience. For example, when a wheel of a car is out of balance, the resulting

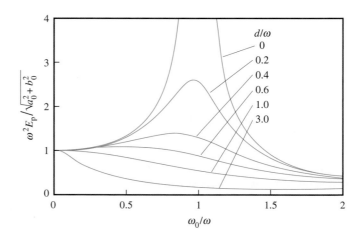

Figure 21.16

Amplitude of the particular (steady-state) solution as a function of the frequency of the forcing function.

vibrations are noticed when the car is moving at a certain speed. At that speed, the wheel rotates at the resonant frequency of the car's suspension. Resonance is of practical importance in many applications, because relatively small oscillatory forces can result in large vibrational amplitudes that may cause damage or interfere with the functioning of a system. The classic example is soldiers marching across a bridge. If their steps in unison coincide with one of the bridge's resonant frequencies, they may damage the bridge even though it can safely support their weight.

Polynomial Forcing Function

Suppose that the forcing function $a(t)$ in Eq. (21.26) is a polynomial function of time; that is,

$$a(t) = a_0 + a_1 t + a_2 t^2 + \cdots + a_N t^N,$$

where $a_1, a_2, \ldots, a_N$ are given constants. This forcing function is important in applications because many smooth functions can be approximated by polynomials over a given interval of time. In this case we can obtain the particular solution of Eq. (21.26) by seeking a solution of the same form, namely,

$$x_p = A_0 + A_1 t + A_2 t^2 + \cdots + A_N t^N, \tag{21.32}$$

where $A_0, A_1, A_2, \ldots, A_N$ are constants to be determined.

For example, if $a(t) = a_0 + a_1 t$, Eq. (21.26) becomes

$$\frac{d^2 x}{dt^2} + 2d \frac{dx}{dt} + \omega^2 x = a_0 + a_1 t, \tag{21.33}$$

and we seek a particular solution of the form $x_p = A_0 + A_1 t$. Substituting this solution into Eq. (21.33), we can write the resulting equation as

$$(2dA_1 + \omega^2 A_0 - a_0) + (\omega^2 A_1 - a_1)t = 0.$$

This equation can be satisfied over an interval of time only if

$$2dA_1 + \omega^2 A_0 - a_0 = 0$$

and

$$\omega^2 A_1 - a_1 = 0.$$

Solving these two equations for A_0 and A_1, we obtain the particular solution:

$$x_p = \frac{a_0 - 2da_1/\omega^2 + a_1 t}{\omega^2}.$$

It can be confirmed that this is a particular solution of Eq. (21.33) by substituting it into that equation.

Study Questions

1. How do you determine the homogeneous solution of Eq. (21.26)?
2. Why is the particular solution for the motion of a damped vibrating system subjected to an oscillatory external force also called the steady-state solution?
3. What is the resonant frequency? Why is it significant?
4. If a vibrating system is subjected to a polynomial forcing function, how do you determine the particular solution of the equation describing the system's motion?

Example 21.5

Oscillatory Forcing Function

An engineer designing a vibration isolation system for an instrument console models the console and isolation system by the damped spring–mass oscillator in Fig. 21.17 with mass $m = 2$ kg, spring constant $k = 8$ N/m, and damping constant $c = 1$ N-s/m. To determine the system's response to external vibration, she assumes that the mass is initially stationary with the spring unstretched, and at $t = 0$ a force

$$F(t) = 20 \sin 4t \text{ N}$$

is applied to the mass.
(a) What is the amplitude of the particular (steady-state) solution?
(b) What is the position of the mass as a function of time?

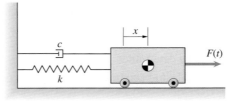

Figure 21.17

Strategy

The forcing function is $a(t) = F(t)/m = 10 \sin 4t$ m/s^2, which is an oscillatory function of the form of Eq. (21.27) with $a_0 = 10$ m/s^2, $b_0 = 0$, and $\omega_0 = 4$ rad/s. The amplitude of the particular solution is given by Eq. (21.31), and the particular solution itself is given by Eq. (21.30). We must also determine whether the damping is subcritical, critical, or supercritical and choose the appropriate form of the homogeneous solution.

Solution

(a) The frequency of the undamped system is $\omega = \sqrt{k/m} = 2$ rad/s and the constant $d = c/2m = 0.25$ rad/s. Therefore, the amplitude of the particular solution is

$$E_{\text{p}} = \frac{a_0}{\sqrt{\left(\omega^2 - \omega_0^2\right)^2 + 4d^2\omega_0^2}} = \frac{10}{\sqrt{\left[(2)^2 - (4)^2\right]^2 + 4(0.25)^2(4)^2}}$$

$$= 0.822 \text{ m.}$$

(b) Since $d < \omega$, the system is subcritically damped and the homogeneous solution is given by Eq. (21.19). The frequency of the damped system is $\omega_d = \sqrt{\omega^2 - d^2} = 1.98$ rad/s, so the homogeneous solution is

$$x_{\text{h}} = e^{-0.25t}(A \sin 1.98t + B \cos 1.98t).$$

From Eq. (21.30), the particular solution is

$$x_p = -0.811 \sin 4t - 0.135 \cos 4t,$$

and the complete solution is

$$x = x_h + x_p$$

$$= e^{-0.25t}(A \sin 1.98t + B \cos 1.98t) - 0.811 \sin 4t - 0.135 \cos 4t.$$

At $t = 0$, $x = 0$ and $dx/dt = 0$. Using these conditions to determine the constants A and B, we obtain $A = 1.651$ m and $B = 0.135$ m. The position of the mass as a function of time is

$$x = e^{-0.25t}(1.651 \sin 1.98t + 0.135 \cos 1.98t)$$

$$- 0.811 \sin 4t - 0.135 \cos 4t \text{ m}.$$

Figure 21.18 shows the homogeneous, particular, and complete solutions for the first 25 s of motion of the system. The complete solution has an initial "transient" phase due to the homogeneous part of the solution. As the homogeneous solution attenuates, the complete solution approaches the particular, or steady-state, solution.

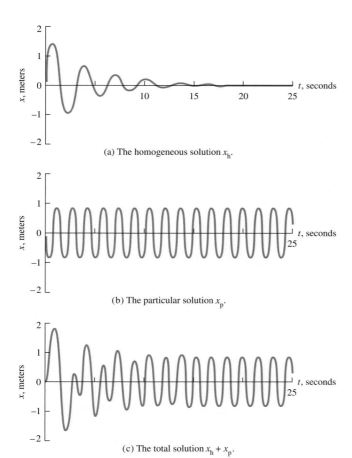

(a) The homogeneous solution x_h.

(b) The particular solution x_p.

(c) The total solution $x_h + x_p$.

Figure 21.18
The homogeneous, particular, and complete solutions.

Example 21.6

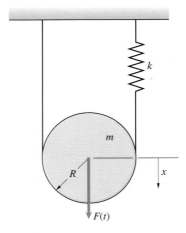

Figure 21.19

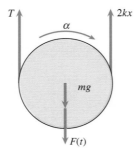

(a) Free-body diagram of the disk.

Polynomial Forcing Function

The homogeneous disk in Fig. 21.19 has radius $R = 2$ m and mass $m = 4$ kg. The spring constant is $k = 30$ N/m. The disk is initially stationary in its equilibrium position, and at $t = 0$ a downward force $F(t) = 12 + 12t - 0.6t^2$ N is applied to the center of the disk. Determine the position of the center of the disk as a function of time.

Strategy

The force $F(t)$ is a polynomial, so we can seek a particular solution of the form of Eq. (21.32).

Solution

Let x be the displacement of the center of the disk relative to its position when the spring is unstretched. We draw the free-body diagram of the disk in Fig. (a), where T is the tension in the cable on the left side of the disk. From Newton's second law,

$$F(t) + mg - 2kx - T = m\frac{d^2x}{dt^2}. \tag{21.34}$$

The angular acceleration of the disk in the clockwise direction is related to the acceleration of the center of the disk by $\alpha = (d^2x/dt^2)/R$. Using this expression, we can write the equation of angular motion of the disk as

$$\Sigma M = I\alpha:$$

$$T R - 2kxR = \left(\frac{1}{2} m R^2\right)\left(\frac{1}{R}\frac{d^2x}{dt^2}\right).$$

Solving this equation for T and substituting the result into Eq. (21.34), we obtain the equation of motion:

$$\frac{3}{2} m \frac{d^2x}{dt^2} + 4kx = F(t) + mg. \tag{21.35}$$

Setting $d^2x/dt^2 = 0$ and $F(t) = 0$ in this equation, we find that the equilibrium position of the disk is $x = mg/4k$. In terms of the position of the center of the disk relative to its equilibrium position $\tilde{x} = x - mg/4k$, the equation of motion is

$$\frac{d^2\tilde{x}}{dt^2} + \frac{8k}{3m} \tilde{x} = \frac{2F(t)}{3m}.$$

This equation is identical in form to Eq. (21.26). Substituting the values of k and m and the polynomial function $F(t)$, we obtain

$$\frac{d^2\tilde{x}}{dt^2} + 20\tilde{x} = 2 + 2t - 0.1t^2. \tag{21.36}$$

Comparing this equation with Eq. (21.26), we see that $d = 0$ (there is no damping) and $\omega^2 = 20$ (rad/s)^2. From Eq. (21.19), the homogeneous solution is

$$\widetilde{x}_h = A \sin 4.472t + B \cos 4.472t.$$

To obtain the particular solution, we seek a solution in the form of a polynomial of the same order as $F(t)$—that is,

$$\widetilde{x}_p = A_0 + A_1 t + A_2 t^2,$$

where A_0, A_1, and A_2 are constants we must determine. We substitute this expression into Eq. (21.36) and collect terms of equal powers in t:

$$\left(2A_2 + 20A_0 - 2\right) + \left(20A_1 - 2\right)t + \left(20A_2 + 0.1\right)t^2 = 0.$$

This equation is satisfied if the coefficients multiplying each power of t equal zero, which yields

$$2A_2 + 20A_0 = 2,$$

$$20A_1 = 2,$$

and

$$20A_2 = -0.1.$$

Solving these three equations for A_0, A_1, and A_2, we obtain the particular solution:

$$\widetilde{x}_p = 0.101 + 0.100t - 0.005t^2.$$

The complete solution is

$$\widetilde{x} = \widetilde{x}_h + \widetilde{x}_p$$

$$= A \sin 4.472t + B \cos 4.472t + 0.101 + 0.100t - 0.005t^2.$$

At $t = 0$, $\widetilde{x} = 0$ and $d\widetilde{x}/dt = 0$. Using these conditions to determine A and B, we obtain the position of the center of the disk (in meters) as a function of time:

$$\widetilde{x} = -0.022 \sin 4.472t - 0.101 \cos 4.472t + 0.101 + 0.100t - 0.005t^2.$$

The position is shown for the first 25 seconds of motion in Fig. 21.20. The undamped, oscillatory homogeneous solution is superimposed on the slowly varying particular solution.

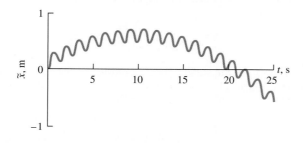

Figure 21.20
Position of the center of the disk as a function of time.

Example 21.7

Application to Engineering:

Displacement Transducers

A damped spring–mass oscillator, or a device that can be modeled as a damped spring–mass oscillator, can be used to measure an object's displacement. Suppose that the base of the spring–mass oscillator in Fig. 21.21 is attached to an object and the coordinate x_i is a displacement to be measured relative to an inertial reference frame. The coordinate x measures the displacement of the mass relative to the base. When $x = 0$, the spring is unstretched. Suppose that the system is initially stationary and at $t = 0$ the base undergoes the oscillatory motion

$$x_i = a_i \sin \omega_i t + b_i \cos \omega_i t. \tag{21.37}$$

If $m = 2$ kg, $k = 8$ N/m, $c = 4$ N-s/m, $a_i = 0.1$ m, $b_i = 0.1$ m, and $\omega_i = 10$ rad/s, what is the resulting steady-state amplitude of the displacement of the mass relative to the base?

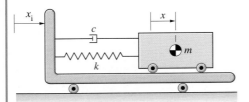

Figure 21.21

Solution

The acceleration of the mass relative to the base is d^2x/dt^2, so its acceleration relative to the inertial reference frame is $(d^2x/dt^2) + (d^2x_i/dt^2)$. Newton's second law for the mass is

$$-c\frac{dx}{dt} - kx = m\left(\frac{d^2x}{dt^2} + \frac{d^2x_i}{dt^2}\right).$$

We can write this equation as

$$\frac{d^2x}{dt^2} + 2d\frac{dx}{dt} + \omega^2 x = a(t),$$

where $d = c/2m = 1$ rad/s, $\omega = \sqrt{k/m} = 2$ rad/s, and the function

$$a(t) = -\frac{d^2x_i}{dt^2} = a_i\omega_i^2 \sin\omega_i t + b_i\omega_i^2 \cos\omega_i t. \tag{21.38}$$

Thus, we obtain an equation of motion identical in form to that for a spring–mass oscillator subjected to an oscillatory force. Comparing Eq. (21.38) with Eq. (21.27), we can obtain the amplitude of the particular (steady-state) solution from Eq. (21.31) by setting $a_0 = a_i\omega_i^2$, $b_0 = b_i\omega_i^2$, and $\omega_0 = \omega_i$:

$$E_p = \frac{\omega_i^2\sqrt{a_i^2 + b_i^2}}{\sqrt{(\omega^2 - \omega_i^2)^2 + 4d^2\omega_i^2}}. \tag{21.39}$$

Therefore, the steady-state amplitude of the displacement of the mass relative to its base is

$$E_p = \frac{(10)^2\sqrt{(0.1)^2 + (0.1)^2}}{\sqrt{[(2)^2 - (10)^2]^2 + 4(1)^2(10)^2}} = 0.144 \text{ m}.$$

𝒟esign Issues

A microphone transforms sound waves into a varying voltage that can be recorded or transformed back into sound waves by a speaker. A device that transforms a mechanical input into an electromagnetic output, or an electromagnetic input into a mechanical output, is called a transducer. Transducers can be used to measure displacements, velocities, and accelerations by transforming them into measurable voltages or currents.

In Fig. 21.21, the coordinate x_i is the displacement to be measured (the input). To use the spring–mass oscillator as a transducer, it would be designed to produce a voltage or current (the output) proportional to the displacement x. If the relationship between the input and output is known, the displacement x_i can be determined. Some seismographs (Fig. 21.22) measure motions of the earth in this way.

If the input is an oscillatory displacement given by Eq. (21.37), the amplitude of the output is given by Eq. (21.39). We can write the latter equation as

$$\frac{E_p}{E_i} = \frac{(\omega_i/\omega)^2}{\sqrt{[1 - (\omega_i/\omega)^2]^2 + 4(d/\omega)^2(\omega_i/\omega)^2}},$$

where $E_i = \sqrt{a_i^2 + b_i^2}$ is the amplitude of the input. In Fig. 21.23, we show the ratio E_p/E_i as a function of the ratio of the input frequency to the frequency of the undamped system (ω_i/ω) for several values of d/ω. If the parameters of the spring–mass oscillator are known, a graph of this type can be used to determine the amplitude of the input by measuring the amplitude of the output.

In practice the input displacement does not usually have a single frequency, but consists of a combination of different frequencies or even a continuous *spectrum* of frequencies. For example, the displacements resulting from earthquakes have a spectrum of frequencies. In that case, it is desirable for the ratio of the output amplitude to the input amplitude to be approximately constant over the range of the input frequencies. The response of the instrument is then said to be "flat." In Fig. 21.23, the response is approximately flat for frequencies ω_i greater than about 2ω if the damping of the system is chosen so that d/ω is in the range 0.6–0.7. Also, notice that making the frequency ω small increases the range of input frequencies over which the response of the instrument is flat. For that reason, seismographs are often designed with large masses and relatively weak springs.

Figure 21.22
A seismograph that measures the local displacement of the earth.

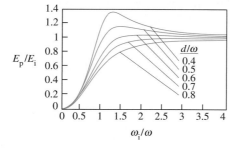

Figure 21.23
Ratio of the output amplitude to the input amplitude.

Chapter Summary

Conservative Systems

Small vibrations of many one-degree-of-freedom conservative systems relative to an equilibrium position are governed by the equation

$$\frac{d^2x}{dt^2} + \omega^2 x = 0,$$

Eq. (21.4)

where ω is a constant determined by the properties of the system. The general solution of this equation is

$$x = A \sin \omega t + B \cos \omega t$$

Eq. (21.6)

where A and B are constants. The general solution can also be expressed in the form

$$x = E \sin(\omega t - \phi),$$

Eq. (21.7)

where the constants E and ϕ are related to A and B by

$$A = E \cos \phi \quad \text{and} \quad B = -E \sin \phi.$$

Eq. (21.8)

The *amplitude* of the vibration is

$$E = \sqrt{A^2 + B^2}.$$

Eq. (21.9)

The *period* τ of the vibration is the time required for one complete oscillation, or *cycle*. The *frequency* f is the number of cycles per unit time. The period and frequency are related to ω by

$$\tau = \frac{2\pi}{\omega}$$

Eq. (21.10)

and

$$f = \frac{\omega}{2\pi}.$$

Eq. (21.11)

The term $\omega = 2\pi f$ is also a measure of the frequency and is expressed in rad/s.

Damped Vibrations

Small vibrations of many damped one-degree-of-freedom systems relative to an equilibrium position are governed by the equation

$$\frac{d^2x}{dt^2} + 2d \frac{dx}{dt} + \omega^2 x = 0.$$

Eq. (21.16)

Subcritical Damping If $d < \omega$, the system is said to be *subcritically damped*. In this case, the general solution of Eq. (21.16) is

$$x = e^{-dt}\left(A \sin \omega_d t + B \cos \omega_d t\right),$$

Eq. (21.19)

where A and B are constants and

$$\omega_d = \sqrt{\omega^2 - d^2}.$$

Eq. (21.18)

The period and frequency of the damped vibrations are

$$\tau_d = \frac{2\pi}{\omega_d} \quad \text{and} \quad f_d = \frac{\omega_d}{2\pi}.$$

Eq. (21.20)

Critical and Supercritical Damping If $d > \omega$, the system is said to be *supercritically damped*. The general solution is

$$x = Ce^{-(d-h)t} + De^{-(d+h)t},$$ Eq. (21.24)

where C and D are constants and

$$h = \sqrt{d^2 - \omega^2}.$$ Eq. (21.23)

If $d = \omega$, the system is said to be *critically damped*. The general solution is

$$x = Ce^{-dt} + Dte^{-dt},$$ Eq. (21.25)

where C and D are constants.

Forced Vibrations

The forced vibrations of many damped one-degree-of-freedom systems are governed by the equation

$$\frac{d^2x}{dt^2} + 2d\frac{dx}{dt} + \omega^2 x = a(t),$$ Eq. (21.26)

where $a(t)$ is the *forcing function*. The general solution of Eq. (21.26) is the sum of the homogeneous and particular solutions:

$$x = x_h + x_p.$$

The *homogeneous solution* x_h is the general solution of Eq. (21.26) with the right side set equal to zero, and the *particular solution* x_p is a solution that satisfies Eq. (21.26).

Oscillatory Forcing Function If $a(t)$ is an oscillatory function of the form

$$a(t) = a_0 \sin \omega_0 t + b_0 \cos \omega_0 t,$$

where a_0, b_0, and ω_0 are constants, the particular solution is

$$x_p = \left[\frac{(\omega^2 - \omega_0^2)a_0 + 2d\omega_0 b_0}{(\omega^2 - \omega_0^2)^2 + 4d^2\omega_0^2} \right] \sin \omega_0 t$$

$$+ \left[\frac{-2d\omega_0 a_0 + (\omega^2 - \omega_0^2)b_0}{(\omega^2 - \omega_0^2)^2 + 4d^2\omega_0^2} \right] \cos \omega_0 t,$$ Eq. (21.30)

and its amplitude is

$$E_p = \frac{\sqrt{a_0^2 + b_0^2}}{\sqrt{(\omega^2 - \omega_0^2)^2 + 4d^2\omega_0^2}}.$$ Eq. (21.31)

The particular solution for the motion of a damped vibrating system subjected to an oscillatory external force is also called the *steady-state solution*. The motion approaches the steady-state solution with increasing time.

Polynomial Forcing Function If $a(t)$ is a polynomial of the form

$$a(t) = a_0 + a_1 t + a_2 t^2 + \cdots + a_N t^N,$$

where $a_1, a_2, \ldots, a_N$ are constants, the particular solution can be obtained by seeking a solution of the same form—that is,

$$x_p = A_0 + A_1 t + A_2 t^2 + \cdots + A_N t^N,$$ Eq. (21.32)

where $A_0, A_1, A_2, \ldots, A_N$ are constants that must be determined.

Review Problems

21.1 The position of a vibrating system is

$$x = 2 \sin(\omega t + 0.4) \text{ m.}$$

(a) If you express the position of the system in the form of Eq. (21.6), what are the constants A and B?
(b) Draw a graph of x for values of ωt from zero to 4π radians.

21.2 The coordinate x measures the displacement of the mass relative to the position in which the spring is unstretched. The mass is given the initial conditions

$$t = 0 \begin{cases} x = 0.1 \text{ m,} \\ \dfrac{dx}{dt} = 0. \end{cases}$$

(a) Determine the position of the mass as a function of time.
(b) Draw graphs of the position and velocity of the mass as functions of time for the first 5 s of motion.

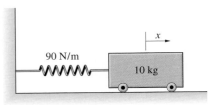

P21.2

21.3 When $t = 0$, the mass in Problem 21.2 is in the position in which the spring is unstretched and has a velocity of 0.3 m/s to the right. Determine the position of the mass as a function of time and the amplitude of the vibration

(a) by expressing the solution in the form given by Eq. (21.6) and
(b) by expressing the solution in the form given by Eq. (21.7)

21.4 A homogeneous disk of mass m and radius R rotates about a fixed shaft and is attached to a torsional spring with constant k. (The torsional spring exerts a restoring moment of magnitude $k\theta$, where θ is the angle of rotation of the disk relative to its position in which the spring is unstretched.) Show that the period of rotational vibrations of the disk is

$$\tau = \pi R \sqrt{2m/k}.$$

P21.4

21.5 Assigned to determine the moments of inertia of astronaut candidates, an engineer attaches a horizontal platform to a vertical steel bar. The moment of inertia of the platform about L is 7.5 kg-m², and the frequency of torsional oscillations of the unloaded platform is 1 Hz. With an astronaut candidate in the position shown, the frequency of torsional oscillations is 0.520 Hz. What is the candidate's moment of inertia about L?

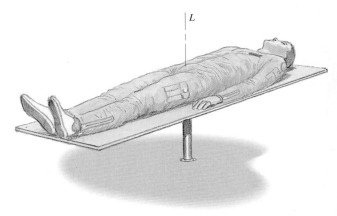

P21.5

21.6 The 22-kg platen P rests on four roller bearings that can be modeled as 1-kg homogeneous cylinders with 30-mm radii. The spring constant is $k = 900$ N/m. What is the frequency of horizontal vibrations of the platen relative to its equilibrium position?

P21.6

21.7 At $t = 0$, the platen described in Problem 21.6 is 0.1 m to the left of its equilibrium position and is moving to the right at 2 m/s. What are the platen's position and velocity at $t = 4$ s?

21.8 The moments of inertia of gears A and B are $I_A = 0.014$ slug-ft² and $I_B = 0.100$ slug-ft². Gear A is connected to a torsional spring with constant $k = 2$ ft-lb/rad. What is the frequency of angular vibrations of the gears relative to their equilibrium position?

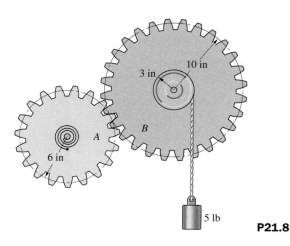

P21.8

21.12 The frequency of the spring–mass oscillator is measured and determined to be 4.00 Hz. The oscillator is then placed in a barrel of oil, and its frequency is determined to be 3.80 Hz. What is the logarithmic decrement of vibrations of the mass when the oscillator is immersed in oil?

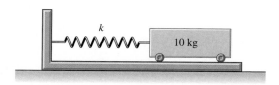

P21.12

21.9 The 5-lb weight in Problem 21.8 is raised 0.5 in. from its equilibrium position and released from rest at $t = 0$. Determine the counterclockwise angular position of gear B relative to its equilibrium position as a function of time.

21.10 The mass of the slender bar is m. The spring is unstretched when the bar is vertical. The light collar C slides on the smooth vertical bar so that the spring remains horizontal. Determine the frequency of small vibrations of the bar.

21.13 Consider the oscillator immersed in oil described in Problem 21.12. If the mass is displaced 0.1 m to the right of its equilibrium position and released from rest, what is its position relative to the equilibrium position as a function of time?

21.14 The stepped disk weighs 20 lb, and its moment of inertia is $I = 0.6$ slug-ft². It rolls on the horizontal surface. If $c = 8$ lb-s/ft, what is the frequency of vibration of the disk?

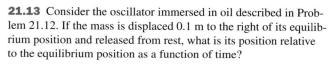

P21.14

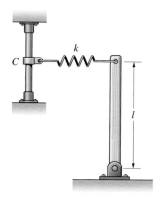

P21.10

21.11 A homogeneous hemisphere of radius R and mass m rests on a level surface. If you rotate the hemisphere slightly from its equilibrium position and release it, what is the frequency of its vibrations?

21.15 The stepped disk described in Problem 21.14 is initially in equilibrium, and at $t = 0$ it is given a clockwise angular velocity of 1 rad/s. Determine the position of the center of the disk relative to its equilibrium position as a function of time.

21.16 The stepped disk described in Problem 21.14 is initially in equilibrium, and at $t = 0$ it is given a clockwise angular velocity of 1 rad/s. Determine the position of the center of the disk relative to its equilibrium position as a function of time if $c = 16$ lb-s/ft.

P21.11

21.17 The 22-kg platen P rests on four roller bearings that can be modeled as 1-kg homogeneous cylinders with 30-mm radii. The spring constant is $k = 900$ N/m. The platen is subjected to a force $F(t) = 100 \sin 3t$ N. What is the magnitude of the platen's steady-state horizontal vibration?

P21.17

21.18 At $t = 0$, the platen described in Problem 21.17 is 0.1 m to the right of its equilibrium position and is moving to the right at 2 m/s. Determine the platen's position relative to its equilibrium position as a function of time.

21.19 The moments of inertia of gears A and B are $I_A = 0.014$ slug-ft^2 and $I_B = 0.100$ slug-ft^2. Gear A is connected to a torsional spring with constant $k = 2$ ft-lb/rad. The bearing supporting gear B incorporates a damping element that exerts a resisting moment on gear B of magnitude $1.5(d\theta_B/dt)$ ft-lb, where $d\theta_B/dt$ is the angular velocity of gear B in rad/s. What is the frequency of angular vibration of the gears?

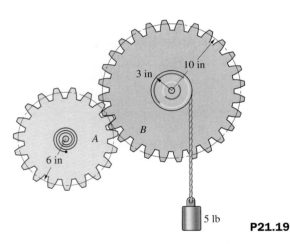

P21.19

21.20 The 5-lb weight in Problem 21.19 is raised 0.5 in. from its equilibrium position and released from rest at $t = 0$. Determine the counterclockwise angular position of gear B relative to its equilibrium position as a function of time.

21.21 The base and mass m are initially stationary. The base is then subjected to a vertical displacement $h \sin \omega_i t$ relative to its original position. What is the magnitude of the resulting steady-state vibration of the mass m *relative to the base*?

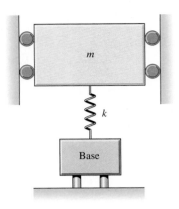

P21.21

21.22 The mass of the trailer, not including its wheels and axle, is m, and the spring constant of its suspension is k. To analyze the suspension's behavior, an engineer assumes that the height of the road surface relative to its mean height is $h \sin(2\pi x/\lambda)$. Assume that the trailer's wheels remain on the road and its horizontal component of velocity is v. Neglect the damping due to the suspension's shock absorbers.

(a) Determine the magnitude of the trailer's vertical steady-state vibration *relative to the road surface*.
(b) At what velocity v does resonance occur?

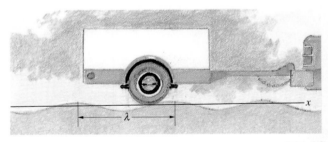

P21.22

21.23 The trailer in Problem 21.22, not including its wheels and axle, weighs 1000 lb. The spring constant of its suspension is $k = 2400$ lb/ft, and the damping coefficient due to its shock absorbers is $c = 200$ lb-s/ft. The road surface parameters are $h = 2$ in. and $\lambda = 8$ ft. The trailer's horizontal velocity is $v = 6$ mi/hr. Determine the magnitude of the trailer's vertical steady-state vibration relative to the road surface, (a) neglecting the damping due to the shock absorbers and (b) not neglecting the damping.

21.24 A disk with moment of inertia I rotates about a fixed shaft and is attached to a torsional spring with constant k. The angle θ measures the angular position of the disk relative to its position when the spring is unstretched. The disk is initially stationary with the spring unstretched. At $t = 0$, a time-dependent moment $M(t) = M_0(1 - e^{-t})$ is applied to the disk, where M_0 is a constant. Show that the angular position of the disk as a function of time is

$$\theta = \frac{M_0}{I} \left[-\frac{1}{\omega(1 + \omega^2)} \sin \omega t - \frac{1}{\omega^2(1 + \omega^2)} \cos \omega t \right.$$
$$\left. + \frac{1}{\omega^2} - \frac{1}{(1 + \omega^2)} e^{-t} \right].$$

Strategy: To determine the particular solution, seek a solution of the form

$$\theta_p = A_p + B_p e^{-t},$$

where A_p and B_p are constants that you must determine.

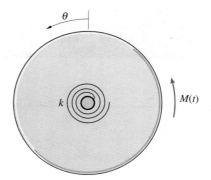

P21.24

Review of Mathematics

A.1 Algebra

Quadratic Equations

The solutions of the quadratic equation

$$ax^2 + bx + c = 0$$

are

$$x = \frac{-b \pm \sqrt{b^2 - 4ac}}{2a}.$$

Natural Logarithms

The natural logarithm of a positive real number x is denoted by $\ln x$. It is defined to be the number such that

$$e^{\ln x} = x,$$

where $e = 2.7182\ldots$ is the base of natural logarithms.
Logarithms have the following properties:

$$\ln(xy) = \ln x + \ln y,$$

$$\ln(x/y) = \ln x - \ln y,$$

$$\ln y^x = x \ln y.$$

A.2 Trigonometry

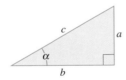

The trigonometric functions for a right triangle are

$$\sin\alpha = \frac{1}{\csc\alpha} = \frac{a}{c}, \qquad \cos\alpha = \frac{1}{\sec\alpha} = \frac{b}{c}, \qquad \tan\alpha = \frac{1}{\cot\alpha} = \frac{a}{b}.$$

The sine and cosine satisfy the relation

$$\sin^2\alpha + \cos^2\alpha = 1,$$

and the sine and cosine of the sum and difference of two angles satisfy

$$\sin(\alpha + \beta) = \sin\alpha\cos\beta + \cos\alpha\sin\beta,$$

$$\sin(\alpha - \beta) = \sin\alpha\cos\beta - \cos\alpha\sin\beta,$$

$$\cos(\alpha + \beta) = \cos\alpha\cos\beta - \sin\alpha\sin\beta,$$

$$\cos(\alpha - \beta) = \cos\alpha\cos\beta + \sin\alpha\sin\beta.$$

The **law of cosines** for an arbitrary triangle is

$$c^2 = a^2 + b^2 - 2ab\cos\alpha_c,$$

and the **law of sines** is

$$\frac{\sin\alpha_a}{a} = \frac{\sin\alpha_b}{b} = \frac{\sin\alpha_c}{c}.$$

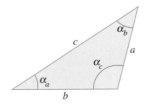

A.3 Derivatives

$$\frac{d}{dx}x^n = nx^{n-1}$$

$$\frac{d}{dx}\sin x = \cos x$$

$$\frac{d}{dx}\sinh x = \cosh x$$

$$\frac{d}{dx}e^x = e^x$$

$$\frac{d}{dx}\cos x = -\sin x$$

$$\frac{d}{dx}\cosh x = \sinh x$$

$$\frac{d}{dx}\ln x = \frac{1}{x}$$

$$\frac{d}{dx}\tan x = \frac{1}{\cos^2 x}$$

$$\frac{d}{dx}\tanh x = \frac{1}{\cosh^2 x}$$

A.4 Integrals

$$\int x^n \, dx = \frac{x^{n+1}}{n+1} \quad (n \neq -1)$$

$$\int \frac{dx}{(1 - a^2x^2)^{1/2}} = \frac{1}{a}\arcsin ax, \quad \text{or} \quad -\frac{1}{a}\arccos ax$$

$$\int x^{-1} \, dx = \ln x$$

$$\int \sin x \, dx = -\cos x$$

$$\int (a + bx)^{1/2} \, dx = \frac{2}{3b}(a + bx)^{3/2}$$

$$\int \cos x \, dx = \sin x$$

$$\int x(a + bx)^{1/2} \, dx = -\frac{2(2a - 3bx)(a + bx)^{3/2}}{15b^2}$$

$$\int \sin^2 x \, dx = -\frac{1}{2}\sin x \cos x + \frac{1}{2}x$$

$$\int (1 + a^2x^2)^{1/2} \, dx = \frac{1}{2}\left\{x(1 + a^2x^2)^{1/2} + \frac{1}{a}\ln\left[x + \left(\frac{1}{a^2} + x^2\right)^{1/2}\right]\right\}$$

$$\int \cos^2 x \, dx = \frac{1}{2}\sin x \cos x + \frac{1}{2}x$$

$$\int x(1 + a^2x^2)^{1/2} \, dx = \frac{a}{3}\left(\frac{1}{a^2} + x^2\right)^{3/2}$$

$$\int \sin^3 x \, dx = -\frac{1}{3}\cos x(\sin^2 x + 2)$$

$$\int x^2(1 + a^2x^2)^{1/2} \, dx = \frac{1}{4}ax\left(\frac{1}{a^2} + x^2\right)^{3/2} - \frac{1}{8a^2}x(1 + a^2x^2)^{1/2}$$

$$\int \cos^3 x \, dx = \frac{1}{3}\sin x(\cos^2 x + 2)$$

$$- \frac{1}{8a^3}\ln\left[x + \left(\frac{1}{a^2} + x^2\right)^{1/2}\right]$$

$$\int \cos^4 x \, dx = \frac{3}{8}x + \frac{1}{4}\sin 2x + \frac{1}{32}\sin 4x$$

$$\int \sin^n x \cos x \, dx = \frac{(\sin x)^{n+1}}{n+1} \quad (n \neq -1)$$

$$\int (1 - a^2x^2)^{1/2} \, dx = \frac{1}{2}\left[x(1 - a^2x^2)^{1/2} + \frac{1}{a}\arcsin ax\right]$$

$$\int \sinh x \, dx = \cosh x$$

$$\int x(1 - a^2x^2)^{1/2} \, dx = -\frac{a}{3}\left(\frac{1}{a^2} - x^2\right)^{3/2}$$

$$\int \cosh x \, dx = \sinh x$$

$$\int x^2(a^2 - x^2)^{1/2} \, dx = -\frac{1}{4}x(a^2 - x^2)^{3/2}$$

$$\int \tanh x \, dx = \ln \cosh x$$

$$+ \frac{1}{8}a^2\left[x(a^2 - x^2)^{1/2} + a^2 \arcsin\frac{x}{a}\right]$$

$$\int e^{ax} \, dx = \frac{e^{ax}}{a}$$

$$\int \frac{dx}{(1 + a^2x^2)^{1/2}} = \frac{1}{a}\ln\left[x + \left(\frac{1}{a^2} + x^2\right)^{1/2}\right]$$

$$\int xe^{ax} \, dx = \frac{e^{ax}}{a^2}(ax - 1)$$

A.5 Taylor Series

The Taylor series of a function $f(x)$ is

$$f(a + x) = f(a) + f'(a)x + \frac{1}{2!}f''(a)x^2 + \frac{1}{3!}f'''(a)x^3 + \cdots,$$

where the primes indicate derivatives.

Some useful Taylor series are

$$e^x = 1 + x + \frac{x^2}{2!} + \frac{x^3}{3!} + \cdots,$$

$$\sin(a + x) = \sin a + (\cos a)x - \frac{1}{2}(\sin a)x^2 - \frac{1}{6}(\cos a)x^3 + \cdots,$$

$$\cos(a + x) = \cos a - (\sin a)x - \frac{1}{2}(\cos a)x^2 + \frac{1}{6}(\sin a)x^3 + \cdots,$$

$$\tan(a + x) = \tan a + \left(\frac{1}{\cos^2 a}\right)x + \left(\frac{\sin a}{\cos^3 a}\right)x^2$$

$$+ \left(\frac{\sin^2 a}{\cos^4 a} + \frac{1}{3\cos^2 a}\right)x^3 + \cdots.$$

A.6 Vector Analysis

Cartesian Coordinates

The gradient of a scalar field ψ is

$$\nabla\psi = \frac{\partial\psi}{\partial x}\mathbf{i} + \frac{\partial\psi}{\partial y}\mathbf{j} + \frac{\partial\psi}{\partial z}\mathbf{k}.$$

The divergence and curl of a vector field $\mathbf{v} = v_x\mathbf{i} + v_y\mathbf{j} + v_z\mathbf{k}$ are

$$\nabla \cdot \mathbf{v} = \frac{\partial v_x}{\partial x} + \frac{\partial v_y}{\partial y} + \frac{\partial v_z}{\partial z},$$

$$\nabla \times \mathbf{v} = \begin{vmatrix} \mathbf{i} & \mathbf{j} & \mathbf{k} \\ \dfrac{\partial}{\partial x} & \dfrac{\partial}{\partial y} & \dfrac{\partial}{\partial z} \\ v_x & v_y & v_z \end{vmatrix}.$$

Cylindrical Coordinates

The gradient of a scalar field ψ is

$$\nabla\psi = \frac{\partial\psi}{\partial r}\mathbf{e}_r + \frac{1}{r}\frac{\partial\psi}{\partial\theta}\mathbf{e}_\theta + \frac{\partial\psi}{\partial z}\mathbf{e}_z.$$

The divergence and curl of a vector field $\mathbf{v} = v_r\mathbf{e}_r + v_\theta\mathbf{e}_\theta + v_z\mathbf{e}_z$ are

$$\nabla \cdot \mathbf{v} = \frac{\partial v_r}{\partial r} + \frac{v_r}{r} + \frac{1}{r}\frac{\partial v_\theta}{\partial\theta} + \frac{\partial v_z}{\partial z},$$

$$\nabla \times \mathbf{v} = \frac{1}{r}\begin{vmatrix} \mathbf{e}_r & r\mathbf{e}_\theta & \mathbf{e}_z \\ \dfrac{\partial}{\partial r} & \dfrac{\partial}{\partial\theta} & \dfrac{\partial}{\partial z} \\ v_r & rv_\theta & v_z \end{vmatrix}.$$

APPENDIX B   Properties of Areas and Lines

B.1 Areas

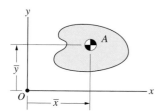

The coordinates of the centroid of the area A are

$$\bar{x} = \frac{\displaystyle\int_A x\,dA}{\displaystyle\int_A dA}, \qquad \bar{y} = \frac{\displaystyle\int_A y\,dA}{\displaystyle\int_A dA}.$$

The moment of inertia about the x axis I_x, the moment of inertia about the y axis I_y, and the product of inertia I_{xy} are

$$I_x = \int_A y^2\,dA, \qquad I_y = \int_A x^2\,dA, \qquad I_{xy} = \int_A xy\,dA.$$

The polar moment of inertia about O is

$$J_O = \int_A r^2\,dA = \int_A (x^2 + y^2)\,dA = I_x + I_y.$$

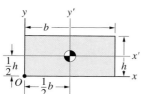

Rectangular area

Area $= bh$

$$I_x = \frac{1}{3}bh^3, \qquad I_y = \frac{1}{3}hb^3, \qquad I_{xy} = \frac{1}{4}b^2h^2$$

$$I_{x'} = \frac{1}{12}bh^3, \qquad I_{y'} = \frac{1}{12}hb^3, \qquad I_{x'y'} = 0$$

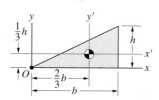

Triangular area

Area $= \dfrac{1}{2}bh$

$$I_x = \frac{1}{12}bh^3, \qquad I_y = \frac{1}{4}hb^3, \qquad I_{xy} = \frac{1}{8}b^2h^2$$

$$I_{x'} = \frac{1}{36}bh^3, \qquad I_{y'} = \frac{1}{36}hb^3, \qquad I_{x'y'} = \frac{1}{72}b^2h^2$$

Triangular area

Area $= \dfrac{1}{2}bh$ $I_x = \dfrac{1}{12}bh^3, \qquad I_{x'} = \dfrac{1}{36}bh^3$

Circular area

Area $= \pi R^2$ $I_{x'} = I_{y'} = \dfrac{1}{4}\pi R^4, \qquad I_{x'y'} = 0$

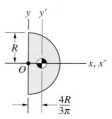

Semicircular area

Area $= \dfrac{1}{2}\pi R^2$ $I_x = I_y = \dfrac{1}{8}\pi R^4, \quad I_{xy} = 0$

$$I_{x'} = \frac{1}{8}\pi R^4, \qquad I_{y'} = \left(\frac{\pi}{8} - \frac{8}{9\pi}\right)R^4, \qquad I_{x'y'} = 0$$

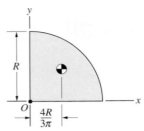

Quarter-circular area

$$\text{Area} = \frac{1}{4}\pi R^2 \qquad I_x = I_y = \frac{1}{16}\pi R^4, \qquad I_{xy} = \frac{1}{8}R^4$$

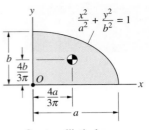

Quarter-elliptical area

$$\text{Area} = \frac{1}{4}\pi ab$$

$$I_x = \frac{1}{16}\pi ab^3, \qquad I_y = \frac{1}{16}\pi a^3 b, \qquad I_{xy} = \frac{1}{8}a^2 b^2$$

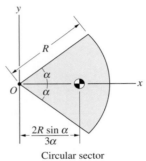

Circular sector

$$\text{Area} = \alpha R^2$$

$$I_x = \frac{1}{4}R^4\left(\alpha - \frac{1}{2}\sin 2\alpha\right), \qquad I_y = \frac{1}{4}R^4\left(\alpha + \frac{1}{2}\sin 2\alpha\right),$$

$$I_{xy} = 0$$

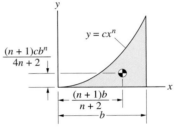

Spandrel

$$\text{Area} = \frac{cb^{n+1}}{n+1}$$

$$I_x = \frac{c^3 b^{3n+1}}{9n+3}, \qquad I_y = \frac{cb^{n+3}}{n+3}, \qquad I_{xy} = \frac{c^2 b^{2n+2}}{4n+4}$$

B.2 Lines

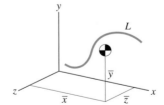

The coordinates of the centroid of the line L are

$$\bar{x} = \frac{\int_L x\,dL}{\int_L dL}, \qquad \bar{y} = \frac{\int_L y\,dL}{\int_L dL}, \qquad \bar{z} = \frac{\int_L z\,dL}{\int_L dL}.$$

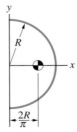

Semicircular arc

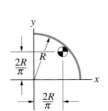

Quarter-circular arc

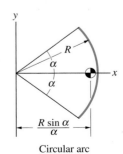

Circular arc

APPENDIX C

Properties of Volumes and Homogeneous Objects

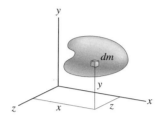

The moments and products of inertia of the object in terms of the xyz coordinate system are

$$I_{(x\text{ axis})} = I_{xx} = \int_m (y^2 + z^2)\, dm,$$

$$I_{(y\text{ axis})} = I_{yy} = \int_m (x^2 + z^2)\, dm,$$

$$I_{(z\text{ axis})} = I_{zz} = \int_m (x^2 + y^2)\, dm,$$

$$I_{xy} = \int_m xy\, dm, \qquad I_{yz} = \int_m yz\, dm,$$

$$I_{zx} = \int_m zx\, dm.$$

Slender bar

$$I_{(x\text{ axis})} = 0, \qquad I_{(y\text{ axis})} = I_{(z\text{ axis})} = \frac{1}{3} ml^2,$$

$$I_{xy} = I_{yz} = I_{zx} = 0.$$

$$I_{(x'\text{ axis})} = 0, \qquad I_{(y'\text{ axis})} = I_{(z'\text{ axis})} = \frac{1}{12} ml^2,$$

$$I_{x'y'} = I_{y'z'} = I_{z'x'} = 0.$$

Thin circular plate

$$I_{(x'\text{ axis})} = I_{(y'\text{ axis})} = \frac{1}{4} mR^2, \qquad I_{(z'\text{ axis})} = \frac{1}{2} mR^2,$$

$$I_{x'y'} = I_{y'z'} = I_{z'x'} = 0.$$

Thin rectangular plate

$$I_{(x\text{ axis})} = \frac{1}{3} mh^2, \qquad I_{(y\text{ axis})} = \frac{1}{3} mb^2, \qquad I_{(z\text{ axis})} = \frac{1}{3} m(b^2 + h^2),$$

$$I_{xy} = \frac{1}{4} mbh, \qquad I_{yz} = I_{zx} = 0.$$

$$I_{(x'\text{ axis})} = \frac{1}{12} mh^2, \qquad I_{(y'\text{ axis})} = \frac{1}{12} mb^2, \qquad I_{(z'\text{ axis})} = \frac{1}{12} m(b^2 + h^2),$$

$$I_{x'y'} = I_{y'z'} = I_{z'x'} = 0.$$

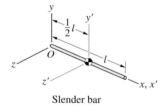

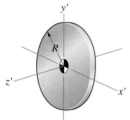

399

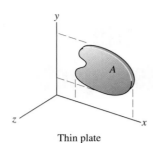

Thin plate

$$I_{(x \text{ axis})} = \frac{m}{A} I_x^A, \qquad I_{(y \text{ axis})} = \frac{m}{A} I_y^A, \qquad I_{(z \text{ axis})} = I_{(x \text{ axis})} + I_{(y \text{ axis})},$$

$$I_{xy} = \frac{m}{A} I_{xy}^A, \qquad I_{yz} = I_{zx} = 0.$$

(The superscripts A denote moments of inertia of the plate's cross-sectional area A).

Rectangular prism

Volume $= abc$

$$I_{(x' \text{ axis})} = \frac{1}{12} m(a^2 + b^2), \qquad I_{(y' \text{ axis})} = \frac{1}{12} m(a^2 + c^2),$$

$$I_{(z' \text{ axis})} = \frac{1}{12} m(b^2 + c^2), \qquad I_{x'y'} = I_{y'z'} = I_{z'x'} = 0.$$

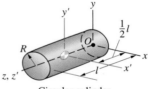

Circular cylinder

Volume $= \pi R^2 l.$

$$I_{(x \text{ axis})} = I_{(y \text{ axis})} = m\left(\frac{1}{3} l^2 + \frac{1}{4} R^2\right), \qquad I_{(z \text{ axis})} = \frac{1}{2} mR^2,$$

$$I_{xy} = I_{yz} = I_{zx} = 0.$$

$$I_{(x' \text{ axis})} = I_{(y' \text{ axis})} = m\left(\frac{1}{12} l^2 + \frac{1}{4} R^2\right), \qquad I_{(z' \text{ axis})} = \frac{1}{2} mR^2,$$

$$I_{x'y'} = I_{y'z'} = I_{z'x'} = 0.$$

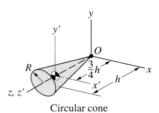

Circular cone

Volume $= \frac{1}{3} \pi R^2 h.$

$$I_{(x \text{ axis})} = I_{(y \text{ axis})} = m\left(\frac{3}{5} h^2 + \frac{3}{20} R^2\right), \qquad I_{(z \text{ axis})} = \frac{3}{10} mR^2,$$

$$I_{xy} = I_{yz} = I_{zx} = 0.$$

$$I_{(x' \text{ axis})} = I_{(y' \text{ axis})} = m\left(\frac{3}{80} h^2 + \frac{3}{20} R^2\right), \qquad I_{(z' \text{ axis})} = \frac{3}{10} mR^2,$$

$$I_{x'y'} = I_{y'z'} = I_{z'x'} = 0.$$

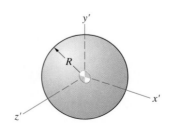

Sphere

Volume $= \frac{4}{3} \pi R^3.$

$$I_{(x' \text{ axis})} = I_{(y' \text{ axis})} = I_{(z' \text{ axis})} = \frac{2}{5} mR^2,$$

$$I_{x'y'} = I_{y'z'} = I_{z'x'} = 0.$$

This appendix summarizes the equations of kinematics and vector calculus in spherical coordinates.

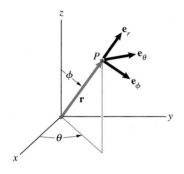

The position vector, velocity, and acceleration are

$$\mathbf{r} = r\mathbf{e}_r,$$

$$\mathbf{v} = \frac{dr}{dt}\mathbf{e}_r + r\frac{d\phi}{dt}\mathbf{e}_\phi + r\frac{d\theta}{dt}\sin\phi\,\mathbf{e}_\theta,$$

$$\mathbf{a} = \left[\frac{d^2r}{dt^2} - r\left(\frac{d\phi}{dt}\right)^2 - r\left(\frac{d\theta}{dt}\right)^2\sin^2\phi\right]\mathbf{e}_r$$

$$+ \left[r\frac{d^2\phi}{dt^2} + 2\frac{dr}{dt}\frac{d\phi}{dt} - r\left(\frac{d\theta}{dt}\right)^2\sin\phi\cos\phi\right]\mathbf{e}_\phi$$

$$+ \left[r\frac{d^2\theta}{dt^2}\sin\phi + 2\frac{dr}{dt}\frac{d\theta}{dt}\sin\phi + 2r\frac{d\phi}{dt}\frac{d\theta}{dt}\cos\phi\right]\mathbf{e}_\theta.$$

The gradient of a scalar field ψ is

$$\nabla\psi = \frac{\partial\psi}{\partial r}\mathbf{e}_r + \frac{1}{r}\frac{\partial\psi}{\partial\phi}\mathbf{e}_\phi + \frac{1}{r\sin\phi}\frac{\partial\psi}{\partial\theta}\mathbf{e}_\theta.$$

The divergence and curl of a vector field $\mathbf{v} = v_r\mathbf{e}_r + v_\theta\mathbf{e}_\theta + v_\phi\mathbf{e}_\phi$ are

$$\nabla\cdot\mathbf{v} = \frac{1}{r^2}\frac{\partial}{\partial r}\left(r^2 v_r\right) + \frac{1}{r\sin\phi}\frac{\partial}{\partial\phi}\left(v_\phi\sin\phi\right) + \frac{1}{r\sin\phi}\frac{\partial v_\theta}{\partial\theta},$$

$$\nabla\times\mathbf{v} = \frac{1}{r^2\sin\phi}\begin{vmatrix} \mathbf{e}_r & r\mathbf{e}_\phi & r\sin\phi\,\mathbf{e}_\theta \\ \dfrac{\partial}{\partial r} & \dfrac{\partial}{\partial\phi} & \dfrac{\partial}{\partial\theta} \\ v_r & rv_\phi & r\sin\phi\,v_\theta \end{vmatrix}.$$

D'Alembert's Principle

In this appendix we describe an alternative approach for obtaining the equations of planar motion for a rigid body. By writing Newton's second law as

$$\Sigma \mathbf{F} + (-m\mathbf{a}) = \mathbf{0}, \tag{E.1}$$

we can regard it as an "equilibrium" equation stating that the sum of the forces, including an *inertial force* $-m\mathbf{a}$, equals zero (Fig. E.1). To state the equation of angular motion in an equivalent way, we use Eq. (18.19), which relates the total moment about a fixed point O to the acceleration of the center of mass and the angular acceleration in general planar motion:

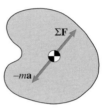

Figure E.1
The sum of the external forces and the inertial force is zero.

$$\Sigma M_O = (\mathbf{r} \times m\mathbf{a}) \cdot \mathbf{k} + I\alpha.$$

We write this equation as

$$\Sigma M_O + \left[\mathbf{r} \times (-m\mathbf{a}) \right] \cdot \mathbf{k} + (-I\alpha) = 0. \tag{E.2}$$

The term $\left[\mathbf{r} \times (-m\mathbf{a}) \right] \cdot \mathbf{k}$ is the moment about O due to the inertial force $-m\mathbf{a}$. We can therefore regard this equation as an "equilibrium" equation stating that the sum of the moments about any fixed point, including the moment due to the inertial force $-m\mathbf{a}$ acting at the center of mass and an *inertial couple* $-I\alpha$, equals zero.

Stated in this way, the equations of motion for a rigid body are analogous to the equations for static equilibrium: The sum of the forces equals zero and the sum of the moments about any fixed point equals zero when we properly account for inertial forces and couples. This is called *D'Alembert's principle*.

If we define ΣM_O and α to be positive in the counterclockwise direction, the unit vector $\mathbf{k}$ in Eq. (E.2) points out of the page and the term $\left[\mathbf{r} \times (-m\mathbf{a}) \right] \cdot \mathbf{k}$ is the counterclockwise moment due to the inertial force. This vector operation determines the moment, or we can evaluate it by using the fact that its magnitude is the product of the magnitude of the inertial force and the perpendicular distance from point O to the line of action of the force (Fig. E.2a). The moment is positive if it is counterclockwise, as in Fig. E.2a, and negative if it is clockwise. Notice that the sense of the inertial couple is opposite to that of the angular acceleration (Fig. E.2b).

As an example, consider a disk of mass m and moment of inertia I that is rolling on an inclined surface (Fig. E.3). We can use D'Alembert's principle to determine the disk's angular acceleration and the forces exerted on it by the surface. The angular

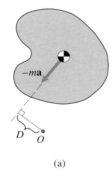

(a)

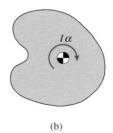

(b)

Figure E.2
(a) The magnitude of the moment due to the inertial force is $|-m\mathbf{a}|D$.
(b) A clockwise inertial couple results from a counterclockwise angular acceleration.

Figure E.3
Disk rolling on an inclined surface.

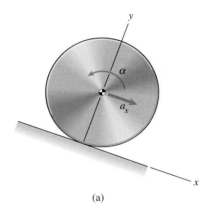

(a)

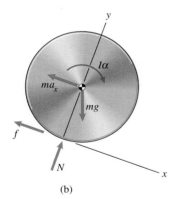

(b)

Figure E.4
(a) Acceleration of the center of the disk and its angular acceleration.
(b) Free-body diagram including the inertial force and couple.

acceleration of the disk and the acceleration of its center are shown in Fig. E.4a. In Figure E.4b, we draw the free-body diagram of the disk showing its weight, the normal and friction forces exerted by the surface, and the inertial force and couple. Equation (E.1) is

$$\Sigma \mathbf{F} + (-m\mathbf{a}) = 0:$$

$$(mg\sin\beta - f)\mathbf{i} + (N - mg\cos\beta)\mathbf{j} - ma_x\mathbf{i} = 0.$$

From this vector equation we obtain the equations

$$mg\sin\beta - f - ma_x = 0,$$
$$N - mg\cos\beta = 0. \tag{E.3}$$

We now apply Eq. (E.2), evaluating moments about the point where the disk is in contact with the surface to eliminate f and N from the resulting equation:

$$\Sigma M_0 + \left[\mathbf{r} \times (-m\mathbf{a})\right] \cdot \mathbf{k} + (-I\alpha) = 0:$$

$$-R(mg\sin\beta) + R(ma_x) - I\alpha = 0. \tag{E.4}$$

The acceleration of the center of the rolling disk is related to the counterclockwise angular acceleration by $a_x = -R\alpha$. Substituting this relation into Eq. (E.4) and solving for the angular acceleration, we obtain

$$\alpha = -\frac{mgR\sin\beta}{mR^2 + I}.$$

From this result we also know a_x and can solve Eqs. (E.3) for the normal and friction forces, obtaining

$$N = mg\cos\beta, \qquad f = \frac{mgI\sin\beta}{mR^2 + I}.$$

In Eq. (E.4) we evaluated the moment due to the inertial force by simply multiplying the magnitude of the force and the perpendicular distance from O to its line of action, but we could have used the vector expression:

$$\left[\mathbf{r} \times (-m\mathbf{a})\right] \cdot \mathbf{k} = \left[(R\mathbf{j}) \times (-ma_x\mathbf{i})\right] \cdot \mathbf{k} = R(ma_x).$$

This example demonstrates an advantage of D'Alembert's principle. Because the moments in Eq. (E.2) can be evaluated about any fixed point, we chose a point that eliminated f and N from the resulting equation and were able to evaluate α directly.

Answers to Even-Numbered Problems

Chapter 13

13.2 (a) 7.10 m. (b) 2.22 s. (c) 11.8 m/s.
13.4 $v = 42.3$ m/s.
13.6 13.1 s.
13.8 68.6 ft/s.
13.10 $|\mathbf{v}| = 2.19$ m/s, $|\mathbf{a}| = 5.58$ m/s^2.
13.12 $\mathbf{a} = -2.75\mathbf{e}_r - 4.86\mathbf{e}_\theta$ (m/s^2).
13.14 (a) $\mathbf{v} = -2.13\mathbf{e}_r + 6.64\mathbf{e}_\theta$ (m/s).
    (b) $\mathbf{v} = -5.90\mathbf{i} + 3.71\mathbf{j}$ (m/s).

Chapter 14

14.2 (a) 34.6 ft/s^2, 3490 lb. (b) $\mu_s = 1.07$.
14.4 (a) $F_1 = 63$ kN, $F_2 = 126$ kN, $F_3 = 189$ kN.
    (b) $F_1 = 75$ kN, $F_2 = 150$ kN, $F_3 = 225$ kN.
14.6 Deceleration is 1.49 m/s^2, compared with 8.83 m/s^2
    on a level road.
14.8 14,300 ft (2.71 mi).
14.10 10.5 lb.
14.12 $a_A = 4.02$ ft/s^2, $T = 17.5$ lb.
14.14 29.7 ft.
14.16 9.30 N.
14.18 $\tan\alpha = v^2/\rho g$.
14.20 $\Sigma\mathbf{F} = -10.7\mathbf{e}_r + 2.55\mathbf{e}_\theta$ (N).

Chapter 15

15.2 (a), (b) 193 ft.
15.4 He should choose (b). Impact velocity is 11.8 m/s,
    work is −251 kN-m. In (a), impact velocity is
    13.9 m/s, work is −119 kN-m.
15.6 $v = 2.08$ m/s.
15.8 (a) 14.3×10^7 ft-lb. (b) $v = 88\left[1 - e^{-(F_0/88m)t}\right]$.
15.10 $k = 163$ lb/ft.
15.12 (a) $k = 809$ N/m. (b) $v_2 = 6.29$ m/s.
15.14 4.39 ft/s.
15.16 $h = 0.179$ m.
15.18 (a) $\theta = 27.9°$. (b) 159 lb. (c) 222 lb.
15.20 $v_1 = 4.73$ ft/s.
15.22 1.02 m.
15.24 2.00 m/s.
15.26 24.8 ft/s.
15.28 (a) $v = 0$.
    (b) $v = \sqrt{gR_E/2} = 18,300$ ft/s, or 12,500 mi/hr.
15.30 (a) 11.3 MW (megawatts). (b) 9.45 MW.

Chapter 16

16.2 (a) $180\mathbf{i} + 360\mathbf{j}$ (lb-s). (b) $12\mathbf{i} + 44\mathbf{j}$ (ft/s).
16.4 877 kN.
16.6 (a) 34.9 ft/s. (b) 15,700 ft-lb.

16.8 (a) 4.80 ft/s. (b) 415 ft.
16.10 (a) 3.45 m/s. (b) 18.7 kN.
16.12 (a) 8.23 kN. (b) 16.5 kN.
16.14 $v_A\sqrt{mk/2}$.
16.16 1.57 ft.
16.18 $e = 0.304$.
16.20 2.30 m.
16.22 $x = -11.13$ in., $y = 6.42$ in.
16.24 $|v_r| = 11,550$ ft/s, $|v_\theta| = 9430$ ft/s.
16.26 867 lb (including the weight of the drum).
16.28 (a) 30.1 ft/s. (b) 46.8 ft/s.

Chapter 17

17.2 OQ: 9.29 rad/s counterclockwise;
    PQ: 2.92 rad/s clockwise.
17.4 $\mathbf{v}_C = -32.0\mathbf{j}$ (ft/s).
17.6 $\alpha_{AB} = 13.60 \times 10^3$ rad/s^2 clockwise,
    $\alpha_{BC} = 8.64 \times 10^3$ rad/s^2 counterclockwise.
17.8 $\mathbf{a}_D = -3490\mathbf{i}$ (in./s^2).
17.10 $\mathbf{a}_G = -8.18\mathbf{i} - 26.4\mathbf{j}$ (in./s^2).
17.12 $\mathbf{v}_C = -1.48\mathbf{i} + 0.79\mathbf{j}$ (m/s).
17.14 $\mathbf{a}_C = -2.99\mathbf{i} - 1.40\mathbf{j}$ (m/s^2).
17.16 Velocity is 55.9 in./s to the right;
    acceleration is 910 in./s^2 to the left.
17.18 $\omega_{BD} = 0.733$ rad/s counterclockwise.
17.20 $\omega_{AB} = 0.261$ rad/s counterclockwise,
    $\omega_{BC} = 2.80$ rad/s counterclockwise.
17.22 $\omega_{BC} = 1.22$ rad/s clockwise, 18.0 m/s from B
    toward C.
17.24 Velocity = 6.89 ft/s upward;
    acceleration = 169 ft/s^2 upward.
17.26 5.66 N.
17.28 (a) $\mathbf{v}_{A\,\text{rel}} = 5\mathbf{i}$ m/s, $\mathbf{a}_{A\,\text{rel}} = \mathbf{0}$.
    (b) $\mathbf{v}_{A/B} = 5\mathbf{i} + 2\mathbf{j}$ (m/s), $\mathbf{a}_{A/B} = -4\mathbf{i} + 20\mathbf{j}$ (m/s^2).

Chapter 18

18.2 (a) 12.1 s. (b) 144 lb.
18.4 (a) 20 m/s^2. (b) $c \leq 49.1$ mm.
18.6 $I = 2.05$ kg-m^2.
18.8 40.2 kN.
18.10 $\alpha = -0.420$ rad/s^2, $F_x = 336$ N, $F_y = 1710$ N.
18.12 $\alpha = (g/l)\left[3\left(1 - \mu^2\right)\sin\theta - 6\mu\cos\theta\right]/\left(2 - \mu^2\right)$
    counterclockwise.
18.14 $B_x = -1959$ N, $B_y = 1238$ N, $C_x = 2081$ N,
    $C_y = -922$ N.
18.16 $\alpha_{OA} = 0.425$ rad/s^2 counterclockwise,
    $\alpha_{AB} = 1.586$ rad/s^2 clockwise.

18.18 $\alpha_{HP} = 5.37$ rad/s^2 clockwise.

18.20 208 m/s^2 to the left.

Chapter 19

19.2 $v = 2.05$ m/s.

19.4 $v = \sqrt{Fb/[\frac{1}{2}m_c + 2(m + I/R^2)]}$.

19.6 $\omega_S = 12.5$ rad/s.

19.8 (a) 1.13 ft. (b) 1.09 ft/s.

19.10 $P_{max} = 580$ ft-lb/s (1.05 hp),
$P_{av} = 290$ ft-lb/s (0.53 hp).

19.12 11.1 rad/s.

19.14 1.77 rad/s counterclockwise.

19.16 $N = 1.433 \, mg$.

19.18 $\omega' = 0.721$ rad/s.

19.20 $\omega' = (\frac{1}{3} + \frac{2}{3}\cos\beta)\omega$.

19.22 $b = 2$ ft.

19.24 $\omega = 8.51$ rad/s.

19.26 $\theta = 54.7°$, $\omega = 10.7$ rad/s counterclockwise.

19.28 $0.0822\mathbf{k}$ (rad/s).

19.30 0.641 rad/s clockwise, $\mathbf{v}' = -2.24\mathbf{i}$ (ft/s).

Chapter 20

20.2 (a) $\boldsymbol{\omega} = \omega_d\mathbf{i} + \omega_0\mathbf{j}$.
(b) $\mathbf{v}_A = -R\omega_0\cos\theta\mathbf{i} + R\omega_d\cos\theta\mathbf{j} + (R\omega_d\sin\theta - b\omega_0)\mathbf{k}$.

20.4 (a) $\boldsymbol{\omega} = 20\mathbf{i} - 5\mathbf{j}$ (rad/s).
(b) $\mathbf{v}_A = 0.25\mathbf{i} + 1.00\mathbf{j} + 3.73\mathbf{k}$ (m/s).

20.6 $\mathbf{a}_B = 166\mathbf{j}$ (ft/s^2),
$\boldsymbol{\alpha} = 7.31\mathbf{i} - 5.52\mathbf{j} + 22.89\mathbf{k}$ (rad/s^2).

20.8 $\mathbf{H} = 25\mathbf{i} + 50\mathbf{k}$ (kg-m^2/s).

20.10 $I_{xx} = 0.0398$ slug-ft^2, $I_{yy} = 0.0373$ slug-ft^2,
$I_{zz} = 0.0772$ slug-ft^2, $I_{xy} = I_{yz} = I_{zx} = 0$.

20.12 $\Sigma\mathbf{M} = -250\mathbf{i} - 250\mathbf{j} + 125\mathbf{k}$ (N-m).

20.14 $\omega_0 = [g\sin\beta/(\frac{2}{3}l\sin\beta\cos\beta + b\cos\beta)]^{1/2}$.

20.16 (b) If $\omega_0 = 0$, the plate is stationary. The solution of the equation $2\cos\beta - \sin\beta = 0$ is the value of β for which the center of mass of the plate is directly above point O; the plate is balanced on one corner.

20.18 $\mathbf{F}_B = 52.72\mathbf{i} + 97.35\mathbf{j} + 9.26\mathbf{k}$ (N),
$\mathbf{M}_B = 0.05\mathbf{i} - 10.25\mathbf{j} + 30.63\mathbf{k}$ (N-m).

20.20

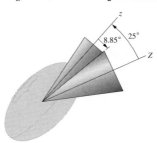

20.22 $\Sigma\mathbf{M} = -283\mathbf{i} - 2546\mathbf{j} - 800\mathbf{k}$ (ft-lb).

20.24 $\boldsymbol{\alpha} = 0.0535\mathbf{i} + 0.0374\mathbf{j} + 0.0160\mathbf{k}$ (rad/s^2)

Chapter 21

21.2 (a) $x = (0.1)\cos 3t$ m.
(b)

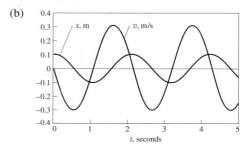

21.6 $f = 0.985$ Hz.

21.8 1.03 Hz.

21.10 $f = (1/2\pi)\sqrt{3[(k/m) - (g/2l)]}$.

21.12 $\delta = 2.07$.

21.14 $f_d = 0.714$ Hz.

21.16 $x = 0.118e^{-2.68t} - 0.118e^{-14.01t}$ ft.

21.18 $\theta_B = e^{-5.05t}(0.244\sin 3.45t + 0.167\cos 3.45t)$ rad.

21.20 $x = 0.253\sin 6.19t + 0.100\cos 6.19t + 0.145\sin 3.00t$ m.

21.22 (a) $E_p = (2\pi v/\lambda)^2 h/[(k/m) - (2\pi v/\lambda)^2]$.
(b) $v = \lambda\sqrt{k/m}/2\pi$.

Index

S

Properties of Areas and Lines

Areas

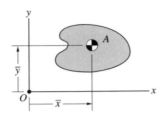

The coordinates of the centroid of the area A are

$$\bar{x} = \frac{\int_A x\, dA}{\int_A dA}, \qquad \bar{y} = \frac{\int_A y\, dA}{\int_A dA}.$$

The moment of inertia about the x axis I_x, the moment of inertia about the y axis I_y, and the product of inertia I_{xy} are

$$I_x = \int_A y^2\, dA, \qquad I_y = \int_A x^2\, dA, \qquad I_{xy} = \int_A xy\, dA.$$

The polar moment of inertia about O is

$$J_O = \int_A r^2\, dA = \int_A (x^2 + y^2)\, dA = I_x + I_y.$$

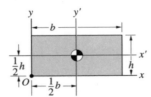

Rectangular area

$$\text{Area} = bh$$

$$I_x = \frac{1}{3} bh^3, \qquad I_y = \frac{1}{3} hb^3, \qquad I_{xy} = \frac{1}{4} b^2 h^2$$

$$I_{x'} = \frac{1}{12} bh^3, \qquad I_{y'} = \frac{1}{12} hb^3, \qquad I_{x'y'} = 0$$

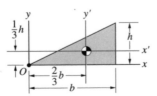

Triangular area

$$\text{Area} = \frac{1}{2} bh$$

$$I_x = \frac{1}{12} bh^3, \qquad I_y = \frac{1}{4} hb^3, \qquad I_{xy} = \frac{1}{8} b^2 h^2$$

$$I_{x'} = \frac{1}{36} bh^3, \qquad I_{y'} = \frac{1}{36} hb^3, \qquad I_{x'y'} = \frac{1}{72} b^2 h^2$$

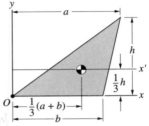

Triangular area

$$\text{Area} = \frac{1}{2} bh \qquad I_x = \frac{1}{12} bh^3, \qquad I_{x'} = \frac{1}{36} bh^3$$

Circular area

$$\text{Area} = \pi R^2 \qquad I_{x'} = I_{y'} = \frac{1}{4} \pi R^4, \qquad I_{x'y'} = 0$$

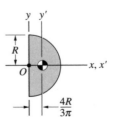

Semicircular area

$$\text{Area} = \frac{1}{2} \pi R^2 \qquad I_x = I_y = \frac{1}{8} \pi R^4, \qquad I_{xy} = 0$$

$$I_{x'} = \frac{1}{8} \pi R^4, \qquad I_{y'} = \left(\frac{\pi}{8} - \frac{8}{9\pi} \right) R^4, \qquad I_{x'y'} = 0$$

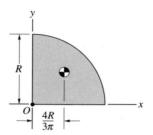

Quarter-circular area

$$\text{Area} = \frac{1}{4}\pi R^2 \qquad I_x = I_y = \frac{1}{16}\pi R^4, \qquad I_{xy} = \frac{1}{8}R^4$$

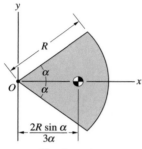

Circular sector

$$\text{Area} = \alpha R^2$$

$$I_x = \frac{1}{4}R^4\left(\alpha - \frac{1}{2}\sin 2\alpha\right), \qquad I_y = \frac{1}{4}R^4\left(\alpha + \frac{1}{2}\sin 2\alpha\right),$$

$$I_{xy} = 0$$

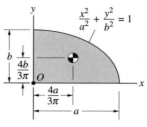

Quarter-elliptical area

$$\text{Area} = \frac{1}{4}\pi ab$$

$$I_x = \frac{1}{16}\pi ab^3, \qquad I_y = \frac{1}{16}\pi a^3 b, \qquad I_{xy} = \frac{1}{8}a^2 b^2$$

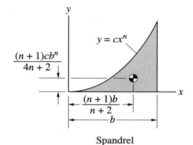

Spandrel

$$\text{Area} = \frac{cb^{n+1}}{n+1}$$

$$I_x = \frac{c^3 b^{3n+1}}{9n+3}, \qquad I_y = \frac{cb^{n+3}}{n+3}, \qquad I_{xy} = \frac{c^2 b^{2n+2}}{4n+4}$$

Lines

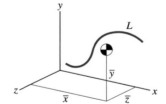

The coordinates of the centroid of the line L are

$$\bar{x} = \frac{\int_L x\, dL}{\int_L dL}, \qquad \bar{y} = \frac{\int_L y\, dL}{\int_L dL}, \qquad \bar{z} = \frac{\int_L z\, dL}{\int_L dL}.$$

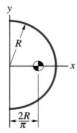

Semicircular arc

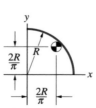

Quarter-circular arc

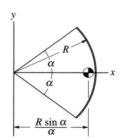

Circular arc